Handbook of Geometric Computing

Eduardo Bayro Corrochano

Handbook of Geometric Computing

Applications in Pattern Recognition, Computer Vision, Neuralcomputing, and Robotics

With 277 Figures, 67 in color, and 38 Tables

Prof. Dr. Eduardo Bayro Corrochano
Cinvestav
Unidad Guadalajara
Ciencias de la Computación
P. O. Box 31-438
Plaza la Luna, Guadalajara
Jalisco 44550
México
edb@gdl.cinvestav.mx

Library of Congress Control Number: 2004118329

ACM Computing Classification (1998): I.4, I.3, I.5, I.2, F. 2.2

ISBN-10 3-540-20595-0 Springer Berlin Heidelberg New York
ISBN-13 978-3-540-20595-1 Springer Berlin Heidelberg New York

This work is subject to copyright. All rights are reserved, whether the whole or part of the material is concerned, specifically the rights of translation, reprinting, reuse of illustrations, recitation, broadcasting, reproduction on microfilm or in any other way, and storage in data banks. Duplication of this publication or parts thereof is permitted only under the provisions of the German Copyright Law of September 9, 1965, in its current version, and permission for use must always be obtained from Springer. Violations are liable for prosecution under the German Copyright Law.

Springer is a part of Springer Science+Business Media

springeronline.com

© Springer-Verlag Berlin Heidelberg 2005
Printed in Germany

The use of general descriptive names, registered names, trademarks, etc. in this publication does not imply, even in the absence of a specific statement, that such names are exempt from the relevant protective laws and regulations and therefore free for general use.

Cover design: KünkelLopka, Heidelberg
Production: LE-TeX Jelonek, Schmidt & Vöckler GbR, Leipzig
Typesetting: by the author
Printed on acid-free paper   45/3142/YL - 5 4 3 2 1 0

Preface

One important goal of human civilization is to build intelligent machines, not necessarily machines that can mimic our behavior perfectly, but rather machines that can undertake heavy, tiresome, dangerous, and even inaccessible (for man) labor tasks. Computers are a good example of such machines. With their ever-increasing speeds and higher storage capacities, it is reasonable to expect that in the future computers will be able to perform even more useful tasks for man and society than they do today, in areas such as health care, automated visual inspection or assembly, and in making possible intelligent man–machine interaction. Important progress has been made in the development of computerized sensors and mechanical devices. For instance, according to Moore's law, the number of transistors on a chip roughly doubles every two years – as a result, microprocessors are becoming faster and more powerful and memory chips can store more data without growing in size.

Developments with respect to concepts, unified theory, and algorithms for building intelligent machines have not occurred with the same kind of lightning speed. However, they should not be measured with the same yardstick, because the qualitative aspects of knowledge development are far more complex and intricate. In 1999, in his work on building anthropomorphic motor systems, Rodney Brooks noted: "A paradigm shift has recently occurred – computer performance is no longer a limiting factor. We are limited by our knowledge of what to build." On the other hand, at the turn of the twenty-first century, it would seem we collectively know enough about the human brain and we have developed sufficiently advanced computing technology that it should be possible for us to find ways to construct real-time, high-resolution, verifiable models for significant aspects of human intelligence.

Just as great strides in the dissemination of human knowledge were made possible by the invention of the printing press, in the same way modern scientific developments are enhanced to a great extent by computer technology. The Internet now plays an important role in furthering the exchange of information necessary for establishing cooperation between different research groups. Unfortunately, the theory for building intelligent machines or perception-and-action systems is still in its infancy. We cannot blame a lack of commitment on the part of researchers or the absence of revolutionary concepts for this state of affairs. Remarkably useful ideas were proposed as early as the mid-nineteenth century, when Babbage was building his first calculating engines. Since then, useful concepts have emerged in mathematics, physics, electronics, and mechanical engineering – all basic fields for the development of intelligent machines. In its time, classical mechanics offered many of the necessary conceptual tools. In our own time, Lie group theory and Riemann differential geometry play a large role in modern mathematics and physics. For instance, as a representation tool, symmetry, a visual primitive probably unattentively encoded, may provide an important avenue for helping us understand perceptual processes. Unfortunately, the application of these concepts in current work on image processing, neural computing, and robotics is still somewhat limited. Statistical physics and optimization theory have also proven to be useful in the fields of numerical analysis, nonlinear dynamics, and, recently, in neural computing. Other approaches for computing under conditions of uncertainty, like fuzzy logic and tensor voting, have been proposed in recent years. As we can see, since Turing's pioneering 1950 work on determining whether machines are intelligent, the development of computers for enhanced intelligence has undergone great progress.

This new handbook takes a decisive step in bringing together in one volume various topics highlighting the geometric aspects necessary for image analysis and processing, perception, reasoning, decision making, navigation, action, and autonomous learning. Unfortunately, even with growing financial support for research and the enhanced possibilities for communication brought about by the Internet, the various disciplines within the research community are still divorced from one another, still working in a disarticulated manner. Yet the effort to build perception–action systems requires flexible concepts and efficient algorithms, hopefully developed in an integrated and unified manner. It is our hope that this handbook will encourage researchers to work together on proposals and methodologies so as to create the necessary synergy for more rapid progress in the building of intelligent machines.

Structure and Key Contributions

The handbook consists of nine parts organized by discipline, so that the reader can form an understanding of how work among the various disciplines is contributing to progress in the area of geometric computing. Understanding in each individual field is a fundamental requirement for the development of perception-action systems. In this regard, a tentative list of relevant topics might include:

- brain theory and neuroscience
- learning
- neurocomputing, fuzzy computing, and quantum computing

- image analysis and processing
- geometric computing under uncertainty
- computer vision
- sensors
- kinematics, dynamics, and elastic couplings
- fuzzy and geometric reasoning
- control engineering
- robot manipulators, assembly, MEMS, mobile robots, and humanoids
- path planning, navigation, reaching, and haptics
- graphic engineering, visualization, and virtual reality
- medical imagery and computer-aided surgery

We have collected contributions from the leading experts in these diverse areas of study and have organized the chapters in each part to address low-level processing first before moving on to the more complex issues of decision making. In this way, the reader will be able to clearly identify the current state of research for each topic and its relevance for the direction and content of future research. By gathering this work together under the umbrella of building perception–action systems, we are able to see that efforts toward that goal are flourishing in each of these disciplines and that they are becoming more interrelated and are profiting from developments in the other fields. Hopefully, in the near future, we will see all of these fields interacting even more closely in the construction of efficient and cost-effective autonomous systems.

Part I Neuroscience

In **Chapter 1** Haluk Öğmen reviews the fundamental properties of the primate visual system, highlighting its maps and pathways as spatio-temporal information encoding and processing strategies. He shows that retinotopic and spatial-frequency maps represent the geometry of the fusion between structure and function in the nervous system, and that magnocellular and parvocellular pathways can resolve the trade-off between spatial and temporal deblurring.

In **Chapter 2** Hamid R. Eghbalnia, Amir Assadi, and Jim Townsend analyze the important visual primitive of symmetry, probably unattentively encoded, which can have a central role in addressing perceptual processes. The authors argue that biological systems may be hardwired to handle filtering with extreme efficiency. They believe that it may be possible to approximate this filtering, effectively preserving all the important temporal visual features, by using current computer technology. For learning, they favor the use of bidirectional associative memories, using local information in the spirit of a local-to-global approach to learning.

Part II Neural Networks

In **Chapter 3** Hyeyoung Park, Tomoko Ozeki, and Shun-ichi Amari choose a geometric approach to provide intuitive insights on the essential properties of neural networks and their performance. Taking into account Riemann's structure of the manifold of multilayer perceptrons, they design gradient learning techniques for avoiding algebraic singularities that have a great negative influence on trajectories of learning. They discuss the singular structure of neuromanifolds and pose an interesting problem of statistical inference and learning in hierarchical models that include singularities.

In **Chapter 4** Gerhard Ritter and Laurentiu Iancu present a new paradigm for neural computing using the lattice algebra framework. They develop morphological auto-associative memories and morphological feed-forward networks based on dendritic computing. As opposed to traditional neural networks, their models do not need hidden layers for solving non-convex problems, but rather they converge in one step and exhibit remarkable performance in both storage and recall.

In **Chapter 5** Tijl De Bie, Nello Cristianini, and Roman Rosipal describe a large class of pattern-analysis methods based on the use of generalized eigenproblems and their modifications. These kinds of algorithms can be used for clustering, classification, regression, and correlation analysis. The chapter presents all these algorithms in a unified framework and shows how they can all be coupled with kernels and with regularization techniques in order to produce a powerful class of methods that compare well with those of the support-vector type. This study provides a modern synthesis between several pattern-analysis techniques.

Part III Image Processing

In **Chapter 6** Jan J. Koenderink sketches a framework for image processing that is coherent and almost entirely geometric in nature. He maintains that the time is ripe for establishing image processing as a science that departs from fundamental principles, one that is developed logically and is free of hacks, unnecessary approximations, and mere showpieces on mathematical dexterity.

In **Chapter 7** Alon Spira, Nir Sochen, and Ron Kimmel describe image enhancement using PDF-based geometric diffusion flows. They start with variational principles for explaining the origin of the flows, and this geometric approach results in some nice invariance properties. In the Beltrami framework, the image is considered to be an embedded manifold in the space-feature manifold, so that the required geometric filters for the flows in gray-level and color images or texture will take into account the induced metric. This chapter presents numerical schemes and kernels for the flows that enable an efficient and robust implementation.

In **Chapter 8** Yaobin Mao and Guanrong Chen show that chaos theory is an excellent alternative for producing a fast, simple, and reliable image-encryption scheme that has a high degree of security. The chapter describes a practical and efficient chaos-based stream-cipher scheme for still images. From an engineer's perspective, the chaos image-encryption technology is very promising for the real-time image transfer and handling required for intelligent discerning systems.

Part IV Computer Vision

In **Chapter 9** Kalle Åström is concerned with the geometry and algebra of multiple one-dimensional projections in a 2D environment. This study is relevant for 1D cameras, for understanding the projection of lines in ordinary vision, and, on the application side, for understanding the ordinary vision of vehicles undergoing planar motion. The structure-of-motion problem for 1D cameras is studied at length, and all cases with non-missing data are solved. Cases with missing data are more difficult; nevertheless, a classification is introduced and some minimal cases are solved.

In **Chapter 10** Anders Heyden describes in-depth, n-view geometry with all the computational aspects required for achieving stratified reconstruction. He starts with camera modeling and a review of projective geometry. He describes the multi-view tensors and constraints and the associated linear reconstruction algorithms. He continues with factorization and bundle adjustment methods and concludes with auto-calibration methods.

In **Chapter 11** Amnon Shashua and Lior Wolf introduce a generalization of the classical collineation of $\mathcal{P}^n$. The m-view tensors for $\mathcal{P}^n$ referred to as homography tensors are studied in detail for the case $n=3,4$ in which the individual points are allowed to move while the projective change of coordinates takes place. The authors show that without homography tensors a recovering of the alignment requires statistical methods of sampling, whereas with the tensor approach both stationary and moving points can be considered alike and part of a global transformation can be recovered analytically from some matching points across m views. In general, the homography tensors are useful for recovering linear models under linear uncertainty.

In **Chapter 12** Abhijit Ogale, Cornelia Fermüller and Yiannis Aloimonos examine the problem of instantaneous finding of objects moving independently in a video obtained by a moving camera with a restricted field of view. In this problem, the image motion is caused by the combined effect of camera motion, scene depth, and the independent motions of objects. The authors present a classification of moving objects and discuss detection methods; the first class is detected using motion clustering, the second depends on ordinal depth from occlusions and the third uses cardinal knowledge of the depth. Robust methods for deducing ordinal depth from occlusions are also discussed.

Part V Perception and Action

In **Chapter 13** Eduardo Bayro-Corrochano presents a framework of conformal geometric algebra for perception and action. As opposed to standard projective geometry, in conformal geometric algebra, using the language of spheres, planes, lines, and points, one can deal simultaneously with incidence algebra operations (meet and join) and conformal transformations represented effectively using bivectors. This mathematical system allows us to keep our intuitions and insights into the geometry of the problem at hand and it helps us to reduce considerably the computational burden of the related algorithms. Conformal geometric algebra, with its powerful geometric representation and rich algebraic capacity to provide a unifying geometric language, appears promising for dealing with kinematics, dynamics, and projective geometry problems without the need to abandon a mathematical system. In general, this can be a great advantage in applications that use stereo vision, range data, lasers, omnidirectionality, and odometry-based robotic systems.

Part VI Uncertainty in Geometric Computations

In **Chapter 14** Kenichi Kanatani investigates the meaning of "statistical methods" for geometric inference on image points. He traces back the origin of feature uncertainty to image-processing operations for computer vision, and he discusses the implications of asymptotic analysis with reference to "geometric fitting" and "geometric model selection." The author analyzes recent progress in geometric fitting techniques for linear constraints and semiparametric models in relation to geometric inference.

In **Chapter 15** Wolfgang Förstner presents an approach for geometric reasoning in computer vision performed under uncertainty. He shows that the great potential of projective geometry and statistics can be integrated easily for propagating uncertainty through reasoning chains. This helps to make decisions on uncertain spatial relations and on the optimal estimation of geometric entities and transformations. The chapter discusses the essential link between statistics and projective geometry, and it summarizes the basic relations in 2D and 3D for single-view geometry.

In **Chapter 16** Gérard Medioni, Philippos Mordohai, and Mircea Nicolescu present a tensor voting framework for computer vision that can address a wide range of middle-level vision problems in a unified way. This framework is based on a data representation formalism that uses second-order symmetric tensors and an information propagation mechanism that uses a tensor voting scheme. The authors show that their approach is suitable for stereo and motion analysis because it can detect perceptual structures based solely on the smoothness constraint without using any model. This property allows them to treat the arbitrary surfaces that are inherent in non-trivial scenes.

Part VII Computer Graphics and Visualization

In **Chapter 17** Lawrence H. Staib and Yongmei M. Wang present two robust methods for nonrigid image registration. Their methods take advantage of differences in available information: their surface warping approach uses local and global surface properties, and their volumetric deformation method uses a combination of shape and intensity information. The authors maintain that, in nonrigid images, registration is desirable for designing a match metric that includes as much useful information as possible, and that such a transformation is tailored to the required deformability, thereby providing an efficient and reliable optimization.

In **Chapter 18** Alyn Rockwood shows how computer graphics indicates trends in the way we think about and represent technology and pursue research, and why we need more visual geometric languages to represent technology in a way that can provide insight. He claims that visual thinking is key for the solution of problems. The author investigates the use of implicit function modeling as a suitable approach for describing complex objects with a minimal database. The author interrogates how general implicit functions in non-Euclidean spaces can be used to model shape.

Part VIII Geometry and Robotics

In **Chapter 19** Neil White utilizes the Grassmann–Cayley algebra framework for writing expressions of geometric incidences in Euclidean and projective geometry. The shuffle formula for the meet operation translates the geometric conditions into coordinate-free algebraic expressions. The author draws our attention to the importance of the Cayley factorization process, which leads to the use of symbolic and coordinate-free expressions that are much closer to the human thinking process. By taking advantage of projective invariant conditions, these expressions can geometrically describe the realizations of a non-rigid, generically isostatic graph.

In **Chapter 20** Jon Selig employs the special Clifford algebra $\mathcal{G}_{0,6,2}$ to derive equations for the motion of serial and parallel robots. This algebra is used to represent the six component velocities of rigid bodies. Twists or screws and wrenches are used for representing velocities and force/torque vectors, respectively. The author outlines the Lagrangian and Hamiltonian mechanics of serial robots. A method for finding the equations of motion of the Stewart platform is also considered.

In **Chapter 21** Calin Belta and Vijay Kumar describe a modern geometric approach for designing trajectories for teams of robots maintaining rigid formation or virtual structure. The authors consider first the problem of generating minimum kinetic energy motion for a rigid body in a 3D environment. Then they present an interpolation method based on embedding SE(3) into a larger manifold for generating optimal curves and projecting them back to SE(3). The novelty of their approach relies on the invariance of the produced trajectories, the way of defining and inheriting physically significant metrics, and the increased efficiency of the algorithms.

Part IX Reaching and Motion Planning

In **Chapter 22** J. Michael McCarthy and Hai-Jun Su examine the geometric problem of fitting an algebraic surface to points generated by a set of spatial displacements. The authors focus on seven surfaces that are traced by the center of the spherical wrist of an articulated chain. The algebraic equations of these reachable surfaces are evaluated on each of the displacements to define a set of polynomial equations which are rich in internal structure. Efficient ways to find their solutions are highly dependent upon the complexity of the problem, which increases greatly with the number of parameters that specify the surface.

In **Chapter 23** Seth Hutchinson and Peter Leven are concerned with planning collision-free paths, one of the central research problems in intelligent robotics. They analyze the probabilistic roadmap (PRM) planner, a graph search in the configuration space, and they discuss its design choices. These PRM planners are confronted with narrow corridors, the relationship between the geometry of both obstacles and robots, and the geometry of the free configuration space, which is still not well understood, making a thorough analysis of the method difficult. PRM planners tend to be easy to implement; however, design choices have considerable impact on the overall performance of the planner.

Guadalajara, Mexico *Eduardo Bayro-Corrochano*
December 2004

Acknowledgments

I am very thankful to CINVESTAV Unidad Guadalajara, and to CONACYT for the funding for Projects 43124 and Fondos de Salud 49, which gave me the freedom and the time to develop this original handbook. This volume constitutes a new venue for bringing together new perspectives in geometric computing that will be useful for building intelligent machines. I would also like to express my thanks to the editor Alfred Hofmann and the associate editor Ingeborg Mayer from Springer for encouraging me to pursue this project. I am grateful for the assistance of Gabi Fischer, Ronan Nugent and Tracey Wilbourn for their LaTeX expertise and excellent copyediting. And finally, my deepest thanks go to the authors whose work appears here. They accepted the difficult task of writing chapters within their respective areas of expertise but in such a manner that their contributions would integrate well with the main goals of this handbook.

Contents

Part I Neuroscience

1 Spatiotemporal Dynamics of Visual Perception Across Neural Maps and Pathways
Haluk Öğmen 3

2 Symmetry, Features, and Information
Hamid R. Eghbalnia, Amir Assadi, Jim Townsend 31

Part II Neural Networks

3 Geometric Approach to Multilayer Perceptrons
Hyeyoung Park, Tomoko Ozeki, Shun-ichi Amari 69

4 A Lattice Algebraic Approach to Neural Computation
Gerhard X. Ritter, Laurentiu Iancu 97

5 Eigenproblems in Pattern Recognition
Tijl De Bie, Nello Cristianini, Roman Rosipal 129

Part III Image Processing

6 Geometric Framework for Image Processing
Jan J. Koenderink 171

7 Geometric Filters, Diffusion Flows, and Kernels in Image Processing
Alon Spira, Nir Sochen, Ron Kimmel 203

8 Chaos-Based Image Encryption
Yaobin Mao, Guanrong Chen 231

Part IV Computer Vision

9 One-Dimensional Retinae Vision
Kalle Åström 269

10 Three-Dimensional Geometric Computer Vision
Anders Heyden 305

11 Dynamic $\mathcal{P}^n$ to $\mathcal{P}^n$ Alignment
Amnon Shashua, Lior Wolf 349

12 Detecting Independent 3D Movement
Abhijit S. Ogale, Cornelia Fermüller, Yiannis Aloimonos 383

Part V Perception and Action

13 Robot Perception and Action Using Conformal Geometric Algebra
Eduardo Bayro-Corrochano 405

Part VI Uncertainty in Geometric Computations

14 Uncertainty Modeling and Geometric Inference
Kenichi Kanatani 461

15 Uncertainty and Projective Geometry
Wolfgang Förstner 493

16 The Tensor Voting Framework
Gérard Medioni, Philippos Mordohai, Mircea Nicolescu 535

Part VII Computer Graphics and Visualization

17 Methods for Nonrigid Image Registration
Lawrence H. Staib, Yongmei Michelle Wang 571

18 The Design of Implicit Functions for Computer Graphics
Alyn Rockwood 603

Part VIII Geometry and Robotics

19 Grassmann–Cayley Algebra and Robotics Applications
Neil L. White 629

20 Clifford Algebra and Robot Dynamics
J. M. Selig 657

21 Geometric Methods for Multirobot Optimal Motion Planning
Calin Belta, Vijay Kumar 679

Part IX Reaching and Motion Planning

22 The Computation of Reachable Surfaces for a Specified Set of Spatial Displacements
J. Michael McCarthy, Hai-Jun Su 709

23 Planning Collision-Free Paths Using Probabilistic Roadmaps
Seth Hutchinson, Peter Leven 737

Index 769

Part I

Neuroscience

1

Spatiotemporal Dynamics of Visual Perception Across Neural Maps and Pathways

Haluk Öğmen

Department of Electrical and Computer Engineering
Center for Neuro–Engineering and Cognitive Science
University of Houston
Houston, TX 77204–4005 USA
ogmen@uh.edu

1.1 Introduction

The relationship between geometry and brain function presents itself as a dual problem: on the one hand, since the basis of geometry is in brain function, especially that of the visual system, one can ask what the brain function can tell us about the genesis of geometry as an abstract form of human mental activity. On the other hand, one can also ask to what extent geometry can help us understand brain function. Because the nervous system is interfaced to our environment by sensory and motor systems and because geometry has been a useful language in understanding our environment, one might expect some convergence of geometry and brain function at least at the peripheral levels of the nervous system. Historically, there has been a close relationship between geometry and theories of vision starting as early as Euclid. Given light sources and an environment, one can easily calculate the corresponding images on our retinae using basic physics and geometry. This is usually known as the "forward problem" [41]. A straightforward approach would be then to consider the function of the visual system as the computation of the inverse of the transformations leading to image formation. However, this "inverse optics" approach leads to ill-posed problems and necessitates the use of a priori assumptions to reduce the number of possible solutions. The use of a priori assumptions in turn makes the approach unsuitable for environments that violate the assumptions. Thus, the inverse optics formulation fails to capture the robustness of human visual perception in complex environments. On the other hand, visual illusions, i.e. discrepancies between the physical stimuli and the corresponding percepts, constitute examples of the limitations of the human visual system. Nevertheless, these illusions do not affect significantly the overall performance of the system, as most people operate succesfully in the environment without even noticing these illusions. The illusions are usually discovered by scientists, artists, and philosophers who scrutinize deeply the relation between the physical and psychological world. These illusions are often used by vision scientists as "singular points" to study the visual system.

How the inputs from the environment are transformed into our conscious percepts is largely unknown. The goals of this chapter are twofold: first, it provides a brief review of the basic neuroanatomical structure of the visual system in primates. Second, it outlines a theory of how neural maps and pathways can interact in a dynamic system, which operates principally in a transient regime, to generate a spatiotemporal neural representation of visual inputs.

1.2 The Basic Geometry of Neural Representation: Maps and Pathways

The first stage of input representation in the visual system occurs in the retina. The retina is itself a complex structure comprising five main neuronal types organized in direct and lateral structures (Fig. 1). The "direct structure"

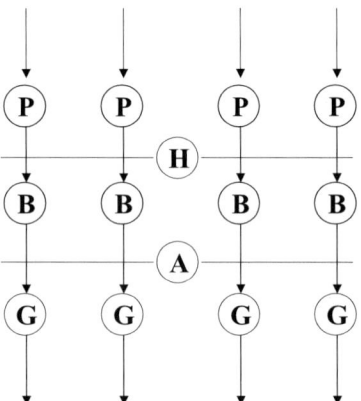

Fig. 1.1. The general architecture of the retina. P, photoreceptor; B, bipolar cell; G, ganglion cell; H, horizontal cell; A, amacrine cell. The *arrows on top* show the light input coming from adjacent spatial locations in the environment, and the *arrows at the bottom* represent the output of the retina, which preserves the two-dimensional topography of the inputs. This gives rise to "retinotopic maps" at the subsequent processing stages

consists of signal flow from the photoreceptors to bipolar cells, and finally to retinal ganglion cells, whose axons constitute the output of the retina. This direct pathway is repeated over the retina and thus constitutes an "image plane" much like the photodetector array of a digital camera. In addition to the cells in the direct pathway, horizontal and amacrine cells carry out signals laterally and contribute to the spatiotemporal processing of the signals. Overall, the three-dimensional world is projected to a two-dimensional *retinotopic map* through the optics of the eye, the two-dimensional sampling by the receptors, and the spatial organization of the post-receptor direct pathway. The parallel fibres from the retina running to the visual cortex via the lateral geniculate nucleus (LGN) preserve the retinal topography, and the early visual representation in the visual cortex maintains the retinotopic map.

In addition to this spatial coding, retinal ganglion cells can be broadly classified into three types: P, M, and K [15, 27]. The characterization of the K type is not fully detailed, and our discussion will focus on the M and P types. These two cell types can be distinguished on the basis of their anatomical and response characteristics; for example, M cell responses have shorter latencies and are more transient than P cell responses [16, 33, 36, 42]. Thus the information from the retina is not carried out by a single retinotopic map, but by three maps that form *parallel pathways*. Moreover, different kinds of information are carried out along these pathways. The pathway originating from P cells is called the parvocellular pathway, and the pathway originating from M cells is called the magnocellular pathway.

The signals that reach the cortex are also channeled into maps and pathways. Two major cortical pathways, the dorsal and the ventral, have been identified (Fig. 1.2) [35]. The dorsal pathway, also called the "where pathway", is specialized in processing information about the position of objects. On the other hand, the ventral pathway, also called the "what pathway", has been implicated in the processing of object identities [35]. Another related functional interpretation of these pathways is that the dorsal pathway is specialized for action, while the ventral pathway is specialized for perception [34]. This broad functional specialization is supplemented by more specialized pathways dedicated to the processing of motion, color, and form [32, 59]. Within these pathways, the cortical organization contains maps of different object attributes. For example, neurons in the primary visual cortex respond preferentially to the orientations of edges. Spatially, neurons that are sensitive to adjacent orientations tend to be located in adjacent locations forming a "map of orientation" on the cortical space [30]. This is shown schematically in Fig. 1.3. Similar maps have been observed for location (retinotopic map) [30], spatial frequency [19], color [52, 58], and direction of motion [2].

Maps build a relatively continuous and periodic topographical representation of stimulus properties (e.g., spatial location, orientation, color) on cortical space. What is the goal of such a representation? In neural computation, in addition to the processing at each neuron, a significant amount of processing takes place at the synapses. Because synapses represent points of connection between neurons, functionally both the development and the processing characteristics of the synapses are often specialized based on processing and encoding characteristics of both pre- and post-synaptic cells. Consequently, map representations in the nervous system appear to be correlated with the

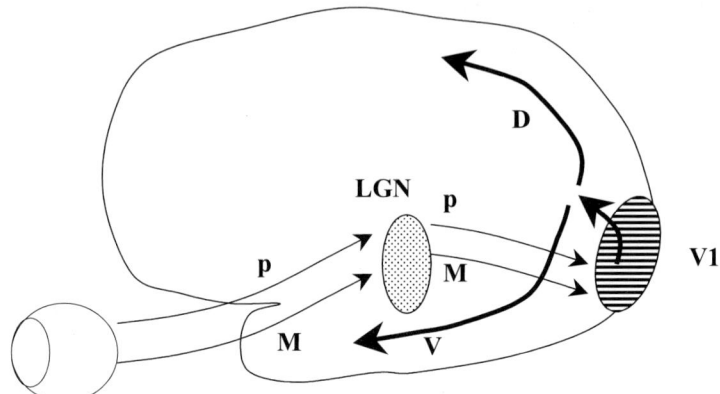

Fig. 1.2. Schematic depiction of the parvocellular (*P*), magnocellular (*M*), and the cortical dorsal (*D*), ventral (*V*) pathways. *LGN*, lateral geniculate nucleus; *V1*, primary visual cortex

Fig. 1.3. Depiction of how orientation columns form an orientation map. Neurons in a given column are tuned to a specific orientation depicted by an *oriented line segment* in the figure. Neurons sensitive to similar orientations occupy neighboring positions on the cortical surface

geometry of synaptic development as well as with the geometry of synaptic patterns as part of information processing. According to this perspective, maps represent the geometry of the fusion between structure and function in the nervous system.

On the other hand, pathways possess more discrete, often dichotomic, representation. What is more important, pathways represent a cascade of maps that share common functional properties. From the functional point of view, pathways can be viewed as complementary systems adapted to conflicting but complementary aspects of information processing. For example, the magnocellular pathway is specialized for processing high-temporal low-spatial frequency information, whereas the parvocellular system is specialized for processing low-temporal and high-spatial frequency information. From the evolutionary point of view, pathways can be viewed as new systems that emerge as the interactions between the organism and the environment become more sophisticated. For example, for a simple organism the localization of stimuli without complex recognition of its figural properties can be sufficient for survival. Thus a basic pathway akin to the primate where/action pathway would be sufficient. On the other hand, more evolved animals may need to recognize and categorize complex aspects of stimuli, and thus an additional pathway specialized for conscious perception may develop.

In the next section, these concepts will be illustrated by considering how the visual system can encode object boundaries in real-time.

1.3 Example: Maps and Pathways in Coding Object Boundaries

1.3.1 The Problem of Boundary Encoding

Under visual fixation conditions, the retinal image of an object boundary is affected by the physical properties of light, the optics of the human eye, the neurons and blood vessels in the eye, eye movements, and the dynamics of the accommodation system [19]. Several studies show that processing time on the order of 100 ms is required in order to reach "optimal" form and sharpness discrimination [4, 11, 29, 55] as well as more veridical perception of the sharpness of edges [44].

A boundary consists of a change of a stimulus attribute, typically luminance, over space. Because this change can occur rapidly for sharp boundaries and gradually for blurred boundaries, measurements at multiple scales are needed to detect and code boundaries and their spatial profile. The visual system contains neurons that respond preferentially to different spatial frequency bands. Moreover, as mentioned in the previous section, these neurons are organized as a "spatial frequency map" [19, 51]. The rate of change of a boundary's spatial profile also depends on the contrast of the boundary as shown in Fig. 1.4. For a fixed boundary transition width (e.g. w_1 in Fig.

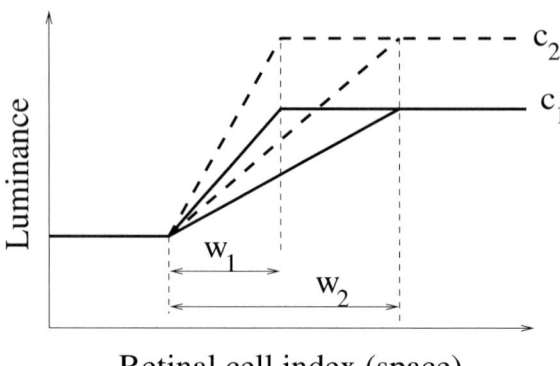

Fig. 1.4. The relationship between contrast and blur for boundaries. Boundary transition widths w_1 and w_2 for boundaries at a low contrast level c_1 (*solid lines*) and a high contrast level c_2 (*dashed lines*)

1.4), the slope of the boundary increases with increasing contrast (c_1 to c_2 in Fig. 1.4). The human visual system is capable of disambiguating the effects of blur and contrast, thereby generating conrast-independent perception of blur [23]. On the other hand, *discrimination* of edge blur depends on contrast, suggesting that the visual system encodes the blur of boundaries at least at two levels, one of which is contrast dependent, and one of which is contrast independent.

1.3.2 A Theory of Visual Boundary Encoding

How does the visual system encode object boundaries and edge blur in real-time? We will present a model of retino-cortical dynamics (RECOD) [37, 44] to suggest (i) how maps can be used to encode the position, blur, and contrast of boundaries; and (ii) how pathways can be used to overcome the real-time dynamic processing limitations of encoding across the maps. The fundamental equations of the model and their neurophysiological bases are given in the Appendix. Detailed and specialized equations of the model can be found in [44].

Figure 1.5 shows a diagrammatic representation of the general structure of RECOD. The lower two populations of neurons correspond to retinal ganglion cells with slow-sustained (parvo) and fast-transient (magno) response properties [16, 33, 36, 42]. Each of these populations contains cells sampling different retinal positions and thus contains a spatial (retinotopic) map. Two pathways, parvocellular (P pathway) and magnocellular (M pathway), emerge from these populations. These pathways provide inputs to post-retinal areas. The model also contains reciprocal inhibitory connections between post-retinal areas that receive their main inputs from P and M pathways. Figure 1.6 shows a more detailed depiction of the model. Here, circular symbols depict neurons whose

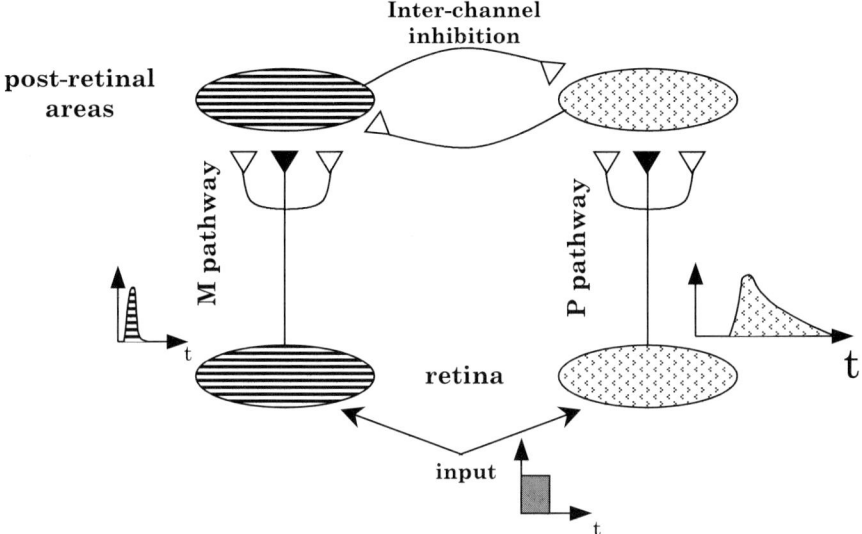

Fig. 1.5. Schematic representation of the major pathways in the RECOD model. *Filled and open synaptic symbols* depict excitatory and inhibitory connections, respectively

spatial relationship follows a retinotopic map. In this figure, the post-retinal area that receives its major input from the P pathway is decomposed into two layers. Both layers preserve the retinotopic map and add a spatial-frequency map (composed of the spatial-frequency channels). For simplicity, only three elements of the spatial-frequency map ranging from the highest spatial frequency class (H) to the lowest spatial frequency class (L) are shown. The M pathway sends a retinotopically organized inhibitory signal to cells in the first post-retinal layer. The direct inhibitory connection from retinal transient cells to post-retinal layers is only for illustrative purpose; in vivo the actual connections are carried out by local inhibitory networks. The first post-retinal layer cells receive center-surround connections from the sustained cells (parvocellular pathway). The rows indicated by H, M, and L represent elements with high, medium, and low spatial frequency tuning in the spatial frequency map, respectively. Each of the H, M, and L rows in the first post-retinal layer receive independent connections from the retinal cells, and there are no interactions between the rows. Cells in the second post-retinal layer receive center-surround connections from the H, M, and L rows of the first post-retinal layer. They also receive center-surround feedback. Sample responses of model

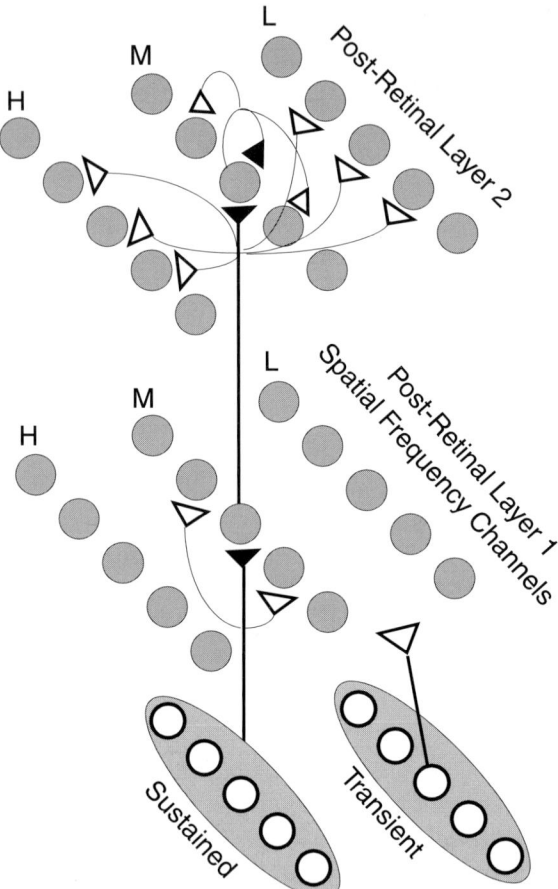

Fig. 1.6. A more detailed depiction of the RECOD model. *Filled and open synaptic symbols* depict excitatory and inhibitory connections, respectively. To avoid clutter, only a representative set of neurons and connections are shown. From [44]

neurons tuned to low spatial frequencies and to high spatial frequencies are shown for sharp and blurred edge stimuli in Fig. 1.7. As one can see in the left panel of this figure, for a sharp edge neurons in the high spatial-frequency channel respond more strongly (dashed curve) compared to neurons in the low spatial-frequency channel (solid curve). Moreover, neurons tuned to low spatial-frequencies tend to blur sharp edges. This can be seen by comparing the spread of activity shown by the dashed and solid curves in the left panel. The right panel of the figure shows the responses of these two channels to a blurred edge. In this case, neurons in the low spatial-frequency channel respond more strongly (solid curve) compared to neurons in the high spatial-frequency channel. Overall, the peak of activity across the spatial-frequency

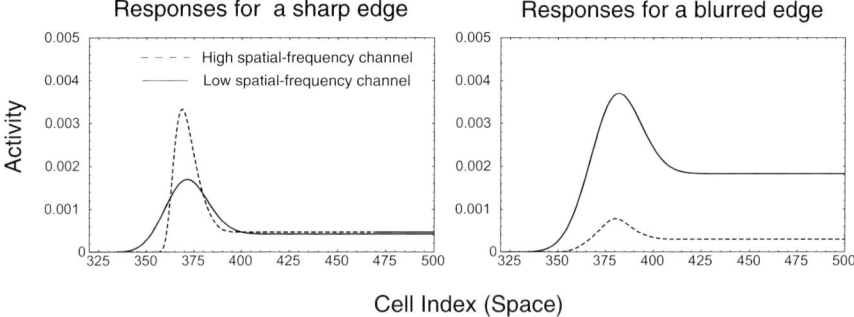

Fig. 1.7. Effect of edge blur on model responses: model responses in the first post-retinal layer for sharp (*left*) and blurred (*right*) edges at high spatial-frequency (*dotted line*) and low spatial-frequency (*continuous line*) loci of the spatial-frequency map. From [44]

map will indicate which neuron's spatial frequency matches best the sharpness of the input edge, and the level of activity for each neuron for a given edge will provide a measure of the level of match. Thus the *distribution* of activity across the spatial-frequency map provides a measure of edge blur. Even though the map is discrete in the sense that it contains a finite set of neurons, the distribution of activity in the map can provide the basis for a fine discrimination and perception of edge blur. This is similar to the encoding of color, where the distributed activities of only three primary components provide the basis for a fine discrimination and perception of color.

The model achieves the spatial-frequency selectivity by the strength and spatial distribution of synaptic connections from the retinal network to the first layer of the post-retinal network. A neuron tuned to high spatial frequencies receives excitatory and inhibitory inputs from a small retinotopic neighborhood, while a neuron tuned to low spatial frequencies receives excitatory and inhibitory inputs from a large retinotopic neighborhood (Fig. 1.8). Thus the retinotopic map allows the simple geometry of neighborhood and the resulting connectivity pattern to give rise to spatial-frequency selectivity. By smoothly changing this connectivity pattern across cortical space, one obtains a spatial-frequency map (e.g. L, M, and H in Fig. 1.6), which in turn, as mentioned above, can relate the geometry of neural activities to the fine coding of edge blur.

The left panel of Fig. 1.9 shows the activities in the first post-retinal layer of the model for a low (dashed curve) and a high (solid curve) contrast input. The response to the high contrast input is stronger. The first post-retinal layer in the model encodes edge blur in a contrast-dependent manner. The second post-retinal layer of cells achieves contrast-independent encoding of edge blur. Contrast independence is produced through connectivity patterns that exploit retinotopic and spatial-frequency maps. The second post-retinal layer implements retinotopic center-surround shunting between the cells in

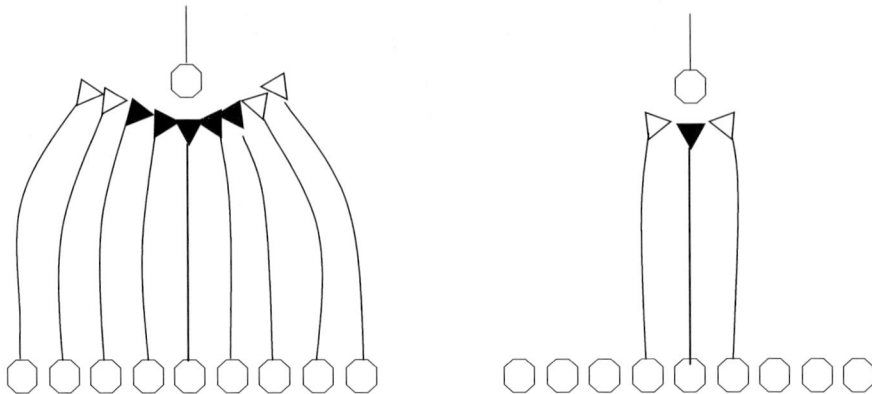

Fig. 1.8. The connectivity pattern on the *left* produces low spatial-frequency selectivity because of the convergence of inputs from an extended retinotopic area. The connectivity pattern on the *right* produces a relatively higher spatial frequency selectivity

the spatial frequency map. Each cell in this layer receives center excitation from the cell at its retinotopic location and only one of the elements in the map below it. However, it receives surround inhibition from all the elements in the map in a retinotopic manner, from a neighborhood of cells around its retinotopic location [12, 18, 20, 49, 50]. In other words, excitation from the bottom layer is one-to-one whereas inhibition is many-to-one pooled activity. This shunting interaction transforms the input activity $p1_i$ for the ith element in the spatial frequency map into an output activity $p2_i = p1_i/(A1 + \sum_i p1_i)$, where $A1$ is the time constant of the response [12, 25]. Therefore, when the total input $\sum_i p1_i$ is large compared to to $A1$, the response of each element in the spatial frequency map is contrast-normalized across the retinotopic map, resulting in contrast-constancy. This is shown in the right panel of Fig. 1.9: the responses to low contrast (dashed curve) and high contrast (solid curve) are identical.

In order to compensate the blurring effects introduced at the retinal level, the RECOD model uses a connectivity pattern across retinotopic maps, but instead of being feedforward as those giving rise to spatial-frequency selectivity, these connections are *feedback* (or re-entrant), as illustrated at the top of Fig. 1.6. Note that, for simplicity, in this figure only the connections for the medium spatial frequencies (M) are shown. Because of these feedback connections and the dynamic properties of the network, the activity pattern is "sharpened" in time to compensate for the early blurring effects. [25, 37]. In

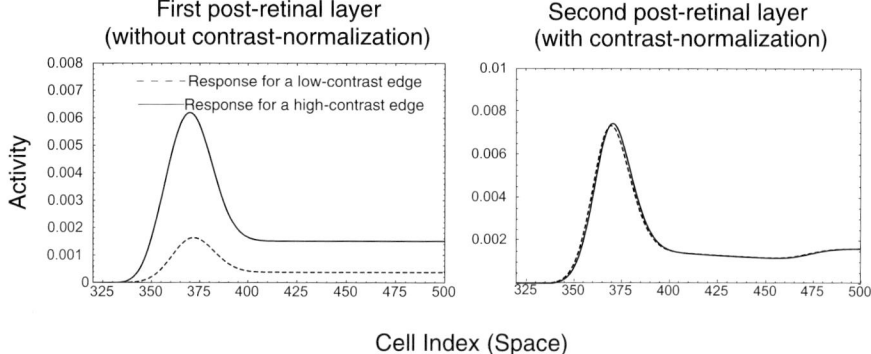

Fig. 1.9. Effect of contrast on model responses: Model responses for a high-contrast edge (*solid curve*) and a low-contrast edge (*dashed curve*) of 2 arcmin blur in the first post-retinal layer (*left*) and the second post-retinal layer (*right*). From [44]

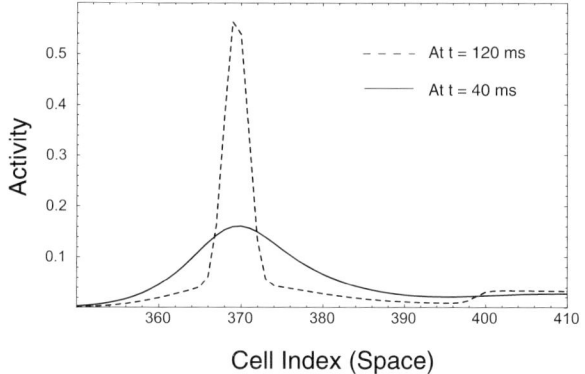

Fig. 1.10. Temporal sharpening of model responses to a blurred edge in the second post-retinal layer: responses at 40 ms (*continuous line*) and 120 ms (*dashed line*) are shown superimposed. From [44]

Fig. 1.10, the response of the model neurons in the second post-retinal layer to an edge stimulus with 2 arcmin base blur at 40 ms after stimulus onset is shown by the dashed curve. The response at 120 ms after stimulus onset is shown by the solid curve. Comparing the width of these activities, one can see that the neural encoding of the edge is initially (at 40 ms) blurred but becomes sharper with more processing time (at 120 ms).

1.3.3 Perception and Discrimination of Edge Blur

The proposed encoding scheme across retinotopic and spatial-frequency maps has been tested by comparing model predictions to a wide range of experimental data [44]. For example, Fig. 1.11 provides a comparison of model predictions to experimental data on the effect of exposure duration on perceived blur for base blurs of 0, 2, and 4 arcmin. The model has been also

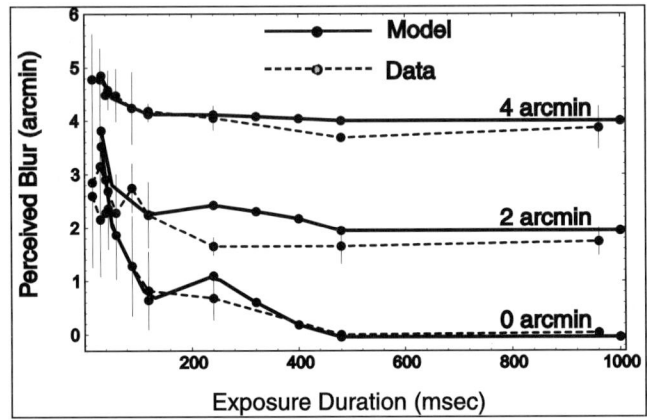

Fig. 1.11. Model predictions (*solid lines*) and data (*dashed lines*) for the effect of exposure duration on perceived blur for base blurs of 0, 2, and 4 arcmin. From [44]

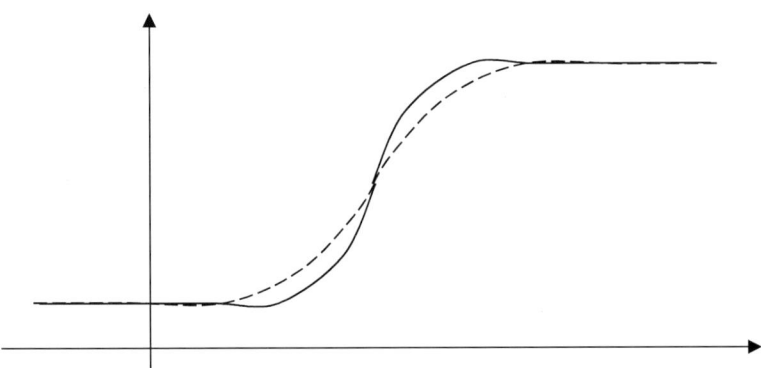

Fig. 1.12. To measure the blur discrimination threshold, first a base blur is chosen (*solid curve*). The ability of the observer to tell apart slightly more blurred edges (*dashed line*) in comparison to this base blur is quantified by psychophysical methods

tested for blur *discrimination* thresholds, i.e. the ability of the observer to tell apart two slightly different amounts of edge blur. As shown in Fig. 1.12, first a base blur (solid curve) is chosen, and the ability of the observer to tell apart slightly more blurred edges (dashed line) in comparison to this base blur is quantified by psychophysical methods. Figure 1.13 compares model predictions and data from [55] for the effect of exposure duration on blur discrimination thresholds. For both blur perception and discrimination, one

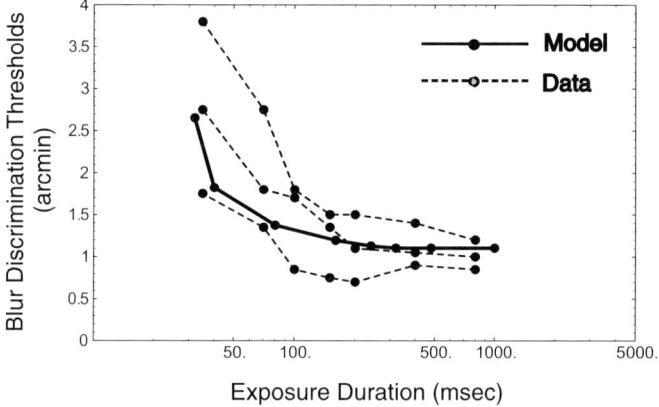

Fig. 1.13. Model predictions (*solid line*) and data (*dashed lines*) of three observers from [55] for blur discrimination threshold as a function of exposure duration. From [44]

observes that an exposure duration on the order of 100 ms is required to reach veridical perception and optimal discrimination of edge blur, and that a good agreement between experimental data and model predictions is found.

Figure 1.14 compares model predictions and data for blur discrimination as a function of base blur. Discrimination thresholds follow a U-shaped function with a minimum value around 1 arcmin. The optics of the eye limits performance for base blurs less than 1 arcmin. For base blurs larger than 1 arcmin, neural factors limit performance.

1.3.4 On and Off Pathways and Edge Localization

Receptive fields of retinal ganglion cells can also be classified as on-center off-surround (Fig. 1.15, left) and off-center on-surround (Fig. 1.15, right). These receptive fields contain two concentric circular regions, called the center and the surround. If a stimulus placed in the center of the receptive field excites the neuron, then a stimulus placed in the surround will inhibit the neuron. Thus the center and the surround of the receptive field have *antagonistic* effects on the neuron. A receptive field whose center is excitatory is called on-center off-surround. Similarly, a receptive field whose center is inhibitory is called off-center on-surround. The outputs of the on-center off-surround cells give rise to the on pathway, and the outputs of the off-center on-surround cells

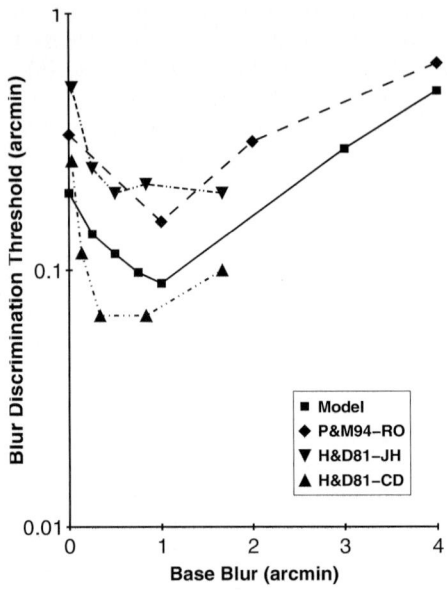

Fig. 1.14. Model predictions and data from [26] (for observers JH and CD) and from [40] (for observer RO) plotting blur discrimination thresholds as a function of base blur. From [44]

give rise to the off pathway. Because the spatial integration of inputs for the P cells is linear, the signals generated by an edge in the on and off pathways will exhibit an odd-symmetry; and their point of balance would correspond to the location of the edge. It has been shown that a contrast-dependent asymmetry exists between the on and off pathways in the human visual system [53]. An implication of this asymmetry is that, if edges are localized based on a comparison of activities in the on and off channels then a systematic mislocalization of the edge should be observed as the contrast of the edge is increased. Indeed, Bex and Edgar [5] showed that the perceived location of an edge shifts towards the darker side of the edge as the contrast is increased. Their data are shown in Fig. 1.16. Negative values on the y-axis indicate that the perceived edge location is shifted towards the darker side of the edge. For a sharp edge (0 arcmin blur), no mislocalization is observed for contrasts ranging from 0.1 to 0.55. However, as the edge blur is increased a systematic shift towards the darker side of the edge is observed. To estimate quantitatively this effect in the model, we introduced an off pathway whose activities consisted of negatively scaled version of the activities in the on pathway. This scaling took into account the aforementioned asymmetry. As a result, as contrast is increased above approximately 0.2, the activities in the off pathway increased slightly more than those in the on pathway. The quantitative predictions of

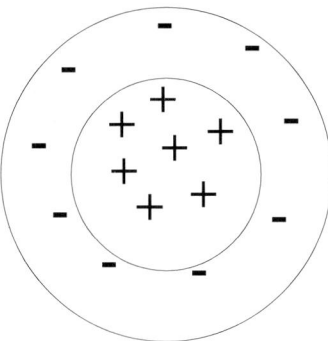 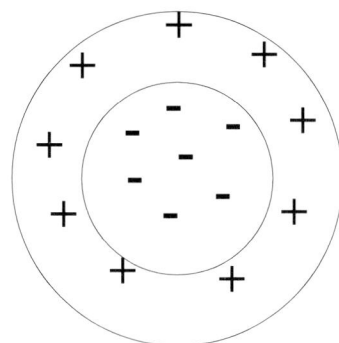

Fig. 1.15. *Left*: On-center off-surround receptive field; *right*: off-center on-surround receptive field. *Plus and minus symbols* indicate excitatory and inhibitory regions of the receptive field, respectively

the model are superimposed on the data in Fig. 1.16. Overall, one can see a good quantitative agreement between the model and the data.

1.3.5 Trade-off Between Spatial and Temporal Deblurring

The aforementioned simulations studied model behavior under the conditions of visual fixation for a static boundary, i.e. when the position of the boundary remains fixed over retinotopic maps. Under these conditions, feedforward retino-cortical signals send blurred boundary information, and gradually post-retinal feedback signals become dominant and construct sharpened representation of boundaries. However, because post-retinal signalling involves positive feedback, at least two major problems need to be taken into consideration:

1) When the positive feedback signals become dominant, the system loses its sensitivity to changes in the input. For example, if the input moves spatially, the signals at the previous location of the input will persist through positive feedback loops and the resulting perception would be highly smeared, similar to pictures of moving objects taken by a camera at long exposure duration. Thus, within a single pathway spatial sharpening comes at the cost of temporal blurring.

2) If left uncontrolled, positive feedback can make the system unstable.

We suggest that the complementary magnocellular pathway solves these problems by rapidly "resetting" the parts of retinotopic map where changes in the input are registered. Accordingly, the real-time operation of the RECOD model unfolds in three phases:

(i) Reset phase: Assume that the post-retinal network has some residual persistent activity due to a previous input. When a new input is applied to

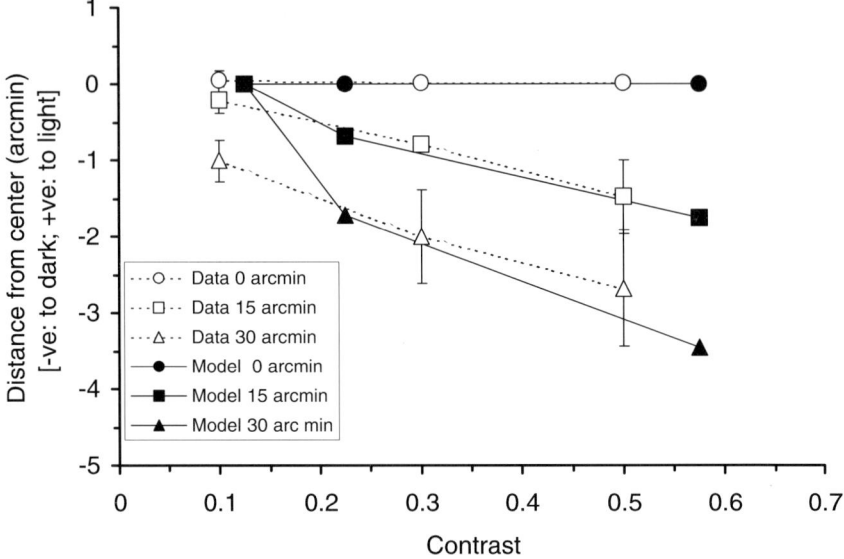

Fig. 1.16. Model predictions and data showing the effect of contrast on the perceived mislocalization of edges with different amounts of blur. The data points are digitized from [5] and represent the mean and the standard error of the mean computed from two observers. From [44]

the RECOD model, the fast-transient neurons respond first. This transient activity inhibits the post-retinal network and removes the persisting residual activity.

(ii) Feedforward dominant phase: The slow-sustained neurons respond next to the applied input and drive the post-retinal network with excitatory inputs.

(iii) Feedback dominant phase: When the activity of the sustained neurons decays from their peak to a plateau, the feedback becomes dominant compared to the sustained feedforward input. This results in the sharpening of the input spatial pattern. Thus, the feedforward reset mode achieves temporal deblurring, and the feedback mode achieves spatial deblurring.

According to the three-phase operation of the model, a single continuous presentation of a blurred edge is necessary for the feedback to sufficiently sharpen the neural image across the retinotopic map. Multiple short exposures cannot achieve the same amount of sharpening as a single long exposure since the post-retinal feedback is reset by the retinal transients. Westheimer [55] measured blur discrimination thresholds for an edge whose blur was temporally modulated in different ways. The reference stimulus was a sharp edge. In the first experiment, the test stimulus was a blurred edge presented alone for durations of 30 ms and 130 ms. Next, the test stimulus was presented as a combination of (i) a sharp edge for 100 ms and a blurred edge for the next 30 ms, (ii) a blurred edge for the first 30 ms and a sharp edge for the next 30 ms, and (iii) a blurred edge for 100 ms and a sharp edge for the next 100 ms. As shown in Table 1, the RECOD model predicts lower differences in the luminance gradients between the test and reference stimuli for conditions (i) and (ii) above than for a 30 ms presentation of a blurred edge. This gives higher blur discrimination thresholds. Similarly, condition (iii) above yields a lower difference in the luminance gradients between the test and reference stimuli than when the test stimuli is a blurred edge presented for 130 ms.

Table 1.1. Model and data from Westheimer [55] for blur discrimination thresholds (arcmin) obtained with hybrid presentations

	30 ms	130 ms	(i)	(ii)	(iii)
Data	3.8	1.43	7.17	8.56	2.06
Model	2.6	1.2	5.33	5.33	1.44

1.3.6 Perceived Blur for Moving Stimuli

Another way to test the proposed reset phase is to compare model predictions with data on the perception of blur for *moving* stimuli. In normal viewing conditions, moving objects do not appear blurred. Psychophysical studies showed that perceived blur for moving objects depends critically on the exposure duration of stimuli. For example, moving targets appear less blurred than predicted from the visual persistence of static targets when the exposure duration is longer than about 40 ms [10, 28]. This reduction of perceived blur for moving targets was named "motion deblurring" [10].

Model predictions for motion deblurring were tested using a "two-dot paradigm", where the stimulus consisted of two horizontally separated dots moving in the horizontal direction, as shown in the top panel of Fig. 1.17. The middle panel of the figure shows a space-time diagram of the dots' trajectories. The afferent short-latency-transient and long-latency-sustained signals are depicted in the bottom panel of Fig. 1.17 by dashed lines and the gray region, respectively. The sustained activity corresponding to both dots are highly spread over space. However, at the post-retinal level, the interaction between the transient activity generated by the trailing dot and the sustained activity generated by the leading dot results in a substantial decrease of the spatial spread of the activity generated by the leading dot. From Fig. 1.17, one can see that the exposure duration needs to be long enough for the transient activity conveyed by the magnocellular pathway for the trailing dot to spatiotemporally overlap with the sustained activity conveyed by the parvocellular pathway for the leading dot.

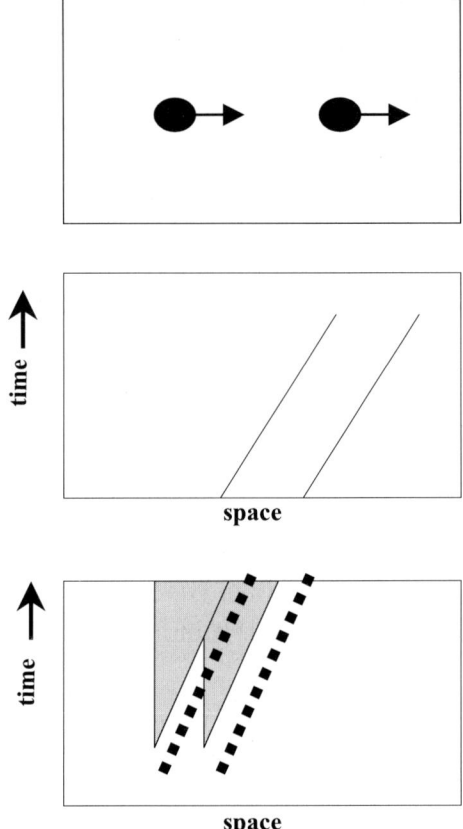

Fig. 1.17. *Top*: Two-dimensional representation of the input. *Arrows* indicate motion. *Middle*: spatiotemporal representation of the input. *Bottom*: superimposed afferent transient and sustained signals

In order to compare model predictions quantitatively with data, Fig. 1.18 plots the duration of perceived blur (calculated as the ratio of the length of perceived blur to the speed) for the leading and the trailing dot, respectively, for two dot-to-dot separations along with the corresponding experimental data [14].

In all cases, when the exposure duration is shorter than 60 msec, no significant reduction of blur is observed and the curves for the leading and trailing dots for both separations largely overlap. The mechanistic explanation of this effect in our model is as follows: due to the relative delay between transient and sustained activities, no spatial overlap is produced when the exposure duration is short. When the moving dots are exposed for a longer duration,

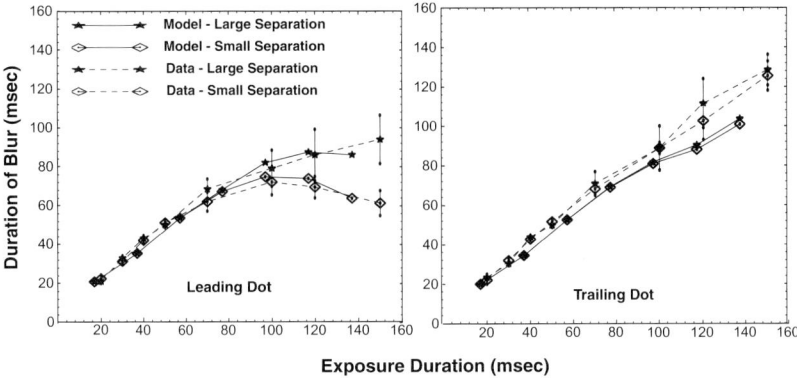

Fig. 1.18. Duration of blur as a function of exposure duration for the leading (left) and trailing (right) dots in the two-dot paradigm for two dot-to-dot separations. From [43]

these two activities overlap and the inhibitory effect of the transient activity on the sustained one reduces the persistent activity from the leading dot. A significant reduction of perceived blur is observed for the leading dot when the dot-to-dot distance is small both in the model and in data. When the dot-to-dot separation is larger, the spatiotemporal overlap of transient and sustained activities is reduced, thereby decreasing the effect of deblurring in agreement with data (Fig. 1.18). For the trailing dot, dot-to-dot separation has no effect on post-retinal activities, and no significant reduction in perceived blur is observed. Quantitatively, the model is in very good agreement with data with the exception of some underestimation for long exposure duration in the case of the trailing dot.

1.3.7 Dynamic Viewing as a Succession of Transient Regimes

Under normal viewing conditions, our eyes move from one fixation point to another, remaining at each fixation for a few hundred milliseconds. Our studies show that a few hundred milliseconds is the time required to attain an "optimal" encoding of object boundaries (Figs. 1.11, 1.13, and 1.18). Therefore, the timing of eye movements correlates well with the timing of boundary analysis. We also suggest that these frequent changes in gaze help the visual system remain mainly at its transient regime and thus avoid unstable behavior that would otherwise result from extensive positive feedback loops observed in the post-retinal areas. Within our theoretical framework, the visual and the oculomotor system together "reset" the activities in the positive feedback loops by using the inhibitory fast transient signals originating from the magnocellular pathway.

1.3.8 Trade-off Between Reset and Persistence

If the system is reset by exogenous signals, as suggested above, one needs to consider the problem that may arise because of internal noise: internal noise in the M pathway could cause frequent resets of information processing in areas that compute object boundaries and form. In addition, such rapid undesirable reset cycles may also occur because of small involuntary eye movements as well as because of small changes in the inputs. We suggest that the inhibition from the P-driven system on the M-driven system prevents these resets through a competition between the two systems (see Fig. 1.5). In our simulations reported in the previous sections, for simplicity we did not include sustained on transient inhibition, for both the inputs and the neural activities were noise-free. The proposed competition between the M-driven and the P-driven systems can be tested by using stimuli that activate successively in time spatially nonoverlapping but adjacent regions. The perceptual correlates for such stimuli have been studied extensively in the masking literature [3, 6, 8]. If we label the stimulus whose perceptual and/or motor effects are measured as the "target" stimulus and the other stimulus as the "mask" stimulus (Fig. 1.19), then the condition where the mask is presented in time before the target is called paracontrast. The condition where the mask is presented after the target is called metacontrast [3, 6, 8]. Based on a broad range of masking

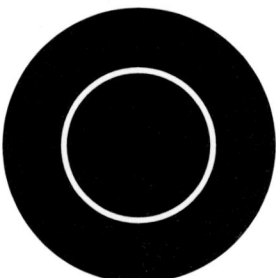

Fig. 1.19. A typical stimulus configuration used in masking experiments. The central disk serves as the target stimulus and the surrounding ring serves as the mask stimulus

data Breitmeyer [7, 6] proposed reciprocal inhibition between sustained and transient channels, and this reciprocal inhibition is also an essential part of the RECOD model. Consider metacontrast: here the aftercoming mask would reset the activity related to the processing of the target. Indeed, a typical metacontrast function is a U-shaped function suggesting that the maximum suppression of target processing occurs when the mask is delayed so that the fast transient activity generated by the mask overlaps in time with the slower sustained activity generated by the target. If the transient activity generated by the mask can be suppressed by sustained activity, then it should be possible to introduce a second mask (Fig. 1.20) whose sustained activity can suppress the transient activity of the primary mask. This in turn results in the disinhibition of the target stimulus. In support of this prediction, several studies

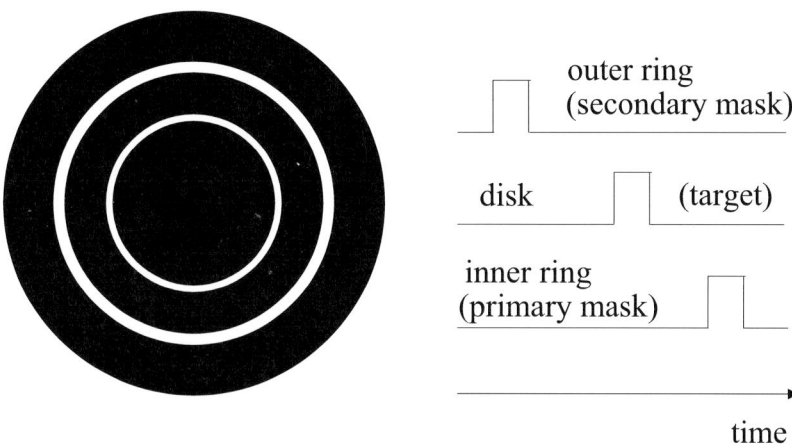

Fig. 1.20. *Left*: modification of the stimulus configuration shown in Fig. 1.19. The second outer ring serves as the secondary mask. *Right*: The temporal order of the stimuli

showed that the second mask allows the recovery of an otherwise suppressed target (e.g. [17]). Furthermore, Breitmeyer et al. [9] showed that the effect of the secondary mask in producing the disinhibition (or recovery) of the target starts when it is presented at about 180 ms prior to the target and gradually increases until it becomes simultaneous with the primary mask. This relatively long range of target recovery provides a time window during which sustained mechanisms can exert their inhibitory influence so as to prevent reset signals generated by noise.

1.3.9 Attention: Real-time Modulation of the Balance Between Reset and Persistence

Having a mechanism to reduce reset signals opens another possibility: modulatory mechanisms can bias the competition in favor of the sustained mechanisms and thereby allow a more persistent and enhanced registration and analysis of stimuli. We suggest that attention serves that purpose. Although a universally adopted definition of attention does not exist, it is often defined as a selection mechanism whereby resources are focused on certain item(s), location(s), etc. Within the framework of the RECOD model, the reset mechanism curtails cortical activity and therefore attention necessitates a reduction of the reset signals for the attended locations, features, objects, and so on. Similarly, attention can also increase the gain of reset signals for unattended locations and objects. A simple way to achieve this in RECOD is to bias the competition between transient and sustained systems in favor of the sustained system for attended locations, features, and objects; and bias the competition in favor of the transient system for unattended locations, features, and objects. For example, assume that attention primes part of the retinotopic map as illustrated in Fig. 1.21. The model then predicts in agreement with experimental data that attention should increase visible persistence [54], decrease temporal sensitivity [57], increase spatial sensitivity [56], and decrease masking [21, 45, 48]. Similarly, it is predicted that attention should enhance

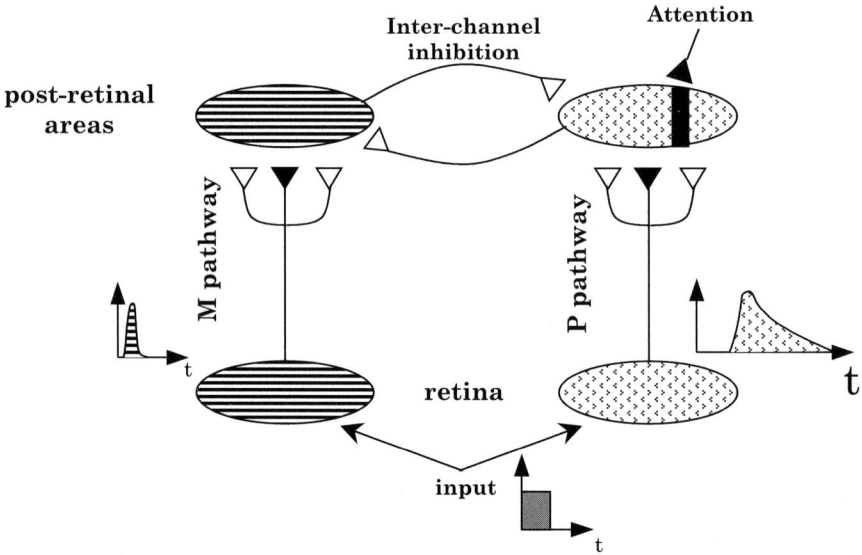

Fig. 1.21. Illustration of attention in RECOD. Priming the activation of the cells in the P pathway biases the competition between sustained and transient systems in favor of the sustained system

target recovery, should increase reaction times to a target in paracontrast, and increase motion blur. These predictions have not been tested.

1.4 Summary

In this chapter we reviewed some fundamental properties of the primate visual system and highlighted maps and pathways as spatiotemporal information encoding and processing strategies. We suggest that maps represent the geometry of the fusion between structure and function in the nervous system, and that the pathways represent complementary aspects of processing whose interactions can solve conflicting requirements arising within a single processing stream. The use of retinotopic and spatial-frequency maps was illustrated by considering the problem of object boundary encoding. The use of parallel, complementary pathways was illustrated by considering how the interactions between magnocellular and parvocellular pathways can resolve the trade-off between spatial and temporal deblurring. We suggested that the interactions between magnocellular and parvocellular pathways play a fundamental role in keeping the system in a succession of transient regimes, thereby avoiding unstable behavior that would result from complex feedback loops that include extensive positive feedback. Finally, we suggested that attention can be viewed as a modulation of the dynamic balance between sustained and transient systems.

Appendix: Fundamental Equations of the Model and Their Neurophysiological Bases

The first type of equation used in the model has the form of a generic Hodgkin–Huxley equation:

$$\frac{dV_m}{dt} = -(E_p + V_m)g_p + (E_d - V_m)g_d - (E_h + V_m)g_h, \tag{1.1}$$

where V_m represents the membrane potential; g_p, g_d, g_h are the conductances for passive, depolarizing, and hyperpolarizing channels, respectively; with E_p, E_d, E_h representing their Nernst potentials. This equation has been used extensively in neural modeling to characterize the dynamics of membrane patches, single cells, as well as networks of cells (rev. [25, 31]). For simplicity, we will assume $E_p = 0$ and use the symbols B, D, and A for E_d, E_h, g_p, respectively, to obtain the generic form for multiplicative or shunting equation (rev. [25]):

$$\frac{dV_m}{dt} = -AV_m + (B - V_m)g_d - (D + V_m)g_h. \tag{1.2}$$

The depolarizing and hyperpolarizing conductances are used to represent the excitatory and inhibitory inputs, respectively. The second type of equation is a simplified version of Eq. (2), called the additive model, or the leaky-integrator model, where the external inputs influence the activity of the cell not through conductance changes but directly as depolarizing I_d and hyperpolarizing I_h currents yielding the form:

$$\frac{dV_\mathrm{m}}{dt} = -AV_\mathrm{m} + I_\mathrm{d} - I_\mathrm{h}. \tag{1.3}$$

Mathematical analyses showed that, with appropriate connectivity patterns, shunting networks can automatically adjust their dynamic range to process small and large inputs (rev. [25]). Accordingly, we use shunting equations when we have interactions among a large number of neurons so that a given neuron can maintain its sensitivity to a small subset of its inputs without running into saturation when a large number of inputs become active. We use the simplified additive equations when the interactions involve few neurons. Finally, a third type of equation is used to express biochemical reactions of the form

$$S + Z \rightarrow Y \rightarrow X \rightarrow S + Z, \tag{1.4}$$

where a biochemical agent S, activated by the input, interacts with a transducing agent Z (e.g. a neurotransmitter) to produce an active complex Y that carries the signal to the next processing stage. This active complex decays to an inactive state X, which in turn dissociates back into S and Z. It can be shown that (see Appendix in Sarikaya et al. [47]), when the active state X decays very fast, the dynamics of this system can be written as:

$$\frac{1}{\tau}\frac{dz}{dt} = \alpha(\beta - z)\gamma S z, \tag{1.5}$$

with the output given by $y(t) = \frac{\gamma}{\delta} S(t) z(t)$, where s, z, y represent the concentrations of S, Z, and Y, respectively, and γ, δ, α denote rates of complex formation, decay to inactive state, and dissociation, respectively. This equation has been used in a variety of neural models, in particular to represent temporal adaptation, or gain control property, occurring, for example, through synaptic depression (e.g. [1, 13, 22, 24, 37, 38]).

Acknowledgements
This study is supported by NIH grant R01–MH49892.

References

1. Abbott L. F., Varela K., Sen K., Nelson S.B. (1997) Synaptic depression and cortical gain control. *Science* **275**:220–223

2. Albright T.D., Desimone R., Gross C.G. (1984) Columnar organization of directionally selective cells in visual area MT of the macaque. *J. Neurophysiol.* **51**:16–31
3. Bachmann T. (1994) *Psychophysiology of Visual Masking: The Fine Structure of Conscious Experience.* Nova Science, New York
4. Baron M., Westheimer, G. (1973) Visual acuity as a function of exposure duration. *J. Opt. Soc. Am.* **63**:212–219
5. Bex P.J., Edgar G.K. (1996) Shifts in perceived location of a blurred edge increase with contrast. *Perception and Psychophysics* **58**:31–33
6. Breitmeyer B.G. (1984) *Visual masking: An Integrative Approach.* Oxford University Press, Oxford
7. Breitmeyer B.G., Ganz, L. (1976) Implications of sustained and transient channels for theories of visual pattern masking, saccadic suppression, and information processing. *Psychological Rev.* **83**:1–36
8. Breitmeyer B.G., Öğmen H. (2000) Recent models and findings in visual backward masking: A comparison, review, and update. *Perception and Psychophysics* **62**:1572–1595
9. Breitmeyer B.G., Rudd M., Dunn K. (1981) Metacontrast investigations of sustained-transient channel inhibitory interactions. *J. of Exp. Psych: Human Perception and Performance* **7**:770–779
10. Burr D. (1980) Motion smear. *Nature* **284**:164–165
11. Burr D.C., Morgan, M.J. (1997) Motion deblurring in human vision. *Proc. R. Soc. Lond. B* **264**:431–436
12. Carandini M., Heeger D.J. (1994) Summation and division by neurons in primate visual cortex. *Science* **264**:1333–1336
13. Carpenter G. A., Grossberg S. (1981) Adaptation and transmitter gating in vertebrate photoreceptors. *J. of Theor. Neurobiology* **1**:1–42
14. Chen S., Bedell H.E., Öğmen H. (1995) A target in real motion appears blurred in the absence of other proximal moving targets. *Vision Res.* **35**:2315–2328
15. Croner L.J., Kaplan E. (1995) Receptive fields of P and M ganglion cells across the primate retina. *Vision Res.* **35**:7–24
16. De Monasterio F.M. (1978) Properties of concentrically organized X and Y ganglion cells of macaque retina. *J. Neurophysiol.* **41**:1394–1417
17. Dember W.N., Purcell D.G. (1967) Recovery of masked visual targets by inhibition of the masking stimulus. *Science* **157**:1335–1336
18. De Valois K.K. (1977) Spatial frequency adaptation can enhance contrast sensitivity. *Vision Res.* **17**:209–215
19. De Valois R.L., De Valois K.K. (1990) *Spatial Vision.* Oxford University Press, New York
20. De Valois K.K., Switkes E. (1980) Spatial frequency specific interaction of dot patterns and gratings. *Proc. Nat. Acad. Sci. USA* **77**:662–665
21. Enns J.T., DiLollo V. (1997) Object substitution: A new form of masking in unattended visual locations. *Psychological Science* **8**:135–139
22. Gaudiano P. (1992) A unified neural network of spatio-temporal processing in X and Y retinal ganglion cells. 2: Temporal adaptation and simulation of experimental data. *Biol. Cybern.* **67**:23–34
23. Georgeson M.A. (1994) From filters to features: location, orientation, contrast and blur. *CIBA Foundation Symposia* **184**:147–169
24. Grossberg S. (1972) A neural theory of punishment and avoidance, II: Quantitative theory. *Mathematical Biosciences* **15**:253–285

25. Grossberg S. (1988) Nonlinear neural networks: Principles, mechanisms and architectures. *Neural Networks* **1**:17–61
26. Hamerly J.R., Dvorak, C.A. (1981) Detection and discrimination of blur in edges and lines. *J. Opt. Soc. Am.* **71**:448–452
27. Hendry S.H.C., Reid, R.C. (2000) The koniocellular pathway in primate vision. *Annu. Rev. Neurosci.* **23**:127–153
28. Hogben J.H., Di Lollo V. (1985) Suppression of visible persistence in apparent motion. *Perception and Psychophysics* **38**:450–460
29. Hood D. (1973) The effects of edge sharpness and exposure duration on detection threshold. *Vision Res.* **13**:759–766
30. Hubel D.H., Wiesel T.N. (1968) Receptive fields and functional architecture of monkey striate cortex. *J. Physiol. London* **195**:215–243
31. Koch C., Segev I. (1989) *Methods in Neuronal Modeling.* MIT Press, Cambridge, MA
32. Livingstone M., Hubel, D. (1988) Segregation of form, color, movement, and depth: Anatomy, physiology, and perception. *Science* **240**:740–749
33. Maunsell J.H.R., Gibson J.R. (1992) Visual response latencies in striate cortex of the macaque monkey. *J. Neurophysiol.* **68**:1332–1344
34. Milner A.D., Goodale M.A. (1995) *The Visual Brain in Action.* Oxford University Press, Oxford
35. Mishkin M., Ungerleider L.G., Macko, K.A. (1983) Object vision and spatial vision: Two cortical pathways. *Trends in Neurosciences* **6**:414–417
36. Nowak L.G., Munk M.H.J., Girard P., Bullier J. (1995) Visual latencies in areas V1 and V2 of the macaque monkey. *Visual Neuroscience* **12**:371–384
37. Öğmen H. (1993) A neural theory of retino-cortical dynamics. *Neural Networks* **6**:245–273
38. Öğmen H., Gagné S. (1990) Neural models for sustained and on-off units of insect lamina. *Biol. Cybern.* **63**:51–60
39. Öğmen H., Breitmeyer B.G., Melvin R. (2003) The what and where in visual masking. *Vision Res.* **43**:1337–1350
40. Pääkkönen A.K., Morgan M.J. (1994) Effect of motion on blur discrimination. *J. Opt. Soc. Am. A* **11**:992–1002
41. Pizlo Z. (2001) Perception viewed as an inverse problem. *Vision Res.* **41**:3145–3161
42. Purpura K., Tranchina D., Kaplan E., Shapley R.M. (1990) Light adaptation in primate retina: Analysis of changes in gain and dynamics of monkey retinal ganglion cells. *Visual Neuroscience* **4**:75–93
43. Purushothaman G., Öğmen H., Chen S., Bedell H.E. (1998) Motion deblurring in a neural network model of retino-cortical dynamics. *Vision Res.* **38**:1827–1842
44. Purushothaman G., Lacassagne D., Bedell H.E., Öğmen H. (2002) Effect of exposure duration, contrast, and base blur on coding and discrimination of edges. *Spatial Vision* **15**:341–376
45. Ramachandran V.S., Cobb S. (1995) Visual attention modulates metacontrast masking. *Nature* **373**:66–68
46. Salin P.-A., and Bullier J. (1995) Corticocortical connections in the visual system: structure and function. *Physiological Reviews* **75**:107–154
47. Sarikaya M., Wang W., Öğmen H. (1998) Neural network model of on-off units in the fly visual system: simulations of dynamic behavior. *Biol. Cybern.* **78**:399–412

48. Shelley-Tremblay J., Mack A. (1999) Metacontrast masking and attention. *Psychological Science* **10**:508–515
49. Stecher S., Sigel C., Lange R.V. (1973) Composite adaptation and spatial frequency interactions. *Vision Res.* **13**:2527–2531
50. Tolhurst D.J. (1972) Adaptation to square-wave gratings: Inhibition between spatial frequency channels in the human visual system. *J. Physiol.* **226**:231–248
51. Tootell R.B.H., Silverman M.S., De Valois R.L. (1981) Spatial frequency columns in primary visual cortex. *Science* **214**:813–815
52. Tootell R.B.H., Silverman M.S., Hamilton S.L., De Valois R.L., Switkes E. (1988) Functional anatomy of macaque striate visual cortex. 3. Color. *J. Neurosci.* **8**:1569–1593
53. Virsu V., Laurinen, P. (1977) Long-lasting afterimages caused by neural adaptation. *Vision Res.* **17**:853–860
54. Visser T.A., Enns J.T. (2001) The role of attention in temporal integration. *Perception* **30**:135–145
55. Westheimer G. (1991) Sharpness discrimination for foveal targets. *J. Opt. Soc. Am.* **8**:681–685
56. Yeshurun Y., Carrasco M. (1998) Attention improves or impairs visual performance by enhancing spatial resolution. *Nature* **396**:72–75
57. Yeshurun Y., Levy L. (2003) Transient spatial attention degrades temporal resolution. *Psychological Science* **14**:225–231
58. Youping X., Yi W., Felleman D.J. (2003) A spatially organized representation of colour in macaque cortical area V2. *Nature* **421**:535–539
59. Zeki S. (1997) The color and motion systems as guides to conscious visual perception. *Cerebral Cortex* **12**:777–809

2

Symmetry, Features, and Information

Hamid R. Eghbalnia,[1] Amir Assadi,[1] Jim Townsend[2]

(1) Department of Mathematics - University of Wisconsin-Madison
    480 Lincoln Dr., Madison, WI 53706
    `eghbalni@nmrfam.wisc.edu`, `ahassadi@facstaff.wisc.edu`
(2) Department of Psychology - Indiana University
    Bloomington, IN 47405
    jtownsen@indiana.edu

To Bill Browder on His Birthday
With Admiration and Friendship

2.1 Introduction

There is growing evidence that a number of problems in perception and perceptual geometry, for example, the problems of figure-ground separation and scene segmentation could be formulated in terms of structural regularity of regions of images in statistical and information theoretic terms. Intuitively, as well as in psychophysical studies performed by cognitive scientists, perception of local structural regularity is fundamentally correlated with perception of local symmetry of surfaces, and under parallel projection of planar surfaces, with local symmetries of their images [50, 14, 37]. In other words, such local symmetries distinguish prevalent regularity of common surfaces in the environment from randomness in arbitrary composition of colored dots; or what is the same, they distinguish meaningful images from a generic pattern of a totally random selection of light intensities in matrices encoding local incoherence in optical properties. Are there features that encode, in a sparse format, such local regularities?

Consider being given an image described by a set of values at a discrete set of points on a finite grid. The task is to find some specific "feature" of the image and to do so in a finite amount of time and with finite computational resources. For example, if the image is that of a rug, the task may be the determination of whether the rug pattern has any symmetric structure. Or, the task may simply be recognition of whether an object in the image is a face or not. This search task may need multiple queries or "looks" at the image in order to obtain pertinent information. What is the pertinent or right structure that needs to be learned, what is the learning? What is to be learned [46]?

Shepherd [38] argues that principles of invariance and elegance are natural principles for formulation of representations of biologically significant objects. Using arguments based on symmetry principles, Palmer [33] develops the "isomorphism constraint", a principled distinction between what can and cannot be determined about the nature of color experience by objective behavioral means. From a mathematical point of view, it can be shown that in the space of all possible patterns of light (i.e. all large matrices of same size with non-negative coefficients), the set of possible images of natural scenes is a very small subset.

Symmetry is widely believed to be an important visual primitive that is probably encoded without the need for attention. The definition of effortless perception proposed by Julesz [20] states that any stimulus property perceived for exposure durations of 160 msec or less is detected preattentively. This notion of effortless perception has contributed greatly to views regarding preattentive symmetry detection. However, there is ample psychophysical data to support the view that a number of symmetry detection tasks require selective attention and may be spatially imprecise, although some grouping or segmentation tasks may operate preattentively [32, 50, 18]. A number of computational approaches to symmetry detection have been proposed. An overview of the various approaches, a discussion of symmetry groups in the context of crystallographic groups, and many references to the computational literature can be found in Liu [24].

Symmetry as a representation tool can take on a central role in addressing perceptual processes. The channel theory in processing of early parts of visual input can be seen as an example of this role. Fourier transform is the most widely known form of symmetry operation arising from group representation. In this new role, symmetry principles are the driving force in obtaining mathematical models for filters (or channels) as well as describing how their information content should be analyzed. In this chapter, we will motivate our approach in the first section and give a brief review of the structure of data flow in the human visual system and the psycophysical evidence for the possible forms of representations used for search tasks. Inspired by the prominent role of symmetry in the description of our physical world as well as its perception through our cognitive processes, our goal in the next section is to use symmetry in a systematic way to formulate the representation and detection of learned signals. This formulation leads to the notion of overcomplete representations [9]. We call this formulation a dynamic search and focus our attention on the central problem of addressing the mathematical formulation for the machinery necessary to solve the dynamic search problem. Next, we describe a measure for the information content of the signals that accounts for the symmetries of our dynamic search representation. We coin the terminology information dynamics, referring to our formulation of search that exploits the idea of overcomplete representations. In the final section, we give a brief overview of the computational implementation and results for the proposed model. A more detailed discussion of the model and the results is found in [7].

2.2 Motivations From the Visual System

The current weight of experimental evidence to support a high-fidelity internal representation for search tasks is at best equivocal. What appears to be a dominant factor is what is referred to as "features". Furthermore, search tasks appear to be segmented to solve simpler problems first, rejecting features irrelevant to the tasks and seeking information for features relevant to the task. If search tasks are based on finding features that involve a decision based on the tasks at hand, in what way are these features compared? It seems likely that a task-dependent feature representation map may exist. What is a likely model of feature comparison? Clues from the theory of guided search combined with the ideas of search for similarity provide the motivation for mathematical models and computational methods presented in the forthcoming sections.

2.2.1 A Quick Review of Retinal Data Flow

The structure of the primate retina is highly inhomogeneous with an extremely high density of receptor and ganglion cells in the center, a specialized fovea, and a rapid decline of the cell densities to the periphery. The lattice spacing of ganglion cell receptive field centers limits certain forms of visual acuity, according to the Nyquist limit. Spatial resolution of ganglion cells and the spacing of receptive field centers appear approximately equal. Given this observed relationship, the spatial low-pass filtering action of ganglion cell receptive fields would appear to attenuate periodic patterns at or above the Nyquist limit by about a factor of five or more, thus reducing the likelihood of detecting potentially confusing higher spatial frequencies with potential for aliasing. However, small offsets or stimulus movements much finer than any retinal cell mosaic spacing can be detected by ganglion cells. This is referred to as visual hyperacuity.

Ganglion cell fields partially overlap in visual space. Moreover, ganglion cells share an overlapping neural substrate composed of retinal interneurons and circuitry. As a result, the physiological signals they transmit are often highly correlated among neighboring cells. Retinal ganglion cells have no true threshold for detection of dim stimuli [3]. When many responses are averaged, signals can be seen for light stimuli as dim as desired, and this includes stimuli so dim that only a few physical quanta of light are delivered to the surface of the eye at the cornea. Of course in total darkness retinal ganglion cells still exhibit a maintained but variable spontaneous firing rate, and it is against this background noise that quantal responses must be detected. Barlow et al. [3] concluded that a single quantum absorption resulted in the firing of 2 or 3 extra ganglion cell nerve impulses. Because of fluctuations in spontaneous background firing, single quantal events might not be readily detected in a single cell. However 3-4 quantal absorptions within the cell's receptive field would be easily noticed in the discharge pattern.

Ganglion cell axons terminate in brain visual centers, principally the lateral geniculate nucleus (LGN) and the superior colliculus. The LGN neurons mainly project to the primary visual cortex, also known as area V1. There is a qualitative difference between neurons in the LGN and those of area V1. For example, while most V1 neurons do not have circularly symmetric receptive fields, most LGN neurons do. Cortical neurons mainly fall into two major categories, simple and complex cells (see Fig. 2.1.).

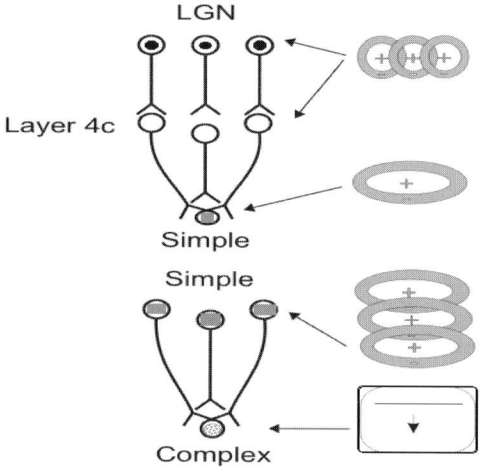

Fig. 2.1. A cartoon presentation of simple and complex cells and their receptive fields.

The sampling of the visual field in V1 is known to be sparse with approximately 5-6 samples of scale and 16–18 samples of orientation spread in a 100-by-100 hypercolumn. The so-called pinwheel structures arise from the attempt to map the external world onto the two-dimensional surface of the cortex parameterized by scale and orientation [34]. This structure of the retina suggests that moving the fovea to different positions is a requirement of having a homogeneous and simultaneous percept of the total visual field. This movement should provide the means for successive "looks" used for acquiring and integrating information. The existence of a fovea requires both, eye movements and periods of fixation, as seen in human saccades. The close relationship of visual perception, cognition, movements of the retinal image, and eye movements has made the independent study of each field without the influence of other effects a difficult task.

2.2.2 Search Tasks

Yarbus [56] was among the first to demonstrate that eye movements reflect cognitive events. However, understanding the relationship between eye movements and cognitive events remained largely unexplored [49]. In his thesis, using the block-copying task, Pelz [35] demonstrated a number of findings that corroborated a number of prior experiments.

Subjects in Pelz's studies made frequent eye movements, returning to inspect the model pattern again and again while copying the eight colored blocks. In essence, they used eye movements to serialize the task into simpler subtasks that were executed sequentially. The constraints of the task and the subjects' common, stereotyped behavior led to a relatively small number of strategies used to copy each block. The modal strategy was the "model-pickup-model-drop" (MPMD) strategy [35]. The subject looked first to the model, then to the resource area (guiding the block pickup), returned gaze to the model, and finally on to the workspace to guide the block drop. Because subjects were given no direction on how to perform the task, other than to complete the copy as quickly as possible without making errors, it is important that subjects chose to complete the task by referring to the model so frequently. None of the subjects in the study used the alternative strategy of first locating and identifying several objects in the scene, then moving a number of blocks without fixating the model again. At least in the context of this study, subjects' use of temporary, task-specific visual representations suggested that vision may be much more "top-down" than was previously thought.

Pelz's work challenged the idea that the visual system is tasked to gather information for integration into a high-fidelity, general-purpose representation of the environment without regard to the immediate task. In comparison, in the classical view of visual perception (also embraced by traditional computer vision approaches (e.g., Marr [26]), planning and cognition was performed by referencing the internal representation. The frequent eye movements used by subjects in these experiments suggest that in real tasks, humans apparently maintain only sparse, transient representations of task-relevant information.

These experiments suggest that the role of perception may be to create descriptions that are relevant to the immediate task. To the extent that manipulations on a given block are largely independent of the information acquired in previous views, performance in this task suggests that it is unnecessary to construct an elaborate scene description to perform the task, and that there is only minimal processing of unattended information. These observations support the suggestion made previously that only minimal information about a scene is represented at any given time, and that the scene can be used as a kind of "external" memory [31, 30, 19] (see also related suggestions by Nakayama [28]).

Research in human visual search over the past 20 years has established three factors that are powerful determinants of search speed and accuracy.

These determinants are the number of objects to be searched for, the degree of target and background similarity, and training [16, 39, 53]. Targets that differ from their backgrounds by a single feature (e.g., a red circle among green circles) usually lead to "pop-out" in that search is fast and relatively independent of the number of distracter objects.

The pop-out effect is in contrast to searching for a conjunction of features (e.g., a red vertical line in a background of red horizontals and green verticals). This type of search is generally slower and more error-prone. Some researchers [42] have suggested that pop-out effects are mediated by preattentive, parallel stages in the visual system, while conjunction searches engage a covert visual attention system that is directed at each display item in turn. This is commonly referred to as feature integration theory. Feature integration theory has been amended to address such variations in search efficiency by suggesting that features are coarsely coded, that the attentional focus may vary in size, and that the representations of distractor features not shared with the target can be inhibited [43, 44, 45]. Independent of the validity of this theory, there is clear evidence that similarity of target and background exerts a powerful influence on search efficiency [53].

Another theory that attempts to account for variations in search efficiency is the guided search theory proposed by Wolfe and coworkers [4, 53, 54, 52]. According to this theory, during visual search, preattentive processes guide shifts of attention by pinpointing stimulus locations likely to contain the target. This preattentive information encompasses both "bottom-up" (i.e., how closely a particular item resembles other items in the display) and "top-down" (i.e., how closely the features of a given item match those characterizing the target) influences. In guided search theory, the sources of information combine to create an "activation map". The activation map contains peaks of activity at likely target locations. The activation map is used during the search process to focus attention on the stimulus location showing the most activity.

A common assumption for explaining the effects of target/background similarity as well as its interaction with the number of objects in the display and training, is based on the idea of a two stage search. The initial stage in visual search is a parallel stage that locates likely target candidates. In a second serial stage, attention is allocated to the target object resulting in its identification [16, 17, 53]. These models have been simulated in computer programs in simple settings and appear to provide a good fit to a variety of search data. However, more realistic situations in which search is accomplished with eye movements and observers must search through complex scenes have not been studied using these models.

The evidence discussed in this section motivates a view of visual processing in which short-term, task-specific, and sparse information plays an important role in natural human tasks. Top-down and bottom-up processing form a synergistic process that creates a map used to guide further visual processing. The sparse and transient nature of this map suggests a feature-based construct that may be coded at multiple scales. Coding the percept using features further suggests an invariant coding to avoid the computation of rotation deconvolution. A more detailed source of the material presented in this section can be found in [7].

2.3 Overview of the Model

To pursue the ideas of search in perceptual mechanisms expressed above, we need to propose a model that has as its main ingredients the element we have called features as well as an invariant measure for comparison of features. In our framework, the analysis of a signal proceeds in three steps that could be potentially coupled. The first stage involves the generation of an overcomplete representation of the signal. Our condition on the generation of the overcomplete representation is the existence of a parameterizing system that can be used to structure the resulting representation. Here, our parameterizing system is the Lie group of symmetries that parameterizes a set of wavelet coefficients obtained by analysis using a mother wavelet. More specifically, a search structure for an image is a collection of elements g belonging to the Lie group of transformations G of the image.[1] The search implementation applies these transformations g to a 2D mother wavelet (or filter), called the *initial search filter*, to commence the process of feature extraction. The role of dynamics is to select an ordering for search transformations g belonging to G. This ordering is dependent on the search objective. The ordering would ideally describe the time parametrization for the path in analogy with attention shifts in the primate visual system.

To select the set of initial filters, we have relied on a physical interpretation of a family of distributions known as Lorentz–Cauchy distributions. These distributions give rise to a family of filters that retain the optimal uncertainty bound under all transformations of the group of rotations, scaling and translations. This property turns out to be unique *within this class* of wavelets and offers an additional reason for selection of these wavelets as feature processors. These wavelets act as projection operators providing a set of coefficients that describe a multiscale and multiresolution, local-to-global version of the object or signal under consideration.

Another set of projection operators, called *decision operators*, is used to model the search objective. Decision operators act on the multiparameter signal obtained from the filtering stage to obtain task-specific feature information. One decision operator we focus on is based on detection of nondegenerate local extrema. We remark that ideally such an operator should be learned using methods of statistical learning theory using data obtained from experiments. The resulting features are points in a compact subset of the plane

---

[1] More generally, a search structure can be any parameterized family of probes where the parameterizing set has a structure with the required properties defined by the problem.

that are encoded in a normalized complex vector called the *feature*. Thus G, in combination with the decision operators, can be thought of as inducing a map from the signal s to a set of feature vectors parameterized by elements g in G. The variation, measured by the Fubini–Study metric on the resulting complex projective space of feature vectors [36], as g varies in G gives rise to *information dynamics*.

An unstructured search among the set of possible matches can lead to computationally intractable problems. Matching one face against a collection of faces at the finest level of detail requires a great deal of computational power. This would require evaluating some form of distance comparison in a large class of objects. Because of the variability of conditions in the collection of pictures, as well as variabilities in the picture to be matched, the comparison of results may be difficult to evaluate. A potential result could be that many faces fall within *"the same distance"* from the item to be matched. This collection of partial matches might not be particularly useful since it is not given any helpful structure. Is there a method of search that in the case of potential rejection, discriminates efficiently, while at the same time the novelty that led to the rejection is captured at its most robust form? A biologically plausible solution of the problem must deal with relevant constraints posed by the human visual system.

It is well known that from the earliest stage of the processing determined by the retinal mosaic, the system is subject to resolution limits. Additionally, the computational machinery of vision is highly evolved and is designed to be particularly efficient for tasks crucial to the ability of humans to reliably perform normal tasks in their environments. Consequently, there are limits imposed on both the computation time and the resolving power of the system for optimal decision. Therefore, it is reasonable to assume that our mechanism is constrained by a level of granularity and to seek solutions that are optimal with respect to time and resources for computation. Not only in the visual system, but also in other systems, such optimal search mechanisms are subject to comparable constraints. Accordingly, we introduce a granularity parameter during the extraction of features in order to accurately represent the limits of computation offered by this machinery.

The second step involves the action of a decision operator.[2] A decision operator is a task-specific projection operator that acts on the overcomplete representation to extract features of significance specific to the signal. We propose to use the nondegenerate zero sets of the derivative operator to construct the decision operator. By appealing to stability theorems [27] we guarantee that the resulting signal is discrete, isolated, and thus the problem now involves a finite set of feature points. We note that in the setting of a learning system the appropriate decision operator should be learned.

---

[2] In the language of cognitive science, our choice of a decision operator is a top-down process.

In the final step, we require that the features selected above interact with the learning system in order to either reject or accept a hypothesis defined by the task, or to identify specific needs for further information. Within the general framework of dynamics search, this step can be coupled to any learning algorithms. For example, support vector machines (SVM), multiscale entropy minimization, and principal component analysis (PCA) methods [1, 2, 47] can be utilized. Our approach is to define a space of features and the corresponding metric in the setting of complex projective spaces. Many methods of learning theory become immediately available to be applied to this problem. We apply learning theoretic estimates to perform PCA on the feature vectors in $\mathbb{C}P^n$. This is the complex analog of PCA that is normally performed in $\mathbb{R}^n$.

In the course of presenting this material we employ a number of results and conditions relevant to the task at hand. To find a suitable representation, we use a condition on the optimality of the overcomplete basis that arises from non-commutativity of operators. We also use the topological stability properties of transversality theory as developed by Thom, Sard, and Smale [27] to extract combinatorics in a signal in a form that can be used for encoding and representation. This discretization process applied to an ensemble of signals is therefore suitable for statistical analysis. To this end, we further transform the ensemble of combinatorial data by eliminating the effects of location, scale, and orientation of features. We call this the ensemble of standardized features. The ensemble of features has a natural metric that arises as a natural analog of the Fisher information metric. The existence of the metric along with the ensemble of data means that methods of statistics and learning theory are available for implementation and computation of parameters. In addition, we note that since our metric is defined in terms of an inner product, nonlinear kernel methods based on reproducing kernel Hilbert spaces (RHKS) can be brought to bear on the problems at hand.

2.4 The Structure of Light as a Signal

Filtering theory has historical ties to statistics and probability. In recent years, in the context of signal processing, other ties such as those to wavelet theory have been explored [41]. Through filtering theory, wavelets share some ties to statistics, and in some rough sense, wavelets can be viewed as filters that estimate statistics for various source signals. We start with a description for the source signal of our interest in terms of a random process and its Fourier transform and ask the following question: "what is the probability distribution of photon detection from a given source incident on a set of detectors ?" We will start with a 1D signal and present the 2D case as an extension of the 1D arguments. This approach establishes some properties of interest and opens the door for further analysis of filters of the following type:

Theorem 1. *There exists a 2D mother wavelet ψ, arising from a probability distribution $p(x, y)$ describing a physical process with the following properties:*

1. The restriction of the probability distribution to the principal coordinate axis is given by the Lorentz–Cauchy distribution that we denote by the LC distribution for short.
2. The probability distribution is related to the physical process of a source with emissions within an angular cone.
3. The characteristic function of the process is described by an exponential decay.
4. The mean has the distribution of any random variable of the process.

More specifically, the following propositions can be proven by direct computation.

Proposition 1. *Let O be a source located at point $(0,1)$ in the plane that emits particles with uniform probability at angle $\theta \in (-\pi/2, \pi/2)$ to the y-axis. Let $(t,0)$ be the point where this particle is detected on the x-axis. Then, the probability distribution of the detected signal is given by:*

$$p(x) = \frac{d}{dt}(\frac{1}{2} + \frac{\tan^{-1} x}{\pi}) = \frac{1}{\pi(1+x^2)} \qquad (2.1)$$

Proposition 2. *The Fourier transform of $p(x)$ is given by $\exp(-|k|)$.*

Proposition 3. *Let $x_1, \ldots, x_n$ be independent random variables drawn from the LC distribution. Then, $P(<x> <\epsilon)$ does not approach 1 as n increases.*

The Fourier transform of a distribution is referred to as the characteristic function in the probability literature. Characteristic functions are particularly useful since their derivatives at zero give the moments of the probability distribution. Viewed as a random process, it is well-known that for the LC process the average of n independent trials does not settle down. This is in contrast to, for example, the Gaussian distribution. However, if the light is detected by a detector of finite width, such as a photoreceptor, an LC process has better estimation accuracy than a Gaussian process. One can show that the efficient equivariant estimator for an LC process is determined by the extreme values of the sample with accuracy of the order of the reciprocal of number of quanta of light hitting the photoreceptor. For a Gaussian process, the accuracy is of the order of the reciprocal of the square root of the number of samples, and therefore the estimate improves more slowly.

These observation can be generalized in a number of ways to develop the 2D analogs of this idea. One possible generalization is to consider the same source now at $(0,0,1)$ emanating at an angle θ to the z-axis along the y-direction as well as along the x-direction and striking the x-y plane with the product probability distribution with Fourier transform of $\exp(-|k_x| - |k_y|)$.

Note that the shape of the 2D filter in the spatial domain is symmetric. However, simple modifications can yield more directionality of shape in the spatial domain, for example, one with Fourier transform of $\exp(-a|k_x| - b|k_y|)$. These distributions, viewed as filters, have a directional feature in the spatial

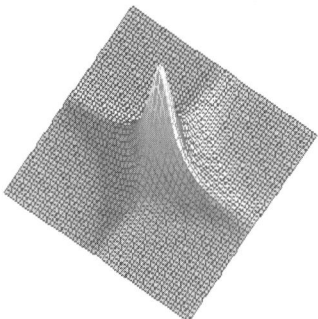

Fig. 2.2. A plot of a 2D LC distribution

domain and allow for immediate generalizations where one can easily construct the filter in a nonorthogonal system. We will see that this will lead to consideration of filters with Fourier transform of the form $F(x, y) \exp(-a|k_x| - b|k_y|)$, where F is in a restricted class of polynomials in x and y that will be defined later. Since multiplication in the Fourier domain corresponds to differentiation in the spatial domain, this filter corresponds to differentials of different orders for the filter above. Therefore, not only directions can be biased but also the rate of change can be biased as well. In the following sections we will investigate these generalizations and additional properties in the wavelet setting.

2.5 Symmetry and Signal Representations

Using the action of groups as generators of infinitesimal symmetries, we proceed to use symmetry in a systematic way to derive our filters. We begin by introducing the basic notation and the ideas necessary for developing the representation of signals of interest. Let $\mathcal{H}$ be a Hilbert space, and $\mathcal{U}(\mathcal{H})$ be the group of unitary operators acting on $\mathcal{H}$.

Definition 1. *A unitary representation is a homomorphism*

$$U : G \to \mathcal{U}(\mathcal{H}) \qquad x \to U(x) \tag{2.2}$$

from the group G to the group of unitary operators.

The set of vectors that can be reached by the action of the representation on the vector $g \in \mathcal{H}$ is called the orbit of g. A subspace of $\mathcal{H}$ is called an invariant subspace iff it is mapped into itself under the action of $U(x)$ for all $x \in G$.

Definition 2. *A representation is called irreducible iff the only closed invariant subspaces are the trivial one and the whole space.*

Consider two unitary representations U_1 and U_2 of the same group G in Hilbert spaces $\mathcal{H}_1$, and $\mathcal{H}_2$, respectively.

Definition 3. *A bounded operator $\mathcal{D} : \mathcal{H}_1 \to \mathcal{H}_2$ is called an intertwining operator iff $\mathcal{D}U_1(x) = U_2(x)\mathcal{D}$ for all $x \in G$.*

Denote the set of all intertwining operators as $R(U_1, U_2)$. For U an irreducible representation, the following important fact [25] is a generalization of Schur's lemma and assures us that our unitary representations on the same Hilbert space differ only by a scalar multiple of identity (see Warner [51] for a proof).

Theorem 2. *$R(U, U)$ is one dimensional and consists of scalar multiples of identity.*

Let G be a group and S an arbitrary set. We say that G is acting on S on the left iff for every $x \in G$ there is a transformation of S into itself, $A_x : S \to S$, such that $A_{xy} = A_x A_y$ for all $x, y \in G$. A right action can be similarly defined. Let G be a locally compact topological group acting on the set S equipped with a Borel measure μ. Let $J \subset S$ be a measurable subset with 'volume' $\mu(J)$. The left action of G on S transforms J into GJ. We say that μ is a left- invariant (or simply invariant) measure if $\mu(GJ) = \mu(J)$. Right-invariant measures may be equivalently defined. In general, such invariant measures need not always exist. However, when $S = G$, invariant measures always exist and are referred to as Haar measures denoted by $\mathrm{d}\mu$. With these definitions at hand, the notion of scalar product, square integrability and convolution in the Hilbert space $L^2(G, \mathrm{d}\mu)$ can be defined.

Definition 4. *A representation U of a group G in a Hilbert space $\mathcal{H}$ is said to be square integrable if there exists a nonzero $g \in \mathcal{H}$ such that*

$$\int_G \mathrm{d}\mu| <U(x)g|g>|^2 < \infty. \qquad (2.3)$$

We can now define the wavelet transform for a locally compact group acting in a Hilbert space [13].

Definition 5. *Let $g \in \mathcal{H}$ be a wavelet and $s \in \mathcal{H}$ a function. The (left) transform over G of s with respect to g is given by*

$$\mathcal{W}_g s(x) = <U(x)g|s>_\mathcal{H}, \qquad x \in G \qquad (2.4)$$

The left transform maps a vector in the Hilbert space $\mathcal{H}$ to a function over the group G.

2.5.1 Gabor and Wavelet Analysis

Gabor analysis can be put in the group theoretical setting. This setting will be useful to establish the link between our analogous use of the uncertainty principle for our group of interest in the next section. The group-theoretical view has been advanced in the atomic decomposition theory of Feichtinger and Gröchenig [9, 10], which applies to general representations on Banach spaces. A survey specifically of Gabor and wavelet analysis on $L^2(\mathbb{R})$ from the group viewpoint can be found in [15]. For example, the group for Gabor analysis is the Heisenberg group $\mathbf{H} = \mathbf{T} \times \mathbb{R} \times \mathbb{R}$, where $\mathbb{R}$ is the real line and $\mathbf{T}$ is the unit circle. The group action is induced by the representation ρ of $\mathbf{H}$ on $L^2(\mathbb{R})$ [12] and is defined by

$$\rho(z,a,b)f(t) = z\,e^{\pi i a b}\,e^{2\pi i b t}\,f(t+a) \quad \text{for} \quad (z,a,b) \in \mathbf{H}\, f \in L^2(\mathbb{R}). \tag{2.5}$$

Then, one way to analyze a function $g \in L^2(\mathbb{R})$ is in terms of the inner products $\{\langle g, \rho(a,b)f \rangle : (a,b) \in \Omega\}$ for $\Omega \subset \mathbb{R}^2$ with $f \in L^2(\mathbb{R})$ fixed. The collection of inner products $\langle g, \rho(a,b)f \rangle$ are termed *Gabor coefficients*, and the mapping $g \mapsto \{\langle g, \rho(a,b)f \rangle : (a,b) \in \mathbb{R}^2\}$ is called the *continuous Gabor transform* of g by f. Stable reconstruction by using all possible Gabor coefficients can be obtained for any g. By selecting a discrete subset Ω of $\mathbb{R}^2$, less redundant representations can be obtained. However, for stable reconstruction, and for g to be completely determined by the Gabor coefficients, the collection

$$S(f, \Lambda) = \{\rho(a,b)f : (a,b) \in \Omega\} \tag{2.6}$$

must be complete in $L^2(\mathbb{R})$ and form a frame for $L^2(\mathbb{R})$.

By replacing the Heisenberg group with the affine group $\mathcal{A} = (\mathbb{R}\setminus\{0\}) \times \mathbb{R}$ we can investigate the parallels between wavelet and Gabor analysis. Then, *time-scale* replace time-frequency translates. Let U be the representation of $\mathcal{A}$ on $L^2(\mathbb{R})$ defined by $U_{(a,b)}\psi(at-b)$ for $(a,b) \in \mathcal{A}$ and $\psi \in L^2(\mathbb{R})$. We may define a continuous transform as above, or seek frames or bases of the form

$$T(\psi, \Omega) = \{U_{(a,b)}\psi : (a,b) \in \Omega\}. \tag{2.7}$$

For Ω that are "sufficiently dense", Feichtinger–Gröchenig theory guarantees the existence of frames. However, the considerable difference in the structure of the two groups is of significant consequence. For example, there are several significant differences in the properties of $S(f, \Lambda)$ versus $T(\psi, \Omega)$. One of the most important results is that $T(\psi, \Omega)$ can form an orthonormal basis with smooth, well-localized ψ. A typical choice for Ω is the "regular" discrete subset $\Omega = \{(a^n, mb) : m, n \in \mathbb{Z}\}$, with $a = 2$ and $b = 1$ [6].

To study local properties of objects at any orientation, location and scale, we focus on the group $G = E(2)$, the Euclidean group of motions and dilations in the plane $\mathbb{R}^2$. We will further focus our discussion to analysis of square integrable functions of two variables $s(x,y)$. The elements of the group can be identified with the vector $v = [t_x, t_y, \theta, a]$, where t_x, t_y parameterize the translations, θ parameterizes the rotations, and a parameterizes the (positive) dilations in the plane. The group $E(2)$ contains the translation group $T \equiv \mathbb{R}^2$, the dilation group $D \equiv \mathbb{R}^+$ and the rotation group $R \equiv S^1$. There is a left Haar measure on this group and each of the subgroups above have the corresponding left action on $L^2(\mathbb{R}^2)$. The unitary representation of $E(2)$ in terms of T, R and D can be written as

$$U : E(2) \to \mathcal{U}(L^2(\mathbb{R}^2)), \qquad [t_x, t_y, \theta, a] \to U(t_x, t_y, \theta, a) \qquad (2.8)$$

Note that $E(2)$ is $\mathbb{R}^2 \bowtie (SO(2) \times \mathbb{R}_+)$, where $\bowtie$ is the semidirect product. Suppose we are given the signal $s \in L^2(\mathbb{R}^2, d^2\mathbf{x})$, then s has the continuous transform with respect to the wavelet ψ:

$$S = \langle \psi_{t_x,t_y,a,\theta} | s \rangle = a^{-1} \int \overline{\psi(a^{-1} r_\theta(x - t_x, y - t_y))}\, s(x,y)\, \mathrm{d}x \mathrm{d}y \qquad (2.9)$$

$$= a \int e^{it_x k_x + it_y k_y} \overline{\widehat{\psi}(ar_\theta(k_x, k_y))}\, \widehat{s}(k_x, k_y)\, \mathrm{d}k_x \mathrm{d}k_y. \qquad (2.10)$$

The wavelet ψ generates, by translation, rotation and dilation, the dictionary $\mathcal{H} = \{\psi_{t_x,t_y,a,\theta}\}$, indexed by $a > 0, \theta \in [0, 2\pi), (t_x, t_y) \in \mathbb{R}^2$. The projections of s onto this family generates the family of wavelet coefficients of the signal s. For G to be a wavelet analysis on $\mathbb{R}^2$ we must require that:
1. U to be irreducible
2. U to be square integrable

This can be shown for unitary group representations [13]. Therefore, we have:

Proposition 4. *U is square-integrable and irreducible.*

Note that although scale and orientation are not specified to be functions of time, global changes of scale and orientation, which may occur as a function of time, are represented by the action of G. In other words, when a natural ordering among scales and orientations is given, one can view scale and orientation change as occurring in a flow of time.

2.5.2 LC Wavelets

In the construction of filters, the properties of signals under study and the desired filter output drive the construction of filters. For example, band-limited signals are typically analyzed by low-pass, band-pass or high-pass filters and support-limited filters operate in the spatial domain. In addition to limiting the support of the function in either the frequency or the spatial domain, there is another method, used by Gabor, which borrows from the notions of simultaneous localization bounds introduced in quantum mechanics. The bounds in quantum mechanics are known as the Heisenberg Uncertainty Principle and define how well the position operator (x) and the momentum operator $(-i\frac{d}{dx})$ can be simultaneously localized by bounding the product of the variance of the operators[3]. In this section, we show how to derive filters that satisfy a bound similar to the uncertainty principal. The derivation shows that LC wavelets arise naturally as the solution to problem of satisfying a form of uncertainty principle with respect to rotation, scaling and translation operators. This property is important to our goal of shifting our focus from perfect reconstruction to efficient feature detection and search. Aside from significance in quantum mechanics, the choice of bounding the product of variances has a rationale of its own in our context. This bound guarantees that, to the first approximation, and over a set of repeated identical observations (or measurements), variance is controlled in a prescribed way.

Let M and P be two self-adjoint operators in a Hilbert space $\mathcal{H}$ with inner product $<\cdot|\cdot>$. Let ψ be a vector in the domain of both operators above with norm equal to one ($||\psi|| = 1$).

Definition 6. *The average of the operator M in the state ψ is defined as*

$$<M> = <\psi|M\psi>. \qquad (2.11)$$

Definition 7. *The variance of the operator in the sate ψ is defined as*

$$\Delta M = \left(<M^2> - <M>^2\right)^{\frac{1}{2}}. \qquad (2.12)$$

Let $\bar{M} = M - <M>$. The following theorem, proved by Gabor using analytic methods, was instrumental in showing that in the case of short-time Fourier transform, Gaussian envelopes saturate the time-frequency localization bounds[4].

Proposition 5. *With the above definitions for M, P and ψ,*

$$2\Delta M \cdot \Delta P \geq |<[M,P]>|. \qquad (2.13)$$

Furthermore, the equality holds iff

$$i\bar{M}\psi = \lambda \bar{P}\psi, \qquad (2.14)$$

where $\lambda > 0$ is a positive constant.

We wish to study other properties of LC distributions within the framework of the uncertainty principle outlined above. These ideas were first utilized in [40] in the context of 2D Gabor wavelets. The use of LC distributions was also investigated[5] in the context of quantum mechanics. We will show that the statistical filters defined above through the LC process have an analogous simultaneous localization property. In fact, more is true and we can prove the following.

---

[3] see [11] pp. 16-20 for a good discussion of this topic and later chapters for a discussion of alternative strategies.
[4] A simple proof based on the algebraic properties of operators can be given.
[5] Reference to Paul's dissertation in French can be found in [11].

Theorem 3. *The LC filter*

$$\hat{\psi}(k_x, k_y) = |(c_1 k_1 + c_2 k_2)| \exp(-|(m_1 k_1 + m_2 k_2)|). \quad (2.15)$$

achieves the simultaneous localization bounds for constants c_1, c_2, m_1, m_2.

The factor for the exponential can be any polynomial H satisfying the proper conditions.

To give an outline of the derivation of the above, let $g(\eta)$ be a one-parameter group of transformations with parameter $\eta \in \mathbb{R}$. Let $\bar{s}(\bar{x}, \bar{y}) = g(\eta) s(x, y)$ be the function obtained from the action of g on the square integrable function $s(x, y)$. The Leibnitz chain rule defines the infinitesimal transformation of s about the identity

$$\frac{d}{dx}(g(\eta))s|_{\eta=0} = \left(\frac{\partial x}{\partial \eta} \frac{\partial}{\partial x} + \frac{\partial y}{\partial \eta} \frac{\partial}{\partial y} + \frac{\partial}{\partial \eta} \right) |_{\eta=0} \bar{s} = D\bar{s}. \quad (2.16)$$

The differential operator D is known as the infinitesimal generator of the transformation. The following propositions follow from computations using the Leibnitz rule and the use of commutation relations.

Proposition 6. *The infinitesimal operator T_x for the x-translation is $-\frac{\partial}{\partial x}$.*

The infinitesimal operator T_y for the y-translation is $-\frac{\partial}{\partial y}$.

The infinitesimal operator R_θ for the rotation through the angle θ is $x\frac{\partial}{\partial y} - y\frac{\partial}{\partial x}$.

The infinitesimal operator D for the scaling is $-x\frac{\partial}{\partial x} - y\frac{\partial}{\partial y} + 1$.

Proposition 7. *For the operators R, D, T_x, and T_y, the following relations hold:*

$$[T_x, T_y] = [D, R] = 0, \quad [R, T_x] = [D, T_y] = T_y, \quad [D, T_x] = -[R, T_y] = T_x. \quad (2.17)$$

We can now apply the conditions of Proposition 5 to obtain a set of differential equations. We will consider the pair $[D, T_x] = -[R, T_y] = T_x$ - symmetry considerations yields the other pair. From the proposition 5, to achieve the bound, we must have:

$$(\bar{D} - i\lambda \bar{T}_x)\psi = 0 \quad (2.18)$$

$$(\bar{R} - i\lambda^* \bar{T}_y)\psi = 0 \quad (2.19)$$

Proposition 8. *The filter described by 3 satisfies the pair of differential equations.*

One can obtain the solutions directly by first using the integrability condition requiring that $\lambda = \lambda^*$. Then, follow by direct substitution and note that it is in the form of the general solution below. To obtain the complete form, we need to compute the solution for the other pair of operators and combine the results to obtain the form of the filter discussed above.

2.6 The Space of Features

In the sections above, we have described how an overcomplete representation of the signal s can be obtained. Our next task is to extract a set of stable features based on our representation. In this section we show how such a set of features can be extracted, and in what sense they can said to be stable. The propositions and statements of this section will lead to the proof of the following theorem:

Theorem 4. *Existence of stable features. There exists a parameterized set of scale, translation, and rotation-invariant stable and finite features associated with the signal s.*

Recall that our representation assigns to every function s from a set of signals $\{s_\alpha\}$ a function on the group G based on the definitions given above, therefore, we have an overcomplete representation. In this section, we set out to define methods that

1. distinguishes the class of signals under consideration efficiently through a sparse representation.
2. gives information as to how additional information about the signal can be obtained when potential ambiguities arise.

The first item above amounts to the determination of a decision operator discussed earlier. The notion of sparse or minimal representation will be defined through features shortly. In this representation, features will be parameterized by angle and scale. It is important to note that a key property used in the succeeding analysis is the ability to parameterize the response of a set of filters to a given signal, which in this case is achieved by the role of the action of G. It is possible to apply the ideas in the remainder of this analysis to the action of another group or another parametrization.

We start with parameterizing the rotations by the angle θ in our fixed coordinates. Our intention is to find the information content carried by the wavelet transform of the signal s at a fixed scale d with respect to all local rotations. Since our rotations are parameterized in a fixed global coordinate system and since translations and rotations in the global system do not commute (unlike a local coordinate system), we must use the appropriate language of semidirect products. Namely, the subset $H = D \bowtie S^1$ of the group G, operate on the wavelet ψ by rotations and translations, denoted by $\psi_{\theta,(u,v)}$ in which θ is the global rotation and (u,v) is a point in $D \subset \mathbb{R}^2$. D parameterizes the translation of signal by the corresponding point. We associate with ψ the wavelet operator $\psi^*_{\theta,(u,v)}$.

Consider $\Psi(\theta, u, v) = <s, \psi^*_{\theta,(u,v)}>$ that associates to each $\psi^*_{\theta,(u,v)}$ a complex number. Denote by $\varphi_\theta(u,v) = |\psi^*_{t,(u,v)}|$ the modulus of complex wavelet coefficient. $\varphi_t(u,v)$ defines a real two-dimensional surface denoted by M_θ over the subset D that measures the information content of the signal s with respect to the wavelet for each angle θ. We call the isolated extrema of this surface for which the Gaussian curvature is non-zero, *non-degenerate*. We wish to define a finite feature set for each θ associated with D. To do so, we introduce a granular resolution δ that defines the resolution of the system under consideration. Then, the maxima and minima as well as the Gaussian curvature of the surface are defined within the said resolution δ. In the neighborhood of a point (u_1, v_1), δ allows us to replace the set of all point with values within δ with a single point at the center of gravity of the said neighborhood. We call these points $\delta - maximum$ and $\delta - minimum$ points.

Definition 8. *Let $F_\theta : M_\theta \to \{-1, 0, 1\}$ be the function that sends every nondegenerate local $\delta - maximum$ to 1, $\delta - minimum$ to -1 and all other points to 0. The set of coordinates $C_\theta = (x_1, ..., x_n)$, for which F_θ assigns the value 1 or -1 is called the **feature set** parameterized by θ.*

Proof of the following proposition follows using Thom's transversality theorem and Sard–Smale stability theorem ([27], p. 10) and justifies our finite point list above.

Proposition 9. *Existence of stable nondegenerate features. With the notation and hypothesis above, there exists a small perturbation η of the signal such that the resulting surface $\bar{s}$ will have only nondegenerate features. Further, η can be arbitrarily small, so that for a choice $\bar{\eta}$, all features in any finite set of scales and orientations will be nondegenerate.*

Numerically we ensure stability and nondegeneracy by introducing a method of differentiation that we coin *noise-enhanced differentiation*. Heuristically, a noise-enhanced differentiation achieves the small perturbation η. Furthermore, control over noise allows us to ignore all spurious features that could be theoretically nondegenerate and stable but fall within the same granularity, or disappear with possibly large perturbations but are still bounded by the granularity. In particular, control over noise allows us to ignore all artifacts of numerical computation that may potentially give rise to accidental features that should not be counted due to any errors tolerated by granularity. Further, the ultimate objective is to perform statistics on ensembles of such features. As a result, a number of potentially error prone features that might have escaped the above-mentioned computation due to special circumstances of the image (as we must be prepared for a great deal of variable conditions within the images of faces with possible outliers in one case or another) will not have influence on the statistical analysis of the ensemble, being filtered by the approximations. This indicates that numerical results depend on the size of the data set for images. If the data set is not too large, then the statistics are affected by numerical outliers. On the other hand, for a large data set we have a stable statistical feature set.

We point out that the introduction of δ makes this definition system dependent. Additionally, we note that a number of other system dependent definitions for the choice of a discrete set of features is possible. For example, one may choose the sparse points obtained from using a support vector machine

(SVM) at a given resolution defined by ϵ-sensitivity of SVM [48]. Additionally, system-independent definitions based on probabilistic definitions may be possible. Our choice for this set of features is motivated by observations (explained below) that may be replaced with a more definite theory. The remainder of this work only assumes that a finite discrete set of features depending on the search task is available. In the definition above, θ parameterizes a collection of sets, each of which contains a finite set of points. Note that for each C_θ, n may be a different number. However, for a finite parameter set θ, the following proposition allows us to view all coordinates in a suitably high-dimensional space (maximum dimension of the coordinates in the coordinate set) by ordering the coordinates.

Proposition 10. *If θ is a finite set, the coordinates describing the feature set can be ordered in a consistent manner.*

It is also important to see how these features vary under other transformations of interest such as the addition of a constant.

Proposition 11. a. *C_θ is invariant with respect to addition of a constant*
b. *If we apply a translation in the plane to s, then C_θ changes with the same translation in the plane.*

We need a few words about the motivation for the above choice of features. Zero-crossings (edges) are well-known for the information they carry about images. Viewing the image as a function, these points are the points where the function vanishes. On the other hand, proof of the Shannon Sampling Theorem for band-limited signals in one dimension results from the statement that the zeros of an entire holomorphic function are isolated. Shannon discovered that the highest rate of zero-crossings of the Fourier transform of the signal can be used to define a sampling frequency for perfect reconstruction of the signal, and he gave a specific formula for its reconstruction. Also, Logan's theorem shows how a band-limited signal of bandwidth less than an octave can be reconstructed from its zero-crossings in the Fourier domain. For holomorphic functions of two variables, the zero set is no longer discrete. In fact, it is a one dimensional analytic submanifold. Therefore, analogous properties in two dimensions are not immediately obvious nor may they exist.

However, we propose an alternative approach in which we simply restrict the class of signals that have a given extremal set. For a sampled signal, the Fourier representation of the signal is a weighted sum of frequency components. For any signal, the extrema in the spatial domain yield restrictions on frequencies in the Fourier domain, since the derivative of the signal in the spatial domain corresponds to multiplication by frequency in the Fourier domain. Representing a band-limited signal via extrema hints at the feasibility of classifying a signal s from the knowledge of the frequencies not present (it is only up to amplitude modulation). From a biological point of view, the selection of extrema can be likened to a neuronal winner-take-all strategy by the neural substrate sensitive to a specific direction. The following lemma is the formal statement of the above strategy.

Lemma 1. *Let s be a smooth band-limited (finite energy) signal whose derivative is also band-limited. Let the Fourier transform of s be denoted by $\tilde{s}$. Then, in the quantized frequency plane, the extrema of s specify the class of signals to which s can belong within the quantization resolution and amplitude modulation.*

2.6.1 Information Content of Features

Our goal now is to show how the points in C_θ can be used as features for representation of signals by introducing an information theoretic measure of distinguishability among signals. The approach to deriving this measure has a distinct geometric character and may hide the information theoretic context. The first subsection is meant to give a brief overview of the relationship between information theory and geometry. A more detailed and in-depth discussion of these connections can be found in [1, 5, 21].

One objective of most learning systems can be formulated as a mechanism to use the given sample for extraction of information about the true distribution. Therefore, we need to state what information is and how its is measured. This problem was considered by Fisher, who provided a solution as follows. Fisher considered the likelihood function of θ given x, $p(x|\theta)$. One observes the random variable x, and uses θ as parameters that represent the unknown distribution from which x has been observed. Let $x|n$ denote n samples of events for a process. Now, suppose that we are given a sample $x|n = [x_1, \ldots, x_n]$, and we obtain an estimate θ_e for θ. Then, for large samples, a reasonable estimate θ_e would tend to a normal distribution in many cases (The Central Limit Theorem). Thus, in such cases and at least for large samples, estimates are characterized by their variance and bias. This idea is often referred to as *asymptotic normality*.

The intuitive description of how to approach the characterization of information based on the observation above is as follows. Since bias can be removed beforehand, it does not affect the amount of information. However, when two estimates approach θ, the estimate with the smaller variance preserves more information compared to the estimate with the larger variance. This again can be intuitively explained by noting that the estimate with the smaller variance can always be made to look similar to the estimate with the larger variance by adding appropriate noise to the one with smaller variance. Using these ideas, Fisher derived a lower bound for the variance of any estimate of the true parameters.

In information theory, Fisher information plays an important role in describing the local statistical properties of an object under study. The geometric interpretation of Fisher information can be given as follows: The state space of a classical n-point system given by all probability distributions is the simplex of probability distribution on the n-point space and is seen to be an $(n-1)$-simplex, which is an $(n-1)$-dimensional manifold. Let P_i denote the probability of the ith event. Introduce the parameterizations $Z_i = 2\sqrt{P_i}$. Since $\sum P_i = 1$, we have $\sum Z_i^2 = 4$. In this way, the probability simplex is parameterized with a portion of the $(n-1)$-sphere. Let $x(t)$ be a curve on the sphere, then the square of the tangent length to $x(t)$ is given by:

$$<\partial_t x, \partial_t x> = \sum_i (\partial_t x_i)^2 = \sum_i P_i(t)[\partial_t log P_i(t)]^2 \qquad (2.20)$$

Observe that this is the Fisher information. Recall that for two matrices A and B of the same dimension, the relation $A > B$ is defined as $A - B$ being positive definite. Then, the following theorem shows the utility of the Fisher information.

Theorem 5. *Let $V(\theta_e)$ be the covariance matrix of the estimator, then*

$$<\partial_i l \partial_j l> \geq [V(\theta_e)]^{-1},$$

where $V(\theta_e)$ is the covariance matrix of any unbiased estimate.

The quantity $g_{ij} = <\partial_i l \partial_j l>$ is called the *Fisher information matrix* and is considered a measure of information. It is easy to see that if $x|n = [x_1, \ldots, x_n]$ is a sample of independent events, then $g_{ij}(x|n) = n g_{ij}(x)$. This simply follows from the observation that for independent events, the probability distribution is the product of distributions and the log of the product distribution is the sum of the log of each distribution. For example, for a binomial distribution with parameter p (probability of heads), $g_{ij}(x|n) = n g_{ij}(x) = n/p(1-p)$.

Suppose now that we have a small sample coming from $p(x|\theta)$ that is not itself normal. Can we find a suitable measure of information for this case? Fisher answered this affirmatively using the same measure above. To see this consider a small sample $x|k$ and a large sample $x|m$. Note that $x|m$ and $[x|k, x|m]$ satisfy the asymptotic normality and thus carry m and $m+k$ units of information, respectively. Then, $x|k$ must carry k units of information. Therefore, g_{ij} provides a suitable measure of information regardless of sample size. Rao's suggestion to consider g_{ij} as defining a Riemannian metric on the parameterized statistical model set the groundwork for the introduction of many concepts of geometry into the arena of statistical inference. In this setting a distance element ds specifying distance between nearby distributions, which is invariant under coordinate transformation, is defined:

$$\mathrm{d}s^2 = \sum_{i,j} g_{ij} \mathrm{d}\theta_i \mathrm{d}\theta_j$$

To achieve the utmost generality, Cencov [5] wrote his theory in the language of modern differential geometry. In this setting, rather than treating the properties of specific distributions or samples, invariant properties of families were treated. Using these tools Cencov was able to prove a far-reaching theorem regarding the uniqueness of the Riemannian metric and invariant affine connections for any finite sample space.

2.6.2 Features and Information

Suppose we are given n points $(x_1, ..., x_n)$ in the plane where each point is specified by Cartesian coordinates. View $(x_1, ..., x_n)$ as a point in $(\mathbb{R}^{2n})$, which provides a $2n$-dimensional summary of major geometric characteristics (features) of an object, including location, orientation, scale, and shape information. To analyze these features in the context of specifying a shape encoding, we must determine the class of all functions of the vector $(x_1, ..., x_n)$ that measure its shape invariants while at the same time eliminating information in $(x_1, ..., x_n)$ that describes the location, scale, or orientation of the features. That is, our encoding must be invariant under $E(2)$ symmetries. Standard statistical tools can be applied to describe the location and scale statistics of a set of points. The location of a data set $(x_1, ..., x_n)$ can be described by its sample mean, or centroid, given by

$$\bar{x} = \frac{1}{n} \sum_{i=1}^{n} x_i. \tag{2.21}$$

The size or scale of our features can be described by a variety of statistics. Let (x_{j1}, x_{j2}) for $j = 1, ..., n$, be two features (feature vectors), centered about their mean, and consider the matrix:

$$X = \begin{pmatrix} x_1 x_1^T & x_2 x_1^T \\ x_2 x_1^T & x_2 x_2^T \end{pmatrix}, \tag{2.22}$$

where T denotes the transpose operation. The trace of X, given by

$$\text{tr}(X) = \sum \text{diag}(X) = \sum_{i=1}^{n} ||x_{ji} - \bar{x}_i||^2. \tag{2.23}$$

is an invariant of the matrix and is therefore a natural measure of the size of the set of features. The matrix X can then be standardized to have trace equal to one. This eliminates location and size information in a data set, which we then call the standardized feature set:

$$\chi = (x_1, \ldots, x_n) = \frac{1}{\sqrt{\text{tr}(X)}} (x_{11} - \bar{x}_1, \ldots, x_{n1} - \bar{x}_1). \tag{2.24}$$

We abuse the notation and write the new coordinates also as $(x_1, \ldots, x_n)$. One point must be noted here. In order for this representation to be meaningful, the features defined above by χ must not all be collinear. However, this does not present a problem for our application. In general, a set of features that are all coincident will be treated separately. Henceforth, we assume that this degeneracy does not occur, as is the case for natural images such as faces.

To eliminate dependence on orientation,[6] we first make the following observation. The vector χ lies in a constrained subset of the original Euclidean space $\mathbb{R}^{2n}$, which is just a lower dimensional sphere,

$$S^{2n-3} = V^{2n-2} \cap S^{2n-1}, \tag{2.25}$$

where

$$V^{2n-2} = \{(x_1, \ldots, x_n) \in \mathbb{R}^{2n} \; : \; \sum x_i = 0\}, \tag{2.26}$$

and S^{2n-1} is the standard $(2n-1)$-dimensional sphere.

Therefore, this subset is represented by the intersection of a subspace of $(2n-2)$-dimensional space with the unit sphere. To eliminate dependence on orientation angle in the plane, consider the following "function of coordinates" from the feature space S^{2n-3} to the circle S^1, heuristically defined via:

$$\vartheta : S^{2n-3} \to S^1. \tag{2.27}$$

ϑ can be used to eliminate orientation by forming the orbit space

$$\Sigma = \{o(\chi) : \chi \in S^{2n-3}\}. \tag{2.28}$$

Note that standardizing for the location and scale of χ does not disturb its orientation. More rigorously, we have an action of the rotation group identified with

$$S^1 : S^1 \times S^{2n-3} \to S^{2n-3} \tag{2.29}$$

whose orbits coincide with the representation of the same point under transformation by all rotations. For every representation, we have also its rotations. Therefore, the group action is well defined, provided we allow the images that are transformations of our original image and may potentially lie not entirely within the domain D. This only adds extra representation points to the ensemble. This does not affect the statistics, since we shall only consider the set of orbits, thus eliminating the spurious additions of these points of orbits.

To compare features we need to devise a metric. The natural metric we consider measures the geodesic distance between two feature sets specified by χ up to the action of S^1 (orientation). For spheres this is easy to compute:

$$\delta(x_1, x_2) = \cos^{-1}(<x_1, x_2>), \tag{2.30}$$

with the induced metric

$$\mathtt{d}(x_1, x_2) = \inf\{\delta(z, w) : z \in o(x_1), w \in o(x_2)\}, \tag{2.31}$$

---

[6] Short for orientation angle with the plane.

where the notion $o(.)$ means the orbit space of $(.)$, so the inf is computed over the orbit of the action of S^1. One can show that this definition satisfies the properties of a metric (see e.g. [36]) . With the metric defined above we now have a manifold that is also known as complex projective space together with the *Fubini–Study metric* [36]. Since we intend to encode features with complex coordinates, we restate the above features making use of the algebraic properties of the complex plane. Once again, consider the features represented by $\chi = (x_1, \ldots, x_n)$, with $x_i \in \mathbb{R}^2$. Consider x_i to be elements of the complex plane by identification of the complex numbers $\mathbb{C}$ with $\mathbb{R}^2$. Given two features we may choose representatives up to the action of the circle. Let us denote complex conjugation by the symbol $*$. Then, it is a standard argument to show that the minimum distance can be obtained as:

$$\mathsf{d}(x_1, x_2) = \cos^{-1}\left(|\sum_{j=1}^{n} x_{1j} x_{2j}^*|\right). \tag{2.32}$$

This definition is independent of the orientation since multiplying both arguments by an element of $U(1)$ does not change the distance. Starting from any point, the direction of maximal information lies along the curve $x(t)$ where $x(t)$ is a geodesic – that is, the information rate is maximal. The geodesic distance between two probability distributions is computed along the great circle of the $(n-1)$-sphere. In complete analogy, the Fubini–Study metric (FS) introduced earlier is the minimal (geodesic) distance between two great circles for an n-level system in the complex projective Hilbert space described by the feature vectors earlier.

Definition 9. *Let t be a parameter of a system and $x(t)$ be the state vector of the system parameterized by t. Let $x(t+\mathsf{d}t)$ denote a small change in the system parameter t by the amount $\mathsf{d}t$. The FS distance between $x(t)$ and $x(t+\mathsf{d}t)$ is called* information distance.

This definition can be understood by considering the analogy to the standard definition of Fisher information. Let q_i be the estimate for the probability P_i on the n-point probability distribution above obtained from the observed frequency after drawing N random samples. For large N it is standard to estimate the frequency distributions given by the multinomial distribution for P_i using the Gaussian $\exp[-N(q_i - P_i)^2/(2P_i)]$. When the exponent of the Gaussian is small, nearby distributions are easily distinguishable. Thus, we have a natural distinguishability described by:

$$\mathsf{d}p^2 = \sum \frac{\mathsf{d}P_i^2}{P_i} = \sum P_i (\mathsf{d}\ln P_i)^2. \tag{2.33}$$

This is the same Fisher information. For the state vector x in $\mathbb{C}P^n$, we can write:

$$|x> = \sum \rho_j^{1/2} \exp(i\varphi_j)|b_j>, \tag{2.34}$$

where b_j is an orthonormal basis for $\mathcal{H}$. A small change for x will result in the vector $\hat{x} = x(t + dt)$, which we can write as:

$$|x> = \sum (\rho_j + d\rho_j)^{1/2} \exp i(\varphi_j + d\varphi_j)|b_j>. \tag{2.35}$$

Then, it is easy to show that $FS[x(t), x(t + dt)]$ is given by:

$$FS[x(t), x(t + dt)] = dp^2 + \text{Var}(\varphi), \tag{2.36}$$

where

$$\text{Var}(\varphi) = \sum \rho_j d\varphi_j^2 - (\sum \rho_j d\varphi_j)^2. \tag{2.37}$$

Therefore, FS is a metric of distinguishability that has the additional term introduced by the variance of phase. FS as an information measure describes the change of the state vector along the geodesic between the two states as the process is evolving, much in the same way that Fisher information achieves the same. When the parameter describing the process is periodic, $x(t)$ traces out a closed loop, and this loop encodes the information dynamics of the process. Note that since a global change of angle does not affect the FS, two periodic processes with the same information evolution and related by a rotation are indistinguishable.

2.7 Information Dynamics

Learning theory, in the mathematical setting, is the study and development of models and the behavior of these models with respect to "learnability" issues [1, 22, 29, 47]. We refer the reader to these references for an in-depth discussion. Our goal here is limited to describing the setup of a dynamics process that can be used to distinguish sets of features. We will first discuss how the dynamics of one set of features parameterized by the angle θ is set up. Next, we show that rigorous computational methods can be applied to establish the dynamics of an ensemble based on their features. We will note that our view of learning here is that of function approximation. This is in line with the views of learning in support vector machines (SVM) and probably approximately correct (PAC) learning.

As discussed earlier, the parametrization of the underlying topological group G gives a natural parameterization to analyze signals. We used this parameterization earlier to define and parameterize a map at a fixed scale d from each angle parameter θ to the corresponding feature set C_θ. We then showed how each C_θ defines a vector in the complex Hilbert space specifying a feature. Next, we defined a metric on this space in a manner analogous to the Fisher information metric.

Similar to a quantized n-state system, where the magnitude of state changes (*with respect to the probes*) provides information about the system, we consider state changes with respect to the family of probes generated from

ψ under the group action. It is important to emphasize that one specific set of probes may provide little or no information under the same dynamics while other probes may. In other words, the choice of probes may be critical in obtaining information about specific processes.[7]

To discuss dynamics search using the features and the metric above we proceed as follows. First, since our features are defined up to the action of S^1, we do not need to find a method to register features corresponding to the same parameter value. Before we proceed, we need to explore the nature of our filter (or probe) and understand its behavior with respect to a given signal s. The next subsection will show that features will vary. The following subsection will address the registration problem. Finally, we deal with learning of the features in the context of principal components analysis (PCA) for vector in $\mathbb{C}P^n$. This will set up all the machinery necessary to address the problem posed at the beginning of this chapter.

2.7.1 Feature Variation

We have established our filter through the action of the elements of the group of Euclidean motion. Specifically, at every fixed scale we have established our probe parameterized by the rotation angle θ as a function of location. To see what information is provided by our probe, we need to examine how feature vectors change with θ. The next proposition tells us that the generic (common) feature coordinates recorded by the directional wavelet $h_t\psi$ as h_t take values in $SO(2)$ are features of s that remain approximately invariant under the local rotation of the wavelet.

Proposition 12. *Let h_t vary in the rotation subgroup of the group G. Consider the set of nondegenerate extrema P of s regarded as $h_t\psi$ at $t = 0$. Assume all such P are fixed under $h_t\psi$ as h_t varies in the rotation subgroup of the group G. Then, P forms a subspace of feature coordinates locally invariant to rotation.*

If the feature vector for a signal s does not vary under action of the rotation group under a particular rotational wavelet, then that group has local rotational symmetry about the "mean" of the wavelet. If this holds for all points in the signal, then the signal is featureless in all directions and thus constant. Therefore, generic signals have feature vectors that vary under the action of the rotation group on a directional wavelet. Roughly speaking, this shows that the feature coordinates highlight locally inhomogeneous structures (with respect to rotation) against a background of rotationally symmetric structures. When a directionally nonhomogeneous signal (defined as a signal that does

---

[7] In fact, it may be necessary to recruit the machinery of learning theory to address the problem of determining probes most suitable to the problem. This is work in progress.

not remain invariant under the action of a group of interest) is transformed via a directional wavelet, the expansion coefficients do not remain constant.

The FS metric is a natural Riemannian metric on the complex projective space that parallels the notion of the Fisher information. For a fixed scale, a vector describing the features of the subset of the Euclidean plane for all orientations is called an information loop. We will view an information loop as a probabilistic indicator for the signal class from which the signals arise. By varying the scale parameter and computing new feature vectors as a function of scale, we have constructed a structure that measures feature variations across scales. Can we associate an information loop to a class of signals that encodes statistical information about the collection? This is the generalization from a single signal to an ensemble of signals that is our learning portion of the problem.

2.7.2 Learning

The methods of learning theory can be applied to learn the information loop and the corresponding feature vector. Given a training set of samples, an information structure can be constructed describing the signal at different scales.

Our approach to build an information structure and its associated loop for the class of signals is to use the eigenstructure of feature vector correlations for a set of signals. Given a data set in $\mathbb{R}^n$, this eigenstructure can be obtained using the PCA method. However, it is not immediately obvious if this method (in its standard formulation in $\mathbb{R}^n$) can be applied to vectors in $\mathbb{C}P^n$. However, the space of features, namely $\mathbb{C}P^n$, has the associated FS metric. Therefore, arguments based on minimal reconstruction error in $\mathbb{R}^n$ can be applied to $\mathbb{C}P^n$ using the FS metric. Another approach with a direct computational implementation is based on the use of the Gram matrix. Given a set of observation vectors x_i, the ijth element of the Gram matrix is given by $W_{ij} =< x_i, x_j >$. The eigenvector of W corresponding to the largest eigenvalue gives the solution of the least square problem. Using the inner product structure we have described, we can show that:

Proposition 13. *The eigenvalues and eigenvectors of the Gram matrix are the same as those of the autocovariance matrix.*

In fact, since the modified Gram–Schmidt orthogonalization process gives an iterative algorithm for this computation, a straightforward implementation of this approach with tolerances within the granularity of the system is available. Furthermore, methods of RKHS apply whenever the properties of interest are defined using inner products. In particular, extensions of existing methods to the complex setting for many approaches such as nonlinear kernels and SVM can be considered.

To finalize the role of the scale parameter, we remark on the following hueristic reasoning. We note that in our formulation of the problem, the search and detection dynamics occur simultaneously. If the information loop at the fine scale is compatible with any element in the learned class of templates, then the dynamics proceeds. However, it is possible that the dynamics at the fine scale detect an incompatibility. In this case, at a finite number of values t, the fine scale dynamics is incompatible (as defined by the FS metric) with all templates in the learned class. Using the FS metric, there is at least one element in the learned class that best matches the feature dynamics at the current scale. If there is more than one, we may choose one selected at random. For this element of the learned class, there is a feature vector that is maximally different from the corresponding feature vector of the element of the learned class. We use this pair of feature vectors to determine the novelty point, as this maximal difference corresponds to the direction of maximal information gain.

These features have emerged as a result of scale change and describe the elements of the underlying signal that emerge in the scale transition from scale $d_2 > d_1$ at the angular parameter t. The subset of features where the largest changes occurs are optimal search locations. The word optimal is used in the information theoretic sense in that this is the point at which our measure of information defined by the FS metric makes the largest contribution to change. This subset can be found as the difference of the two feature vectors at the same angular parameter. Then, the largest components of this vector define locations, or possibly, clusters of maximal change.

If dynamics at all scales agree, how likely is it that the signals are different? To answer this question we need to consider error probabilities of testing a hypothesis H. For example, if our goal is to find a yes or no answer to a detection problem, then our hypothesis space is the binary decision. A direct argument yields the following proposition.

Proposition 14. *Let $|H|$ be the cardinality of the hypothesis space. If the probability of error at each scale and orientation is p_e (which may be a function of granularity), then the hypothesis $h \in H$ can be determined with probability*

$$p > 1 - |H| p_e^{m-1}, \tag{2.38}$$

where m is the number of scale and orientation angle comparisons.

2.8 Application and Computation

Our goal in this section is to give visual examples demonstrating the application of certain aspects of the work we have described, and to remark on future directions of our work. The examples presented pertain to the detection of human faces from a computational perspective. We refer to [7] for an extended discussion regarding computational issues.

Detection of faces and recognition of their identity has long been an important problem in computational vision [34]. Among the reasons that make face detection hard, one can point to both intrinsic and extrinsic effects. Extrinsic effects arise from the context in which faces are viewed, or specifically, from changes in the light intensity distribution that arrives at the detecting instrument (e.g. the eye). For example, illumination conditions can vary from a dimly lit room to sunlight conditions, faces can be viewed from a variety of viewpoints, and the existence of shadows or occlusions can cause local changes in the intensity map detected by the eye. Intrinsic effects also introduce changes. Among them, we can point to color of skin, facial gestures, and changes in appearance by wearing glasses, growing beards, or wearing makeup. Our computational experiments so far have not explicitly addressed variations of facial expression or occlusions and shadows. Although we believe that a learning machine based on these principles can address a number of these issues, we have concentrated on a more realistic goal of obtaining results under more restricted conditions.

The input to our algorithm consists of grayscale images of faces selected from the ATT database of faces. We preprocess each image by normalizing and compensating for the effect of "boundaries" or "edges" of an image. We then apply the LC wavelets to an oversampled representation of the image and find local maxima and minima based on stability criteria derived using our earlier theoretical work on the existence of such points. Figure 2.3 shows the phase and amplitude component of a filtered image at three scales, illustrating the information recovered by the phase component of the LC filters. Figure 2.4 shows the points detected as the extrema by the detection procedure, where we note that the detected extrema in filtered faces carry a good deal of information despite their sparse representation (see Fig. 2.5)

The learned feature vectors were obtained by applying the singular value decomposition (SVD) method using the FS metric. Feature vectors (maxima and minima) for all faces at a single scale and orientation are used to build a single matrix. In this case, since the entries are complex, it may be natural to refer to them as density matrices. Application of singular value decomposition gives a set of eigenvectors sorted by decreasing eigenvalue. After training, we used our method to match faces and nonfaces from a random training set against the learned set. Our test case showed a 94% correct detection against a test set that was not used in the training trials.

Our promising results for the small sample data set motivate the need for a larger study to obtain further experimental verification. Guided by our mathematical framework, we find it reasonable to expect good results with a larger data set as well. A larger experimental setup requires addressing some limitations in our implementation. Biological systems may be hardwired to handle filtering with extreme efficiency, while our software filtering in the image domain has a large computational cost. We have considered two changes to improve the efficiency of this approach. The first consideration is to use a parallel and distributed approach, essentially following a model of the brain circuits. In addition, we believe that an approximation to the filtering approach where all important features are effectively preserved is possible. A

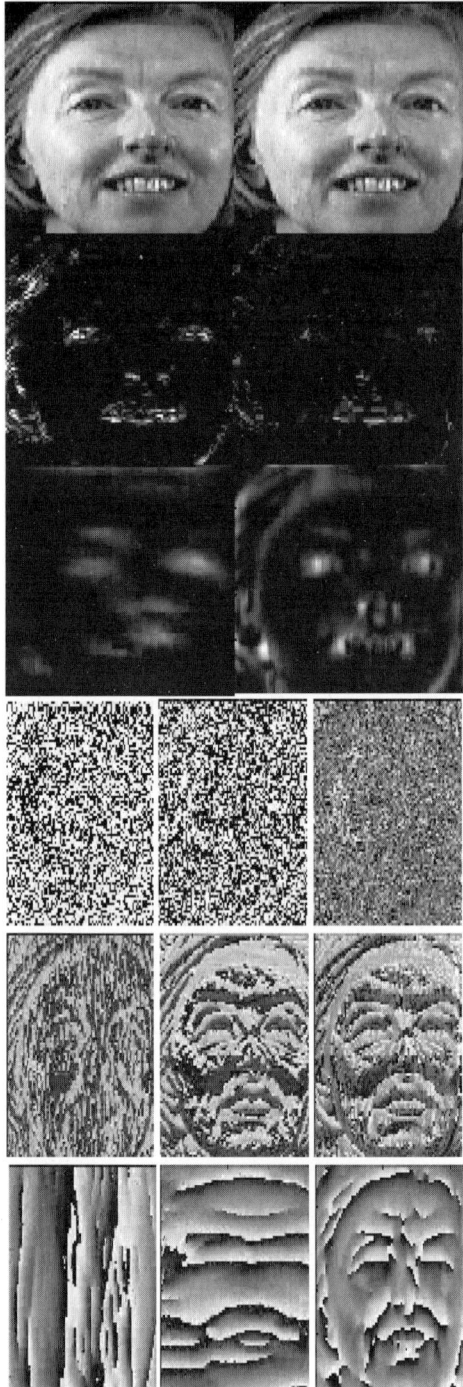

Fig. 2.3. The amplitude (*top*) and phase of wavelet coefficients at various scales and angles for an example face

Fig. 2.4. *Top three rows* represent Maxima while the *bottom three* represent minima. From *left to right* and *top to bottom* of each set angles vary at 10-degree increments. This image represents projection at the coarsest scale

good approximation would require a detailed study of the numerical behavior of the filters.

The form of the learning algorithm used in the current formulation of our simulation needs modification in order to become biologically more plausible. A straightforward modification of the learning algorithm to deal with access to feature vectors of all faces is to use a neural network approach, much in the spirit of bidirectional associative memories or Hopfield's model. However, these models are not entirely satisfactory either. An online version using local information based on approaches of local to global methods [8] and references therein) seems to be a more natural approach for this form of learning.

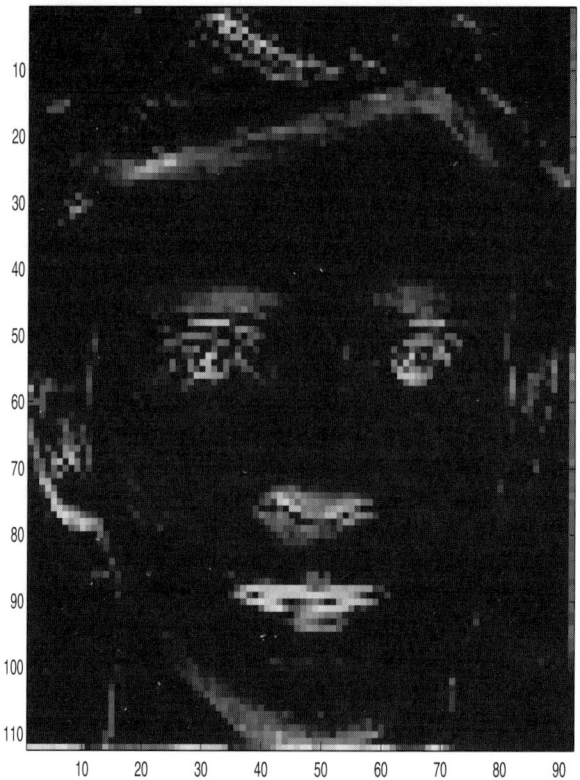

Fig. 2.5. The reconstruction of basic face features from extrema by summing values at corresponding grid points at different scales. The extrema representation is approximately 50 times more compressed than the corresponding GIF image.

References

1. Amari S.I. (1987) *Differential geometry in statistical inference.* Lecture notes-monograph series; v10. Institute of Mathematical Statistics.
2. Anthony M, Biggs M. (1992) *Computational Learning Theory.* Cambridge Tracts in Theoretical Computer Science.
3. Barlow H.B., Levick W.R. , Yoon M. (1971) Responses to single quanta of light in retinal ganglion cells of the cat, *Vision Research Supplement* 3:87–101.
4. Cave K.R. , Wolfe J.M. (1990) Modeling the role of parallel processing in visual search. *Cognitive Psychology*, 22:225 – 271.
5. Chentsov N.N. (1982) Statistical decision rules and optimal inference / N. N. Cencov; (translated from Russian) *Translations of mathematical monographs*; v.53, American Mathematical Society.
6. Daubechies I. (1992) *Ten Lectures on Wavelets*, SIAM Press, Philadelphia.
7. Eghbalnia H.R. (2000) *A Complex-valued, Overcomplete representation of signal, PhD. Thesis* University of Wisconsin-Madison.

8. Eghbalnia H. R., Assadi A. (2001) Visual target selection employing local-to-global strategies for support vector machines. Proc. SPIE Vol. 4055, pp. 130 –139, Applications and Science of Computational Intelligence III; Kevin L. Priddy, Paul E. Keller, David B. Fogel; Eds.
9. Feichtinger H. G., Gröchenig K. (1988) *A unified approach to atomic decompositions through integrable group representations, Function Spaces and Applications*, Lecture Notes in Math. (M. Cwikel et al., eds.), vol. 1302, Springer, New York, pp. 52–73.
10. Feichtinger H. G., Gröchenig K. (1989) Banach spaces related to integrable group representations and their atomic decompositions, I, *J. Funct. Anal.* 86:307–340; Banach spaces related to integrable group representations and their atomic decompositions, II, *Monatshefte für Mathematik* 108:129–148.
11. Flanders P. (1999) Time-Frequency, *Time-Scale Analysis. Elsevier*, New York.
12. Folland G. B. (1989) *Harmonic Analysis in Phase Space*, Princeton University Press, Princeton, NJ.
13. Grossmann A., Morlet J. (1984) *SIAM J. Math. Anal.* 15, 723.
14. Hargittai I., Harrgittai M. (2000) *In Our Own Image: Personal Symmetry in Discovery*, Kluwer Academin/Plenum Publishers, New York.
15. Heil C. and Walnut D. (1989) Continuous and discrete wavelet transforms, *SIAM Review*, 31:628–666.
16. Hoffman J.E. (1979) A two-stage model of visual search. *Perception and Psychophysics*, 25:319–327.
17. Hoffman J.E., Nelson B. (1981) Spatial selectivity in visual search. *Perception and Psychophysics*, 30:283–290.
18. Huang L., Pashler H. (2002) Symmetry detection and visual attention: a "binary-map" hypothesis. *Vision Res* 42:1421–1430.
19. Irwin D.E. (1992) Visual memory within and across fixations. In: *Eye Movements and Visual Cognition; Scene Perception and reading*. Raynor, K. Ed. Springer-Verlag, New York.
20. Julesz B. (1981). Textons, the elements of texture perception, and their interactions. *Nature*, 290:91–97.
21. Kass R. E., Voss P. W. (1997) *Geometrical Foundations of Asymptotic Inference*. Wiley, New York.
22. Kearns M., Vazirani U. (1994) *An Introduction to Computational Learning Theory*, pp. 52–64.
23. Land H., McCann J.J. (1971) Lightness and retinex theory. *Journal of the Optical Society of America*, 61:1–11.
24. Liu Y. (2002). Computational Symmetry, In: *Proceedings of Symmetry 2000*, v. 1888 Portland, London, Vol. 80(1):231– 245.
25. Mackey G.W. (1952) Induced representations of locally compact groups, I. *Ann. of Math.* 55:101–139.
26. Marr, D.C. (1982) *Vision*. W.H. Freeman, New York.
27. Milnor J. (1965) *Topology From a differentiable viewpoint*. University Press of Virginia, Virginia.
28. Nakayama K. (1990) The iconic bottleneck and the tenuous link between early visual processing and perception. In: *Vision: Coding and Efficiency*, C. Blakemore (Ed.), Cambridge Univ. Press, Cambridge, pp. 411– 422.
29. Natarajan B.K. (1991) *Machine Learning: A theoretical approach*. Morgan-Kauffman, San Mateo.

30. O'Regan J.K. (1992) Solving the "real" mysteries of visual perception: the world as an outside memory. *Canadian Journal of Psychology*, 46:461– 488.
31. O'Regan J.K., Lèvy-Schoen A. (1983) Integrating visual information from successive fixations: Does trans-saccadic fusion exist? *Vision Research*, 23:765 –769.
32. Olivers C.N., Van der Helm P.A. (1998). Symmetry and selective attention: a dissociation between effortless perception and serial search. *Percept Psychophys* 60:1101–1116.
33. Palmer S.E. (1999) Color, consciousness, and the isomorphism constraint. *Behav Brain Sci.* 22:923 – 943.
34. Palmer S. (1999) *Vision Science: Photons to Phenomenology*. MIT Press, Cambridge, MA.
35. Pelz B. (1995). *Visual representations in a natural visuo-motor task*. PhD thesis, Carlson Center for Imaging Science, Rochester Institute of Technology.
36. Sakai T. (1996) *Riemannian Geometry*, American Mathematical Scociety, Providence.
37. D. Schattschneider. (1998) *Visions of Symmetry*, W.H. Freeman, New York..
38. Shepard R.N. (2001) Perceptual-cognitive universals as reflections of the world. *Behav Brain Sci.* 24:581– 601.
39. Shiffrin R.M., Schneider W. (1977). Controlled and automatic human information processing: II. Perceptual learning, automatic attending and a general theory. *Psychological Review*, 84:127–190.
40. Simoncelli P., Freeman W.T., Adelson E.H., Heeger D.J. (1992). Shiftable Multiscale Transforms. *IEEE transactions on Information Theory*, 38(2):587– 607.
41. Strang G.W. (1988) . *Linear Algebra and Its applications*, HBJ, San Diego.
42. Treisman A., Gelade G. (1980) A feature integration theory of attention. *Cognitive Psychology*, 12:97– 136.
43. Treisman A. (1991a) Search, similarity, and integration of features between and within dimensions. *Journal of Experimental Psychology: Human Perception and Performance*, 17:652– 676.
44. Treisman A. (1991b) Representing visual objects. In: *Attention and Performance*, D. Meyer S. Kornblum (Eds.), XIV, Hillsdale, N.J.: Lawrence Erlbaum Associates, pp. 163 –175.
45. Treisman A. (1993) The perception of features and objects. In: *Attention: Selection, awareness, and control: A tribute to Donald Broadbent*, A. Baddeley L. Weiskrantz (Eds.), Clarendon Press, Oxford, pp. 5 – 35.
46. Valiant L.G. (1984). A theory of the learnable. *Communications of the ACM*, 27(11):1134 –1142.
47. Vapnik V.N. (1995). *The Nature of Statistical Learning Theory*. Springer, New York.
48. Vapnik V.N., Chervonenkis A.Y. (1971) On the uniform covergence of relative frequences of events to their probabilities. *Theory of Probability and its Applications*, 16(2):264-280.
49. Viviani P. (1990) Eye movements in visual search: cognitive, perceptual, and motor control aspects. In: *Eye movements and their role in visual and cognitive processes*, Kowler, E. (ed.). Reviews of Oculomotor Research V4, Elsevier, Amsterdam, pp. 353–383.
50. Wagemans J. (1995) Detection of visual symmetries. *Spat. Vis.*, 9, pp. 9 – 32.
51. Warner F. W. (1994) *Foundations of Differentiable Manifolds and Lie Groups*. Springer, New York.

52. Wolfe J.M., Cave K.R., Franzel S.L. (1989) Guided search: An alternative to the feature integration model for visual search. *Journal of Experimental Psychology: Human Perception and Performance*, 15:419 – 433.
53. Wolfe J.M. (1994) Guided search 2.0: A revised model of visual search. *Psychonomic Bulletin Review*, 1:202 – 238.
54. Wolfe J.M. (1996) Extending guided search: Why guided search needs a preattentive "item map." In: *Converging operations in the study of visual attention*, A.F. Kramer, M.G.H. Coles, G.D. Logan (Eds), Washington, DC: American Psychological Association, pp. 247 – 270.
55. Wolfe J.M. (1998) Visual search. In: *Attention*, H. Pashler (Ed.), Hove, England UK, pp. 13 – 73.
56. Yarbus A.F. (1967) *Eye Movements and Vision*. Plenum, New York.

Part II

Neural Networks

3

Geometric Approach to Multilayer Perceptrons

Hyeyoung Park[1], Tomoko Ozeki[2], and Shun-ichi Amari[2]

[1] Dept. of Computer Science, Kyungpook National University
  Sangyuk-dong, Buk-gu, Daegu, 702-701, Korea `hypark@knu.ac.kr`
[2] RIKEN Brain Science Institute
  2-1 Hirosawa, Wako, Saitama, 351-0198, Japan
  `{tomoko,amari}@brain.riken.go.jp`

3.1 Introduction

When we use a multilayer perceptron (MLP) model with a fixed architecture, its functional behavior is determined by the values of weight parameters. We can modify the parameter values by learning to obtain an optimal network for our purpose. It is possible to consider the space of all the multilayer perceptrons, in which the set of modifiable parameters plays the role of coordinates. The parameter space of a MLP is called the neuromanifold. Learning takes place in the parameter space, forming a trajectory that approaches the optimal point. Such a neuromanifold has rich geometrical structures that are responsible for various phenomena observed in practical applications of MLP. By considering those geometrical properties, it is possible to find some solutions to improve the performance as well as some explanations to those phenomena.

On the other hand, a multilayer perceptron works in a stochastic environment, and its output may be corrupted by random noise. In this sense, the behavior of a MLP can be represented by a probability distribution, e.g., a conditional probability density of the output conditioned on the input. Therefore, the space of the MLP, the neuromanifold, can also be identified with the set of conditional probability distributions specified by the weight parameters of the MLP.

Based on these statistical and geometrical viewpoints on neural networks, we use information geometrical approaches in order to investigate the properties of the MLP. Information geometry [2, 3, 5] is a powerful tool to study the intrinsic geometry of parameter spaces related to probability distributions. It introduces an invariant Riemannian metric and a dual pair of affine connections. It is useful for studying statistical inference, system theory, information theory and many others having stochastic natures as well as nonstochastic ones such as optimization. In the 1990s, information geometry was also applied successfully to neural networks. It was applied to Boltzmann machines, higher-order neurons, and Expectation-Maximization (EM) algorithms [4].

Based on information geometry, we have succeeded in obtaining a new efficient learning algorithm called the adaptive natural gradient method [7]. It gives ideal performances of convergence for learning of multilayer perceptrons. Natural gradient learning and its adaptive version take the entire geometrical structure into account and use the Riemannian metric so as to accelerate the convergence. It is also confirmed that the natural gradient learning method can avoid or alleviate plateaus, which are known to be the main cause of slow convergence of the conventional gradient descent learning method [25].

Recently, it was also found that the plateaus in learning dynamics are closely related to the hierarchical structure and permutation symmetries of MLP [17, 29]. The hierarchical structure brings complicated singularities in the space of MLP. At the singularities, the Riemannian metric degenerates and the conventional statistical theory in the Cramér–Rao paradigm does not hold. In the theoretical sense, this is an important problem for constructing theories of nonregular statistical models. At the same time, in the practical sense, it is also related to necessity of developing a new criterion for model selection as an alternative to the Akaike information criteria (AIC) [1] and minimum description length (MDL) [26], which were developed for regular models.

In the present chapter, we first review the natural gradient learning method and its adaptive versions, which were derived by using the Riemannian metric in the space of MLP. Next, we go into the problem of singularities. Using simple toy models, we analyze the explicit properties at singularities. We also review some recent results related to the singularities. Finally, we discuss the influence of singular structures on the learning dynamics of MLP, and show how the natural gradient method can solve the problem.

3.2 Space of Neural Networks

3.2.1 Stochastic Multilayer Perceptrons

By taking the information geometrical approach, we consider a MLP as a stochastic model, and then investigate the geometrical properties of the space of the stochastic MLP. We will first describe the concept of stochastic multilayer perceptrons.

In many practical problems such as regressions and time series predictions, learning data include noises so that the input-output relation is described stochastically in terms of the conditional probability density $p(\boldsymbol{y}|\boldsymbol{x})$ of the output $\boldsymbol{y}$ when an input $\boldsymbol{x}$ is given. In other practical applications such as classifications, on the other hand, the output has discrete values so that the continuous values of output nodes of neural networks can be considered as representing a conditional probability $P(C|\boldsymbol{x})$ of class C given input $\boldsymbol{x}$. Thus, even though most neural network models are deterministic in their nature, it can be regarded as a stochastic system estimating a true probability density function from which the input-output data are generated. This approach leads to a stochastic model of neural networks.

We will explain this concept more clearly using a simple three-layer MLP with one output node. However, its generalization can be easily done. The network calculates an input-output function $f(\boldsymbol{x}, \boldsymbol{\theta})$ that is determined by the network structure and is written by

$$f(\boldsymbol{x}, \boldsymbol{\theta}) = \varphi_\mathrm{o} \left(\sum_j v_j \varphi_\mathrm{h}(\boldsymbol{w}_j^T \boldsymbol{x} + b_j) + b_\mathrm{o} \right), \qquad (3.1)$$

where $v_j, \boldsymbol{w}_j, b_j, b_\mathrm{o}$ are the weight parameters of the network, and are summarized to $\boldsymbol{\theta}$. The functions φ_o and φ_h are the activation functions of output and hidden nodes, respectively. When we treat a neural network from the stochastic viewpoint, we say that the final output y is emitted through a stochastic operation $\mathcal{S}$ operated on the deterministic function f,

$$y = \mathcal{S}\{f(\boldsymbol{x}; \boldsymbol{\theta})\}. \qquad (3.2)$$

The stochastic operation can be defined properly for given applications. For regression problems, $\mathcal{S}$ is usually defined by a Gaussian additive random noise, so that y can be written by a sum of the deterministic function and a random noise:

$$y = f(\boldsymbol{x}; \boldsymbol{\theta}) + n, \qquad n \sim \mathcal{N}(0, \sigma^2). \qquad (3.3)$$

From the assumption that the random noise is subject to Gaussian distribution, the corresponding conditional probability density function is given by

$$p(y|\boldsymbol{x}; \boldsymbol{\theta}) = \frac{1}{\sqrt{2\pi}\sigma} \exp\left\{-\frac{1}{2}[y - f(\boldsymbol{x}; \boldsymbol{\theta})]^2\right\}. \qquad (3.4)$$

For classification problems, we can use the coin-flipping process for the stochastic operation $\mathcal{S}$, so that the binary output y representing class C_1 and C_0 is determined by

$$y = \begin{cases} 1 & (\boldsymbol{x} \in C_1), \\ 0 & (\boldsymbol{x} \in C_0), \end{cases} \quad \text{with probability} \quad \begin{matrix} f(\boldsymbol{x}, \boldsymbol{\theta}), \\ 1 - f(\boldsymbol{x}, \boldsymbol{\theta}). \end{matrix} \qquad (3.5)$$

Then the corresponding conditional probability density function for y given input $\boldsymbol{x}$ is written as

$$p(y|\boldsymbol{x}; \boldsymbol{\theta}) = f(\boldsymbol{x}, \boldsymbol{\theta})^y [1 - f(\boldsymbol{x}, \boldsymbol{\theta})]^{(1-y)}. \qquad (3.6)$$

Using this stochastic viewpoint, we can describe the behavior of the network with the probability density function $p(y|\boldsymbol{x}; \boldsymbol{\theta})$, and the properties of the space of MLP can be explained by investigating the space of corresponding probability density functions.

3.2.2 Metric of Neuromanifold

When we consider the space of MLP, the neuromanifold, the most basic property is its metric. Since the neuromanifold can be considered as a space of probability density functions, we can exploit the Kullback–Leibler divergence, which is often used to show the discrepancy between two probability distributions specified by $p(\boldsymbol{y}|\boldsymbol{x},\boldsymbol{\theta})$ and $p(\boldsymbol{y}|\boldsymbol{x},\boldsymbol{\theta}')$. It is defined by

$$D\left[\boldsymbol{\theta}:\boldsymbol{\theta}'\right] = \int p(\boldsymbol{y}|\boldsymbol{x},\boldsymbol{\theta})q(\boldsymbol{x})\log\frac{p(\boldsymbol{y}|\boldsymbol{x},\boldsymbol{\theta})}{p\left(\boldsymbol{y}|\boldsymbol{x},\boldsymbol{\theta}'\right)}\mathrm{d}\boldsymbol{x}\mathrm{d}\boldsymbol{y}, \qquad (3.7)$$

where $q(\boldsymbol{x})$ is the probability density function of input $\boldsymbol{x}$ [14].

When the two distributions are infinitesimally close, $\boldsymbol{\theta}' = \boldsymbol{\theta} + \mathrm{d}\boldsymbol{\theta}$, and the divergence between two nearby distributions is expanded as

$$D\left[\boldsymbol{\theta}:\boldsymbol{\theta}+\mathrm{d}\boldsymbol{\theta}\right] = \frac{1}{2}\mathrm{d}\boldsymbol{\theta}^\tau G(\boldsymbol{\theta})\mathrm{d}\boldsymbol{\theta}, \qquad (3.8)$$

where $G(\boldsymbol{\theta})$ is the Fisher information matrix defined by

$$G(\boldsymbol{\theta}) = E_{\boldsymbol{x},\boldsymbol{y}}\left[\frac{\partial \log p(\boldsymbol{y}|\boldsymbol{x},\boldsymbol{\theta})}{\partial \boldsymbol{\theta}}\frac{\partial \log p(\boldsymbol{y}|\boldsymbol{x},\boldsymbol{\theta})^\tau}{\partial \boldsymbol{\theta}}\right]. \qquad (3.9)$$

$E_{\boldsymbol{x},\boldsymbol{y}}[\cdot]$ denotes the expectation with respect to $p(\boldsymbol{y}|\boldsymbol{x},\boldsymbol{\theta})q(\boldsymbol{x})$, and superfix τ is transposition of a column vector.

A manifold is said to be Riemannian when it is locally Euclidean, which means that the metric $G(\boldsymbol{\theta})$ is defined at every point $\boldsymbol{\theta}$ such that the square of the distance between two nearby points $\boldsymbol{\theta}$ and $\boldsymbol{\theta} + \mathrm{d}\boldsymbol{\theta}$ is defined by the quadratic form in Eq. (3.8). When $G(\boldsymbol{\theta})$ does not depend on $\boldsymbol{\theta}$ and is equal to the identity matrix, the manifold is Euclidean and the corresponding coordinate system $\boldsymbol{\theta}$ is orthonormal. When $G(\boldsymbol{\theta})$ cannot be reduced to the identity matrix whatever coordinate system is used, the manifold is intrinsically Riemannian, and $G(\boldsymbol{\theta})$ is called a Riemannian metric. Especially, the Fisher information matrix (3.9) is called a Fisher metric.

Since the Fisher metric $G(\boldsymbol{\theta})$ is a unique one that is invariant over the choice of coordinate system, we can say that it is the most appropriate Riemannian metric for the neuromanifold (see [5] for details). The manifold of the previous two examples of stochastic MLP models are all Riemannian. In the case of Gaussian additive noise model defined in Eq. (3.4), the explicit form of $G(\boldsymbol{\theta})$ is given by

$$G(\boldsymbol{\theta}) = E_{\boldsymbol{x}}\left[E_{\boldsymbol{y}|\boldsymbol{x};\boldsymbol{\theta}}\left[\{y - f(\boldsymbol{x},\boldsymbol{\theta})\}^2\frac{\partial f(\boldsymbol{x},\boldsymbol{\theta})}{\partial \boldsymbol{\theta}}\frac{\partial f(\boldsymbol{x},\boldsymbol{\theta})^\tau}{\partial \boldsymbol{\theta}}\right]\right], \qquad (3.10)$$

$$= \sigma^2 E_{\boldsymbol{x}}\left[\frac{\partial f(\boldsymbol{x},\boldsymbol{\theta})}{\partial \boldsymbol{\theta}}\frac{\partial f(\boldsymbol{x},\boldsymbol{\theta})^\tau}{\partial \boldsymbol{\theta}}\right], \qquad (3.11)$$

where $E_{\boldsymbol{x}}[\cdot]$ and $E_{y|\boldsymbol{x};\boldsymbol{\theta}}[\cdot]$ denote the expectations with respect to $q(\boldsymbol{x})$ and $p(y|\boldsymbol{x},\boldsymbol{\theta})$, respectively. In the case of coin-flipping model defined in Eq. (3.6), the explicit form of $G(\boldsymbol{\theta})$ is given by

$$G(\boldsymbol{\theta}) = E_{\boldsymbol{x}}\left[E_{y|\boldsymbol{x},\boldsymbol{\theta}}\left[\frac{(y-f)^2}{f^2(1-f)^2}\frac{\partial f(\boldsymbol{x},\boldsymbol{\theta})}{\partial \boldsymbol{\theta}}\frac{\partial f(\boldsymbol{x},\boldsymbol{\theta})}{\partial \boldsymbol{\theta}}^T\right]\right], \tag{3.12}$$

$$= E_{\boldsymbol{x}}\left[\frac{1}{f(1-f)}\frac{\partial f(\boldsymbol{x},\boldsymbol{\theta})}{\partial \boldsymbol{\theta}}\frac{\partial f(\boldsymbol{x},\boldsymbol{\theta})}{\partial \boldsymbol{\theta}}^T\right]. \tag{3.13}$$

By using this metric for the space of MLP, we can derive a new gradient descent learning method, which is different from the conventional ones, as we discuss in Sect. 3.

3.2.3 Hierarchical Structure

Another important characteristic of the space of MLP is the hierarchical structures such that the space of MLP with a smaller number of hidden nodes are included in the one with a larger number of hidden nodes. The hierarchical systems have an interesting geometry, which has some influence on the dynamical behaviors of learning of the systems. We give a simple example of such hierarchies.

Let us consider a simple MLP, which receives an input vector signal $\boldsymbol{x}$ and emits a scalar output signal y. Let h be the number of hidden units, and let $\boldsymbol{w}_i$ be the weight vectors of the ith hidden unit, $i=1\cdots h$. Let φ be a sigmoidal activation function such as the hyperbolic tangent, and let v_i be the weight from the ith hidden unit to the output unit. We assume that the output unit is linear and is disturbed by Gaussian noise n with mean 0 and variance 1. The input-output relation of the simple MLP is then represented as

$$y = f(\boldsymbol{x},\boldsymbol{\theta}) + n, \tag{3.14}$$

where

$$f(\boldsymbol{x},\boldsymbol{\theta}) = \sum_{i=1}^{h} v_i \varphi(\boldsymbol{w}_i \cdot \boldsymbol{x}), \tag{3.15}$$

and $\boldsymbol{\theta}$ is the vector parameter summarizing all the modifiable parameters $\boldsymbol{w}_1 \cdots \boldsymbol{w}_h$ and $v_1 \cdots v_h$.

Let us denote by $M(h)$ the set of all such multilayer perceptrons which forms a space with coordinates $\boldsymbol{\theta}$. One can easily see that $M(h)$ includes $M(h-1), M(h-2), \ldots$ as subspaces. For example, when

$$v_i \|\boldsymbol{w}_i\| = 0 \tag{3.16}$$

holds, the ith hidden node does not play any role, so it can be removed. Hence, the subspace defined by Eq. (3.16) in $M(h)$ corresponds to $M(h-1)$. In addition, when $\boldsymbol{w}_i = \boldsymbol{w}_j$, the ith and jth hidden neurons play the same role so that they can be merged into one neuron. Hence, the subspace defined by $\boldsymbol{w}_i = \boldsymbol{w}_j$ is also identified with $M(h-1)$. In the same manner, we can see that a hierarchy such as

$$M(h) \supset M(h-1) \supset \cdots \supset M(0) \tag{3.17}$$

exists among the spaces of MLP models with different numbers of hidden nodes. The subspaces $M(h-1)$, $M(h-2)$,... in the space of $M(h)$ make complicated singularities as discussed in the next section.

3.2.4 Singular Structure

A point in a neuromanifold $M(h)$ is represented by a parameter vector $\boldsymbol{\theta}$, and the parameter vector specifies an explicit functional relation of the network. Two points $\boldsymbol{\theta}$ and $\boldsymbol{\theta}'$ are said to be equivalent in $M(h)$ when their corresponding functional relations are the same, that is, the related probability distributions are the same. In the case of the MLP model shown in the previous section, the following three types of equivalence relations are known [13, 22, 27, 30]:

1. When $v_i = 0$, any points $\boldsymbol{\theta}$ and $\boldsymbol{\theta}'$ are equivalent when they differ only by $\boldsymbol{w}_i$.
2. When $\|\boldsymbol{w}_i\| = 0$, any points $\boldsymbol{\theta}$ and $\boldsymbol{\theta}'$ are equivalent when they differ only by v_i.
3. When $\boldsymbol{w}_i = \boldsymbol{w}_j$ (or $\boldsymbol{w}_i = -\boldsymbol{w}_j$) holds, any points $\boldsymbol{\theta}$ and $\boldsymbol{\theta}'$ are equivalent when

$$v_i + v_j = v'_i + v'_j \quad (v_i - v_j = v'_i - v'_j).$$

Each set of equivalent points forms a subset in $M(h)$. One can easily see that the subsets correspond to $M(h-1)$ discussed in the previous section.

Since we are not interested in the individual parameter values but instead in the functional behavior of the network, we regard all the mutually equivalent points (i.e., networks) as one and the same network. Mathematically speaking, we take the residue class by the equivalence relations as defined above. Then, an equivalent subspace reduces to one point, and the resultant space around the point has a cone-type singular structure. Figure 3.1a shows a conceptual illustration of the singular structure. More generally, many cone-type singularities can be connected to form a lower-dimensional subset (Fig. 3.1b). More explicit examples of the singular structure of neuromanifolds are given in [6]. The reduced space has lots of singularities where dimensionalities are reduced. This type of singularity is closely related to the performance of the learning dynamics of MLP, and also poses an important problem in classical statistical theory of inference.

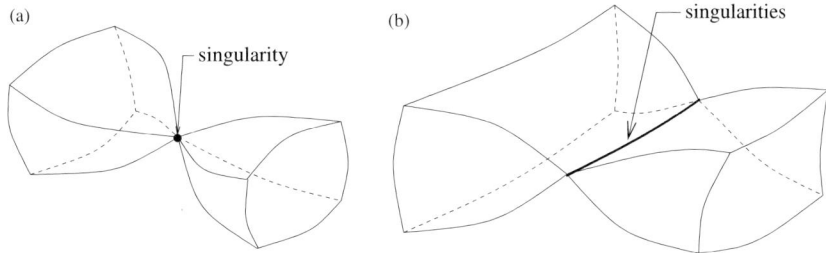

Fig. 3.1. Cone-type singular structures

3.3 Learning in Neuromanifolds

3.3.1 Basic Concepts

For learning, we consider a space of conditional probability density functions $\{p(\boldsymbol{y}|\boldsymbol{x};\boldsymbol{\theta})|\boldsymbol{\theta} \in \Re^M\}$ of stochastic MLP with input $\boldsymbol{x}$, output $\boldsymbol{y}$, and a parameter vector $\boldsymbol{\theta}$. The space of the MLP, the neuromanifold, has a coordinate system $\boldsymbol{\theta}$. The parameter $\boldsymbol{\theta}$ is modified by learning from examples. The set of examples of input-output pairs $\{(\boldsymbol{x}_1, \boldsymbol{y}_1), \ldots, (\boldsymbol{x}_T, \boldsymbol{y}_T)\}$ is called the training set. It is assumed that they are generated from the true conditional probability distribution $p^*(\boldsymbol{y}|\boldsymbol{x})$, which might not be included in the neuromanifold. If we consider the space of all the conditional probability distributions, the neuromanifold is included in it. The goal of learning is to find the optimum $\boldsymbol{\theta}^*$ that minimizes the discrepancy from the true probability density function $p^*(\boldsymbol{y}|\boldsymbol{x})$ to the neuromanifold (Fig. 3.2).

The discrepancy is generally measured by the Kullback–Leibler divergence, which has the form

$$D[p^*(\boldsymbol{y}|\boldsymbol{x}) : p(\boldsymbol{y}|\boldsymbol{x}, \boldsymbol{\theta})] = \int p^*(\boldsymbol{y}|\boldsymbol{x})q(\boldsymbol{x}) \log \frac{p^*(\boldsymbol{y}|\boldsymbol{x})}{p(\boldsymbol{y}|\boldsymbol{x},\boldsymbol{\theta})} \mathrm{d}\boldsymbol{x}\mathrm{d}\boldsymbol{y}, \quad (3.18)$$

$$= E_{p^*}\left[\log \frac{p^*(\boldsymbol{y}|\boldsymbol{x})}{p(\boldsymbol{y}|\boldsymbol{x},\boldsymbol{\theta})}\right], \quad (3.19)$$

$$= E_{p^*}[\log p^*(\boldsymbol{y}|\boldsymbol{x})] - E_{p^*}[\log p(\boldsymbol{y}|\boldsymbol{x},\boldsymbol{\theta})], \quad (3.20)$$

where $E_{p^*}[\cdot]$ denotes the expectation with respect to the true probability density $p^*(\boldsymbol{y}|\boldsymbol{x})q(\boldsymbol{x})$. By neglecting the $\boldsymbol{\theta}$-independent part, we obtain the typical error function, the negative logarithm of the likelihood, which is written as

$$l_{\mathrm{gen}}(\boldsymbol{\theta}) = -E_{p^*}[\log p(\boldsymbol{y}|\boldsymbol{x},\boldsymbol{\theta})]. \quad (3.21)$$

This is called the generalization error. In practical implementation, however, l_{gen} cannot be calculated because the true probability density is unknown. Therefore, the empirical error obtained from an observed data set $D = \{(\boldsymbol{x}_t, \boldsymbol{y}_t)\}_{t=1\cdots T}$ is defined by

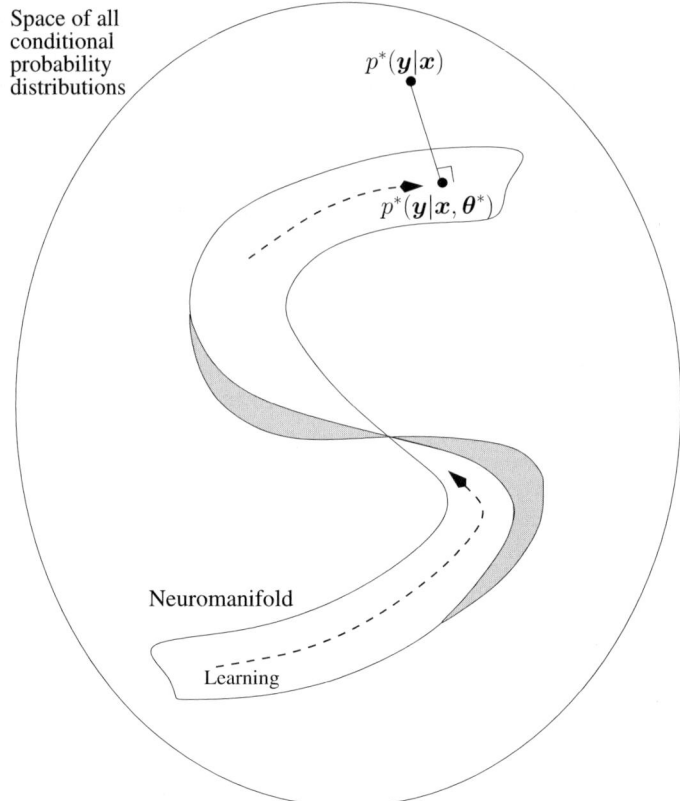

Fig. 3.2. Learning in a neuromanifold

$$l_{\text{train}}(\boldsymbol{\theta}) = -\frac{1}{T}\sum_{t=1}^{T}\log p(\boldsymbol{y}_t|\boldsymbol{x}_t,\boldsymbol{\theta}). \tag{3.22}$$

This is called the training error. Using this error function, one can calculate the standard gradient $\nabla l = \partial l/\partial \boldsymbol{\theta}$, which represents the steepest direction of l when the parameter space is Euclidean. The standard gradient descent learning algorithm is then given by

$$\boldsymbol{\theta}_{t+1} = \boldsymbol{\theta}_t - \eta_t \nabla l_{\text{train}}(\boldsymbol{\theta}_t) = \boldsymbol{\theta}_t - \eta_t \frac{\partial l_{\text{train}}(\boldsymbol{\theta}_t)}{\partial \boldsymbol{\theta}_t}, \tag{3.23}$$

where η_t is a learning rate. This updating rule uses the whole data set at each update, and thus is called batch learning. It is also possible to use only one data in the training data set at each update, giving the updating rule

$$\boldsymbol{\theta}_{t+1} = \boldsymbol{\theta}_t - \eta_t \frac{\partial l(\boldsymbol{y}_t, \boldsymbol{x}_t, \boldsymbol{\theta}_t)}{\partial \boldsymbol{\theta}_t}, \tag{3.24}$$

where $l(\boldsymbol{y}, \boldsymbol{x}, \boldsymbol{\theta}) = -\log p(\boldsymbol{y}|\boldsymbol{x}, \boldsymbol{\theta})$. This is called the online learning scheme.

The online learning scheme is realized by the standard stochastic gradient descent learning method, which uses the standard gradient of the Euclidean space and a scalar learning rate. If the conditional probability $p(\boldsymbol{y}|\boldsymbol{x}; \boldsymbol{\theta})$ is defined as a Gaussian distribution with zero mean and unit variance, then the error function becomes the squared error function and the learning rule given in Eq. (3.23) has the same form as the standard back-propagation learning method [28], which is the most popular learning method for neural networks. However, we can obtain many variations of the gradient learning algorithm by using different forms of $p(\boldsymbol{y}|\boldsymbol{x}; \boldsymbol{\theta})$. In this sense, the stochastic gradient descent learning method is a generalization of standard back-propagation. It is also obvious that we can expect better performance by using a more appropriate form of $p(\boldsymbol{y}|\boldsymbol{x}; \boldsymbol{\theta})$ for given applications rather than using the standard one.

The gradient is closely related to the metric, and standard gradient descent learning uses the Euclidean metric. As discussed in Sect. 2, however, the neuromanifold has a Riemannian metric. By using the Riemannian metric, we developed a new gradient called a natural gradient.

3.3.2 Natural Gradient Learning

Based on the fact that the space of the MLP is Riemannian, the natural gradient learning method was developed as the true steepest descent learning method. As discussed in Sect. 2.2, the Riemannian metric of the space is given by the Fisher information matrix $G(\boldsymbol{\theta})$ defined in Eq. (3.9). Using this Fisher information matrix, we can obtain the steepest direction $\tilde{\nabla} l$,

$$\tilde{\nabla} l(\boldsymbol{x}, \boldsymbol{y}, \boldsymbol{\theta}) = G^{-1}(\boldsymbol{\theta}) \nabla l(\boldsymbol{x}, \boldsymbol{y}, \boldsymbol{\theta}) = G^{-1}(\boldsymbol{\theta}) \frac{\partial l(\boldsymbol{x}, \boldsymbol{y}, \boldsymbol{\theta})}{\partial \boldsymbol{\theta}}, \quad (3.25)$$

in the Riemannian space. The $\tilde{\nabla} l$ is called the natural gradient, and the related learning algorithm is given by

$$\boldsymbol{\theta}_{t+1} = \boldsymbol{\theta}_t - \eta_t \tilde{\nabla} l(\boldsymbol{x}_t, \boldsymbol{y}_t, \boldsymbol{\theta}_t). \quad (3.26)$$

Even though it has been proved that the natural gradient learning algorithm gives a Fisher efficient online estimator [3], there are some problems in implementation of this method. First, we have to know the probability density function $q(\boldsymbol{x})$ to get an explicit form of $G(\boldsymbol{\theta})$, but this information is hardly given in practical applications. Second, even if we can get the explicit form of $G(\boldsymbol{\theta})$, the inversion of $G(\boldsymbol{\theta})$ is necessary in order to get the natural gradient at each learning step, which is very time consuming. To solve these difficulties, Amari et al. [7] proposed an adaptive method of directly obtaining an estimate of $G^{-1}(\boldsymbol{\theta})$ for the stochastic neural network having one output node with Gaussian random noise. Subsequently, Park et al. [24] developed explicit algorithms of adaptive natural gradient learning for various stochastic models and error functions. These algorithms are widely used in practical applications.

In order to implement the adaptive natural gradient explicitly, we need to define the explicit form of $p(y|x;\theta)$ as discussed in Sect. 2. In this section, we briefly review two explicit forms of adaptive natural gradient learning developed in Park et al. [24]: one for regression problems and the other for classification problems.

For regression problems, such as function approximation, time series prediction, and nonlinear system identification, we can use the following type of the stochastic network:

$$y = f(x, \theta) + n, \tag{3.27}$$

with an additive noise $n \in \mathcal{R}^M$. Note that this model is an extension of the simple Gaussian noise model defined in Eq. (3.3) to multivariate outputs with M output nodes. The value of each output node y_i is decided by the sum of the output of deterministic function $f_i(x, \theta)$ and additional random noise n_i. Assuming that each noise element n_i is independent and subject to the standard normal distribution, we can get the conditional probability density function of output y given input x, which can be written by

$$p(y|x;\theta) = \prod_{i=1}^{M} \frac{1}{\sqrt{2\pi}} \exp\left\{\frac{1}{2}[y_i - f_i(x,\theta)]^2\right\}. \tag{3.28}$$

The negative of log-likelihood gives an error function for this model, which is written by

$$l(x, y; \theta) = -\frac{1}{2} \sum_{i=1}^{M} [y_i - f_i(x, \theta)]^2. \tag{3.29}$$

This is the same form as the standard sum of squares error function. Using this stochastic model, we can obtain the Fisher information matrix $G(\theta)$ of the form,

$$G(\theta) = E_x \left[\nabla F(x, \theta) \nabla F(x, \theta)^T\right], \tag{3.30}$$

where

$$\nabla F(x, \theta) = \left(\frac{\partial f_1(x, \theta)}{\partial \theta}, \cdots, \frac{\partial f_M(x, \theta)}{\partial \theta}\right). \tag{3.31}$$

Since we do not know the distribution $q(x)$, we estimate the matrix $G(\theta)$ adaptively at each step t using

$$\hat{G}_{t+1} = (1 - \epsilon_t)\hat{G}_t + \epsilon_t \nabla F(x_t, \theta_t) \nabla F(x_t, \theta_t)^T, \tag{3.32}$$

where ε_t is a small time-dependent constant such as $1/t$, and $\hat{G}_0$ at $t = 0$ is an arbitrary nonsingular matrix such as the identity matrix. From this estimation, we can directly get the estimate $\hat{G}_{t+1}^{-1}$ of the inverse of the Fisher information matrix, which is given by

$$\hat{G}_{t+1}^{-1} = \frac{1}{1-\varepsilon_t}\hat{G}_t^{-1}$$
$$- \frac{\varepsilon_t}{(1-\varepsilon_t)}\hat{G}_t^{-1}\nabla F_t \left[(1-\varepsilon_t)I + \varepsilon_t \nabla F_t^\tau \hat{G}_t^{-1}\nabla F_t\right]^{-1} \nabla F_t^\tau \hat{G}_t^{-1}. \quad (3.33)$$

Whereas the outputs for regression problems are in general continuous values, the target output values for classification problems are discrete, representing the classes of patterns. Therefore, the additive noise model and the corresponding squared error function are not appropriate for classification problems. Park et al. [24] used the Bayesian stochastic model for classification problems [10] and gave an explicit form of adaptive natural gradient learning for the model.

In this section, we consider the case of M-class classification problems as the extension of the coin-flipping model defined in Eq. (3.5). We need a network with M output nodes so that the ith output node represents the ith class C_i. We use the target coding scheme, that is, $y_j = \delta_{ij}$ $(j = 1, \ldots, M)$ for class C_i. In this coding scheme, the value of the ith output node $f_i(\boldsymbol{x}, \boldsymbol{\theta})$ can be considered as representing the posterior probability $P(C_i|\boldsymbol{x})$ for class C_i. The conditional distribution can be written as

$$p(\boldsymbol{y}|\boldsymbol{x}; \boldsymbol{\theta}) = \prod_{i=1}^{M}(f_i(\boldsymbol{x}, \boldsymbol{\theta}))^{y_i}. \quad (3.34)$$

Since the output values $f_i(\boldsymbol{x}, \boldsymbol{\theta})$ are interpreted as probabilities, they must lie in the range $[0, 1]$, and their sum must be equal to 1. This can be achieved by using a generalized form of the logistic function for the activation function of output nodes, which is defined as

$$f_i(\boldsymbol{x}, \boldsymbol{\theta}) = \varphi_o(z_i) = \frac{\exp(z_i)}{\sum_{j=1}^{M}\exp(z_j)}. \quad (3.35)$$

Here z_i is the linear sum of hidden outputs given to the ith output node.

The corresponding error function is given by

$$l(\boldsymbol{x}, \boldsymbol{y}; \boldsymbol{\theta}) = -\sum_{i=1}^{M} y_i \log f_i(\boldsymbol{x}, \boldsymbol{\theta}). \quad (3.36)$$

This error function is called the cross-entropy error function. It is known that in the case of classification problems, the cross-entropy function gives better convergence and generalization performance than the sum of squares error function [10].

Using this stochastic model, we can obtain the Fisher information matrix $G(\boldsymbol{\theta})$ and the adaptive estimate $\hat{G}_{t+1}^{-1}$ of the inverse of the Fisher information matrix, which is given by

$$G(\boldsymbol{\theta}) = E_{\boldsymbol{x}} \left[\sum_{i=1}^{M} \frac{1}{f_i(\boldsymbol{x}, \boldsymbol{\theta})} \frac{\partial f_i}{\partial \boldsymbol{\theta}} \left(\frac{\partial f_i}{\partial \boldsymbol{\theta}} \right)^{\tau} \right], \quad (3.37)$$

$$\hat{G}_{t+1}^{-1} = \frac{1}{1-\varepsilon_t} \hat{G}_t^{-1}$$
$$- \frac{\varepsilon_t}{(1-\varepsilon_t)} \hat{G}_t^{-1} \nabla \tilde{F}_t \left[(1-\varepsilon_t) I + \varepsilon_t \nabla \tilde{F}_t^{\tau} \hat{G}_t^{-1} \nabla \tilde{F}_t \right]^{-1} \nabla \tilde{F}_t^{\tau} \hat{G}_t^{-1}, \quad (3.38)$$

where

$$\nabla \tilde{F}_t = \left[\frac{1}{\sqrt{f_1(\boldsymbol{x}_t, \boldsymbol{\theta}_t)}} \frac{\partial f_1(\boldsymbol{x}_t, \boldsymbol{\theta}_t)}{\partial \boldsymbol{\theta}_t}, \ldots, \frac{1}{\sqrt{f_M(\boldsymbol{x}_t, \boldsymbol{\theta}_t)}} \frac{\partial f_M(\boldsymbol{x}_t, \boldsymbol{\theta}_t)}{\partial \boldsymbol{\theta}_t} \right]. \quad (3.39)$$

In general, the natural gradient method is different from second-order methods such as the Newton method, the conjugate gradient method, and the like, which use a Hessian matrix instead of a Fisher information matrix (see [10] for details). However, when log-likelihood is taken as the cost function, it is locally equivalent to the Newton method and the Fisher scoring method. Hence, its convergence is second order. The adaptive method of evaluating G^{-1} is similar in this case to the Gauss–Newton method [12]. However, the natural gradient method is more general and is applicable to various cost functions other than the square loss or the negative log-likelihood [24].

The merit of the natural gradient method is not merely given by its good local convergence property. Its merit lies mostly in the global property of convergence, escaping from plateaus. This is because the Riemannian metric is responsible for the singularities of the space that are related to plateaus. We show this in Sect. 4.

3.3.3 Computational Experiments

In order to see the convergence performance of adaptive natural gradient learning, we show two experimental results on benchmark problems. First is the Mackey–Glass chaotic time series prediction, which is a well-known benchmark problem for neural networks. The time series data were generated from the equation

$$x(t+1) = (1-b)x(t) + a \frac{x(t-\tau)}{1+x(t-\tau)^{10}}, \quad (3.40)$$

where $a = 0.2$, $b = 0.1$, and $\tau = 17$. The input values of the network are given from four previous time series data, i.e., $x(t), x(t-6), x(t-12)$, and $x(t-18)$. The output value of the network is given from one future time series datum, $x(t+6)$. For training, 500 data generated at $t = 200, \ldots, 700$ were used, and 500 other data at $t = 5000, \ldots, 5500$ were used for testing.

Since the output values are continuous, the Gaussian additive noise model was used. The adaptive natural gradient learning was compared to the standard gradient descent learning. Each learning method was tried ten times with different initialization in order to get the average result. The learning process was stopped when the mean square error (MSE) for the training data became smaller than 2×10^{-5}.

Table 3.1. Average results on the Mackey–Glass time series prediction problem (SGL, standard gradient learning; ANGL, adaptive natural gradient learning)

	SGL	ANGL
Learning rate	$\eta = 0.1$	$\eta = 0.005, \varepsilon_t = 1/t$
No. of hidden nodes	10	10
Rate of success	10/10	10/10
Learning cycle for MSE $< 2 \times 10^{-5}$	836,480	502.2
(standard deviation)	(396,320)	(132.8)
MSE for test data	7.6265×10^{-5}	2.4716×10^{-5}
(standard deviation)	(3.0823×10^{-5})	(4.7204×10^{-6})
Processing time (relative to SGL)	1.0	0.064

The average results are shown in Table 3.1. The learning rates η were tuned through the experiments so as to get high rates of success and rapid convergence. For this problem, the adaptive natural gradient converged more than 1600 times faster than the standard gradient learning method in the sense of necessary number of learning cycles for convergence. Regarding the processing time, the proposed learning algorithm was more than 15 times faster than the ordinary one. Figure 3.3 shows the learning curves of the two algorithms. From this figure, one can see that the plateaus that appear in the standard gradient learning method mostly do not exist in the adaptive natural gradient learning method. Note that the network used for training has ten hidden units, which implies that its parameter space has very complex singularities (Sect. 2.4). The plateau observed in standard gradient learning is known to be caused by the singularities [29]. From this experiment, we can say that the adaptive natural gradient can alleviate the plateau remarkably. We will discuss the learning dynamics and singularity in Sect. 4.4 with a simple and visual model.

As an example of a classification problem, the thyroid disease problem was exploited, which is also a well-known benchmark problem. It is known to be hard to train a feedforward neural network by using the standard gradient leaning algorithm with the data set. The task is to determine whether a patient referred to the clinic is hypothyroid. Therefore, three classes are built: normal (not hypothyroid), hyperfunctioning, and subnormal functioning. Each measurement data has 21 attributes. Fifteen of them are binary values, and the others are continuous values. Since 92% of the patients are not hyperthyroid,

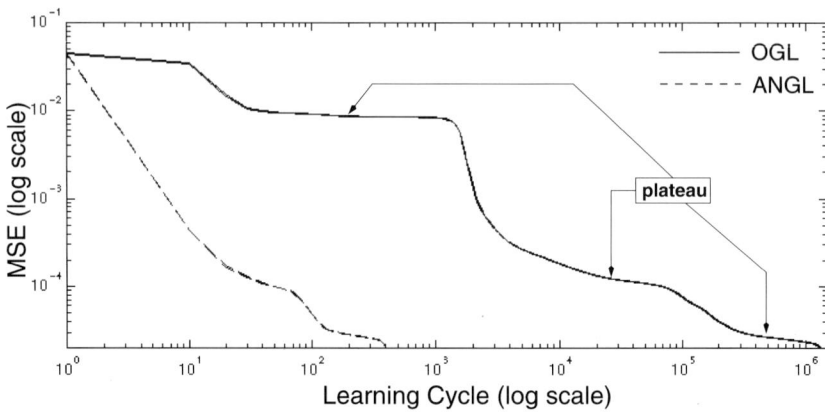

Fig. 3.3. Learning curves for the Mackey–Glass problem (OGL, standard (ordinary) gradient learning; ANGL, adaptive natural gradient learning). Reprinted with permission from [24]

a good classifier must give a significantly better classification rate than 92%. The number of data points in the training set is 3,772, and the number of data points in the test set is 3,428. The network has 21 input nodes, 5 hidden nodes, and 3 output nodes. The learning process was stopped when the MSE became smaller than 10^{-3}.

Table 3.2. Average results on Thyroid disease classification problem (SGL, standard gradient learning; ANGL, adaptive natural gradient learning)

	SGL	ANGL
Learning rate	$\eta = 0.005$	$\eta = 0.0002, \varepsilon_t = 1/t$
No. of hidden nodes	5	5
Rate of success	5/5	5/5
Learning cycle for MSE $< 10^{-3}$	459,900.0	349.0
(standard deviation)	(124,100)	(172.5)
Classification rate (training)	99.81%	99.87%
(standard deviation)	(0.01%)	(0.05%)
Classification rate (test)	97.55%	98.19%
(standard deviation)	(0.26%)	(0.12%)
Processing time (relative to SGL)	1.0	0.24

The average results over five independent runs are shown in Table 3.2. Regarding the learning cycles, the adaptive natural gradient method was more than 1,300 times faster than the standard gradient method. Also, considering the classification performance, the adaptive natural gradient method gave higher classification rates for the test data than the ordinary one. As far as the processing time is concerned, the proposed method was more than four times faster than the ordinary one.

3.4 Problem of Singular Structures

3.4.1 Singularity Problem

The remarkable superiority of natural gradient learning, which was shown in the previous section, is achieved by exploiting an appropriate metric, the Fisher metric, for the neuromanifold. However, in order to gain a deeper understanding of why the plateaus disappear in natural gradient learning, we need to consider the singular structure of the neuromanifold. The singular structure of the neuromanifold seems to be a main cause of various problems in the learning and designing of neural networks. By paying more attention to the singularities, we can expect to get remarkable solutions or understanding of the problems.

The basic problem related to the singular structure is in the case when the optimal network is exactly on the singular point of the neuromanifold, which is called the singularity problem. This situation occurs when we use larger networks than the optimal one; thus we also call it the over-realizable scenario (note that this situation occurs easily because we do not know the optimal size of networks and tend to use sufficiently large networks). When the optimal network is at the singularity, it is impossible to use conventional model selection criteria such as AIC or MDL in order to find an optimal size of network. These criteria are based on the classical asymptotic theory of statistical inference such as the Fisher–Cramér–Rao paradigm, but these theoretical results lose their validity in the neighborhood of singularities. Therefore, it is necessary to build a new theory for singular models and to develop a new criterion for model selection for over-realizable scenarios.

In this section, we present our preliminary approaches toward building a new theory. We define two simple toy models that have intrinsic attributes of singularities, and try to evaluate their estimation ability in over-realizable scenarios. In addition, in the last part of this section, we also show the influence of singularity in the dynamics of standard gradient learning and natural gradient learning using the simple toy model.

3.4.2 Related Works

Before presenting our own results on the singularity problem, we briefly review related works. The singularity problem in hierarchical models has been noted by statisticians. For example, in the case of the Gaussian mixture model, where a probability distribution is represented by a mixture of Gaussian distributions, it has a hierarchical structure determined by the number of Gaussian distributions. Hotelling [21] and Weyl [32] investigated the problem of the degeneracy of the Fisher information matrix by means of a geometrical treatment in the 1930s. In systems theory, these properties were pointed out by Brockett [11]. Recently, Hartigan [19], Dacunha-Castelle and Gassiat [15], and Kuriki and Takemura [23] studied statistical models with nonregularity in the field of statistics.

Related to the neural networks, the problem was first pointed out by Hagiwara [18]. Hagiwara clarified that the AIC does not work on MLP through computer simulations, and suggested that this is caused by the hierarchical structure of the model. Moreover, Hagiwara used simple models to show that the least-square error of the estimator does not asymptotically obey the conventional asymptotic rule of $1/T$ where T is the number of data, but instead $\log T/T$ [18]. This result is consistent with Hartigan's conjecture [19].

Through these results, it was noted that analysis of the error of estimators is important for model selection of neural networks, and various works started thereafter. Watanabe [31] applied algebraic geometry to elucidate the behavior of the Bayesian predictive estimator in MLP, and developed a general framework for obtaining its generalization error by using Hironaka's resolution theorem of singularity [20]. As a result, he showed a sharp difference between regular cases and singular cases and the superiority of the Bayesian predictive distribution for singular models. Fukumizu [16] gave a general analysis of maximum likelihood estimators in singular statistical models including the MLP. He showed that behavior does not obey their regular statistical theory. He also showed that the estimation error has a variance of $\log T/T$ for MLP models with more than two redundant hidden units.

3.4.3 Analytical Solutions in Simple Models

One of the most important issues related to the singularity problem in neural networks is to develop a new criterion for model selection. To do this, one must know the relation between the generalization error and the training error mentioned in Sect. 3. Unfortunately, it is very difficult to find these relations for the general model. Thus, we try to attack this problem using some simple models, and analyze their asymptotic properties.

4.3.1. Asymptotic Statistical Inference: Generalization Error and Training Error

Before going to our results, we briefly review the asymptotic results of statistical estimators in the regular case. Let $D = \{x_1 \cdots x_T\}$ be T independent observations from the true distribution $p^*(x)$. We first consider the maximum-likelihood estimator (MLE), which maximizes the log-likelihood of data D. The estimated distribution from the data is given by $\hat{p}_{\text{MLE}}(x) = p(x; \hat{\theta})$, where

$$\hat{\boldsymbol{\theta}} = \mathrm{argmax}_{\boldsymbol{\theta}} \left\{ \frac{1}{T} \sum_{t=1}^{T} \log p(\boldsymbol{x}_t; \boldsymbol{\theta}) \right\}.$$

While the MLE searches for an asymptotically optimal point estimator in the model, the Bayes paradigm studies a posterior probability of the parameters based on the set of observations D. The posterior probability density is written as

$$p(\boldsymbol{\theta}|D) = c(D)\pi(\boldsymbol{\theta}) \prod_{t=1}^{T} p(\boldsymbol{x}_t|\boldsymbol{\theta}), \qquad (3.41)$$

where $c(D)$ is the normalization factor depending only on data D, and $\pi(\boldsymbol{\theta})$ is the prior distribution on the parameters. The Bayesian predictive distribution $\hat{p}_{\mathrm{Bayes}}(\boldsymbol{x}) = p(\boldsymbol{x}|D)$ is obtained by averaging $p(\boldsymbol{x}|\boldsymbol{\theta})$ with respect to the posterior distribution of $\boldsymbol{\theta}$, and can be written as

$$\hat{p}_{\mathrm{Bayes}}(\boldsymbol{x}) = p(\boldsymbol{x}|D) = \int p(\boldsymbol{x}|\boldsymbol{\theta}) p(\boldsymbol{\theta}|D) \mathrm{d}\boldsymbol{\theta}. \qquad (3.42)$$

These estimators are evaluated by the generalization error defined by the Kullback–Leibler divergence from $p^*(\boldsymbol{x})$ to $\hat{p}(\boldsymbol{x})$ (Sect. 3). Since the individual estimated probability density $\hat{p}(\boldsymbol{x})$ depends on data set D, we need to take the expectation with respect to the distribution of data set D. The expected generalization error is thus defined as

$$L_{\mathrm{gen}} = -E_D \left[E_{p^*} \left[\log \hat{p}(\boldsymbol{x}) \right] \right], \qquad (3.43)$$

where E_D represents the expectation with respect to D. Similarly, the expected training error is defined by using the empirical expectation,

$$L_{\mathrm{train}} = -E_D \left[\frac{1}{T} \sum_{t=1}^{T} \log \hat{p}(\boldsymbol{x}_t) \right]. \qquad (3.44)$$

In order to evaluate the estimator $\hat{p}(\boldsymbol{x})$, it is desirable to use L_{gen}, which is not computable because we do not know $p^*(\boldsymbol{x})$. Instead, we use L_{train}, which is computable. Hence, it is important to see the difference between L_{gen} and L_{train}. This is used as a principle of model selection.

When the statistical model M is regular, or the best approximation to the true distribution $p^*(\boldsymbol{x})$ is at a regular point of M, the estimated probability density $\hat{p}(\boldsymbol{x})$ is known to have the relations [1],

$$L_{\mathrm{train}} \approx H^* - \frac{d}{2T}, \qquad (3.45)$$

$$L_{\mathrm{gen}} \approx H^* + \frac{d}{2T} \approx L_{\mathrm{train}} + \frac{d}{T}, \qquad (3.46)$$

where d is the dimension number of parameter vector $\boldsymbol{\theta}$, and H^* represents the data-independent term, $-E_{p^*}[\log p^*(\boldsymbol{x})]$.

These universal relations are not guaranteed in the singular case. The relation between the generalization and training errors is different, so that we need a different criterion to evaluate the generalization performance of singular models such as the MLP.

4.3.2. Simple Toy Models

For the singular case, we investigated the relation between the generalization and training errors using two simple toy models (see [8, 9] for detailed discussions). One is a very simple multilayer perceptron having only one hidden unit. The other is a simple cone model.

The simple cone model describes a typical singularity in MLP mentioned in Sect. 1. Let x be a Gaussian random variable $x \in R^{d+2}$, with mean μ and identity covariance matrix I, then the probability density function of x is given by

$$p(x|\mu) = \frac{1}{(\sqrt{2\pi})^{d+2}} \exp\left\{-\frac{1}{2}\|x-\mu\|^2\right\}. \tag{3.47}$$

If we consider a set $S_{d+2} = \{\mu | \mu \in \mathcal{R}^{d+2}\}$, the cone model M_{d+1} is a subset of S_{d+2}, in which μ is restricted by

$$M : \mu = \frac{\xi}{\sqrt{1+c^2}} \begin{pmatrix} 1 \\ c\omega \end{pmatrix} = \xi a(\omega). \tag{3.48}$$

Here c is a constant determining the tangent of the vertical angle of the cone ($\tan \vartheta$ in Fig. 3.4), $\|a\|^2 = 1$, $\omega \in U^d$, and U^d is a d-dimensional unit sphere. Then M_{d+1} is a cone having (ξ, ω) as coordinates, where the apex $\xi = 0$ is the singular point. When $d = 1$, U^1 is a circle so that ω is replaced by angle θ, and we have

$$\mu = \frac{\xi}{\sqrt{1+c^2}} \begin{pmatrix} 1 \\ c\cos\theta \\ c\sin\theta \end{pmatrix}. \tag{3.49}$$

Figure 3.4 shows the one-dimensional cone model M_2, which is embedded in $S_3 = \{\mu | \mu \in \mathcal{R}^3\}$ by the condition given by Eq. (3.49).

We also exploit a simple MLP, which has the input-output relation given by

$$y = v\varphi(w \cdot x) + n, \tag{3.50}$$

where n is a Gaussian random noise subject to $\mathcal{N}(0,1)$. When $v = 0$, the behavior is the same whatever the value of w. Let us put $w = \beta\omega$, where $\beta = \|w\|$, ω is on the d-dimensional hypersphere, $\xi = v\|w\|$, and $\psi(x;\beta,\omega) = \varphi\{\beta(\omega \cdot x)\}/\beta$. We then have an alternative expression,

$$y = \xi\psi(x;\beta,\omega) + n, \tag{3.51}$$

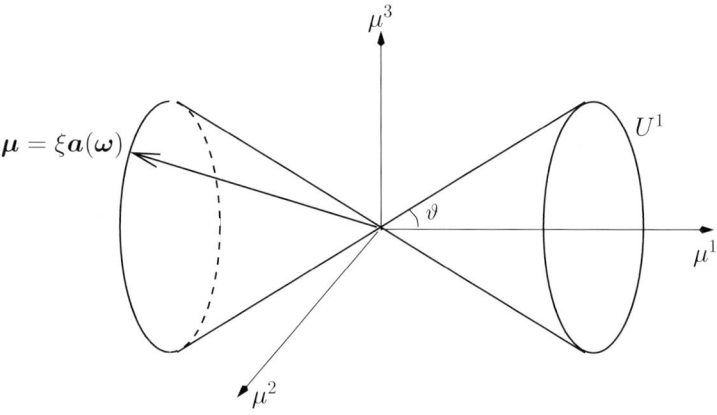

Fig. 3.4. One-dimensional cone model M_2 embedded in $S_3(d=1,\ c=\tan\vartheta)$

which shows the cone structure with apex at $\xi = 0$. In this section, we assume that β is known and does not need to be estimated. The conditional probability density of y given input $\boldsymbol{x}$ and learning parameters ξ and $\boldsymbol{\omega}$ is written as

$$p(y|\boldsymbol{x};\xi,\boldsymbol{\omega}) = \frac{1}{\sqrt{2\pi}}\exp\left\{-\frac{1}{2}[y-\xi\psi(\boldsymbol{x};\beta,\boldsymbol{\omega})]^2\right\}. \tag{3.52}$$

For these toy models, when the true or optimal parameter is at a singular point, the conventional asymptotic results for regular models cannot be applied. In the next section, we gave some analytic results in the singular case, i.e., $\xi^* = 0$.

4.3.3. Analytic Results of MLE

For the cone model defined in Eq. (3.47), the log-likelihood of data $D = \{\boldsymbol{x}_t\}_{t=1\cdots T}$ is written as

$$L(D,\xi,\boldsymbol{\omega}) = -\frac{1}{2}\sum_{t=1}^{T}||\boldsymbol{x}_t - \xi\boldsymbol{a}(\boldsymbol{\omega})||^2. \tag{3.53}$$

The maximum likelihood estimator (MLE) is the one that maximizes $L(D,\xi,\boldsymbol{\omega})$. However, $\partial^k L/\partial\boldsymbol{\omega}^k = 0$ at $\xi = 0$ for any k, so that we cannot analyze the behaviors of the MLE by the Taylor expansion at the optimal point. Therefore, we first fix $\boldsymbol{\omega}$ and search for the value of ξ that maximizes L. This is easy since L is a quadratic function of ξ. The maximum $\hat{\xi}$ is given by a function of $\boldsymbol{\omega}$,

$$\hat{\xi}(\boldsymbol{\omega}) = \operatorname{argmax}_\xi L(D,\xi,\boldsymbol{\omega}) = \frac{1}{\sqrt{T}}Y(\boldsymbol{\omega}), \tag{3.54}$$

where

$$Y(\boldsymbol{\omega}) = \boldsymbol{a}(\boldsymbol{\omega}) \cdot \tilde{\boldsymbol{x}}, \quad \tilde{\boldsymbol{x}} = \frac{1}{\sqrt{T}} \sum_{t=1}^{T} \boldsymbol{x}_t. \tag{3.55}$$

Here, $Y(\boldsymbol{\omega}) = \boldsymbol{a}(\boldsymbol{\omega}) \cdot \tilde{\boldsymbol{x}}$ is subject to the Gaussian distribution depending on $\boldsymbol{\omega}$. More precisely, $Y(\boldsymbol{\omega})$ is a zero-mean Gaussian random field over U^d with covariance $A(\boldsymbol{\omega}, \boldsymbol{\omega}') = \boldsymbol{a}(\boldsymbol{\omega}) \cdot \boldsymbol{a}(\boldsymbol{\omega}')$. By substituting $\hat{\xi}(\boldsymbol{\omega})$ from Eq. (3.54), the log-likelihood function becomes

$$\hat{L}(\boldsymbol{\omega}) = -\frac{1}{2}\sum_{t=1}^{T}||\boldsymbol{x}_t||^2 + \frac{1}{2}Y^2(\boldsymbol{\omega}). \tag{3.56}$$

Therefore, the MLE $\hat{\boldsymbol{\omega}}$ is given by the maximizer of

$$\hat{\boldsymbol{\omega}} = \mathrm{argmax}_{\boldsymbol{\omega}} Y^2(\boldsymbol{\omega}). \tag{3.57}$$

Using the MLE, we obtain the expected generalization and training errors in the following theorem.

Theorem 1. In the case of the cone model, the MLE satisfies

$$L_{\mathrm{gen}} = H^* + \frac{1}{2T} E_D \left[\sup_{\boldsymbol{\omega}} Y^2(\boldsymbol{\omega}) \right], \tag{3.58}$$

$$L_{\mathrm{train}} = H^* - \frac{1}{2T} E_D \left[\sup_{\boldsymbol{\omega}} Y^2(\boldsymbol{\omega}) \right]. \tag{3.59}$$

In the simple cone model, we can obtain the explicit value of $E_D\left[\sup_{\boldsymbol{\omega}} Y^2(\boldsymbol{\omega})\right]$. We show the asymptotic results (the large d limit).

Corollary 1. When d is large, the MLE satisfies

$$L_{\mathrm{gen}} \approx H^* + \frac{1 + 2c\sqrt{2d/\pi} + c^2(d+1)}{(1+c^2)2T} \approx H^* + \frac{c^2}{(1+c^2)} \frac{d}{2T}, \tag{3.60}$$

$$L_{\mathrm{train}} \approx H^* - \frac{1 + 2c\sqrt{2d/\pi} + c^2(d+1)}{(1+c^2)2T} \approx H^* - \frac{c^2}{(1+c^2)} \frac{d}{2T}. \tag{3.61}$$

It should be remarked that the generalization and training errors depend on the shape parameter c as well as on the dimension number d. In the regular case, they depend only on d. As one can easily see, when c is small, the cone looks like a needle, and its behavior resembles a one-dimensional model. When c is large, it resembles two $(d+1)$-dimensional hypersurfaces, so that its behavior is like a $(d+1)$-dimensional regular model.

In the case of the simple MLP defined in Eq. (3.51), the log-likelihood of data set $D = \{(\boldsymbol{x}_t, y_t)\}_{t=1\cdots T}$ is written as

$$L(D; \xi, \boldsymbol{\omega}) = -\frac{1}{2}\sum_{t=1}^{T} \{y_t - \xi\varphi_\beta(\boldsymbol{\omega}\cdot\boldsymbol{x}_t)\}^2. \tag{3.62}$$

Let us define two random variables that depend on D and $\boldsymbol{\omega}$:

$$Y(\boldsymbol{\omega}) = \frac{1}{\sqrt{T}} \sum_{t=1}^{T} y_t \varphi_\beta(\boldsymbol{\omega} \cdot \boldsymbol{x}_t), \tag{3.63}$$

$$A_D(\boldsymbol{\omega}) = \frac{1}{n} \sum_{t=1}^{T} \varphi_\beta^2(\boldsymbol{\omega} \cdot \boldsymbol{x}_t). \tag{3.64}$$

Note that $A_D(\boldsymbol{\omega})$ converges to $A(\boldsymbol{\omega}) = E_{\boldsymbol{x}}[\varphi_\beta^2(\boldsymbol{\omega} \cdot \boldsymbol{x})]$ as T goes to infinity, and $Y(\boldsymbol{\omega})$ defines asymptotically a Gaussian random field with mean 0 and covariance $A(\boldsymbol{\omega}, \boldsymbol{\omega}') = E_{\boldsymbol{x}}[\varphi_\beta(\boldsymbol{\omega} \cdot \boldsymbol{x})\varphi_\beta(\boldsymbol{\omega}' \cdot \boldsymbol{x})]$. From the same approach as the cone model case, we obtain

$$\hat{\xi}(\boldsymbol{\omega}) = \mathrm{argmax}_\xi L(D; \xi, \boldsymbol{\omega}) = \frac{1}{\sqrt{n}} \frac{Y(\boldsymbol{\omega})}{A_D(\boldsymbol{\omega})}, \tag{3.65}$$

$$L(\hat{\xi}(\boldsymbol{\omega}), \boldsymbol{\omega}) = -\frac{1}{2} \sum_{t=1}^{T} y_t^2 + \frac{1}{2} \frac{Y^2(\boldsymbol{\omega})}{A_D(\boldsymbol{\omega})}, \tag{3.66}$$

$$\hat{\boldsymbol{\omega}} = \mathrm{argmax}_{\boldsymbol{\omega}} \frac{1}{2} \frac{Y^2(\boldsymbol{\omega})}{A_D(\boldsymbol{\omega})}. \tag{3.67}$$

Using the mle, we get the following theorem.

Theorem 2. For the MLE of the simple MLP, we have

$$L_{\mathrm{gen}} = H^* + \frac{1}{2T} E_D \left[\sup_{\boldsymbol{\omega}} \frac{Y(\boldsymbol{\omega})^2}{A(\boldsymbol{\omega})} \right], \tag{3.68}$$

$$L_{\mathrm{train}} = H^* - \frac{1}{2T} E_D \left[\sup_{\boldsymbol{\omega}} \frac{Y(\boldsymbol{\omega})^2}{A(\boldsymbol{\omega})} \right]. \tag{3.69}$$

There is a nice correspondence between the cone model and the simple MLP. However, because of the existence of nonlinearity of the hidden unit, it is not easy to get the explicit relation between the error and the dimension number of parameters like that of the cone in Corollary 1.

We checked the analytic results using computer simulations. The cone model and MLP were trained by the gradient method to get the MLE. For each d ($d = 1, \ldots, 25$), we generated 100 training sets with $10 \times d$ data from the true distribution, and repeated the training processes with the sets to get the average generalization and training errors. We set $c = 1, \beta = 1$ in the simulations. The learning procedure was stopped when the decrease in the training error was smaller than 10^{-10}. In Figures 3.5 and 3.6, one can see the symmetry of L_{gen} and L_{train}, which is obtained in Theorems 1 and 2. For the cone model, we confirmed that the errors approach the approximate values of Corollary 1. In addition, it is possible from the figures to suppose that a similar relationship is also sustained between d and the errors for the MLP.

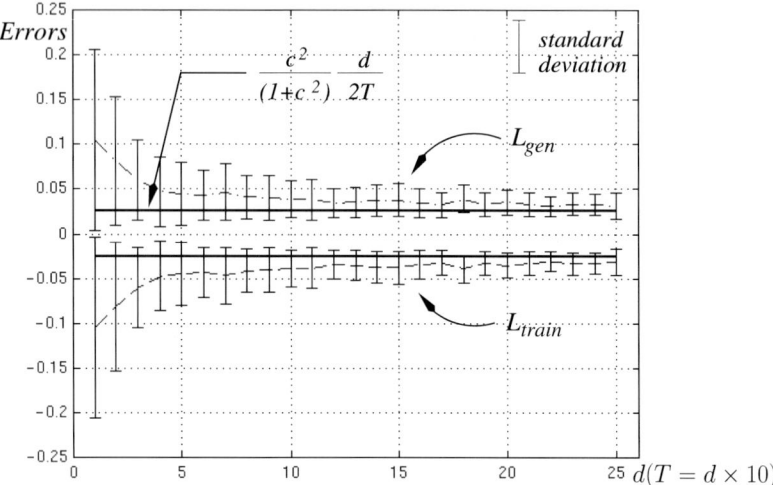

Fig. 3.5. Simulation results on cone model

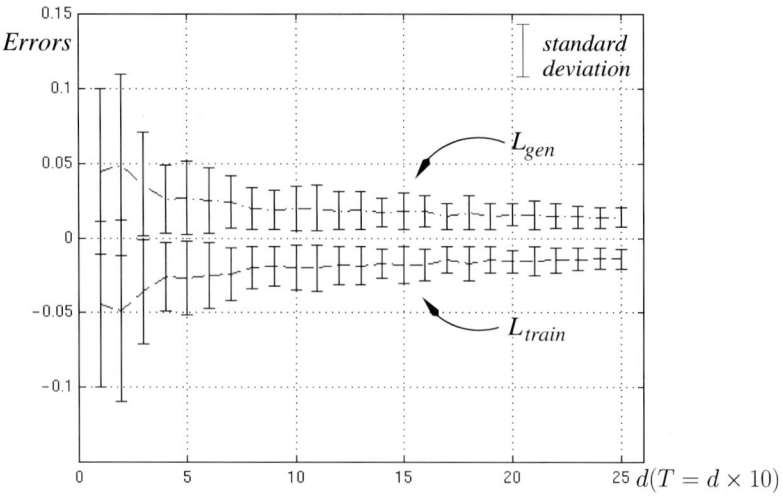

Fig. 3.6. Simulation results on simple MLP

4.3.4. Analytic Results of Bayesian Predictive Distribution

For the Bayesian predictive distribution, we show some results from the cone model here. For the simple MLP model defined in Eq. (3.51), we can also apply the same approach and get similar results [9].

Unlike the regular case, the asymptotic behavior of the Bayesian predictive distribution depends on the prior. Let us define the prior as $\pi(\xi, \boldsymbol{\omega})$. Then the probability density of observed samples is given by

$$Z_T = p(D) = \int \pi(\xi, \boldsymbol{\omega}) \prod_{t=1}^{T} p(\boldsymbol{x}_t|\xi, \boldsymbol{\omega}) \mathrm{d}\xi \mathrm{d}\boldsymbol{\omega}. \tag{3.70}$$

When a new data $\boldsymbol{x}_{T+1}$ is given, we can similarly obtain the joint probability density $p(\boldsymbol{x}_{T+1}, D)$ as

$$Z_{T+1} = p(\boldsymbol{x}_{T+1}, D) = \int \pi(\xi, \boldsymbol{\omega}) \prod_{t=1}^{T+1} p(\boldsymbol{x}_t|\xi, \boldsymbol{\omega}) \mathrm{d}\xi \mathrm{d}\boldsymbol{\omega}. \tag{3.71}$$

From the Bayesian theorem, we can easily see that the Bayes predictive distribution is given by

$$\hat{p}_{\text{Bayes}}(\boldsymbol{x}|D) = \frac{Z_{T+1}}{Z_T}. \tag{3.72}$$

When the prior is uniform, i.e., $\pi(\xi, \boldsymbol{\omega}) = $ constant, we can calculate Z_T explicitly, and obtain

$$\hat{p}_{\text{Bayes}}(\boldsymbol{x}|D) = \frac{1}{\sqrt{2\pi}^{d+2}} \sqrt{\frac{T}{T+1}} \exp\left\{-\frac{||x||^2}{2}\right\} \frac{P_d(\tilde{\boldsymbol{x}}_{T+1})}{P_d(\tilde{\boldsymbol{x}})}, \tag{3.73}$$

where

$$\tilde{\boldsymbol{x}}_{T+1} = \frac{1}{\sqrt{T+1}}(\boldsymbol{x} + \sqrt{T}\tilde{\boldsymbol{x}}), \tag{3.74}$$

$$P_d(\tilde{\boldsymbol{x}}) = \int \exp\left\{\frac{1}{2}Y^2(\boldsymbol{\omega})\right\} \mathrm{d}\boldsymbol{\omega}. \tag{3.75}$$

From the fact that $Y(\boldsymbol{\omega}) = \boldsymbol{a}(\boldsymbol{\omega}) \cdot \tilde{\boldsymbol{x}}$ and $Y_{T+1}(\boldsymbol{\omega}) = \boldsymbol{a}(\boldsymbol{\omega}) \cdot \tilde{\boldsymbol{x}}_{T+1}$ are subject to the same probability distribution, the generalization error can be obtained as

$$L_{\text{gen}} \approx H^* + \frac{1}{2n}. \tag{3.76}$$

Furthermore, using the Edgeworth expansion, we can also have

$$\hat{p}_{\text{Bayes}}(\boldsymbol{x}|D) \approx \frac{1}{\sqrt{2\pi}^{d+2}} \exp\left\{-\frac{||x||^2}{2}\right\}$$
$$\left\{1 + \frac{1}{\sqrt{T}} \nabla \log P_d(\tilde{\boldsymbol{x}}) \cdot \boldsymbol{x} + \frac{1}{2T} \mathrm{tr}\left(\frac{\nabla \nabla P_d}{P_d} H_2(\boldsymbol{x})\right)\right\}, \tag{3.77}$$

where ∇ is the gradient, and $H_2(\boldsymbol{x})$ is the Hermite polynomial. Then, the training error is given by

$$L_{\text{train}} \approx L_{\text{gen}} - \frac{1}{T} E_D\left[\nabla \log P_d(\tilde{\boldsymbol{x}}) \cdot \tilde{\boldsymbol{x}}\right]. \tag{3.78}$$

In the case of Jeffreys' prior, i.e., $\pi(\xi, \boldsymbol{\omega}) \propto |\xi|^d$, we can do similar analysis, and obtain

$$L_{\text{gen}} \approx H^* + \frac{d+1}{2n}. \tag{3.79}$$

One can refer [9] for detailed results.

These results are rather surprising: under the uniform prior, the generalization error is constant and does not depend on d. This is completely different from the regular case. However, this striking result is given rise to by the uniform prior on ξ. The uniform prior puts strong emphasis on the singularity, showing that one should be very careful in choosing a prior when the model includes singularities. In the case of the Jeffreys' prior, the generalization error increases in proportion to d, which is the same result as the regular case. In addition, the symmetric duality between L_{gen} and L_{train} does not hold for both the uniform prior and the Jeffreys' prior.

3.4.4 Learning and Singularities

It has been elucidated in the statistical mechanics framework that the learning trajectory is ubiquitously attracted by singularities, and that plateaus are the result of this singular structure [29]. The natural gradient learning method takes this structure into account by using the Riemannian metric, so that it works efficiently in learning. It was also confirmed that the natural gradient can avoid plateaus through statistical mechanical approaches [25].

In this chapter, we show the effect of singularities in learning dynamics of the simple cone model. We exploited the one-dimensional cone model defined in Eq. (3.49) with $c = 1$. The true model generating training data was set to $(\xi^*, \theta^*) = (1, 0)$. Starting from $(\xi_o, \theta_o) = (1, \frac{11}{12}\pi)$, we calculated the average trajectories of standard gradient learning and natural gradient learning in order to show how the learning parameters (ξ, θ) approach the optimal point $(1, 0)$. Here, the average trajectories of standard gradient learning given in Eq. (3.24) are given by using

$$\begin{pmatrix} \Delta\xi(t) \\ \Delta\theta(t) \end{pmatrix}_{\text{SGD}} = -\eta_t \left(\begin{matrix} \left\langle \frac{\partial l(\boldsymbol{x}_t, \xi_t, \theta_t)}{\partial \xi} \right\rangle \\ \left\langle \frac{\partial l(\boldsymbol{x}_t, \xi_t, \theta_t)}{\partial \theta} \right\rangle \end{matrix} \right), \tag{3.80}$$

and those of the natural gradient learning in Eq. (3.26) are given by

$$\begin{pmatrix} \Delta\xi(t) \\ \Delta\theta(t) \end{pmatrix}_{\text{NGD}} = -\eta_t G^{-1}(\xi, \theta) \left(\begin{matrix} \left\langle \frac{\partial l(\boldsymbol{x}_t, \xi_t, \theta_t)}{\partial \xi} \right\rangle \\ \left\langle \frac{\partial l(\boldsymbol{x}_t, \xi_t, \theta_t)}{\partial \theta} \right\rangle \end{matrix} \right), \tag{3.81}$$

where $<\cdot>$ denotes the expectation with respect to the true distribution of $\boldsymbol{x}$.

As shown in Fig. 3.7, the dynamics of standard gradient learning is attracted to the singular point, and it takes a long time to go out from the trap, making a plateau. On the contrary, Fig. 3.8 shows that the dynamics of the natural gradient is not affected by the singular point. The differences of

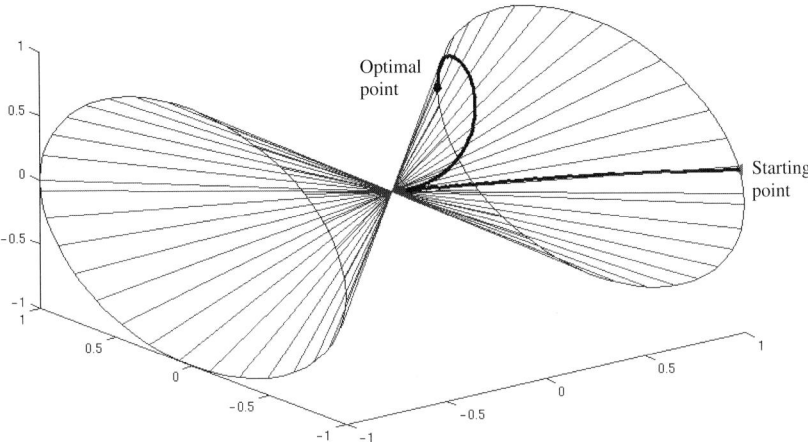

Fig. 3.7. Standard gradient descent learning dynamics in cone model

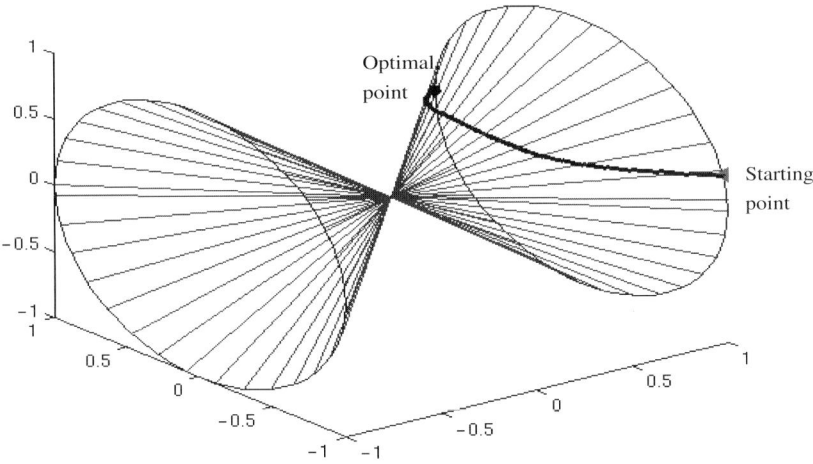

Fig. 3.8. Natural gradient descent learning dynamics in cone model

the two learning algorithms are clearly shown in the learning curves of Fig. 3.9. The plateaus shown in standard gradient learning correspond to the part of the trajectories trapped in singularity. The plateau has completely disappeared in the natural gradient. This phenomenon commonly occurs in the learning of general MLP with much more complicated singularities as shown in Fig. 3.3.

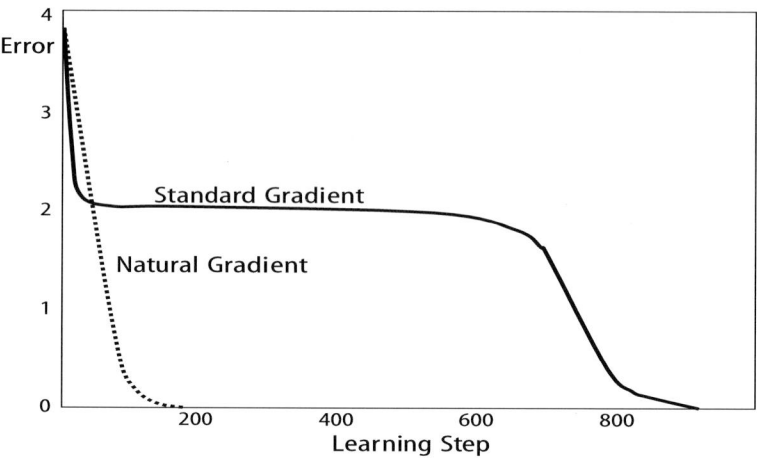

Fig. 3.9. Standard gradient learning (*solid line*) versus natural gradient descent learning (*dashed line*) in the learning of cone model

3.5 Discussion and Conclusions

We have investigated the geometrical structures of the space of MLP, the neuromanifold, by using information geometry. There are a number of interesting characteristics that should be taken into account to design a good network.

First, the parameter space of MLP is not Euclidean, having an intrinsic Riemannian metric, which is given by the Fisher information matrix of the corresponding probability density function. By using the metric, we developed a new learning method, called natural gradient learning. The performance of natural gradient learning was confirmed through computational experiments on benchmark data as well as through statistical mechanical analysis.

Second, the neuromanifold includes complicated algebraic singularities, which are due to its hierarchical structure. At the singularities, the conventional model selection criteria cannot be applied, and we need a new criterion for the over-realizable scenario. Using simple toy models, we showed some preliminary results toward a new criterion and also briefly reviewed other recent works related to this problem. The singular structure is also closely related to the dynamics of the learning method, and we gave a simple example showing the effect of a singular point on the learning trajectories and the plateau phenomenon.

These singularities are ubiquitous in hierarchical models such as the Gaussian mixture model and the ARMA model. However, the current results are very preliminary, and there exist a large number of open problems to be solved.

References

1. Akaike H. (1974) A new look at the statistical model identification. *IEEE Trans.*, AC-19:716–723.
2. Amari S. (1985) *Differential-Geometrical Method in Statistics*. Springer, Berlin Heidelberg New York
3. Amari S. (1998) Natural gradient works efficiently in learning. *Neural Computation* 10:251–276
4. Amari S., Kurata K. and Nagaoka H. (1992). Information geometry of Boltzmann machines. *IEEE Trans. on Neural Networks* 3:260–271
5. Amari S., Nagaoka H. (2000) *Methods of Information Geometry*. AMS and Oxford University press, New York
6. Amari S., Ozeki, T. (2001) Differential and algebraic geometry of multilayer perceptrons. *IEICE Trans. Fundamentals* E84-A:31–38
7. Amari S., Park H., Fukumizu K. (2000) Adaptive method of realizing natural gradient learning for multilayer perceptrons. *Neural Computation* 12:1399–1409
8. Amari S., Park H., Ozeki, T. (2000) Statistical inference in non-identifiable and singular statistical models. *Journal of Korean Statistical Society* 30:179–192
9. Amari S., Park H., Ozeki, T. (2003) Geometrical singularities in the neuromanifold of multilayer perceptrons. *Advances in NIPS* 14:343–350
10. Bishop, C. (1995) *Neural Networks for Pattern Recognition* Oxford University Press, Oxford
11. Brockett, R. W. (1976) Some geometric questions in the theory of linear systems. *IEEE Trans. on Automatic Control* 21:449–455
12. Bottou, L. (1998) Online Algorithms and Stochastic Approximations. In Saad D. ed., *Online Learning and Neural Networks*. Cambridge University Press, Cambridge, UK
13. Chen A.M., Lu H., Hecht-Nielsen R. (1993) On the geometry of feedforward neural network error surfaces. *Neural Computations* 5:910–927
14. Cover, T.M., Thomas, J.A. (1991) *Elements of Information Theory* Wiley, New York
15. Dacunha-Castelle, D., Gassiat, É. (1997) Testing in locally conic models, and application to mixture models. *Probability and Statistics* 1:285–317
16. Fukumizu, K. (2003) Likelihood ratio of unidentifiable models and multilayer neural networks. *The Annals of Statistics* 31:833-851
17. Fukumizu, K., Amari, S. (2000) Local minima and plateaus in hierarchical structures of multilayer perceptrons. *Neural Networks* 13:317–327
18. Hagiwara, K. (2002) On the problem in model selection of neural network regression in overrealizable scenario. *Neural Computation* 14:1979–2002
19. Hartigan, J. A. (1985) A failure of likelihood asymptotics for normal mixtures. *Proc. Berkeley Conf. in Honor of J. Neyman and J. Kiefer* 2:807–810
20. Hironaka, H. (1964) Resolution of singularities of an algebraic variety over a field of characteristic zero. *Annals of Mathematics* 79:109–326
21. Hotelling, H. (1939) Tubes and spheres in n-spaces, and a class of statistical problems. *Amer. J. Math.* 61:440–460
22. Kurkova V., Kainen P.C. (1994) Functionally equivalent feedforward neural networks. *Neural Computation* 6:543–558
23. Kuriki S., Takemura, A. (2001) Tail probabilities of the maxima of multilinear forms and their applications. *The Annals of Statistics* 29:328–371

24. Park H., Amari S., Fukumizu K. (2000) Adaptive natural gradient learning algorithms for various stochastic models. *Neural Networks* 13:755-764
25. Rattray M., Saad D., Amari S. (1998) Natural gradient descent for on-line learning. *Physical Review Letters* 81:5461–5464
26. Rissanen J. (1978) Modelling by shortest data description, *Automatica* 14:465-471
27. Rüger S.M., Ossen A. (1997) The metric of weight space. *Neural Processing Letters* 5:63–72
28. Rumelhart D.E., McClelland J.L. (1986) *Parallel Distributed Processing: Explorations in the Microstructure of Cognition, vol. 1*. MIT Press, Cambridge, MA
29. Saad, D. and Solla, A. (1995) On-line learning in soft committee machines. *Physical Review E* 52:4225–4243
30. Sussmann H.J. (1992) Uniqueness of the weights for minimal feedforward nets with a given input–output map. *Neural Networks* 5:589–593
31. Watanabe, S. (2001) Algebraic analysis for non-identifiable learning machines. *Neural Computation* 13:899–933
32. Weyl, H. (1939) On the volume of tubes. *Amer. J. Math.* 61:461–472

4

A Lattice Algebraic Approach to Neural Computation

Gerhard X. Ritter and Laurentiu Iancu

Computer and Information Science and Engineering Department,
University of Florida, P.O. Box 116120, Gainesville, FL 32611-6120, USA
{ritter,liancu}@cise.ufl.edu

4.1 Introduction

During the past decade lattice algebraic operations have been used extensively to derive a variety of novel artificial neural network models. Many of these models have been successfully employed to solve real-world problems. Among the different models based on lattice algebra are the following: morphological associative memories [1, 2, 3, 4, 5], shared-weight neural networks [6, 7], regularization neural networks [8], hybrid morphological-rank-linear neural networks [9], min-max neural networks [10, 11, 12], morphological perceptrons [13, 14], fuzzy lattice networks [15, 16], and adaptive logic networks, which combine linear functions by a tree expression of maximum and minimum operations [17]. In this treatise we restrict our discussion to those neural networks whose computational basis is strictly limited to lattice-ordered groups (ℓ-groups) and bounded lattice-ordered groups (blogs). These types of networks have become known as *morphological neural networks* (MNNs). The name morphological neural networks stems from the fact that lattice-ordered groups and bounded ℓ-groups also provide for the mathematical foundation of the area of image processing known as mathematical morphology.

Our particular focus will be on matrix memories and feedforward types of networks. Such a narrow focus allows for a deeper immersion into the fundamental computational principles of morphological neural networks, which is a main goal of this treatise. Before discussing the two very distinct networks we first provide a quick review of lattice algebra as related to ℓ-groups and blogs.

4.2 Some Elementary Concepts From Lattice Algebra

The concept of lattices was formed with a view to generalize and unify certain relationships between subsets of a set, between substructures of an algebraic structure such as groups, and between geometric structures such as topological spaces. The development of the theory of lattices started about 1930 and was influenced by the work of Garrett Birkhoff [18].

In the subsequent discussion we assume that the reader is familiar with the basic concepts of lattice theory, in particular those that apply to the real numbers. The real numbers $\mathbb{R}$ together with the relation of less or equal ($\leq$) is a *totally ordered* set; i.e., given any pair $x, y \in \mathbb{R}$, then either $x \leq y$ or $y \leq x$. If $x \vee y = \max\{x, y\}$ and $x \wedge y = \min\{x, y\}$ $\forall x, y \in \mathbb{R}$, then $\mathbb{R}$ together with the operations of $\vee$ and $\wedge$ is a lattice. However, $(\mathbb{R}, \vee, \wedge)$ is not a *complete* lattice as there is no largest and smallest number. A complete lattice can be obtained by extending the real numbers to include the symbols $-\infty$ and ∞ by setting $\mathbb{R}_{\pm\infty} = \mathbb{R} \cup \{-\infty, \infty\}$ and defining $-\infty < x < \infty$ $\forall x \in \mathbb{R}$ and $-\infty \leq x \leq \infty$ $\forall x \in \{-\infty, \infty\}$. The extended structure $(\mathbb{R}_{\pm\infty}, \vee, \wedge)$ is now a complete lattice with largest element ∞ and smallest element $-\infty$. The dual of this lattice is obtained by replacing $\leq$ with the relation of greater or equal $\geq$. Obviously, $(\mathbb{R}, \vee, \wedge)$ is a sublattice of $(\mathbb{R}_{\pm\infty}, \vee, \wedge)$. Similarly, if $\mathbb{R}^+$ denotes the set of positive numbers, then the set $\mathbb{R}_\infty^{\geq 0} = \mathbb{R}^+ \cup \{0, \infty\}$ with the relation $\leq$ is a complete lattice. Here 0 is the smallest element and ∞ the largest element. $(\mathbb{R}_\infty^{\geq 0}, \vee, \wedge)$ is a sublattice of $(\mathbb{R}_{\pm\infty}, \vee, \wedge)$ but not of $(\mathbb{R}, \vee, \wedge)$.

The lattices $(\mathbb{R}_{\pm\infty}, \vee, \wedge)$ and $(\mathbb{R}, \vee, \wedge)$ are *distributive lattices* because the distributive law $x \wedge (y \vee z) = (x \wedge y) \vee (x \wedge z)$ holds. This equation expresses the similarity with the distributive law $x(y + z) = xy + xz$ of ordinary arithmetic. Note also that by duality we have that $x \vee (y \wedge z) = (x \vee y) \wedge (x \vee z)$.

In addition to being a distributive lattice, the set of real numbers is also a group under addition, and our early experience in elementary algebra has taught us the useful properties

- P_1 $x \geq y \Rightarrow z + x \geq z + y$
- P_2 $x \geq y \Rightarrow z + x + w \geq z + y + w$
- P_3 $z + (x \vee y) = (z + x) \vee (z + y)$ and $z + (x \wedge y) = (z + x) \wedge (z + y)$,

where $x, y, z, w \in \mathbb{R}$. These properties exhibit the interplay between the lattice and group operations. A lattice that is also a group and satisfies property P_2 is called a *lattice-ordered group* or *ℓ-group*.

It is often convenient to deal with only one of the operations $\vee$ or $\wedge$. Every partially ordered set $\mathbb{F}$ for which the operation $x \vee y$ (or $x \wedge y$) is associative and is defined for each pair $x, y \in \mathbb{F}$ is called a *semilattice* whenever $x \vee x = x$. For example, $(\mathbb{R}, \vee)$ is a semilattice with *dual* $(\mathbb{R}, \wedge)$. If $\mathbb{R}_\infty = \mathbb{R} \cup \{\infty\}$ and $\mathbb{R}_{-\infty} = \mathbb{R} \cup \{-\infty\}$, then $(\mathbb{R}_{-\infty}, \vee)$ is a semilattice with *dual* $(\mathbb{R}_\infty, \wedge)$. Note that the semilattice $(\mathbb{R}_{-\infty}, \vee)$ is a monoid with zero element $-\infty$ since $r \vee (-\infty) = (-\infty) \vee r$ $\forall r \in \mathbb{R}_{-\infty}$. Similarly, $(\mathbb{R}_\infty, \wedge)$ is a monoid with zero element ∞. If a semilattice is also a group, then it is called a *semilattice-ordered group* or *$s\ell$-group*. If $(\mathbb{F}, \vee, +)$ is an $s\ell$-group and $(\mathbb{F}, \wedge, +')$ is also an $s\ell$-group with semilattice operation $\wedge$ and group operation $+'$ and satisfies $a \vee (b \wedge a) = a \wedge (b \vee a) = a$ $\forall a, b \in \mathbb{F}$, then we say that $\mathbb{F}$ is an *$s\ell$-group with duality*. If the operations $+$ and $+'$ coincide, then the operation $+$ is called *self-dual*. If $(\mathbb{F}, +)$ and $(\mathbb{F}, +')$ are only semigroups, then $\mathbb{F}$ is called an *$s\ell$-semigroup with duality* or *$s\ell$-semigroup*. The ℓ-group $(\mathbb{R}, \vee, \wedge, +)$ has the identity element 0 but no null element, as there is no "smallest" element in $\mathbb{R}$. As another example, the ℓ-semigroup $(\mathbb{R}_{-\infty}, \vee, \wedge, +)$, where $+$ denotes the extended real addition $a + (-\infty) = -\infty + a = -\infty \ \forall a \in \mathbb{R}_{-\infty}$, has the null element $-\infty$ but has no identity element, as $-\infty$ has no inverse under extended real addition. Similarly, if we adjoin the element ∞, then the ℓ-group $\mathbb{R}$ degenerates into the ℓ-semigroup $(\mathbb{R}_\infty, \vee, \wedge, +)$ as ∞ has no inverse under the addition $\infty + a = \infty \ \forall a \in \mathbb{R}_\infty$. This is not really as much of a disadvantage as it seems. We can extend the ℓ-group $\mathbb{R}$ to include the elements ∞ and $-\infty$ in a well-defined manner as follows. If $a, b \in \mathbb{R}$, then $a + b$ is already defined. Let $+' = +$ be the self-dual addition of elements of $\mathbb{R}$. For $a \in \mathbb{R}$, define

$$a + -\infty = -\infty + a = a +' -\infty = -\infty +' a = -\infty, \text{ and}$$
$$a + \infty = \infty + a = a +' \infty = \infty +' a = \infty.$$

If we now define $-\infty + \infty = \infty + -\infty = -\infty$ and $-\infty +' \infty = \infty +' -\infty = \infty$, then the resultant structure $(\mathbb{R}_{\pm\infty}, \vee, \wedge, +, +')$ is a distributive lattice which is called a *bounded ℓ-group* or *blog*.

In recent years, lattice-based matrix operations have found widespread applications in the engineering sciences. In these applications, the usual matrix operations of addition and multiplication are replaced by corresponding lattice operations. For example, given the bounded ℓ-group $(\mathbb{R}_{\pm\infty}, \vee, +)$ and $A = (a_{ij})$, $B = (b_{ij})$ two $m \times n$ matrices with entries in $\mathbb{F}_{\pm\infty}$, then the *pointwise maximum*, $A \vee B$, of A and B, is the $m \times n$ matrix C defined by $A \vee B = C$, where $c_{ij} = a_{ij} \vee b_{ij}$. If A is $m \times p$ and B is $p \times n$, then the *max product* of A and B is the matrix $C = A \boxvee B$, where $c_{ij} = \bigvee_{k=1}^{p}(a_{ik} + b_{kj})$. Observe that this product is analogous to the usual matrix product $c_{ij} = \sum_{k=1}^{p}(a_{ik} \times b_{kj})$, with the symbol $\sum$ replaced by $\bigvee$. Since $\bigvee$ replaces $\sum$ in our definition, the pointwise maximum can be thought of as matrix addition.

Example 1. An illustration of the max product of a 5×4 and a 4×3 matrix with entries from $\mathbb{R}_{\pm\infty}$ is the following:

$$\begin{bmatrix} -\infty & 6 & -2 & 2 \\ 7 & -5 & 10 & -4 \\ 8 & 4 & 11 & 9 \\ -3 & +\infty & 1 & -7 \\ -1 & 1 & 0 & 5 \end{bmatrix} \boxvee \begin{bmatrix} -\infty & 6 & -2 \\ 7 & -5 & 10 \\ 8 & 4 & 11 \\ -1 & 1 & 0 \end{bmatrix} = \begin{bmatrix} 13 & 3 & 16 \\ 18 & 14 & 21 \\ 19 & 15 & 22 \\ +\infty & +\infty & +\infty \\ 8 & 6 & 11 \end{bmatrix}.$$

The *min product* of A and B in the dual structure $(\mathbb{R}_{\pm\infty}, \wedge, +')$ is the matrix $C = A \boxwedge B$, where $c_{ij} = \bigwedge_{k=1}^{p}(a_{ik} +' b_{kj})$. Similarly, the *pointwise minimum* $A \wedge B$ of two matrices of the same size is defined as $A \wedge B = C$, where $c_{ij} = a_{ij} \wedge b_{ij}$.

Lattice-induced matrix operations lead to an entirely different perspective of a class of nonlinear transformations. These ideas were applied by Shimbel

[19] to communications networks, and to machine scheduling by Cuninghame-Green [20, 21] and Giffler [22]. Others have discussed their usefulness in applications to shortest path problems in graphs [23, 24, 25, 26]. Additional examples are given in [27], primarily in the field of operations research. Another useful application to image processing was developed by Ritter and Davidson [28, 29].

While lattice theory and lattice-ordered groups have only marginal connections to the computational aspects of linear algebra, Cuninghame-Green developed a novel nonlinear matrix calculus based on the min and max product, called *minimax algebra*, which is very reminiscent of linear algebra [27]. Problems notated using the *minimax products* take on the flavor of problems in linear algebra. By allowing for the minimax matrix products to take on the character of the familiar matrix products, concepts analogous to those in linear algebra, such as solutions to systems of equations, linear dependence and independence, rank, seminorms, eigenvalues and eigenvectors, spectral inequalities, and invertible and equivalent matrices, can be formulated.

Originally, many of these concepts were developed primarily to help solve operations research types of problems. Our interest in these notions is due to their applicability to artificial neural networks. For instance, we will view associative matrix memories as transforms $\mathbb{R}^n \to \mathbb{R}^m$. For such memories the notions of independence, dependence, and fixed points play a major role in the development of memories that are robust in the presence of noise. All these notions resemble their analogues encountered in linear algebra. For example, given a matrix transform $M : \mathbb{R}^n \to \mathbb{R}^n$ and $\mathbf{x} \in \mathbb{R}^n$, then $\mathbf{x}$ is a *fixed point* of M if and only if $M \triangledown \mathbf{x} = \mathbf{x}$ (or $M \triangle \mathbf{x} = \mathbf{x}$).

Property P_3 also holds for the $s\ell$-semigroups $(\mathbb{R}_{-\infty}, \vee, +)$ and $(\mathbb{R}_\infty, \wedge, +')$. It follows that each of the structures is also a semiring. These two semirings are in a one-to-one correspondence given by $r^* = -r$, where $\infty^* = -\infty$, $(-\infty)^* = \infty$, and $r \in \mathbb{R}_{\pm\infty}$. That is, r^* is the *dual* of r as $(r^*)^* = r$ and $r \wedge u = (r^* \vee u^*)^* \; \forall r, u \in \mathbb{R}_{\pm\infty}$.

This conjugacy extends to matrices with entries from $\mathbb{R}_{\pm\infty}$. Here the conjugate of a matrix $A = (a_{ij})$ with entries in $\mathbb{R}$ or $\mathbb{R}_{\pm\infty}$ is the matrix $A^* = (b_{ij})$, where $b_{ij} = [a_{ji}]^*$ and $[a_{ji}]^*$ is the additive conjugate of a_{ji} defined earlier. The notions of pointwise maximum and dual product could have been defined in terms of conjugation since $A \wedge B = (A^* \vee B^*)^*$ and $A \triangle B = (B^* \triangledown A^*)^*$ for appropriately sized matrices.

In the next two sections we will employ the lattice theoretic notions developed in this section to serve as the underlying mathematical foundation for the theory of morphological neural networks.

4.3 Morphological Associative Memories

One of the first goals achieved in the development of morphological neural networks was the establishment of a morphological associative memory network (MAM). In its basic form, this model of an associative memory resembles the well-known correlation memory or linear associative memory [30]. As in correlation encoding, the morphological associative memory provides a simple method to add new associations. A weakness in correlation encoding is the requirement of orthogonality of the key vectors in order to exhibit perfect recall of the fundamental associations. The morphological autoassociative memory does not restrict the domain of the key vectors in any way. Thus, as many associations as desired can be encoded into the memory [1, 31]. In the real number case, the capacity for a memory of length n can be as large as desired. That is, if k denotes the number of distinct patterns of length n to be encoded, then k is allowed to be any integer, no matter how large. Of course, in the binary case, the limit is $k = 2^n$, as this is the maximum number of distinct patterns of length n. In comparison, McEliece et al. showed that the asymptotic limit capacity of the Hopfield associative memory is $n/2\log n$ if with high probability the unique fundamental memory is to be recovered, except for a vanishingly small fraction of fundamental memories [32]. Likewise, the information storage capacity (number of bits that can be stored and recalled associatively) of the morphological autoassociative memory also exceeds the respective number of certain linear matrix associative memories which was calculated by Palm [33] and Willshaw et al. [34].

Among the various autoassociative networks the Hopfield network is the most widely known today [35, 36, 37]. A large number of researchers have exhaustively studied this network, its variations, and generalizations [32, 38, 39, 40, 41, 42, 43, 44, 45]. Hardware implementation issues of various associative memories have also been extensively studied [46, 47, 48, 49, 50]. Unlike the Hopfield network, which is a recurrent neural network, the morphological model provides the final result in one pass through the network without any significant amount of training.

To begin our discussion on associative memories, let $(\mathbf{x}^1, \mathbf{y}^1), \ldots, (\mathbf{x}^k, \mathbf{y}^k)$ be k vector pairs with $\mathbf{x}^\xi = (x_1^\xi, \ldots, x_n^\xi)' \in \mathbb{R}^n$ and $\mathbf{y}^\xi = (y_1^\xi, \ldots, y_m^\xi)' \in \mathbb{R}^m$ for $\xi = 1, \ldots, k$. For a given set of pattern associations $\{(\mathbf{x}^\xi, \mathbf{y}^\xi) : \xi = 1, \ldots, k\}$ we define a pair of associated pattern matrices (X, Y), where $X = (\mathbf{x}^1, \ldots, \mathbf{x}^k)$ and $Y = (\mathbf{y}^1, \ldots, \mathbf{y}^k)$. Thus, X is of dimension $n \times k$ with i, jth entry x_i^j and Y is of dimension $m \times k$ with i, jth entry y_i^j.

The earliest neural network approach to associative memories was the linear associative memory or correlation memory [30]. In this approach the goal is to store k vector pairs $(\mathbf{x}^1, \mathbf{y}^1), \ldots, (\mathbf{x}^k, \mathbf{y}^k)$ in an $m \times n$ *associative memory* W such that for any given input vector $\mathbf{x}^\xi$, the associative memory W recalls the output vector $\mathbf{y}^\xi = W\mathbf{x}^\xi, \forall \xi = 1, \ldots, k$. The simplest solution for this goal is to set

$$W = \sum_{\xi=1}^{k} \mathbf{y}^\xi (\mathbf{x}^\xi)' . \qquad (4.1)$$

In this case, the i,jth entry of W is given by $w_{ij} = \sum_{\xi=1}^{k} y_i^\xi x_j^\xi$. If the input patterns $\mathbf{x}^1,\ldots,\mathbf{x}^k$ are orthonormal, that is, $(\mathbf{x}^j)' \cdot \mathbf{x}^i = \begin{cases} 1 \text{ if } i = j \\ 0 \text{ if } i \neq j \end{cases}$, then $W\mathbf{x}^\xi = \mathbf{y}^\xi \left((\mathbf{x}^\xi)' \cdot \mathbf{x}^\xi\right) + \sum_{\gamma \neq \xi} \mathbf{y}^\gamma \left((\mathbf{x}^\gamma)' \cdot \mathbf{x}^\xi\right) = \mathbf{y}^\xi$.

Thus, we have *perfect recall* of the output patterns $\mathbf{y}^1,\ldots,\mathbf{y}^k$. If $\mathbf{x}^1,\ldots,\mathbf{x}^k$ are not orthonormal (as in most realistic cases), then filtering processes using activation functions become necessary in order to retrieve the desired output pattern.

Morphological associative memories are surprisingly similar to these classical correlation memories. With each pair of pattern associations (X,Y) we associate two natural morphological $m \times n$ memories W_{XY} and M_{XY} defined by

$$W_{XY} = \bigwedge_{\xi=1}^{k} \left[\mathbf{y}^\xi \times (-\mathbf{x}^\xi)'\right], \text{ and } M_{XY} = \bigvee_{\xi=1}^{k} \left[\mathbf{y}^\xi \times (-\mathbf{x}^\xi)'\right], \quad (4.2)$$

where the *morphological outer product* is defined as

$$\mathbf{y} \times \mathbf{x}' = \begin{pmatrix} y_1 + x_1 & \cdots & y_1 + x_n \\ \vdots & \ddots & \vdots \\ y_m + x_1 & \cdots & y_m + x_n \end{pmatrix}.$$

It is worthwhile to note that $\mathbf{y} \times \mathbf{x}' = \mathbf{y} \boxtimes \mathbf{x}' = \mathbf{y} \boxtimes \mathbf{x}'$.

Note the similarities between the definition of the memory given by Eq. (4.1) and those defined by Eq. (4.2). Also, a consequence of Eq. (4.2) is that $W_{XY} \boxtimes X \leq Y \leq M_{XY} \boxtimes X$. Here we use the notion that matrix A is *less or equal* than a matrix B of the same dimension, denoted by $A \leq B$, and A is *strictly less than* B, denoted by $A < B$, if and only if for each corresponding entry of these matrices we have that $a_{ij} \leq b_{ij}$ and $a_{ij} < b_{ij}$, respectively.

A fundamental relationship between the canonical MAMs and other morphological associative memories is given by the next theorem, which was proved in [1].

Theorem 1. *Let (X,Y) denote the associate sets of pattern vector pairs. Whenever there exist perfect recall memories A and B such that $A \boxtimes \mathbf{x}^\xi = \mathbf{y}^\xi$ and $B \boxtimes \mathbf{x}^\xi = \mathbf{y}^\xi$ for $\xi = 1,\ldots,k$, then $A \leq W_{XY} \leq M_{XY} \leq B$ and $\forall \xi$, $W_{XY} \boxtimes \mathbf{x}^\xi = \mathbf{y}^\xi = M_{XY} \boxtimes \mathbf{x}^\xi$.*

Hence, W_{XY} is the least upper bound of all perfect recall memories involving the $\boxtimes$ operation, and M_{XY} is the greatest lower bound of all perfect memories involving the $\boxtimes$ operation. Furthermore, if there exist perfect recall memories, then the canonical memories are also perfect recall memories.

If $X = Y$ (i.e., $\forall \xi$, $\mathbf{x}^\xi = \mathbf{y}^\xi$), then we obtain the morphological autoassociative memories W_{XX} and M_{XX}. In [1] we proved that $W_{XX} \boxtimes X = X =$

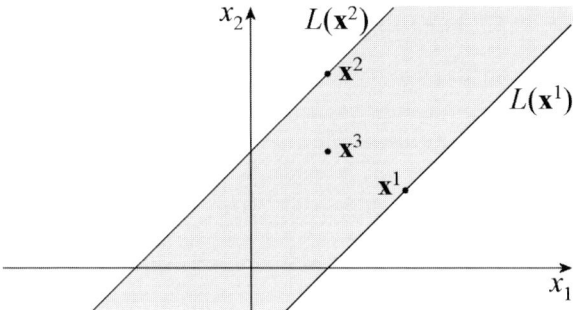

Fig. 4.1. The *shaded area* including the boundaries $L(\mathbf{x}^1)$ and $L(\mathbf{x}^2)$ constitutes the set of fixed points of W_{XX}, where $X = \{\mathbf{x}^1, \mathbf{x}^2\}$

$M_{XX} \boxtimes X$, where X can consist of any arbitrarily large number of pattern vectors.

Example 2. Let $\mathbf{x}^1 = \begin{pmatrix} 4 \\ 2 \end{pmatrix}$, $\mathbf{x}^2 = \begin{pmatrix} 2 \\ 5 \end{pmatrix}$, $\mathbf{x}^3 = \begin{pmatrix} 2 \\ 3 \end{pmatrix}$; then $X = \{\mathbf{x}^1, \mathbf{x}^2, \mathbf{x}^3\}$, $W_{XX} = \begin{pmatrix} 0 & -3 \\ -2 & 0 \end{pmatrix}$, and $W_{XX} \boxtimes \mathbf{x}^\xi = \mathbf{x}^\xi \ \forall \xi = 1, 2, 3$.

It is interesting to note that when taking the subset $Y = \{\mathbf{x}^1, \mathbf{x}^2\} \subset X$, then $W_{YY} \boxtimes \mathbf{x}^3 = \mathbf{x}^3$. In fact, we have that $W_{XX} \boxtimes \mathbf{x} = \mathbf{x} \iff W_{YY} \boxtimes \mathbf{x} = \mathbf{x}$. That is, W_{XX} and W_{YY} have identical fixed point sets. The reason is that $W_{XX} = W_{YY}$ and $\mathbf{x}^3$ is *lattice dependent* on the set $\{\mathbf{x}^1, \mathbf{x}^2\}$. More precisely, $\mathbf{x}^3$ is a "linear" combination of $\mathbf{x}^1$ and $\mathbf{x}^2$, namely $\mathbf{x}^3 = (\alpha + \mathbf{x}^1) \vee (\beta + \mathbf{x}^2)$, where $\alpha = \beta = -2$. Here addition is again pointwise; i.e., if $\mathbf{x} = (x_1, \ldots, x_n)'$ and $\alpha \in \mathbb{R}_{\pm\infty}$, then $\alpha + \mathbf{x} = (\alpha + x_1, \ldots, \alpha + x_n)$.

Lattice dependency and the set of fixed points of the transformation W_{XX} turn out to be equivalent notions. Figure 4.1 illustrates this for the set $X = \{\mathbf{x}^1, \mathbf{x}^2\}$, where $\mathbf{x}^1 = (4, 2)'$ and $\mathbf{x}^2 = (2, 5)'$.

The lines $L(\mathbf{x}^\xi) = \{\alpha + \mathbf{x}^\xi : \alpha \in \mathbb{R}\}$, $\xi = 1, 2$, passing through the points $\mathbf{x}^\xi$ form the boundary of an infinite strip as shown. Every point $\mathbf{x} \in \mathbb{R}^2$ in this strip is of form

$$\mathbf{x} = (\alpha_1 + \mathbf{x}^1) \vee (\alpha_2 + \mathbf{x}^2), \tag{4.3}$$

as well as

$$\mathbf{x} = (\beta_1 + \mathbf{x}^1) \wedge (\beta_2 + \mathbf{x}^2), \tag{4.4}$$

for some scalars $\alpha_1, \alpha_2, \beta_1, \beta_2 \in \mathbb{R}_{\pm\infty}$. Any point satisfying Eq. (4.3) or Eq. (4.4) is also a fixed point of W_{XX}. In fact, we have the following:

Theorem 2. *Suppose* $X = \{\mathbf{x}^1, \mathbf{x}^2, \ldots, \mathbf{x}^K\} \subset \mathbb{R}^n$, $A(X) = \{\mathbf{x} \in \mathbb{R}^n : \mathbf{x} = \bigvee_{\xi=1}^{K}(\alpha_\xi + \mathbf{x}^\xi), \alpha_\xi \in \mathbb{R}\}$, *and* $B(X) = \{\mathbf{x} \in \mathbb{R}^n : \mathbf{x} = \bigwedge_{\xi=1}^{K}(\alpha_\xi + \mathbf{x}^\xi), \alpha_\xi \in \mathbb{R}\}$. *If* $\mathbf{x} \in A(X) \cup B(X)$, *then* $W_{XX} \boxtimes \mathbf{x} = \mathbf{x} = M_{XX} \boxtimes \mathbf{x}$.

Proof. Suppose $\mathbf{x} \in A(X)$. Then $\mathbf{x} = \bigvee_{\xi=1}^{K} (\alpha_\xi + \mathbf{x}^\xi)$, and hence, $x_j = \bigvee_{\xi=1}^{K} \left(\alpha_\xi + x_j^\xi\right) \forall j = 1, \ldots, n$. Thus, for each $j = 1, \ldots, n$, there exists an index $k \in \{1, \ldots, K\}$ *depending on* j such that $x_j = \alpha_k + x_j^k$. We denote this k by $k(j)$. Since W_{XX} is a *dilative* transform, $W_{XX} \boxtimes \mathbf{x} \geq \mathbf{x} \ \forall \mathbf{x} \in \mathbb{R}^n$. Therefore,

$$(W_{XX} \boxtimes \mathbf{x})_i \geq \mathbf{x}_i \quad \forall i = 1, \ldots, n. \tag{4.5}$$

On the other hand, $(W_{XX} \boxtimes \mathbf{x})_i = \bigvee_{j=1}^{n} (w_{ij} + x_j) = \bigvee_{j=1}^{n} \left[\bigwedge_{\xi=1}^{K} \left(x_i^\xi - x_j^\xi\right) + x_j\right] = \bigwedge_{\xi=1}^{K} \left(x_i^\xi - x_\ell^\xi\right) + x_\ell$, for some $\ell \in \{1, \ldots, n\}$. Thus, $(W_{XX} \boxtimes \mathbf{x})_i = \bigwedge_{\xi=1}^{K} \left(x_i^\xi - x_\ell^\xi\right) + x_\ell \leq \left(x_i^{k(\ell)} - x_\ell^{k(\ell)}\right) + x_\ell = \left(x_i^{k(\ell)} - x_\ell^{k(\ell)}\right) + \left(\alpha_{k(\ell)} + x_\ell^{k(\ell)}\right) = \alpha_{k(\ell)} + x_i^{k(\ell)} \leq \bigvee_{\xi=1}^{K} \left(\alpha_\xi + x_i^\xi\right) = x_i$. It follows that

$$(W_{XX} \boxtimes \mathbf{x})_i \leq \mathbf{x}_i \quad \forall i = 1, \ldots, n. \tag{4.6}$$

Equations (4.5) and (4.6) imply that $W_{XX} \boxtimes \mathbf{x} = \mathbf{x}$.

As an easy consequence of Theorem 1, we now also have that $M_{XX} \boxtimes \mathbf{x} = \mathbf{x}$. The case $\mathbf{x} \in B(X)$ is handled in an analogous manner. Q.E.D. □

As a consequence, it follows that every point in the shaded area of Fig. 4.1 can be perfectly recalled by the memories W_{XX} and M_{XX}, where $X = \{\mathbf{x}^1, \mathbf{x}^2\}$. Thus, the memories not only store the vectors $\mathbf{x}^1$ and $\mathbf{x}^2$, but also an infinite number of patterns, namely those that *depend* on $\mathbf{x}^1$ and $\mathbf{x}^2$.

Obviously, $A(X) = B(X)$ in Theorem 2, and it is not difficult to ascertain that if $\mathbf{x}^1, \mathbf{x}^2 \in \mathbb{R}^2$, with $\mathbf{x}^1 \neq \alpha + \mathbf{x}^2 \ \forall \alpha \in \mathbb{R}$, then $\mathbf{x}$ is a fixed point of W_{XX} if and only if $\mathbf{x} = (\alpha_1 + \mathbf{x}^1) \vee (\alpha_2 + \mathbf{x}^2)$ for some $\alpha_1, \alpha_2 \in \mathbb{R}$. Although the condition $\mathbf{x} \in A(X)$ implies that $W_{XX} \boxtimes \mathbf{x} = \mathbf{x}$, the converse, however, does not hold if the dimension $n > 2$.

Example 3. Let $X = \{\mathbf{x}^1, \mathbf{x}^2\} \subset \mathbb{R}^3$, where $\mathbf{x}^1 = \begin{pmatrix} 0 \\ 0 \\ 0 \end{pmatrix}$ and $\mathbf{x}^2 = \begin{pmatrix} 2 \\ 1 \\ 0 \end{pmatrix}$. If $\mathbf{x} = \begin{pmatrix} 1 \\ 1 \\ 0 \end{pmatrix}$, then $W_{XX} \boxtimes \mathbf{x} = \begin{pmatrix} 0 & 0 & 0 \\ -1 & 0 & 0 \\ -2 & -1 & 0 \end{pmatrix} \boxtimes \begin{pmatrix} 1 \\ 1 \\ 0 \end{pmatrix} = \begin{pmatrix} 1 \\ 1 \\ 0 \end{pmatrix}$. Thus $\mathbf{x}$ is a fixed point for W_{XX}. If $\mathbf{x} = (\alpha_1 + \mathbf{x}^1) \vee (\alpha_2 + \mathbf{x}^2)$, then

$$(\alpha_1 + x_1^1) \vee (\alpha_2 + x_1^2) = 1, \tag{4.7}$$

$$(\alpha_1 + x_2^1) \vee (\alpha_2 + x_2^2) = 1, \text{ and} \tag{4.8}$$

$$\alpha_1 \vee \alpha_2 = 0. \tag{4.9}$$

It follows from Eq. (4.9) that $\alpha_1 \leq 0$ and $\alpha_2 \leq 0$ and that at least one of $\alpha_1 = 0$ or $\alpha_2 = 0$. By Eq. (4.7) we have $\alpha_1 \vee (\alpha_2 + 2) = 1$. Since $\alpha_1 \leq 0$, we must have $\alpha_2 = -1$. But according to Eq. (4.8), $\alpha_1 \vee (-1 + 1) = 1$, which is impossible. Therefore, $\mathbf{x} \neq (\alpha_1 + \mathbf{x}^1) \vee (\alpha_2 + \mathbf{x}^2) \ \forall \alpha_1, \alpha_2 \in \mathbb{R}$.

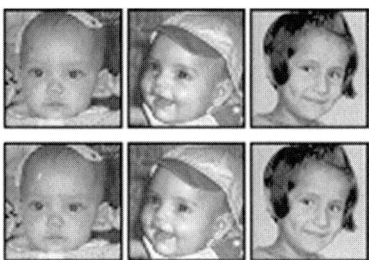

Fig. 4.2. The three patterns in the *top row* were used in constructing the morphological autoassociative memories W_{XX} and M_{XX} (of size 2500×2500). The *bottom row* shows the perfect output of these memories when presented with the respective patterns from the *top row*

The above example shows that a fixed point of W_{XX} need not be an element of $A(X)$. This raises the question as to the structure of the set of fixed points of W_{XX}. Before answering this question, observe that $\mathbf{x} = \bigvee_{j=1}^{3} \bigwedge_{\xi=1}^{2} \left(\alpha_{\xi,j} + \mathbf{x}^{\xi} \right)$, where $\alpha_{1,1} = \alpha_{1,2} = 1$, $\alpha_{1,3} = \alpha_{2,2} = \alpha_{2,3} = 0$, and $\alpha_{2,1} = -1$. In terms of lattice theory this means that $\mathbf{x}$ is lattice dependent on X. More precisely, we have the following:

Definition 1. *Suppose* $X = \{\mathbf{x}^1, \ldots, \mathbf{x}^k\} \subset \mathbb{R}^n$. *A vector* $\mathbf{x} \in \mathbb{R}^n$ *is* lattice dependent *on* X *if and only if* $\mathbf{x} = p$ *for some lattice polynomial p over X.*

The vector $\mathbf{x}$ is said to be lattice independent *of X if and only if $\mathbf{x}$ is not lattice dependent on X.*

The set X is said to be lattice independent *if and only if* $\forall \lambda \in \{1, \ldots, k\}$, $\mathbf{x}^{\lambda}$ *is lattice independent of* $X \setminus \{\mathbf{x}^{\lambda}\}$.

Recall that a lattice polynomial over X is any finite expression involving the symbols $\wedge$ and $\vee$, and letters, of form $a + \mathbf{x}^{\xi}$, where $\mathbf{x}^{\xi} \in X$ and $a \in \mathbb{R}$. Such a polynomial is also referred to as a polynomial of degree one.

It is not difficult to show that if $\mathbf{x}$ and $\mathbf{y}$ are fixed points of W_{XX} or M_{XX}, then so are $(a + \mathbf{x}) \vee (b + \mathbf{y})$ and $(a + \mathbf{x}) \wedge (b + \mathbf{y})$ for any pair $a, b \in \mathbb{R}$. This observation implies the following:

Theorem 3. *Suppose* $X = \{\mathbf{x}^1, \ldots, \mathbf{x}^k\} \subset \mathbb{R}^n$. *If* $\mathbf{x} \in \mathbb{R}^n$, *then* $W_{XX} \boxvee \mathbf{x} = \mathbf{x}$ *if and only if $\mathbf{x}$ is lattice dependent on X.*

The notion of lattice independence is important in the recall of noisy patterns. It is well known that the memories W_{XX} and M_{XX} are extremely robust in the presence of erosive and dilative noise, respectively [1, 3]. Given a pattern $\mathbf{x}^{\gamma}$ we say that a distorted version $\tilde{\mathbf{x}}^{\gamma}$ of the pattern $\mathbf{x}^{\gamma}$ has undergone an *erosive change* whenever $\tilde{\mathbf{x}}^{\gamma} \leq \mathbf{x}^{\gamma}$ and a *dilative change* whenever $\tilde{\mathbf{x}}^{\gamma} \geq \mathbf{x}^{\gamma}$. As an example, consider the three pattern images $\mathbf{p}^1$, $\mathbf{p}^2$, and $\mathbf{p}^3$ shown in Fig. 4.2. Each $\mathbf{p}^{\xi}$ is a 50×50-pixel 256 grayscale image. For uncorrupted input, perfect recall is guaranteed if we use either memory W_{XX} or M_{XX}. Using the standard row-scan method, each pattern image $\mathbf{p}^{\xi}$ can be converted

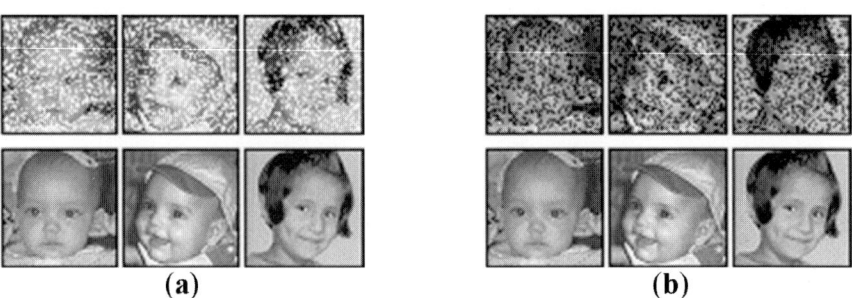

Fig. 4.3. The *top row* shows the input patterns corrupted with dilative (**a**) and erosive (**b**) noise. The *bottom row* shows the corresponding recalled patterns using the morphological memory M_{XX} (**a**) and, respectively, W_{XX} (**b**)

into a pattern vector $\mathbf{x}^\xi = (x_1^\xi, \ldots, x_{2500}^\xi)$ by defining $x_{50(r-1)+c}^\xi = p^\xi(r,c)$ for $r,c = 1,\ldots,50$.

Corrupting the patterns $\mathbf{x}^\xi$ with 30% randomly generated erosive and dilative noise with an intensity level of 128 results in almost perfect recall (NMSE[1] $< 10^{-3}$) when using the memory W_{XX} and M_{XX}, respectively. Figure 4.3 provides for a visual example of this experiment.

The reason for the robustness of associative memories in the presence of erosive or dilative noise is a consequence of a sequence of theorems, which are given in [1]. These theorems provide necessary and sufficient conditions for the bounds of the corruption of a pattern $\mathbf{x}^\gamma$ that guarantees perfect recall; they also imply that W_{XX} will fail miserably if dilative noise not satisfying these bounds is present. Our experiments have shown that insertion of only minute amounts of dilative noise, often in only one vector component, can result in complete recall failure. Similar comments hold for the memory M_{XX} and erosive noise. Hence, neither memory W_{XX} or M_{XX} is useful in the presence of random noise, which, generally, consists of both erosive as well as dilative noise.

The kernel method proposed in [1] suggests a solution to this dilemma. However, it became clear that finding an algorithmic method for selecting an optimal set of proper kernels was not going to be an easy task. Part of the difficulty is due to the fact that the existence of proper kernels for a given set of pattern vectors remains an unsolved problem if the definition of kernels proposed in [1] is used. Additionally, to be useful for pattern recognition in the presence of large amounts of noise, such kernel patterns need to represent greatly reduced (eroded) versions of the exemplar patterns. However, simply eroding exemplar patterns will, generally, not result in kernel vectors. Before addressing solutions to the problem of random noise, it is necessary to gain

---

[1] Normalized mean-square error, computed as $\sum_j \left(\tilde{x}_j^\xi - x_j^\xi\right)^2 \Big/ \sum_j \left(x_j^\xi\right)^2$ for each ξ.

an understanding of the kernel method and its relationship to the notion of morphological independence.

4.4 Kernels and Morphological Independence

Since W_{XX} is suitable for recognizing patterns corrupted by erosive noise and M_{XX} is suitable for recognizing patterns corrupted by dilative noise, an intuitive idea is to process a noisy version $\tilde{\mathbf{x}}^\gamma$ of $\mathbf{x}^\gamma$ containing both erosive and dilative noise through a combination of W_{XX} and M_{XX}. Sussner proved that passing the output of $M_{XX} \boxtimes \tilde{\mathbf{x}}^\gamma$ through the memory W_{XX} or, dually, the output of $W_{XX} \boxdot \tilde{\mathbf{x}}^\gamma$ through M_{XX} will, generally, not result in $\mathbf{x}^\gamma$ [51]. Nevertheless, the modified kernel approach proposed by Ritter et al. [3] is based on this intuitive idea using the memories M_{XX} and W_{XX} in sequence in order to create a morphological memory that is robust in the presence of random noise, even in the general situation where $X \neq Y$ and X and Y are not Boolean [1]. The underlying idea is to define a memory M that associates with each input pattern $\mathbf{x}^\gamma$ an intermediate pattern $\mathbf{z}^\gamma$. Another associative memory W is defined that associates each pattern $\mathbf{z}^\gamma$ with the desired output pattern $\mathbf{y}^\gamma$. In terms of min-max products, one obtains the equation $W \boxdot (M \boxtimes \mathbf{x}^\gamma) = \mathbf{y}^\gamma$.

If the $n \times k$ matrix $Z = (\mathbf{z}^1, \ldots, \mathbf{z}^k)$ satisfies certain conditions, then the matrices M_{ZZ} and W_{ZY} can serve as M and W, respectively. Furthermore, if Z is properly chosen, then $W \boxdot (M_{ZZ} \boxtimes \tilde{\mathbf{x}}^\gamma) = \mathbf{y}^\gamma$ for most corrupted versions $\tilde{\mathbf{x}}^\gamma$ of $\mathbf{x}^\gamma$. If Z satisfies these basic properties, then Z is called a *kernel* for the associated pair (X, Y). The following formal definition of a kernel was proposed in [3]:

Definition 2. *Let $Z = (\mathbf{z}^1, \ldots, \mathbf{z}^k)$ be an $n \times k$ matrix. We say that Z is a kernel for (X, Y) if and only if $Z \neq X$ and there exists a memory W such that $W \boxdot (M_{ZZ} \boxtimes \mathbf{x}^\gamma) = \mathbf{y}^\gamma$. If $Y = X$, then we say that Z is a kernel for X.*

For kernels to be effective in recognizing patterns that are severely corrupted by random noise, they need not only represent eroded subsets of X but should also be extremely sparse; i.e., for each γ, $\mathbf{z}^\gamma$ consists of mostly zero entries. The reason for sparseness is roughly based on the following observation. If Z is sparse, then the corrupted version $\tilde{\mathbf{x}}^\gamma$ of $\mathbf{x}^\gamma$ will generally be able to afford a high degree of erosive noise and still satisfy the inequality $\mathbf{z}^\gamma \leq \tilde{\mathbf{x}}^\gamma$. As M_{ZZ} is robust in the presence of dilative noise, $\tilde{\mathbf{x}}^\gamma$ will be conceived as a dilated version of $\mathbf{z}^\gamma$ by the memory M_{ZZ}. On the other hand, if $\mathbf{z}^\gamma$ is not sparse and $\tilde{\mathbf{x}}^\gamma$ contains large amounts of erosive noise, then it is far more likely that $\mathbf{z}^\gamma \not\leq \tilde{\mathbf{x}}^\gamma$ and M_{ZZ} will have difficulty in recognizing $\tilde{\mathbf{x}}^\gamma$. Ideally, we would like that for each γ, $z_j^\gamma = x_j^\gamma$ for exactly one $j \in \{1, \ldots, n\}$ and $z_j^\gamma = 0 \; \forall i \neq j$. If Z results in a kernel under these conditions, then we are guaranteed the recovery of $\mathbf{x}^\gamma$ from $\tilde{\mathbf{x}}^\gamma$ as long as $\mathbf{z}^\gamma \leq M_{ZZ} \boxtimes \tilde{\mathbf{x}}^\gamma \leq \mathbf{x}^\gamma$. These loose concepts lead to the definition of minimal representations of a pattern set X.

Definition 3. A set of patterns $Z \leq X$ is said to be a *minimal representation* of X if and only if for $\gamma = 1, \ldots, k$,

1. $\mathbf{z}^\gamma \wedge \mathbf{z}^\xi = \mathbf{0} \ \forall \xi \neq \gamma$,
2. $\mathbf{z}^\gamma$ contains at most one nonzero entry, and
3. $W_{ZX} \boxvee \mathbf{z}^\gamma = \mathbf{x}^\gamma$.

Condition 1 of this definition satisfies part of the following equation from Sussner's theorem [51], which provides for *binary* associative memories that are robust in the presence of random bit reversals: $\mathbf{z}^\gamma \wedge \mathbf{z}^\xi = \mathbf{0}$ and $\mathbf{z}^\xi \not\leq \mathbf{x}^\gamma \ \forall \gamma$ and $\forall \xi$ with $\gamma \neq \xi$. Condition 2 assures sparsity, while condition 3 simply says that X can be reconstructed from Z. In this sense Z acts as an orthogonal basis within the lattice algebra underlying the morphological operations.

The connection between kernels and minimal representations is given by the following theorems, which are proven in [3].

Theorem 4. *If X is lattice independent, then there exists a set of patterns $Z \leq X$ with the property that for $\gamma = 1, \ldots, k$,*

1. $\mathbf{z}^\gamma \wedge \mathbf{z}^\xi = \mathbf{0} \ \forall \xi \neq \gamma$,
2. $\mathbf{z}^\gamma$ contains at most one nonzero entry, and
3. $W_{XX} \boxvee \mathbf{z}^\gamma = \mathbf{x}^\gamma$.

Corollary 1. *If X and Z are as in Theorem 4, then Z is a minimal representation of X.*

Corollary 2. *If X and Z are as in Theorem 4, then Z is a kernel for X.*

According to Corollary 1, a minimal representation is also a kernel. Hence, for a set of patterns X to be reducible to a kernel, it is sufficient that X is lattice independent. Furthermore, if X is lattice independent, then in order to obtain a kernel one simply selects a minimal representation Z of X using the constructive method given in the proof of Theorem 4.

Given a minimal representation Z that is also a kernel for X and a noisy version $\tilde{\mathbf{x}}^\gamma$ of the pattern $\mathbf{x}^\gamma$ having the property that $\mathbf{z}^\gamma \leq \tilde{\mathbf{x}}^\gamma$ and $M_{ZZ} \boxwedge \tilde{\mathbf{x}}^\gamma \leq \mathbf{x}^\gamma$, then it must follow that $W_{XX} \boxvee (M_{ZZ} \boxwedge \tilde{\mathbf{x}}^\gamma) = \mathbf{x}^\gamma$.

Using the method of proof of Theorem 4, we constructed a kernel matrix Z for the pattern images $\mathbf{p}^1$, $\mathbf{p}^2$, and $\mathbf{p}^3$ shown in the top row of Fig. 4.2. The patterns are lattice independent and Fig. 4.4a depicts the patterns and associated minimal representation $(\mathbf{z}^1, \mathbf{z}^2, \mathbf{z}^3)$. Randomly corrupting the patterns with 30% of noise with an intensity level of 128, and using the minimal representation Z as our kernel set, we obtained the perfect recall $W_{XX} \boxvee (M_{ZZ} \boxwedge \tilde{\mathbf{x}}^\gamma) = \mathbf{x}^\gamma$ for $\gamma = 1, \ldots, 3$, as illustrated in Fig. 4.4b.

It is important to note that minimal representations are not unique. However, given two minimal representations Z and $\bar{Z}$ of X, then statistically either representation performs as well as the other. More precisely, if $z_i^\gamma = 1 = \bar{z}_j^\gamma$, where $\mathbf{z}^\gamma \in Z$ and $\bar{\mathbf{z}}^\gamma \in \bar{Z}$, then $p(z_i^\gamma | \tilde{x}_i^\gamma) = p(\bar{z}_j^\gamma | \tilde{x}_j^\gamma)$, where $p(z_i^\gamma | \tilde{x}_i^\gamma)$ denotes

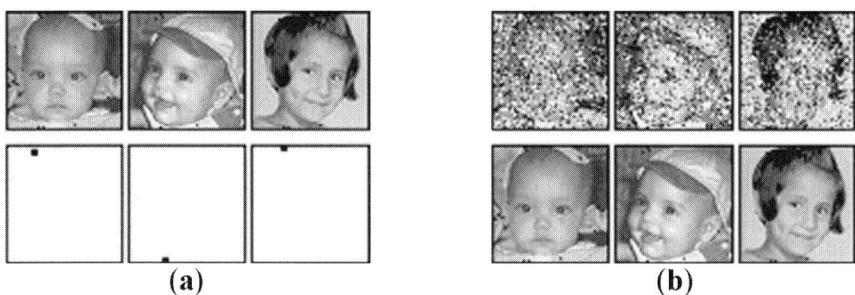

Fig. 4.4. In (**a**) the *top row* shows the lattice-independent patterns and the *bottom row* shows the corresponding nonzero entries in matrix Z used as a minimal representation or kernel. In (**b**) the *top row* contains the input patterns corrupted with random noise, while the *bottom row* illustrates perfect recall using the kernel Z for the memory scheme $M_{ZZ} \to W_{XX}$ (also the output of memory W_{ZX})

the probability that $\tilde{x}_i^\gamma < \bar{z}_i^\gamma$. It follows that there is no *optimal* minimal representation unless some a priori knowledge of noise characteristics is available.

The problem of kernels for pattern pairs (X, Y) where $X \neq Y$ follows from the results established in this section. The following theorem is an easy consequence of Theorem 4 and its corollaries.

Theorem 5. *If X and Z are as in Theorem 4 and W_{XY} is a perfect associative recall memory, then Z is a kernel for (X, Y).*

In order to verify this theorem, simply let $W = W_{XY} \boxtimes W_{XX}$. Then, for all $\gamma = 1, \ldots, k$, $W \boxtimes (M_{ZZ} \boxtimes \mathbf{x}^\gamma) = (W_{XY} \boxtimes W_{XX}) \boxtimes (M_{ZZ} \boxtimes \mathbf{x}^\gamma) = W_{XY} \boxtimes [W_{XX} \boxtimes (M_{ZZ} \boxtimes \mathbf{x}^\gamma)] = W_{XY} \boxtimes \mathbf{x}^\gamma = \mathbf{y}^\gamma$. The sequence of this associative feedforward network is given by $\mathbf{x}^\gamma \to M_{ZZ} \to W_{XX} \to W_{XY} \to \mathbf{y}^\gamma$ or, simply, $\mathbf{x}^\gamma \to M_{ZZ} \to W \to \mathbf{y}^\gamma$ where $W = W_{XY} \boxtimes W_{XX}$.

As a final observation, it has been our experience that the method described above increases in robustness for noisy pattern recall as the dimension n of pattern size increases. There are some probabilistic reasons for supporting this observation [3].

The ring $(\mathbb{R}, +, \times)$ forms the computational basis for the traditional artificial neural model. The matrix associative memories presented in this section amount to a simple reformulation of the classical matrix associative memories based on $(\mathbb{R}, +, \times)$ in terms of the algebras $(\mathbb{R}_{-\infty}, \vee, +)$ and $(\mathbb{R}_\infty, \wedge, +')$. Considering the fact that the latter two lattice algebras are only semirings, it is somewhat surprising that these *weaker* algebras provide for more robust memories with larger storage capacities. Similar observations hold for morphological feedforward networks discussed in the following sections. However, in contrast to traditional feedforward networks, the morphological counterparts also include dendritic computing.

4.5 Dendritic Computation Based on Lattice Algebra

In the classical theory of artificial neural networks (ANNs), the main processing element is the neuron. Computation at a neuron N is performed within the context of the ring of real numbers $(\mathbb{R}, +, \times)$ by summing the products of neural values and connection weights from all neurons in the network connected to N. Generally, a neural activation function is applied to the sum, which provides for nonlinearity of the neural output. Application of the activation function is the only nonlinear component in this model; all other operations carried out at a neuron are linear algebraic operations.

Morphological neural networks (MNNs) represent the counterpart of the above model, obtained when lattice algebra is employed instead of linear algebra. Computation here is performed within the general context of the bounded ℓ-group $(\mathbb{R}_{\pm\infty}, \vee, \wedge, +, +')$ that was defined in Sect. 4.2, or within context of the particular $s\ell$-semigroups $(\mathbb{R}_{-\infty}, \vee, +)$ or $(\mathbb{R}_{\infty}, \wedge, +')$. Thus, the total net input at a neuron M_j is computed as the maximum (or minimum) of the sums of neural values and corresponding synaptic weights. It is apparent that, since the maximum (or minimum) of sums is used instead of the sum of products, the lattice algebraic model is nonlinear *before* the application of an activation function.

The previous sections have discussed one category of MNNs, namely morphological associative memories. In the remaining part of this exposition, we will examine another MNN category, morphological perceptrons (MPs) with dendritic structures, and will show that this novel model has greater computational capability and pattern discrimination power than traditional perceptrons. Besides being based on lattice algebra, MPs with dendritic structures bear a closer resemblance to biological neural networks. Neurons in the mammalian brain have two important processes, *dendrites* and *axons*. The axon is the principal output fiber that branches toward its end into an *axonal tree*. Its tips synapse with the dendritic structures of other neurons at *synaptic sites*. Dendrites create large and complicated trees, and the number of synapses on a *single* cortical neuron typically ranges between 500 and 200,000. Synapses are of two types, *excitatory* and *inhibitory*.

In biological neural networks, dendrites make up the largest component in both surface area and volume of the brain, and span all cortical layers in all regions of the cerebral cortex [52, 53, 54]. Thus, when attempting to model artificial brain networks, one cannot ignore dendrites, which make up more than 50% of the neuron's membrane. This is especially true in light of the fact that some researchers have proposed that dendrites, and not the neurons, are the elementary computing devices of the brain, capable of implementing such logical functions as AND, OR, and NOT [52, 53, 54, 55, 56, 57, 58, 59, 60].

Current ANN models, and in particular perceptrons, do not include dendritic structures. As a result, problems occur that may be easily preventable when employing dendritic computing. For example, M. Gori and F. Scarselli have shown that multilayer perceptrons (MLPs) are not adequate for pattern recognition and verification [61]. Specifically, they proved that multilayer perceptrons with sigmoidal units and a number of hidden units, less than or equal to the number of input units, are unable to model patterns distributed in typical clusters. The reason is that these networks draw open separation surfaces in pattern space. In this case, all patterns not members of the cluster but contained in an open area determined by the separation surfaces will be misclassified. When using more hidden units than input units, closed surfaces *may* result but, unfortunately, determining whether or not the perceptron draws closed separation surfaces in pattern space is NP-hard. This is quite opposite to what is commonly believed and reported in the literature. We have proven that MPs with dendritic structures do not suffer from these problems, because the separation surfaces are guaranteed to be closed [13, 62].

Let $N_1, \ldots, N_n$ denote a collection of neurons with dendritic structures, whose morphology is based on the biological model described earlier. Suppose these neurons provide synaptic input to another collection $M_1, \ldots, M_m$ of neurons having the same processes. The value of a neuron N_i ($i = 1, \ldots, n$) propagates through its axonal tree all the way to the terminal branches that make contact with the neuron M_j ($j = 1, \ldots, m$). The weight of an axonal branch of neuron N_i terminating on the kth dendrite of M_j is denoted by w_{ijk}^ℓ, where the superscript $\ell \in \{0, 1\}$ distinguishes between *excitatory* ($\ell = 1$) and *inhibitory* ($\ell = 0$) input to the dendrite. The kth dendrite of M_j will respond to the total input received from the neurons $N_1, \ldots, N_n$ and will either accept or inhibit the received input. The computation of the kth dendrite of M_j, denoted by D_{jk}, is given by

$$\tau_k^j(\mathbf{x}) = p_{jk} \bigwedge_{i \in I(k)} \bigwedge_{\ell \in L(i)} (-1)^{1-\ell} \left(x_i + w_{ijk}^\ell \right) , \qquad (4.10)$$

where $\mathbf{x} = (x_1, \ldots, x_n)$ denotes the input value of the neurons $N_1, \ldots, N_n$ with x_i representing the value of N_i; $I(k) \subseteq \{1, \ldots, n\}$ corresponds to the set of all input neurons with terminal fibers that synapse on the kth dendrite of M_j; $L(i) \subseteq \{0, 1\}$ corresponds to the set of terminal fibers of N_i that synapse on the kth dendrite of M_j; and $p_{jk} \in \{-1, 1\}$ denotes the excitatory ($p_{jk} = 1$) or inhibitory ($p_{jk} = -1$) response of the kth dendrite of M_j to the received input.

It follows from the formulation $L(i) \subseteq \{0, 1\}$ that the ith neuron N_i can have at most two synapses on a given dendrite k. Also, if the value $\ell = 1$, then the input $\left(x_i + w_{ijk}^1 \right)$ is excitatory, and inhibitory for $\ell = 0$ since in this case we have $-\left(x_i + w_{ijk}^0 \right)$.

The value $\tau_k^j(\mathbf{x})$ is passed to the cell body, and the state of M_j is a function of the input received from all its dendrites. The total value received by M_j is given by

$$\tau^j(\mathbf{x}) = p_j \bigwedge_{k=1}^{K_j} \tau_k^j(\mathbf{x}) , \qquad (4.11)$$

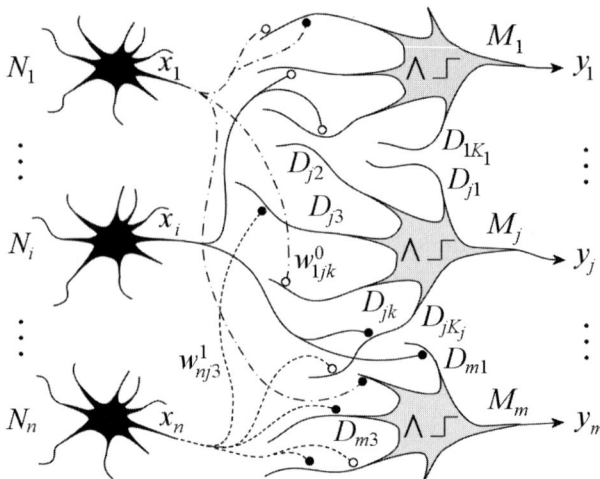

Fig. 4.5. Single-layer morphological perceptron with dendritic structures. Here D_{jk} denotes the kth dendrite of neuron M_j, and K_j the number of dendrites of M_j. An input neuron can make synapse on a dendrite with excitatory and/or inhibitory fibers, e.g., w^1_{nj3} is the weight of an excitatory fiber from neuron N_n to dendrite D_{j3}, while w^0_{1jk} is the weight of an inhibitory fiber coming from input neuron N_1 to dendrite D_{jk}

where K_j denotes the total number of dendrites of M_j, and $p_j = \pm 1$ denotes the response of the cell body to the received dendritic input. Here again, $p_j = 1$ means that the input is accepted, while $p_j = -1$ means that the cell rejects the received input. The *next* state of M_j is then determined by an activation function f, namely $y_j = f\left(\tau^j(\mathbf{x})\right)$. In this exposition we restrict our discussion to the hard-limiter

$$f\left(\tau^j(\mathbf{x})\right) = \begin{cases} 1 \text{ if } \tau^j(\mathbf{x}) \geq 0 \,, \\ 0 \text{ if } \tau^j(\mathbf{x}) < 0 \,. \end{cases} \qquad (4.12)$$

A *single-layer morphological perceptron* (SLMP) is a special case of this model. Here the neurons $N_i, \ldots, N_n$ would denote the input neurons, and the neurons $M_1, \ldots, M_m$ the output neurons. For SLMPs we allow $\mathbf{x} = (x_1, \ldots, x_n) \in \mathbb{R}^n$. That is, the value x_i of the ith input neuron N_i need not be binary. The structure of a single-layer morphological perceptron is illustrated in Fig. 4.5.

4.6 Computational Capability of Perceptrons Based on the Dendritic Model

The dendritic computing framework introduced in the previous section allows the construction of novel models of artificial neural networks, which share common characteristics with classical models, such as basic architecture, but exhibit significantly different capabilities. Based on the dendritic model we can design a single-layer morphological perceptron (SLMP) as an artificial neural network that is similar in structure to the classical single-layer perceptron (SLP), but incorporates dendritic structures and operates in terms of lattice algebraic operations. In this section we will discuss the specific characteristics of morphological perceptrons with dendritic structures.

Analogous to the classical SLP with one output neuron, an SLMP with one output neuron consists of a finite number of input neurons that are connected via axonal fibers to the output neuron. However, in contrast to an SLP, the output neuron of an SLMP has a dendritic structure and performs the lattice computation embodied by Eqs. (4.10) and (4.11). The computational capability of an SLMP is vastly different from that of an SLP as well as that of classical perceptrons in general. For example, no hidden layers are necessary to solve the XOR problem with an SLMP or to specify the points of a nonconvex region in pattern space. The specific computational capability of an SLMP with one output neuron is governed by the following:

Theorem 6. *If $X \subset \mathbb{R}^n$ is compact and $\varepsilon > 0$, then there exists a single-layer morphological perceptron that assigns every point of X to class C_1 and every point $\mathbf{x} \in \mathbb{R}^n$ to class C_0 whenever $\mathrm{d}(\mathbf{x}, X) > \varepsilon$.*

The expression $\mathrm{d}(\mathbf{x}, X)$ in the statement of Theorem 6 refers to the distance of the point $\mathbf{x} \in \mathbb{R}^n$ to the set X. As a consequence, any compact configuration, as the one shown in Fig. 4.6a, whether it is convex or nonconvex, connected or not connected, contains a finite or infinite number of points, can be approximated within any desired degree of accuracy $\varepsilon > 0$ by an SLMP with one output neuron.

The proof of Theorem 6 requires tools from elementary point set topology and is given in [13]. Although the proof is an existence proof, part of it is constructive and provides the basic idea for the training algorithms that we developed.

An SLMP can be extended to multiple output neurons in order to handle multiclass problems, just like its classical counterpart. However, unlike the SLP, which is a linear discriminator, the SLMP with multiple outputs can solve multiclass nonlinear classification problems. This computational capability of an SLMP with multiple output neurons is attested by Theorem 7 below, which is a generalization of Theorem 6 to multiple sets.

Suppose $X_1, X_2, \ldots, X_m$ denotes a collection of disjoint compact subsets of $\mathbb{R}^n$. The goal is to classify, $\forall j = 1, \ldots, m$, every point of X_j as a point belonging to class C_j and not belonging to class C_i whenever $i \neq j$. For each $p \in \{1, \ldots, m\}$, define $Y_p = \bigcup_{j=1, j\neq p}^{m} X_j$. Since each Y_p is compact and $Y_p \cap X_p = \emptyset, \varepsilon_p = \mathrm{d}(X_p, Y_p) > 0 \; \forall p = 1, \ldots, m$. Let $\varepsilon_0 = \frac{1}{2} \min\{\varepsilon_1, \ldots, \varepsilon_p\}$.

Theorem 7. *If $\{X_1, X_2, \ldots, X_m\}$ is a collection of disjoint compact subsets of $\mathbb{R}^n$ and ε a positive number with $\varepsilon < \varepsilon_0$, then there exists a single-layer*

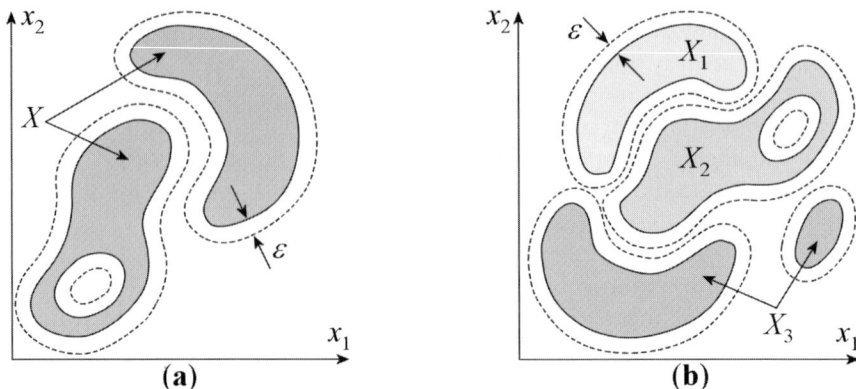

Fig. 4.6. a Compact set X, and b collection of disjoint compact sets X_1, X_2, X_3 and banded region of thickness ε (*dashed*). Theorems 6 and 7 guarantee the existence of SLMPs able to classify sets X and, respectively, X_1, X_2, X_3, within desired ε accuracy

morphological perceptron that assigns each point $\mathbf{x} \in \mathbb{R}^n$ *to class* C_j *whenever* $\mathbf{x} \in X_j$ *and* $j \in \{1, \ldots, m\}$, *and to class* $C_0 = \neg \bigcup_{j=1}^{m} C_j$ *whenever* $\mathrm{d}(\mathbf{x}, X_i) > \varepsilon$, $\forall i = 1, \ldots, m$. *Furthermore, no point* $\mathbf{x} \in \mathbb{R}^n$ *is assigned to more than one class.*

Figure 4.6b illustrates the conclusion of Theorem 7 for the case $m = 3$. The proof of this theorem is somewhat lengthy and is not included here; it is given in [62]. Based on the proofs of these two theorems, we constructed training algorithms for SLMPs [13, 14, 62, 63, 64]. During the learning phase, the output neurons grow new dendrites while the input neurons expand their axonal branches to terminate on the new dendrites. The algorithms always converge and have rapid convergence rate when compared to back-propagation learning in traditional perceptrons.

These training algorithms are similar in that they all dynamically grow dendrites and axonal fibers during the learning phase, which will use the patterns of the training set in just one iteration (one *epoch*). Thus, the architecture of the network is *not* predetermined beforehand. It is during training that the network grows new structures as necessary to learn the training patterns. The algorithms differ in the strategy of partitioning the pattern space. They either reduce an initial large box through elimination of foreign patterns and smaller regions that enclose them, or grow a class region by merging smaller hyperboxes and taking their union when they remain disconnected. In either case, the separation surfaces drawn in pattern space during training are always closed, making the SLMPs immune to the problems that MLPs suffer from due to open separation surfaces [61]. Also, as a consequence of the aforementioned theorems on which the algorithms are based, the trained SLMPs will always correctly recognize 100% of the patterns in the training set.

4.7 Training Algorithms for SLMPs with Dendritic Structures

The mathematical results provided in Sect. 4.6 and proved in [13] and [62] established that, for any collection of m compact sets, there exists an SLMP with dendritic structures than can classify the sets as m distinct classes to within any desired degree of accuracy. Although these results are existence theorems, their proofs provide the main ideas for developing training methods for the MP with dendritic structures.

Training can be realized in one of two main strategies, which differ in the way the separation surfaces in pattern space are determined. One strategy is based on elimination, whereas the other is based on merging. In the former approach, a hyperbox is initially constructed large enough to enclose all patterns belonging to the same class, possibly including foreign patterns from other classes. This large region is then carved to eliminate the foreign patterns. Training completes when all foreign patterns in the training set have been eliminated. The elimination is performed by computing the intersection of the regions recognized by the dendrites, as expressed in Eq. (4.11) for some neuron M_j: $\tau^j(\mathbf{x}) = p_j \bigwedge_{k=1}^{K_j} \tau_k^j(\mathbf{x})$.

The latter approach starts by creating small hyperboxes around individual patterns or small groups of patterns all belonging to the same class. Isolated boxes that are identified as being close according to a distance measure are then merged into larger regions that avoid including patterns from other classes. Training is completed after merging the hyperboxes for all patterns of the same class. The merging is performed by computing the union of the regions recognized by the dendrites. Thus, the total net value received by output neuron M_j is computed as:

$$\tau^j(\mathbf{x}) = p_j \bigvee_{k=1}^{K_j} \tau_k^j(\mathbf{x}) \ . \tag{4.13}$$

The two strategies are equivalent in the sense that they are based on the same mathematical framework and they both result in closed separation surfaces around patterns. The equivalence can be attested by examining the equations employed to compute the total net values in the two approaches given in Eqs. (4.11) and (4.13), and remarking that the maximum of any K values $a_1, a_2, \ldots, a_K$, can be equivalently written as a minimum: $\bigvee_{k=1}^{K} a_k = -\bigwedge_{k=1}^{K}(-a_k)$. Thus, if the output value y_j at neuron M_j is computed in terms of minimum as $y_j = f\left(p_j \bigwedge_{k=1}^{K_j} \tau_k^j(\mathbf{x})\right)$, then y_j can be equivalently computed in terms of maximum as $y_j = f\left(-p_j \bigvee_{k=1}^{K} -\tau_k^j(\mathbf{x})\right)$.

The major difference between the two approaches is in the shape of the separation surface that encloses the patterns of a class, and in the number of dendrites that are grown during training to recognize the region delimited by that separation surface. Since the elimination strategy involves removal of pieces from an originally large hyperbox, the resulting region is bigger than the one obtained with the merging strategy. The former approach is thus more general, while the latter is more specialized. This observation can guide the choice of the method for solving a particular problem.

Figure 4.7 illustrates two possible partitionings of the pattern space $\mathbb{R}^2$ in terms of intersection (Fig. 4.7a) and, respectively, union (Fig. 4.7b), in order to recognize the solid circles ($\bullet$) as one class C_1. In Fig. 4.7a the C_1 region is determined as the intersection of three regions, each identified by a corresponding dendrite. The rectangular region marked D_1 is intersected with the complement of the region marked D_2 and the complement of the region marked D_3. An excitatory dendrite recognizes the interior of an enclosed region, whereas an inhibitory dendrite recognizes the exterior of a delimited region. Thus, the corresponding dendrite D_1 is excitatory, while dendrites D_2 and D_3 are inhibitory. If we assign $j = 1$ in Eqs. (4.10) and (4.11) as representing the index of the output neuron for class C_1, then the output value is computed as $y_1 = f\left(\tau^1(\mathbf{x})\right) = f\left(p_1 \bigwedge_{k=1}^{3} \tau_k^1(\mathbf{x})\right)$, where $\tau_k^1(\mathbf{x})$ is computed as in Eq. (4.10). There, the responses p_{1k} are $p_{11} = 1$ and $p_{12} = p_{13} = -1$. The response of the cell body is $p_1 = 1$.

In Fig. 4.7b the C_1 region is determined as the union of four regions, each identified by a corresponding dendrite. This time, all dendrites $D_1, \ldots, D_4$ are excitatory, so their responses will be $p_{1k} = 1$, $k = 1, \ldots, 4$. Using Eqs. (4.10) and (4.13) we obtain the output value $y_1 = f\left(\tau^1(\mathbf{x})\right) = f\left(p_1 \bigvee_{k=1}^{4} \tau_k^1(\mathbf{x})\right)$, where $p_1 = 1$. As noted above, the output value can be equivalently computed with minimum instead of maximum as $y_1 = f\left(\tau^1(\mathbf{x})\right) = f\left(-p_1 \bigwedge_{k=1}^{4} -\tau_k^1(\mathbf{x})\right)$, i.e., by conjugating the responses p_1 and p_{1k} and using the minimum operator.

4.7.1 Training Algorithm Based on Elimination

A training algorithm that employs elimination is discussed in [13]. The algorithm constructs and trains an SLMP with dendritic structures with a single output neuron, able to recognize the training patterns as either belonging to the class of interest C_1 or not belonging to it. Thus, it solves a one-class problem, but can be generalized as discussed in Sect. 4.7.3.

In its first step, the algorithm creates the first dendrite, which encloses the entire training set belonging to class C_1 within a single hyperbox. Subsequent steps carve out regions containing points belonging to class C_0 from the hyperbox. Thus, the final class C_1 region will be contained within the original hyperbox, hence, making it a bounded region. This is in contrast to the MLP, which often creates open regions as illustrated by the following example.

Example 4. To illustrate the results of an implementation of the training algorithm based on elimination, we employed a data set from [65], where it was

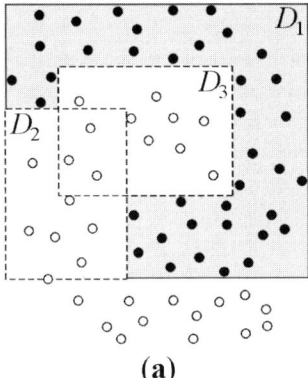 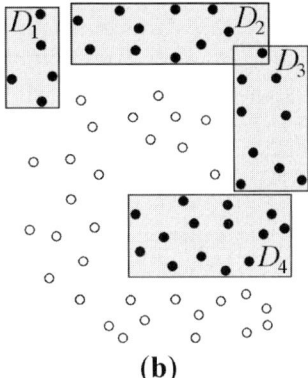

Fig. 4.7. Two partitionings of the pattern space $\mathbb{R}^2$ in terms of intersection (**a**) and union (**b**), respectively. The *solid circles* (•) belong to class C_1, which is recognized as the *shaded area*. *Solid and dashed lines* enclose regions learned by excitatory and, respectively, inhibitory dendrites

used to test a simulation of a radial basis function network (RBFN). The data set consists of two nonlinearly separable classes of ten patterns each, where the class of interest C_1 comprises the patterns depicted with solid circles (•). All patterns were used for both training and test. Figure 4.8 compares the class C_1 regions learned by an SLMP with dendritic structures using the elimination-based algorithm (Fig. 4.8a) and, respectively, by a back-propagation MLP (Fig. 4.8b).

The first step of the algorithm creates the first dendrite, which sends an excitatory message to the cell body of the output neuron if and only if a point of $\mathbb{R}^2$ is in the rectangle (solid lines) shown in Fig. 4.8a. This rectangle encloses the entire training set of points belonging to class C_1. Subsequent steps of the algorithm create two more dendrites having inhibitory responses. These dendrites will inhibit responses to points in the carved out region of the rectangle as indicated by the dashed lines in Fig. 4.8a. The only "visible" region for the output neuron will now be the dark shaded area of Fig. 4.8a.

The three dendrites grown in a single epoch during training of the SLMP are sufficient to partition the pattern space. In contrast, the MLP created the open surface in Fig. 4.8b using 13 hidden units and 2000 epochs. The RBFN also required 13 basis functions in its hidden layer [65]. The separation surfaces drawn by the SLMP are closed, which is not the case for the MLP, and classification is 100% correct, as guaranteed by the theorems in Sect. 4.6.

4.7.2 Training Algorithm Based on Region Merging

A second class of training algorithms for SLMPs is based on merging rather than elimination. This strategy provides for the design of various algorithms

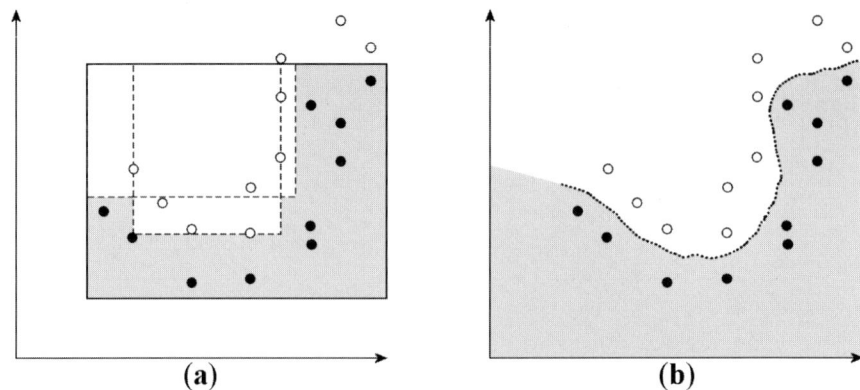

Fig. 4.8. The closed class C_1 region (*shaded*) learned by an SLMP with dendritic structures using the elimination algorithm (**a**), in comparison to the open region learned by an MLP (**b**), both applied to the data set from [65]. During training, the SLMP grows only three dendrites, one excitatory and two inhibitory (*dashed*). Compare (**a**) to the output in Fig. 4.9 of the merging version of the SLMP training algorithm

depending on the manner in which small hyperboxes can be enlarged to incorporate neighboring patterns, or individual boxes can be connected together by creating new boxes that partially overlap. Customized versions of the training algorithm can be devised for specific types of problems, e.g., problems where the patterns are known to be be arranged on curves in 2-D or 3-D space. Such a tailored, merging-based algorithm was applied to solve the embedded spirals problem in [66] with 100% correct classification results.

A training algorithm based on region merging for SLMPs is outlined below and discussed in detail in [62]. The algorithm constructs and trains an SLMP with dendritic structures to recognize the patterns belonging to the class of interest C_1. The remaining patterns in the training set are labeled as belonging to class $C_0 = \neg C_1$. In the form provided below, the algorithm will construct an SLMP comprising one dendrite either per pattern in class C_1 or per unique pair of patterns in class C_1.

Step 1. Compute $d_{\min}$ as the minimal Chebyshev interset distance between classes C_1 and C_0. The Chebyshev distance between two n-dimensional patterns $\mathbf{x}^\xi$ and $\mathbf{x}^\gamma$ is defined as $\mathrm{d}(\mathbf{x}^\xi, \mathbf{x}^\gamma) = \max_{i=1,\dots,n} |x_i^\xi - x_i^\gamma|$.

Step 2. Initialize a dendrite counter $K = 0$ and set all patterns in C_1 as unmarked.

Step 3. Select an unmarked pattern $\mathbf{x}^\xi$ belonging to class C_1 and mark it. (Two immediate options are to pick the first pattern encountered in the training set, or to pick one at random.)

Step 4. For each unmarked pattern $\mathbf{x}^\zeta$ situated in the vicinity of $\mathbf{x}^\xi$, i.e., for each $\mathbf{x}^\zeta$ such that $\mathrm{d}(\mathbf{x}^\zeta, \mathbf{x}^\xi) < d_{\min} + d_\varepsilon$, do step 5. The term d_ε is a tolerance parameter that controls how close two class C_1 patterns must be in order to qualify for merging. A possible value is $d_\varepsilon = \frac{1}{2}d_{\min}$.

Step 5. Identify a region in pattern space that would connect patterns $\mathbf{x}^\xi$ and $\mathbf{x}^\zeta$. (This involves several computations that are not detailed here.)
- If the identified merging region is free of foreign (class C_0) patterns, then increment K and grow a new excitatory dendrite D_K, and assign weights to make D_K recognize that region.
- Otherwise, i.e., if there exists at least one close foreign pattern $\mathbf{x}^\gamma$, do not grow a dendrite in this step.

Step 6. If no dendrite was grown in step 5, i.e., no merging occurred, then increment K and grow an excitatory dendrite that recognizes an isolated region around $\mathbf{x}^\xi$. The size of this region must be less than $d_{\min}$ in each coordinate $i = 1, \ldots, n$ to ensure that it will not touch patterns from class C_0.

Step 7. If there are unmarked class C_1 patterns remaining, repeat from step 3; otherwise, stop.

We need to point out that this training algorithm as well as our previously mentioned algorithm starts with the creation of hyperboxes enclosing training points. In this sense there is some similarity between our algorithms and those established in the fuzzy min-max neural networks approach, which also uses hyperboxes [10, 11]. However, this is also where the similarity ends, as all subsequent steps are completely different. Furthermore, our approach does not employ fuzzy set theory.

Example 5. Figure 4.9 illustrates the results of an implementation of the SLMP training algorithm based on merging, applied to the same data set as in Example 4. Again, all patterns were used for both training and test. During training 19 excitatory dendrites are grown, 10 for regions around each pattern from class C_1, and 9 more to merge the individual regions. The separation surface is closed and recognition is 100% correct, as expected.

There is one more region in Fig. 4.9, drawn in dashed line, corresponding to an inhibitory dendrite, and its presence is explained as follows. The merging algorithm outlined above creates regions that are sized to avoid touching patterns belonging to class C_0 or approaching them closer than a certain distance. In a more general version, larger hyperboxes are allowed to be constructed. In this case, an additional step would identify foreign patterns that are approached or touched, and create inhibitory dendrites to eliminate those patterns. This more general approach is being used in the experiment of Fig. 4.9 and explains the presence of the inhibitory dendrite whose region is depicted with dashed line.

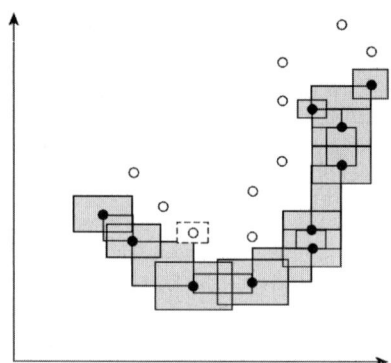

Fig. 4.9. The class C_1 region (*shaded*) learned by an SLMP with dendritic structures using the merging-based algorithm, applied to the data set from [65]. During training, the SLMP grows 20 dendrites, 19 excitatory and 1 inhibitory (*dashed*). Compare to the results in Fig. 4.8a obtained with the elimination version of the algorithm

4.7.3 Generalization of the Training Algorithms to Multiple Classes

For better clarity of the description, the training algorithms described so far were limited to a single nonzero class, which corresponds to a single output neuron of the SLMP with dendritic structures. Following we present a straightforward generalization to multiple classes, which will invoke either one of the procedures in Sects. 4.7.1 or 4.7.2 as a subroutine.

The generalized algorithm consists of a main loop that is iterated m times, where m represents the number of nonzero classes and also the number of output neurons of the resulting SLMP. Within the loop, the single-class procedure is invoked. Thus, one output neuron at a time is created and trained to classify the patterns belonging to its corresponding class. The algorithm proceeds as follows:

Step 1. For each nonzero class index $j = 1, \ldots, m$, do steps 2 through 4.
Step 2. Create a new output neuron M_j.
Step 3. For each pattern $\mathbf{x}^\xi$ of the training set:
- If $\mathbf{x}^\xi$ is labeled as belonging to class C_j, then temporarily reassign $\mathbf{x}^\xi$ as belonging to C_1.
- Otherwise, temporarily reassign $\mathbf{x}^\xi$ to class C_0.

The assignment is for this iteration only. The original pattern labels are needed in subsequent iterations.
Step 4. Invoke the single-class procedure to train output neuron M_j on the training set modified to contain patterns of only one nonzero class.

The straightforward generalization presented above suffers from a potential problem. The resulting SLMP partitions the pattern space in regions that might partially overlap. It is desirable that the learned regions be disjoint. Otherwise, a test pattern located in an area of overlap will erroneously be classified as belonging to more than one class. Theorem 7 guarantees the existence of an SLMP with multiple output neurons that is able to classify m classes disjointly. Therefore, the generalization of the algorithm can be modified to prevent overlap between classes.

One way of modifying the algorithm would consist of taking into account current information during training about the shape of the regions learned so far, and using this information when growing new dendrites and assigning synaptic weights. An alternative would be to draw regions of controlled size based on minimum interset distance, in such a way that two regions that belong to different classes cannot touch. A similar idea was mentioned in the training algorithm based on merging (Sect. 4.7.2).

Yet another approach to prevent overlap of different class regions would involve the augmentation of the SLMP with an additional layer of morphological neurons. The former output layer of the SLMP will thus become the hidden layer of a two-layer morphological perceptron with dendritic structures. The role of the supplemental layer is basically to change the neural values of the hidden nodes, where several can be active simultaneously, into values where at most one output neuron may be active (may fire) at a time. This approach is discussed in detail in [62].

It is worthwhile mentioning that multiple layers are not required to solve a nonlinear problem with a morphological perceptron, as is the case for classical perceptrons. The theorems in Sect. 4.6 prove that a single layer is sufficient. The two-layer MP described in the previous paragraph simply provides a conceivable manner to prevent ambiguous classification in the straightforward generalization of the training algorithm. Existence of single-layer MPs that are able to solve a multiclass problem with no class overlap is guaranteed.

4.8 SLMPs with Fuzzy-Valued Outputs

In our SLMP model the values of the output neurons are always crisp, i.e., having either value 1 or 0. In many application domains it is often desirable to have fuzzy-valued outputs in order to describe such terms as very tall, tall, fairly tall, somewhat tall, and not tall at all. Obviously, the boundaries between these linguistic concepts cannot be exactly quantified. In particular, we would like to have output values $y_j(\mathbf{x})$ such that $0 \leq y_j(\mathbf{x}) \leq 1$, where $y_j(\mathbf{x}) = 1$ if $\mathbf{x}$ is a clear member of class C_j, and $y_j(\mathbf{x}) = 0$ whenever $\mathbf{x}$ has no relation to class C_j. However, we would like to say that $\mathbf{x}$ is close to full membership of class C_j the closer the value of $y_j(\mathbf{x})$ is to value 1.

In order to extend the SLMP to accommodate fuzzy outputs, we redefine the activation function in Eq. (4.12) from a hard-limiter to a ramp:

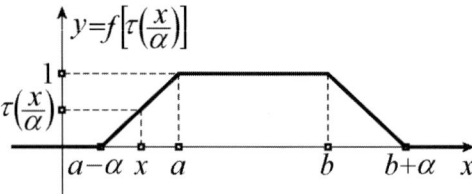

Fig. 4.10. Computing fuzzy output values with an SLMP using the ramp activation function given by Eq. (4.14)

$$f(z) = \begin{cases} 1 & \text{if } z \geq 1, \\ z & \text{if } 0 \leq z \leq 1, \\ 0 & \text{if } z \leq 0. \end{cases} \tag{4.14}$$

The following example illustrates an SLMP extended to produce fuzzy outputs that employs the ramp activation function given by Eq. (4.14).

Example 6. Suppose we would like to have every point in the interval $[a, b] \subset \mathbb{R}$ to be classified as belonging to class C_1 and every point outside the interval $[a - \alpha, b + \alpha]$ as having no relation to class C_1, where $\alpha > 0$ is a specified fuzzy boundary parameter. For a point $x \in [a - \alpha, a]$ or $x \in [b, b + \alpha]$ we would like $y(x)$ to be close to 1 when x is close to a or b, and $y(x)$ close to 0 whenever x is close to $a - \alpha$ or $b + \alpha$. In this case we simply convert the input $x \in \mathbb{R}$ to a new input format $\frac{x}{\alpha}$. If $w_1^0 = -b$ and $w_1^1 = -a$ denote the weights found either by inspection or the aforementioned algorithms for input x, then we set $v_1^0 = -\frac{w_1^0}{\alpha} - 1$ and $v_1^1 = -\frac{w_1^1}{\alpha} + 1$ for the weights of the new input $\frac{x}{\alpha}$ and use the ramp activation function in Eq. (4.14).

Computing $\tau\left(\frac{x}{\alpha}\right)$ we obtain $\tau\left(\frac{x}{\alpha}\right) = \left(\frac{x}{\alpha} + v_1^1\right) \wedge -\left(\frac{x}{\alpha} + v_1^0\right) = \left[\frac{1}{\alpha}(x - a) + 1\right] \wedge \left[-\frac{1}{\alpha}(x - b) + 1\right]$. Thus,

$$f\left[\tau\left(\frac{x}{\alpha}\right)\right] = \begin{cases} 1, & \text{if } x \in [a, b], \\ 0 \leq \tau\left(\frac{x}{\alpha}\right) < 1, & \text{if } x \notin [a, b], \\ 0, & \text{if } x \notin [a - \alpha, b + \alpha]. \end{cases} \tag{4.15}$$

Equation (4.15) is illustrated in Fig. 4.10. By choosing fuzzy factors α_i for each x_i, it is intuitively clear how this example generalizes to n-dimensional pattern vectors.

4.9 Conclusions

We presented a new paradigm of neural computation based on lattice algebra. After a brief introduction to lattice algebra, we focused on two types of neural networks, namely morphological associative memories and morphological feedforward networks based on dendritic computing. These networks have proven to be radically different in behavior than traditional neural networks. In contrast to traditional associative memories, morphological associative memories converge in one step! Thus, convergence problems do not exist. Morphological analogues to the Hopfield network not only proved to be far more robust in the presence of noise, but have also unlimited storage capacity for perfect inputs. For noisy inputs and carefully chosen kernels, morphological autoassociative memories again exhibit superior performance in both recall and storage [3].

For feedforward neural nets, our paradigm takes into account the dendritic processes of neurons. The theorems, algorithms, and examples presented make it obvious that the feedforward neural network presented here has several advantages over both traditional single-layer perceptrons and perceptrons with hidden layers. For instance, our network needs no hidden layers for solving nonconvex problems. Also, it does not suffer from the problem of misclassification due to open separation surfaces, which can happen in multilayer perceptrons as was shown in [61]. Based on the proofs of the two theorems in Sect. 4.6, we developed training algorithms which *always* draw closed regions around pattern clusters [13, 62].

Questions may be raised as to whether dendrites merely represent hidden layers in disguise. Such questions are valid in light of the fact that, theoretically, a two-hidden-layer perceptron can also classify any compact region in $\mathbb{R}^n$. However, there are some major differences between the model presented here and hidden-layer perceptrons. In comparison to hidden-layer neurons, which generally use sigmoidal activation functions, dendrites have no activation functions. They only compute the basic logic functions of AND, OR, and NOT. Activation takes place only within the neuron via the hard-limiter function. Also, with hidden layers the number of neurons within a hidden layer is predetermined before training of weights, which traditionally involves backpropagation methods. In our model, dendrites are grown automatically as the neuron learns its specific task. Furthermore, *no* error remains after training. *All* pattern vectors of the training set will *always* be correctly identified after the training stops [13, 62].

Since the model presented is still in its infancy, general performance and comparison to traditional feedforward neural networks need further investigation. The training algorithms constructed thus far are only a first attempt and will, no doubt, need further refinements.

References

1. Ritter GX, Sussner P, Diaz de Leon JL (1998) Morphological associative memories. *IEEE Trans. on Neural Networks* 9(2):281–293
2. Ritter GX, Diaz de Leon JL, Sussner P (1999) Morphological bidirectional associative memories. *Neural Networks* 12:851–867
3. Ritter GX, Urcid G, Iancu L (2003) Reconstruction of noisy patterns using morphological associative memories. *J. of Mathematical Imaging and Vision* 19(2):95–111
4. Graa M, Raducanu B (2001) On the application of morphological heteroassociative neural networks. In: *Proc. of Intern. Conf. on Image Processing ICIP'01*, Thessaloniki, Greece, pp. 501–504
5. Sussner P (2003) Generalizing operations of binary autoassociative morphological memories using fuzzy set theory. *J. of Mathematical Imaging and Vision* 19(2):81–93
6. Khabou MA, Gader PD, Keller JM (2000) LADAR target detection using morphological shared-weight neural networks. *Machine Vision and Applications* 11(6):300–305
7. Won Y, Gader PD, Coffield PC (1997) Morphological shared-weight networks with applications to automatic target recognition. *IEEE Trans. on Neural Networks* 8(5):1195–1203
8. Gader PD, Won Y, Khabou MA (1994) Image algebra networks for pattern classification. In: Dougherty ER, Gader PD, Schmitt M (eds) *Image Algebra and Morphological Image Processing V.* San Diego, California, pp. 157–168
9. Pessoa LFC, Maragos P (2000) Neural networks with hybrid morphological/rank/linear nodes: a unifying framework with applications to handwritten character recognition. *Pattern Recognition* 33(6):945–960
10. Simpson PK (1992) Fuzzy min–max neural networks. 1. Classification. *IEEE Trans. on Neural Networks* 3:776–786
11. Simpson PK (1993) Fuzzy min–max neural networks. 2. Clustering. *IEEE Trans. on Neural Networks* 4:32–45
12. Zhang X, Hang CC, Tan S, Wang PZ (1996) The min–max function differentiation and training of fuzzy neural networks. *IEEE Trans. on Neural Networks* 7(5):1139–1150
13. Ritter GX, Urcid G (2003) Lattice algebra approach to single neuron computation. *IEEE Trans. on Neural Networks* 14(2):282–295
14. Ritter GX, Iancu L, Urcid G (2003) Morphological perceptrons with dendritic structure. In: *Proc. of IEEE Intern. Conf. on Fuzzy Systems FUZZ-IEEE'03*, St. Louis, Missouri, pp. 1296–1301
15. Petridis V, Kaburlasos VG (1998) Fuzzy lattice neural network (FLNN): a hybrid model for learning. *IEEE Trans. on Neural Networks* 9:877–890
16. Kaburlasos VG (2003) Improved fuzzy lattice neurocomputing (FLN) for semantic neural computing. In: *Proc. of Intern. Joint Conf. on Neural Networks IJCNN'03*, Portland, Oregon, pp. 1850–1855
17. Armstrong WW, Thomas MM (1996) Adaptive logic networks. In: Fiesler E, Beale R (eds) *Handbook of Neural Computation.* Oxford University, Oxford C1.8:1–14
18. Birkhoff G (1984) *Lattice Theory.* American Mathematical Society, Providence, Rhode Island

19. Shimbel A (1954) Structure in communication nets. In: *Proc. of Symposium on Information Networks*. Polytechnic Institute of Brooklyn, New York, pp. 119–203
20. Cuninghame-Green RA (1960) Process synchronization in steelworks—a problem of feasibility. In: Banbury, Maitland (eds) *Proc. of 2nd Intern. Conf. on Operations Research*. English University, London, pp. 323–328
21. Cuninghame-Green RA (1962) Describing industrial processes with interference and approximating their steady-state behaviour. *Oper. Research Quart.* 13:95–100
22. Giffler B (1960) *Mathematical Solution of Production Planning and Scheduling Problems*. Tech. Rep., IBM ASDD, Yorktown Heights, NY
23. Peteanu V (1967) An algebra of the optimal path in networks. *Mathematica* 9:335–342
24. Benzaken C (1968) Structures algébra des cheminements. In: Biorci G (ed) *Network and Switching Theory*. Academic, New York Boston, pp. 40–57
25. Carré B (1971) An algebra for network routing problems. *J. Inst. Math. Appl.* 7:273–294
26. Backhouse RC, Carré B (1975) Regular algebra applied to path-finding problems. *J. Inst. Math. Appl.* 15:161–186
27. Cuninghame-Green RA (1979) *Minimax Algebra: Lecture Notes in Economics and Mathematical Systems* 166. Springer, Berlin Heidelberg New York
28. Ritter GX, Davidson JL (1987) *The Image Algebra and Lattice Theory*. Tech. Rep. TR 87-09. University of Florida CIS Dept., Gainesville, Florida
29. Davidson JL (1989) *Lattice Structures in the Image Algebra and Applications to Image Processing*. PhD Thesis, University of Florida, Gainesville, Florida
30. Kohonen T (1972) Correlation matrix memory. *IEEE Trans. on Computers* C-21:353–359
31. Ritter GX, Sussner P (1997) Associative memories based on lattice algebra. In: *Proc. of IEEE Intern. Conf. on Systems, Man, and Cybernetics*, Orlando, Florida, pp. 3570–3575
32. McEliece R, et al. (1987) The capacity of Hopfield associative memory. *Trans. Information Theory* 1:33–45
33. Palm G (1980) On associative memory. *Biological Cybernetics* 36:19–31
34. Willshaw DJ, Buneman OP, Longuet-Higgins HC (1969) Non-holographic associative memories. *Nature* 222:960–962
35. Hopfield JJ (1982) Neural networks and physical systems with emergent collective computational abilities. In: *Proc. of National Academy of Sciences*, USA, 79:2554–2558
36. Hopfield JJ (1984) Neurons with graded response have collective computational properties like those of two state neurons. In: *Proc. of National Academy of Sciences*, USA, 81:3088–3092
37. Hopfield JJ, Tank DW (1986) Computing with neural circuits. *Science* 233:625–633
38. Abu-Mostafa Y, St. Jacques J (1985) Information capacity of the Hopfield model. *IEEE Trans. on Information Theory* 7:1–11
39. Chen HH, et al. (1986) Higher order correlation model for associative memories. In: Denker JS (ed) *Neural Networks for Computing*. AIP Proceedings 151:398–403
40. Denker JS (1986) Neural network models of learning and adaption. *Physica* 22D:216–222

41. Gimenez-Martinez V (2000) A modified Hopfield autoassociative memory with improved capacity. *IEEE Trans. on Neural Networks* 11(4):867–878
42. Keeler JD (1986) Basins of attraction of neural network models. In: Denker JS (ed) *Neural Networks for Computing.* AIP Proceedings 151:259–265
43. Kosko B (1987) Adaptive bidirectional associative memories. *IEEE Trans. Systems, Man, and Cybernetics* pp. 124–136
44. Li WJ, Lee T (2001) Hopfield neural networks for affine invariant matching. *IEEE Trans. on Neural Networks* 12(6): 1400–1410
45. Munehisa T, Kobayashi M, Yamazaki H (2001) Cooperative updating in the Hopfield model. *IEEE Trans. on Neural Networks* 12(5):1243–1251
46. Akyama MT, Kikuti M (2001) Recognition of character using morphological associative memory. In: *Proc. of 14th Brazilian Symposium on Computer Graphics and Image Processing*, Florianópolis, Brazil, p. 400
47. Annovi A, Bagliesi MG, et al. (2001) A pipeline of associative memory boards for track finding. *IEEE Trans. on Nuclear Science* 48(3) Part 1:595–600
48. Kung SY, Zhang X (2001) An associative memory approach to blind signal recovery for SIMO/MIMO systems. In: *Proc. of 2001 IEEE Signal Processing Society Workshop.* Neural Networks for Signal Processing XI:343–362
49. Matsuda S (2001) Theoretical limitations of a Hopfield network for crossbar switching. *IEEE Trans. on Neural Networks* 12(3):456–462
50. Mihu IZ, Brad R, Breazu M (2001) Specifications and FPGA implementation of a systolic Hopfield-type associative memory. In: *Proc. of Intern. Joint Conference on Neural Networks IJCNN'01* 1:228–233
51. Sussner P (2000) Observations on morphological associative memories and the kernel method. *Elsevier Neurocomputing* 31:167–183
52. Eccles JC (1977) *The Understanding of the Brain.* McGraw-Hill, New York
53. Koch C, Segev I (eds) (1989) *Methods in Neuronal Modeling: From Synapses to Networks.* MIT, Boston
54. Segev I (1998) Dendritic processing. In: Arbib M (ed) *The Handbook of Brain Theory and Neural Networks.* MIT, Boston, pp. 282–289
55. Arbib MA (ed) (1998) *The Handbook of Brain Theory and Neural Networks.* MIT, Boston
56. Holmes WR, Rall W (1992) Electronic models of neuron dendrites and single neuron computation. In: McKenna T, Davis J, Zornetzer SF (eds) *Single Neuron Computation.* Academic, San Diego, California, pp. 7–25
57. McKenna T, Davis J, Zornetzer SF (eds) (1992) *Single Neuron Computation.* Academic, San Diego, California
58. Mel BW (1993) Synaptic integration in excitable dendritic trees. *J. of Neurophysiology* 70:1086–1101
59. Rall W, Segev I (1987) Functional possibilities for synapses on dendrites and dendritic spines. In: Edelman GM, Gall EE, Cowan WM (eds) *Synaptic Function.* Wiley, New York, pp. 605–636
60. Shepherd GM (1992) Canonical neurons and their computational organization. In: McKenna T, Davis J, Zornetzer SF (eds) *Single Neuron Computation.* Academic, San Diego, California, pp. 27–55
61. Gori M, Scarselli F (1998) Are multilayer perceptrons adequate for pattern recognition and verification? *IEEE Trans. on Pattern Analysis and Machine Intelligence* 20(11):1121–1132
62. Ritter GX, Iancu L (2003) Morphological perceptrons. Submitted to *IEEE Trans. on Neural Networks*

63. Ritter GX, Iancu L (2003) Single layer feedforward neural network based on lattice algebra. In: *Proc. of Intern. Joint Conf. on Neural Networks IJCNN'03*, Portland, Oregon, pp. 2887–2892
64. Ritter GX, Iancu L (2003) Lattice algebra approach to neural networks and pattern classification. In: *Proc. of 6th Open German–Russian Workshop on Pattern Recognition and Image Understanding*, Katun Village, Altai Region, Russian Federation, pp. 18–21
65. Wasnikar VA, Kulkarni AD (2000) Data mining with radial basis functions. In: Dagli CH, et al. (eds) *Intelligent Engineering Systems Through Artificial Neural Networks*. ASME, New York, 10:443–448
66. Ritter GX, Iancu L, Urcid G (2003) Neurons, dendrites, and pattern classification. In: *Proc. of 8th Ibero–American Congress on Pattern Recognition CIARP'03*, Havana, Cuba, pp. 1–16

5

Eigenproblems in Pattern Recognition

Tijl De Bie[1], Nello Cristianini[2], and Roman Rosipal[3]

[1] K.U.Leuven ESAT-SCD/SISTA
   Kasteelpark Arenberg 10
   3001 Leuven, Belgium
   `tijl.debie@esat.kuleuven.ac.be`

[2] U.C. Davis Department of Statistics
   360 Kerr Hall One Shields Ave.
   Davis, CA 95616
   `nello@support-vector.net`

[3] NASA Ames Research Center
   Computational Sciences Division
   Moffett Field, CA 94035
   `rrosipal@mail.arc.nasa.gov`

5.1 Introduction

The task of studying the properties of configurations of points embedded in a metric space has long been a central task in pattern recognition, but has acquired even greater importance after the recent introduction of kernel-based learning methods. These methods work by virtually embedding general types of data in a vector space, and then analyzing the properties of the resulting data cloud. While a number of techniques for this task have been developed in fields as diverse as multivariate statistics, neural networks, and signal processing, many of them show an underlying unity. In this chapter we describe a large class of pattern analysis methods based on the use of generalized eigenproblems, which reduce to solving the equation $\mathbf{Aw} = \lambda \mathbf{Bw}$ with respect to $\mathbf{w}$ and λ.

The problems in this class range from finding a set of directions in the data-embedding space containing the maximum amount of variance in the data (principal components analysis), to finding a hyperplane that separates two classes of data minimizing a certain cost function (Fisher discriminant), or finding correlations between two different representations of the same data (canonical correlation analysis). Also some important clustering algorithms can be reduced to solving eigenproblems. The importance of this class of algorithms derives from the facts that generalized eigenproblems provide an efficient way to optimize an important family of cost functions, of the type $f(\mathbf{w}) = \frac{\mathbf{w}'\mathbf{Aw}}{\mathbf{w}'\mathbf{Bw}}$ (known as a *Rayleigh quotient*); they can be studied with very simple linear algebra; and they can be solved or approximated efficiently using a number of well-known techniques from computational algebra.

Their statistical behavior has also been studied to some extent (e.g. [24] and [25]), allowing us to efficiently design regularization strategies in order to reduce the risk of overfitting. However, methods limited to detecting linear relations among vectors could hardly be considered to constitute state-of-the-art technology, given the nature of the challenges presented by modern data analysis. Therefore it is crucial that *all such problems* can be cast and solved in a kernel-induced feature space; that is, they only require information about inner products between data points. The entire toolbox of generalized eigenproblems for pattern analysis can then be applied to detection of generalized relations on a wide range of data types, such as sequences, text, images, and so on.

In this chapter we will first review the general theory of eigenvalue problems, then we will give a brief review of kernel methods in general. Finally, we will discuss a number of algorithms based in multivariate statistics: principal components analysis, partial least squares, canonical correlation analysis, Fisher discriminant, and spectral clustering, where appropriate both in their primal and in their dual form, leading to a version involving kernels.

5.1.1 Notation

All matrices are boldface uppercase. Vectors are boldface lowercase. Scalar variables are lowercase. Sets and spaces are denoted with calligraphic letters.

With $(\mathbf{a}\ \mathbf{b}\ \cdots\ \mathbf{z})$, the matrix built by stacking the vectors $\mathbf{a}, \mathbf{b}, \ldots, \mathbf{z}$ next to each other is meant.

The symbols used are:

- The vector containing all ones is denoted by $\mathbf{1}$. The identity matrix is denoted by $\mathbf{I}$. The matrix or vector containing all zeros is denoted by $\mathbf{0}$. Their dimensionality is clear from the context.
- $\mathbf{x}$ or $\mathbf{x}_i$, column vectors represent a vector in the $\mathcal{X}$-space. When we have n samples, the matrix $\mathbf{X}$ is built up as $\mathbf{X} = (\mathbf{x}_1\ \mathbf{x}_2\ \cdots\ \mathbf{x}_n)'$.
- Similarly, $\mathbf{y}$ or $\mathbf{y}_i$ are sample vectors from the $\mathcal{Y}$-space. The matrix $\mathbf{Y}$ containing samples $\mathbf{y}_1$ through $\mathbf{y}_n$ is built up as $\mathbf{Y} = (\mathbf{y}_1\ \mathbf{y}_2\ \cdots\ \mathbf{y}_n)'$.
- When $\mathcal{Y}$ is one-dimensional, a sample from this space is denoted by y or y_i, and the vector containing all samples is $\mathbf{y} = (y_1\ y_2\ \cdots\ y_n)'$.
- Unless stated differently, *all data are assumed to be centered (have zero mean) throughout this chapter.* This means that $\mathbf{1}' \cdot \mathbf{X} = \mathbf{0}'$, $\mathbf{1}' \cdot \mathbf{Y} = \mathbf{0}'$, or when $\mathcal{Y}$ is one-dimensional, $\mathbf{1}' \cdot \mathbf{y} = 0$.
- $\mathbf{K}_\mathbf{X}$ and $\mathbf{K}_\mathbf{Y}$ are the so-called *kernel* or *Gram matrices* corresponding to $\mathbf{X}$ and $\mathbf{Y}$. They are the inner product matrices $\mathbf{K}_\mathbf{X} = \mathbf{X}\mathbf{X}'$ and $\mathbf{K}_\mathbf{Y} = \mathbf{Y}\mathbf{Y}'$. When it is clear from the context which data the kernel is built from, we just use $\mathbf{K}$. When we want to stress the kernel is centered we use $\mathbf{K}_c$.
- For centered data matrices $\mathbf{X}$ and $\mathbf{Y}$, the matrices $\mathbf{S}_{\mathbf{XX}} = \mathbf{X}'\mathbf{X}$, $\mathbf{S}_{\mathbf{XY}} = \mathbf{X}'\mathbf{Y}$, $\mathbf{S}_{\mathbf{YY}} = \mathbf{Y}'\mathbf{Y}$, and $\mathbf{S}_{\mathbf{YX}} = \mathbf{S}_{\mathbf{XY}}'$ are the *scatter matrices*.

- $\boldsymbol{\alpha}, \boldsymbol{\alpha}_{\mathbf{X}}, \boldsymbol{\alpha}_{\mathbf{Y}}, \boldsymbol{\alpha}_i, \boldsymbol{\alpha}_{\mathbf{X},i}$, and $\boldsymbol{\alpha}_{\mathbf{Y},i}$ will be referred to as *dual vectors* and their respective ith coordinates. When an index i is used as a subscript after a boldface $\boldsymbol{\alpha}$, this refers to a dual vector indexed by i, and not to the ith coordinate.
- $\mathbf{w}, \mathbf{w}_{\mathbf{X}}, \mathbf{w}_{\mathbf{Y}}$ will be referred to as *weight vectors*. Their respective ith coordinates are denoted by $w_i, w_{\mathbf{X},i}, w_{\mathbf{Y},i}$. When an index i is used as a subscript after a boldface $\mathbf{w}$, this refers to a weight vector indexed by i, and not to the ith coordinate.
- The feature map from the input space to the feature space is denoted with $\phi(\mathbf{x}_i)$.
- $d, n, m, \ldots$ are scalar integers; d is used for indicating dimensionality.

5.2 Linear Algebra

In this section we will review some basic properties of linear algebra that will prove useful in this chapter. We use the standard linear algebra notation in the beginning and translate the important results to the kernel methods conventions afterwards. Extensive references for matrix analysis can be found in [12] and [13].

5.2.1 Symmetric (Generalized) Eigenvalue Problems

Notation. In this introductory section, we will use a notation that is to be distinguished from the notation in the remainder of the chapter:

- $\mathbf{A} \in \mathcal{R}^{n \times m}$, a general matrix.
- $\mathbf{M}, \mathbf{N} \in \mathcal{R}^{n \times n}$, symmetric matrices. $\mathbf{N}$ is invertible.
- $\boldsymbol{\Lambda}, \mathbf{S} \in \mathcal{R}^{n \times n}$, diagonal matrices.
- $\mathbf{U}, \mathbf{V} \in \mathcal{R}^{n \times n} : \mathbf{U}\mathbf{U}' = \mathbf{I} = \mathbf{U}'\mathbf{U}, \mathbf{V}\mathbf{V}' = \mathbf{I} = \mathbf{V}'\mathbf{V}$, orthogonal matrices.
- $\mathbf{W} \in \mathcal{R}^{n \times n}$, a matrix orthogonal in the metric defined by $\mathbf{N}$: $\mathbf{w}'\mathbf{N}\mathbf{w} = \mathbf{I}$.
- λ or λ_i, an eigenvalue.
- σ or σ_i, a singular value.

Variational Characterization

The optimization problems we are concerned with in this chapter are all basically of the form (we assume $\mathbf{N}$ is invertible)

$$\max_{\mathbf{w}} \frac{\mathbf{w}'\mathbf{M}\mathbf{w}}{\mathbf{w}'\mathbf{N}\mathbf{w}}.$$

This is an optimization of a Rayleigh quotient. One can see the norm of $\mathbf{w}$ does not matter: scaling $\mathbf{w}$ does not change the value of the object function. Thus, one can impose an additional scalar constraint on $\mathbf{w}$ and optimize the object function without losing any solutions. This constraint is chosen to be

$\mathbf{w}'\mathbf{N}\mathbf{w} = 1$. Then the optimization problem becomes a constrained optimization problem of the form:

$$\max_{\mathbf{w}} \mathbf{w}'\mathbf{M}\mathbf{w} \quad \text{s.t.} \quad \mathbf{w}'\mathbf{N}\mathbf{w} = 1,$$

or by using the Lagrangian $\mathcal{L}(\mathbf{w})$:

$$\max_{\mathbf{w}} \mathcal{L}(\mathbf{w}) = \max_{\mathbf{w}} \mathbf{w}'\mathbf{M}\mathbf{w} - \lambda \mathbf{w}'\mathbf{N}\mathbf{w}.$$

Equating the first derivative to zero leads to

$$\mathbf{M}\mathbf{w} = \lambda \mathbf{N}\mathbf{w}. \tag{5.1}$$

The optimal value reached by the object function is equal to the maximal eigenvalue, the Lagrange multiplier λ. This is the symmetric generalized eigenvalue problem that will be studied here.

Note that the vector $\mathbf{w}$ with the scalar λ leading to the optimum of the Rayleigh quotient is not the only solution of the generalized eigenvalue problem given by Eq. (5.1). There exist other eigenvector-eigenvalue pairs that do not correspond to the optimum of the Rayleigh quotient. For any pair $(\mathbf{w}, \lambda)$ that is a solution of Eq. (5.1), $\mathbf{w}$ is called a (generalized) eigenvector and λ is called a (generalized) eigenvalue. In many cases several of these eigenvector-eigenvalue pairs are of interest.

Symmetric Eigenvalue Problems

For the ordinary symmetric eigenvalue problem (where $\mathbf{N} = \mathbf{I}$):

$$\mathbf{M}\mathbf{w} = \lambda \mathbf{w}.$$

Eigenvectors $\mathbf{w}_i$ corresponding to different eigenvalues λ_i are orthogonal to each other. Furthermore, the eigenvalues of symmetric matrices are real, and a real eigenvector corresponds to them.

Proof. For $\lambda_i \neq \lambda_j$,

$$\mathbf{M}\mathbf{w}_i = \lambda_i \mathbf{w}_i,$$
$$\Rightarrow \lambda_i(\mathbf{w}_j'\mathbf{w}_i) = \mathbf{w}_j'\mathbf{M}\mathbf{w}_i = \mathbf{w}_i'\mathbf{M}'\mathbf{w}_j = \mathbf{w}_i'\mathbf{M}\mathbf{w}_j,$$
$$= \lambda_j(\mathbf{w}_i'\mathbf{w}_j),$$
$$\Rightarrow \mathbf{w}_j'\mathbf{w}_i = 0.$$

Thus, eigenvectors corresponding to different eigenvalues λ_i and λ_j are orthogonal. Furthermore, with $\cdot^*$ the adjoint operator:

$$\mathbf{M}\mathbf{w}_i = \lambda_i \mathbf{w}_i \quad \text{and} \quad \mathbf{M} = \mathbf{M}' = \mathbf{M}^* \quad (\mathbf{M} \text{ is real symmetric}),$$
$$\Rightarrow \lambda_i^* \mathbf{w}_i' \mathbf{w}_i^* = (\lambda_i \mathbf{w}_i^{*\prime} \mathbf{w}_i)^* = (\mathbf{w}_i^{*\prime} \mathbf{M} \mathbf{w}_i)^* = \mathbf{w}_i' \mathbf{M}^* \mathbf{w}_i^* = \mathbf{w}_i' \mathbf{M}' \mathbf{w}_i^*,$$
$$= \lambda_i \mathbf{w}_i' \mathbf{w}_i^*,$$
$$\Rightarrow \lambda_i = \lambda_i^*.$$

Therefore the eigenvalues of a real symmetric matrix are real. Then also the eigenvectors are real up to a complex scalar (and can thus be made real by scalar multiplication), since if they were not, we could take the real part and the imaginary part separately, and both would be eigenvectors corresponding to the same eigenvalue.

When eigenvalues are *degenerate*, that is, they are equal but correspond to a different eigenvector, then these eigenvectors can be chosen to be orthogonal to each other. This follows from the fact that they are in a subspace orthogonal to the space spanned by all eigenvectors corresponding to the other eigenvalues. In this subspace an orthogonal basis can be found. The number of eigenvalues and corresponding orthogonal eigenvectors of a real symmetric matrix thus is equal to the dimensionality n of $\mathbf{M}$.

If we normalize all eigenvectors $\mathbf{w}_i$ to unit length and choose them to be orthogonal to each other, they are said to form an *orthonormal* basis. For $\mathbf{W}$ being the matrix built by stacking these normalized eigenvectors $\mathbf{w}_i$ next to each other, we have

$$\mathbf{W}\mathbf{W}' = \mathbf{W}'\mathbf{W} = \mathbf{I},$$

that is, the matrix $\mathbf{W}$ is *orthogonal*.

Since then $\mathbf{M}\mathbf{w}_i = \mathbf{w}_i \lambda_i$ for all i, we can state that

$$\mathbf{M}\mathbf{W} = \mathbf{W}\mathbf{\Lambda},$$

where $\mathbf{\Lambda}$ contains the corresponding eigenvalues λ_i on its diagonal. Then, taking into account that $\mathbf{W}^{-1} = \mathbf{W}'$, we can express the matrix $\mathbf{M}$ as:

$$\mathbf{M} = \mathbf{W}\mathbf{\Lambda}\mathbf{W}' = \sum_i \lambda_i \mathbf{w}_i \mathbf{w}_i'.$$

This is called the *eigenvalue decomposition* of the matrix $\mathbf{M}$, also known as the *spectral decomposition* of $\mathbf{M}$.

Symmetric Generalized Eigenvalue Problems

In general, we will deal with generalized eigenvalue problems of the form

$$\mathbf{M}\mathbf{w} = \lambda \mathbf{N}\mathbf{w}.$$

This could be solved as an ordinary but nonsymmetric eigenvalue problem (by multiplying with $\mathbf{N}^{-1}$ on the left-hand side). We can also convert it to a symmetric eigenvalue problem by defining $\mathbf{v} = \mathbf{N}^{1/2}\mathbf{w}$:

$$\mathbf{M}\mathbf{N}^{-1/2}\mathbf{N}^{1/2}\mathbf{w} = \lambda \mathbf{N}^{1/2}\mathbf{N}^{1/2}\mathbf{w},$$

and thus by left multiplication with $\mathbf{N}^{-1/2}$:

$$(\mathbf{N}^{-1/2}\mathbf{M}\mathbf{N}^{-1/2})\mathbf{v} = \lambda \mathbf{v}.$$

For this type of problem, we know that the different eigenvectors $\mathbf{v}$ can be chosen to be orthogonal and of unit length, thus:

$$\mathbf{V}'\mathbf{V} = \mathbf{I} = \mathbf{W}'\mathbf{N}\mathbf{W},$$

which means that the generalized eigenvectors $\mathbf{w}_i$ of a symmetric eigenvalue problem are orthogonal in the metric defined by $\mathbf{N}$.

5.2.2 Singular Value Decompositions, Duality

The singular value decomposition of a general real matrix $\mathbf{A}$ is defined as

$$\mathbf{A} = \begin{pmatrix} \mathbf{U} & \mathbf{U}_0 \end{pmatrix} \begin{pmatrix} \mathbf{S} & \mathbf{0} \\ \mathbf{0} & \mathbf{0} \end{pmatrix} \begin{pmatrix} \mathbf{V} & \mathbf{V}_0 \end{pmatrix}' = \mathbf{U}\mathbf{S}\mathbf{V}',$$

where $\mathbf{S}$ contains the *singular values* s_i in decreasing order (by convention) on the diagonal, and dimensions of all blocks are compatible. The matrices $\begin{pmatrix} \mathbf{U} & \mathbf{U}_0 \end{pmatrix}$ and $\begin{pmatrix} \mathbf{V} & \mathbf{V}_0 \end{pmatrix}$ are orthogonal matrices, respectively containing the *left* and the *right singular vectors* as their columns. This decomposition can be calculated for any real matrix.

One can see that multiplying $\mathbf{A}$ on the left with a column of $\mathbf{U}_0$ gives zero: $\mathbf{U}_0'\mathbf{A} = \mathbf{0}'$. Therefore $\mathbf{U}_0$ is said to span the left *null space* of $\mathbf{A}$. Similarly, $\mathbf{V}_0$ is a basis for the right null space of $\mathbf{A}$. On the other hand, $\mathbf{U}$ and $\mathbf{V}$ respectively span the column and the row space of $\mathbf{A}$.

Note that $\mathbf{A}\mathbf{A}'$ and $\mathbf{A}'\mathbf{A}$ are symmetric, and their eigenvalue decompositions are:

$$\mathbf{A}\mathbf{A}' = \mathbf{U}\mathbf{S}^2\mathbf{U}',$$
$$\mathbf{A}'\mathbf{A} = \mathbf{V}\mathbf{S}^2\mathbf{V}'.$$

Another important property of singular value decompositions is that the nonzero singular values and corresponding singular vectors are the nonzero eigenvalues and corresponding eigenvectors of the matrix $\begin{pmatrix} \mathbf{0} & \mathbf{A} \\ \mathbf{A}' & \mathbf{0} \end{pmatrix}$:

$$\begin{pmatrix} \mathbf{0} & \mathbf{A} \\ \mathbf{A}' & \mathbf{0} \end{pmatrix} \begin{pmatrix} \mathbf{u}_i \\ \mathbf{v}_i \end{pmatrix} = s_i \begin{pmatrix} \mathbf{u}_i \\ \mathbf{v}_i \end{pmatrix}, \tag{5.2}$$

the solution of which leads to the singular value decomposition of $\mathbf{A} = \mathbf{USV}'$.

In a pattern recognition problem, the rows of the matrix $\mathbf{A}$ may consist of different data vectors. Above, we used the standard linear algebra notation. In pattern recognition, the matrix $\mathbf{A}$ will then correspond to $\mathbf{X}$, the columns of $\mathbf{V}$ to $\mathbf{w}$ being the weight vectors, and the columns of $\mathbf{U}$ to $\boldsymbol{\alpha}$, being the dual vectors. Thus, in the notation we adopt in this chapter:

$$\mathbf{X}'\boldsymbol{\alpha}_i = s_i \mathbf{w}_i,$$
$$\mathbf{X}\mathbf{w}_i = s_i \boldsymbol{\alpha}_i.$$

When the norm is not an issue, which is often the case, the factor s_i can be omitted, so up to a scaling factor:

$$\mathbf{X}'\boldsymbol{\alpha}_i = \mathbf{w}_i, \tag{5.3}$$
$$\mathbf{X}\mathbf{w}_i = \boldsymbol{\alpha}_i.$$

The matrix $\mathbf{X}'\mathbf{X} = \mathbf{S_{XX}}$ will be called a scatter matrix. Since the samples making up the rows of $\mathbf{X}$ are assumed to have zero mean, it is proportional to the finite sample covariance matrix $\mathbf{C_{XX}} = \frac{1}{n}\mathbf{S_{XX}}$. On the other hand, $\mathbf{XX}' = \mathbf{K_X}$ is a Gram or kernel matrix. (Note that element (i, j) corresponds to the inner product of samples $\mathbf{x}_i$ and $\mathbf{x}_j$.) Thus, the weight vectors are the eigenvectors of the scatter matrix, and the dual vectors are the eigenvectors of the kernel matrix. Given the dual vectors, the weight vectors can be found by multiplication with the data matrix $\mathbf{X}'$, and vice versa. *This type of relation between primal and dual variables forms the basis of the duality and enables the use of kernels.*

5.3 Kernel Methods

Kernel methods (KMs) [7, 21, 23, 27, 29] are a relatively new family of algorithms that presents a series of useful features for pattern analysis in data sets. In recent years, their simplicity, versatility, and efficiency have made them a standard tool for practitioners, and a fundamental topic in many data analysis courses. We will outline some of their important features, referring the interested reader to more detailed articles and books for a deeper discussion (see, for example, [23] and references therein).

KMs combine the simplicity and computational efficiency of linear algorithms, such as the perceptron algorithm or ridge regression, with the flexibility of nonlinear systems, such as, for example, neural networks, and the rigor of statistical approaches, such as regularization methods in multivariate statistics. As a result of the special way they represent functions, these algorithms typically reduce the learning step to a simple optimization problem that can always be solved in polynomial time, avoiding the problem of local minima typical of neural networks, decision trees, and other nonlinear approaches.

Their foundation in the principles of statistical learning theory makes them remarkably resistant to overfitting especially in regimes where other methods are affected by the 'curse of dimensionality'. Another important feature for applications is that they can naturally accept input data that are not in the form of vectors, such as, for example, strings, trees, and images. Their characteristically modular design makes them amenable to theoretical analysis, but also makes them well suited to a software engineering approach in which a general-purpose learning module is combined with a data-specific 'kernel function' that provides the interface with the data and incorporates domain knowledge.

Many learning modules can be used, depending on whether the task is one of classification, regression, clustering, novelty detection, ranking, and so on. At the same time, many kernel functions have been designed, for example, for protein sequences, for text and hypertext documents, for images, time series, etc. As a result, this method can be used for dealing with rather exotic tasks, such as ranking strings, or clustering graphs, in addition to such classical tasks as classifying vectors. In the remainder of this section, we will briefly describe theory behind kernel methods, followed by a brief example of how this can be used in practice: kernelizing least squares regression and ridge regression.

5.3.1 Theory

Kernel-based learning algorithms work by embedding the data into a Hilbert space and searching for linear relations in such space. The embedding is performed implicitly, that is, by specifying the inner product between each pair of points, rather than by giving their coordinates explicitly. This approach has several advantages, the most important being the observation that often the inner product in the embedding space can be computed much more easily than the coordinates of the points themselves.

Given an input set $\mathcal{X}$ and an embedding vector space $\mathcal{F}$ (often called the feature space), we consider a map $\phi : \mathcal{X} \to \mathcal{F}$ (often called the feature map). The function that, given two points $\mathbf{x}_i \in \mathcal{X}$ and $\mathbf{x}_j \in \mathcal{X}$, returns the inner product between their images in the space $\mathcal{F}$ is known as *kernel function*.

Definition 1. *A kernel is a function k, such that for all $\mathbf{x}, \mathbf{z} \in \mathcal{X}$, $k(\mathbf{x}, \mathbf{z}) = \langle \phi(\mathbf{x}), \phi(\mathbf{z}) \rangle$, where ϕ is a mapping from $\mathcal{X}$ to a Hilbert space $\mathcal{F}$, and $\langle \cdot, \cdot \rangle$ denotes the inner product.*

We also consider the matrix $\mathbf{K}_{ij} = k(\mathbf{x}_i, \mathbf{x}_j)$, called the *kernel matrix* or the *Gram matrix*. Thanks to the fact it is built from inner products it is always a symmetric, positive semidefinite matrix, and since it specifies the inner products between all pairs of points, it completely determines the relative positions between those points in the embedding space. For example, given such information, it is trivial to recover all the pairwise distances between them.[1]

The solutions sought by kernel-based algorithms are linear functions in the feature space:

$$f(\mathbf{x}) = \mathbf{w}'\phi(\mathbf{x}),$$

for some weight vector $\mathbf{w}$. The kernel can be exploited whenever the weight vector can be expressed as a linear combination of the training points, $\mathbf{w} = \sum_{i=1}^n \alpha_i \phi(\mathbf{x}_i)$, implying that we can express f as follows:

$$f(\mathbf{x}) = \sum_{i=1}^n \alpha_i k(\mathbf{x}_i, \mathbf{x}).$$

This will be the case for any of the algorithms considered in this chapter.

5.3.2 Example: Least Squares and Ridge Regression

We consider the well-known problem of least squares regression to start with and derive a *kernelized* version for it. Consider the vector $\mathbf{y} \in \mathcal{R}^n$ and the data points $\mathbf{X} \in \mathcal{R}^{n \times d}$. We want to find the weight vector $\mathbf{w} \in \mathcal{R}^d$ that minimizes $\|\mathbf{y} - \mathbf{X}\mathbf{w}\|^2$. Taking the gradient of this cost function with respect to $\mathbf{w}$ and equating to zero leads to:

$$\begin{aligned}
\nabla_{\mathbf{w}} \|\mathbf{y} - \mathbf{X}\mathbf{w}\|^2 &= \nabla_{\mathbf{w}} (\mathbf{y}'\mathbf{y} + \mathbf{w}'\mathbf{X}'\mathbf{X}\mathbf{w} - 2\mathbf{w}'\mathbf{X}'\mathbf{y}), \\
&= 2\mathbf{X}'\mathbf{X}\mathbf{w} - 2\mathbf{X}'\mathbf{y}, \\
&= 0, \\
\Rightarrow \mathbf{w} &= (\mathbf{X}'\mathbf{X})^{-1}\mathbf{X}'\mathbf{y}.
\end{aligned}$$

This is the well-known least squares solution.

However, least squares is highly sensitive to overfitting. Especially when $\mathbf{X}$ lives in a high-dimensional (feature) space, care needs to be taken (ultimately, when the dimensionality $d > n$, regression can always be carried out exactly, which means that any noise sequence could be fit by the model). In order to avoid overfitting, a standard approach is to reduce the *capacity* of the learner, or the *effective number of degrees of freedom*, by imposing a prior on the solution, thus introducing a bias. In the case of regression, for example, one usually prefers a weight vector with small norm. This is taken into account by introducing an additional term $\gamma \|\mathbf{w}\|^2$ in the cost function, with γ the *regularization parameter*. Minimizing leads to the ridge regression estimate:

---

[1] Notice that we do not really need $\mathcal{X}$ to be a vector space; in fact, $\mathcal{X}$ can be a generic finite set. This is because we are guaranteed that the data are *implicitly* mapped to some Hilbert space by simply checking that the kernel matrix $\mathbf{K}$ satisfies the conditions above.

$$\nabla_{\mathbf{w}}\left[\|\mathbf{y}-\mathbf{Xw}\|^2+\gamma\|\mathbf{w}\|^2\right] = \nabla_{\mathbf{w}}\left[(\mathbf{y}'\mathbf{y}+\mathbf{w}'\mathbf{X}'\mathbf{Xw}-2\mathbf{w}'\mathbf{X}'\mathbf{y})+\gamma(\mathbf{w}'\mathbf{w})\right],$$
$$= 2(\mathbf{X}'\mathbf{X}+\gamma\mathbf{I})\mathbf{w}-2\mathbf{X}'\mathbf{y},$$
$$= 0,$$
$$\Rightarrow \mathbf{w} = (\mathbf{X}'\mathbf{X}+\gamma\mathbf{I})^{-1}\mathbf{X}'\mathbf{y}.$$

To evaluate the regression function in a new test point, it can simply be projected on the weight vector:

$$y_{\text{test}} = \mathbf{x}'_{\text{test}}\mathbf{w}.$$

So far we have discussed the primal version of the ridge regression method. The dual version can be derived by noting that the minimum norm weight vector will always be in the span of the data $\mathbf{X}$. This can be seen by replacing $(\mathbf{X}'\mathbf{X}+\gamma\mathbf{I})^{-1}$ with $(\mathbf{V}\mathbf{\Lambda}\mathbf{V}'+\gamma\mathbf{I})^{-1} = (\mathbf{V}(\mathbf{\Lambda}+\gamma\mathbf{I})\mathbf{V}')^{-1} = \mathbf{V}(\mathbf{\Lambda}+\gamma\mathbf{I})^{-1}\mathbf{V}'$, where the columns of $\mathbf{V}$ are the right singular vectors of $\mathbf{X}$ and are thus a basis for the row space of $\mathbf{X}$. Thus the weight vector $\mathbf{w} = \mathbf{V}\left[(\mathbf{\Lambda}+\gamma\mathbf{I})^{-1}\mathbf{V}'\mathbf{X}'\mathbf{y}\right]$ lies in the column space of $\mathbf{V}$, or equivalently in the row space of $\mathbf{X}$, and can thus be expressed as $\mathbf{w} = \mathbf{X}'\boldsymbol{\alpha}$ (cf. Eq. (5.3)). Here $\boldsymbol{\alpha} \in \mathcal{R}^n$ is called the dual vector. Plugging this into the equations leads to:

$$\nabla_{\boldsymbol{\alpha}}\left[\|\mathbf{y}-\mathbf{XX}'\boldsymbol{\alpha}\|^2+\gamma\|\mathbf{X}'\boldsymbol{\alpha}\|^2\right] = 2(\mathbf{XX}'\mathbf{XX}')\boldsymbol{\alpha}-2\mathbf{XX}'\mathbf{y}+2\gamma\mathbf{XX}'\boldsymbol{\alpha},$$
$$= 2(\mathbf{K}^2+\gamma\mathbf{K})\boldsymbol{\alpha}-2\mathbf{K}\mathbf{y},$$
$$= \mathbf{0},$$
$$\Rightarrow \mathbf{K}(\mathbf{K}+\gamma\mathbf{I})\boldsymbol{\alpha} = \mathbf{K}\mathbf{y}. \tag{5.4}$$

In the second step, $\mathbf{XX}'$, which is the matrix containing the inner products between any two points as its elements, is replaced by the kernel matrix $\mathbf{K}$. Since the inner products in $\mathbf{K}$ can be inner products in a feature space, they can in fact be a nonlinear function of the data points, namely the kernel function. In this way, nonlinearities can be dealt with in a very natural way. This is the essence of the 'kernel trick'. A general solution for Eq. (5.4) is given by:

$$\boldsymbol{\alpha} = (\mathbf{K}+\gamma\mathbf{I})^{-1}\mathbf{y}+\boldsymbol{\alpha}_0,$$

where $\boldsymbol{\alpha}_0$ is any vector in the null space of $\mathbf{K}$: $\mathbf{X}'\boldsymbol{\alpha}_0 = \mathbf{K}\boldsymbol{\alpha}_0 = \mathbf{0}$.

The projection of a test point $\mathbf{x}_{\text{test}}$ onto the weight vector $\mathbf{w} = \mathbf{X}'\boldsymbol{\alpha} = \mathbf{X}'\left[(\mathbf{K}+\gamma\mathbf{I})^{-1}\mathbf{y}+\boldsymbol{\alpha}_0\right] = \mathbf{X}'(\mathbf{K}+\gamma\mathbf{I})^{-1}\mathbf{y}$, can be written as $y_{\text{test}} = \mathbf{x}'_{\text{test}}\mathbf{X}'\boldsymbol{\alpha}$ (as one can see, the actual value of $\boldsymbol{\alpha}_0$ does not matter). Written in terms of kernel evaluations, this becomes:

$$y_{\text{test}} = \sum_{i=1}^{n}\alpha_i k(\mathbf{x}_i, \mathbf{x}_{\text{test}}).$$

This is indeed the standard form.

5.3.3 Kernels in This Chapter

In this chapter, we will aim at deriving primal and dual versions of spectral algorithms in pattern recognition. Whereas the primal formulation is usually the standard form in which algorithms are known, the dual form is formulated in terms of inner products only.[2] This is important, since then the kernel trick can be used in any algorithm where such a dual version can be derived, very much in the same way as shown in the example above: by replacing the matrix containing inner products with the kernel matrix. The inner products are considered to be carried out implicitly between nonlinear mappings of the points in a feature space.

As mentioned before, we will assume all data are centered. In primal space, this centering is a trivial operation, as it is done by simply subtracting the mean of each of the coordinates (n is the number of samples): $\mathbf{X}_c = \left(\mathbf{X} - \frac{\mathbf{1}\mathbf{1}'}{n}\mathbf{X}\right)$. However, centering in feature space deserves some attention since we do not compute the feature vectors explicitly, but only the inner products between them. Thus we have to compute the centered kernel matrix based on the uncentered kernel matrix.

For an uncentered $\mathbf{K}$ corresponding to uncentered $\mathbf{X}$, the centered version $\mathbf{K}_c$ can be computed as the product of the centered matrices $\mathbf{X}_c = \left(\mathbf{X} - \frac{\mathbf{1}\mathbf{1}'}{n}\mathbf{X}\right)$, where $\mathbf{1} \in \mathcal{R}^n$ is the column vector containing n ones:

$$\mathbf{K}_c = \left(\mathbf{X} - \frac{\mathbf{1}\mathbf{1}'}{n}\mathbf{X}\right)\left(\mathbf{X} - \frac{\mathbf{1}\mathbf{1}'}{n}\mathbf{X}\right)'$$
$$= \mathbf{K} - \frac{\mathbf{1}\mathbf{1}'}{n}\mathbf{K} - \mathbf{K}\frac{\mathbf{1}\mathbf{1}'}{n} + \frac{\mathbf{1}\mathbf{1}'}{n}\mathbf{K}\frac{\mathbf{1}\mathbf{1}'}{n}. \quad (5.5)$$

In this chapter, unless stated otherwise, we assume all kernel matrices are centered as such. Therefore, the subscript c will be omitted for brevity, wherever this does not cause confusion.

Similarly, a test sample $\mathbf{x}_{\text{test}}$ should be centered accordingly. Let $\mathbf{k}_{\text{test}} = [k(\mathbf{x}_{\text{test}}, \mathbf{x}_i)]_{i=1:n}$ be the vector containing the kernel evaluations of $\mathbf{x}_{\text{test}}$ with all n training samples $\mathbf{x}_i$. Then again, we can do the centering implicitly: the properly centered version (in correspondence with the centering of Eq. (5.5)) of this vector can be shown to be

$$\mathbf{k}_{\text{test},c} = \mathbf{k}_{\text{test}} - \mathbf{K}\frac{\mathbf{1}}{n} - \frac{\mathbf{1}\mathbf{1}'}{n}\mathbf{k}_{\text{test}} + \frac{\mathbf{1}\mathbf{1}'}{n}\mathbf{K}\frac{\mathbf{1}}{n}.$$

In this chapter we assume all test samples are already centered in this way as well. Again, the subscript c will be omitted wherever this does not cause confusion.

---
[2] In many if not all practical cases, the dual can be motivated using an optimization perspective. The reader is referred to [27] for an in-depth treatment.

5.4 Dimensionality Reduction: PCA, (R)CCA, PLS

The general philosophy that motivates dimensionality reduction techniques is the fact that real-life data contain redundancies and noise. Dimensionality reduction is often a good way to deal with this: by using a low-dimensional approximate representation, noise can be suppressed and redundancies removed. The data are replaced by a summary that still captures as much information as possible. All methods described in this section can be useful as a preprocessing step for other algorithms like clustering, classification, regression, and so on.

We will discuss various ways to perform dimensionality reduction. They all share the property that they rely on inner products and on eigenproblems. This has as a consequence that they can easily be made nonlinear using the kernel trick, and that they are efficiently solved. The difference between them lies in the cost function they optimize.

Therefore, each of the subsections will be structured as follows: first the different cost functions leading to the algorithm are described, subsequently the primal is derived and some properties are given, and finally the dual formulation is presented. For a previous treatment of these algorithms in their primal version, we refer to [6].

5.4.1 PCA

Cost Function

The motivation for performing *principal component analysis* (PCA) [16] is often the assumption that directions of high variance will contain more information than directions of low variance. The rationale behind this could be that the noise can be assumed to be uniformly spread. Thus, directions of high variance will have a higher signal-to-noise ratio. Mathematically:

$$\begin{aligned} \mathbf{w} &= \operatorname{argmax}_{\|\mathbf{w}\|=1} \mathbf{w}'\mathbf{X}'(\mathbf{w}'\mathbf{X}')', \\ &= \operatorname{argmax}_{\|\mathbf{w}\|=1} \mathbf{w}'\mathbf{X}'\mathbf{X}\mathbf{w}, \\ &= \operatorname{argmax}_{\|\mathbf{w}\|=1} \mathbf{w}'\mathbf{S_{XX}}\mathbf{w}. \end{aligned} \quad (5.6)$$

Or, for $\mathbf{w}$ not normalized this can be written as:

$$\mathbf{w} = \operatorname{argmax}_{\mathbf{w}} \frac{\mathbf{w}'\mathbf{S_{XX}}\mathbf{w}}{\mathbf{w}'\mathbf{w}}.$$

The solution of Eq. (5.6) is also equivalent to minimizing the 2-norm of the residuals. This can be seen by projecting all samples $\mathbf{X}$ on the subspace orthogonal to $\mathbf{w}$ (by left multiplication with $(\mathbf{I} - \mathbf{ww}')$), and computing the Frobenius norm:

$$\begin{aligned}
\mathbf{w} &= \operatorname{argmin}_{\|\mathbf{w}\|=1} \|\mathbf{X}(\mathbf{I}-\mathbf{w}\mathbf{w}')\|_F^2, \\
&= \operatorname{argmin}_{\|\mathbf{w}\|=1} \operatorname{trace}\left([\mathbf{X}(\mathbf{I}-\mathbf{w}\mathbf{w}')]'[\mathbf{X}(\mathbf{I}-\mathbf{w}\mathbf{w}')]\right), \\
&= \operatorname{argmin}_{\|\mathbf{w}\|=1} \operatorname{trace}\left(\mathbf{X}'\mathbf{X} + \mathbf{w}\mathbf{w}'\mathbf{X}'\mathbf{X}\mathbf{w}\mathbf{w}' - 2\mathbf{X}'\mathbf{X}\mathbf{w}\mathbf{w}'\right), \\
&= \operatorname{argmin}_{\|\mathbf{w}\|=1} \operatorname{trace}(\mathbf{S}_{\mathbf{X}\mathbf{X}}) + \|\mathbf{w}\|^2 \mathbf{w}'\mathbf{S}_{\mathbf{X}\mathbf{X}}\mathbf{w} - 2\mathbf{w}'\mathbf{S}_{\mathbf{X}\mathbf{X}}\mathbf{w}, \\
&= \operatorname{argmin}_{\|\mathbf{w}\|=1} -\mathbf{w}'\mathbf{S}_{\mathbf{X}\mathbf{X}}\mathbf{w}.
\end{aligned}$$

Primal

Differentiating the Lagrangian $\mathcal{L}(\mathbf{w}, \lambda) = \mathbf{w}'\mathbf{S}_{\mathbf{X}\mathbf{X}}\mathbf{w} - \lambda \mathbf{w}'\mathbf{w}$ corresponding to Eq. (5.6) with respect to $\mathbf{w}$ and equating to zero leads to

$$\begin{aligned}
\nabla_\mathbf{w} \mathcal{L}(\mathbf{w}, \lambda) &= \nabla_\mathbf{w}(\mathbf{w}'\mathbf{S}_{\mathbf{X}\mathbf{X}}\mathbf{w} - \lambda \mathbf{w}'\mathbf{w}) = 0, \\
\Leftrightarrow \mathbf{S}_{\mathbf{X}\mathbf{X}}\mathbf{w} &= \lambda \mathbf{w}.
\end{aligned}$$

This is a symmetric eigenvalue problem as presented in Sect. 5.2. Such an eigenvalue problem has d eigenvectors. All are called principal directions, corresponding to their variance λ.

Properties

- All principal directions are orthogonal to each other.
- The principal directions can all be obtained by optimizing the same cost function, where the above property is explicitly imposed.
- The projections of the data onto different principal directions are *uncorrelated*: $(\mathbf{X}\mathbf{w}_i)'\mathbf{X}\mathbf{w}_j = 0$ for $i \neq j$. Note that one could as well say the projections are *orthogonal*. This is equivalent, but we will use the notion of correlation when we are talking about projections of data onto a weight vector. Because of this property of PCA, it is sometimes called *linear decorrelation*.
- The PCA solution is equivalent to, and can thus be obtained by computing, the singular value decomposition of $\mathbf{X}$.

Dual

To derive the dual, we use the key fact that $\mathbf{w}$ will always be a linear combination of the columns of $\mathbf{X}'$ (to see this, note that $\mathbf{w} = \frac{1}{\lambda}\mathbf{S}_{\mathbf{X}\mathbf{X}}\mathbf{w} = \mathbf{X}'\frac{\mathbf{X}\mathbf{w}}{\lambda}$). We can thus replace $\mathbf{w}$ with $\mathbf{X}'\boldsymbol{\alpha}$, where $\boldsymbol{\alpha}$ are the dual variables. The dual problem is then:

$$\begin{aligned}
\mathbf{S}_{\mathbf{X}\mathbf{X}}\mathbf{X}'\boldsymbol{\alpha} &= \lambda \mathbf{X}'\boldsymbol{\alpha}, \\
\Rightarrow \mathbf{X}\mathbf{S}_{\mathbf{X}\mathbf{X}}\mathbf{X}'\boldsymbol{\alpha} &= \lambda \mathbf{X}\mathbf{X}'\boldsymbol{\alpha}, \\
\Rightarrow \mathbf{K}_\mathbf{X}^2 \boldsymbol{\alpha} &= \lambda \mathbf{K}_\mathbf{X} \boldsymbol{\alpha}. \quad (5.7)
\end{aligned}$$

When $\mathbf{K_X}$ has full rank, we can multiply Eq. (5.7) by $\mathbf{K_X}^{-1}$ on the left-hand side, leading to:

$$\mathbf{K_X}\alpha = \lambda\alpha. \qquad (5.8)$$

On the other hand, when $\mathbf{K_X}$ is rank deficient, a solution for Eq. (5.7) is not always a solution for Eq. (5.8) anymore (however, the converse is still true). Then for α_0 lying in the null space of $\mathbf{K_X}$, and α a solution of Eq. (5.8) (and thus also of Eq. (5.7)), also $\alpha + \alpha_0$ is a solution of Eq. (5.7) but generally not of Eq. (5.8). But, since $\mathbf{K_X}\alpha_0 = \mathbf{0}$ and thus $\mathbf{X}'\alpha_0 = \mathbf{0}$, the component α_0 will have no effect on $\mathbf{w} = \mathbf{X}'(\alpha + \alpha_0) = \mathbf{X}'\alpha$ anyway, and we can ignore the null space of $\mathbf{K_X}$ by simply solving Eq. (5.8) also in the case $\mathbf{K_X}$ is rank deficient.

Since $\mathbf{K_X}$ is a symmetric matrix, the dual eigenvectors will be orthogonal to each other. The projections of the training samples onto the weight vector $\mathbf{w}$ are $\mathbf{Xw} = \mathbf{XX}'\alpha = \lambda\alpha$. Thus, the vector α is proportional with (and thus up to a normalization equal to) the projections of the training samples onto this weight vector. The fact that different dual vectors are orthogonal is thus equivalent to the observation that the projections of the data onto different weight vectors is uncorrelated.

Projection of a test point onto the PCA direction found can be carried out as

$$y_{\text{test}} = \sum_{i=1}^{n} \alpha_i k(\mathbf{x}_i, \mathbf{x}_{\text{test}}).$$

5.4.2 Canonical Correlation Analysis (CCA) and Regularized CCA

While PCA deals with only one data space $\mathcal{X}$ where it identifies directions of high variance, *canonical correlation analysis* (CCA, first introduced in [15]) proposes a way for dimensionality reduction by taking into account relations between samples coming from *two* spaces $\mathcal{X}$ and $\mathcal{Y}$. The assumption is that the data points coming from these two spaces contain some joint information that is reflected in correlations between them. Directions along which this correlation is high are thus assumed to be relevant directions when these relations are to be captured.

Again a primal and a dual form are available. The dual form makes it possible to capture nonlinear correlations as well, thanks to the kernel trick [1, 3, 11].

When data are scarce as compared to the dimensionality of the problem, it is important to regularize the problem in order to avoid overfitting. This is provided in the *regularized CCA* (RCCA) algorithm.

A Small Example

To make things more concrete, consider the following example described in [31]. Suppose we have two text corpora, one containing English texts, and another one containing the same texts but translated in French. The text corpora can be represented by the matrices $\mathbf{X}$ and $\mathbf{Y}$ containing vectors that are the bag of words representations of the texts as its rows. Now, since we know that the same basic semantic information must be present in both the English text and the French translation, we must be able to extract some information from every row of $\mathbf{X}$ that is similar to information extracted from the rows of $\mathbf{Y}$. If we do this in a linear way, this would mean that $\mathbf{X}\mathbf{w_X}$ and $\mathbf{Y}\mathbf{w_Y}$ are similar in a way, for some $\mathbf{w_X}$ and $\mathbf{w_X}$ representing a certain semantic meaning. This could be: $\mathbf{X}\mathbf{w_X}$ and $\mathbf{Y}\mathbf{w_Y}$ are correlated, thus motivating the cost function introduced below. In [31], it is pointed out that many of the $\mathbf{w_X}$-$\mathbf{w_X}$ pairs found can indeed be related to an intuitively satisfying semantic meaning. Other examples are available in literature, notably in bioinformatics [30, 35].

Cost Function

We thus want to maximize the correlation between a projection $\mathbf{X}\mathbf{w_X}$ of $\mathbf{X}$ and a projection $\mathbf{Y}\mathbf{w_Y}$ of $\mathbf{Y}$. Or, another geometrical interpretation is: find directions $\mathbf{X}\mathbf{w_X}, \mathbf{Y}\mathbf{w_Y}$ in the column space of $\mathbf{X}$ and $\mathbf{Y}$ with a minimal angle between each other (we will use the notation $\mathbf{S_{XY}} = \mathbf{X}'\mathbf{Y}$, the *cross-scatter* matrix):

$$\begin{aligned}
\{\mathbf{w_X}, \mathbf{w_Y}\} &= \operatorname{argmax}_{\mathbf{w_X}, \mathbf{w_Y}} \cos\left(\angle(\mathbf{X}\mathbf{w_X}, \mathbf{Y}\mathbf{w_Y})\right), \\
&= \operatorname{argmax}_{\mathbf{w_X}, \mathbf{w_Y}} \frac{(\mathbf{X}\mathbf{w_X})'(\mathbf{Y}\mathbf{w_Y})}{\sqrt{(\mathbf{X}\mathbf{w_X})'(\mathbf{X}\mathbf{w_X})}\sqrt{(\mathbf{Y}\mathbf{w_Y})'(\mathbf{Y}\mathbf{w_Y})}}, \\
&= \operatorname{argmax}_{\mathbf{w_X}, \mathbf{w_Y}} \frac{\mathbf{w}'_\mathbf{X}\mathbf{S_{XY}}\mathbf{w_Y}}{\sqrt{\mathbf{w}'_\mathbf{X}\mathbf{S_{XX}}\mathbf{w_X}}\sqrt{\mathbf{w}'_\mathbf{Y}\mathbf{S_{YY}}\mathbf{w_Y}}}.
\end{aligned}$$

Since the norm of the weight vectors does not matter, we can maximize correlation along the weight vectors, or 'fit' subject to constraints fixing the value of these weight vectors:

$$\begin{aligned}
\{\mathbf{w_X}, \mathbf{w_Y}\} &= \operatorname{argmax}_{\mathbf{w_X}, \mathbf{w_Y}} \mathbf{w}'_\mathbf{X}\mathbf{S_{XY}}\mathbf{w_Y} \\
&\text{s.t. } \|\mathbf{X}\mathbf{w_X}\|^2 = \mathbf{w}'_\mathbf{X}\mathbf{S_{XX}}\mathbf{w_X} = 1, \|\mathbf{Y}\mathbf{w_Y}\|^2 - \mathbf{w}'_\mathbf{Y}\mathbf{S_{YY}}\mathbf{w_Y} = 1.
\end{aligned}$$

This is equivalent to the minimization of a 'misfit' subject to these constraints:

$$\begin{aligned}
\{\mathbf{w_X}, \mathbf{w_Y}\} &= \operatorname{argmin}_{\mathbf{w_X}, \mathbf{w_Y}} \|\mathbf{X}\mathbf{w_X} - \mathbf{Y}\mathbf{w_Y}\|^2 \\
&\text{s.t. } \|\mathbf{X}\mathbf{w_X}\|^2 = 1, \|\mathbf{Y}\mathbf{w_Y}\|^2 = 1.
\end{aligned}$$

Primal

We solve the second formulation of the problem. Differentiating the Lagrangian $\mathcal{L}(\mathbf{w}_X, \mathbf{w_Y}, \lambda_\mathbf{X}, \lambda_\mathbf{Y}) = \mathbf{w}'_\mathbf{X}\mathbf{S_{XY}}\mathbf{w_Y} - \lambda_\mathbf{X}\mathbf{w}'_\mathbf{X}\mathbf{S_{XX}}\mathbf{w_X} - \lambda_\mathbf{Y}\mathbf{w}'_\mathbf{Y}\mathbf{S_{YY}}\mathbf{w_Y}$ with respect to $\mathbf{w_X}$ and $\mathbf{w_Y}$ and equating to 0, gives

$$\begin{cases} \frac{\partial}{\partial \mathbf{w_X}} \mathcal{L}(\mathbf{w_X}, \mathbf{w_Y}, \lambda_X, \lambda_Y) = 0, \\ \frac{\partial}{\partial \mathbf{w_Y}} \mathcal{L}(\mathbf{w_X}, \mathbf{w_Y}, \lambda_X, \lambda_Y) = 0, \end{cases}$$

$$\Rightarrow \begin{cases} \mathbf{S_{XY} w_Y} = \lambda_X \mathbf{S_{XX} w_X}, \\ \mathbf{S_{YX} w_X} = \lambda_Y \mathbf{S_{YY} w_Y}. \end{cases}$$

Now, since from this

$$\lambda_X \mathbf{w'_X S_{XX} w_X} = \mathbf{w'_X S_{XY} w_Y} = \mathbf{w'_Y S_{YX} w_X} = \lambda_Y \mathbf{w'_Y S_{YY} w_Y},$$

and since $\mathbf{w'_X S_{XX} w_X} = \mathbf{w'_Y S_{YY} w_Y} = 1$, we find that $\lambda_X = \lambda_Y = \lambda$, and thus

$$\begin{cases} \mathbf{S_{XY} w_Y} = \lambda \mathbf{S_{XX} w_X}, \\ \mathbf{S_{YX} w_X} = \lambda \mathbf{S_{YY} w_Y}. \end{cases} \tag{5.9}$$

Or, stated in another way as a generalized eigenvalue problem,

$$\begin{pmatrix} \mathbf{0} & \mathbf{S_{XY}} \\ \mathbf{S_{YX}} & \mathbf{0} \end{pmatrix} \begin{pmatrix} \mathbf{w_X} \\ \mathbf{w_Y} \end{pmatrix} = \lambda \begin{pmatrix} \mathbf{S_{XX}} & \mathbf{0} \\ \mathbf{0} & \mathbf{S_{YY}} \end{pmatrix} \begin{pmatrix} \mathbf{w_X} \\ \mathbf{w_Y} \end{pmatrix}. \tag{5.10}$$

This generalized eigenvalue problem has $2d$ eigenvalues. But, for each positive eigenvalue λ and corresponding eigenvector $\begin{pmatrix} \mathbf{w_X} \\ \mathbf{w_Y} \end{pmatrix}$, $-\lambda$ is an eigenvalue too with corresponding eigenvector $\begin{pmatrix} \mathbf{w_X} \\ -\mathbf{w_Y} \end{pmatrix}$. Thus, we get all the information by only looking at the d positive eigenvalues. The largest one with its eigenvector corresponds to the optimum of the cost function described earlier. The weight vectors making up the other eigenvectors will be referred to as other canonical directions, corresponding to a smaller canonical correlation quantized by their corresponding eigenvalue.

Properties

- CCA not only finds pairs of directions that capture maximal correlations between each other. Projections onto canonical directions corresponding to a different canonical correlation are *uncorrelated*:

$$\begin{aligned} \lambda_i \mathbf{w'_{Y,j}}(\mathbf{S_{YY} w_{Y,i}}) &= \mathbf{w'_{Y,j}}(\mathbf{S_{YX} w_{X,i}}), \\ &= \mathbf{w'_{X,i}}(\mathbf{S_{XY} w_{Y,j}}), \\ &= \lambda_j \mathbf{w'_{X,i}}(\mathbf{S_{XX} w_{X,j}}), \\ &= \lambda_j \mathbf{w'_{X,j}}(\mathbf{S_{XX} w_{X,i}}). \end{aligned}$$

And similarly,

$$\lambda_i \mathbf{w'_{X,j}}(\mathbf{S_{XX} w_{X,i}}) = \lambda_j \mathbf{w'_{Y,j}}(\mathbf{S_{YY} w_{Y,i}}).$$

So for $\lambda_i \neq \lambda_j$, the projection of $\mathbf{Y}$ onto $\mathbf{w_{Y,j}}$ is uncorrelated with the projection of $\mathbf{X}$ onto $\mathbf{w_{X,i}}$: $\mathbf{w'_{Y,j} S_{YX} w_{X,i}} = 0$. Similarly, $\mathbf{w'_{X,j} S_{XX} w_{X,i}} =$

0, and $\mathbf{w}'_{\mathbf{Y},j}\mathbf{S}_{\mathbf{YY}}\mathbf{w}_{\mathbf{Y},i} = 0$. Another way to state this is to say that $\mathbf{w}_{\mathbf{X},i}$ is orthogonal to $\mathbf{w}_{\mathbf{X},j}$ in the metric defined by $\mathbf{S}_{\mathbf{XX}}$; similarly, $\mathbf{w}_{\mathbf{Y},i}$ is orthogonal to $\mathbf{w}_{\mathbf{Y},j}$ in the metric defined by $\mathbf{S}_{\mathbf{YY}}$.

- All canonical directions can be captured by a constrained optimization problem in which the above property is explicitly imposed:

$$\{\mathbf{w}_{\mathbf{X},i}, \mathbf{w}_{\mathbf{Y},i}\} = \operatorname{argmax}_{\mathbf{w}_{\mathbf{X},i}, \mathbf{w}_{\mathbf{Y},i}} \mathbf{w}'_{\mathbf{X},i} \mathbf{S}_{\mathbf{XY}} \mathbf{w}_{\mathbf{Y},i}$$
$$\text{s.t. } \|\mathbf{X}\mathbf{w}_{\mathbf{X},i}\| = \mathbf{w}'_{\mathbf{X},i}\mathbf{S}_{\mathbf{XX}}\mathbf{w}_{\mathbf{X},i} = 1$$
$$\|\mathbf{Y}\mathbf{w}_{\mathbf{Y},i}\| = \mathbf{w}'_{\mathbf{Y},i}\mathbf{S}_{\mathbf{YY}}\mathbf{w}_{\mathbf{Y},i} = 1$$
$$\text{and for } j < i : \begin{array}{l} \mathbf{w}'_{\mathbf{X},j}\mathbf{S}_{\mathbf{XX}}\mathbf{w}_{\mathbf{X},i} = 0, \\ \mathbf{w}'_{\mathbf{Y},j}\mathbf{S}_{\mathbf{YY}}\mathbf{w}_{\mathbf{Y},i} = 0. \end{array}$$

- The CCA problem can be reformulated as an ordinary eigenvalue problem:

$$\begin{pmatrix} 0 & \mathbf{S}_{\mathbf{XX}}^{-1}\mathbf{S}_{\mathbf{XY}} \\ \mathbf{S}_{\mathbf{YY}}^{-1}\mathbf{S}_{\mathbf{YX}} & 0 \end{pmatrix} \begin{pmatrix} \mathbf{w}_{\mathbf{X}} \\ \mathbf{w}_{\mathbf{Y}} \end{pmatrix} = \lambda \begin{pmatrix} \mathbf{w}_{\mathbf{X}} \\ \mathbf{w}_{\mathbf{Y}} \end{pmatrix}.$$

This eigenvalue problem can be made symmetric by introducing $\mathbf{v}_{\mathbf{X}} = \mathbf{S}_{\mathbf{XX}}^{1/2}\mathbf{w}_{\mathbf{X}}$ and $\mathbf{v}_{\mathbf{Y}} = \mathbf{S}_{\mathbf{YY}}^{1/2}\mathbf{w}_{\mathbf{Y}}$:

$$\begin{pmatrix} 0 & \mathbf{S}_{\mathbf{XX}}^{-1/2}\mathbf{S}_{\mathbf{XY}}\mathbf{S}_{\mathbf{YY}}^{-1/2} \\ \mathbf{S}_{\mathbf{YY}}^{-1/2}\mathbf{S}_{\mathbf{YX}}\mathbf{S}_{\mathbf{XX}}^{-1/2} & 0 \end{pmatrix} \begin{pmatrix} \mathbf{v}_{\mathbf{X}} \\ \mathbf{v}_{\mathbf{Y}} \end{pmatrix} = \lambda \begin{pmatrix} \mathbf{v}_{\mathbf{X}} \\ \mathbf{v}_{\mathbf{Y}} \end{pmatrix}.$$

Note that this eigenvalue problem is of the form of Eq. (5.2), so here $\mathbf{v}_{\mathbf{X}}$ and $\mathbf{v}_{\mathbf{Y}}$ are the left and right singular vectors of $\mathbf{S}_{\mathbf{XX}}^{-1/2}\mathbf{S}_{\mathbf{XY}}\mathbf{S}_{\mathbf{YY}}^{-1/2}$. The weight vectors can be retrieved as $\mathbf{w}_{\mathbf{X}} = \mathbf{S}_{\mathbf{XX}}^{-1/2}\mathbf{v}_{\mathbf{X}}$ and $\mathbf{w}_{\mathbf{Y}} = \mathbf{S}_{\mathbf{YY}}^{-1/2}\mathbf{v}_{\mathbf{Y}}$.

By the orthogonality of the singular vectors, we can derive in an alternative way that projections onto noncorresponding canonical directions are uncorrelated: $0 = \mathbf{v}'_{\mathbf{X},i}\mathbf{v}_{\mathbf{X},j} = \mathbf{w}'_{\mathbf{X},i}\mathbf{S}_{\mathbf{XX}}\mathbf{w}_{\mathbf{X},j}$, and $0 = \mathbf{v}'_{\mathbf{Y},i}\mathbf{v}_{\mathbf{Y},j} = \mathbf{w}'_{\mathbf{Y},i}\mathbf{S}_{\mathbf{YY}}\mathbf{w}_{\mathbf{Y},j}$. Also, we find that $0 = \mathbf{v}'_{\mathbf{X},i}\mathbf{S}_{\mathbf{XX}}^{-1/2}\mathbf{S}_{\mathbf{XY}}\mathbf{S}_{\mathbf{YY}}^{-1/2}\mathbf{v}_{\mathbf{Y},j} = \mathbf{w}'_{\mathbf{X},i}\mathbf{S}_{\mathbf{XY}}\mathbf{w}_{\mathbf{Y},j}$.

- As a last remark, we note that CCA where one of both data spaces is one-dimensional is equivalent to least squares regression (LSR).

Dual

To derive the dual, again note that the (minimum norm[3]) $\mathbf{w}_{\mathbf{X}}$ and $\mathbf{w}_{\mathbf{Y}}$ will lie in the column space of $\mathbf{X}$ and $\mathbf{Y}$, respectively (thus, analogously to Eq. (5.3),

---

[3] The motivation for taking the minimum norm solution is as follows: first of all, we need to make a choice in cases where there is an indeterminacy as is when the rows of $\mathbf{X}$ and/or $\mathbf{Y}$ do not span the whole space. And a component of the weight vectors orthogonal to the data would never contribute to the correlation of a projection of the data onto this weight vector anyway; the projection onto this orthogonal direction would be zero. We do not get any information concerning the orthogonal subspace, and thus do not want $\mathbf{w}$ to make any unmotivated predictions on this. In this chapter we always look for minimum norm solutions.

$\mathbf{w_X} = \mathbf{X}'\boldsymbol{\alpha_X}$ and $\mathbf{w_Y} = \mathbf{Y}'\boldsymbol{\alpha_Y}$; see also [3] for a more detailed explanation). Thus we can write

$$\begin{pmatrix} 0 & \mathbf{S_{XY}} \\ \mathbf{S_{YX}} & 0 \end{pmatrix} \begin{pmatrix} \mathbf{X}'\boldsymbol{\alpha_X} \\ \mathbf{Y}'\boldsymbol{\alpha_Y} \end{pmatrix} = \lambda \begin{pmatrix} \mathbf{S_{XX}} & 0 \\ 0 & \mathbf{S_{YY}} \end{pmatrix} \begin{pmatrix} \mathbf{X}'\boldsymbol{\alpha_X} \\ \mathbf{Y}'\boldsymbol{\alpha_Y} \end{pmatrix}$$

$$\Downarrow \text{ multiplying left with } \begin{pmatrix} \mathbf{X} & 0 \\ 0 & \mathbf{Y} \end{pmatrix}$$

$$\begin{pmatrix} 0 & \mathbf{X S_{XY} Y}' \\ \mathbf{Y S_{YX} X}' & 0 \end{pmatrix} \begin{pmatrix} \boldsymbol{\alpha_X} \\ \boldsymbol{\alpha_Y} \end{pmatrix} = \lambda \begin{pmatrix} \mathbf{X S_{XX} X}' & 0 \\ 0 & \mathbf{Y S_{YY} Y}' \end{pmatrix} \begin{pmatrix} \boldsymbol{\alpha_X} \\ \boldsymbol{\alpha_Y} \end{pmatrix}$$

$$\Downarrow$$

$$\begin{pmatrix} 0 & \mathbf{K_X K_Y} \\ \mathbf{K_Y K_X} & 0 \end{pmatrix} \begin{pmatrix} \boldsymbol{\alpha_X} \\ \boldsymbol{\alpha_Y} \end{pmatrix} = \lambda \begin{pmatrix} \mathbf{K_X^2} & 0 \\ 0 & \mathbf{K_Y^2} \end{pmatrix} \begin{pmatrix} \boldsymbol{\alpha_X} \\ \boldsymbol{\alpha_Y} \end{pmatrix}.$$

Projections of test points $\mathbf{x}_{\text{test}}$ and $\mathbf{y}_{\text{test}}$ onto the CCA directions corresponding to $\boldsymbol{\alpha_X}$ and $\boldsymbol{\alpha_Y}$ can then be carried out as

$$\sum_{i=1}^{n} \alpha_{\mathbf{X},i} k(\mathbf{x}_i, \mathbf{x}_{\text{test}}), \text{ and } \sum_{i=1}^{n} \alpha_{\mathbf{Y},i} k(\mathbf{y}_i, \mathbf{y}_{\text{test}}). \qquad (5.11)$$

Regularization

Primal problem

Regularization is often necessary in doing CCA for the following reason. The scatter matrices $\mathbf{S_{XX}}$ and $\mathbf{S_{YY}}$ are proportional to finite sample estimates of the covariance matrices. This generally leads to poor performance in case of small eigenvalues of these covariances. Remember the generalized eigenvalue problem is (theoretically) equivalent with a standard eigenvalue problem where the right-hand side matrix containing the scatter matrices is inverted. Any fluctuation of the smallest eigenvalue will thus be blown up in the inverse. To counteract this effect, one often adds a diagonal to the scatter matrices, or equivalently to each of their eigenvalues [3]. In this way, a bias is introduced, but it is hoped that for a certain bias, the total variance will be lower than the case when no bias is present.

An equivalent way to view this is, as presented above in the ridge regression derivation, by interpreting the regularization as a reduction of the effective number of degrees of freedom. Generalization will be more likely to be good.

The primal regularized problem is thus

$$\begin{pmatrix} 0 & \mathbf{S_{XY}} \\ \mathbf{S_{YX}} & 0 \end{pmatrix} \begin{pmatrix} \mathbf{w_X} \\ \mathbf{w_Y} \end{pmatrix} = \lambda \begin{pmatrix} \mathbf{S_{XX}} + \gamma\mathbf{I} & 0 \\ 0 & \mathbf{S_{YY}} + \gamma\mathbf{I} \end{pmatrix} \begin{pmatrix} \mathbf{w_X} \\ \mathbf{w_Y} \end{pmatrix}.$$

Intuitively, this type of regularization boils down to trusting correlations along high-variance directions more than along low-variance directions. Or, equivalently, it corresponds to a modified optimization problem where the constraints contain an additional term constraining the norm of $\mathbf{w_X}$ and $\mathbf{w_Y}$, similarly to the ridge regression cost function.

Note that RCCA with one of both spaces one-dimensional is equivalent to ridge regression (RR).

Dual problem

The dual of this generalized eigenvalue problem can be derived in the same way as the unregularized problem, leading to:

$$\begin{pmatrix} 0 & \mathbf{K_X K_Y} \\ \mathbf{K_Y K_X} & 0 \end{pmatrix} \begin{pmatrix} \alpha_X \\ \alpha_Y \end{pmatrix} = \lambda \begin{pmatrix} \mathbf{K_X^2} + \gamma \mathbf{K_X} & 0 \\ 0 & \mathbf{K_Y^2} + \gamma \mathbf{K_Y} \end{pmatrix} \begin{pmatrix} \alpha_X \\ \alpha_Y \end{pmatrix} \quad (5.12)$$

In the dual case, the need for regularization is often even stronger than in the primal case. This is because the feature space is often infinite-dimensional, so that the freedom to find correlations is much too high. All correlations would be equal to 1, which means no generalization is possible at all. Penalizing a large weight vector as above thus makes sense to improve generalization.

When both the kernels have full rank, left-multiplication on both sides of Eq. (5.12) with $\begin{pmatrix} \mathbf{K_X^{-1}} & 0 \\ 0 & \mathbf{K_Y^{-1}} \end{pmatrix}$ reveals that this generalized eigenvalue problem is equivalent with

$$\begin{pmatrix} 0 & \mathbf{K_Y} \\ \mathbf{K_X} & 0 \end{pmatrix} \begin{pmatrix} \alpha_X \\ \alpha_Y \end{pmatrix} = \lambda \begin{pmatrix} \mathbf{K_X} + \gamma \mathbf{I} & 0 \\ 0 & \mathbf{K_Y} + \gamma \mathbf{I} \end{pmatrix} \begin{pmatrix} \alpha_X \\ \alpha_Y \end{pmatrix}. \quad (5.13)$$

Kernel matrices are often rank deficient, however (e.g. when they are centered). In that case the solutions of Eq. (5.13) are still solutions for Eq. (5.12), but the converse is no longer always true. The reason is that for any generalized eigenvector $\begin{pmatrix} \alpha_X \\ \alpha_Y \end{pmatrix}$ of Eq. (5.13) and thus of Eq. (5.12), $\begin{pmatrix} \alpha_X + \alpha_{X0} \\ \alpha_Y + \alpha_{Y0} \end{pmatrix}$, where α_{X0} and α_{Y0} are arbitrary vectors lying respectively in the null spaces of $\mathbf{K_X}$ and $\mathbf{K_Y}$, is also an eigenvector with the same eigenvalue of Eq. (5.12) but generally not of Eq. (5.13). However, similarly as in the ridge regression derivation, it can be seen that these components α_{X0} and α_{Y0} play no role in the calculation of Eq. (5.11). This is because the weight vectors $\mathbf{w_X} = \mathbf{X}'(\alpha_X + \alpha_{X0}) = \mathbf{X}'\alpha_X$ and $\mathbf{w_Y} = \mathbf{Y}'(\alpha_Y + \alpha_{Y0}) = \mathbf{Y}'\alpha_Y$ are unaffected by the components in the null spaces of $\mathbf{K_X}$ and $\mathbf{K_Y}$. Therefore, we can choose to solve either Eq. (5.12) or Eq. (5.13).

5.4.3 Partial Least Squares

Partial least squares (PLS, introduced in [33, 34]; see also [14] for a good review) can be interpreted in two ways. The first PLS component is the maximally regularized version of the first CCA component (the case where $\gamma \to \infty$, after rescaling the eigenvalues by multiplying them with γ). Another view is as a covariance maximizer instead of a correlation maximizer, this again for the first PLS component. Whereas all PLS formulations compute the first component in the same way, there is no one way to compute the other components. We will present two variants: so-called EZ-PLS, which consists of only one eigenvalue decomposition (or a singular value decomposition) and which is used mainly for exploratory purposes (similar to CCA), and regression-PLS which is a more involved version that is most widely used in (multivariate) regression applications.

Because of the iterative way PLS components are computed in, and because of the fact that there exist several variants of PLS, the discussion is somewhat more involved. We will first give a general discussion on the cost function optimized in all PLS formulations, followed by the eigenproblem optimizing this cost function. Next, we will shortly go into some computational aspects. Finally, we will show the particularities of the two PLS formulations EZ-PLS and regression-PLS, followed by a discussion of the regression step in regression-PLS. Again, a primal and a dual (see [19] where this was first derived) formulation will be provided.

Cost Function

Maximize the sample *covariance*[4] between a projection of $\mathbf{X}$ and a projection of $\mathbf{Y}$:

$$\{\mathbf{w_X}, \mathbf{w_Y}\} = \operatorname{argmax}_{\mathbf{w_X}, \mathbf{w_Y}} \frac{(\mathbf{X}\mathbf{w_X})'(\mathbf{Y}\mathbf{w_Y})}{\sqrt{\mathbf{w_X'}\mathbf{w_X}}\sqrt{\mathbf{w_Y'}\mathbf{w_Y}}},$$

$$= \operatorname{argmax}_{\mathbf{w_X}, \mathbf{w_Y}} \frac{\mathbf{w_X'}\mathbf{S_{XY}}\mathbf{w_Y}}{\sqrt{\mathbf{w_X'}\mathbf{w_X}}\sqrt{\mathbf{w_Y'}\mathbf{w_Y}}}.$$

This is equivalent to maximizing the sample covariance, or the 'fit' subject to constraints:

$$\{\mathbf{w_X}, \mathbf{w_Y}\} = \operatorname{argmax}_{\mathbf{w_X}, \mathbf{w_Y}} \mathbf{w_X'}\mathbf{S_{XY}}\mathbf{w_Y}$$
$$\text{s.t. } \|\mathbf{w_X}\|^2 = \mathbf{w_X'}\mathbf{w_X} = 1, \|\mathbf{w_Y}\|^2 = \mathbf{w_Y'}\mathbf{w_Y} = 1,$$

and equivalent to minimizing the misfit subject to these constraints:

$$\{\mathbf{w_X}, \mathbf{w_Y}\} = \operatorname{argmin}_{\mathbf{w_X}, \mathbf{w_Y}} \|\mathbf{X}\mathbf{w_X} - \mathbf{Y}\mathbf{w_Y}\|^2$$
$$\text{s.t. } \|\mathbf{w_X}\|^2 = 1, \|\mathbf{w_Y}\|^2 = 1.$$

Primal

We solve the second formulation of the problem. Differentiating the Lagrangian $\mathcal{L}(\mathbf{w_X}, \mathbf{w_Y}, \lambda_\mathbf{X}, \lambda_\mathbf{Y}) = \mathbf{w_X'}\mathbf{S_{XY}}\mathbf{w_Y} - \lambda_\mathbf{X}\mathbf{w_X'}\mathbf{w_X} - \lambda_\mathbf{Y}\mathbf{w_Y'}\mathbf{w_Y}$ with respect to $\mathbf{w_X}$ and $\mathbf{w_Y}$ and equating to 0 gives

---
[4] Note the difference between CCA where *correlation* was maximized.

$$\begin{cases} \frac{\partial}{\partial \mathbf{w_X}}\mathcal{L}(\mathbf{w_X}, \mathbf{w_Y}, \lambda_X, \lambda_Y) = 0, \\ \frac{\partial}{\partial \mathbf{w_Y}}\mathcal{L}(\mathbf{w_X}, \mathbf{w_Y}, \lambda_X, \lambda_Y) = 0, \end{cases}$$

$$\Rightarrow \begin{cases} \mathbf{S_{XY} w_Y} = \lambda_X \mathbf{w_X}. \\ \mathbf{S_{YX} w_X} = \lambda_Y \mathbf{w_Y}. \end{cases}$$

Since from this

$$\lambda_X \mathbf{w'_X w_X} = \mathbf{w'_X S_{XY} w_Y} = \mathbf{w'_Y S_{YX} w_X} = \lambda_Y \mathbf{w'_Y w_Y},$$

and since $\mathbf{w'_X w_X} = \mathbf{w'_Y w_Y} = 1$, we find that $\lambda_X = \lambda_Y = \lambda$. Thus

$$\begin{cases} \mathbf{S_{XY} w_Y} = \lambda \mathbf{w_X}, \\ \mathbf{S_{YX} w_X} = \lambda \mathbf{w_Y}. \end{cases} \qquad (5.14)$$

Or, stated in another way as an eigenvalue problem,

$$\begin{pmatrix} 0 & \mathbf{S_{XY}} \\ \mathbf{S_{YX}} & 0 \end{pmatrix} \begin{pmatrix} \mathbf{w_X} \\ \mathbf{w_Y} \end{pmatrix} = \lambda \begin{pmatrix} \mathbf{w_X} \\ \mathbf{w_Y} \end{pmatrix}. \qquad (5.15)$$

This eigenvalue problem has d eigenvalues, corresponding to a covariance between projections onto $\mathbf{w_X}$ and $\mathbf{w_Y}$. The largest one with its eigenvector corresponds to the optimum of the cost function described earlier.

Note that Eq. (5.15) is of the form of Eq. (5.2). Thus the EZ-PLS problem can be solved by calculating the singular value decomposition of $\mathbf{S_{XY}}$.

Dual

The dual problem can easily be found by using Eq. (5.3):

$$\begin{pmatrix} 0 & \mathbf{K_X K_Y} \\ \mathbf{K_Y K_X} & 0 \end{pmatrix} \begin{pmatrix} \alpha_X \\ \alpha_Y \end{pmatrix} = \lambda \begin{pmatrix} \mathbf{K_X} & 0 \\ 0 & \mathbf{K_Y} \end{pmatrix} \begin{pmatrix} \alpha_X \\ \alpha_Y \end{pmatrix},$$

which includes all solutions of

$$\begin{pmatrix} 0 & \mathbf{K_Y} \\ \mathbf{K_X} & 0 \end{pmatrix} \begin{pmatrix} \alpha_X \\ \alpha_Y \end{pmatrix} = \lambda \begin{pmatrix} \alpha_X \\ \alpha_Y \end{pmatrix} \qquad (5.16)$$

as its solutions as well. Similarly, as in CCA this is the formulation of the dual problem that is solved, since it does not suffer from indeterminacies.

Projections of test points $\mathbf{x}_{\text{test}}$ and $\mathbf{y}_{\text{test}}$ onto the PLS directions corresponding to α_X and α_Y can then be computed as

$$\sum_{i=1}^{n} \alpha_{X,i} k(\mathbf{x}_i, \mathbf{x}_{\text{test}}), \text{ and } \sum_{i=1}^{n} \alpha_{Y,i} k(\mathbf{y}_i, \mathbf{y}_{\text{test}}).$$

It is important to note that the first component corresponds to maximally regularized RCCA. Taking more than one component lessens this regularization in an alternative way in comparison to RCCA. This will be the subject of the remainder of this section on PLS.

Nonlinear Iterative Partial Least Squares and Primal-Dual Symmetry in PLS

A straightforward way to solve for the largest eigenvector of Eq. (5.15) could be by using the power method. However, thanks to the structure of the eigenvalue problems at hand, it can be solved by using the so-called nonlinear iterative partial least squares (NIPALS) method [33]. Note that, from Eqs. (5.15) and (5.16):

- $\mathbf{YY'XX'}\alpha_\mathbf{X} = \lambda^2 \alpha_\mathbf{X}$.
- $\mathbf{X'YY'X}\mathbf{w_X} = \lambda^2 \mathbf{w_X}$.
- $\mathbf{XX'YY'}\alpha_\mathbf{Y} = \lambda^2 \alpha_\mathbf{Y}$.
- $\mathbf{Y'XX'Y}\mathbf{w_Y} = \lambda^2 \mathbf{w_Y}$.

Thus it follows that both the primal and the dual eigenvalue problem are actually solved at the same time, using the following 'power' method:

0. Fix initial value $\mathbf{w_Y}$, normalize. Then iterate over steps 1-4.
1. $\alpha_\mathbf{X} = \mathbf{Y}\mathbf{w_Y}$.
2. $\mathbf{w_X} = \mathbf{X'}\alpha_\mathbf{X}$, normalize $\mathbf{w_X}$ to unit length.
3. $\alpha_\mathbf{Y} = \mathbf{X}\mathbf{w_X}$.
4. $\mathbf{w_Y} = \mathbf{Y'}\alpha_\mathbf{Y}$, normalize $\mathbf{w_Y}$ to unit length.

After convergence, the normalizations carried out in steps 2 and 4 both amount to a division by λ; then $\mathbf{w_X} = \frac{1}{\lambda}\mathbf{X'}\alpha_\mathbf{X}$ and $\mathbf{w_Y} = \frac{1}{\lambda}\mathbf{Y'}\alpha_\mathbf{Y}$.

In case the feature vectors $\mathbf{X}$ are only implicitly determined by a kernel function, steps 2 and 3 must be combined in one step:

2,3. $\alpha_\mathbf{Y} = \mathbf{K_X}\alpha_\mathbf{X}$, normalize.

It can be seen that each of these weight vectors or dual vectors converge to the eigenvector of the above four eigenvalue problems (combining four steps following each other gives the power method for one of these four eigenvalue problems). Since these are equivalent with Eqs. (5.15) and (5.16), they converge to the PLS weight vectors and dual vectors.

In this way, we can solve efficiently for the largest singular value and singular vectors. Only this one component is not enough to solve most practical problems, however. We discuss two ways to extract more information present in the data: what we call EZ-PLS and regression-PLS. For both methods first the primal versions will be discussed, then afterwards the dual.

EZ-PLS

Primal

In EZ-PLS, the other PLS directions are the other eigenvectors corresponding to a different covariance (eigenvalue) λ. This can be accomplished by using an iterative deflation scheme:

1. Initialize: $\mathbf{S_{XY}}^0 \leftarrow \mathbf{S_{XY}}$.
2. Compute the largest singular value of $\mathbf{S_{XY}}^i$ with NIPALS. This gives the ith PLS component. Normalize so that $\|\mathbf{w}_{\mathbf{X},i}\| = \|\mathbf{w}_{\mathbf{Y},i}\| = 1$.
3. Deflate the scatter matrices:

$$\mathbf{S_{XY}}^{i+1} \leftarrow \mathbf{S_{XY}}^i - \lambda_i \mathbf{w}_{\mathbf{X},i} \mathbf{w}'_{\mathbf{Y},i}.$$

The rank of $\mathbf{S_{XY}}^{i+1}$ is 1 less than the rank of $\mathbf{S_{XY}}^i$.
4. When the number of desired components (necessarily lower than the rank of $\mathbf{S_{XY}}$) is not yet reached, go to step 2.

The deflation of the $\mathbf{X}^i$ matrix for EZ-PLS, in order to get the desired deflation of the cross-scatter matrix, is

$$\mathbf{X}^{i+1} \leftarrow \mathbf{X}^i - \mathbf{X}^i \mathbf{w}_{\mathbf{X},i} \mathbf{w}'_{\mathbf{X},i}.$$

Similarly, one could do the deflation of the $\mathbf{Y}^i$ matrix

$$\mathbf{Y}^{i+1} \leftarrow \mathbf{Y}^i - \mathbf{Y}^i \mathbf{w}_{\mathbf{Y},i} \mathbf{w}'_{\mathbf{Y},i},$$

also leading to the same desired deflation of the cross-scatter matrix.

Dual

Taking Eq. (5.3) or equivalently the NIPALS iteration into account, the deflation of the kernel matrices corresponding to the EZ-PLS deflation is found to be

$$\mathbf{K}_{\mathbf{X}}^{i+1} \leftarrow \mathbf{K}_{\mathbf{X}}^i - \frac{1}{\lambda_i^2} \mathbf{K}_{\mathbf{X}}^i \alpha_{\mathbf{X},i} \alpha'_{\mathbf{X},i} \mathbf{K}_{\mathbf{X}}^i = \mathbf{K}_{\mathbf{X}}^i - \alpha_{\mathbf{Y},i} \alpha'_{\mathbf{Y},i}.$$

Properties

- Since the $\mathbf{w}_{\mathbf{X},i}$ and the $\mathbf{w}_{\mathbf{Y},i}$ are the left and right singular vectors of $\mathbf{S_{XY}}$, all $\mathbf{w}_{\mathbf{X},i}$ are orthogonal to each other, and all $\mathbf{w}_{\mathbf{Y},i}$ are orthogonal to each other.
- For the same reason, if $i \neq j$: $\mathbf{w}'_{\mathbf{X},i} \mathbf{S_{XY}} \mathbf{w}_{\mathbf{Y},j} = 0$. In other words, projections onto noncorresponding $\mathbf{w}_{\mathbf{X},i}$ and $\mathbf{w}_{\mathbf{Y},j}$ are uncorrelated.
- All EZ-PLS components can be calculated at once by optimizing the same cost function as for the first component, taking the first (orthogonality) property into account as an additional constraint.

The EZ-PLS form is the easiest, in the sense that because of the nature of the deflation, it is in fact not more than solving for the most important singular vectors of $\mathbf{S_{XY}}$. That is why it is discussed here; it is less useful in practice.

Regression-PLS

Whereas EZ-PLS is not often used for regression (note that it is entirely symmetric between $\mathbf{X}$ and $\mathbf{Y}$, whereas regression is not; it is rather used for modelling though), regression-PLS is the PLS formulation that is generally preferred for (multivariate) regression (see [14]). We will first discuss the deflations that are characteristic for regression-PLS. Further on we will explain how regression can be carried out using the results from these deflations.

Primal

The difference between EZ-PLS and Regression-PLS lies in the way the deflation is carried out. Regression-PLS has the intention of modelling one (possibly) vectorial variable $\mathbf{Y}$ with the other vectorial variable $\mathbf{X}$, hence the name.[5] It is thus asymmetric between the two spaces, which is expressed in the deflation step:

2,4. Deflate by orthogonalizing $\mathbf{X}^i$ to its projection onto the weight vector $\mathbf{w}_{\mathbf{X},i}$, $\mathbf{X}^i \mathbf{w}_{\mathbf{X},i}$, and recomputing the scatter matrix:

$$\mathbf{X}^{i+1} \leftarrow \left(\mathbf{I} - \frac{\mathbf{X}^i \mathbf{w}_{\mathbf{X},i} \mathbf{w}'_{\mathbf{X},i} \mathbf{X}^{i'}}{\mathbf{w}'_{\mathbf{X},i} \mathbf{X}^{i'} \mathbf{X}^i \mathbf{w}_{\mathbf{X},i}}\right) \mathbf{X}^i = \mathbf{X}^i - \frac{\mathbf{X}^i \mathbf{w}_{\mathbf{X},i} \mathbf{w}'_{\mathbf{X},i} \mathbf{X}^{i'}}{\mathbf{w}'_{\mathbf{X},i} \mathbf{X}^{i'} \mathbf{X}^i \mathbf{w}_{\mathbf{X},i}} \mathbf{X}^i \quad (5.17)$$

$$= \left(\mathbf{I} - \frac{\alpha_{\mathbf{Y},i} \alpha'_{\mathbf{Y},i}}{\alpha'_{\mathbf{Y},i} \alpha_{\mathbf{Y},i}}\right) \mathbf{X}^i. \quad (5.18)$$

Finally (see later, Eq. (5.28)) we will perform a regression of $\mathbf{Y}$ based on the $\alpha_{\mathbf{Y},i}$. (The $\alpha_{\mathbf{Y},i}$ can be computed from $\mathbf{X}$ as will become clear later, see Eq. (5.27).) Therefore, we also deflate $\mathbf{Y}^i$ with $\alpha_{\mathbf{Y},i}$ to remove the information captured by the ith iteration:

$$\mathbf{Y}^{i+1} \leftarrow \left(\mathbf{I} - \frac{\alpha_{\mathbf{Y},i} \alpha'_{\mathbf{Y},i}}{\alpha'_{\mathbf{Y},i} \alpha_{\mathbf{Y},i}}\right) \mathbf{Y}^i. \quad (5.19)$$

This boils down to the following deflation of the scatter matrix:

$$\mathbf{S_{XY}}^{i+1} \leftarrow \mathbf{S_{XY}}^i - \frac{\lambda_i}{\mathbf{w}'_{\mathbf{X},i} \mathbf{S_{XX}}^i \mathbf{w}_{\mathbf{X},i}} \mathbf{S_{XX}}^i \mathbf{w}_{\mathbf{X},i} \mathbf{w}'_{\mathbf{Y},i}.$$

The philosophy behind this kind of deflation is as follows: after step i, part of the information in $\mathbf{X}^i$, namely its projection $\alpha_{\mathbf{Y},i}$ onto the ith PLS direction $\mathbf{w}_{\mathbf{X},i}$, is captured already: the component $\frac{\alpha_{\mathbf{Y},i} \alpha'_{\mathbf{Y},i}}{\alpha'_{\mathbf{Y},i} \alpha_{\mathbf{Y},i}} \mathbf{X}^i$ of $\mathbf{X}^i$ (along $\alpha_{\mathbf{Y},i}$) perfectly models the component $\frac{\alpha_{\mathbf{Y},i} \alpha'_{\mathbf{Y},i}}{\alpha'_{\mathbf{Y},i} \alpha_{\mathbf{Y},i}} \mathbf{Y}^i$ of $\mathbf{Y}^i$. This information should not be used or modelled again in next steps, so it is 'subtracted' from both $\mathbf{X}^i$ and $\mathbf{Y}^i$. In the next step, the direction of maximal covariance between the remaining information $\mathbf{X}^{i+1}$ and $\mathbf{Y}^{i+1}$ is found, and so on.

---

[5] In literature this form of PLS is best known as PLS2, or PLS1 for the case where $\mathbf{Y}$ is one-dimensional.

Dual

Using Eqs. (5.18) and (5.19), the deflation of the kernel matrices corresponding to the regression-PLS deflation can be shown to be

$$\mathbf{K}_\mathbf{X}^{i+1} \leftarrow \left(\mathbf{I} - \frac{\alpha_{\mathbf{Y},i}\alpha'_{\mathbf{Y},i}}{\alpha'_{\mathbf{Y},i}\alpha_{\mathbf{Y},i}}\right) \mathbf{K}_\mathbf{X}^i \left(\mathbf{I} - \frac{\alpha_{\mathbf{Y},i}\alpha'_{\mathbf{Y},i}}{\alpha'_{\mathbf{Y},i}\alpha_{\mathbf{Y},i}}\right).$$

Analogously,

$$\mathbf{K}_\mathbf{Y}^{i+1} \leftarrow \left(\mathbf{I} - \frac{\alpha_{\mathbf{Y},i}\alpha'_{\mathbf{Y},i}}{\alpha'_{\mathbf{Y},i}\alpha_{\mathbf{Y},i}}\right) \mathbf{K}_\mathbf{Y}^i \left(\mathbf{I} - \frac{\alpha_{\mathbf{Y},i}\alpha'_{\mathbf{Y},i}}{\alpha'_{\mathbf{Y},i}\alpha_{\mathbf{Y},i}}\right).$$

Properties

- The different weight vectors $\mathbf{w}_{\mathbf{Y},i}$ are *not* orthogonal (it is even possible that they are all collinear, e.g. in the case where $\mathbf{Y}$ is one-dimensional). The different weight vectors $\mathbf{w}_{\mathbf{X},i}$, however, are orthogonal. Using Eq. (5.17),

$$\mathbf{w}'_{\mathbf{X},i}\mathbf{S}_{\mathbf{XY}}^{i+1} = \mathbf{w}'_{\mathbf{X},i}\left(\left(\mathbf{I} - \frac{\mathbf{X}^i\mathbf{w}_{\mathbf{X},i}\mathbf{w}'_{\mathbf{X},i}\mathbf{X}^{i'}}{\mathbf{w}'_{\mathbf{X},i}\mathbf{X}^{i'}\mathbf{X}^i\mathbf{w}_{\mathbf{X},i}}\right)\mathbf{X}^i\right)' \mathbf{Y}^{i+1} = \mathbf{0},$$

so that $\mathbf{w}_{\mathbf{X},i}$ is in the left null space of $\mathbf{S}_{\mathbf{XY}}^{i+1}$. Since $\mathbf{w}_{\mathbf{X},i+1}$ is a left singular vector of $\mathbf{S}_{\mathbf{XY}}^{i+1}$ this means that $\mathbf{w}_{\mathbf{X},i+1}$ will be orthogonal to $\mathbf{w}_{\mathbf{X},i}$. By replacing the left-most $\mathbf{X}^i$ in the above equation by $\left(\mathbf{I} - \frac{\mathbf{X}^{i-1}\mathbf{w}_{\mathbf{X},i-1}\mathbf{w}'_{\mathbf{X},i-1}\mathbf{X}^{i-1'}}{\mathbf{w}'_{\mathbf{X},i-1}\mathbf{X}^{i-1'}\mathbf{X}^{i-1}\mathbf{w}_{\mathbf{X},i-1}}\right)\mathbf{X}^{i-1}$, and so on for $\mathbf{X}^{i-1},\ldots$, one can see that also for $j < i$, $\mathbf{w}_{\mathbf{X},j}$ is orthogonal to $\mathbf{w}_{\mathbf{X},i}$. Thus, all $\mathbf{w}_{\mathbf{X},i}$ are mutually orthogonal:

$$\mathbf{W}'_\mathbf{X}\mathbf{W}_\mathbf{X} = \mathbf{I},$$

where $\mathbf{W}_\mathbf{X}$ represents the matrix built by stacking the vectors $\mathbf{w}_{\mathbf{X},i}$ next to each other.

- The vectors $\alpha_{\mathbf{Y},i}$ are mutually orthogonal. Using Eq. (5.18), for $i \leq j$ one has:

$$\mathbf{X}^{j'}\alpha_{\mathbf{Y},i} = \mathbf{X}^{i'}\left(\mathbf{I} - \frac{\alpha_{\mathbf{Y},i}\alpha'_{\mathbf{Y},i}}{\alpha'_{\mathbf{Y},i}\alpha_{\mathbf{Y},i}}\right)\cdots\left(\mathbf{I} - \frac{\alpha_{\mathbf{Y},j-1}\alpha'_{\mathbf{Y},j-1}}{\alpha'_{\mathbf{Y},j-1}\alpha_{\mathbf{Y},j-1}}\right)\alpha_{\mathbf{Y},i}.$$

For $j = i+1$, this is immediately proven to be zero. When this product is zero for all $j : i < j < j*$, $\alpha'_{\mathbf{Y},j}\alpha_{\mathbf{Y},i} = \mathbf{w}'_{\mathbf{X},j}\mathbf{X}^{j'}\alpha_{\mathbf{Y},i} = 0$, and the matrices between brackets in the above product commute. Since this is indeed true for $j = i+1$, by induction it is proved for all $i < j$ that:

$$\mathbf{X}^{j'}\alpha_{\mathbf{Y},i} = \mathbf{0}, \qquad (5.20)$$

and thus by left multiplication with $\mathbf{w}_{\mathbf{X},j}$

$$\alpha'_{\mathbf{Y},j}\alpha_{\mathbf{Y},i} = 0. \tag{5.21}$$

Note that since $\alpha_{\mathbf{Y},i} = \mathbf{X}^i\mathbf{w}_{\mathbf{X},i}$, this means that the projections $\alpha_{\mathbf{Y},i}$ of $\mathbf{X}^i$ onto their weight vectors $\mathbf{w}_{\mathbf{X},i}$ are uncorrelated with each other. This property may remind you of CCA.

- This orthogonality property in Eq. (5.21) of the $\alpha_{\mathbf{Y},i}$ leads to the fact that

$$\mathbf{w}_{\mathbf{Y},i} = \mathbf{Y}^{i'}\alpha_{\mathbf{Y},i} = \mathbf{Y}'\left(\mathbf{I} - \frac{\alpha_{\mathbf{Y},1}\alpha'_{\mathbf{Y},1}}{\alpha'_{\mathbf{Y},1}\alpha_{\mathbf{Y},1}}\right)\cdots\left(\mathbf{I} - \frac{\alpha_{\mathbf{Y},i-1}\alpha'_{\mathbf{Y},i-1}}{\alpha'_{\mathbf{Y},i-1}\alpha_{\mathbf{Y},i-1}}\right)\alpha_{\mathbf{Y},i}$$

$$\Rightarrow \mathbf{w}_{\mathbf{Y},i} = \mathbf{Y}'\alpha_{\mathbf{Y},i}, \tag{5.22}$$

up to a normalization.

- Furthermore, one finds that for $i < j$:

$$\mathbf{X}^j\mathbf{w}_{\mathbf{X},i} = \left(\mathbf{I} - \frac{\alpha_{\mathbf{Y},j-1}\alpha'_{\mathbf{Y},j-1}}{\alpha'_{\mathbf{Y},j-1}\alpha_{\mathbf{Y},j-1}}\right)\cdots\left(\mathbf{I} - \frac{\alpha_{\mathbf{Y},i}\alpha'_{\mathbf{Y},i}}{\alpha'_{\mathbf{Y},i}\alpha_{\mathbf{Y},i}}\right)\mathbf{X}^i\mathbf{w}_{\mathbf{X},i},$$

$$= \left(\mathbf{I} - \frac{\alpha_{\mathbf{Y},j-1}\alpha'_{\mathbf{Y},j-1}}{\alpha'_{\mathbf{Y},j-1}\alpha_{\mathbf{Y},j-1}}\right)\cdots\left(\mathbf{I} - \frac{\alpha_{\mathbf{Y},i}\alpha'_{\mathbf{Y},i}}{\alpha'_{\mathbf{Y},i}\alpha_{\mathbf{Y},i}}\right)\alpha_{\mathbf{Y},i},$$

$$= \mathbf{0}. \tag{5.23}$$

This generally does not hold for $i \geq j$.

- Another consequence of Eq. (5.21) is, for $i < j$:

$$\mathbf{Y}^{j'}\alpha_{\mathbf{Y},i} = \mathbf{Y}^{i'}\left(\mathbf{I} - \frac{\alpha_{\mathbf{Y},i}\alpha'_{\mathbf{Y},i}}{\alpha'_{\mathbf{Y},i}\alpha_{\mathbf{Y},i}}\right)\cdots\left(\mathbf{I} - \frac{\alpha_{\mathbf{Y},j-1}\alpha'_{\mathbf{Y},j-1}}{\alpha'_{\mathbf{Y},j-1}\alpha_{\mathbf{Y},j-1}}\right)\alpha_{\mathbf{Y},i},$$

$$= \mathbf{0}. \tag{5.24}$$

- And thus also, for $i < j$:

$$\alpha'_{\mathbf{X},j}\alpha_{\mathbf{Y},i} = \mathbf{w}_{\mathbf{X},j}\mathbf{Y}^{j'}\alpha_{\mathbf{Y},i},$$

$$= 0. \tag{5.25}$$

- From this it follows that

$$\mathbf{w}_{\mathbf{X},i} = \mathbf{X}^{i'}\alpha_{\mathbf{X},i} = \mathbf{X}'\left(\mathbf{I} - \frac{\alpha_{\mathbf{Y},1}\alpha'_{\mathbf{Y},1}}{\alpha'_{\mathbf{Y},1}\alpha_{\mathbf{Y},1}}\right)\cdots\left(\mathbf{I} - \frac{\alpha_{\mathbf{Y},i-1}\alpha'_{\mathbf{Y},i-1}}{\alpha'_{\mathbf{Y},i-1}\alpha_{\mathbf{Y},i-1}}\right)\alpha_{\mathbf{X},i}$$

$$\Rightarrow \mathbf{w}_{\mathbf{X},i} = \mathbf{X}'\alpha_{\mathbf{X},i}, \tag{5.26}$$

up to a normalization factor.

Thus as a summary:

$$\mathbf{w}_{\mathbf{X},i} \propto \mathbf{X}'\alpha_{\mathbf{X},i},$$
$$\mathbf{w}_{\mathbf{Y},i} \propto \mathbf{Y}'\alpha_{\mathbf{Y},i},$$
$$\mathbf{w}'_{\mathbf{X},j}\mathbf{w}_{\mathbf{X},i} = 0,$$
$$\alpha'_{\mathbf{Y},j}\alpha_{\mathbf{Y},i} = 0,$$
$$\alpha'_{\mathbf{X},j}\alpha_{\mathbf{Y},i} = \mathbf{0} \text{ for } i < j,$$
$$\mathbf{X}^{j}\mathbf{w}_{\mathbf{X},i} = \mathbf{0} \text{ for } i < j,$$
$$\mathbf{Y}^{j'}\alpha_{\mathbf{Y},i} = \mathbf{0} \text{ for } i < j.$$

Final Regression in Regression-PLS

Primal

The entire regression-PLS algorithm is composed of a (generally noninvertible) linear mapping of $\mathbf{X}$ towards k so-called *latent variables* (in the current context we would rather call them dual variables) $\alpha_{\mathbf{Y},i} = \mathbf{X}^i \mathbf{w}_{\mathbf{X},i}$, followed by a regression of $\mathbf{Y}$ on $\mathbf{A}_{\mathbf{Y}}$, where $\mathbf{A}_{\mathbf{Y}}$ contains $\alpha_{\mathbf{Y},i}$ as its columns.

The part of $\mathbf{X}$ that has been deflated and thus will be used for regression is equal to the sum $\sum_{i=1}^{k} \frac{\alpha_{\mathbf{Y},i}\alpha'_{\mathbf{Y},i}}{\alpha'_{\mathbf{Y},i}\alpha_{\mathbf{Y},i}} \mathbf{X}^i = \mathbf{A}_{\mathbf{Y}}\mathbf{P}'$, where the vectors $\mathbf{p}_i = \mathbf{X}^{i'}\frac{\alpha_{\mathbf{Y},i}}{\alpha'_{\mathbf{Y},i}\alpha_{\mathbf{Y},i}}$ make up the columns of $\mathbf{P}$. Analogously, define $\mathbf{c}_i = \mathbf{Y}^{i'}\frac{\alpha_{\mathbf{Y},i}}{\alpha'_{\mathbf{Y},i}\alpha_{\mathbf{Y},i}}$ making up the columns of $\mathbf{C}$.

Now, if we go on with the deflations until the rank of $\mathbf{X}^i$ is zero,[6] the space spanned by the orthogonal vectors $\alpha_{\mathbf{Y},i}$ is complete and we have that

$$\mathbf{X} = \mathbf{A}_{\mathbf{Y}}^{\text{tot}}\mathbf{P}^{\text{tot}'} = \mathbf{A}_{\mathbf{Y}}\mathbf{P}' + \mathbf{A}_{\mathbf{Y}}^{\text{rem}}\mathbf{P}^{\text{rem}'} = \mathbf{A}_{\mathbf{Y}}\mathbf{P}' + \mathbf{E}_{\mathbf{X}},$$

with $\mathbf{E}_{\mathbf{X}}$ the part of $\mathbf{X}$ that is not used in regression when the components corresponding to $\mathbf{A}_{\mathbf{Y}}^{\text{rem}}$ are not kept. Also, because of Eq. (5.23) and the definition of $\mathbf{P}$: $\mathbf{p}_j'\mathbf{w}_{\mathbf{X},i} = 0$ for $i < j$, and thus:

$$\mathbf{P}^{\text{rem}'}\mathbf{W}_{\mathbf{X}} = \mathbf{0}.$$

This leads to the linear mapping from $\mathbf{X}$ to $\mathbf{A}_{\mathbf{Y}}$:

$$\mathbf{A}_{\mathbf{Y}}\mathbf{P}'\mathbf{W}_{\mathbf{X}} = \mathbf{X}\mathbf{W}_{\mathbf{X}}$$
$$\Rightarrow \mathbf{A}_{\mathbf{Y}} = \mathbf{X}\mathbf{W}_{\mathbf{X}}\left(\mathbf{P}'\mathbf{W}_{\mathbf{X}}\right)^{-1}, \tag{5.27}$$

where the matrix to be inverted is lower triangular (again because $\mathbf{p}_j'\mathbf{w}_{\mathbf{X},i} = 0$ for $i < j$), so the inversion can be carried out efficiently.

---
[6] Note that the number of deflations k will always be smaller (or equal, in full LSR) than the rank of $\mathbf{X}$. This results in matrices $\mathbf{W}_{\mathbf{X}}, \mathbf{W}_{\mathbf{Y}}, \mathbf{A}'_{\mathbf{X}}, \mathbf{A}'_{\mathbf{Y}}, \mathbf{P}$, and $\mathbf{C}$ all having k columns.

The regression from the latent variables α_Y towards Y is given by

$$Y = \sum_{i=1}^{k} \frac{\alpha_{Y,i}\alpha'_{Y,i}}{\alpha'_{Y,i}\alpha_{Y,i}} Y^i + Y^{k+1} = A_Y C' + E_Y, \qquad (5.28)$$

where $E_Y = Y^{k+1}$ is the part of Y that is not predicted by the first k PLS components (the misfit).

Thus, the entire PLS regression formula is given by

$$y_{\text{pred}} = \left[W_X (P'W_X)^{-1} C'\right]' x_{\text{pred}} = \left[C (W'_X P)^{-1} W'_X\right] x_{\text{pred}}.$$

Dual

Let us define A_X as the matrix containing $\alpha_{X,i}$ as its columns. Now we use the properties in Eqs. (5.26) and (5.23), showing that $W_X = X'A_X$ and $X^{k+1}W_X = 0$ leading to $W'_X P \propto W'_X X'A_Y = A'_X K_X A_Y$, where the proportionality is an equality up to a diagonal normalization matrix $A'_Y A_Y$ on the right-hand side. Furthermore, using Eq. (5.24), it is seen that $E'_Y A_Y = 0$ and thus (from Eq. (5.28)) that with the same diagonal normalization matrix as proportionality factor (which will thus be cancelled out), $C \propto CA'_Y A_Y = Y'A_Y$. This leads to the complete dual form of regression-PLS:

$$y_{\text{pred}} = \left[Y'A_Y (A'_X K_X A_Y)^{-1} A'_X X\right] x_{\text{pred}}.$$

Note that the entire algorithm only requires the evaluation of kernel functions, since Xx_{pred} also consists of inner products only (or equivalently kernel evaluations $k(\cdot,\cdot)$). Using this fact, the solution can be cast in the standard form of kernel-based pattern recognition algorithms:

$$y_{\text{pred}} = \sum_i \beta_i k(x_i, x_{\text{pred}}), \qquad (5.29)$$

where β_i are the columns of $\beta = Y'A_Y (A'_X K_X A_Y)^{-1} A'_X$.

5.5 Classification: Fisher Discriminant Analysis (FDA)

Definitions

We first define some symbols necessary to develop the theory. Since these quantities are defined in general for uncentered data, first this general definition is given. Afterwards, when appropriate the simplified formula will be provided for centered data. The latter formulas are the ones used in this section.

- Mean (n is the total number of samples $\mathbf{x}_i$)

$$\mathbf{m} = \frac{1}{n} \sum_i \mathbf{x}_i.$$

- Class mean ($\mathcal{S}_k$ is the set of samples belonging to cluster k, and $n_k = |\mathcal{S}_k|$, the number of samples in cluster k; thus $n = \sum_k n_k$)

$$\mathbf{m}_k = \frac{1}{n_k} \sum_{i:\mathbf{x}_i \in \mathcal{S}_k} \mathbf{x}_i.$$

- Total scatter matrix

$$\mathbf{S}_T = \sum_k \sum_{\mathbf{x}_i \in \mathcal{S}_k} (\mathbf{x}_i - \mathbf{m})(\mathbf{x}_i - \mathbf{m})'.$$

- Within-class k scatter matrix

$$\mathbf{S}_k = \sum_{\mathbf{x}_i \in \mathcal{S}_k} (\mathbf{x}_i - \mathbf{m}_k)(\mathbf{x}_i - \mathbf{m}_k)'.$$

- Within-class scatter matrix

$$\mathbf{S}_\mathrm{W} = \sum_k \mathbf{S}_k. \tag{5.30}$$

- Between-class scatter matrix

$$\mathbf{S}_\mathrm{B} = \sum_k n_k (\mathbf{m}_k - \mathbf{m})(\mathbf{m}_k - \mathbf{m})'.$$

For centered data (as we will assume in the remainder of this section), we get:

$$\mathbf{m} = \mathbf{0},$$

$$\frac{1}{n} \sum_k n_k \mathbf{m}_i = \mathbf{0},$$

$$\mathbf{S}_T = \sum_k \sum_{\mathbf{x}_i \in \mathcal{S}_k} \mathbf{x}_i \mathbf{x}_i' = \mathbf{X}'\mathbf{X} = \mathbf{S}_{\mathbf{XX}},$$

$$\mathbf{S}_\mathrm{B} = \sum_k n_k \mathbf{m}_k \mathbf{m}_k'.$$

Finally, the following properties hold:

- $\mathbf{S}_T = \mathbf{S}_\mathrm{B} + \mathbf{S}_\mathrm{W}$.
- When the number of classes is 2, they can be indexed as $+$ and $-$, and:

$$\mathbf{S}_\mathrm{B} = \frac{n_+ n_-}{n} (\mathbf{m}_+ - \mathbf{m}_-)(\mathbf{m}_+ - \mathbf{m}_-)'. \tag{5.31}$$

5.5.1 Cost Function

Fisher discriminant analysis (FDA) [10] is designed for discrimination between two classes, indexed by $+$ and $-$. It finds the direction $\mathbf{w}$ along which the between-class variance divided by within-class variance is maximized:

$$\mathbf{w} = \operatorname{argmax}_\mathbf{w} \frac{\mathbf{w}'\mathbf{S}_B\mathbf{w}}{\mathbf{w}'\mathbf{S}_W\mathbf{w}}. \tag{5.32}$$

Note that when $\mathbf{w}$ is a solution, $c\mathbf{w}$ with c a real number is a solution too. In fact, we are not interested in the norm of $\mathbf{w}$, but only in the direction it is pointing at. Thus, equivalently, we could optimize the constrained optimization problem

$$\mathbf{w} = \operatorname{argmax}_\mathbf{w} \mathbf{w}'\mathbf{S}_B\mathbf{w} \tag{5.33}$$
$$\text{s.t. } \mathbf{w}'\mathbf{S}_W\mathbf{w} = 1.$$

5.5.2 Primal

This optimization problem can be solved by differentiating the Lagrangian $\mathcal{L}(\mathbf{w},\mu) = \mathbf{w}'\mathbf{S}_B\mathbf{w} - \mu\mathbf{w}'\mathbf{S}_W\mathbf{w}$ with respect to $\mathbf{w}$ and equating to zero:

$$\nabla_\mathbf{w}\mathcal{L}(\mathbf{w},\mu) = \mathbf{0}$$
$$\Rightarrow \mathbf{S}_B\mathbf{w} = \mu\mathbf{S}_W\mathbf{w}. \tag{5.34}$$

This is again a generalized eigenvalue problem, with both $\mathbf{S}_B$ and $\mathbf{S}_W$ symmetric and positive semidefinite. We are interested in the dominant eigenvector.

Another way to get the same result is by maximizing the correlation between the data projected on a weight vector $\mathbf{w}$ with the labels $\mathbf{y}$ (for each sample being 1 or -1, depending on the class the sample belongs to) of the corresponding data points. This is in fact CCA, applied on the data vectors on the one hand, and the labels on the other hand:

$$\begin{pmatrix} \mathbf{0} & \mathbf{S}_{\mathbf{X}\mathbf{y}} \\ \mathbf{S}_{\mathbf{y}\mathbf{X}} & \mathbf{0} \end{pmatrix} \begin{pmatrix} \mathbf{w}_\mathbf{X} \\ w_\mathbf{y} \end{pmatrix} = \lambda \begin{pmatrix} \mathbf{S}_{\mathbf{X}\mathbf{X}} & \mathbf{0} \\ \mathbf{0} & \mathbf{S}_{\mathbf{y}\mathbf{y}} \end{pmatrix} \begin{pmatrix} \mathbf{w}_\mathbf{X} \\ w_\mathbf{y} \end{pmatrix},$$

from which $\mathbf{w}_\mathbf{X}$ can be solved as

$$\mathbf{S}_{\mathbf{X}\mathbf{X}}^{-1}\mathbf{S}_{\mathbf{X}\mathbf{y}}\mathbf{S}_{\mathbf{y}\mathbf{y}}^{-1}\mathbf{S}_{\mathbf{y}\mathbf{X}}\mathbf{w}_\mathbf{X} = \lambda^2\mathbf{w}_\mathbf{X}.$$

To see that $\mathbf{w}_\mathbf{X} = \mathbf{w}$, note that for centered data $\mathbf{X}$ (so $\mathbf{m}$ is made equal to $\mathbf{0}$ by centering), $\mathbf{S}_{\mathbf{X}\mathbf{X}} = \mathbf{S}_T = \mathbf{S}_B + \mathbf{S}_W$, $\mathbf{S}_{\mathbf{y}\mathbf{y}} = n$ is a scalar, and $\mathbf{S}_{\mathbf{X}\mathbf{y}} = \mathbf{X}'\mathbf{y} = n_+\mathbf{m}_+ - n_-\mathbf{m}_-$. One can then show that $\mathbf{S}_{\mathbf{X}\mathbf{y}}\mathbf{S}_{\mathbf{y}\mathbf{y}}^{-1}\mathbf{S}_{\mathbf{y}\mathbf{X}} = \frac{4n_+n_-}{n^2}\mathbf{S}_B$, and thus

$$\frac{4n_+n_-}{n^2}\mathbf{S}_B\mathbf{w}_\mathbf{X} = \lambda^2(\mathbf{S}_B + \mathbf{S}_W)\mathbf{w}_\mathbf{X}$$

$$\Rightarrow \mathbf{S}_B\mathbf{w}_\mathbf{X} = \frac{\lambda^2}{\frac{4n_+n_-}{n^2} - \lambda^2}\mathbf{S}_W\mathbf{w}_\mathbf{X}.$$

This is exactly the Fisher discriminant generalized eigenvalue problem, with $\mu = \frac{\lambda^2}{\frac{4n_+ n_-}{n^2} - \lambda^2}$ and $\mathbf{w} = \mathbf{w_X}$.

5.5.3 Dual

Define $\mathbf{y}_+$ as $(\mathbf{y}_+)_i = \delta_{y_i, 1}$ and $\mathbf{y}_-$ as $(\mathbf{y}_-)_i = \delta_{y_i, -1}$ (where we use the Dirac delta $\delta_{i,j}$, which is equal to 1 if $i = j$ and to 0 if $i \neq j$). The dual can again be derived by using $\mathbf{w} = \mathbf{X}'\boldsymbol{\alpha}$:

$$\mathbf{S}_B \mathbf{w} = \mu \mathbf{S}_W \mathbf{w}$$
$$\Downarrow \qquad \text{Eqs. (5.30), (5.31)}$$
$$\frac{n_+ n_-}{n} \mathbf{X}(\mathbf{m}_+ - \mathbf{m}_-)(\mathbf{m}_+ - \mathbf{m}_-)' \mathbf{X}' \boldsymbol{\alpha}$$
$$= \mu \mathbf{X} \sum_{k=+,-} \sum_{\mathbf{x}_i \in \mathcal{S}_k} (\mathbf{x}_i - \mathbf{m}_k)(\mathbf{x}_i - \mathbf{m}_k)' \mathbf{X}' \boldsymbol{\alpha}$$
$$\Downarrow$$
$$\frac{n_+ n_-}{n} \mathbf{K_X} \left(\frac{\mathbf{y}_+}{n_+} - \frac{\mathbf{y}_-}{n_-} \right) \left(\frac{\mathbf{y}_+}{n_+} - \frac{\mathbf{y}_-}{n_-} \right)' \mathbf{K_X} \boldsymbol{\alpha}$$
$$= \mu \mathbf{K_X} \left(\mathbf{I} - \frac{1}{n_+} \mathbf{y}_+ \mathbf{y}'_+ - \frac{1}{n_-} \mathbf{y}_- \mathbf{y}'_- \right) \mathbf{K_X} \boldsymbol{\alpha}$$
$$\Downarrow$$
$$\mathbf{M}\boldsymbol{\alpha} = \mu \mathbf{N}\boldsymbol{\alpha},$$

where we substituted $\mathbf{M} = \frac{n_+ n_-}{n} \mathbf{K_X} \left(\frac{\mathbf{y}_+}{n_+} - \frac{\mathbf{y}_-}{n_-} \right) \left(\frac{\mathbf{y}_+}{n_+} - \frac{\mathbf{y}_-}{n_-} \right) \mathbf{K'_X}$, and $\mathbf{N} = \mathbf{K_X} \left(\mathbf{I} - \frac{1}{n_+} \mathbf{y}_+ \mathbf{y}'_+ - \frac{1}{n_-} \mathbf{y}_- \mathbf{y}'_- \right) \mathbf{K_X}$.

For centered data as is assumed here, the projection of a test point $\mathbf{x}_{\text{test}}$ onto the FDA direction corresponding to $\boldsymbol{\alpha}$ can again be computed as

$$\sum_{i=1}^{n} \alpha_i k(\mathbf{x}_i, \mathbf{x}_{\text{test}}).$$

5.5.4 Multiple Discriminant Analysis (MDA)

While Fisher discriminant analysis is originally designed for the two-class problem, optimization of the very same cost function (Eqs. (5.32) and (5.33)) leading to the same generalized eigenvalue problem in Eq. (5.34) can be used for solving the multiclass problem (e.g. [9]). In that case, a few generalized eigenvector may be necessary to do the classification (typically the number of clusters minus one).

The intuition behind this is to maximize the total between-class covariance for a certain amount of within-class covariance. This amounts to maximizing the signal-to-noise ratio present in the projections of the samples onto the discriminant directions. Here, the distance between the projected clusters is the signal one is interested in, and the variance in the projections of the clusters is the noise. Interestingly, it has been shown that PLS also maximizes the between-class covariance when computed on a class indicator matrix $\mathbf{Y}$, however, this is done without considering the within-class covariance [4, 20]. Deriving the dual version of MDA can be done in a similar way as for FDA.

5.6 Spectral Methods for Clustering

Clustering is a standard problem in pattern recognition: identify groups of samples that supposedly belong to the same class, without any information on the class labels (unsupervised). The problem is often solved with classical algorithms of which the K-means algorithm is the best known. Most of these algorithms are designed for data with Gaussian class distributions. In many cases, however, this is an oversimplification. Furthermore, many well-known algorithms are based on a nonconvex optimization problem.

Therefore in recent years a significant amount of research has been carried out in the field of *spectral clustering* (SC) [2, 5, 8, 17, 18, 22, 26, 32]. The clustering problem is relaxed or restated, leading to efficient algorithms with a simple eigenvalue problem at the core. Furthermore, in general no Gaussianity assumptions are made.

Spectral clustering algorithms generally consist of three components: the computation of a suitable *affinity matrix*, expressing the similarities between the samples; an *eigenvalue problem* based on this affinity matrix, returning (eigen)vectors that reflect the cluster structure in the data; and a final step performing the *actual clustering*, based on these eigenvectors. In the next three subsections we will briefly go into each of these aspects.

5.6.1 The Radial Basis Function as the Kernel

Whereas standard clustering methods assume Gaussian class distributions (or make similar assumptions on the distribution), spectral clustering methods intend not to do this. In order to achieve this goal, the use of the Euclidian inner product as a similarity measure between the samples is avoided. Instead, the kernel trick can be used to implicitly compute an inner product between feature maps of the samples. More specifically, in spectral clustering algorithms, most often the radial basis kernel function (*RBF kernel*) is used as similarity measure:

$$k(\mathbf{x}_i, \mathbf{x}_j) = \exp\left(-\frac{\|\mathbf{x}_i - \mathbf{x}_j\|^2}{2\sigma^2}\right).$$

Note that for $\|\mathbf{x}_i - \mathbf{x}_j\| \ll \sigma$, the RBF kernel is $k(\mathbf{x}_i, \mathbf{x}_j) \simeq 1 - \frac{\|\mathbf{x}_i - \mathbf{x}_j\|^2}{2\sigma^2}$. Thus, locally, the RBF kernel is related to the Euclidian metric. On the other hand, for two points at a farther Euclidian distance from each other (that is, $\|\mathbf{x}_i - \mathbf{x}_j\| \gg \sigma$), we have that $k(\mathbf{x}_i, \mathbf{x}_j) \simeq 0$. The result is that the algorithm will not see if a group of points with a 'diameter' considerably larger than σ is Gaussianly distributed or not. Only for samples that are relatively close to each other, it will give an indication of how close exactly they are. This is desirable: it allows us to cluster samples that are stretched out in a nonlinear shape.

Even though, in spectral clustering methods, very often an RBF kernel is used, it is important to know that the similarity measure does not have to be positive definite; however, for most spectral clustering variants (such as the ones described in Sects. 5.6.2 and 5.6.2), it has to be *nonnegative* (which is indeed true for the RBF kernel). Because of the absence of the positive definiteness requirement, the matrix containing the similarities between the samples is usually called the *affinity matrix* in this context, instead of the *kernel matrix*. Besides the RBF kernel matrix, other affinity matrices are used in literature, such as the k-nearest neighbor affinity matrix. However, for uniformity in this chapter, here we will continue to use the term kernel matrix instead of affinity matrix, and denote it by **K**.

As opposed to the techniques discussed in the previous sections, *in spectral clustering, usually the kernel/affinity matrix is* not *centered*. In case it is centered, we will denote this explicitly, here, by using $\mathbf{K}_c$.

5.6.2 Which Eigenvectors?

We will only give a brief overview of the methods available in the literature. All of them compute the eigenvectors of a (generalized) eigenproblem involving **K**. We will outline two methods that represent a relaxation of a discrete optimization problem on a graph, and another method based on the alignment between two matrices. Every method described is derived for the two-cluster case. However, they appear to be extendible towards multicluster problems, by taking more than one eigenvector (often $k-1$ when there are k clusters).

Normalized Cut Cost

Shi and Malik [26] start from graph theoretic concepts. They relax the problem of finding the minimal *normalized cut cost (NCut)* of the graph, where nodes of the graph correspond to samples and the (positive) kernel entries are the weights (*affinities*) of the edges in between the nodes. Intuitively, an NCut is the total affinity between the clusters, normalized by the total affinity of each cluster with the entire sample. Mathematically, this is

$$\mathrm{NCut}(\mathbf{K},\mathbf{y}) = \frac{\sum_{i,j:y_i=-y_j=1} K_{ij}}{\sum_{i:y_i=1}\sum_j K_{ij}} + \frac{\sum_{i,j:y_i=-y_j=-1} K_{ij}}{\sum_{i:y_i=-1}\sum_j K_{ij}}.$$

Thus, one looks for a label assignment $y_i \in \{1,-1\}$ such that $\mathrm{NCut}(\mathbf{K},\mathbf{y})$ is minimized.

This problem can be proven to be equivalent to minimizing $\frac{\tilde{\mathbf{y}}'(\mathbf{D}-\mathbf{K})\tilde{\mathbf{y}}}{\tilde{\mathbf{y}}'\mathbf{D}\tilde{\mathbf{y}}}$ subject to $\tilde{y}_i \in \{1, -\tilde{y}\}$, and $\tilde{\mathbf{y}}'\mathbf{D}\mathbf{1} = 0$, for some $\tilde{y}$ and for $\mathbf{D} = \text{diag}(\mathbf{K}\mathbf{1})$. When the discrete vector $\tilde{\mathbf{y}}$ is replaced by a continuous vector α_i, so the problem is relaxed, an *approximation* for the unrelaxed problem solution can be found by solving the generalized eigenvalue equation:

$$(\mathbf{D} - \mathbf{K})\alpha = \lambda \mathbf{D}\alpha \quad \text{s.t.} \quad \alpha'\mathbf{D}\mathbf{1} = 0,$$

where one is interested in the vector α corresponding to the smallest eigenvalue λ while satisfying the constraint. One can show, however, that the constraint is satisfied for all of the generalized eigenvectors except for the one with smallest eigenvalue $\lambda = 0$ with corresponding generalized eigenvector $\alpha = \mathbf{1}$. Thus, one searches for the eigenvector with the smallest nonzero eigenvalue.

Average Cut Cost

Another approach discussed in [26] is based on a relaxation of the minimum *average cut cost (ACut)* problem. The ACut cost is the sum of the (positive) kernel entries corresponding to pairs of points belonging to different classes, normalized by the number of samples in both classes:

$$\text{ACut}(\mathbf{K}, \mathbf{y}) = \frac{\sum_{i,j:y_i=-y_j=1} K_{ij}}{\sum_{i:y_i=1} 1} + \frac{\sum_{i,j:y_i=-y_j=-1} K_{ij}}{\sum_{i:y_i=-1} 1},$$

where again $y_i \in \{1, -1\}$. This is similar to the NCut problem, and gives rise to a similar eigenvalue problem to be solved after relaxation:

$$(\mathbf{D} - \mathbf{K})\alpha = \lambda \alpha.$$

The eigenvector α corresponding to the smallest nonzero eigenvalue will reflect the cluster structure of the data.

Alignment-Based Approach

The alignment-based method (proposed in [8]) is a relaxation of the problem to find a label assignment that maximizes the alignment between the label matrix and the centered kernel matrix $\mathbf{K}_c$:

$$\max_{\mathbf{y}} \ \mathbf{y}'\mathbf{K}_c\mathbf{y} \quad \text{s.t.} \quad y_i \in \{1, -1\}.$$

Since this problem would be combinatoric again, it is relaxed by replacing the discrete vector $\mathbf{y}$ with a continuous vector α

$$\max_{\alpha} \ \alpha'\mathbf{K}_c\alpha \quad \text{s.t.} \quad \|\alpha\| = n$$

for n samples. This corresponds to solving the eigenvalue problem:

$$\mathbf{K}_c\alpha = \lambda\alpha.$$

Here the *dominant* eigenvector contains the relaxed labels as its entries.

5.6.3 What to do With the Eigenvectors?

We have now discussed how to compute eigenvectors that reflect the clustering in some way. There are different methods to extract the final clustering from these eigenvectors. In general, one constructs a matrix $\mathbf{A} = (\alpha_1 \alpha_2 \cdots \alpha_k)$ containing the eigenvectors as its columns. Then some traditional distance-based clustering is performed on the rows of $\mathbf{A}$ in this k-dimensional space, sometimes after normalizing all rows of $\mathbf{A}$ to unit length. For further reading on different possible approaches we refer to the literature, see e.g. [18, 22, 36].

5.7 Summary

Table 5.1 contains the cost functions optimized for most of the algorithms described in this chapter. Tables 5.2 and 5.3 give the primal and the dual eigenproblems to be solved in order to optimize these cost functions. These tables contain columns $\mathbf{M}$, $\mathbf{N}$, and $\mathbf{v}$, each indicating which matrices and eigenvector to use in the generalized eigenproblem of the form $\mathbf{Mv} = \lambda \mathbf{Nv}$.

Given this, we still need to know how to project test data on the directions found by solving these generalized eigenproblems. This is summarized as:

- projection of a test sample onto weight vector in primal space $\mathbf{w}$: $\mathbf{w}'\mathbf{x}_{\text{test}}$.
- projection of a test sample onto weight vector in feature space corresponding to the dual vector α: $\sum_{i=1}^{n} \alpha_i k(\mathbf{x}_i, \mathbf{x}_{\text{test}})$.

5.8 Conclusions

Among the algorithms discussed in this chapter, there are a number of classic methods from multivariate statistics, such as PCA and CCA; some methods that are virtually unknown in that field but are hugely popular in specific application domains, such as PLS; and finally some methods that are typically the product of the machine learning community, such as the clustering methods presented here, and all the extensions based on the use of kernels. Despite coming from so many different fields, the algorithms clearly display their common features, and we have emphasized them by casting them in a common notation and with a common language. From those comparisons, and from the comparison with the family of kernel methods based on quadratic programming, it is clear that this approach based on spectral methods can be considered another major branch of the KM family. The duality that emerges here from SVD approaches naturally matches the duality derived by the Kuhn–Tucker Lagrangian theory developed for those methods, and the statistical study demonstrates similar properties as shown in [27] and [28].

Some properties of this class of algorithms are already extremely appealing to machine learning practitioners, while others still need research attention.

Table 5.1. Cost functions optimized by the different methods

	Maximize variance	$\frac{\mathbf{w'S_{XX}w}}{\mathbf{w'w}}$
PCA		$\mathbf{w'S_{XX}w}$ s.t. $\|\mathbf{w}\|^2 = 1$
	Minimize residuals	$\|(\mathbf{I} - \mathbf{ww'})\mathbf{X}\|_F^2$
	Maximize correlation	$\frac{\mathbf{w'_X S_{XY} w_Y}}{\sqrt{\mathbf{w'_X S_{XX} w_X}}\sqrt{\mathbf{w'_Y S_{YY} w_Y}}}$
CCA	Maximize fit	$\mathbf{w'_X S_{XY} w_Y}$ s.t. $\|\mathbf{Xw_X}\|^2 = \|\mathbf{Yw_Y}\|^2 = 1$
	Minimize misfit	$\|\mathbf{w'_X X} - \mathbf{w'_Y Y}\|^2$ s.t. $\|\mathbf{Xw_X}\|^2 = \|\mathbf{Yw_Y}\|^2 = 1$
	Maximize covariance	$\frac{\mathbf{w'_X S_{XY} w_Y}}{\sqrt{\mathbf{w'_X w_X}}\sqrt{\mathbf{w'_Y w_Y}}}$
PLS	Maximize fit	$\mathbf{w'_X S_{XY} w_Y}$ s.t. $\|\mathbf{w_X}\|^2 = \|\mathbf{w_Y}\|^2 = 1$
	Minimize misfit	$\|\mathbf{w'_X X} - \mathbf{w'_Y Y}\|^2$ s.t. $\|\mathbf{w_X}\|^2 = \|\mathbf{w_Y}\|^2 = 1$
	Maximize between-class to	$\frac{\mathbf{w'S_B w}}{\mathbf{w'S_W w}}$
FDA	within-class covariance	$\mathbf{w'S_B w}$ s.t. $\mathbf{w'S_W w}$
SC1	Normalized cut cost	$\frac{\sum_{i,j:y_i=-y_j=1} K_{ij}}{\sum_{i:y_i=1} \sum_j K_{ij}} + \frac{\sum_{i,j:y_i=-y_j=-1} K_{ij}}{\sum_{i:y_i=-1} \sum_j K_{ij}}$
SC2	Average cut cost	$\frac{\sum_{i,j:y_i=-y_j=1} K_{ij}}{\sum_{i:y_i=1} 1} + \frac{\sum_{i,j:y_i=-y_j=-1} K_{ij}}{\sum_{i:y_i=-1} 1}$
SC3	Alignment	$\mathbf{K}_c$

Table 5.2. Primal forms (not for spectral clustering algorithms)

	$\mathbf{M}$	$\mathbf{N}$	$\mathbf{v}$
PCA	$\mathbf{S_{XX}}$	$\mathbf{I}$	$\mathbf{w}$
RCCA	$\begin{pmatrix} 0 & \mathbf{S_{XY}} \\ \mathbf{S_{YX}} & 0 \end{pmatrix}$	$\begin{pmatrix} \mathbf{S_{XX}}+\gamma\mathbf{I} & 0 \\ 0 & \mathbf{S_{YY}}+\gamma\mathbf{I} \end{pmatrix}$	$\begin{pmatrix} \mathbf{w_X} \\ \mathbf{w_Y} \end{pmatrix}$
PLS	$\begin{pmatrix} 0 & \mathbf{S_{XY}} \\ \mathbf{S_{YX}} & 0 \end{pmatrix}$	$\begin{pmatrix} \mathbf{I} & 0 \\ 0 & \mathbf{I} \end{pmatrix}$	$\begin{pmatrix} \mathbf{w_X} \\ \mathbf{w_Y} \end{pmatrix}$
FDA	$\mathbf{S}_B$	$\mathbf{S}_W$	$\mathbf{w}$

PLS, for example, is designed precisely to operate with input data that are high-dimensional and present highly correlated features, exactly the situation created by the use of kernel functions. The match between the two concepts is perfect, and in a way PLS can be better suited to the use of kernels than maximal-margin methodologies. Furthermore it is easily extendible towards multivariate regression. On the other hand, one of the major properties of support vector machines is not naturally present in eigenalgorithms: sparseness. Deliberate design choices can be made in order to enforce it, but the optimal way to include sparseness in this class of methods still remains an

Table 5.3. Dual forms

	M	N	v
PCA	$\mathbf{K}$	$\mathbf{I}$	α
RCCA	$\begin{pmatrix} \mathbf{0} & \mathbf{K_X K_Y} \\ \mathbf{K_Y K_X} & \mathbf{0} \end{pmatrix}$	$\begin{pmatrix} \mathbf{K_X}^2 + \gamma \mathbf{K_X} & \mathbf{0} \\ \mathbf{0} & \mathbf{K_Y}^2 + \gamma \mathbf{K_Y} \end{pmatrix}$	$\begin{pmatrix} \alpha_X \\ \alpha_Y \end{pmatrix}$
PLS	$\begin{pmatrix} \mathbf{0} & \mathbf{K_X K_Y} \\ \mathbf{K_Y K_X} & \mathbf{0} \end{pmatrix}$	$\begin{pmatrix} \mathbf{I} & \mathbf{0} \\ \mathbf{0} & \mathbf{I} \end{pmatrix}$	$\begin{pmatrix} \alpha_X \\ \alpha_Y \end{pmatrix}$
FDA	$\frac{n_+ n_-}{n} \mathbf{K_X} \left(\frac{\mathbf{y_+}}{n_+} - \frac{\mathbf{y_-}}{n_-} \right) \cdot \left(\frac{\mathbf{y_+}}{n_+} - \frac{\mathbf{y_-}}{n_-} \right)' \mathbf{K_X}$	$\mathbf{K_X} \left(\mathbf{I} - \frac{\mathbf{y_+ y_+'}}{n_+} - \frac{\mathbf{y_- y_-'}}{n_-} \right) \mathbf{K_X}$	α
SC1	$\mathbf{D} - \mathbf{K}$	$\mathbf{D}$	α
SC2	$\mathbf{D} - \mathbf{K}$	$\mathbf{I}$	α
SC3	$\mathbf{K}_c$	$\mathbf{I}$	α

open question. Another important point of research is the stability and statistical convergence of general eigenproblems for finite sample sizes. For work on the stability of the spectrum of Gram matrices, we refer to [24] and [25].

The synthesis offered by this unified view has immediate practical consequences, allowing for unified statistical analysis and for unified implementation strategies.

Acknowledgements

Tijl De Bie is a research assistant with the Fund for Scientific Research Flanders (F.W.O.–Vlaanderen). Furthermore, his research is supported by: the Research Council KUL: GOA-Mefisto-666, GOA-Ambiorics; the FWO: G.0240.99 (multilinear algebra), G.0407.02 (support vector machines); the Belgian Federal Government: Belgian Federal Science Policy Office, IUAP V-22 (Dynamical Systems and Control: Computation, Identification, Modelling, 2002-2006). Roman Rosipal's research was supported by funding from the NASA CICT/ITSR/NeMC and IS/HCC programs.

References

1. S. Akaho. A kernel method for canonical correlation analysis. In: *Proceedings of the International Meeting of the Psychometric Society (IMPS2001)*. Springer, Berlin Heidelberg New York, Osaka, July 2001
2. Y. Azar, A. Fiat, A. Karlin, F. McSherry, and J. Saia. Spectral analysis of data. In: *Proceedings of The 42nd Annual Symposium on Foundations of Computer Science (FOCS2001)*, Las Vegas, October 2001
3. F. R. Bach and M. I. Jordan. Kernel independent component analysis. *Journal of Machine Learning Research*, 3:1–48, 2002
4. M. Barker and W.S. Rayens. Partial least squares for discrimination. *Journal of Chemometrics*, 17:166–173, 2003
5. Y. Bengio, P. Vincent, and J.F. Paiement. Learning eigenfunctions of similarity: Linking spectral clustering and kernel PCA. Technical Report 1232, Département d'informatique et recherche opérationnelle, Université de Montréal, 2003
6. M. Borga, T. Landelius, and H. Knutsson. A Unified Approach to PCA, PLS, MLR and CCA. Report LiTH-ISY-R-1992, ISY, SE-581 83 Linköping, Sweden, November 1997
7. N. Cristianini and J. Shawe-Taylor. *An Introduction to Support Vector Machines*. Cambridge University Press, Cambridge, 2000
8. N. Cristianini, J. Shawe-Taylor, and J. Kandola. Spectral kernel methods for clustering. In: T. G. Dietterich, S. Becker, and Z. Ghahramani (eds.), *Advances in Neural Information Processing Systems 14*. MIT Press, Cambridge, MA, 2002
9. R. O. Duda, P. E. Hart, and D. G. Stork. *Pattern Classification,* 2nd edn, Wiley, New York, 2001
10. R. A. Fisher. The use of multiple measurements in taxonomic problems. *Annals of Eugenics*, 7, Part II:179–188, 1936
11. C. Fyfe and P. L. Lai. ICA using kernel canonical correlation analysis. In: *International workshop on Independent Component Analysis and Blind Signal Separation (ICA2000)*, Helsinki, June 2000
12. R. A. Horn and C. R. Johnson. *Matrix Analysis*. Cambridge University Press, Cambridge, 1985
13. R. A. Horn and C. R. Johnson. *Topics in Matrix Analysis*. Cambridge University Press, Cambridge, 1991
14. A. Höskuldsson. Pls regression methods. *Journal of Chemometrics*, 2:211–228, 1988
15. H. Hotelling. Relations between two sets of variables. *Biometrika*, 28:321–377, 1936
16. I. T. Jolliffe. *Principal Component Analysis*. Springer, Berlin Heidelberg New York, 1986
17. R. Kannan, S. Vempala, and A. Vetta. On clusterings: good, bad and spectral. In: *Proc. of the 41st Foundations of Computer Science (FOCS2000)*, Redondo Beach, November 2000
18. A. Ng, M. I. Jordan, and Y. Weiss. On spectral clustering: Analysis and an algorithm. In: T. G. Dietterich, S. Becker, and Z. Ghahramani (eds.), *Advances in Neural Information Processing Systems 14*. MIT Press, Cambridge, MA, 2002
19. R. Rosipal and L. J. Trejo. Kernel partial least squares regression in reproducing kernel hilbert space. *Journal of Machine Learning Research*, 2:97–123, 2001

20. R. Rosipal, L.J. Trejo, and B. Matthews. Kernel PLS-SVC for linear and nonlinear classification. In: *Proceedings of the Twentieth International Conference on Machine Learning (ICML-2003)*, pp. 640–647, Washington DC, 2003
21. B. Schölkopf and A. Smola. *Learning with Kernels*. MIT Press, Cambridge, MA, 2002
22. G. Scott and H. Longuet-Higgins. An algorithm for associating the features of two patterns. In: *Proceedings of the Royal Society London B*, 224:21–26, 1991
23. J. Shawe-Taylor and N. Cristianini. *Kernel methods for Pattern Analysis*. Cambridge University Press, Cambridge, 2004
24. J. Shawe-Taylor, N. Cristianini, and J. Kandola. On the concentration of spectral properties. In T. G. Dietterich, S. Becker, and Z. Ghahramani (eds.), *Advances in Neural Information Processing Systems 14*. MIT Press, Cambridge, MA, 2002
25. J. Shawe-Taylor, C. Williams, N. Cristianini, and J. S. Kandola. On the eigenspectrum of the gram matrix and its relationship to the operator eigenspectrum. In: *Proceedings of the 13th International Conference on Algorithmic Learning Theory (ALT2002)*, pp. 23–40, 2002
26. J. Shi and J. Malik. Normalized cuts and image segmentation. *IEEE Transactions on Pattern Analysis and Machine Intelligence*, 22(8):888–905, 2000
27. J. A. K. Suykens, T. Van Gestel, J. De Brabanter, B. De Moor, and J. Vandewalle. *Least Squares Support Vector Machines*. World Scientific, Singapore, 2002
28. J. A. K. Suykens, T. Van Gestel, J. Vandewalle, and B. De Moor. A support vector machine formulation to PCA analysis and its kernel version. *IEEE Transactions on Neural Networks*, 14(2):447–450, 2003
29. V. N. Vapnik. *The Nature of Statistical Learning Theory*, 2nd edn., Springer, Berlin Heidelberg New York, 1999
30. J.-P. Vert and M. Kanehisa. Graph-driven features extraction from microarray data using diffusion kernels and kernel CCA. In: S. Becker, S. Thrun, and K. Obermayer (eds.), *Advances in Neural Information Processing Systems 15*. MIT Press, Cambridge, MA, 2003
31. A. Vinokourov, N. Cristianini, and J. Shawe-Taylor. Inferring a semantic representation of text via cross-language correlation analysis. In: T. G. Dietterich, S. Becker, and Z. Ghahramani (eds.) *Advances in Neural Information Processing Systems 14*. MIT Press, Cambridge, MA, 2002
32. Y. Weiss. Segmentation using eigenvectors: A unifying view. In: *Proceedings of the 7th International Conference on Computer Vision (ICCV1999)*, pp. 975–982, Kerkyra, September 1999
33. H. Wold. Path models with latent variables: The NIPALS approach. In: H.M. Blalock et al. (eds.), *Quantitative Sociology: International Perspectives on Mathematical and Statistical Model Building*. pp. 307–357. Academic, NY, 1975
34. H. Wold. Partial least squares. In S. Kotz and N.L. Johnson (eds.), *Encyclopedia of the Statistical Sciences*, vol. 6. pp. 581–591. Wiley, New York, 1985
35. Y. Yamanishi, J.-P. Vert, A. Nakaya, and M. Kanehisa. Extraction of correlated gene clusters from multiple genomic data by generalized kernel canonical correlation analysis. *Bioinformatics*, 19:323i–330i, 2003
36. H. Zha, C. Ding, M. Gu, X. He, and H. Simon. Spectral relaxation for k-means clustering. In T. G. Dietterich, S. Becker, and Z. Ghahramani (eds.), *Advances in Neural Information Processing Systems 14*. MIT Press, Cambridge, MA, 2002

Part III

Image Processing

6

Geometric Framework for Image Processing

Jan J. Koenderink

Universiteit Utrecht, Department of Physics and Astronomy
Princetonplein 5, 3508TA Utrecht,The Netherlands
j.j.koenderink@phys.uu.nl

6.1 Geometrical Aspects of Image Processing

"Image processing" is ill defined, but is perhaps best understood as a discipline defined by the totality of operations people apply to images that yield again images as their result. This conveniently distinguishes image processing from "image analysis", "image interpretation", "object recognition" and so forth. Images are thus understood only in a structural, syntactical, but not semantic sense. Structural analysis is a sine qua non for semantic analysis or interpretation.

6.1.1 What are "Images"?

There exist two major definitions of images in the literature. In the pragmatist's view images are discrete data structures. A rectangular matrix of nonnegative numbers (typically "bytes", i.e., in the range $[0, 255]$) is the prototype. Such images tend to be squarish, and typical sizes are 32×32 (an "icon") to 4096×4096 (presently a very "large" image). In the more sophisticated view images are fields over some base space. For instance, everything visible from a point is a radiance (in the simplest case a nonnegative radiant power per area and per solid angle) seen in some direction (base space $\mathbb{S}^2$). The field could be more complicated (we might include spectral density and state of polarization, for instance) but typically is a scalar physical quantity with a dimension that is incompatible with the dimensions of the base space. A common definition of an image is as a cross section $I(x,y)$ of $\mathbb{E}^2 \times \mathbb{R}^+$, the nonnegative reals over the Euclidian plane. Here $\mathbb{E}^2$ is conventionally known as "the image plane", and I as "the intensity".

The former definition makes sense in many settings, but I will not consider it image processing proper ("pixel pushing" is a term one sometimes hears). My reason is that "images" in common understanding have nothing to do with discrete data structures. *One* image may be viewed on a CRT (at 72 ppi, the "pixels" having byte values) or on paper from a laser printer (at 600 dpi, the "pixels" being black (pigment dot) or white (paper base) as the case may be), and these representations [41] may be considered "the same". Pixels are, or at least should be, irrelevant. It is like arithmetic on a computer: The user should be oblivious of the actual representation of numbers in the machine. Likewise, *image processing is the domain where the pixels are irrelevant.*

The latter definition at least does not mention pixels at all. Although this is as it should be, there are other severe problems with it. One is that the physical dimensions of the base space and the intensity domain are incommensurable. This means that the group of Euclidian similarities cannot possibly be the group of similarities of image space $\mathbb{I}$. Thus the use of Euclidian differential invariants in image processing must be considered nonsensical, albeit very common in the literature. Another problem is that the image in this definition has infinite resolution and thus cannot be known. Anything real (I mean knowable by man) has a finite number of degrees of freedom. Thus the definition assumes a God's Eye perspective. We need an operational definition: The intensity at a given point of the base space can be sampled at a certain (arbitrary) resolution, and only such samples are real. The resolution may be as high as you like, but it necessarily has to be finite. The resolution (I will speak of the "inner scale" of the image) is an essentially arbitrary choice on the part of the image processor, arbitrary in the sense of not given by nature. This has numerous important consequences, although this goes often unrecognized. In the literature one often considers the resolution as given by the image (then pixel pushing turns into a vice), or one uses a finite resolution because forced to (in order to "regularize" the "ill-posed" computations [39]) but tries to get at the "real" (at infinite resolution) structure. A common example is the differentiation of images. It is not so much that images are nondifferentiable functions, it is that they are not functions at all. Only derivatives at a certain scale make sense (and then differentiability is not an issue), the scale being arbitrary, that is to say, your choice. At any finite resolution the image (as an *ideal*, i.e., before being sampled and turned into something real) *is* a differentiable function by design (see below).

6.1.2 What is "Image Processing"?

In image processing one operates on images to produce other images. Some operations are so extremely common that they often escape notice. For instance, when you print an image seen on the screen, when you view an image on different screens, and so forth, you perform image processing whether you know it or not. Sometimes you perform operations on the image that somehow do not seem to change it. For instance, when you adjust the position of an image on the CRT screen or adjust the brightness of the display you merely "translate" the image in $\mathbb{I}$. Likewise, when you print an image at a different scale or use "gamma transformations", such as different grades of photo paper, "contrast" control of your display, you merely "scale" the image in $\mathbb{I}$, and so forth. When you use "edge burning" in the darkroom [1] you apparently "rotate" the image in $\mathbb{I}$, and so forth. In all those cases it seems that only "congruences" or "similarities" (both to be defined formally) are concerned.

Cases of "true" image processing are also common in the visual arts just think of solarization, psychedelic coloring, and so forth. In scientific work one might produce "edginess" images, "zero-crossingness" images and the like.

6.1.3 What Has Geometry To Do With It?

Structural analysis of multiply extended continua is synonymous with geometry. The image space $\mathbb{I}$ is intuitively a homogeneous space because we can easily imagine the "free movement" of configurations in the space. In order to understand the structure of $\mathbb{I}$ we have to formally define its groups of congruences and similarities. An exhaustive study of global and local invariants then goes a long way towards understanding image structure. Of course, this is a chore of pure geometry.

Global Geometry

Certain global entities are immediately implicated as invariants because they are invariant under manipulations of the size, position, rotation, brightness and gamma controls of your display unit. The images $I(x,y) = I_0 \exp(\alpha x + \beta y)$ are examples. They are "intensity gradient" images, and fiddling with the controls merely changes their steepness, brightness and direction. Other examples are images of the kind $I(x,y) = I_0 \exp[(x-a)(y-b)]$. These are remarkable in many respects because of their behavior not only under fiddling with the aforementioned knobs of the display, but also the *focus* control: These images are also invariant under changes of inner scale (see below).

I will show that the gradients are to be considered "planes" in the geometry of $\mathbb{I}$, whereas the latter type of images are "Clifford planes", that is, surfaces that contain two mutually transverse pencils of metrically parallel (to be explained below) lines. This is typical for the geometry of $\mathbb{I}$; in $\mathbb{E}^3$ such surfaces would have to be generic planes.

Geometry of the "Deep Structure"

With "deep structure" I mean the structure as a function of resolution. This is an important problem in practice because many structural elements of images only occur at certain limited ranges of inner scales. This is intuitively clear when you consider a "powers of ten" type of scenario [24], where you see a distant scene, zoom in to see a forest, a treetop, a leaf cluster on a branch, a leaf, and so forth. Sometimes you "don't see the forest for the trees"; in order to see something you have to decide on the relevant inner scale. As you change your focus (inner scale) the image structure changes and the structure of these changes is itself a matter of great interest. Its study is the topic of "deep structure", another intrinsically geometric topic.

Human observers are uncannily apt at focussing on the right scales, although this is a moot point as we obviously have an incomplete idea of what we miss. Increasing resolution is generally considered beneficial (witness the popularity of microscopes and telescopes), whereas *decreasing* resolution is mainly practiced by visual artists in order to better see the composition of a scene. Indeed, a minifying glass is just as revealing as a magnifying glass. The image processing literature has only scratched the surface of the topic of "selecting the right scale" [23].

Local Geometry

Any point of an image is an *edge* point (or "corner" point, or ..., you name it) when defined as the finite activity of edge operators, although some points carry more "edginess" than others. Thus the very epithet "edge detector" is void: *At any generic point the image is anything you want it to be* (i.e., you run an operator for). Thus feature detectors have no proper place in image processing. Features are semantic entities (interpretations), not syntactic ones. Here, (in contradistinction to the bulk of the literature), I talk of "local operators" instead of feature detectors. The "meaning" of the output of a local operator is not intrinsic to it, but is dependent upon the context and the task at hand.

Multilocal Geometry, "Outer Scale"

In "multilocal" geometry it is convenient to define an "outer scale" to complement the notion of "inner scale". The inner scale is the resolution and is intuitively thought of as the "size of the points". Many algorithms imply a "region of interest" in the sense of the smallest region such that only points within the region of interest enter the computation. The size of that region of interest is the outer scale. Here is an example: We think of the points as circular disks in a densest, thus hexagonal, packing. In order to estimate the Laplacian of the intensity we would use a region of interest composed of seven points, the point at the fiducial location and its six nearest neighbors. We estimate the Laplacian as proportional with the difference of the intensity at the fiducial point and the average intensity over the region. Thus the outer scale would be three times the inner scale. Now suppose that we had a second-order jet of differential operators at any point, then we could perform the computation at a single point! Thus one may trade the complexity of the structure of a point (the order of the jet of differential operators) with the extent of the outer scale. For a base space $\mathbb{E}^2$ the nth-order jet has $(n+1)(n+2)/2$ degrees of freedom, thus the diameter of the region in the purely multilocal (zeroth-order jet) representation is asymptotically roughly the order times the inner scale.

Fig. 6.1. A point (*left*) and another point (*center*). At *right* is a locally disordered region that implements two points at the same location in base space but different intentities ("parallel points", see below)

There are many other useful forms of multilocal geometry, of which we single out one other example [12,13]. Consider a circular region centered at a given point. Let the diameter of the region (the outer scale) significantly exceed the inner scale. Then the area contains many independent samples of intensity. Suppose we ignore their spatial order, then we are left with the local intensity histogram. When we repeat this for all points we have a histogram-valued image. Since histograms are invariant against permutation of the samples all these histogram valued images agree up to local scrambling, which is why I refer to them as "locally disorderly". Such structures have many important applications. Since the histograms are density distributions in the intensity domain, such images do not correspond to surfaces $z(x, y)$, but to three-dimensionally extended "clouds" $P(x, y, z)$ in image space. See Fig. 6.1 for the simplest possible and Fig. 6.2 for a more interesting example.

The histogram itself is, of course, also a "deep structure", since there exist a continuous family of histograms at different bin widths. In changing the bin width all the aforementioned constraints that should apply to a proper deep structure apply. What is especially interesting about histogram-valued images is that a single location can represent more than one value of the intensity (because a "point" may contain many "pixels"). Thus it may make sense to say that the image is both white and black at a certain location, and a strongly bimodal histogram indeed suggests such an abuse of language. This is similar to human amblyopic or eccentric vision in which observers can easily distinguish between average gray surfaces and black and white striped surfaces, but are at a loss to make out whether the stripes are horizontal or vertical (or whether the surface is checkered, rather than striped, etc.).

With this kind of structure we arrive at a representation of the image that is a number of intricately interlocked deep structures, often called "scale spaces". The "scales" involved are of various nature: the classical *inner scale*, that is, the rock-bottom spatial resolution; the *bin width*, that is, the resolution in the intensity domain; and the *outer scale* of the disorderly representation, which is the inner scale of the histogram-valued image. In this representation

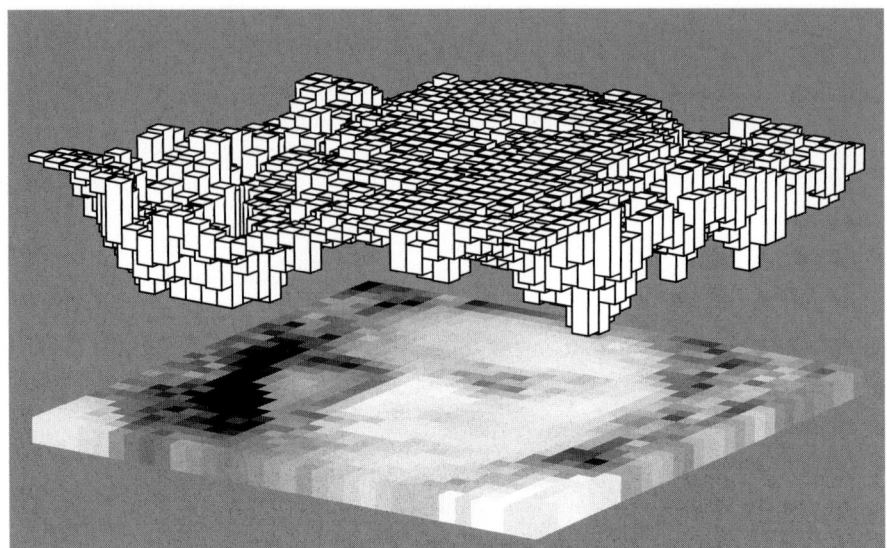

Fig. 6.2. A histogram-valued image. The total range of the histogram is represented by means of cuboids whose tops and bottoms are located at the local maximum and minimum, respectively, of the intensity

the extremes are "the" image with a definite intensity at every point (then the histogram degenerates into a single bin!) and "the" histogram of the whole image (then the "image" has been shrunk to a single point!). In general one has an image with a certain inner scale (the ultimate spatial resolution), a certain outer scale (the size of the region over which local histograms are collected) and a certain bin width (which sets the inner scale, or resolution, for the local histograms). In the general theory all these entities are described within a *single, unified formalism*.

6.2 Theory

Many threads of theory are already well established, but an overall framework is still sadly lacking. Here I will not spend much discussion on known topics, I will merely highlight the key principles (from a fundamental perspective, that is, but the literature often focusses on details that might well be ignored at a first, global investigation) and indicate how the various threads hang together. A few topics are much less well known, and I will devote proportionally more discussion on them.

6.2.1 Image Space

Image space $\mathbb{I}$ can hardly be studied without specifying the inner scale in advance. In this section I assume the inner scale has been settled. Then the images are differentiable (see below) cross sections of the "intensity" dimension over the base space (*by construction*).

The Base Space

The nature of the base space is not too important here. Just for ease of exposition I will consider two instances, the Euclidian line and the Euclidian plane. Both are very important in practice. The base space $\mathbb{E}^1$ occurs with linear CCD arrays in scanners and many industrial imaging applications, as well as in spectroscopy, and so forth. The base space $\mathbb{E}^2$ is, of course the prototypical example (although an "image" typically covers only a rectangular region $\mathbb{I}^2 \subset \mathbb{E}^2$).

The Intensity Domain

The "intensity domain" can be of very various nature. For this chapter I only consider a scalar, nonnegative quality; thus the intensity domain is $\mathbb{R}^+$, at least if one abstracts from the physical dimension. However, that the intensity domain is incommensurable with the base space is *crucial* [27]. The literature is rife with nonsensical instances due to neglecting this fact.

Can we find a "natural" representation of the intensity domain? Indeed we can. Such a representation should (at the very least!) be invariant against changes of the physical units. For a nonnegative entity this implies that one should use a logarithmic representation, $z = \log(I/I_0)$, and consider the z-dimension to be an *affine line*. Notice that z is dimensionless because I/I_0 is. A change of units would imply $I' = \lambda I$, thus $z' = \log(I'/I_0') = z$. A change of the reference intensity I_0 to I_0' would imply $z' = \log(I/I_0') = \log(I/I_0 \times I_0/I_0') = z + a$, which is a mere shift of origin. Moreover, one would like to admit of a uniform prior probability density in the absence of all prior knowledge. Suppose we sample from a two-parameter distribution $P(\mathrm{d}x|\mu,\sigma) = h((x-\mu)/\sigma)\,\mathrm{d}x/\sigma$ and are given a sample $\{x_1,\ldots,x_n\}$ (for μ a location, σ a scale parameter), and we have to estimate μ and σ. What is the prior $f(\mu,\sigma)\,\mathrm{d}\mu\,\mathrm{d}\sigma$ that expresses total ignorance? Clearly complete ignorance implies invariance with respect to arbitrary scalings and shifts. Consider the priors f, g assigned by two observers. You have $f(\mu,\sigma) = a\,g(\mu,\sigma)$ (a a scaling factor), whereas objectivity implies that $f = g$. The general solution is $f(\mu,\sigma) = (const)/\sigma$, a result derived by Jaynes [18] that is known in statistics as "Jeffreys Rule" [19]. This latter finding fits in very well with our earlier considerations based on the physics. Notice that a gamma transformation merely scales the z-representation, for $z' = [\log(I/I_0)]' = \log(I^\gamma/I_0^\gamma) = \gamma z$, and thus conserves constant probability densities. When the z-dimension is treated as the affine line $\mathbb{A}^1$ (no point singled out as an "origin", and no "unit" distance) we have solved the problem of representation. In this chapter I will use z for the "log-intensity". (Notice that we now consider points on the full line $\mathbb{A}^1$ instead of the halfline $\mathbb{R}^+$.)

6.2.2 Scale Space

An image at different scales is like an atlas [29]: When you page through the atlas you see the (same) city at various scales. On a world map the city may fail to be indicated at all, on a finer scale it may be indicated by a conventional sign (a circle, say) without internal detail, on a still finer map you may see the structure indicated in a "generalized" manner, on a still finer scale the generalization becomes less severe and the structure splits up into substructures, and on the finest scale you may be able to see individual blocks, parks and major throughways, and, of course, we can image this process to continue indefinitely. A few general properties are immediately evident. For instance, everything that is on a coarse map can be traced to a finer map but not *vice versa* (see Fig. 6.3). In order to make a coarse map; you need not even refer to the finest map, you can start at any finer one.

A totally general "scale space" [40] for images should be shift, scale and rotation invariant (no place, size or orientation being "special"), it should be build by a linear process (nonlinearity implies specialization too), and the operation of generalization (called "blurring" here) should yield a semigroup. Thus one arrives at convolution with a rotationally symmetric kernel that reproduces itself under convolution. Clearly the kernel should be nonnegative throughout. It is tricky to formalize the notion that "blurring should only destroy, not generate detail". The best way is to insist that the height of maxima should decrease, while that of minima should increase under blurring. This guarantees that extrema are not generated by blurring for linear images, though not for images with a higher dimensional base space. Pairs of critical points [26] can be generated by blurring, though they tend to live a short life (blurring even more kills them). It can be shown that the unique kernel with the desired properties is the Gaussian kernel [9,14].

Since the Gaussian kernel is the Green's function of the diffusion equation, the blurring can be described as a diffusion process $\Delta z(x, y; t) = z_t$, where t denotes the scale parameter. This neatly illustrates the semigroup property. Notice that (with $t_2 > t_1$) one has $z(x, y, t_1) - z(x, y; t_2) = \int_{t_1}^{t_2} \Delta z(x, y; t)\,dt$, in particular $z(x, y, t) = \int_t^\infty \Delta z(x, y; t)\,dt + z(x, y; \infty)$. Thus all the information (except for $z(x, y; \infty)$, which is typically a constant) is already contained in the Laplacian, see Fig. 6.4). The Laplacian of an image can be thought of as an "infinitesimally thin slice" of scale space. This is of interest because (as I will show below) there is no intrinsic (local) information in the orders below the second, as both the level and the gradient can be changed by congruences of image space at any given point.

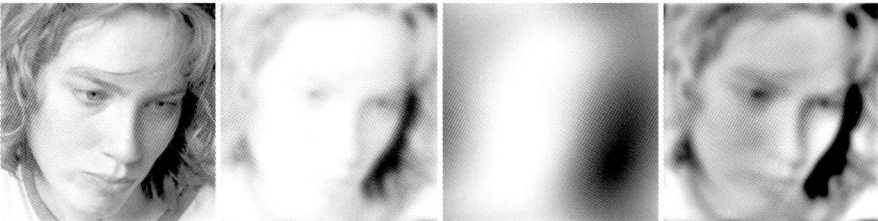

Fig. 6.3. An image (*left*) and two representations at two different inner scales (second and thirdly from left). The scales differ by a factor of 8. Notice that you can find corresponding locations in these images, but only one way (the "atlas principle"). On the *right* a three-octave scale space slice between the inner scales of the examples

One may link points at different scales by linking two points at infinitesimal close scales when they agree in their log-intensity and are as close together (in the base space) as possible. This generates a field of directions whose integral curves form the desired linking structure. It can be shown that this structure has singularities [6] such that structures collapse at some critical scale the linkage containing branching points. This is what we saw in the atlas example, when the scale becomes too coarse the structure of the city collapses to a point and has to be indicated with a conventional sign. For instance, look for the eyes in the third figure from the left in Fig. 6.3: There are none! Yet there are on the finer scale. They simply collapsed at some scale.

6.2.3 Image Tangent Spaces

The "tangent space" of the "surface" $\{x, y, z(x,y)\}$ (I take the example $\mathbb{I} = \mathbb{E}^2 \times \mathbb{A}^1$) at the point $\{x_0, y_0, z_0\}$ is spanned by the tangent vectors in the coordinate directions, that is to say, $\{1, 0, z_x(x_0, y_0)\}$ and $\{0, 1, z_y(x_0, y_0)\}$. Thus we have a natural interest in derivatives of $z(x,y)$. When we are interested in higher order properties (like curvature) we have to consider higher order derivatives too. In general we need to consider the "nth-order jet" at any point.

A tangent vector like $\{1, 0, z_x\}$ is a "first-order directional derivative" (in the x-direction). It can be understood geometrically as the "velocity vector" of the equivalence class of all arc-length parameterized curves that are tangent to $\{1, 0, z_x\}$ at the fiducial point (as I will show later x is the arc-length). Alternatively, it can be understood as a "bilocal" operator as follows: A "point" is a local operator that acts upon a scalar field to produce a scalar, the value of the field at the point. A point operator is conceived of as an aperture (slit of a monochromator, photosensitive area of a CCD array, etc.) that collects a "flux" in order to produce a sample. The term point is very apt because there is no way to differentiate spatially between localities inside the aperture. In Euclid's terms "*a point is that which has no parts*". Notice that this in no way constrains the *size* of the point. Points come in all conceivable sizes. The size is a choice of you. A bilocal operator can be constructed from two point operators separated at some finite distance. We wire them up such that the pair yields a single scalar that is the difference of the point samples divided by their separation. When the separation is very small its actual value turns out to be irrelevant. In the limit we obtain a bilocal operator no larger than the point itself, and we may as well call it a "first-order directional derivative operator", or a "structured point". It is best to think of such operators as little machines whose internal structures are hidden from us. They take a bite from the image and spew out a number, which is the value of the directional derivative of the image at that location in that direction.

Since these operators implement directional derivatives *exactly* [33], they span a linear space (the tangent space at the point). Thus arbitrary linear combinations of outputs of such operators are equivalent to the output of some single operator. In $\mathbb{I}^3$ a basis of just two operators suffices to mimic operators in arbitrary directions. The literature has picked up on that with the notion of "steerable filters" [10]. However, this notion is far more limited in its conceptual scope, it is essentially little more than an engineer's trick to compute arbitrary directions cheaply. The geometrical notion of the tangent space at a point is far richer in its implications. It indicates an *exact* isomorphism between machine implementation and abstract differential geometry.

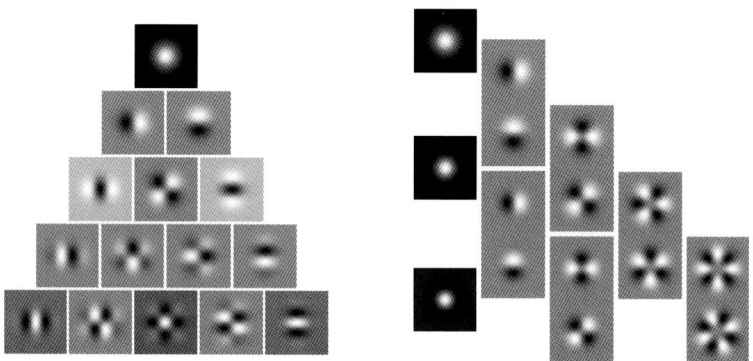

Fig. 6.4. Two different representations of the fourth-order jet. On the *left* is a Cartesian representation, on the *right* is a polar representation. The polar representation can be obtained by linear combination of the Cartesian representation of that order and *vice versa*. Many different representations are possible and sometimes convenient. All representations span the same space. The purely directional operators (first items of the rows in the Cartesian representation) also span the space if one selects a sufficient (order plus one) number of orientations. Then the number of distinct types equals the order of the jet. Such (nearly) "Gabor filters" are popular in many circles [7], though for irrelevant reasons

It works exactly similarly for the higher order differential operators. The nth-order jet is a bunch of such machines, all stationed at the same location, all with the same resolution (the inner scale), only with various orders and directions (Fig. 6.4). For the nth-order jet we need $(n+1)(n+2)/2$ independent machines. In a typical image processing application one limits oneself to $n = 2$ and thus has a six degrees of freedom representation of the image at any location, roughly a "2×2 to 3×3 pixels" approximation to the image structure at that location, that is pretty coarse. However, there is no obvious reason not to include higher orders. The endemic fear of differentiation in the image processing community derives from an abortive understanding of what differentiation and inner scale mean. When you represent an image through local jets you may subsample it because each "point" is structured and represents not a single but roughly $n^2/2$ degrees of freedom. (Namely $\sum_0^n (i+1) = \frac{1}{2}(n+1)(n+2)$ degrees of freedom, where $(i+1)$ is the number of terms of order i in a Taylor development of the intensity at a point.)

6.2.4 Definition of Geometrical Loci

Geometrical loci are commonly defined through constraints, e.g., the constraint $x = 0$ defines the y-axis of $\mathbb{E}^2$ fitted out with a Cartesian frame. Another way to define loci is the ostensive one: to draw a line, etc. Clearly, the ostensive manner will not work in image processing (although often done!) because it means the *ex machina* introduction of an additional image instead of processing the given image. There are problems with the former method too though, for $x = 0$ would define an entity known at infinite resolution, which clearly cannot be. In fact, the constraint $x = 0$ itself is only ideal. Instead, one might specify $z(x,y) = \exp{-(x^2/2\sigma^2)}$ as an alternative. The log-intensity of this "image" is almost everywhere near zero, except near the points $x = 0$ (the original constraint!), though with a finite width given by the inner scale (of the constraint) σ. Such a definition has the additional advantage that the constraint itself is an image. The major point is that the unrealistic infinite resolution has been circumvented. The image may be considered as a "soft" or "blurry" constraint, and its log-intensity specifies the degree (on a zero-to-one scale) to which the constraint is satisfied.

Many entities of interest in image processing are points, curves or areas. At finite resolution these must appear as blobs, ribbons and regions with blurry boundaries. This means that their nature as submanifolds also changes. A "curve" may become really indistinguishable from a blurry region, although there will also be clearcut cases, where the curve is a ribbon of about the width of the inner scale. In this representations *there are no points, curves or areas*, instead, they are all *images* of some kind. This is similar to the problem of "features". There are no feature detectors in image processing proper and there are no "curve finders" (or what have you) either. Features and geometrical loci are the domain of image interpretation. In the simplest case they are found through thresholding followed by skeletonization. This is a nonlinear form of image processing, and thus is admissable up to the point of interpretation, which involves idiosyncratic selection procedures, etc.

When we consider submanifolds like surfaces, curves or points in $\mathbb{I}$ we discuss *ideal* geometrical entities that are often useful in the analysis (like real numbers are). They do not immediately correspond to anything real though. Only images are real.

6.2.5 Definition of "Isophotes"

A prototypical example of a geometrical locus is the "isophote" [4] for some fiducial log-intensity of, say, $z(x,y) = z_0$. The appropriate soft constraint is $\exp[-(z(x,y) - z_0)^2 / 2\Delta z_0^2]$. Here the new parameter Δz_0 is a resolution in the log-intensity domain. This resolution must be sharply distinguished from the inner scale (which is the resolution in the base space). One may naturally conceive of Δz as the "bin width" of the log-intensity histogram of the image.

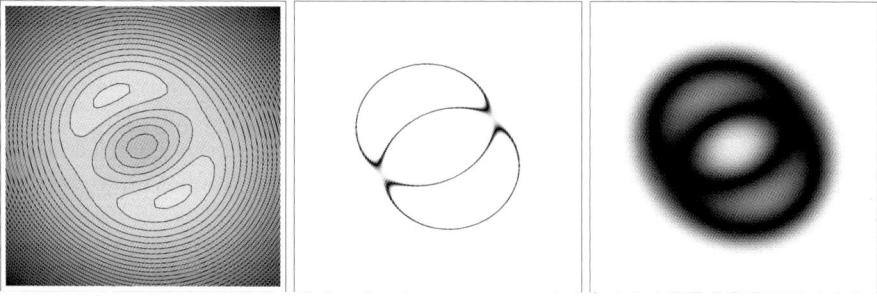

Fig. 6.5. An image and two "isophote images". The images are for the same level, but have been computed for different histogram bin width

These isophotes are "blurry curves" with a width that is inversely proportional to the width of the surface (Fig. 6.5). They actually look much more reasonable than the "isophotes of infinite precision", which often are broken up in many disconnected components and are almost fractallike. Even when the surface is as steep as possible the width of the blurry isophotes is finite, because it is limited by the inner scale.

In this representation the resolution of the log-intensity is a vital parameter. It links the log-intensity representation to more general "disorderly" or "histogram-valued" representations in which the log-intensity resolution is understood as a (local) histogram "bin width".

In any real application there exist physical and/or technical limits to the inner and outer scales. For instance, the spatial inner scale is limited by either the size of the photosensitive elements (pixel limited devices) or by the physical optics of the imaging optics (diffraction-limited devices). The outer scale is limited by the size of the sensor (CCD chip, photographic plate) or by the useful field of view of the imaging optics. The inner scale involved in the log-intensity dimension is typically limited by the photon shot noise, the thermal noise of the photoelectric sensors or by the discretization (bit depth), while the outer scale is limited by saturation effects in the sensor or arbitrary upper limits set by the encoding. This is similar to the representation of the (ideal) natural numbers in computer languages through a finite set of bit patterns. In this contribution I completely abstract from such (very real) limits, and I concentrate upon the ideal case. Although the ideal does not exist, I believe that any real system should be gauged against the ideal and against the fundamental physical limits. If one is satisfied with the mere description of actual systems one has moved from science to engineering. In that case no general theory is possible in principle.

6.3 Image Processing

In this section I begin with the theory of image processing proper. The first thing to do is to identify the group of motions. Because image space is a *homogeneous* space it "looks the same" from any of its points, the group of motions defines the geometry. In fact, we are bound to arrive at one of the classical Cayley–Klein geometries [3,20–22]. That $\mathbb{I}$ is indeed homogeneous is a matter of definition if you want: that is, in the absence of any prior knowledge we have to assume that the space is the same everywhere, that no location, direction or log-intensity level is in any way singled out. But then the same configuration should be realizable everywhere, in short, the space should admit the free movement of rigid configurations.

6.3.1 Operations in the Intensity Domain and in the Base Space

In this section I restrict the discussion to either $\mathbb{I}^2 = \mathbb{E}^1 \times \mathbb{A}^1$ ("linear images") or $\mathbb{I}^2 = \mathbb{E}^2 \times \mathbb{A}^1$ ("planar images"), although one could easily generalize to $\mathbb{I} = \mathbb{E}^n \times \mathbb{A}^1$. Extensions to cinematic images are less immediate (though possible) because the time dimension is not commensurable with either the dimensions of the base space nor the intensity dimension. Extensions to color images, etc., are possible but not immediate.

The base space is either the Euclidian line or plane, thus the movements are translations and rotations. Similarities have a single modulus because all dimensions have to be scaled equally due to the freedom of rotation.

The log-intensity dimension is the affine line, and the movements are simply translations. Scalings are possible too and are similarities with a modulus unrelated to the modulus of scalings in the base space.

Movements in $\mathbb{I}$ involve simultaneously the base space and the log-intensity domain. They are constrained because–due to reasons of simple physics–no log-intensity difference may ever end up as a stretch in the base space. This means that the lines $\{x, y, \lambda z\}$, with λ variable, must be invariant under movements. We can set up such a geometry starting from the projective space $\mathbb{P}^3$

with coordinates $\{x_0, x_1, x_2, x_3\}$ such that $x_0 = 0$ denotes the plane at infinity, which we assume fixed. Since we need a fixed point for the group of similarities, we may assume the absolute conic to be of the form $x_1^2 + x_2^2 = 0$, this can be written $(x_1 + ix_2)(x_1 - ix_2) = 0$, and thus represents two complex lines $x_1 = \pm ix_2$ (first and second absolute lines) which intersect in the point $\mathcal{F} = \{0, 0, 0, 1\}$, the absolute point. General similarities conserve the absolute line pair, and consequently $\mathcal{F}$ is a fixed point. It represents a pencil of parallel real lines, exactly what we set out to construct.

Thus the group of movements conserves a pencil of parallel lines. This completely fixes the geometry, and we obtain the singly isotropic Cayley–Klein plane and the singly isotropic Cayley–Klein space. For $\mathbb{I}^2$ we obtain one out of the nine, and for $\mathbb{I}^3$ one out of the twenty seven possible Cayley–Klein geometries.

There exist two types of similarities that leave absolute lines invariant, we single out those that leave each one individually invariant. The similarities that interchange the absolute lines correspond to inversions, for which we have no use.

6.3.2 Global Operations

Consider the linear images first. The geometry is the same as that induced by the kinematical Galilean group [43] on the line (where time corresponds with log-intensity).

A generic similarity is [30]

$$x' = e^{\sigma_1} x + \tau_x,$$
$$z' = \rho x + e^{\sigma_2} z + \tau_z,$$

where we have proper movements (a typical one shown in Fig. 6.6) for $\sigma_{1,2} = 0$. Notice that it is a five-parameter group, whereas the group of Euclidian similarities is only a four-parameter group (rotation angle, scaling factor and two components of translation). The reason is that similarities of $\mathbb{I}^2$ have two (not one) moduli: Both the distances and the angles can be scaled (see Fig. 6.7), whereas in $\mathbb{E}^2$ only the distances can be scaled. While $\mathbb{E}^2$ has a parabolic distance measure and an elliptic angle measure, in $\mathbb{I}^2$ both measures are parabolic.

I will use the following terms quite frequently: A point in image space has a certain log-intensity and a certain *trace* in the base space (where "it is at"). Points with the same trace lie on a *normal line*. Planes that contain a normal line are *normal planes*. Lines and planes that are not normal are *generic*, though I will seldom use the term explicitly. In image processing one is only interested in generic lines and planes. Thus lines and planes have well-defined slopes in all cases. Although I will not make use of it here, there exists a full metrical duality betweeen the generic lines and the points of $\mathbb{I}^2$ and between the generic planes and the points of $\mathbb{I}^3$. Image space is far more elegant and

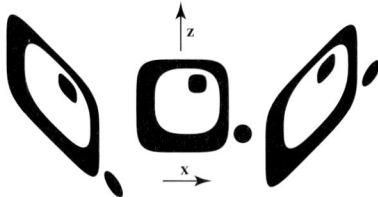

Fig. 6.6. Three phases of a rigid rotation in $\mathbb{I}^2$. Both distances and angles are conserved, though it may take some time to notice

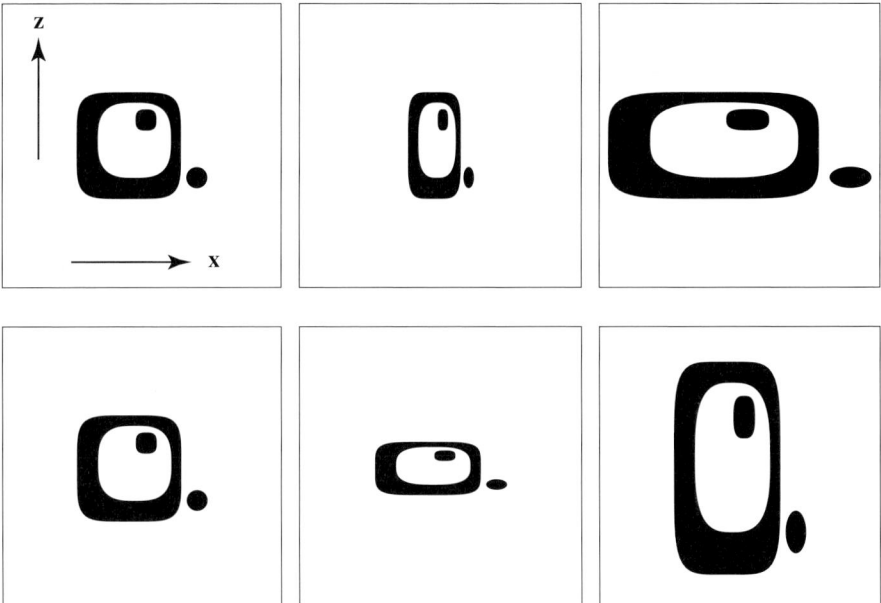

Fig. 6.7. Two types of similaries in $\mathbb{I}^2$, in the *top row* those "of the first-kind" which affect distances and in the *bottom row* those those "of the second-kind" which affect the angles

symmetric than Euclidian space in this respect, because both the distance and the angle metric are parabolic.

One easily checks that a motion leaves the length of the trace of the linear segment defined by two points invariant, thus the length of the trace is the "proper distance". In the case of $\mathbb{I}^2$ it is a signed distance. When the distance of two points vanishes they may still be distinct. If so, they lie on a normal line. Such points are called *parallel*. The log-intensity difference of parallel points is not changed by a motion. Thus parallel points (and only those) can be assigned a "special distance". "The" distance is defined as either the proper or the special distance. All pairs of points have a distance and this distance is invariant under arbitrary motions. When we write the metric of a plane as

$ds^2 = dx^2 + \mu dz^2$, we obtain the Euclidian plane for $\mu = +1$, the Minkowski plane for $\mu = -1$, and $\mathbb{I}^2$ for $\mu = 0$. Thus the metric of $\mathbb{I}^2$ appears as a limiting case of both Euclidean and Minkowski space. Indeed, both views are useful. As a limit of Minkowski space we see that the normal lines are degenerated light cones. Thus we have order on the normal lines, but points on different normal lines are "elsewhere" and their log-intensities cannot be compared. Indeed, it is easy to find motions that equate their log-intensities. As a limit of Euclidian space, we see that the "rotations" of $\mathbb{I}^2$ appear to our Euclidean eyes as *shears* about normal lines or planes (Fig. 6.6). Thus you cannot make a full turn and turn the image surface upside down. Just try to imagine what that would mean! If you actually *can* you should consider to stop reading on.

Notice that the motions leave the area element $dx \wedge dz$ invariant, thus it makes sense to speak of the area of regions in the normal planes. Physically the area represents a flux, the construction is dimensionally consistent since it involves no "mixture" of incommensurable quantities.

The geometry of $\mathbb{I}^2$ is perhaps most easily developed in terms of the dual numbers $\mathbb{D}$, which are a kind of imaginary numbers [5,42] of the form $a + \varepsilon b$ with nilpotent imaginary unit, thus $\varepsilon^2 = 0$. In terms of these numbers the motions are simply linear transformations. Consider the point $\{x, z\}$ which we represent as the dual number $w = x + \varepsilon z$. The linear transformation $w' = aw + b$ where $a = a_1 + \varepsilon a_2$ and $b = b_1 + \varepsilon b_2$ yields $w' = (a_1 x + b_1) + \varepsilon(a_2 x + a_1 z + b_2) = \{a_1 x + b_1, a_2 x + a_1 z + b_2\}$, i.e., for $a_1 = 1$ proper motions, otherwise similarities. All of the standard formulas from the algebra of complex numbers carry over (though the nilpotency of the imaginary unit often leads to surprises), so we immediately gain great power over the geometry of $\mathbb{I}^2$.

Instead of this *analytic* model of $\mathbb{I}^2$ we may construct a *geometric* model using multivector algebra [15] by introducing the orthogonal basis $\{\boldsymbol{e}_x, \boldsymbol{e}_z\}$ such that $\boldsymbol{e}_x \boldsymbol{e}_x = 1$, $\boldsymbol{e}_z \boldsymbol{e}_z = 0$. With $\omega = \boldsymbol{e}_x \wedge \boldsymbol{e}_z$ as the bivector (oriented area) and the unit scalar 1 we have the multiplication table

$$\begin{array}{llll} 1^2 = 1, & 1\boldsymbol{e}_x = \boldsymbol{e}_x, & 1\boldsymbol{e}_z = \boldsymbol{e}_z, & 1\omega = \omega, \\ \boldsymbol{e}_x 1 = \boldsymbol{e}_x, & \boldsymbol{e}_x^2 = 1, & \boldsymbol{e}_x \boldsymbol{e}_z = \omega, & \boldsymbol{e}_x \omega = \boldsymbol{e}_z, \\ \boldsymbol{e}_z 1 = \boldsymbol{e}_z, \, \boldsymbol{e}_z \boldsymbol{e}_x = -\omega, & \boldsymbol{e}_z^2 = 0, & \boldsymbol{e}_z \omega = 0, \\ \omega 1 = \omega, & \omega \boldsymbol{e}_x = -\boldsymbol{e}_z, & \omega \boldsymbol{e}_z = 0, & \omega^2 = 0. \end{array}$$

The zeros signal the presence of divisors of zero, which slightly complicates the algebra. Notice that $\boldsymbol{e}_x(x\boldsymbol{e}_x + z\boldsymbol{e}_z) = x + \omega z$, and because $\omega^2 = 0$ we regain the dual number representation. Thus the two models are fully isomorphic. In the case of planar images one should use the geometric model of course, but the analytic model is still applicable (and convenient) for the restriction to normal planes.

Notice that for an analytic real function of a real variable you may write the Taylor series $f(x+a) = f(x) + af'(x) + a^2/2! f''(x) + \ldots$, and on setting $a = \varepsilon dx$, $f(x + \varepsilon dx) = f(x) + \varepsilon dx f'(x)$ (exactly). Thus the dual numbers "implement the infinitesimal domain"; that is, they are a kind of non-standard reals [28]. As a consequence we have $\exp(\varepsilon x) = 1 + \varepsilon x$, $\sin(\varepsilon x) = \varepsilon x$, $\cos(\varepsilon x) = $

1. We can write the dual number $a + \varepsilon b$ in the polar representation $r \exp(\varepsilon \varphi)$, with $r = a$ (a signed quantity) and $\varphi = b/a$. This immediately reveals the angle measure as b/a. This is also intuitively obvious when you think of the circle $x^2 = a^2$. The arc subtended by the angle is b, and dividing by the radius a we arrive again at φ. Notice that the circle $x^2 = a^2$ expresses constant distance from the normal line $x = 0$, all of whose points are centers of the circle (thus any generic line passes through the center!). Notice also that equal arcs subtend equal angles as should be, and that motions conserve angular differences and turn all generic lines over the same angle. Finally, notice that the normal lines subtend infinite angle ("are normal to", hence the name) with *any* generic line.

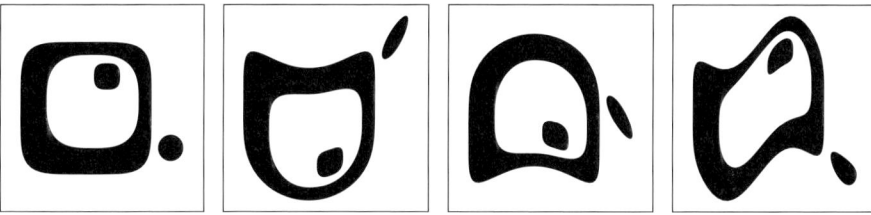

Fig. 6.8. On the *left* a gauge figure in $\mathbb{I}^2$. The *next two figures* show the effect of a nonlinear transformation, that is, the inversion in spheres of the second-kind with positive and negative radius, which are conformal transformations. The *rightmost* figure shows the effect of a conformal transformation involving an arbitrary function (in this case a sinewave). The normal direction is the vertical in these figures

The dual number plane as a model of $\mathbb{I}^2$ is convenient in many respects. For instance, one can easily develop the homographies parallel to the conventional (imaginary unit $i^2 = -1$) case [25] (Fig. 6.8), complex function theory [32], etc. Especially important for the image domain are the conformal mappings. Here image space is more flexible than the Euclidian case for one may also have nontrivial conformal maps in $\mathbb{I}^3$ (Fig. 6.8). The examples from Fig. 6.8 are locally rotations (no scaling involved), thus these are still rather special conformal mappings. In the case of images these correspond to multiplicative combination of images ("sandwiching negatives" in the photographic darkroom). Human observers cheerfully "read" such combinations, which perhaps finds a partial explanation in the fact that this involves a conformal mapping via an arbitrary additional function.

A circle like $(x-a)^2 = b^2$ is known as a "circle of the first-kind", the circles "of the second-kind" are parabolæ with normal lines as axes. The latter (like the former) can be rotated such as to shift along themselves. Such a circle $z(x) = \kappa(x-a)^2/2$ has a curvature κ (radius $1/\kappa$), which is a signed quantity. This curvature equals z_{xx}, which is the lowest order and simplest differential invariant of $\mathbb{I}^2$. The circle $z(x) = \kappa x^2/2$ is shifted over a distance d along itself by the transformation

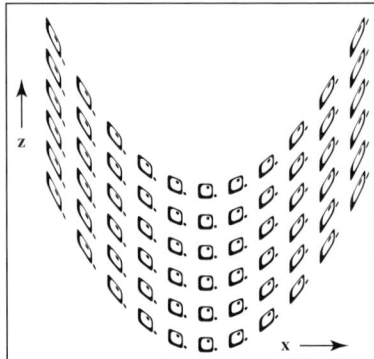

Fig. 6.9. A "parabolic limit rotation" in $\mathbb{I}^2$. This is the equivalent of a rigid rotation of $\mathbb{E}^2$. However, the "spokes" are the normal lines and the "hub" is their vanishing point at infinity. Notice that this is an isometry: All the gauge figures have the same shape. If it does not look that way to you then adjust your mental eye to the group of congruences of image space

$$x' = x + d,$$
$$z' = 2\kappa dx + z + \kappa d^2,$$

and so are all circles $z(x) = \kappa x^2/2 + z_0$ concentric with it (Fig. 6.9). The transformation is thus a *rigid rotation* of $\mathbb{I}^2$ where the normal lines appear as the "spokes" of the wheel with "hub" at infinity (at the vanishing point of the normal lines).

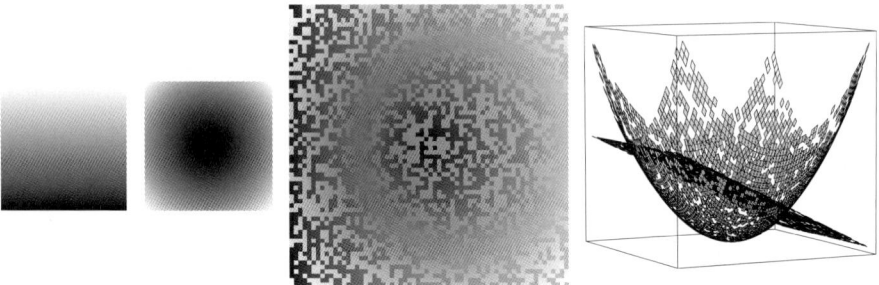

Fig. 6.10. A generic plane intersects a sphere of the second-kind. At *left* the images, at the *center* a disorderly representation, at *right* the geometry in $\mathbb{I}^3$

The case of geometry of $\mathbb{I}^3$ is not that different from that of $\mathbb{I}^2$, although richer because of the higher dimensionality. Notice that the geometry in the normal planes of $\mathbb{I}^3$ simply repeats the geometry of $\mathbb{I}^2$. Instead of circles we are mainly interested in spheres (Fig. 6.10), which (to our Euclidian eyes) appear as paraboloids with normal axes. The generic planes are the duals of the points.

6 Geometric Framework for Image Processing   189

Fig. 6.11. A screw movement in $\mathbb{I}^3$. This is a nonperiodic movement that corresponds to a Euclidian rotation in the trace. however, a translation is involved in the log-intensity direction

Fig. 6.12. A pure rotation in $\mathbb{I}^3$. This example shows a "rotation about a horizontal axis", which is actually a shear in the normal planes with vertical traces

Fig. 6.13. A parabolic limit rotation in $\mathbb{I}^3$. This is the equivalent of the transformation illustrated earlier in Fig. 6.9

The groups of similarities and motions for planar images are quite similar to that of linear images (it contains the latter, of course, when you restrict yourself to the normal planes), but it is richer because of the higher dimensionality. (Figs 6.11-6.15.) One obtains a seven-parameter group [31] (against the six-dimensional group of similarities of $\mathbb{E}^3$):

Fig. 6.14. A Clifford shift in $\mathbb{I}^3$. As the image translates towards the right it progressively "rotates about the horizontal". For the other type of Clifford shift this rotation would be in the opposite sense

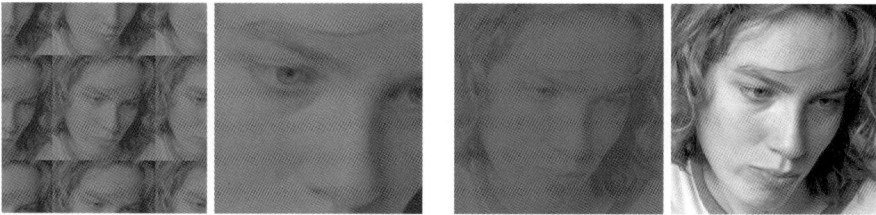

Fig. 6.15. Similarities of the first (*left*) and of the second-kind (*right*) in $\mathbb{I}^3$. The similarities of the first-kind are the familiar Euclidian ones. The similarities of the second-kind scale the isotropic angles and appear as contrast changes or "gamma transformations"

$$x' = e^{k_1}(x\cos\varphi - y\sin\varphi) + \tau_x,$$
$$y' = e^{k_1}(x\sin\varphi + y\cos\varphi) + \tau_y,$$
$$z' = \sigma_x x + \sigma_y y + e^{k_2} z + \tau_z.$$

When you consider only the trace, the groups appear as the group of similarities and motions (for $k_{1,2} = 0$) of $\mathbb{E}^2$, that is, the base space. There are two distinct motions that correspond with rotations in the base space though, one a screw motion with normal axis (which is not periodic) and one that transforms each one of a family of parallel generic planes in itself. To the identity in the base space there corresponds a translation into the normal direction ("brightness adjustment") and a shear that conserves a family of parallel normal planes (an "additive plane" adjustment). The latter is a pure rotation in $\mathbb{I}^3$ (Fig. 6.12). The case of translations in the base space is especially interesting. One has shifts in a generic direction, and a motion that is the equivalent of the parabolic rotation discussed for the linear images (Fig. 6.13), neither of them very surprising. The interesting transformations [35-38], are the so-called "left and right Clifford shifts" (Fig. 6.14). These transformations shift certain surfaces (Clifford planes) within themselves. The Clifford planes are denoted "planes" because they are covered with two mutually transverse families of *metrically* parallel generic lines (which would render them proper planes in Euclidian geometry). The Clifford planes are surfaces of the type $z(x,y) = xy + \mu x^2$. Notice that the blur-invariant images $z(x,y) = a + bx + cy + dxy + e(x^2 - y^2)$ (with $\Delta z(x,y) = 0$, hence $z_t = 0$, thus blur invariant) are a special type of Clifford planes.

Fig. 6.16. Example of a conformal transformation in $\mathbb{I}^3$ involving an arbitrary additional image. Notice that the composite image "reads" without problems

One has conformal mappings in $\mathbb{I}^3$ (Fig. 6.16) much like in $\mathbb{I}^2$, this is in contradistinction with the Euclidian case where the conformal mappings in the plane are much more various than those in space. For image processing purposes the conformal transformations that appear as identities in the trace and involve an arbitrary additional image are perhaps the most interesting.

Generic planes $z = ux + vy + w$ are defined to have "plane coordinates" $\{u, v, w\}$; indeed, the triple $\{u, v, w\}$ defines the plane uniquely. It can be shown that the plane coordinates transform just like the point coordinates [31] and that the description can naturally be extended to normal planes and the plane at infinity. Thus there exists a full metric duality between lines and planes in $\mathbb{I}^3$ (and, likewise, between points and lines in $\mathbb{I}^2$), something lacking in Euclidean space. Geometricaly one defines the plane $\{u, v, w\}$ dual to the point $\{x, y, z\}$ as the polar plane of the in-the-image plane reflected point $\{x, y, -z\}$ with respect to the sphere $z = \frac{1}{2}(x^2 + y^2)$, and vice versa. This correlation is equivalent to the algebraic definition [31]. The duality makes it trivial to interpret the point invariants in terms of planes and thus define the angle between planes. The result is what one expects *more geometrico*: The angle between two normal planes is the Euclidean angle between their traces or their normal distance of the traces (in case they are parallel), and the angle between two generic planes is the isotropic angle in the normal plane whose trace is perpendicular to the trace of the intersection of the planes, or the normal separation in the case of parallel planes. In a similar manner one defines invariants for a plane and a line, two lines, etc. This is often important because planes appear as local surface elements (tangent planes), that is to say, local (linear) approximations to images.

The subgroup of "unimodular limit movements"

$$x' = x + \tau_x,$$
$$y' = y + \tau_y,$$
$$z' = \sigma_x x + \sigma_y y + z + \tau_z,$$

is arguably very relevant to image processing. It is easily shown that each such movement can be obtained as the commutative product of a Clifford left shift

$$x' = x + \alpha,$$
$$y' = y + \beta,$$
$$z' = -\beta x + \alpha y + z + \gamma,$$

and a Clifford right shift

$$x' = x + A,$$
$$y' = y + B,$$
$$z' = Bx - Ay + z + C,$$

modulo a pure isotropic shift $x' = x$, $y' = y$, $z' = z + \gamma$. The group of unimodular limit movements is a normal quotient group of the group of movements, and the Clifford shifts are three-parameter simply transitive subgroups of it.

There exists an algebraic description that superficially resembles the dual number model, but is really in a different spirit. We write a point of $\mathbb{I}^3$, say $\boldsymbol{x} = \lambda\{1, x, y, z\} = \{x_0, x_1, x_2, x_3\}$ in homogeneous coordinates, as a certain hypercomplex number $x_0 + x_1\boldsymbol{i} + x_2\boldsymbol{j} + x_3\boldsymbol{k}$, with $\boldsymbol{i}^2 = \boldsymbol{j}^2 = \boldsymbol{k}^2 = 0$, $\boldsymbol{ij} = -\boldsymbol{ji} = \boldsymbol{k}$, $\boldsymbol{ik} = \boldsymbol{ki} = \boldsymbol{jk} = \boldsymbol{kj} = 0$. We conceive of $\mathbb{I}^3$ as extended with ideal elements $\{0, x_1, x_2, x_3\}$ "at infinity". Multiplication of these hypercomplex numbers is not commutative, and we have to reckon with divisors of zero. We have $[\boldsymbol{x}, \boldsymbol{y}] = \boldsymbol{xy} - \boldsymbol{yx} = 2(x_1 y_2 - x_2 y_1)\boldsymbol{k}$, that is, the oriented area in the trace. Only elements with vanishing trace or collinear traces commute. Multiplication is associative though. Defining conjugation through $\overline{\boldsymbol{x}} = x_0 - x_1\boldsymbol{i} - x_2\boldsymbol{j} - x_3\boldsymbol{k}$, we find that $\boldsymbol{x}\overline{\boldsymbol{x}} = \overline{\boldsymbol{x}}\boldsymbol{x} = x_0^2$ is real, thus we may define it as the norm $|\boldsymbol{x}|$. (Notice that this "norm" has nothing to do with the metric, in practice it will be zero for elements at infinity and one for generic points.) One has $\overline{\boldsymbol{xy}} = \overline{\boldsymbol{y}}\,\overline{\boldsymbol{x}}$ from which we obtain $|\boldsymbol{xy}| = |\boldsymbol{x}|\,|\boldsymbol{y}|$. The only elements of norm zero are the elements at infinity. For generic elements we may define the inverse as $\boldsymbol{x}^{-1} = \overline{\boldsymbol{x}}/|\boldsymbol{x}|$, for $\boldsymbol{x}\boldsymbol{x}^{-1} = 1$. Notice that the inverse is the additive inverse, just like the product is much like vector addition, except for an additional purely isotropic shift. When we have a pair of generic elements $|\boldsymbol{a}| \neq 0$, $|\boldsymbol{b}| \neq 0$ we have that $\boldsymbol{ab} = 0$ when and only when their corresponding points at infinity $\mathcal{A} = a_1\boldsymbol{i} + a_2\boldsymbol{j} + a_3\boldsymbol{k}$, $\mathcal{B} = b_1\boldsymbol{i} + b_2\boldsymbol{j} + b_3\boldsymbol{k}$ are collinear or coincident with the absolute point $\mathcal{F}$. In this algebraic system we may write the Clifford left shifts as $\boldsymbol{x}' = \boldsymbol{ax}$ and the Clifford right shifts as $\boldsymbol{x}' = \boldsymbol{xb}$ where $|\boldsymbol{a}| \neq 0$, $|\boldsymbol{b}| \neq 0$. Moreover, $\boldsymbol{c}^{-1}\boldsymbol{xc}$ generates a pure isotropic shift. Consequently, all unimodular movements can be cast in the form $\boldsymbol{x}' = \boldsymbol{a}\boldsymbol{x}\boldsymbol{b}$, and we obtain a coherent and very convenient algebraic representation.

Finally, the subgroup

$$x' = x, \quad y' = y,$$
$$z' = \sigma_x x + \sigma_y y + \gamma z + \tau,$$

of the similarities is the one that is perhaps most pervasive in image processing since in the trace (the image plane) it boils down to the identity. I will denote these important transformations $\Gamma_\gamma^{\sigma,\tau}$. It is made up of the subgroups of pure isotropic rotations $z' = \sigma_x x + \sigma_y y$, the subgroup of pure isotropic shifts $z' = z + \tau$ and the subgroup of similarities of the second-kind $z' = \gamma z$ (where $\gamma > 0$ for regular transformations, $\gamma < 0$ inducing an inversion). These are the well known *gradients* (or additive planes, or edge burnings), the *intensity adjustments* (or lightenings and darkenings) and *contrast adjustments* (or gamma transformations, or scale adjustments via paper grades, etc.). The group fails to be communitative if contrast changes are involved for $\Gamma_{\gamma_2}^{\sigma_2,\tau_2}\Gamma_{\gamma_1}^{\sigma_1,\tau_1} = \Gamma_{\gamma_1\gamma_2}^{\sigma_2+\gamma_2\sigma_1,\tau_2+\gamma_2\tau_1}$. The unit element is $\Gamma_1^{0,0}$ and the inverse is $\Gamma_{1/\gamma}^{-\sigma/\gamma,-\tau/\gamma}$. Notice that movements of this type ($\gamma = 1$) suffice to bring any surface element into some canonical position and orientation. The attitude of surface elements is something that can be changed through a movement and hence can have no intrinsic meaning. Consequently, the "local structure" starts only at the second-order, it is the local curvature. The second-order is not influenced by the lower orders, thus the curvature is fully determined by the partial derivatives of log-intensity of the second-order. This makes most formulas of the differential geometry of surfaces in $\mathbb{I}^3$ rather simpler than the similar (and familiar) formulas from differential geometry of $\mathbb{E}^3$.

Concatenation of Processes

Since image processing turns an image into an image, one may "pipeline" image operations. Since images may also be subjected to arbitrary functions of log-luminance ($f(\{x,z\}) = \{x, f(z)\}$), can be added, multiplied, etc., point by point (e.g., $\{x,z\} + \{x,z'\} = \{x, z+z'\}$, etc), we obtain a rich repertoire of operations. In many cases one really requires an image algebra, one example being a change of inner scale.

6.3.3 Local Operations

"Local" operations involve only operations *at a point*. Apart from the obvious case $f(\{x,z\}) = \{x, f(z)\}$, more interesting cases involve intricate structure of the "point" itself. The point operations of interest here involve convolutions with derivatives of the point operation and subsequent algebraic combination of the images. Whereas "differentiation of the image" as such is nonsensical, the point operator can be differentiated because it is not a real, but an ideal entity (an analytic function). The "convolution" of the image with any operator is best understood as the action of some machine whose internal structure is forever beyond our ken. It "just happens" (the photographic plate or a CCD array provide examples). As long as the operation is linear we may interpret convolution with the derivative of the point operator as the convolution of the "derivative of the image" (essentially a nonsensical notion) with the point operator, that is to say, as the derivative of the image *at the given inner scale*.

Differential Invariants

Perhaps the most important instances of local operations are the computation of *differential invariants* [16]. Here we meet with a major cleft between the present treatment and the bulk of the literature. In the literature either of two approaches are most common: One group is thoroughly pragmatic and computes such quantities (typically approximately) as $I_{xx} + I_{yy}$ or $I_{xx}I_{yy} - I_{xy}^2$ because these turn out to be useful. Some token remarks make clear one knows these are not proper differential invariants. Another group is more sophisticated and computes much more complicated "true" (but Euclidian) differential invariants. The results are roughly similar. The irony is that the former group (unknowingly) computes the true invariants (under the group of congruences of $\mathbb{I}^3$; in many cases the "intensity" already is "log-intensity" due to nonlinearities in the imaging device), where the latter group computes nonsensical entities. The Euclidian invariants [8,34] assume that configurations that are related through Euclidian movements are congruent in $\mathbb{I}^2$. This is indeed nonsensical (consider what it means to turn an image over 90°!). The formal expressions for invariants of $\mathbb{I}^3$ are similar, though different and generally rather simpler, than those of $\mathbb{E}^3$, which is the reason (implicit and generally unrecognized though) why one is nevertheless happy to live with such nonsense.

In $\mathbb{I}$ the lowest order differential invariant are second-order (i.e., curvatures). Notice that at any given point one may apply a movement such that $z(x,y) = (z_{xx}x^2 + 2z_{xy}xy + z_{yy}y^2)/2! + \ldots$, and that the coefficients z_{xx}, etc., are not affected by that movement.

The most convenient way to represent the second-order is through the differential invariants $2H = z_{xx} + z_{yy}$ (the mean curvature), $K = z_{xx}z_{yy} - z_{xy}$ (the Gaussian curvature) and the pair (not themselves differential invariants) $A = z_{xy}$, $2B = z_{xx} - z_{yy}$ whose sum of squares $A^2 + B^2$ *is* invariant and whose ratio depends only on the direction of principal curvature. In the principal frame we have $2H = \kappa_{\max} + \kappa_{\min}$, $K = \kappa_{\max}\kappa_{\min}$, $A = 0$ and $2B = \kappa_{\max} - \kappa_{\min}$. For a Gaussian random surface the triple $\{H, A, B\}$ are statistically independent variables.

Notice that the intrinsic curvature [11] of *any* surface $\{x, y, z(x,y)\}$ in $\mathbb{I}^3$ vanishes identically because of the degenerate metric! It is more useful to define the "Gaussian curvature" as the magnification of the Weingarten map, that is, the map from the surface to its Gaussian "spherical" image.

6 Geometric Framework for Image Processing   195

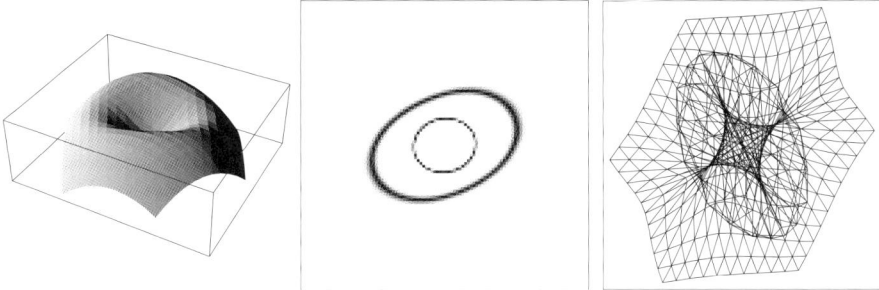

Fig. 6.17. From left to right: an image surface, its parabolic points, and its attitude image. The folds of the attitude image correspond to the parabolic curves

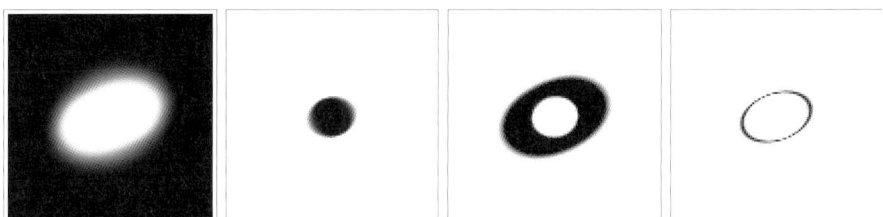

Fig. 6.18. The hills, dales, saddlelike points and minimal points for the surface shown in Fig. 6.17 left. Notice that the minimal points lie in the interior of the region of saddlelike points

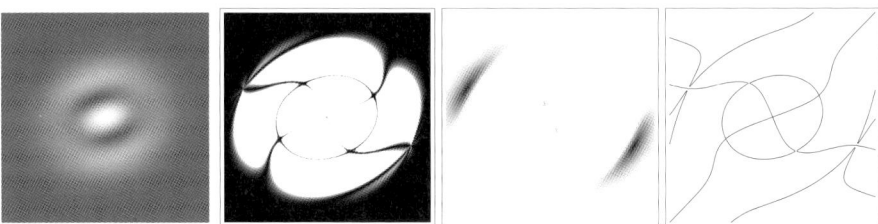

Fig. 6.19. On the *left* is the Casorati curvature, and the second and third are images the ridges and umbilics. For this example the ridges can be found analytically (*rightmost image*). The ridges image (*second from the left*) is more instructive than the analytic result: One sees immediately that virtually all points in the outer region are nearly umbilical (the whole region is nearly spherical). Notice that "the" umbilics in this region are poorly defined and that the umbilics lie on branchings of the ridges (the remaining "branches" are not what they seem, but are "fly overs")

The "Gaussian attitude image" is the map $\{x, y, z(x,y)\} \mapsto \{x, y, (x^2+y^2)/2\}$ (that is, to the unit sphere of $\mathbb{I}^3$) such that the tangent planes at corresponding points are parallel. A stereographic projection from the vanishing points of the normals maps the unit sphere isometrically(!) on $\{z_x, z_y\}$-space (often called the map to "gradient space"). The Jacobian of this mapping is the Hessian

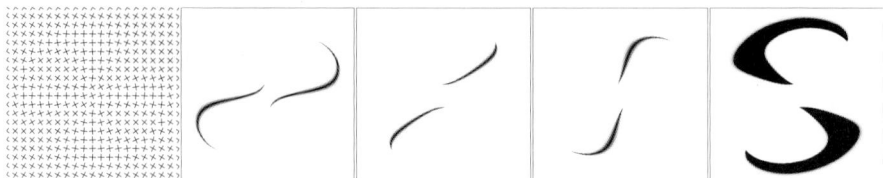

Fig. 6.20. *Left* are the directions of principal curvature for the surface depicted in Fig. 6.17 right. The images are isocline images at 45° intervals. The isocline images have been computed for the same angular width, thus the inherent fuzziness is evident from the widths of the ribbons

of the log-intensity, thus the area magnification is the determinant of the Hessian which equals the Gaussian curvature as defined above. Likewise, the magnification in any direction is the "normal curvature" for that direction. This identification allows one to develop the differential geometry of surfaces in $\mathbb{I}^3$ in analogy to that of $\mathbb{E}^3$, with a great many striking parallels (Fig. 6.17).

The differential invariants $C = \sqrt{(\kappa_{\max}^2 + \kappa_{\min}^2)/2}$ (the Casorati curvature [2] or "curvedness") and $S = \arctan(\kappa_{\max} + \kappa_{\min})/(\kappa_{\max} - \kappa_{\min})$ (the "shape index") with the direction of maximum principal curvature form a system of polar coordinates in $\{H, A, B\}$ space (the space one wants to be in because the coordinates are statistically uncorrelated). The distance to the origin (C) is a very intuitive notion [2] of "curvature" (it vanishes only for planar surfaces, which is the reason Casorati invented it), and the latitude S is an intuitive descriptor of shape (modulo size). The longitude is simply the orientation of the principal frame. This is by far the most convenient and intuitive representation of curvature (Figs. 6.18-6.20).

6.3.4 Multilocal Operations

In most cases one is not simply interested in differential invariants at a point, but in their spatial distribution. From differential geometry we know that such distributions often define submanifolds. For instance, the "umbilicals" define sets of isolated *points*, the "parabolic points" *curves* (typically closed, nonbranching), the "ridges and ruts" *curves* (with characteristic branchings), the "elliptic convex" points *regions* (bounded by parabolic curves), and so forth. Much of the interest is in these geometrical loci and their interrelations. Typically these loci are defined through the vanishing of differential invariants, thus one may easily turn the classical expressions into the "fuzzy" (inner scale) representations needed in image processing.

I illustrate a few of the more interesting examples (Figs 6.21-6.25). Consider what one might call a "hill" or "dale", I mean, can we put limits to them? Where does a hill stop being a hill? I propose defining hills as areas of elliptic curvature ($K > 0$) such that the surface is convex ($H < 0$). Using a suitable motion in $\mathbb{I}^3$ such a point could be turned into a summit. Notice that H cannot change sign in a connected region $K > 0$. Thus we need to find

6 Geometric Framework for Image Processing   197

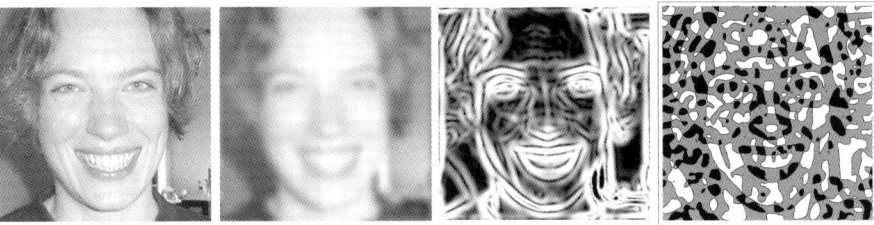

Fig. 6.21. From left to right: the original image, the image at the inner scale used for the calculations, the Casorati curvature and the shape index

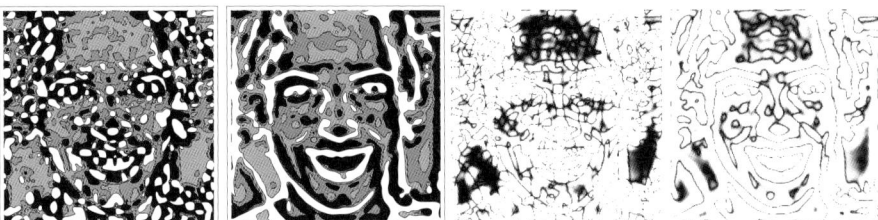

Fig. 6.22. From left to right: the Gaussian curvature, the mean curvature, the parabolic curves and the minimal curves

Fig. 6.23. The sign of the shape index divides the image into predominantly positively (*white*) and predominantly negatively (*black*) curved areas. Such images typically look like "sketches", even more so when the black areas are skeletonized

the the region $K > 0$ and in retrospect sort them with respect to the sign of H (negative hills, positive dales). The image $\{x, y, (1 + \mathrm{erf}(K(x,y)/K_0))/2\}$ (with $\mathrm{erf}(z) = \frac{2}{\sqrt{\pi}} \int_0^z e^{-z^2} \, dz$) expresses this. (Alternatively, one could use the constraint $S > \pi/4$ for hills, $S < -\pi/4$ for dales.) It ranges from zero to one and is only significantly different from zero when $K \gg K_0$.

The boundaries of the elliptic areas (and thus of the hyperbolic areas) are the parabolic curves that are the loci $K(x,y) = 0$. Thus we find them via the image $\{x, y, \exp(-K(x,y)^2/2\Delta K^2)\}$.

Fig. 6.24. From left to right: the ridges and the concave and the convex umbilical points. Notice that the ridge and rut points appear to cluster on curves, whereas the umbilical point are scattered, thin blobs

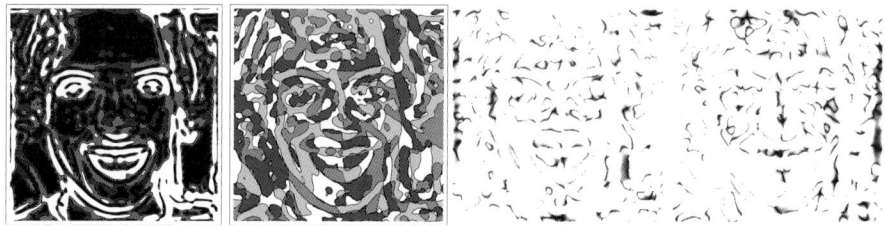

Fig. 6.25. The magnitude of the antisymmetric terms $A^2 + B^2$ (*left*) and the (*third and fourth images*, respectively) isoclines (*second from left*, isoclines binned in 45° increments), as well as isocline images for the horizontal and the vertical orientation of the direction of largest principal curvature

The "minimal curves" are the curves $H(x,y) = 0$, and they are made up of points where the shape is congruent to its mould. We find these interesting curves via the image $\{x, y, \exp(-H(x,y)^2/2\Delta H^2)\}$.

The umbilical points are points where the surface is locally isotropic, i.e., $S = \pm \pi/2$. We find them via the images $\{x, y, \exp(-(S(x,y) \pm \pi/2)^2/2\Delta S^2)\}$.

The isoclines of the curves of principal curvature are the contour lines of A/B. The A/B image is easily computed, but the principal curves themselves have to be determined via numerical integration, a truly multilocal process.

At the ridge and rut points a principal curvature is extremal along the direction of the other principal curvature. The constraint contains cubic terms. Other entities of potential interest involve even higher orders, i.e., the flecnodal points quartic terms. Despite the nearly universal abhorrence of such matters in image processing circles, this really poses no problems. The implementation is immediate (the expressions from differential geometry can be blindly compiled into image operators) and robust.

6.4 Outlook

I have roughly sketched a framework for image processing that is coherent and almost entirely geometrical in nature. Although complicated through the importance of many different spaces (the base space, the image domain, image space, a variety of scale spaces, complicated mixtures in the case of locally disorderly representations) one arrives at a fully coherent view because the geometries of all these spaces are variously interrelated. There is no "adhockery" involved.

Is "image processing" a science? Well, not right now. But there is no reason it could not be. At this moment the field is only defined by what its practitioners do and I consider it as largely a grab bag of hacks (theory is not valued highly by a community mainly interested in applications). However, most of the fundamentals for a principled framework are in place, though these threads are scattered around throughout the literature and are often only partially (or not at all) recognized for what they are. In short, I do not think much fundamental work remains to be done for someone to write a textbook on image processing that departs from first principles, develops the field logically, and steers free of hacks, unnecessary approximations and mere showpieces of mathematical dexterity. All that is needed is "good taste" (in the mathematician's sense) and a solid intuitive feeling for what is conceptually important and what is mere fluff (no matter how well it works or how fast, or how impressively flashy the mathematics). Of course, such a textbook would only serve to establish (or define) the field as a science. Much remains to be done (I am happy to say). Unfortunately, it may be some time before someone takes on this challenge seriously, as the field appears to perceive no need for it.

6.5 References

1. A. Adams, *The Print: Contact Printing and Enlarging*, Basic Photo **3**, Morgan and Lester, New York, 1950.

2. F. Casorati, *Nuova definitione dello curvatura delle superficie e suo confronto con quella di Gauss*, Rend.Inst.Matem.Accad.Lomb. **2**, p. 22, 1867–68.

3. A. Cayley, *Sixth memoir upon the quantics*, Philosophical Transactions of the Royal Society London **149**, pp. 61–70, 1859.

4. A. Cayley, *On contour and slope lines*, The London, Edingburgh and Dublin Philosophical Magazine and Journal of Science **120**, pp. 264–268, 1859.

5. W. K. Clifford, *Preliminary sketch of the biquaternions*, Proceedings of the London Mathematical Society, pp. 381–395, 1873.

6. J. Damon, *Local Morse theory for solutions of the heat equation and Gaussian blurring*, Journal of Differential Equations **115**, pp. 368–401, 1995.

7. J. D. Daugman, *Complete discrete 2–D Gabor transforms by neural networks for image analysis and compression*, IEEE Transactions on Acoustics,, Speech and Signal Processing **36**, pp. 1169–1179, 1988.

8. M. P. do Carmo, *Differential geometry of curves and surfaces*, Prentice Hall, Englewood Cliffs, NJ, 1976.

9. L. Florack, Image structure, Kluwer, Dordrecht, 1997.

10. W. T. Freeman and E. H. Adelson, *The design and use of steerable filters*, IEEE Transactions on Pattern Analysis and Machine Intelligence **13**, pp. 891–906, 1991.

11. C. F. Gauss, *Algemeine Flächentheorie*, German translation of the *Disquisitiones generales circa Superficies Curvas*, Hrsg. A. Wangerin, Ostwald's Klassiker der exakten Wissenschaften **5**, Engelmann, Leipzig, 1889.

12. B. van Ginneken and B. M.ter Haar Romeny, *Applications of locally orderless images*, In: Eds. M. Nielsen, P. Johansen, O. F. Olsen and J. Weickert, *Scale–Space Theories in Computer Vision*, Second International Conference on Scale–Space'99, Lecture Notes in Computer Science **1682**, pp. 10–21, Springer, Berlin, 1999.

13. L. Griffin, *Scale–imprecision space*, Image and Vision Computing **15**, pp. 369–398, 1997.

14. B. M. ter Haar Romeny, *Front-end vision and multi-scale image analysis*, Kluwer, Dordrecht, 2002.

15. D. Hestenes and G. Sobezyk, Clifford algebra to geometric calculus: A unified language for mathematics and physics, D. Reidel, Dordrecht, 1984.

16. D. Hilbert, *Über die vollen Invariantensystemen*, Mathematische Annalen **42**, pp. 313–373, 1893.

17. T. Lindeberg, *Scale–Space theory in computer vision*, Kluwer, Dordrecht, 1994.

18. E. T. Jaynes, *Prior probabilities*, IEEE Transaction on Systems Science and Cybernetics **SSC–4** pp. 227–241, 1968.

19. H. Jeffreys, *Theory of probability*, Clarendon, Oxford, 1939.

20. F. Klein, *Über die sogenannte nicht–Euklidische Geometrie*, Mathematische Annalen **6**, pp. 112–145, 1871.

21. F. Klein, *Vergleichende Betrachtungen über neuere geometrische Forschungen*, Mathematische Annalen **43**, pp. 63–100, 1893.

22. F. Klein, *Vorlesungen über nicht–Euklidische Geometrie*, Springer, Berlin, 1928.

23. T. Lindeberg, *Scale–Space theory in computer vision*, Kluwer, Dordrecht, 1999.

24. P. Morrison and P. Morrison, *Powers of Ten*, Scientific American Library and W. H. Freeman, New York, 1994.

25. T. Needham, *Visual complex analysis*, Clarendon, Oxford, 1997.

26. T. Poston and I. Stewart, *Catastrophy theory and its applications*, Pitman, London, 1978.

27. H. Pottmann and K. Opitz, *Curvature analysis and visualization for functions defined on Euclidean spaces or surfaces*, Computer aided geometric design **11**, pp. 655–674, 1993.

28. A. Robinson, *Non–standard analysis*, North–Holland, Amsterdam, 1974.

29. A. H. Robinson, J. L. Morrison, P. C. Muehrcke, A. J. Kimerling and S. C. Guptill, *Elements of cartography*, Wiley, New York, 1995.

30. H. Sachs, *Ebene isotrope Geometrie*, Friedrich Vieweg & Sohn, Braunschweig, 1987.

31. H. Sachs, *Isotrope Geometrie des Raumes*, Friedrich Vieweg, Braunschweig, 1990.

32. G. Scheffer, *Verallgemeinerung der Grundlagen der gewöhnlichen komplexen funktionen*, Sitz. ber. Sächs. Ges. Wiss., Math.–phys.Klasse, **42**, pp. 828–842, 1893.

33. L. Schwartz, *Théorie des distributions*, Hermann, Paris, 1966.

34. M. Spivak, *Differential geometry*, Publish or Perish, Berkeley, 1975.

35. K. Strubecker, *Differentialgeometrie des isotropen Raumes I*, Sitzungsberichte der Akademie der Wissenschaften Wien **150**, pp. 1–43, 1941.

36. K. Strubecker, *Differentialgeometrie des isotropen Raumes II*, Mathematische Zeitschrift **47**, pp. 743–777, 1942.

37. K. Strubecker, *Differentialgeometrie des isotropen Raumes III*, Mathematische Zeitschrift **48**, pp. 369–427, 1943.

38. K. Strubecker, *Differentialgeometrie des isotropen Raumes IV*, Mathematische Zeitschrift **50**, pp. 1–92, 1945.

39. A. Tikhonov, *Solutions of incorrectly formulated problems and the regularization method*, Soviet. Math. Dokl. **4**, pp. 1035–1038, 1963.

40. A. P. Witkin, *Scale–space filtering*, In: Proceedings of the International Joint Conference on Artificial Intelligence, pp. 1019–1022, Karlsruhe, 1983.

41. R. A. Ulichney, *Digital halftoning*, The M.I.T. Press, Cambridge MA, 1987.

42. I. M. Yaglom, *Complex numbers in geometry*, Academic, New York, 1968.

43. I. M. Yaglom, *A simple non–Euclidean geometry and its physical basis*, Springer, New York, 1979.

7

Geometric Filters, Diffusion Flows, and Kernels in Image Processing

Alon Spira[1], Nir Sochen[2], and Ron Kimmel[1]

[1] Department of Computer Science, Technion, Israel
{salon,ron}@cs.technion.ac.il
[2] Department of Applied Mathematics, University of Tel-Aviv, Israel
sochen@math.tau.ac.il

7.1 Introduction

Diffusion flows are processes applied to digital images in order to enhance or simplify them. These flows are usually implemented by appropriate discretizations of partial differential equations (PDEs). Iteratively applying these discretizations, called also *numerical schemes*, to an image results in a series of images with decreasing detail (Fig. 7.1). Using a suitable flow, one can enhance important image features such as edges and objects while filtering the image from undesired noise. This can be done not only to gray-level and color images but also to textures, movies, volumetric medical images, and so on.

Diffusion flows are important members of the family of methods for image processing, computer vision, and computer graphics based on the numerical solution of PDEs. Other members of the family include active contours/surfaces for image segmentation, reconstruction of three-dimensional scenes from their shading or stereo images, graphic visualization of natural phenomena, and many others. This family of methods has many advantages, among them theoretical origin due to derivation from a minimization of (usually geometric) cost functions, efficiency, and robustness.

7.2 Diffusion Flows and Geometric Filters

Diffusion processes are widely spread in many areas of physics. Naturally, they found their way to the field of image processing. At first, only linear diffusion was used, but gradually also nonlinear diffusions were introduced and geometry-based filters proposed. This section reviews the development of these methods from the early days to the present (mid-2004).

Fig. 7.1. A diffusion flow of a color image. The original image is *top left*

7.2.1 The Heat Equation

The simplest diffusion is the one generated by the two-dimensional heat equation

$$I_t = \Delta I,$$

with $I(x, y)$ the two-dimensional data, I_t its partial derivative according to time, and Δ the Laplacian operator $(\partial_{xx} + \partial_{yy})$. This equation depicts, for instance, the temporal change in the heat profile of a metal sheet. In our case $I(x, y)$ gives the gray-level values of the image.

The heat equation was the first diffusion process applied to images [45]. It was mainly used to create a *scale space* for an image, meaning a three-dimensional volume with a scale coordinate t added to the spatial coordinates x and y. At the origin of t we have the original image as initial condition, and as we advance along t we get smoother versions of it. The idea behind scale space is that important features of the original image should survive the change of scale, and therefore all the scale space of the image should be used to detect these features. Later on, the heat equation was suggested for filtering noise corrupting the image [9]. After applying the heat equation for a short duration, the noise that is of fine resolution disappears.

The heat equation as a diffusion flow generating a scale space has an important attribute, its linearity. It is therefore also referred to as *linear diffusion*. However, it damages the edges of objects in images and does not preserve connected components (Fig. 7.2). This simple example was presented in the introduction of the first collection of papers on this topic [38].

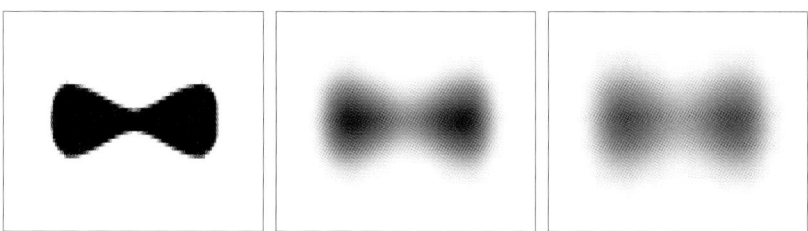

Fig. 7.2. The heat equation damages edges and separates connected components

7.2.2 The Geometric Heat Equation

New flows were suggested to overcome the problem of the change in the number of connected components. One such flow, first introduced by Alvarez et al. [1] in the context of invariant image processing, is the level set curvature flow. *Level set* curves are another way to describe the structure of a gray-level image. Given an image $I(x,y)$, its level set curves are defined as $C(h) = \{(x,y) : I(x,y) = h\}$. See Fig. 7.3 for the level curves of the images in Fig. 7.2. The interior of a closed contour can be considered as a component, and the number of components somehow indicates the complexity of the image [3].

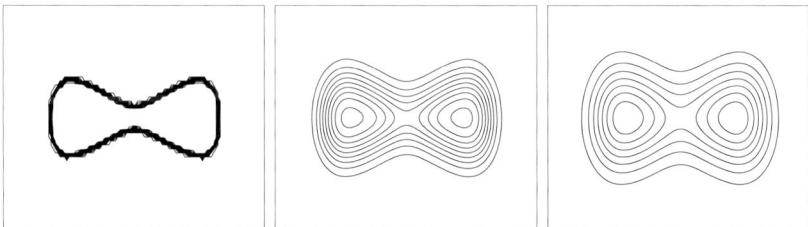

Fig. 7.3. Level set curves of the images in Fig. 7.2

The idea was to use the powerful Grayson theorem [10] for curve evolution via its curvature. The theorem states that the curvature flow

$$C_t = \kappa \mathbf{n},$$

with κ the curvature of the closed planar curve C, and $\mathbf{n}$ the unit normal vector to the curve, results in the convergence of the curve to a point.

Using the Osher–Sethian [21] level set formulation the whole image could be propagated via the curvature flow equation. That is, each and every level set of the original image could be propagated by its curvature flow, and all this process could be described by a single evolution equation for the whole image given by

$$I_t = \operatorname{div}\left(\frac{\nabla I}{|\nabla I|}\right)|\nabla I|.$$

This process is possible due to the Evans–Spruck [7] confirmation that as embedding of such propagating curves is preserved, the level set formulation is indeed valid for the curvature flow. One nice property of this flow is that connected components remain connected until they disappear. Moreover, this flow is invariant to Euclidean transformations in the image plane.

Next came the interesting question of what could be said about more complicated transformations. In [1] the authors also introduced the equi-affine invariant flow given by

$$I_t = \left(\operatorname{div}\left(\frac{\nabla I}{|\nabla I|}\right)\right)^{1/3}|\nabla I|. \tag{7.1}$$

Again, the connection to curve evolution was presented at the same time by Sapiro [26]. First, the curvature flow can be equivalently written by

$$C_t = C_{ss},$$

where $C(s) = \{x(s), y(s)\}$, and s is the Euclidean arc length parameterization. This is why it is also known as the *geometric heat equation*. Using similar writing for the equi-affine flow, that is,

$$C_t = C_{vv},$$

where v is the equi-affine arc length $dv = \kappa^{1/3}ds$, the resulting geometric flow can be written by

$$C_t = \kappa^{1/3}\mathbf{n}.$$

This equation, known as the *affine heat equation*, enjoys some of the nice properties of Grayson's theorem, like preservation of embedding of the propagating contours. It is thus directly related to Eq. (7.1), again via the Osher–Sethian level set formulation. These beautiful relations and geometric properties started a new era in the image processing and analysis field. For example, when smoothing stereo images we would be better off using the affine heat equation, and not the geometric heat equation, which would distort the geometric structure relating the two images. Applications of these operators include computation of geometric signatures [8, 12], and extensions of these ideas to deal with problems like geometric scale space for images painted on surfaces [13, 34].

7.2.3 Isotropic Nonlinear Diffusion

At the other extreme, researchers started to explore the field of variational principles and geometry in image processing. That is, define an integral measure that somehow captures the norm of the image. For example, the *total variation* (TV) norm was a popular selection proposed in [25]. The TV is defined by

$$\iint |\nabla I| dx dy,$$

for which the Euler–Lagrange equation is given by

$$\text{div}\left(\frac{\nabla I}{|\nabla I|}\right) = 0.$$

That is, the level set curvature should be equal to zero. This geometric connection should not come as a surprise, since by the co-area equation we have that

$$\iint |\nabla I| dx dy = \iint_\Omega ds dh$$

where s is the arc length parameter of each and every level set contour, and h is a parameter running over the image intensities I. The zero curvature is indeed the result of minimizing the arc length of all level set contours in the image.

The methods used to denoise an image based on the TV norm usually apply the Euler–Lagrange as a gradient descent via a PDE of the form

$$I_t = \text{div}\left(\frac{\nabla I}{|\nabla I|}\right).$$

Again, the corresponding flow of the image level sets can be written as

$$C_t = \frac{1}{|\nabla I|} \kappa \mathbf{n},$$

[14]. This is nothing but a selective curvature flow, where the flow is enhanced at smooth regions and suppressed near the image edges (where the image gradient is high), so that these important features are preserved.

Another popular filter proposed at the same time is the Perona–Malik [23] *anisotropic diffusion*. Unlike its name, the filter is an inhomogeneous yet locally isotropic flow given by

$$I_t = \text{div}\left(f(|\nabla I|)\nabla I\right).$$

We see that setting $f(s) = s^{-1}$ we are back with the TV flow, while other selections lead to other filters.

The role of the *diffusivity* function f is to control the amount of diffusion according to the gradient of the image. At image edges, where $|\nabla I|$ is large, the diffusion should be minimal, and vice versa at the interior of objects. To accomplish that, f should be monotonically decreasing. A popular choice for f is

$$f(|\nabla I|) = \frac{1}{1 + |\nabla I|^2/\lambda^2}.$$

7.2.4 Anisotropic Nonlinear Diffusion

Gabor [2, 9, 16, 20] was probably the first to consider anisotropic diffusion by smoothing along the edge and inverting the heat operator and thereby generating an unstable enhancing process across the edge. If we write the gradient direction as $\xi = \nabla I/|\nabla I|$ and η as the orthogonal direction (Fig. 7.4), $I_t = I_{\eta\eta}$ is nothing but the curvature flow. Gabor proposed to use one iteration of a discretization of the equation

$$I_t = I_{\eta\eta} - \epsilon I_{\xi\xi},$$

where ϵ determines the amount of inverse diffusion. This simple and nice formulation for image enhancement (which cannot be easily extracted from a variational principle) was rediscovered many times along the evolution of the image processing field.

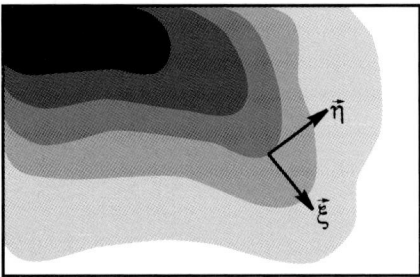

Fig. 7.4. The gradient direction and the tangent direction of the image level sets

A recent interesting anisotropic differential filter for image analysis is Weickert's [43] edge direction sensitive flow. Weickert's idea was to plug a 2×2 symmetric positive definite matrix instead of the scalar function $f(s)$ of the Perona–Malik flow. The orthonormal eigenvectors of the matrix are selected according to the image gradient direction

$$v_1 \parallel \nabla I, \qquad v_2 \perp \nabla I,$$

and their corresponding eigenvalues are taken such that

$$\lim_{|\nabla I| \to \infty} \frac{\lambda_1(|\nabla I|)}{\lambda_2(|\nabla I|)} = 0.$$

This way, the smoothing is mostly along the edges and not across them.

7.2.5 Mean Curvature Flow

Many times the way we represent objects defines the way(s) in which we can manipulate them. Images, for example, were represented traditionally as a matrix of numbers. Many image processing techniques followed this representation. Recently, a more geometric point of view emerged. An image is regarded and represented as a surface. In fact, the graph of the intensity function for gray-valued images is a two-dimensional surface. One may think of it as embedded in $\mathbb{R}^3$ with coordinates x, y, and I. Once described in this way it is natural to ask geometric questions such as about the curvature of the surface at a given point. We may also envisage processes that alter the geometric properties of the surface. Noting that noise is represented in the image as points (or small regions) of high curvature, it is natural to give a smoother version of the image by reducing points with high curvature. One way to achieve this goal is to define an evolution equation that depends on the curvature. We move, at each instant, the image surface in the direction of the normal to the surface. Note that this is the only direction that changes the *shape* of the image. Movement along the other two directions simply causes a reparameterization that does not change the image's gray-value content. The amount of change at each point is proportional to the mean curvature in that point. Denoting the mean curvature H, and the normal to the surface $\mathbf{N}$, we find the following PDE

$$\mathbf{S}_t = H\mathbf{N}.$$

How should we understand this equation? How is it applied to images? In order to answer these questions we go back to the representation of the image as a surface. The graph of the image embedded in $\mathbb{R}^3$ is represented as the ternary $(x, y, I(x, y))$. The two tangent vectors along the canonical coordinates x and y are given by $X_1 = (1, 0, I_x)$ and $X_2 = (0, 1, I_y)$. The normal vector is derived easily as orthogonal to X_1 and X_2. Its form is

$$\mathbf{N} = \frac{1}{\sqrt{1 + |\nabla I|^2}} (-I_x, -I_y, 1).$$

The mean curvature at each point is

$$H(x, y) = \frac{(1 + I_x)^2 I_{yy} - 2 I_x I_y I_{xy} + (1 + I_y^2) I_{xx}}{(1 + I_x^2 + I_y^2)^{\frac{3}{2}}}.$$

It follows that the equation is

$$(x, y, I)_t^T = H(-I_x, -I_y, 1)^T \frac{1}{\sqrt{1 + |\nabla I|^2}}.$$

Since we work in a constant domain and a constant coordinate system, namely the Cartesian x and y coordinates, the only change that actually takes place is the value of the gray value at each pixel. In order to have the required effect while changing the gray values only, we change the gray value at each point such that its projection on the normal has exactly the magnitude of the mean curvature. A simple calculation shows that we need to multiply by a factor of $\sqrt{1+|\nabla I|^2}$ (Fig. 7.5).

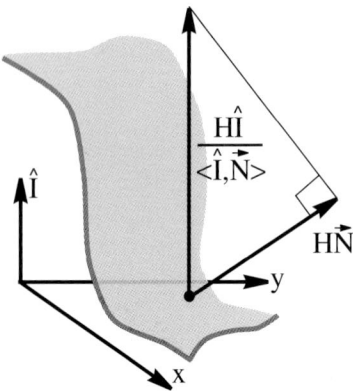

Fig. 7.5. The mean curvature flow for gray level images is accomplished by only changing the intensity component

The final equation is

$$I_t = \frac{(1+I_x)^2 I_{yy} - 2I_x I_y I_{xy} + (1+I_y^2)I_{xx}}{(1+I_x^2+I_y^2)}.$$

7.2.6 Color Images

Color images are the canonical example of vector value images. The light that is reflected from a surface is described by the wavelength spectrum $R(\lambda) = S(\lambda)\rho(\lambda)$, where $S(\lambda)$ is the spectrum of the illumination and $\rho(\lambda)$ is the material reflectance property known as the albedo. Three filters are applied at each spatial point to the spectrum to produce the three channels $I^i = \int d\lambda R(\lambda) f^i(\lambda)$. These three channels are usually called red, green, and blue (RGB) with respect to the regions in spectrum space where the filters extract most of their energy. The information is then encoded in three functions $R(x,y)$, $G(x,y)$, and $B(x,y)$.

There are several approaches in the denoising process of color and other multichannel images. The first and most simple and naive approach is to apply a denoising process to each channel separately. This approach ignores completely the correlation between the different channels. Since the channel edges are not necessarily aligned, an anisotropic channel-by-channel process may blur regions where only one channel has an edge. In case several strong edges in all channels exist with small offsets, artificial colors may appear.

We describe in this section a different approach where the color channels are correlated via the Di Zenzo metric [5]. More elaborate approaches that incorporate perceptual psychophysical data will be discussed below in the context of the Beltrami framework. Here we follow the approach of Sapiro and Ringach [27]. The Di Zenzo metric is defined in the color space. Its explicit form is

$$D = \begin{pmatrix} R_x^2 + G_x^2 + B_x^2 & R_x R_y + G_x G_y + B_x B_y \\ R_x R_y + G_x G_y + B_x B_y & R_y^2 + G_y^2 + B_y^2 \end{pmatrix},$$

where the subscripts x and y mean partial derivation. The elements can be written more simply with the Einstein summation convention: indices that appear twice are summed over. The elements are written as $D_{xx} = I_x^i I_x^i$, where the summation is over the index $i = 1, 2, 3$, and $I^1 = R$, $I^2 = G$, $I^3 = B$. In general $D_{\mu\nu} = I_\mu^i I_\nu^i$, where μ and ν take the values 1 and 2. They stand for x_μ and x_ν, where by convention $x_1 = x$ and $x_2 = y$.

The matrix D is real, and symmetric and it can be diagonalized. Formally, we can write $D = U^T \Lambda U$ where $\Lambda = \text{diag}(\lambda_+, \lambda_-)$. The matrix U is composed of the eigenvectors that give the direction of maximal variation in color space and its perpendicular direction. The λ_i indicate the amount of change in each direction. Sapiro and Ringach suggest in their paper constructing an anisotropic process in the following manner:

$$I_t^i = \text{div}\left(f(\lambda_+ + \lambda_-) \nabla I^i\right).$$

This equation can be derived as a gradient descent of a functional. It is simply $S[I^i] = \int \Psi(\lambda_+ + \lambda_-) dx dy$. A new analysis of this and many other approaches can be found in Tschumperlé's thesis [40].

7.2.7 The Beltrami Flow

In the Beltrami framework [15, 31] the image is regarded as an embedding of the image manifold in the space-feature manifold. In more rigorous terms we describe the image as a *section* of a *fiber bundle*. The fiber bundle is composed of the spatial part, which is usually a rectangle in $\mathbb{R}^2$, called the base manifold, and the fiber that describes the feature space, i.e. intensity, color, texture, and so on. A section of the fiber bundle is a choice of a specific feature from the feature space for every point in the base manifold. The feature space may be a linear space or a more complicated manifold. In the first case we call the section a vector field.

The most simple example is the gray-level image. Denote the embedding map by X. The explicit form of this map for gray-level images is

$$X(u^1, u^2) = (u^1, u^2, I(u^1, u^2)),$$

where u^1, u^2 are the spatial coordinates and I is the intensity component (Fig. 7.6). For color images the embedding map reads:

$$X(u^1, u^2) = (u^1, u^2, I^1(u^1, u^2), I^2(u^1, u^2), I^3(u^1, u^2)),$$

where I^1, I^2, I^3 are the three color components (for instance, red, green, and blue for the RGB color space).

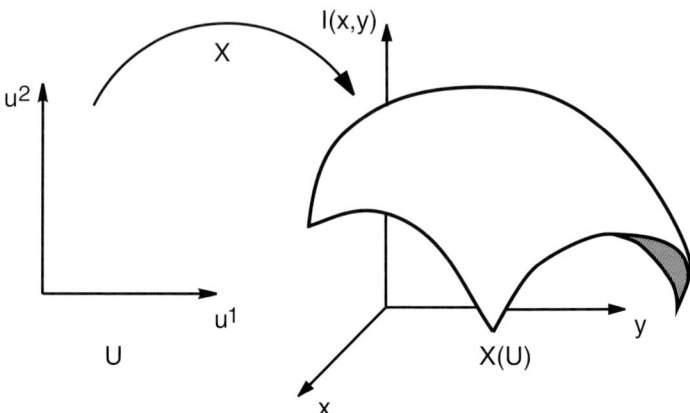

Fig. 7.6. A gray-level image according to the Beltrami framework

The geometry of the image manifold, i.e. the section, is determined according to its metric tensor **G**, which is the result of the metric **H** chosen for the space-feature manifold, i.e. the fiber bundle. A natural choice for gray-level images is a Euclidean space-feature manifold with the metric

$$\mathbf{H} = (h_{ij}) = \begin{pmatrix} 1 & 0 & 0 \\ 0 & 1 & 0 \\ 0 & 0 & \beta^2 \end{pmatrix},$$

where β is the relative scale between the space coordinates and the intensity component. The metric **G** of the image manifold is derived from the metric **H** and the embedding X by the pullback procedure

$$(\mathbf{G})_{ij} = \partial_i X^a \partial_j X^b h_{ab}.$$

Using the explicit form of the embedding map X and the metric of the fiber bundle **H** for gray-level images, we can find the metric **G**:

$$\mathbf{G} = (g_{ij}) = \begin{pmatrix} 1 + \beta^2 I_1^2 & \beta^2 I_1 I_2 \\ \beta^2 I_1 I_2 & 1 + \beta^2 I_2^2 \end{pmatrix},$$

where $I_i \triangleq \frac{\partial I}{\partial u^i}$.

The Euclidean metric **H** of the space-feature manifold for color images is

7 Geometric Filters, Diffusion Flows, and Kernels in Image Processing

$$\mathsf{H} = (h_{ij}) = \begin{pmatrix} 1 & 0 & 0 & 0 & 0 \\ 0 & 1 & 0 & 0 & 0 \\ 0 & 0 & \beta^2 & 0 & 0 \\ 0 & 0 & 0 & \beta^2 & 0 \\ 0 & 0 & 0 & 0 & \beta^2 \end{pmatrix},$$

where the same scaling factor was chosen for the three color channels. The resulting image metric is

$$\mathsf{G} = (g_{ij}) = \begin{pmatrix} 1 + \beta^2 \sum_a (I_1^a)^2 & \beta^2 \sum_a I_1^a I_2^a \\ \beta^2 \sum_a I_1^a I_2^a & 1 + \beta^2 \sum_a (I_2^a)^2 \end{pmatrix}.$$

The Beltrami flow is obtained by minimizing the area of the image manifold

$$S = \iint \sqrt{g}\, du_1 du_2,$$

with respect to the intensity components, where $g = \det(\mathsf{G}) = g_{11}g_{22} - g_{12}^2$. The gradient descent process is given by the corresponding Euler–Lagrange equations

$$X_t^a = -g^{-\frac{1}{2}} h^{ab} \frac{\delta S}{\delta X^b} = g^{-\frac{1}{2}} \partial_i (g^{\frac{1}{2}} g^{ij} \partial_j X^a) + \Gamma_{bc}^a \partial_i X^b \partial_j X^c g^{ij},$$

with g^{ij} the components of the contravariant metric of the image manifold G^{-1} (the inverse of the metric tensor G). The Christoffel symbols (also known as the Levi–Civita coefficients) Γ_{bc}^a are defined in terms of the fiber bundle metric H:

$$\Gamma_{bc}^a = \frac{1}{2} h^{ad} (\partial_b h_{dc} + \partial_c h_{bd} - \partial_d h_{bc}). \tag{7.2}$$

In matrix form it reads

$$X_t^a = \underbrace{\frac{1}{\sqrt{g}} \mathrm{div}\left(\sqrt{g}\mathsf{G}^{-1}\nabla X^a\right)}_{\Delta_g X^a} + Tr(\Gamma^a F),$$

where Γ^a is the matrix whose elements are $(\Gamma^a)_{ab} = \Gamma_{ab}^a$ and $F_{ab} = \partial_i X^a \partial_j X^b g^{ij}$. The symbol Δ_g is the Laplace–Beltrami operator, which is the extension of the Laplacian to manifolds. The resulting diffusion flow for gray-level images is

$$I_t = \Delta_g I = H \langle \hat{I}, \mathbf{N} \rangle,$$

i.e. the image surface moves according to the intensity component of the mean curvature flow (Fig. 7.7). Because we chose a Euclidean feature space the Christoffel symbols are identically zero in this case. They vanish for color images as well. The diffusion equation for each color component reads

$$I_t^i = \Delta_g I^i. \tag{7.3}$$

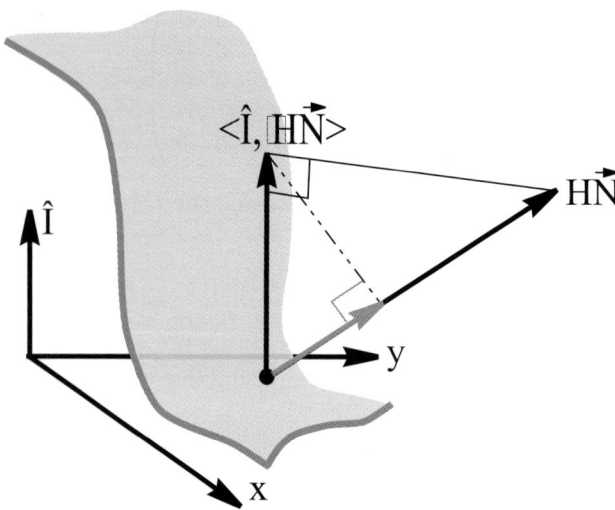

Fig. 7.7. In the Beltrami flow for gray-level images the image surface moves according to the intensity component of the mean curvature flow. Geometrically, only the projection of this movement on the normal to the surface matters

The diffusion process in Fig. 7.1 is actually the Beltrami flow. Figure 7.8 contains a closeup of the images in Fig. 7.1, including the two-dimensional manifolds of the red, green, and blue color components. It is evident that the Beltrami flow filters out the noise while not only preserving the edges, but keeping their location in the three color components aligned.

7.3 Extending the Beltrami Framework

The basic idea of the Beltrami framework of treating the image as a manifold and enhancing it by minimizing its area can be extended in various ways. In this section the framework is extended to higher dimensional spaces (for texture, video, and volumetric data), non-Euclidean feature spaces, and other diffusion directions.

7.3.1 Texture, Video and Volumetric Data

We have discussed color for which researchers try to give a simple geometric interpretation, like an arc length that would capture our visual sensitivity to colors. Next, we claimed that in order to extract technology from such definitions we need to link the color arc length to another measure of distance in the image domain. In this way we came up with the hybrid space idea.

Next come interesting questions of what is texture and how should we treat it? Like color, we try to interpret texture as a region for which homogeneity is

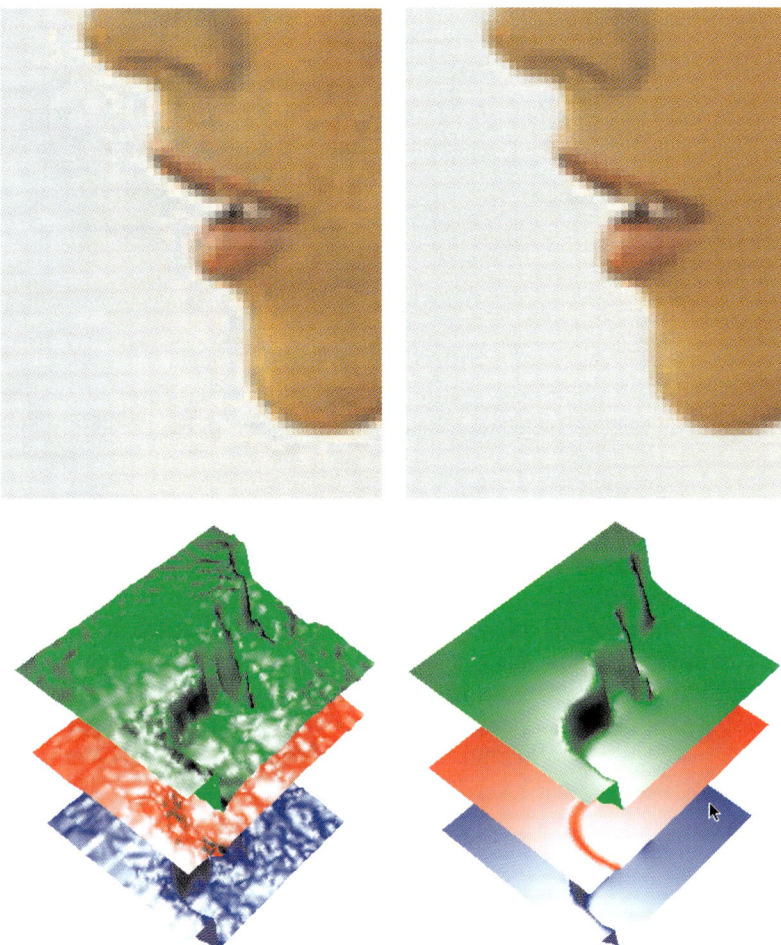

Fig. 7.8. The results of the Beltrami filter. The original image is on the *left* and the filtered one on the *right*

no longer determined by a single constant like color, but rather by repeating patterns in the image domain. Again, we need some sort of measure that defines a distance between different patterns. There are many ways to achieve this goal [24]. Once such an arc length is defined, all we need to do is to plug it into our Beltrami framework and we have a filter for texture.

Such filters are reported in [16], where the texture is represented by using the Gabor–Morlet wavelet transform $W(x, y, \theta, \sigma)$ [19], with x and y the spatial coordinates, θ the wavelet orientation parameter, and σ the wavelet scale parameter. The texture image is the embedding $(x, y, \theta, \sigma) \to (x, y, \theta, \sigma, R, J)$, where $R = \text{real}(W)$ and $J = \text{imag}(W)$. Each scale is considered as a different space, resulting in the metric

$$G = (g_{ij}) = \begin{pmatrix} 1 + R_x^2 + J_x^2 & R_x R_y + J_x J_y & R_x R_\theta + J_x J_\theta \\ R_x R_y + J_x J_y & 1 + R_y^2 + J_y^2 & R_y R_\theta + J_y J_\theta \\ R_x R_\theta + J_x J_\theta & R_y R_\theta + J_y J_\theta & 1 + R_\theta^2 + J_\theta^2 \end{pmatrix},$$

and the Beltrami flow

$$R_t = \Delta_g R,$$
$$J_t = \Delta_g J.$$

Consequently, each scale can be filtered in a different way and to a different extent. See Fig. 7.9 for a demonstration of texture enhancement using the Beltrami flow.

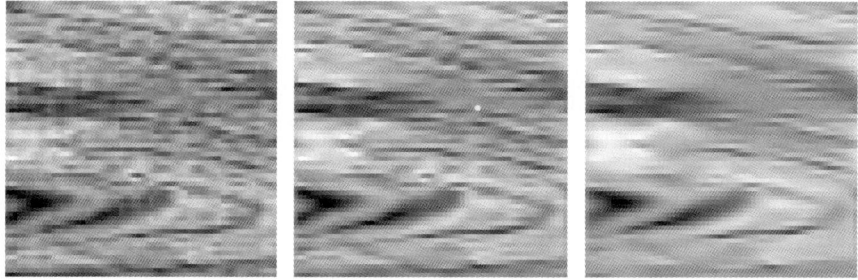

Fig. 7.9. Texture enhancement by using the Beltrami filter on the Gabor–Morlet wavelet transform of the texture image. The original image is on the *left*

The Beltrami filter for gray-level video and volumetric medical data (such as CT or MRI) is accomplished by considering them as the embedding $(x, y, z) \to (x, y, z, I)$, where for video z represents time and for volumetric data the third spatial coordinate. The induced metric in this case is

$$G = (g_{ij}) = \begin{pmatrix} 1 + I_x^2 & I_x I_y & I_x I_z \\ I_x I_y & 1 + I_y^2 & I_y I_z \\ I_x I_z & I_y I_z & 1 + I_z^2 \end{pmatrix},$$

and the Beltrami flow is

$$I_t = \frac{1}{\sqrt{g}} \mathrm{div}\left(\frac{\nabla I}{\sqrt{g}}\right),$$

where $\nabla I = (I_x, I_y, I_z)$, and $g = 1 + I_x^2 + I_y^2 + I_z^2$.

7.3.2 Non-Euclidean Feature Spaces

We have seen above that the image is represented as an embedding of a surface in a spatial-feature space. In the previous subsections we treated many tasks in which the feature space is Euclidean and endowed with a Cartesian coordinate system. There are many instances where the situation is different. We shall present below two such cases: perceptual color denoising and orientation diffusion.

Color Image Denoising

The construction of the RGB color space was described in the section on color images (Sect. 7.2.6). While the coordinates in this color space are perfectly defined from a physical point of view, they are not enough in order to denoise color images that are to be seen by human beings. The most important notion in denoising is distance. What is relevant in denoising color images is to understand how distances between colors are *perceived* by humans. In other words, we treat the perceptual color space as a three-dimensional manifold whose local coordinates are given by the RGB system. What is needed in order to complete the picture is to provide the *metric* on this manifold such that distances between colors can be measured with accordance to perception. This distance cannot be deduced from physics and must be given from psychophysical experiments and considerations. Albeit its modern appearance, this paradigm is more than a century old. The first formulation of the perceptual color space as a Riemannian manifold is due to Helmholtz [11] in 1896! Helmholtz suggested a metric that is based on the famous log response of our senses. While it is good as a first approximation, it was soon realized that his metric is inappropriate and does not describe well the experiments in various regions of the perceptual color space. The experiments are based on the notion of just noticeable differences (JND). In a typical JND experiment two squares of the same color are shown to a subject. One of these squares gradually changes its color until the subject declares that the colors are different. This gives a map of infinitesimal distances in color space and can be compared directly to metrics that model this human color perception. The construction of such metrics captured the interest of prominent scientists such as Helmholtz and Schrödinger [28]. The Helmholtz model is given simply by the following line element:

$$ds^2 = (d \log R)^2 + (d \log G)^2 + (d \log B)^2$$

This equation ignores the dependence of the JND on the overall luminance. Schrödinger tried to rectify this line element and suggested the following model:

$$ds^2 = \frac{1}{R+G+B}\left(\frac{dR^2}{R} + \frac{dG^2}{G} + \frac{dB^2}{B}\right).$$

More recent efforts to model the metric of the perceptual color space include Stiles [37] and Vos and Walraven [42].

We will demonstrate here the denoising with respect to the Helmholtz and Schrödinger metrics only. For a thorough discussion refer to [32]. Let us denote the perceptual color Riemannian manifold by M_C. The Beltrami framework describes a color image as the embedding of a two-dimensional surface in the fiber bundle $\mathbb{R}^2 \times M_C$. The base manifold is $\mathbb{R}^2$. At each point in the base manifold the fiber M_C is attached. A color image is a *section* of this fiber bundle. The metric on the fiber bundle is simply

$$ds^2 = ds^2_{\text{spatial}} + ds^2_{\text{color}} = dx^2 + dy^2 + dI^i dI^j h_{ij},$$

where for the Helmholtz model

$$(h_{ij}) = \begin{pmatrix} \frac{1}{R^2} & 0 & 0 \\ 0 & \frac{1}{G^2} & 0 \\ 0 & 0 & \frac{1}{B^2} \end{pmatrix},$$

and for the Schrödinger model it is

$$(h_{ij}) = \frac{1}{R+G+B} \begin{pmatrix} \frac{1}{R} & 0 & 0 \\ 0 & \frac{1}{G} & 0 \\ 0 & 0 & \frac{1}{B} \end{pmatrix}.$$

The induced metric on the section is simply

$$g_{\mu\nu} = \delta_{\mu\nu} + I^i_\mu I^j_\nu h_{ij},$$

and the Levi–Civita coefficients are given by Eq. (7.2). The Beltrami flow is then

$$I^i_t = \Delta_g I^I + \Gamma^I_{jk} \partial_\mu X^j \partial_\nu X^k g^{\mu\nu}.$$

Orientation Diffusion

Another example of a non-Euclidean feature space is the orientation [18]. In this case the feature manifold is the unit circle $\mathbf{S}^1$. We again construct the fiber bundle $\mathbb{R}^2 \times \mathbf{S}^1$ and regard the orientation vector field as a section of this fiber bundle. In order to express the metric on this fiber bundle we cover $\mathbf{S}^1$ with two coordinate patches. This can be done in various ways. We present here the hemispheric coordinates for simplicity. Embedding the orientation circle in $\mathbb{R}^2$ with Cartesian coordinates u and v we find that $\mathbf{S}^1$ is given by $u^2 + v^2 = 1$. We write the metric on the patch of $\mathbf{S}$ described by u as

$$ds^2 = du^2 + dv^2 = (1 + \frac{u^2}{1-u^2})du^2 = \frac{1}{1-u^2}du^2 = A(u)du^2.$$

Having calculated the metric on the fiber we can now deduce the induced metric on the section

$$ds^2 = dx^2 + dy^2 + A(u)du^2$$
$$= (1 + A(u)u_x^2)dx^2 + 2A(u)u_x u_y dxdy + (1 + A(u)u_y^2)dy^2.$$

Note that the metric on the fiber bundle is given by

$$(h_{ij}) = \begin{pmatrix} 1 & 0 & 0 \\ 0 & 1 & 0 \\ 0 & 0 & \frac{1}{1-u^2} \end{pmatrix}.$$

The Levi–Civita coefficients can be calculated by Eq. (7.2). The Beltrami flow equation reads:

$$u_t = \Delta_g u + \Gamma^u_{jk} \partial_\mu X^j \partial_\nu X^k g^{\mu\nu}.$$

The Beltrami flow modifies the features *in the feature manifold* such that a unit length vector stays always a unit length vector along the flow.

7.3.3 Inverse Diffusion Across Edges

An interesting approach to extend Gabor's original idea [9] for image enhancement via

$$I_t = I_{\eta\eta} - \epsilon I_{\xi\xi},$$

is to try to manipulate the eigenvalues of the inverse metric matrix in the Beltrami operator. If these values are kept positive, the result is a diffusion that can be enhanced in a specific direction, as proposed by Weickert in his 'coherence enhancement' filters [44]. More interesting, yet obviously less stable, is the concept of negative eigenvalues that mimic Gabor's inverse diffusion across the edge. This was first introduced in [16].

The concept is simple. We first extract the inverse metric matrix (g^{ij}) and compute its eigenstructure, $(g^{ij}) = U \Lambda U^T$. Next, manipulate the eigenvalues so that the smaller one gets a negative sign. This way, the inverse diffusion across the edge, because of the negative sign, enhances and sharpens the edges in the image, while the diffusion along the edges (the direction orthogonal to the maximal change direction) smooths the boundaries and adds some control to the process. See Fig. 7.10 for an example of this process. This is an extension to Gabor's original idea from 1965 that exploits the geometric structure of the color image, where there are no level sets or 'isophots' due to its multichannel nature.

7.4 Numerical Schemes

The PDEs describing the diffusion processes are continuous, but they are implemented on discrete digital images by computer algorithms with discrete representations. The means to bridge this gap are the numerical schemes that ensure that the discrete solution will converge to the continuous one as the grid is refined.

Fig. 7.10. Edge enhancement by diffusion along the edge and inverse diffusion across it. The original image is on the *left*

Many numerical schemes are used for the solution of the image diffusion PDEs. Among them are fast Fourier transform (FFT), wavelet transforms, finite element techniques, neural networks, multigrid methods, and many more. However, in most cases *finite difference* schemes are used. In these schemes continuous derivatives are approximated by discrete differences. The parameter domain is covered by a grid with step sizes k in time and h in space, and the variables are discretized. For instance, $u(t,x)$ is replaced by $u_m^n \triangleq u(t=nk, x=mh)$ (Fig. 7.11).

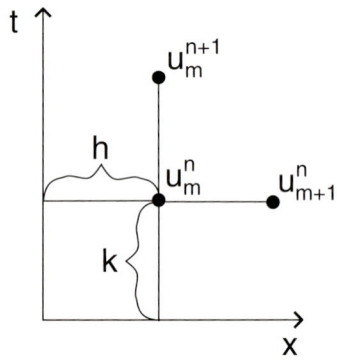

Fig. 7.11. The numerical grid for finite difference schemes

In most cases the design of satisfactory finite difference numerical schemes is quite straightforward. However, because of the size of the data, simplistic schemes might be inefficient and require a long run time. In the following subsections the main principles of the finite difference schemes are presented along with a few more elaborate schemes required to efficiently tackle the more challenging PDEs used for image processing.

7.4.1 Linear Diffusion

For one-dimensional linear diffusion, the first derivative in time of the function $u(t, x)$ can be approximated by the first-order accurate forward difference

$$D_t^+ u_m^n \triangleq \frac{u_m^{n+1} - u_m^n}{k},$$

and the second derivative in space can be approximated by the second-order central difference

$$D_{xx}^0 u_m^n \triangleq \frac{u_{m+1}^n - 2u_m^n + u_{m-1}^n}{h^2}.$$

The resulting numerical scheme for the linear diffusion is

$$\frac{u_m^{n+1} - u_m^n}{k} = \frac{u_{m+1}^n - 2u_m^n + u_{m-1}^n}{h^2},$$

and if we define

$$r \triangleq \frac{k}{h^2},$$

we get

$$u_m^{n+1} = (1 - 2r) u_m^n + r \left(u_{m+1}^n + u_{m-1}^n \right).$$

All we need is to add the initial condition

$$u_m^0 = f_m,$$

and to define the boundary conditions.

This is an *explicit* numerical scheme, because the value of u at iteration $n+1$ is given explicitly by the value of u at previous times (Fig. 7.12). The update step consists of merely additions and multiplications. The problem with explicit schemes is that their time step is limited by reasons of stability. For linear diffusion we require $r \leq 1/2$. Taking a bigger time step may result in an unstable process, whose outcome does not depend on the initial data but on the computation errors. In many equations the allowed time step is rather small and necessitates many iterations till the required output is reached. One solution is *implicit* numerical schemes, where the desired value u_m^{n+1} depends on the value of u at the same time $n+1$ and at other spatial locations (Fig. 7.12). One example is the Crank–Nicolson second order accurate scheme in time and space

$$\frac{u_m^{n+1} - u_m^n}{k} = \frac{1}{2} \left[\frac{u_{m+1}^{n+1} - 2u_m^{n+1} + u_{m-1}^{n+1}}{h^2} + \frac{u_{m+1}^n - 2u_m^n + u_{m-1}^n}{h^2} \right].$$

In this case we need to solve a tridiagonal system of equations in every update step. This can be done efficiently by the Thomas algorithm. A large time step would affect the accuracy of the solution, but it would not generate any instabilities.

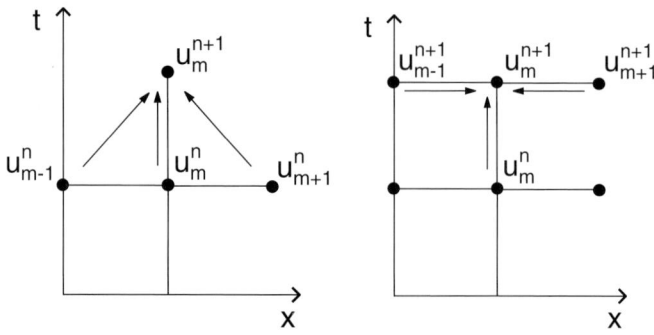

Fig. 7.12. The time and space dependencies of the explicit (*left*) and implicit (*right*) schemes for linear diffusion

For images where the equations have more than one dimension in space, explicit schemes are usually impractical due to the decrease of the bound on the time step. For linear diffusion we have $r \leq 1/(2D)$, with D the spatial dimension of the equation. On the other hand, implicit schemes result in a system of equations that is not tridiagonal and usually cannot be solved efficiently. More elaborate implicit schemes are required.

One such numerical scheme is the alternating direction implicit (ADI) scheme. Peaceman and Rachford's [22] version is

$$\left(\mathbb{I} - \frac{k}{2}A_1\right)\tilde{u}^{n+\frac{1}{2}} = \left(\mathbb{I} + \frac{k}{2}A_2\right)u^n,$$
$$\left(\mathbb{I} - \frac{k}{2}A_2\right)u^{n+1} = \left(\mathbb{I} + \frac{k}{2}A_1\right)\tilde{u}^{n+\frac{1}{2}}, \quad (7.4)$$

with $\mathbb{I}$ the identity matrix, and the operators $A_1 u = u_{xx}$ and $A_2 u = u_{yy}$ replaced by their second-order approximations. It can be seen from Eq. (7.4) that each iteration includes two steps where first the x direction is solved implicitly and the y direction explicitly, and then the opposite. Both steps consist of solving a tridiagonal system of equations, which can be done efficiently by the Thomas algorithm.

7.4.2 Nonlinear Diffusion

The original Perona–Malik filter [23] suffered from instabilities. The regularization presented by Catté et al. [4] consists of replacing $f(|\nabla I|)$ with $f(|\nabla I_\sigma|)$, where I_σ is the convolution of I with a Gaussian kernel with a standard deviation of σ. This smoothing of I eliminates some of the small-scale noise and makes the filter well-posed.

Weickert et al. [46] introduced the first-order accurate additive operator splitting (AOS) scheme to numerically implement this filter. The update step is

7 Geometric Filters, Diffusion Flows, and Kernels in Image Processing

$$I^{n+1} = \frac{1}{m} \sum_{i=1}^{m} (\mathbb{I} - mkA_i(I^n))^{-1} I^n,$$

with m the dimension of the image. The elements of the matrix A_i are given by

$$(A_i)_{pq} = \begin{cases} \frac{f_p + f_q}{2h^2} & q \in N(p) \\ -\sum_{l \in N(p)} \frac{f_p + f_l}{2h^2} & p = q \\ 0 & \text{otherwise} \end{cases}$$

with $N(p)$ the neighbors of the grid point p in the ith direction, and f_p the value of $f(|\nabla I_\sigma^n|)$ at grid point p.

The AOS scheme is semi-implicit and the size of the time step does not affect its stability. The scheme is efficient because it only requires the solution of tridiagonal systems of equations. It creates a discrete scale space [44], and its additivity gives equal importance to all coordinate axes, as opposed to the multiplicative locally one dimensional (LOD) scheme, which uses the update step

$$I^{n+1} = \prod_{i=1}^{m} (\mathbb{I} - kA_i(I^n))^{-1} I^n.$$

The AOS may also be used for some anisotropic nonlinear filters applied to gray level images. For color images and filters like the Beltrami flow, where each color component depends on the value of the others, the splitting is impossible. To date, there is no PDE-based implicit scheme for the color Beltrami. This is one of the main motivations for the construction of numerical kernels, described in the next section.

7.5 Kernels

It was shown in the previous section that the bound on the time step of some of the explicit numerical schemes can be alleviated by the use of implicit schemes. This enables a trade-off between the efficiency of the scheme and its accuracy. Unfortunately, this is not the case in some of the important geometric filters, such as the Beltrami filter. Another approach, namely the use of kernels, is the answer in some of these cases. Moreover, the kernels add a new perspective to these filters and present connections to other existing image-enhancing procedures.

7.5.1 The Gaussian Kernel for the Heat Equation

It can be shown that linear diffusion of an image can be accomplished by convolving it with a Gaussian kernel. Applying the heat equation to the two-dimensional data $I(u^1, u^2, t_0)$ for the duration t is equivalent to the convolution

$$I(u^1, u^2, t_0 + t) = \iint I(\tilde{u}^1, \tilde{u}^2, t_0) K(|u^1 - \tilde{u}^1|, |u^2 - \tilde{u}^2|; t) \mathrm{d}\tilde{u}^1 \mathrm{d}\tilde{u}^2$$
$$= I(u^1, u^2, t_0) * K(u^1, u^2; t) , \tag{7.5}$$

where the kernel is given by

$$K(u^1, u^2; t) = \frac{1}{4\pi t} \exp\left(-\frac{(u^1)^2 + (u^2)^2}{4t}\right) .$$

The use of the kernel enables us to replace the iterative application of the numerical scheme for the PDE with a one-step filter.

7.5.2 One-Dimensional Kernel for Nonlinear Diffusion

A kernel for the nonlinear diffusion of one-dimensional signals was presented in [30]. The nonlinear kernel adapts itself to the local amplitude of the signal. Adaptive filtering has been done before, mainly by using robust estimation techniques. However, the nonlinear kernel relates to the signal as a curve, and its adaptivity originates from the geometry of this curve.

The main idea behind the nonlinear kernel is presented in Fig. 7.13. For the linear kernel the amplitude of the filtered signal at a specific point is the sum of the neighboring points' amplitudes weighted according to their distance along the coordinate axis. For the nonlinear kernel the weighting is according to the distance on the signal itself. The nonlinear kernel 'resides' on the signal, while for the linear kernel the Gaussian 'resides' on the coordinate axis.

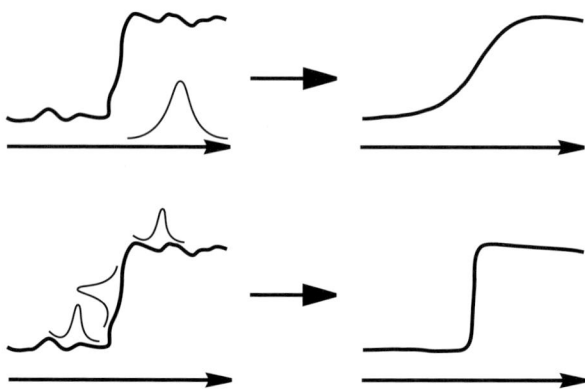

Fig. 7.13. Filtering a signal with a linear Gaussian kernel (*top*) and a nonlinear kernel (*bottom*)

The distance along the signal is calculated using the metric of the curve, which is the signal. Various metrics are possible, and they yield different filtering results. The Euclidean metric, for instance, using the curve representation

$C(p) = (x(p), y(p)) = (x, y(x))$ is $g(x) = 1 + y_x^2$. The kernel is constructed for the one-dimensional Beltrami flow

$$C_t = \Delta_g C.$$

The kernel cannot be global in time due to its nonlinearity (the kernel depends on the signal's local amplitudes, which change in each iteration of the kernel). Therefore the PDE cannot be replaced with a one-step filter like in linear diffusion. Only a short time kernel that is applied iteratively is possible. After each iteration the signal is

$$C(p, t_0 + t) = \int C(\tilde{p}, t_0) K(p, \tilde{p}; t) \mathrm{d}\tilde{p},$$

with the kernel

$$K(p, \tilde{p}; t) = \frac{H(p, \tilde{p}; t)}{\sqrt{t}} \exp\left(-\frac{\psi(p, \tilde{p})}{t}\right).$$

$H(p, \tilde{p}; t)$ can be taken to be a constant [30], and we get

$$\psi(p) = \frac{1}{4} \left(\int_p^{\tilde{p}} \mathrm{d}s\right)^2,$$

where $\mathrm{d}s$ is an arc length element given by $\mathrm{d}s = \sqrt{g(p)} \mathrm{d}p$. Since $\int_p^{\tilde{p}} \mathrm{d}s$ is the distance on the signal from point p to point $\tilde{p}$, the resulting kernel is indeed a Gaussian 'residing' on the signal (Fig. 7.13).

7.5.3 The Short Time Kernel for the Beltrami Flow

A short time kernel for the two-dimensional Beltrami flow was introduced in [36]. If used iteratively, it has an equivalent effect to that of the Beltrami flow. We replace Eq. (7.5) with

$$I^i(u^1, u^2, t_0 + t) = \iint I^i(\tilde{u}^1, \tilde{u}^2, t_0) K(u^1, u^2, \tilde{u}^1, \tilde{u}^2; t) \mathrm{d}\tilde{u}^1 \mathrm{d}\tilde{u}^2,$$

which we denote by

$$I^i(u^1, u^2, t_0 + t) = I^i(u^1, u^2, t_0) *_g K(u^1, u^2; t).$$

This is not a convolution in the strict sense, because K does not depend on the differences $u^i - \tilde{u}^i$. It will be shown later that $*_g$ is the geometric equivalent for manifolds of convolution. The general form of K is

$$K(u^1, u^2; t) = \frac{H(u^1, u^2; t)}{t} \exp\left(-\frac{\psi^2(u^1, u^2)}{t}\right),$$

where we take, without loss of generality, $(\tilde{u}^1, \tilde{u}^2) = (0,0)$ and omit from K the notation of dependency on these coordinates. In order to find K, we use the fact that it should satisfy Eq. (7.3), and after a few mathematical manipulations we get

$$g^{ij}\psi_i\psi_j = \|\nabla_g \psi\|^2 = \frac{1}{4},$$

with ∇_g the extension of the gradient to the manifold. This is the Eikonal equation on the manifold, and its viscosity solution is a geodesic distance map ψ on the manifold. The resulting short time kernel is

$$K(u^1, u^2, \tilde{u}^1, \tilde{u}^2; t) = \frac{H_0}{t} \exp\left(-\frac{\left(\int_{(u^1,u^2)}^{(\tilde{u}^1,\tilde{u}^2)} \mathrm{d}s\right)^2}{4t}\right),$$

$$= \frac{H_0}{t} \exp\left(-\frac{\mathrm{d}_g^2\left((u^1, u^2), (\tilde{u}^1, \tilde{u}^2)\right)}{4t}\right), \quad (7.6)$$

where $\mathrm{d}s$ is an arc length element on the manifold, and $\mathrm{d}_g(p_1, p_2)$ is the geodesic distance between two points, p_1 and p_2, on the manifold. Note that in the Euclidean space with a Cartesian coordinate system $\mathrm{d}_\mathrm{E}(p_1, p_2) = |p_1 - p_2|$. The geodesic distance on manifolds is therefore the natural generalization of the difference between coordinates in the Euclidean space. It is natural then to define the convolution on a manifold by

$$I^i(u^1, u^2) *_g K(u^1, u^2; t) = \iint I^i(\tilde{u}^1, \tilde{u}^2) K\left(\mathrm{d}_g\left((u^1, u^2), (\tilde{u}^1, \tilde{u}^2)\right)\right) \mathrm{d}\tilde{u}^1 \mathrm{d}\tilde{u}^2.$$

The resulting update step for the image is

$$I^i(u^1, u^2, t_0 + t) =$$

$$= \frac{H_0}{t} \iint_{(\tilde{u}^1, \tilde{u}^2) \in N(u^1, u^2)} I^i(\tilde{u}^1, \tilde{u}^2, t_0) \exp\left(-\frac{\left(\int_{(u^1,u^2)}^{(\tilde{u}^1,\tilde{u}^2)} \mathrm{d}s\right)^2}{4t}\right) \mathrm{d}\tilde{u}^1 \mathrm{d}\tilde{u}^2,$$

with $N(u^1, u^2)$ the neighborhood of the point (u^1, u^2), where the value of the kernel is above a certain threshold. Because of the monotonic nature of the fast marching algorithm used for the solution of the Eikonal equation, once a point is reached where the value of the kernel is smaller than the threshold, the algorithm can stop and thereby naturally bound the numerical support of the kernel. The value of the kernel for the remaining points of the manifold would be negligible. Therefore, the Eikonal equation is solved only in a small neighborhood of each image point. H_0 is taken such that integration over the kernel in the neighborhood $N(u^1, u^2)$ of the point equals one.

The short time Beltrami kernel in Eq. (7.6) is very similar to the bilateral filter kernel [6, 39]. The difference between them is that the Beltrami kernel uses geodesic distances on the image manifold, while the bilateral kernel uses

Euclidean distances. As can be seen from the derivation of the Beltrami kernel, the bilateral filter originates from image manifold area minimization. The bilateral filter can actually be viewed as an Euclidean approximation of the Beltrami flow.

The Euclidean distance used in the bilateral filter, while being easier to calculate, does not take into account the image intensity values between two image points. A point can have a relatively high kernel value, although it belongs to a different object than that of the filtered image point. The Beltrami kernel takes this effect into account and penalizes a point that belongs to a different connected component. That is, it is not 'as blind' as the bilateral filter to the spatial structure of the image.

The short time kernel for the Beltrami flow requires the solution of the Eikonal equation on the image manifold. The image manifold is a parametric manifold, where the metric G is given for every point. The solution to the Eikonal equation on parametric manifolds [33, 35] is based on the solution of the same problem on triangulated manifolds [17], which in turn is an extension of Sethian's fast marching method [29]. Another Eikonal solver on flat domains with regular grids was proposed by Tsitsiklis [41].

The original fast marching algorithm [29] solves the Eikonal equation in an orthogonal coordinate system. This is not the case for image manifolds. There $g_{12} \neq 0$ and we get a nonorthogonal coordinate system on the manifold. The solution for that is similar to that of [17], where a preprocessing stage is used to construct a suitable numerical stencil for each grid point. In this case there is no need to perform the unfolding step of [17] because the structure of the nonorthogonal grid on the manifold is given by its metric G. Figure 7.14 demonstrates the solution of the Eikonal equation for the parametric manifold $z = 0.5 \sin(4\pi x) \sin(4\pi y)$.

In order to demonstrate the spatial structure of the kernel, we tested it on the synthetic image in Fig. 7.15. At isotropic areas of the image, the kernel is isotropic, and its weights are determined solely by the spatial distance from the filtered pixel. Across edges the significant change in intensity is translated into a long geodesic distance, which results in negligent kernel weights on the other side of the edge. The filtered pixel is computed as an average of the pixels on the 'right' side of the edge.

7.6 Conclusion

This chapter described image enhancement using PDE based geometric diffusion flows. On the theoretical side, starting with variational principles explains the origin of the flows, and the geometric approach results in some nice invariance properties. On the practical side, using carefully selected numerical schemes and developing kernels for the flows enables an efficient and robust implementation. Combined together, we get a fascinating area of research yielding state-of-the-art algorithms.

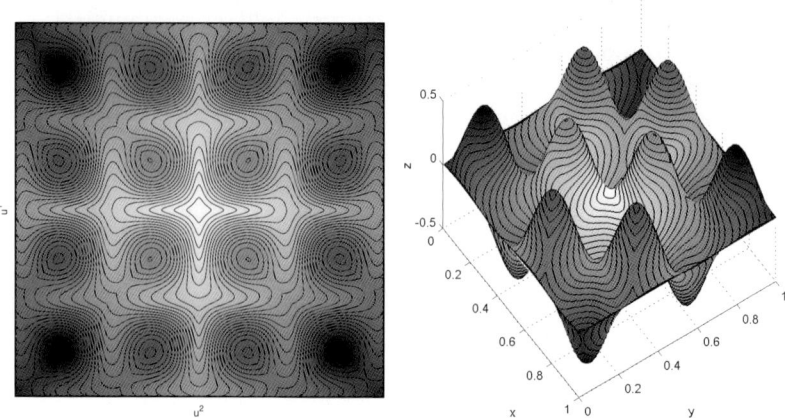

Fig. 7.14. Fast marching on the manifold $z = 0.5\sin(4\pi x)\sin(4\pi y)$. *Left*: implemented on the parameterization plane. *Right*: projected on the manifold. Lower values are assigned brighter colors. The *black curves* are the level curves

Fig. 7.15. Level curves of the kernel at various locations in a synthetic image

References

1. L. Alvarez, F. Guichard, P.L. Lions, J.M. Morel (1993) Axioms and fundamental equations of image processing. *Arch. Rational Mechanics*, 123:199–257
2. L. Alvarez, L. Mazora (1994) Signal and image restoration using shock filters and anisotropic diffusion. *SIAM J. Numer. Anal*, 31:590–605
3. C. Ballester, V. Caselles, P. Monasse (2001) The tree of shapes of an image. In: *Preprint, C.M.L.A, No. 2001-02, Ecole Normale Supérieure de Cachan*
4. F. Catté, P. Lions, J. Morel, T. Coll (1992) Image selective smoothing and edge detection by nonlinear diffusion. *SIAM Journal of Numerical Analysis*, 29:182–193
5. S. Di Zenzo (1986) A note on the gradient of a multi image. *Computer Vision, Graphics, and Image Processing*, 33:116–125

6. M. Elad (2002) On the bilateral filter and ways to improve it. *IEEE Transactions on Image Processing*, 11(10):1141–1151
7. L.C. Evans, J. Spruck (1991) Motion of level sets by mean curvature, I. *J. Diff. Geom.*, 33
8. O. Faugeras, R. Keriven (1995) Scale-space and affine curvature. In: *Proceedings Europe–China Workshp on Geometrical modelling and Invariants for Computer Vision*, pp. 17–24
9. D. Gabor (1965) Information theory in electron microscopy. *Laboratory Investigation*, 14(6):801–807
10. M.A. Grayson (1987) The heat equation shrinks embedded plane curves to round points. *J. Diff. Geom.*, 26:285–314
11. H. Helmholtz (1896) *Handbuch der Psychologischen Optik*. Voss, Hamburg
12. R. Kimmel (1996) Affine differential signatures for gray level images of planar shapes. In: *Proc. of ICPR'96*, Vienna, Austria
13. R. Kimmel (1997) Intrinsic scale space for images on surfaces: The geodesic curvature flow. *Graphics Modeling and Image Processing*, 59(5):365–372
14. R. Kimmel (2003) *Numerical Geometry of Images: Theory, Algorithms, and Applications*. Springer, Berlin Heidelberg New York
15. R. Kimmel, R. Malladi, N. Sochen (1998) Image processing via the Beltrami operator. In: *Proc. of 3-rd Asian Conf. on Computer Vision*, Hong Kong
16. R. Kimmel, R. Malladi, N. Sochen (2000) Images as embedding maps and minimal surfaces: Movies, color, texture, and volumetric medical images. *International Journal of Computer Vision*, 39(2):111–129
17. R. Kimmel and J. Sethian (1998) Computing geodesic paths on manifolds. *Proceedings of National Academy of Sciences*, 95(15):8431–8435
18. R. Kimmel, N. Sochen (2002) Orientation diffusion or how to comb a porcupine? *Special issue on PDEs in Image Processing, Computer Vision, and Computer Graphics, Journal of Visual Communication and Image Representation*, 13:238–248
19. T.S. Lee (1996) Image representation using 2D Gabor-wavelets. *IEEE Trans. on Pattern Analysis and Machine Intelligence*, 18(10):959–971
20. M. Lindenbaum, M. Fischer, A.M. Bruckstein (1994) On Gabor's contribution to image enhancement. *Pattern Recognition*, 27(1):1–8
21. S. Osher, J. Sethian (1988) Fronts propagation with curvature dependent speed: Algorithms based on Hamilton–Jacobi formulations. *J. Comput. Phys.*, 79:12–49
22. D. Peaceman, H. Rachford (1955) The numerical solution of parabolic and elliptic differential equations. *Journal of the Society for Industrial and Applied Mathematics*, 3:28–41
23. P. Perona, J. Malik (1990) Scale-space and edge detection using anisotropic diffusion. *IEEE Trans. on Pattern Analysis and Machine Intelligence*, 12:629–639
24. Y. Rubner, C. Tomasi (2000) *Perceptual Metrics for Image Database Navigation*. Kluwer Academic, Boston, NY
25. L. Rudin, S. Osher, and E. Fatemi (1992) Nonlinear total variation based noise removal algorithms. *Physica D*, 60:259–268
26. G. Sapiro (2001) *Geometric Partial Differential Equations and Image Processing*. Cambridge University Press, Cambridge
27. G. Sapiro, D.L. Ringach (1996) Anisotropic diffusion of multivalued images with applications to color filtering. *IEEE Trans. Image Proc.*, 5:1582–1586

28. E. Schrödinger (1920) Grundlinien einer theorie der farbenmetrik in tagessehen. *Ann. Physik*, 63:481
29. J. Sethian (1996) A fast marching level set method for monotonically advancing fronts. *Proceedings of National Academy of Sciences*, 93(4):1591–1595
30. N. Sochen, R. Kimmel, A.M. Bruckstein (2001) Diffusions and confusions in signal and image processing. *Journal of Mathematical Imaging and Vision*, 14(3):195–209
31. N. Sochen, R. Kimmel, R. Malladi (1998) A general framework for low level vision. *IEEE Trans. on Image Processing*, 7(3):310–318
32. N. Sochen, Y.Y. Zeevi (1998) Representation of colored images by manifolds embedded in higher dimensional non-Euclidean space. In: *Proc. of ICIP98*, pp. 166–170, Chicago, IL
33. A. Spira, R. Kimmel (2004) An efficient solution to the Eikonal equation on parametric manifolds. accepted to *Interfaces and Free Boundaries*
34. A. Spira, R. Kimmel (2002) Geodesic curvature flow on parametric surfaces. In: *Curve and Surface Design: Saint-Malo 2002*, pp. 365–373, Saint-Malo, France
35. A. Spira, R. Kimmel (2003) An efficient solution to the Eikonal equation on parametric manifolds. In: *INTERPHASE 2003 meeting, Isaac Newton Institute for Mathematical Sciences, 2003 Preprints, Preprint no. NI03045-CPD*, UK
36. A. Spira, R. Kimmel, N. Sochen (2003) Efficient Beltrami flow using a short time kernel. In: *Proc. of Scale Spcace 2003, Lecture Notes in Computer Science (vol. 2695)*, pp. 511–522, Isle of Skye, Scotland, UK
37. W.S. Stiles (1946) A modified Helmholtz line element in brightness–colour space. *Proc. Phys. Soc. (London)*, 58:41
38. B.M. ter Haar Romeny (1994) *Geometry-Driven Diffusion in Computer Vision*. Kluwer Academic, Dordrecht
39. C. Tomasi, R. Manduchi (1998) Bilateral filtering for gray and color images. In: *Sixth International Conference on Computer Vision*, Bombay, India
40. D. Tschumperlé (2002) *PDE's based regularization of multivalued images and applications*. Ph.D. thesis
41. J.N. Tsitsiklis (1995) Efficient algorithms for globally optimal trajectories. *IEEE Trans. on Automatic Control*, 40(9):1528–1538
42. J.J. Vos, P.L. Walraven (1972) An analytical desription of the line element in the zone-fluctuation model of colour vision. II. The derivative of the line element. *Vision Research*, 12:1345–1365
43. J. Weickert (1996) Theoretical foundation of anisotropic diffusion in image processing. *Computing, Suppl.*, 11:221–236
44. J. Weickert (1998) *Anisotropic Diffusion in Image Processing*. Teubner, Stuttgart
45. J. Weickert, S. Ishikawa, A. Imiya (1999) Linear scale-space has first been proposed in Japan. *Journal of Mathematical Imaging and Vision*, 10:237–252
46. J. Weickert, B.M. ter Haar Romeny, M.A. Viergever (1998) Efficient and reliable scheme for nonlinear diffusion filtering. *IEEE Trans. on Image Processing*, 7(3):398–410

8

Chaos-Based Image Encryption

Yaobin Mao[1] and Guanrong Chen[2]

[1] Nanjing University of Science and Technology `maoyaobin@163.com`
[2] City University of Hong Kong `gchen@ee.cityu.edu.hk`

8.1 Introduction

With the proliferation of the Internet and maturation of the digital signal processing technology, applications of digital imaging are prevalent and are still continuously and rapidly increasing today. Yet the main obstacle in the widespread deployment of digital image services has been enforcing security and ensuring authorized access to sensitive data. In this regard, a direct solution is to use an encryption algorithm to mask the image data streams, which has led to the celebrated number-theory-based encryption algorithms such as Data Encryption Standard (DES), International Data Encryption Algorithm (IDEA), and the algorithm developed by Rivest, Shamir and Adleman (RSA) [24, 40, 41]. However, these encryption schemes appear not to be ideal for image applications, due to some intrinsic features of images such as bulk data capacity and high redundancy, which are troublesome for traditional encryption. Moreover, these encryption schemes require extra operations on compressed image data, thereby demanding long computational time and high computing power. In real-time communications, because of their low encryption and decryption speeds, they may introduce significant latency.

Compared with text encryption, which most existing encryption standards aim at, image encryption (or more generally, multimedia encryption) has its own characteristics and special features with many unique specifications. In order to develop effective image encryption techniques, these have to be fully understood. In the following section, some basic concepts in cryptography with respect to image encryption are introduced.

8.1.1 Fundamentals of Cryptology

The basic idea of encryption is to modify the message in such a way that its content can be reconstructed only by a legal recipient. A discrete-valued *cryptosystem* can be characterized by [33]

- a set of possible *plaintexts*, $\mathcal{P}$
- a set of possible *ciphertexts*, $\mathcal{C}$
- a set of possible *cipherkeys*, $\mathcal{K}$
- a set of possible encryption and decryption transformations, $\mathcal{E}$ and $\mathcal{D}$

For each key, $k \in \mathcal{K}$, there exists an encryption function $e(k,\cdot) \in \mathcal{E}$ and a corresponding decryption function $d(k,\cdot) \in \mathcal{D}$, such that for each plaintext $p \in \mathcal{P}$ the condition for unique decoding, $d(k, e(k, p)) = p$, is satisfied.

The security of a cryptosystem usually relies on the key only. In other words, it is assumed that the opponent knows the structure of the encryption system, has the ciphering algorithm, and has access to the transmission channel to obtain an arbitrary segment of the ciphertext c. A good cryptosystem always allows for this situation to happen, and this condition or requirement is referred to as *Kerckhoff's principle* [41].

Encryption algorithms, also called *ciphers*, can also be classified with respect to the structures of the algorithms. There are two kinds of ciphers: stream ciphers and block ciphers.

A block cipher is a type of symmetric-key encryption algorithm that transforms a fixed-length block of plain-text data into a block of ciphertext data of the same length. The fixed length is called the block size, and for many block ciphers the block size is 64 or 128 bits. The larger the block size, the more secure the cipher, but the more complex the encipher and decipher algorithms and devices. Typical block ciphers include DES, Triple DES, Blowfish, IDEA, and AES; some of them have become cipher standards lately.

Unlike block ciphers, which operate on large blocks of data, stream ciphers typically operate on smaller units of plaintext, usually bits. So, stream ciphers can be designed to be exceptionally fast, much faster than a typical block cipher. Generally, a stream cipher generates a sequence of bits as a key (called *keystream*), and the encryption is accomplished by combining the keystream with the plaintext. Usually, the bitwise Exclusive-OR (XOR) operation [24] is chosen to perform ciphering, basically for its simplicity. As of today, no stream cipher has emerged as a standard. The most widely used stream cipher is RC4, while RC5 has been incorporated into some major products such as BSAFE, JSAFE, and S/MAIL of the RSA Data Security, Inc. [40]

Traditional cryptology is studied by applying mathematical tools such as number theory, algebra, algebraic geometry, and combinatorics [9]. In the past years, a new approach of constructing cryptosystems based on the theory of chaotic dynamical systems has been gradually developed. The similarity and difference of both traditional cryptology and chaotic cryptology are further expounded in Sect. 8.2.

A good cipher should have strong ability to withstand all kinds of cryptanalysis and attacks that try to break the system. To a certain extent, the resistance against attacks is a good measure of the performance of a cryptosystem; thus, it is often used to evaluate cryptosystems.

According to the method of the opponent's access to additional information, attacks on a cryptosystem may be classified into four classes:

- *Ciphertext-only attack*: Opponent has access to communication channel and can eavesdrop some segments of the ciphertext, encrypted by a certain key. The task of the opponent is to reveal as much plaintext as possible, and even to be able to deduce the cipher key.
- *Known-plaintext attack*: In addition to the obtained ciphertext segments, the opponent knows also an associated piece of plaintext. The task of the opponent is then to deduce the cipher key.
- *Chosen-plaintext attack*: The opponent not only has access to some segments of the cipher and the plaintext, but also can choose plaintext to encrypt and accordingly gets some corresponding ciphertext that he wants for comparison. This kind of attack is more intensive than the known-plaintext attack.
- *Chosen-ciphertext attack*: The opponent can choose different segments of the ciphertext and accordingly get its corresponding plaintext.

Apart from the aforementioned typical attacks, there is a type of attack named *exhaustive key search*, which tries all possibilities for the key in the keyspace to completely decrypt the plain message. If the keyspace of a cipher is relatively small, this exhaustive searching works quite well, given the availability of supercomputing power today.

It should be emphasized that any encryption algorithm, traditional or chaos-based, should obey basic cryptographical principles in order to be able to resist serious attacks.

8.1.2 Particularities of Image Encryption

Unlike text messages, image data have special features such as bulk capacity, high redundancy, and high correlation among pixels, not to mention that they usually are huge in size, which together make traditional encryption methods difficult to apply and slow to process. Sometimes image applications also have their own requirements like real-time processing, fidelity reservation, image format consistence, and data compression for transmission. Simultaneous fulfillments of these requirements, along with high security and high quality demands, have presented great challenges to real-time imaging practice. One example is the case where one needs to manage both encryption and compression. In doing so, if an image is to be encrypted after its format is converted, say from a TIFF file to a GIF file, encryption has to be implemented before compression. However, a conventional encrypted image has very little compressibility. On the other hand, compression will make a correct and lossless decipher impossible, particularly when a highly secure image encryption scheme is used. This conflict between the compressibility and the security is very difficult, if not impossible, to completely resolve.

Particularities of image encryption may be summarized as follows:

1. High redundancy and bulk capacity generally make encrypted image data vulnerable to attacks via cryptanalysis. Based on the bulk capacity, the opponent can gain enough ciphertext samples (even from one picture) for statistical analysis. Meanwhile, since data in images have high redundancy, adjacent pixels likely have similar grayscale values, or image blocks have similar patterns, which usually embed the image with certain patterns that result in secret leakage.
2. Image data have strong correlations among adjacent pixels, which makes fast data-shuffling quite difficult. Statistical analysis on large numbers of images shows that averagely adjacent 8 to 16 pixels are correlative in the horizontal, vertical, and also diagonal directions for both natural and computer-graphical images. According to Shannon's information theory [35], a secure cryptosystem should fulfill a condition on the information entropy, $E(P|C) = E(P)$, where P stands for plain message and C for ciphered message; that is, the ciphered (i.e., encrypted) image should not provide any information about the plain image. To meet this requirement, therefore, the ciphered image should be presented as randomly as possible. Since a uniformly distributed message source has a maximum uncertainty [34], an ideal cipher image should have an equilibrium histogram, and any two adjacent pixels should be uncorrelated statistically. This goal is not easy to achieve under only a few rounds of permutation and diffusion.
3. Bulk capacity of image data also makes real-time encryption difficult. Compared with texts, image data capacity is horrendously large. For example, a common 24-bit true-color image of 512-pixel height and 512-pixel width occupies $512 \times 512 \times 24/8 = 768$ KB in space. Thus, a one-second motion picture will reach up to about 19 MB. Real-time processing constraints are often required for imaging applications, such as video conferencing, image surveillance, and so on. Vast amounts of image data put a great burden on the encoding and decoding processes. Encryption during or after the encoding phase, and decryption during or after the decoding phase, will aggravate the problem. If an encryption algorithm runs very slowly, even with high security, it would have little practical value for real-time imaging applications. That is the reason why current encryption methods such as DES, IDEA, and RSA are not the best candidates for this consideration.
4. Image encryption is often to be carried out in combination with data compression. In almost all cases, the data are compressed before they are stored or transmitted due to the huge amount of image data and their very high redundancy. Thus, directly incorporating security requirements in the data compression system is a very attractive approach. The main challenge is how to ensure reasonable security while reducing the computational cost without downgrading the compression performance.
5. In image usage, file format conversion is a frequent operation. It is desirable that image encryption not affect such an operation. Thus, directly treating image data as ordinary data for encryption will make file format

conversion impossible. In this scenario, content encryption, where only the image data are encrypted, leaving file header and control information unencrypted, is preferable.
6. Human vision has high robustness to image degradation and noise. Only encrypting those data bits tied with intelligibility can efficiently accomplish image protection [47]. However, conventional cryptography treats all image data bits equally in importance, and thus requires a considerable amount of computational power to encrypt all of them, which has often proved unnecessary.
7. In terms of security, image data are not as sensitive as text information. Security of images is largely determined by the real situation in an application. Usually, the value of the image information is relatively low, except in some specific situations like military and espionage applications or video conferencing in business. A very expensive attack of encrypted median data is generally not worthwhile. In practice, many image applications do not have very strict security requirements. Under certain circumstances, protection of the fidelity of an image object is more important than its secrecy. An example is electronic signatures. As another example, in image database applications, only those users who have paid for the service can have access to large-size images with high resolution. Adversaries may be able to get some small-size images with low resolution by attacks based on cryptanalysis, but those images have little business values–and perhaps much cheaper than the cost are of preparing and executing the attacks. In the worst case, possible partial leakage of some secrecy in multi-media, within a certain limitations, is always permitted, while for text information this scenario is largely forbidden because it is then quite easy to predict the entire message based on the obtained information from a partial leakage.

Today, there does not seem to be any image encryption algorithm that can fulfill all the aforementioned specifications and requirements.

Chaos-based image encryption, further described below, cannot solve all these problems either. However, it can provide a class of very promising methods that can partially fulfill many of these requirements and demonstrate superiority over the conventional encryption methods, particularly with a good combination of speed, security, and flexibility. As seen below, through an elaborative design, either chaotic block cipher or chaotic stream cipher can achieve very good overall performance.

8.1.3 Some Existing Image Encryption Schemes

Some image encryption methods have been proposed in the current literature. In order to inspire the development of better chaotic ciphers, this review is not only intended for chaos-based methods, but is also meant for understanding image encryption technology in general.

Image encryption algorithms, which can be classified with respect to the approach in constructing the scheme, are divided into two groups here: *chaos-based methods* and *non-chaos-based methods*. Image encryption also can be divided into *full encryption* and *partial encryption* (also called *selective encryption*) according to the percentage of the data encrypted. Moreover, they can be classified into *compression-combined methods* and *noncompression methods*.

Some existing proposals of chaos-based image encryption algorithms are now introduced. In [13], two kinds of schemes based on higher-dimensional chaotic maps were proposed. By using a discretized chaotic map, pixels in an image are permuted in shuffling after several rounds of operations. Between every two adjacent rounds of permutations, a diffusion process is performed, which can significantly change the distribution of the image histogram that makes statistical attack infeasible. Empirical testing as well as cryptanalysis both demonstrated that the chaotic baker map and cat map are good candidates for this kind of image encryption. Similar thoughts also appeared, e.g., in [31], where a fast bulk data encryption scheme was designed by combining chaotic Kolmogorov flows with an adaption of a very fast shift-register-based pseudorandom number generator.

The aforementioned schemes are block cipher, and they have some prominent merits, including high security and fast processing. However, their defects are also significant since the encrypted image has very little compressibility and is unable to abide any lossy compression (e.g., JPEG). To alleviate the conflict between compressibility and encryption, several suggestions of combining compression and encryption have been proposed. In [47], the so-called MHT scheme was proposed that encrypts image via a manipulation of Huffman coding tables in the image coding system. The MHT scheme chooses several different Huffman tables from a large number of possible candidates, and uses them alternatively to encode the image data. The choice of Huffman tables and the order in which they are used are kept secret as the key. It was advocated that the method requires very little computational overhead and can be applied to MPEG and JPEG/JPEG 2000, but it cannot resist chosen-plaintext attacks [47].

A somewhat different chaos-based image encryption method was proposed in [2] that makes use of the SCAN language. Through substitution of each pixel based on an additive noise vector and scramble scanning patterns, an image can be encrypted and compressed simultaneously. The idea seems to be quite good, but it was pointed out in [6] that this method is weak against exhaustive key searching and chosen-plaintext attacks. In [3], another image compression and encryption algorithm was proposed based on the lossless quadtree image compression scheme. The quadtree data structure is used to represent the image, and the scanning sequences of image data comprise a private key for encryption. Also in [6], numerous attacks on the proposed algorithm were tested and presented, which include keyspace reduction, histogram attack, known-plaintext attack, and chosen-plaintext attack.

In order to speed up encryption processes so as to make them feasible for real-time applications, most of the existing schemes follow the idea of selective encryption. Actually, according to Shannon's theory, both encryption and compression are processes of redundancy reduction [34], but their purposes are different. In [7], several partial encryption schemes were provided. It was reported that by a partial encryption, only 13% to 27% of the output from a quadtree compression algorithm is encrypted for a typical image, and less than 2% is encrypted for a 512×512 image compressed by set-partitioning in the hierarchical trees algorithm.

There are also several proposed schemes that consider matching the compatibility to current international standards. Since many international standards on videos and images use block-based discrete cosine transform (DCT), including the familiar JPEG, MPEG-1, MPEG-2, H.261, and H.263 formats, the current research has been concentrated on selective encryptions within the framework of DCT. However, with the emergence of MPEG-4 and JPEG-2000, research emphasis may soon be redirected to a combination of encryption and wavelet compression. In the following, some proposed schemes are briefly reviewed and commented upon.

To achieve high encryption speed, in the early stage some elementary cryptographic methods using random permutation lists were suggested. Since the operations are simple, the encryption does not require high computational cost. The challenge is how to achieve reasonable security with such simple operations. The method recommended in [44] replaces the zig-zag scan by the random permutation lists of MPEG. In doing so, if the decoder does not know the permutation lists, the DCT coefficients in a block will be in the wrong order although the values are not modified. It is well known that encryption using only permutation is not secure enough, therefore it was pointed out in [45] that the method proposed in [44] and its enhancement version given in [39] are not able to resist known-plaintext attacks.

Another fast encryption scheme was proposed in [37], which encrypts the sign bits of the DCT coefficients (i.e., the sign bits of differential DC values for the DC coefficients). Because DC values significantly affect the quality of an image, changing them will render the whole image unreadable. For the same reason as discussed in [44] and [39], the method proposed in [37] is not secure enough either. Therefore, an enhanced scheme called RVEA was composed in [38], which tried to implement DES or IDEA aiming to strengthen the sign-bit encryption.

Since wavelet-based image compression achieves both high compression rates with reasonably high image quality and low computational complexity, many image compression standards (for moving or still pictures) have selected to use wavelets. Integrating an encryption algorithm with wavelet image coding is reasonable and has great usage potential. In [46], a wavelet-based system combining compression and encryption was recommended. By using Antonini wavelets [1], an image is decomposed into several subbands. In each level of the subbands, encryption is performed using random permutation.

Experiments show that permutation does not affect compression significantly: according to [46], it causes only 2% compression-rate drops. But permutation in different levels of the subbands affects the image quality significantly, so encryption performed on low-pass subbands may render the whole image unreadable, while one performed on high-pass subbands may only create some noiselike spots on the image. This scheme is also unable to resist the known-plaintext attack or chosen-plaintext attack. Knowing this, DES may be used to strengthen its security, but it brings in extra computational loads.

8.2 Chaos-Based Encryption Schemes

Since the demonstration of possibility for self-synchronization of chaotic oscillations [26], a great deal of work on application of chaos to cryptography has been carried out in the last decade. Early works on chaos in cryptography were connected with encrypting messages through modulation of chaotic orbits of continuous-time dynamical systems. These methods are strongly related to the concept of synchronization of two chaotic systems and to chaos control [5]. Several different ways have been proposed to achieve synchronization of chaotic systems, thereby transmitting information on a chaotic carrier signal. Some typical forms have been brought up, which includes chaotic masking, chaotic shift keying, and chaotic modulation using inverse systems [10, 15].

In spite of the fact that many "secure" communication schemes have been proposed based on the use of the chaos synchronization principle, they all suffer from some common weakness. The following technical problems were listed in [23]:

- It is difficult to determine the synchronization time; therefore, the message during the transient period will be lost, sometimes causing fairly long transient times.
- Noise throughout the transmission significantly affects the intended synchronization. This means the synchronization noise intensity should be small compared to the signal level, or the desired synchronization will not be achieved.
- Technically, it is difficult to implement two well-matched analog chaotic systems, which are required in synchronization, and if this is not required (i.e., with certain robustness) then the opponent can also easily achieve the same synchronization for attack.

In contrast to synchronization-based techniques, a direct application of a chaotic transformation to a plaintext, or applying a chaotic signal in the design of an encryption algorithm, seems to be a more promising approach. The sensitivity to initial conditions and parameters as well as the mixing (ergodicity) characteristics of chaos are very beneficial to cryptosystems. The main difference is that cryptosystems are operated on a finite set of integers, while chaotic maps are defined on an infinite set of real numbers. Therefore, how to merge these two kinds of systems so as to take advantage of the good properties of chaos is worthy of further exploration.

In the next section, some basic concepts of chaos are introduced, and the possibility of integrating chaos into the design of better encryption algorithms is investigated.

8.2.1 Basic Features of Chaos

Chaos is a ubiquitous phenomenon existing in deterministic nonlinear systems that exhibit extreme sensitivity to initial conditions and have random-like behaviors. Since its discovery by Edward N. Lorenz in 1963 [22], chaos theory has become a branch of scientific studies today [5]. Since discrete chaotic dynamic systems (i.e., maps) are used in cryptography, this notion is briefly introduced.

Definition of discrete chaos

There are several definitions of chaos, which are similar but are actually not equivalent [4]. Only a textbook definition is introduced here for brevity.

For simplicity, one-dimensional maps are discussed. Consider a discrete dynamical system in the general form of

$$x_{k+1} = f(x_k), \quad f : I \longrightarrow I, \quad x_0 \in I, \tag{8.1}$$

where f is a continuous map on the interval $I = [0,1]$. This system is said to be *chaotic* if the following conditions are satisfied [11]:

1. Sensitive to initial conditions:

$$\exists \delta > 0 \ \forall x_0 \in I, \varepsilon > 0 \ \exists n \in \mathbf{N}, y_0 \in I :$$
$$|x_0 - y_0| < \varepsilon \Rightarrow |f^n(x_0) - f^n(y_0)| > \delta. \tag{8.2}$$

2. Topological transitivity:

$$\forall I_1, I_2 \subset I \ \exists x_0 \in I_1, n \in \mathbf{N} : f^n(x_0) \in I_2. \tag{8.3}$$

3. Density of periodic points in I:
   Let $P = \{p \in I | \exists n \in \mathbf{N} : f^n(p) = p\}$ be the set of periodic points of f. Then P is dense in I: $\overline{P} = I$.

This definition has some redundancy, which is not discussed here.

The sensitivity of chaos to initial conditions is often illustrated as *the butterfly effect*, which is rooted in Lorenz's original wording "Does the flap of a butterfly's wings in Brazil set off a tornado in Texas?" This sensitivity property is commonly utilized for the keys of cryptosystems.

The topological transitivity property ensures the ergodicity of a chaotic map, which means that if we partition the state space into a finite number of regions, no matter how many, any orbit of the map will pass through all these regions. This property is linked to the diffusion feature of cryptosystems.

In chaotic cryptology, the above two properties are often used to construct stream ciphers and block ciphers. Further comparison given in the following section.

8.2.2 Relationships Between Chaos and Cryptography

There have been many discussions in the literature about the relationships between chaotic systems and cryptosystems [10, 16, 17, 18, 32]. As mentioned above, the main difference between chaos theory and cryptography is that cryptosystems work on a finite field, while chaos is meaningful only on a continuum. Nevertheless, these two scientific notions are very closely related. Many fundamental concepts in chaos theory, such as mixing and sensitivity to initial conditions and parameters, actually coincide with those in cryptography.

The following excerpt from Shannon's masterpiece [35] demonstrates that cryptographic algorithms have unconsciously used the mixing property of chaos, even before the dawn of chaos research [12, 13, 16, 17]:

> Good mixing transformations are often formed by repeated products of two simple non-commuting operations. Hopf has shown, for example, that pastry dough can be mixed by such a sequence of operations. The dough is first rolled out into a thin slab, then folded over, then rolled, and the folded again, etc. ... In a good mixing transformation ... functions are complicated, involving all variables in a sensitive way. A small variation of any one (variable) changes (the output) considerably.

The similarities and differences between the two subjects can be listed [16], as shown in Table 8.1. Chaotic maps and cryptographic algorithms have some similar properties: both are sensitive to tiny changes in initial conditions and parameters; both have random like behaviors; and cryptographic algorithms shuffle and diffuse data by rounds of encryption, while chaotic maps spread a small region of data over the entire phase space via iterations. The only difference in this regard is that encryption operations are defined on finite sets of integers while chaos is defined on real numbers.

8.2.3 Chaos for Cryptography

It is natural to apply the discrete chaos theory to cryptography for the following reasons [18]:

- The property of sensitive dependence of orbits on initial conditions makes the nature of encryption very complicated. Suppose that one has the following chaos-based encryption scheme:

Table 8.1. Similarities and differences between chaos and cryptography

Chaotic systems	Cryptographic algorithms
Phase space: set of real numbers	Phase space: finite set of integers
Iterations	Rounds
Parameters	Key
Sensitivity to initial conditions and parameters	Diffusion

For a plaintext P in $(0,1)$, some parameters of a chaotic map are used as the key for encryption. Choose a one-dimensional chaotic dynamical system (I, ϕ) to perform encryption. Then, the encryption procedure is the n-fold iteration of the map ϕ with the initial value P, and the ciphertext C is a result of the encryption:

$$C = \phi^n(P) = (\phi(\phi(\cdots \phi(P)))). \tag{8.4}$$

Since the map is chaotic, it has a positive Lyapunov exponent, namely, at some point $x \in I$, $\lambda_x > 0$; therefore,

$$\forall \varepsilon > 0 \ \exists n_1, n_2 \ \exists U_{n_1,n_2} \ni x, \forall n_1 \leq n \leq n_2, \forall z_1, z_2 \in U_{n_1,n_2}$$

$$e^{(\lambda_x - \varepsilon)n}|z_1 - z_2| < |\phi^n(z_1) - \phi^n(z_2)| < e^{(\lambda_x + \varepsilon)n}|z_1 - z_2|, \tag{8.5}$$

where U_{n_1,n_2} is some neighborhood of x in I. The above expression implies that, after n iterations, the initial distance $|z_1 - z_2|$ between two arbitrarily close but distinct points z_1 and z_2 will increase exponentially as $e^{\lambda_x n}|z_1 - z_2|$. That means if one uses k_1 as the encipher key but the opponent uses k_2 to decipher the message, then even with $|k_1 - k_2| < \varepsilon$, the difference $|z_1 - z_2|$ will be significantly huge. This prevents the system from any brute-force attack.

- A Lebesgue measure μ is said to be *invariant* if and only if it satisfies

$$\forall A \in \sigma(X), \ \mu(A) = \mu(\phi(A)), \tag{8.6}$$

where $\sigma(X)$ is the σ-algebra of all measurable subsets in X. Here, (X, ϕ) is called *ergodic* if and only if it has only trivial invariant sets, i.e., $\phi(B) \subseteq B$ implies either $\mu(B) = 0$ or $\mu(B) = \mu(X)$. The ergodicity implies that the state space cannot be nontrivially divided into several subspaces. So, if some orbit starts from an arbitrary point x, it will then never be restricted within a small region. This property indicates that if a chaotic map is used to compose encryption then the plaintext space will not be restricted to a small subspace. Thus, for ciphertext C, to search for the corresponding plaintext P one must go over the entire state space X.
- The aforementioned system is *mixing* if the following condition is satisfied (assume $\mu(X) = 1$):

$$\lim_{n \to \infty} \frac{\mu\left(\phi^n(B) \bigcap A\right)}{\mu(A)} = \frac{\mu(B)}{\mu(X)}. \tag{8.7}$$

This property implies that after n iterations, part of B will be contained in A, and the percentage of B that are contained in A is asymptotically proportional to the percentage of B in X with respect to the measure μ. Thus, plaintext P is thoroughly contained in its ciphertext C if a chaotic map is used for encryption.

The above-described properties of chaos are the foundations of chaotic crytography. To design secure cryptographic algorithms for both stream and block ciphers, all these properties should be well utilized.

8.3 Chaos-Based Image Encryption

Many chaos-based encryption schemes have been proposed, and some of them have been extended from text encryption to image encryption. A direct extension of a chaos-based text encryption scheme to also work for images is possible, but this simple modification may not provide an efficient solution to these image encryption problems. As pointed out in Sect. 8.1.2, image encryption has its own specifications such as encryption speed, compatibility to image format and compression standards, and real-time implementation, therefore it requires a special design of the encryption algorithm.

Some existing image encryption schemes were briefly reviewed in Sect. 8.1.3. However, with a few exception, dedicated chaos-based image encryption schemes do not often appear in the literature. These exceptions are further discussed here.

In [49], an encryption method called chaotic key-based algorithm (CKBA) was proposed. The algorithm first generates a time series based on a chaotic map, and then uses it to create a binary sequence as a key. According to the binary sequence generated, image pixels are rearranged and XOR or XNOR operated with the selected key. This method is very simple but has defects in security, as pointed out in [20]: this method is very weak to the chosen/known-plaintext attack with only one plain image. Moreover, its security to brute-force attack is also questionable.

In [31], a chaotic Kolmogorov-flow-based image encryption algorithm was designed. In this scheme, the whole image is taken as a single block and permuted through a key-controlled chaotic system based on the Kolmogorov flow. In order to confuse the data, a substitution based on a shift-registered pseudorandom number generator is applied, which alters the statistical property of the cipher image. It was advocated that the scheme is computationally efficient secure, and superior to contemporary bulk encryption systems when aiming at efficient image and video data encryption.

In [13], a systematic method was suggested for adapting an invertible two-dimensional chaotic map on a torus or on a square to create a symmetric block encryption scheme. The approach to constructing the symmetric block cipher consists of three steps: (1) choose a chaotic map and generalize it by introducing some parameters; (2) discretize it to a finite square lattice of points that represent pixels; (3) extend the discretized map to three dimensions and compose it with a simple diffusion mechanism. In this design, an example based on the standard two-dimensional baker map was given to illustrate the construction procedure and to demonstrate the security.

Furthermore, some other two-dimensional chaos-based encryption examples such as the chaotic cat map and standard map have also been used for cipher design. In [21], for example, a chaotic video encryption scheme (CVES) was proposed based on a multiple digital chaotic system. In this scheme, 2^n chaotic maps controlled by another single chaotic map are used to generate pseudorandom signals to mask the video, and to perform pseudorandom permutation of the masked video. It was claimed that the CVES is independent of any video compression algorithms and can provide high security for real-time digital videoing with a fast encryption speed. In [21], the method was also extended to the so-called RRS-CVES, which supports random retrieval of cipher video with maximal time-out.

It seems that most chaotic image encryption methods are concentrated on block ciphers. Actually, both block cipher and stream cipher have their own merits and can be used for different applications under different conditions to meet different requirements. In the next two sections, more general constructive approaches of these two types of chaos-based ciphers are further discussed.

8.4 Chaos-Based Block Ciphers for Image Encryption

Chaos-based block ciphers have excellent flexibility. Applying block ciphering on a whole picture can achieve very fast shuffling. Meanwhile, if a huge-sized image needs to be encrypted by a device with limited memory or computational power, say a single chip or a mobile phone, the image can be split into several small blocks, which are then encrypted in serial. Unlike stream ciphers, block ciphers are suitable for parallel processing. Moreover, using a block cipher it is easy to balance the requirements of encryption intensity and cipher speed by simply controlling the cipher rounds.

8.4.1 Construction of Chaos-Based Block Ciphers

Integrating a chaotic map into a block cipher utilizes chaos properties to rapidly scramble and diffuse data. Two general principles that guide the design of block ciphers are *diffusion* and *confusion*. Diffusion means spreading out the influence of a single plaintext digit over many ciphertext digits, so that the statistical structure of the plaintext becomes unclear. Confusion, on the other hand, means using transformations that complicate the dependence of the statistics of the ciphertext on the statistics of the plaintext [16, 17]. These two principles are closely related to the mixing and ergodicity properties of chaotic maps.

In [17], a general approach to chaos-based block cipher design was provided, which consists of four steps:

1. *Choosing a chaotic map*: one should consider maps with good mixing property, robust chaos, and a large parameter set.
2. *Introducing the parameters.*
3. *Discretization.*
4. *Cryptanalysis and key scheduling.*

This framework is recapitulative and can be used to direct block cipher design. However, to design a fast block cipher applicable to real-time imaging, more considerations are in order. For example, since images are highly correlated, it is better to choose a higher-dimensional chaotic map to speed up the permutation process. A block image cipher design framework, based on higher-dimensional chaotic maps, is recommended here:

1. *Choose a higher-dimensional chaotic map and generalize it by introducing a large number of parameters.* The chaotic map chosen should have a large parameter set and good mixing properties. In addition, the map should be a measure-preserved map, to ensure one-to-one mapping after discretization, which is needed for decryption. Good examples of such maps include the generalized cat map, generalized baker map, and so on, as are further discussed below.
2. *Discretize the map.* Despite the fact that in theory a discretized map is only defined over a finite field and therefore can never to truly chaotic, one should keep certain strong features of chaos, such as mixing and sensitivity to parameters, while keeping the data shuffling speed fast. For practical use, this oftentimes proves sufficient.
3. *Compose a diffusion process.* Although pixel positions of an image were scrambled in the last step, generally the distribution of gray-scales of the image is still unchanged, i.e., the histogram of the plain image is about the same as that of the cipher image. This leaves a door widely open for statistical attack and chosen-plaintext attack. Thus, a diffusion process is necessary to make the spread influence of each single pixel over all of the image. The diffusion process may simply be accomplished using a one-dimensional chaotic map. In addition, one may also introduce an additional substitution procedure to speed up the diffusion process.
4. *Perform security evaluation.* Many cryptanalysis methods that are widely used in traditional cryptography should be applied to analyze the performance of a proposed chaos-based cipher. These methods include keyspace analysis, statistical attack, differential and linear attacks, know-plaintext attack, and chosen-plaintext attack.
5. *Other performance evaluation.* Apart from security analysis, other issues should also be considered for image encryption. These include cipher

speed, cipher image size, deciphered image quality (if data compression is combined), and computational overhead.

Next, the generalized three-dimensional baker map is used as an example to illustrate the composition of a chaos-based image block cipher.

8.4.2 A Fast Image Cipher Based on Chaotic 3D Baker Map

In applications, there are two kinds of methods for constructing secure encryption algorithms. For quite a long time, many now-classic schemes like DES and IDEA put more emphasis on substitution than on permutation. Actually, permutation plus diffusion can also compose good encryption schemes with fast speed and high security. In the last section, some encryption schemes using two-dimensional chaotic maps were discussed that were designed based on this idea [12, 13, 31].

Here, the two-dimensional (2D) baker map is extended to three dimension (3D), and then is used to compose a fast image encryption scheme. Empirical results have shown that, compared with encryption using the 2D baker map, this new 3D version is not only more secure but is also 2 to 3 times faster.

Extending 2D Baker Map to 3D Version

The standard 2D baker map, denoted by B hereafter, is described by [12, 13]

$$B(x) = \begin{cases} (2x, \frac{y}{2}) & 0 \leq x < \frac{1}{2} \\ (2x-1, \frac{y}{2}+\frac{1}{2}) & \frac{1}{2} \leq x \leq 1 \end{cases} \quad (8.8)$$

This 2D baker map is a chaotic bijection of the unit square $I \times I$ onto itself. The generalized baker map [27, 28] is defined as follows: divide the unit square into k vertical rectangles, $[F_{i-1}, F_i) \times [0,1), i = 1, \cdots, k, F_i = p_1 + p_2 + \cdots + p_i$, $F_0 = 0$, such that $p_1 + \cdots + p_k = 1$. The lower right corner of the ith rectangle is located at $F_i = p_1 + \cdots + p_i$. The generalized baker map stretches each rectangle horizontally by a factor of $1/p_i$. At the same time, the rectangle is contracted vertically by a factor of p_i. Formally,

$$B(x, y) = \left(\frac{1}{p_i}(x - F_i), p_i y + F_i\right), \quad (8.9)$$

for $(x, y) \in [F_i, F_i + p_i) \times [0, 1)$.

A direct extension of the 2D baker map to 3D version can be accomplished by the following procedure. First, divide the unit cube into four even, narrow stripes of small cubes, and then press each of them and pile them up one by one to form a new unit cube that has the same volume as the former. Mathematically,

$$B(x,y,z) = \begin{cases} (2x, 2y, \frac{z}{4}) & 0 \le x < \frac{1}{2}, 0 \le y < \frac{1}{2} \\ (2x, 2y-1, \frac{z}{4}+\frac{1}{2}) & 0 \le x < \frac{1}{2}, \frac{1}{2} \le y \le 1 \\ (2x-1, 2y, \frac{z}{4}+\frac{1}{4}) & \frac{1}{2} \le x < 1, 0 \le y < \frac{1}{2} \\ (2x-1, 2y-1, \frac{z}{4}+\frac{3}{4}) & \frac{1}{2} \le x \le 1, \frac{1}{2} \le y \le 1 \end{cases} \quad (8.10)$$

which is illustrated by Fig. 8.1.

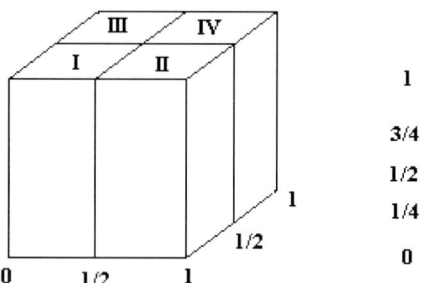
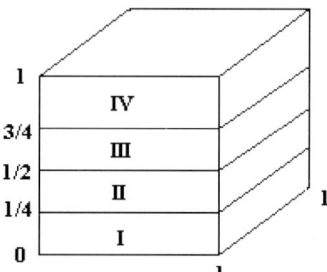

Fig. 8.1. The 3D baker map

Similar to the 2D baker map, the 3D baker map also has its general form. As can be seen from Fig. 8.2, the unit cube is first divided into several small stripes, and then each stripe is pressed and then piled up to form a new unit cube of the same volume. Assume that the unit cube is divided into $k \times t$ blocks, $[W_{i-1}, W_i) \times [H_{j-1}, H_j) \times [0, 1)$, $i = 1, \cdots, k$, $j = 1, \cdots, t$, $W_i = w_1, w_2, \cdots, w_i$, $W_0 = 0$, such that $w_1 + w_2 + \cdots + w_k = 1$, and $H_j = h_1 + h_2 + \cdots + h_j$, $H_0 = 0$, such that $h_1 + h_2 + \cdots + h_t = 1$. The generalized 3D baker map is then defined as follows:

$$B_3(x, y, z) = \left(\frac{1}{w_i}(x - W_i), \frac{1}{h_j}(y - H_j), w_i h_j z + L_{ij} \right), \quad (8.11)$$

for $(x, y, z) \in [W_{i-1}, W_i) \times [H_{j-1}, H_j) \times [0, 1)$, where $L_{ij} = W_i \times h_j + H_j$, $i = 1, \cdots, k$, $j = 1, \cdots, t$.

The continuous 3D baker map is then extended to its discrete version with an arbitrary cube size. Without loss of generality, assume that the cube is $W \times H \times L$, and is split into $k \times t$ blocks. The sequence of k integers, $w_1, w_2, \ldots, w_k$, is chosen such that $W_i = w_1 + w_2 + \cdots + w_i$, $W = w_1 + w_2 + \cdots + w_k$, and $W_0 = 0$. The same for the sequence of t integers, $h_1, h_2, \ldots, h_t$, i.e., $H_j = h_1 + h_2 + \cdots + h_j$, $H = h_1 + h_2 + \cdots + h_t$, and $H_0 = 0$. By using the following formulas:

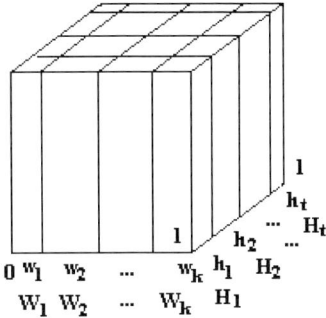 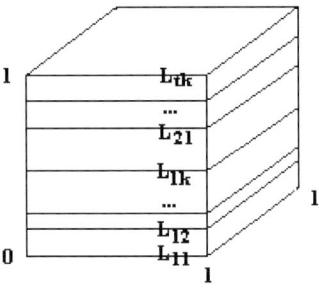

Fig. 8.2. The 3D baker map: general form

$$S = (H_{j-1} \times W + W_{i-1}) \times L + w_i \times h_j \times l +$$
$$(n - H_{j-1}) \times w_i + (m - W_{i-1}),$$
$$(m', n', l') = B_{3D}(m, n, l), \quad (8.12)$$
$$= \left((S \mod (W \times H)) \mod W, \left\lfloor \frac{S \mod (W \times H)}{W} \right\rfloor, \left\lfloor \frac{S}{W \times H} \right\rfloor \right),$$

an arbitrary point (m, n, l) in the original cube is mapped to (m', n', l') in the new cube.

Image Encryption Scheme Based on 3D Baker Map

The discrete 3D baker map, designed in Sect. 8.4.2, is now applied to construct a fast image encryption scheme. As mentioned previously, a secure encryption scheme should have a mechanism of diffusion that makes known-plaintext attack infeasible. In this new image encryption scheme, an XOR plus modulo operation is inserted to each pixel in between every two adjacent rounds of the map used. Below, the diffusion process is first discussed, and then the entire encryption scheme is described in detail.

1. *Diffusion Procedure.* First, choose two numbers: one (denoted L_i) is a float number in (0,1), to be used as initial condition; another (denoted S) is an integer, to be used as seed. Then, use L_i as the initial value to compute the logistic map

$$x(k+1) = 4x(k)[1 - x(k)]. \quad (8.13)$$

If the next value obtained is in (0.2,0.8), then go to the next step; otherwise, the iteration goes on until a desired number in (0.2,0.8) is obtained. Here, notice that the value of 0.5 is a "bad" point, which will lead the iteration being trapped in the fixed point 0. If such a case is encountered, a small perturbation should apply. Once a proper value is obtained from the logistic map, digitize it by amplifying it with a proper scale and then do sampling. The digitized value is designated as $\phi(k)$ and is XOR-ed with the values of the currently operated pixel and the previously operated pixel in the image, according to the following formulas:

$$C(k) = \phi(k) \oplus \{[I(k) + \phi(k)] \bmod N\} \oplus C(k-1), \tag{8.14}$$

where $I(k)$ is the currently operated pixel and $C(k-1)$ is the previously operated pixel in a vector that was strung out from the image, and $C(k)$ is the XOR-ed value. One may set the initial value $I(0) = S$. The inverse transform of the above is simple, and is given by

$$I(k) = \{\phi(k) \oplus C(k) \oplus C(k-1) + N - \phi(k)\} \bmod N. \tag{8.15}$$

Since in step k the previous value $C(k-1)$ is known, the value $C(k)$ can be ciphered out.

2. *Image encryption scheme.* The integrated image encryption scheme, illustrated in Fig. 8.3, consists of five steps of operations:

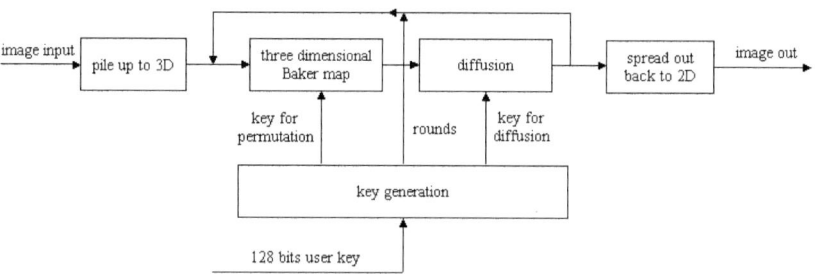

Fig. 8.3. Block diagram of image encryption using the 3D baker map

Step 1. *Key generation.* Select a sequence of 128 bits as the key, and split them into six groups among which the first four groups contain 24 bits each and the last two groups contain 16 bits each. Map these six groups of bits into six numbers, $k_1, k_2, k_3, k_4, k_5,$ and k_6, where $k_1, k_2,$ and k_3 are floating numbers in $(0,1)$, while the rest are integers.

Step 2. *Pile up the two-dimensional image to a three-dimensional one.* Suppose that the image to be encrypted is with W pixels wide and H pixels high. First, one needs to pile up all pixels of the image, to form a cube of size $M \times N \times L$. Since the number of total pixels is unchanged, the integers M, N, and L must be chosen such that $M \times N \times L = W \times H$. The decomposition algorithm for M, N, L is described as follows:

1. Set $T = W \times H$, and then factor out all prime numbers of T and list them out as a sequence $\{p_1, p_2, \ldots, p_n\}$ such that $T = p_1 \times p_2 \cdots \times p_n \times 1$.
2. Permute the sequence $\{p_1, p_2, \ldots, p_n, 1\}$, and then regroup them into three groups. During the permutation process, two integers are needed: one is used as the seed and the other determines the shuffle rounds. Here, k_5 and k_6 are used for these purposes, respectively.

Step 3. *Perform the 3D baker map.* First, select k_1 and k_2 as two initial values to perform the logistic map, respectively. After several rounds of mappings, followed by a floating point to integer transformation, one can select two sequences, $\{m_1, m_2, \ldots, m_k\}$ and $\{n_1, n_2, \ldots, n_t\}$, such that $M = m_1 + m_2 + \cdots + m_k$ and $N = n_1 + n_2 + \cdots + n_t$. Then, perform the discrete 3D baker map (Sect. 8.4.2), on the image cube, to get a shuffled image.

Step 4. *Diffusion process.* Set $k_3 = L_i$ and $k_4 = S$, and then perform the diffusion process once according to the algorithm described in the first part of this subsection.

Step 5. *Reverse process.* Transform the 3D cube back to a 2D image for display or storage.

Note that operations in step 3 and step 4 are often interleaved for several rounds, which depends on the security requirements.

To this end, the decipher procedure is similar to that of the encipher illustrated above, but with the inverse operational sequences to those described in steps 3 and 4. Since decipher and encipher procedures possess similar structures, they have the same algorithmic complexity and time consumption.

Security Analysis and Test Results

Compared with other similar encryption schemes, the new one described above has higher security and can resist all kinds of known attacks, such as the known-plaintext attack, ciphertext-only attack, and so on. Here, some security analysis results on the scheme are described, including the most important ones like keyspace analysis, statistical analysis, and differential analysis.

1. *Keyspace analysis.* A good image encryption algorithm should be sensitive to the cipher key, and the keyspace should be large enough to make brute-force attacks infeasible. For the above-described chaos-based image encryption algorithm based on the generalized 3D baker map, basic analysis and test results are summarized as follows:

- *Number of control parameters.* This algorithm is a 128-bit encryption scheme whose keyspace size is $2^{128} \approx 3.4028 \times 10^{38}$. Since this scheme takes advantage of the 3D baker map, the opponent may try to bypass guessing the key and directly guess all the possible combinations of the sequences $\{m_1, m_2, ..., m_k\}$ and $\{n_1, n_2, ..., n_t\}$, as well as all the possible decomposition of M, N, and L that are used in the 3D baker map. Therefore, the combinations of the baker map control parameters should be large enough to prevent such exhaustive search. In [13], the possible combinations of control parameters for a 2D baker map were estimated. According to the conservative estimate for an $N \times N$ image, the total number of ciphering keys is about $K(N,t) = \binom{N}{t}$, where t is the length of the key sequence $\{n_1, n_2, ..., n_t\}$. For a 2D image, since the key sequences of width and height are different, the size of the keyspace will be twice this estimate. If each ciphering round of the baker map uses different ciphering keys, then an increase in round numbers will also enlarge the keyspace. Compared with the 2D baker map, the keyspace of the 3D one is further enlarged, since the keyspace of the 2D map is just a subspace of the 3D one. For example, suppose that an image size is $W \times H$. In order to perform the 3D baker map, the image must be piled up to a cube with size $M \times N \times L$ such that $W \times H = M \times N \times L$. Among all possible decompositions, $W \times H \times 1$ is a special case that reduces the 3D map to a 2D one. Therefore, it is clear that the 3D baker map has a much larger keyspace than that of the 2D one.

- *Key sensitivity test.* Assume that a 16-character ciphering key is used. This means that the key consists of 128 bits. A typical key sensitivity test is performed according to the following steps:
  1. First, a 512×512 image is encrypted by using the test key "1234567890123456".
  2. Then, the least significant bit of the key is changed, so that the original key becomes "1234567890123457" in this example, which is used to encrypt the same image.
  3. Finally, the above two cipher images, encrypted by the two slightly different keys, are compared.

  As a result: the image encrypted by the key "1234567890123456" has 99.59% difference from the image encrypted by the key "1234567890123457" in terms of pixel gray-scale values, although there is only one bit difference in the keys.

2. *Statistical analysis.* Shannon said, in his masterpiece [35], "It is possible to solve many kinds of ciphers by statistical analysis," and therefore he suggested two methods of diffusion and confusion for frustrating the powerful statistical analysis. Next, it is demonstrated that the above-described image encryption scheme, based on the generalized 3D baker map, has good confusion and diffusion properties. This is shown by a test on the histogram of the cipher images and on the correlations of adjacent pixels in the cipher image.

1. *Histograms of encrypted images.* Select several 256 gray-level images with size of 512×512 that have different contents, and calculate their histograms. One typical example among them is shown in Fig. 8.4. From the

figure, one can see that the histogram of the cipher-image is fairly uniform and is significantly different from that of the original image.

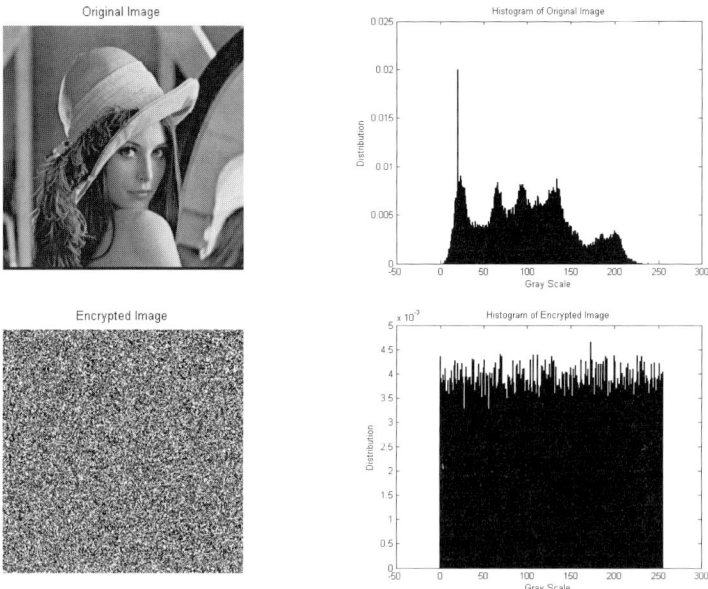

Fig. 8.4. Histograms of the plain image and the cipher image

2. *Correlation of two adjacent pixels.* To test the correlation between two vertically adjacent pixels, two horizontally adjacent pixels, and two diagonally adjacent pixels in a cipher image, respectively, the procedure is as follows: First, randomly select 1000 pairs of adjacent pixels from an image. Then, calculate their correlation coefficient using the following two formulas:

$$cov(x, y) = E(x - E(x))(y - E(y)), \qquad (8.16)$$

$$r_{xy} = \frac{cov(x, y)}{\sqrt{var(x)}\sqrt{var(y)}}, \qquad (8.17)$$

where x and y are gray levels of two adjacent pixels in the image. Figure 8.5 shows the correlations of two horizontally adjacent pixels in the plain image and in the cipher image: the correlation coefficients are 0.96638 and 0.0057765, respectively. Similar results for the diagonal and vertical directions are shown in Table 8.2.

3. *Differential attacks.* Generally, an opponent may make a slight change (e.g., modify only one pixel) of the encrypted image so as to observe the change in the result. In this way, he may be able to find out a meaningful relationship

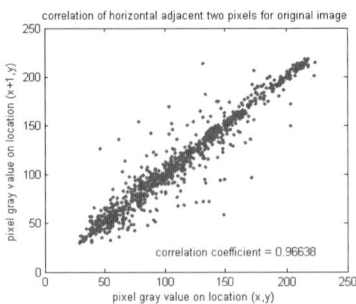 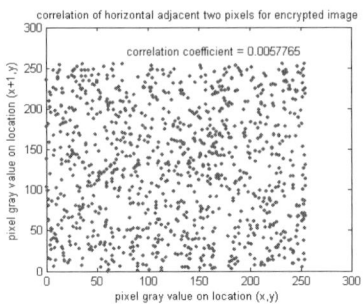

Fig. 8.5. Correlations of two horizontally adjacent pixels in the plain image and in the cipher image

Table 8.2. Correlation coefficients of adjacent pixels in two images

	Plain-image	Cipher-image
Horizontal	0.96638	0.0057765
Vertical	0.97961	0.028434
Diagonal	0.95025	0.020662

between the plain image and the cipher image. This is known as the differential attack. However, if one minor change in the plain image can cause a significant change in the cipher image, with respect to diffusion and confusion, then the differential attack would become very inefficient and useless.

To test the influence of a one-pixel change on the whole image encrypted by the above-described 3D chaos-based algorithm, two common measures can be used: the number of pixels change rate (NPCR) and the unified average changing intensity (UACI). Let two cipher images, whose corresponding plain images have only one pixel difference be C_1 and C_2. Label the gray values of the pixels at grid (i,j) in C_1 and C_2 by $C_1(i,j)$ and $C_2(i,j)$, respectively. Define a bipolar array D with the same size as image C_1 or C_2. Then, $D(i,j)$ is determined by $C_1(i,j)$ and $C_2(i,j)$; namely, if $C_1(i,j) = C_2(i,j)$ then $D(i,j) = 1$, otherwise $D(i,j) = 0$. The NPCR is defined by

$$\text{NPCR} = \frac{\sum_{i,j} D(i,j)}{W \times H} \times 100\%, \tag{8.18}$$

where W and H are the width and height of both C_1 and C_2, and NPCR measures the percentage of different pixel numbers between the two images. The UACI is define by

$$\text{UACI} = \frac{1}{W \times H} \left[\sum_{i,j} \frac{|C_1(i,j) - C_2(i,j)|}{255} \right] \times 100\%, \tag{8.19}$$

which measures the average intensity of differences between the two images.

One performed test is on the one-pixel change influence on a 512 gray-level image of size 512×512. The test results are shown in Fig. 8.6. Generally, with the increase in ciphering rounds, the influence of a one-pixel change is increased. Hence, it is reasonable to increase the ciphering rounds in the test to achieve higher security, yet this is at the expense of processing speed.

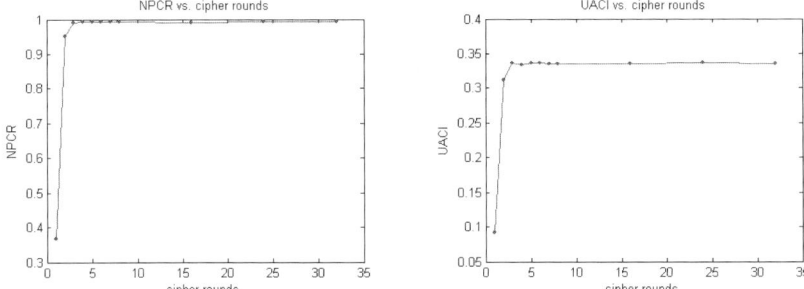

Fig. 8.6. NPCR vs. ciphering rounds and UACI vs. ciphering rounds

4. *Enciphering/deciphering speeds.* The above-described 3D chaos-based image encryption algorithm is very fast. Simulation shows that the average enciphering/deciphering speed is 1.2 MB/s, and the peak speed can reach up to 2.8 MB/s, on a 1-GHz Pentium IV computer. Even the designed cipher based on the 2D baker map in this study is different from that suggested in [13] on the diffusion operation. Taking into account improvements in computers, the speeds of the 3D cipher is slightly faster than that of the 2D cipher. The encryption rate of the algorithm of [13] is only about 1 MB with an unoptimized C code on a 60-MHz Pentium. A comparison between the 2D and 3D chaos-based schemes is shown in Table 8.3.

8.5 Chaos-Based Stream Cipher for Image Encryption

Compared with a block cipher, the main advantage of a chaotic stream cipher is that it can be designed to accommodate image compression. An elaborately designed stream cipher only introduces a small computational overhead in image coding. Moreover, if the image compression algorithm is an embedded one, i.e., the decoding procedure is from coarse to fine progressively, the cipher-image stream can also be truncated at any desired point without influencing the decoding process.

8.5.1 Design of Chaos-Based Stream Ciphers

Encryption by a stream cipher uses a sequence of random numbers to mask a sequence of plaintext of the same length, bit by bit. Although using random

Table 8.3. Comparison of ciphering speed between the 2D baker map and the 3D baker map schemes

Image size (in pixels)	Colors	2D baker map (in seconds)	3D baker map (in seconds)
256×256	2	< 0.3	< 0.3
256×256	16	< 0.3	< 0.3
256×256	256	< 0.3	< 0.3
256×256	16777216	< 0.3	< 0.3
512×512	2	1	< 0.3
512×512	16	1	< 0.3
512×512	256	1.1	< 0.3
512×512	16777216	1	< 0.3
1024×1024	2	3.3	1.0
1024×1024	16	3.3	1.1
1024×1024	256	3.3	1.2
1024×1024	16777216	3.3	1.3
2048×2048	2	13.6	3.4
2048×2048	16	13.5	3.2
2048×2048	256	14.0	3.4
2048×2048	16777216	13.6	4.3

Test Conditions:

(1) The computer used in this test is a Pentium IV, 1-GHz CPU with 256-MB memory and 40-GB hard disk capacity.

(2) Theoretically, both algorithms are symmetric, i.e., both encipher and decipher procedures have the same complexity. But, due to the programming realization issue, the decipher procedure may consume a little more time than enciphering. The time recorded in Table 8.3 is the average time of the encipher and decipher procedures.

numbers to mask a plaintext can achieve theoretical security [35], its practical implementation is impossible. John von Neumann once said, "Any one who considers arithmetical methods of producing random digits is, of course, in a state of sin." The fact is that it is practically very difficult, if not impossible, to generate a truly random number sequence with a deterministic algorithms. In practice, pseudorandom numbers are used instead. Then, the main problem is to generate pseudorandom numbers with "good" properties to meet the need of a key stream. A commonly used pseudorandom number generator (PRNG) is the linear congruential generator (LCG). Since chaotic systems can generate orbits that prove to be nondistinguishable from truly random orbits (e.g., they both have broad power spectra, and they are both extremely sensitive to small changes of initial conditions), chaotic pseudorandom number generators (CPRNG) have attracted more and more attention [19, 42, 43].

Both compressed and uncompressed image data are treated as bitstreams in stream ciphers. It is more interesting to construct pseudorandom bit sequences. Traditionally, linear feedback shift registers (LFSR) are popular generators of pseudorandom bit sequences like the m-sequence. In practice, the Gold sequence is also used frequently. By using a chaotic map, CPRNG is easy to construct. As an example, the method introduced in [19] is employed here to describe the process. Assume that a dynamical system, denoted (X, ϕ), has a normalized invariant measure μ. Divide the state space X into two disjointed parts, X_0 and X_1, such that $\mu(X_0) = \mu(X_1) = 1/2$. Take an initial value $x_0 \in X$ as seed, and start to evolute the system governed by ϕ and x_0. Suppose that after n iterations, a value x_n is obtained. The nth bit b_n of the sequence is then determined by the following formula:

$$b_n = \begin{cases} 0 & \text{if } x_n \in X_0, \\ 1 & \text{if } x_n \in X_1. \end{cases} \qquad (8.20)$$

Thus, one obtains a bit sequence, $\{b_1, b_2, \ldots, b_n, \ldots\}$. Owing to the intrinsic properties of chaos like ergodicity and mixing, the CPRNG has many good features: unique dependence of the sequence on the seed, equiprobable occurrence of "0" and "1," and asymptotic statistical independence of bits.

The assessment of pseudorandom number generators (PRNG) is now discussed. In [14], three postulates concerning properties of periodic PRNG were recommended; to calculate quantities over one complete period of the generator, the following three conditions should be satisfied:

- The number of "0" bits should differ from the number of "1" bits by at most one.
- Among all the runs, half should be of length 1, a quarter should be of length 2, an eighth should be of length 3, and so on, and for each of these lengths there should be equally many runs of "0" bits and runs of "1" bits.
- The value of the autocorrelation function is equivalent to the period of the generator when the offset is 0; otherwise, the value is equal to a certain constant integer.

A practical and widely used test standard is specified by the National Institute of Standards and Technology (NIST) in the United States, called FIPS 140-2 [25]. It consists of 4 tests on a total of 16 aspects. More specifically, a single stream of 20,000 consecutive bits should be subjected to the following 4 tests:

1. *Monobit test.* A monobit test first counts the number of "1" in the 20,000 bitstream. Denote this quantity by X. If $9,725 < X < 10,275$, then the test is passed.
2. *Poker test.* Poker test firstly divides the 20,000 bitstream into 5,000 consecutive 4-bit segments. Count and store the number of occurrences of the 16 possible 4-bit values. Denote $f(i)$ as the number of each 4-bit value i, where $0 \le i \le 15$. Evaluate the following value:

$$X = \frac{16}{5000} \sum_{i=1}^{n} [f(i)]^2 - 5000. \qquad (8.21)$$

The test is passed if $2.16 < X < 46.17$.

3. *Runs test.* A run is defined as a maximal sequence of consecutive bits of either all "1" or all "0", which is part of the 20,000 bitstream. The incidences of runs of all lengths in the bitstream should be counted and stored. The test is passed if the runs that occurred are within the corresponding interval specified in Table 8.4. Note that for the purpose of this test, runs of length greater than 6 are considered to be of length 6.

Table 8.4. Run test specification

Length of run	Required interval
1	2,315—2,685
2	1,114—1,386
3	527—723
4	240—384
5	103—209
6+	103—209

4. *Long run test.* A long run is defined to be a run of length exceeding 25, of either all "0" or all "1". On a sample of 20,000 bits, the test is passed if there are no long runs.

In the next section, as an example, a secure image coding scheme based on the hierarchical trees algorithm is introduced. This scheme is incorporated within a chaos-based stream cipher.

8.5.2 Chaotic Secure Image Coding Based on SPIHT

As mentioned in the previous section, stream ciphers can be incorporated with image compression. Here, a fast chaos-based image encryption scheme is proposed that integrates the encryption with the compression in the image bitstream.

In this design, during the compression process, the set partitioning in hierarchical trees (SPIHT) image coding algorithm [29] is used, which can achieve a reasonably good compression rate. By making the ciphertext be correlated to the plaintext, the encryption scheme can well resist the known-plaintext attack. Furthermore, the algorithm can decode and decrypt ciphertext with an arbitrary bit rate. To introduce the new design, the SPIHT image coding is first reviewed, followed by the chaos-based image cipher scheme, along with some corresponding security analysis and experiment results.

A Brief Review of SPIHT

Among all wavelet-based image compression schemes, the embedded zerotree wavelet (EZW) coding [36] and SPIHT coding [29, 30] show their remarkable performance not only in terms of efficiency but also in their low computational cost and progressive coding characteristics. Progressive coding (also called embedding coding) refers to the way that the most significant bits representing an image are placed at the beginning of the code, and the code bits are arranged according to their importance relative to the representation of the image. A decoder can truncate the code at any position and obtain an estimate of the image, based on the information up to that point. Both EZW and SPIHT algorithms are progressive coding schemes, and SPIHT is a more efficient implementation of the EZW. After the subband decomposition is applied to the image, the main algorithm works by partitioning the subband-decomposed image into significant and insignificant partitions using the following function:

$$S_n(T) = \begin{cases} 1, & \max_{(i,j) \in T}\{|c_{i,j}|\} \geq 2^n, \\ 0, & \text{otherwise,} \end{cases} \quad (8.22)$$

where $S_n(T)$ represents the significance of a set of coordinates T, and $c_{i,j}$ is the coefficient value of coordinate (i,j).

There are two passes in the algorithm: the *sorting pass* and the *refinement pass*. Three lists are defined, which are the list of insignificant sets (LIS), the list of insignificant pixels (LIP), and the list of significant pixels (LSP), respectively. The LIP and the LSP consist of nodes that contain single pixels, while the LIS contains nodes that have descendants. The sorting pass is performed on these three lists and finally makes pixels in the LSP, which is arranged in an order according to the information importance. The maximum number of bits required to represent the largest coefficient in the spatially oriented tree is designated as $n_{\max}$, and is computed by the following formula:

$$n_{\max} = \left\lfloor \log_2 \left(\max_{(i,j)}\{|c_{i,j}|\} \right) \right\rfloor \quad (8.23)$$

During the sorting pass, those coordinates of the pixels that remain in LIP are tested for the significance. The result, $S_n(T)$, is then sent to the output. Those that are significant will be moved to LSP, as well as having their sign bits output. Sets in LIS will also have their significance tested, and, if they are found to be significant, then they will be removed and consequently the result will be partitioned into subsets. Subsets with a single coefficient, if found to be significant, will be added to LSP, or else they will be added to LIP.

During the refinement pass, the nth most significant bit of the coefficients in LSP is output. The value of n is then decreased by 1, and the sorting and refinement passes are repeated.

This continues until either the desired rate is reached, or $n = 0$, and all the nodes in LSP have their bits output. The latter case will result in an almost perfect reconstruction since all the coefficients have been processed completely.

Neither EZW nor SPIHT is noise tolerant, i.e., both methods are sensitive to small modification of bits in their bitstreams. Knowing this, many methods have been introduced to improve the error-prone nature of SPIHT as well as EZW coding schemes [8, 48]. The new design to be introduced below, however, makes use of this error sensitivity that resides in the image source coding.

>From the above discussion, it is clear that there are two kinds of data contained in a SPIHT-coded bitstream, and similar data structure can also be found in an EZW-coded one. They are named *structure bits* and *data bits*, respectively. Structure bits refer to those used for synchronizing the encoding end and the decoding end in the construction of spatially oriented tree. These bits are extremely sensitive to noise, especially the first few bits in the bitstream. Data bits refer to those coding signs of image coefficients or coding values of coefficients generated in the refinement pass. Change of data bits does not seriously affect the reconstruction of the image, but only introduces a small amount of noise to the result.

A large value of the ratio of structure bits to data bits is a basic feature of this designed encryption scheme, and simulation shows that this requirement can be met. Table 8.5 lists some ratios of structure bits to data bits under different coding rates.

Table 8.5. Ratios of structure bits to data bits under different coding rates (test subject is the 512×512 Lena image with 256 gray levels)

Compression rate (bpp)	Number of structure bits (bits)	Number of data bits (bits)	Ratio of structure bits to data bits
0.125	25,692	7,076	3.63:1
0.25	49,524	16,012	3.09:1
0.5	95,936	35,136	2.73:1
1	188,078	74,066	2.54:1
2	367,562	156,726	2.35:1
3	539,624	246,808	2.19:1
4	638,310	410,266	1.56:1
5	638,310	534,826	1.19:1
6	638,310	534,826	1.19:1
7	638,310	534,826	1.19:1
8	638,310	534,826	1.19:1

Chaos-Based Binary Stream Encryption

Given the structure of SPIHT coding, image encryption is incorporated into the SPIHT encoding process. A combination of encoding and encrypting processes is actually very simple. Notice that an SPIHT-encoding bitstream is generated bit by bit, so when each coded bit is output the encryption process can be performed. Three kinds of bitwise operations can be defined, namely, "keep operation", "exclusive or (XOR) operation", and "invert operation", which are denoted by ♯, ⊕, and ¬, respectively. Both ♯ and ¬ are unary operations, while ⊕ is a binary operation. Rules for each operation are listed in Table 8.6, where A and B are operands. Note that two operands of XOR operation are the current-operated bit and its previous-output bit, respectively.

Table 8.6. Rules for ♯,¬ and ⊕ operations

B	A	♯A	¬A	A⊕B
0	0	0	1	0
0	1	1	0	1
1	0	—	—	1
1	1	—	—	0

In order to make even the transfer probabilities of 0 to 0, 0 to 1, 1 to 0, and 1 to 1, which ensures the cipher stream to be more random, operators ♯ and ¬ each should have about 25% operational opportunity, while ⊕ should have 50% operational opportunity. To achieve this goal, the logistic map is employed to determine the kind of operation to be used in each round. The entire encryption procedure is described as follows:

Step 1. Choose two 24-bit-long sequences, $K1 = \{K1_0, K1_1, \ldots, K1_{23}\}$, and $K2 = \{K2_0, K2_1, \ldots, K2_{23}\}$ as the initial key, where $K1_i, K2_j \in \{0, 1\}$. By using the formula

$$x_0 = (K2_0 \times 2^0 + K2_1 \times 2^1 + \cdots + K2_i \times 2^i + \cdots + K2_{23} \times 2^{23})/2^{24} \quad (8.24)$$

sequence $K2$ is assembled into a floating point number x_0, where $x_0 \in (0, 1)$. Suppose that a binary sequence $C = \{C_0, C_1, \ldots, C_{n-1}\}$ is a set of encrypted bitstream, where n is the total number of bits in the set, and is increased as the encryption process goes on. Initially, let $C = K1$, i.e., $C = \{K1_0, K1_1, \cdots, K1_{23}\}$.

Step 2. Let x_0 be the initial value, and then continuously use the logistic map

$$x_k = 4\, x_{k-1}\,(1 - x_{k-1}) \quad (8.25)$$

to get 23 numbers that fall in $(0.2, 0.8)$. Including the initial value x_0, group the 24 numbers in a set denoted by $X = \{x_0, x_1, \ldots, x_{23}\}$. Through X and the formula

$$t_j = \lfloor n \times (x_j - 0.2)/0.6 \rfloor, \quad j = 0, 1, \ldots, 23 \qquad (8.26)$$

the 24 positions in set C are specified, where each position corresponds to either 1 or 0. Subsequently, take out the above 24 binary numbers and regroup them into a new set, $C_n = \{C_{t_0}, C_{t_1}, \ldots, Ct_{23}\}$. This set C_n can be further mapped onto a floating number $y_0 \in (0, 1)$ according to formula

$$y_{n_0} = \left(P_{t_0} \times 2^0 + P_{t_1} \times 2^1 + \cdots + P_{t_{23}} \times 2^{23} \right) / 2^{24}. \qquad (8.27)$$

Then, take y_0 as the initial value, and use the logistic map again to get a new number $y_m \in (0.2, 0.8)$ through iteration. This y_m is the number to be used to determine the bitwise operation type in the next step.

Step 3. Divide the interval $(0.2, 0.8)$ unevenly into the following three nonoverlapping intervals: $(0.2, 0.35]$, $(0.35, 0.65]$, and $(0.65, 0.8)$. The type of bitwise operation in this round is determined according to which interval y_m falls into. If $y_m \in (0.2, 0.35]$, then the operator $\sharp$ is chosen; if $y_m \in (0.35, 0.65]$, operator $\oplus$ is chosen; if $y_m \in (0.65, 0.8)$, operator $\neg$ is chosen. Suppose that the plain bit of the lth round is P_l. According to the above convention, one has $C_l = op(P_l)$, where P_l is the corresponding cipher bit in the same round, and $op \in \{\sharp, \oplus, \neg\}$. The last job in this step is to append C_l to the sequence C and then denote it as C_{l+23}.

Step 4. Set $n = n + 1$ and let $x_0 = x_{23}$ if all bits are encrypted. Then, exit the routine. Otherwise, go to Step 2.

A reversed procedure is performed in the decoding procedure, which decrypts the encrypted bitstream with an arbitrary desired bit rate.

Security Analysis and Experimental Results

The security of the above-described chaos-based encryption scheme is now analyzed. The results show that although this scheme has a moderate size of keyspace, it already has good key sensitivity and can well resist both ciphertext-only attack and known-plaintext attack.

1. *Keyspace analysis.* The keys used in this scheme are composed of several initial values of the logistic map. For 256 gray-scale pictures, two of them are used; for 24-bit true-color pictures, 4 of them are used. Theoretically, since the logistic map in its chaotic phase is very sensitive to initial values, the keyspace of this encryption scheme can be arbitrarily large. However, because of the limitation of numerical precision of a computer, a practical implementation has to be restrained only in a small keyspace. In the simulation of this scheme,

24 bits are used to assemble a floating number. Thus, for gray-scale images, the cipher key is chosen to be 48 bits long, which has about $2^{48} \simeq 2.81475 \times 10^{14}$ possible combinations.

It may seem that the keyspace is not very large, but this does not mean that the designed scheme has little practical value. There are several reasons for this. First, since images have bulk data, the time consumed for one attack will be much more than that to a textfile. In other word, a small keyspace does not mean the total attack time can be reduced for images. Second, because for true-color images there are two more image channels, each of which will occupy an additional initial value, the keyspace is actually enlarged to 96 bits long. Therefore, since one often uses 24-bit true-color images, the keyspace of this scheme is considered to be large enough for practical applications.

2. *Key sensitivity analysis.* The logistic map is sensitive to initial values in its chaotic phase, which ensures the key sensitivity of the encryption scheme.

A test is performed on gray-scale images that randomly changes one of the 48 bits in the key and calculates the percentage of different bits in the encrypted bitstreams. Test results are summarized in Table 8.7. Although, on average, about 52.52% bits are different, the decrypted images with wrong cipher keys are literally unreadable, for the structural bits are extremely sensitive to the key.

Table 8.7. Test results of percentage of different bits in encrypted bitstreams with respect to one-bit change in the key

No.	1	2	3	4	5	Average
Percentage of different bits	36.21%	72.58%	54.40%	45.16%	54.25%	52.52%

3. *Ciphertext-only attack analysis.* Ciphertext-only attack on this encryption scheme is impossible. Suppose that an opponent wants to attack only the first 1000 bits of an encrypted image bitstream by exhaustive searching, where each bit is either 1 or 0. Then, the possible combinations will be a terrible number of $2^{1000} \simeq 1.0715 \times 10^{301}$. Nevertheless, 1000 bits is really trivial as a tiny part of the data set for an image: according to Table 8.5, for a 64:1 compressed 512×512 gray-scale image, the structural bits alone will be 25,692.

Another feature of this encryption scheme is that with the increase of data bits, it is more difficult to break the cryptosystem by brute force attack. Notice that each time before an operation is chosen from among ♯, ⊕, and ¬, a comparison with two thresholds will be performed on a floating number, y_m. The generation of the floating number y_m is based on a 24-bit binary sequence, which is correlated to the preceding encrypted plaintext. Even if the opponent knew the previous n bits, he still could not guess the floating number y_m because the number of all possible choices would be $\binom{n}{24}$. So, with the increase of n, the combinations would increase very significantly.

4. *Known-plaintext attack analysis.* In this encryption scheme, there exist three kinds of operations that make four kinds of bit changes with an equal probability: 1 to 1, 1 to 0, 0 to 1, and 0 to 0. These types of bit operations are not fixed, and they are correlated to the previously encrypted bits, therefore frustrating the known-plaintext attack.

If the operation choosing procedure is removed, and only the XOR operation is used, then the encryption process would be reduced to the way that a compressed image bitstream has XOR operations bit by bit with a binary sequence generated by the logistic map followed by a bi-polarizing operation with a fixed threshold. Here, denote the binary sequence as $\{S_0, S_1, \ldots, S_n, \ldots\}$, where $S_i \in \{0,1\}$, and denote the compressed image bitstream by $\{I_0, I_1, \ldots, I_n, \ldots\}$. The encrypted bitstream is $\{C_0, C_1, \ldots, C_n, \ldots\}$, where $C_i = S_i \oplus I_i$. Once the ciphertext is given, i.e., once $\{I_0, I_1, \ldots, I_n, \ldots\}$ is known, the sequence $\{S_0, S_1, \ldots, S_n, \ldots\}$ can be revealed. Even if the XOR operation is related to the previous data, a problem still exists. For instance, if the enciphering operation is defined as $C_i = S_i \oplus I_i \oplus C_{i-1}$, then since I_1, C_1, and C_0 are known, one can get S_1, then S_2, and so forth, until finally all S_i are obtained iteratively. However, if the operation selection procedure is added into each step, then because this operator is unknown, the sequence $\{S_0, S_1, \ldots, S_n, \ldots\}$ could not be revealed unless the key used for generating it can be obtained.

5. *Histogram analysis of encrypted images.* The histograms of several 512×512 gray-scale images were analyzed with different contents and natures. Test results show that histograms of encipher-images are very uniform, which makes statistical attacks difficult. Figure 8.7 shows one typical result.

8.6 Conclusion: An Engineer's Perspective

The security of an image is very different from that of a textfile. Because of its intrinsic characteristics, the encryption speed and algorithm simplicity are usually considered more important than the "absolute security", – even if that were possible. Chaos theory has proved to be an excellent alternative to provide a fast, simple, and reliable image encryption scheme that has a high enough degree of security. In this article, two chaos-based cipher schemes for still images have been described in detail. Both security analysis and experiments show that, taking into account the trade-off between attack expense and information value as well as other issues such as operational speed, computational cost, and implementation simplicity, these kind of chaos-based

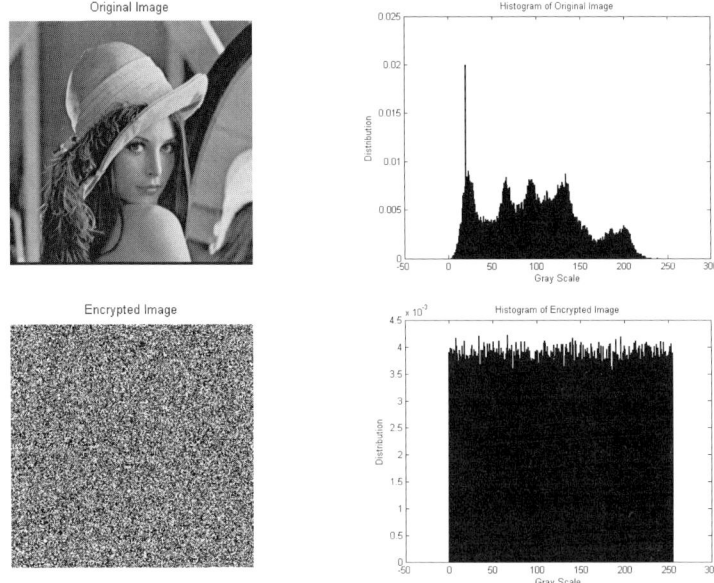

Fig. 8.7. Histograms of plain-image and encrypted image

image encryption schemes are very practical. From an engineer's perspective, chaos-based image encryption technology is very promising for real-time secure image and video communications in military, industrial, and commercial applications.

References

1. Antonini M, Barlaud M, Mathieu P, Daubechies I (1992) Image coding using wavelet transform. *IEEE Trans Image Processing* 1:205 – 220
2. Bourbakis N, Alexopoulos C (1992) Picture data encryption using scan patterns. *Pattern Recognition* 25(6):567 – 581
3. Chang HKC, Liu JL (1997) A linear quadtree compression scheme for image encryption. *Signal Process Image Commun.* 10(4):279 – 290
4. Chen G (2003) Chaotification via feedback: the discrete case. In: Chen G, Yu X (eds) *Chaos Control: Theory and Applications*, Springer, Berlin Heidelberg New York, 159 – 177
5. Chen G, Dong X (198) *From Chaos to Order: Methodologies, Perspectives and Applications.* World Scientific, Singapore
6. Cheng H (1998) Partial Encryption for Image and Video Communication. *M.S. Thesis, Univ. Alberta*, Edmonton, Canada
7. Cheng H, Li XB (2000) Partial encryption of compressed images and videos. *IEEE Trans Signal Processing* 48(8):2439 – 2451
8. Collins T, Atkins P (2001) Error-tolerant SPIHT image compression. *IEE Proc. Image Signal Process* 148(3):182 – 186

9. Coutinho SC (1999) *The Mathematics of Ciphers: Number Theory and RSA Cryptography*, A.K.Peters, Natick, MA, USA
10. Dachselt F, Schwarz W (2001) Chaos and cryptography. *IEEE Trans Circuits and Systems-I* 48(12):1498 – 1509
11. Devaney RL (1989) *An Introduction to Chaotic Dynamical Systems*, (2nd edn.) Addison-Wesley, Reading, MA, USA
12. Fridrich J (1997) *Secure image ciphering based on chaos: Final report for AFRL*. Rome, New York
13. Fridrich J (1998) Symmetric ciphers based on two-dimensional chaotic maps. *Int J Bifurcation and Chaos* 8(6):1259 – 1284
14. Golomb SW (1967) *Shift Register Sequences*. Holden-Day, San Francisco, USA
15. Hasler M (1997) Recent advances in the transmission of information using a chaotic signal. http://icwww.epfl.ch/publications/documents/IC_TECH_REPORT_199729.pdf Dec. 2003
16. Kocarev L (2001) Chaos-based cryptography: a brief overview. *IEEE Circuits and Systems Magazine* 1(3):6 – 21
17. Kocarev L, Jakimovski G (2001) Chaos and cryptography: From chaotic maps to encryption algorithms. *IEEE Trans Circuits and Systems-I* 48(2):163 – 169
18. Kotulski Z, Szczepański J (1997) Discrete chaotic cryptography (DCC): New method for secure communication. Proc NEEDS'97. http://www.ippt.gov.pl/~zkotulsk/kreta.pdf. Dec. 2003
19. Kotulski Z, Szczepański J (2000) On constructive approach to chaotic pseudo-random number generator. *Proc Regional Conference on Military Communication and Information Systems, CIS Solutions for an Enlarged NATO*, RCMIS2000, Zegrze 1:191 – 203
20. Li SJ, Zheng X (2002) Cryptanalysis of a chaotic image encryption method. *IEEE Int Symposium Circuits and Systems*, Scottsdale, AZ, 2:708 – 711
21. Li SJ, Zheng X, Mou X, Cai Y (2002) Chaotic encryption scheme for real-time digital video. *Proc SPIE on Electronic Imaging*, San Jose CA, Real-Time Imaging VI 4666:149 – 160
22. Lorenz EN (1993) *The Essence of Chaos*. University of Washington Press, Seattle, WA
23. Masuda N, Aihara K (2002) Cryptosystems with discretized chaotic maps. *IEEE Trans Circuits and Systems-I* 49(1):28 – 40
24. Menezes AJ, van Oorschot PC, Vanstone SA (1996) *Handbook of Applied Cryptography*. CRC Press, Boca Raton, FL
25. National Institute of Standards and Technology (2001) *Federal Information Processing Standards Publication FIPS PUB 140-2: Security Requirements for Cryptographic Modules*, May 25
26. Pecora LM, Carroll TL (1990) Synchronization in chaotic systems. *Physical Review Letters* 64(8):821 – 824
27. Pichler F, Scharinger J (1995) Ciphering by Bernoulli-shifts in finite Abelian groups. In: Kaiser HK, Muller WB, Pilz GF (eds) *Contributions to General Algebra* 9:249 – 256
28. Pichler F, Scharinger J (1996) Ciphering by Bernoulli shifts in finite Abelian groups. In: *Contributions to General Algebra: Proc Linz-Conference*, 465 – 476
29. Said A, Pearlman WA (1996) A new fast and efficient image codec based on set partitioning in hierarchical trees. *IEEE Trans Circuits and Systems for Video Technology* 6(6):243 – 250

30. Said A, Pearlman WA (1996) An image multiresolution representation for lossless and lossy compression. *IEEE Trans Image Processing* 5(9):1303 – 1310
31. Scharinger J (1998) Fast encryption of image data using chaotic Kolmogorov flows. *J Electronic Imaging* 7(2):318 – 325
32. Schmitz R (2001) Use of chaotic dynamical systems in cryptography. *J Franklin Institute* 338:429 – 441
33. Schneier, B (1995) *Applied Cryptography: Protocols, Algorithms, and Source Code in C* (2nd edn.) Wiley, New York
34. Shannon CE (1948) A mathematical theory of communication. *Bell System Technical Journal* 27:379 – 423,623 – 656
35. Shannon CE (1949) Communication theory of secrecy system. *Bell System Technical Journal* 28:656 – 715
36. Shapiro JM (1993) Embedded image coding using zerotrees of wavelet coefficients. *IEEE Trans Signal Processing* 41(12):3445 – 3462
37. Shi C, Bhargava, B (1998) A fast MPEG video encryption algorithm. *Proc 6th ACM Int Multimedia Conference*, Bristol, UK, Sept. 12–16, 1998:81 – 88
38. Shi C, Wang S, Bhargava B (1999) MPEG video encryption in real-time using secret key cryptography. *Proc Int Conference Parallel and Distributed Processing Techniques and Applications (PDPTA'99)*:191 – 201
39. Shin SU, Sim KS, Rhee HK (1999) A secrecy scheme for MPEG video data using the joint of compression and encryption. *LNCS 1729*:191-201.
40. Stallings W (1999) *Cryptography and Network Security: Principles and Practice*. Prentice-Hall, Upper Saddle River, NJ
41. Stinson DR (2002) *Cryptography: Theory and Practice* (2nd edn.) Chapman and Hall/CRC, Boca Raton, FL
42. Stojanovski T, Kocarev L (2001) Chaos-based random number generators – Part I: Analysis. *IEEE Trans Circuits and Systems-I* 48(3):281 – 288
43. Stojanovski T, Pihl J, Kocarev L (2001) Chaos-based random number generators – Part II: Practical realization. *IEEE Trans Circuits and Systems-I* 48(3):382 – 385
44. Tang L (1996) Methods for encrypting and decrypting MPEG video data efficiently. *Proc ACM Multimedia 96*: 219 – 229.
45. Uehara T, Safavi-Naini R (2000) Chosen DCT coefficients attack on MPEG encryption scheme. *The First IEEE Pacific Rim Conference on Multimedia*, Sydney, Australia, Dec. 13–15, 2000: 316 – 319
46. Uehara T, Safavi-Naini R, Ogunbona P (2000) Securing wavelet compression with random permutations. *The First IEEE Pacific Rim Conference on Multimedia*, Sydney, Australia, Dec. 13–15, 2000:332 – 335
47. Wu CP, Kuo CCJ (2000) Fast encryption methods for audiovisual data confidentiality. *Proc SPIE 4209*:284 – 295
48. Yap CW, Ngan, KN (2001) Error resilient transmission of SPIHT coding images over fading channels. *IEE Proc Image Signal Process* 148(1):59 – 64
49. Yen JC, Guo JI (2000) A new chaotic key-based design for image encryption and decryption. *Proc IEEE Int Symposium Circuits and Systems*, Geneva, Switzerland, May 28-31, 2000, 4:49 – 52

Part IV

Computer Vision

9
One-Dimensional Retinae Vision

Kalle Åström[1]

Center for Mathematical Sciences
Lund University
Sweden
kalle@maths.lth.se

9.1 Introduction

Fig. 9.1. a: A laser guided vehicle. b: A laser scanner or angle meter

Understanding of one-dimensional cameras is important in several applications. In [21] it was shown that the structure and motion problem using line features in the special case of affine cameras can be reduced to the structure and motion problem for points in one dimension less, i.e. one-dimensional cameras. This was used to solve the problem of three views of seven lines. Two solutions were obtained. However, no geometrical interpretation of these two solutions were given.

Another area of application is vision for planar motion. It is shown that ordinary vision (two-dimensional retina) can be reduced to that of one-dimensional cameras if the motion is planar, i.e. if the camera is rotating and translating in one specific plane only, cf. [12]. In another paper the planar motion is used for auto calibration [2]. A typical example is the case where a camera is mounted on a vehicle that moves on a flat plane or flat road.

Our personal motivation, however, stems from **autonomous guided vehicles (AGV)**, which are important components for factory automation. Such vehicles have traditionally been guided by wires buried in the factory floor, gives a very rigid system. Removal and change of wires is cumbersome and costly. The system can be drastically simplified using navigation methods based on laser sensors and computer vision algorithms. With such a system the position of the vehicle can be computed instantly. The vehicle can then be guided along any feasible path in the room. This paper deals with some navigation problems for laser-guided vehicles (LGV). The navigation system uses strips of inexpensive reflector tape (called **reflectors** or **beacons**) which are put on walls or objects along the route of the vehicle [15]. The **laser scanner**, also called the **angle meter** or **meter**, measures the direction from the vehicle to the beacons, but not the distance. This is the information used to calculate the position of the vehicle.

One interesting problem is the so-called **surveying** problem [3, 4]. This is the procedure to obtain a map of the unknown positions of the beacons using images at unknown positions and orientations. This is usually done off line, once and for all, when the system is installed, and then occasionally if there are changes in the environment. High accuracy is needed since the map has to be hard-coded in the system. The performance of the navigation routines depends on the precision of the surveyed map. The surveying problem is in essence a *structure and motion* problem, i.e. one tries to solve for both the structure (the map) and the motion of the vehicle.

Note that the discussion here is focused on finding initial estimates of structure and motion. In practice it is necessary to refine these estimates using non linear optimization or bundle adjustment [5, 23].

The chapter is a collection of results obtained together with Fredrik Kahl, Magnus Oskarsson and Niels Christian Overgaard. The chapter is organized as follows. In Sect. 9.2 a brief introduction to the geometry of the problem is given. Some important notations are introduced, and the structure and motion problem is formalized. In Sect. 3 we solve all structure and motion cases with non missing data. In Sect. 4 we classify similar problems with missing data. Some of the so-called prime problems are also solved. Both Sects. 3 and 4 assume that points and cameras are in general position. In Sect 5 we discuss cases in which there might be ambiguous solutions to the structure and motion problem without missing data.

9.2 Scanner Geometry

A laser-navigated vehicle is shown in Fig. 9.1a The laser scanner, which is shown in detail in Fig. 9.1b, is mounted on the top of the vehicle. A laser beam generated by a vertical laser in the scanner is deflected by a rotating mirror at the top of the scanner. Thus, the laser beam scans the room at a fixed height. When the laser beam hits a beacon (a retroreflective tape, also shown in Fig 9.1a), a large part of the light is reflected back to the scanner. The reflected light is processed to find sharp intensity changes. When this happens the bearing α of the laser beam relative to a fixed direction of the scanner is stored. The time t when the reflection occurs is also stored. All beacons are identical. This means that the identity of a beacon cannot be determined from a single measurement.

We introduce an object coordinate system which will be held fixed with respect to the scene. The bearing α, defined above, depends on the position of the beacon (U_x, U_y) and the position (P_x, P_y) and orientation P_θ of the scanner, according to

$$\alpha(P, U) = \arg[U_x - P_x + i(U_y - P_y)] - P_\theta \,, \tag{9.1}$$

where arg is the complex argument (the angle of the vector $(U_x - P_x, U_y - P_y)$ relative to the positive x-axis). The vector (P_x, P_y, P_θ) is called the **camera state**.

Equation (9.1) for the measured bearing is non linear. A somewhat simpler representation of the same equation can be obtained as follows. The vector between the camera center and the beacon can be written as

$$\lambda \begin{bmatrix} \cos(\alpha + P_\theta) \\ \sin(\alpha + P_\theta) \end{bmatrix} = \begin{bmatrix} U_x - P_x \\ U_y - P_y \end{bmatrix} = \begin{bmatrix} 1 & 0 & -P_x \\ 0 & 1 & -P_y \end{bmatrix} \begin{bmatrix} U_x \\ U_y \\ 1 \end{bmatrix} . \tag{9.2}$$

By multiplying each side with a rotation matrix we obtain

$$\lambda \begin{bmatrix} \cos(\alpha) \\ \sin(\alpha) \end{bmatrix} = \begin{bmatrix} \cos(P_\theta) & \sin(P_\theta) \\ -\sin(P_\theta) & \cos(P_\theta) \end{bmatrix} \begin{bmatrix} 1 & 0 & -P_x \\ 0 & 1 & -P_y \end{bmatrix} \begin{bmatrix} U_x \\ U_y \\ 1 \end{bmatrix} . \tag{9.3}$$

We introduce alternative representations for the **bearing**

$$\alpha \longleftrightarrow \mathbf{u} = \begin{bmatrix} \cos(\alpha) \\ \sin(\alpha) \end{bmatrix} \tag{9.4}$$

for the **beacon position**

$$(U_x, U_y) \longleftrightarrow \mathbf{U} = \begin{bmatrix} U_x \\ U_y \\ 1 \end{bmatrix} , \tag{9.5}$$

and for the **camera state**

$$(P_x, P_y, P_\theta) \longleftrightarrow \mathbf{P} = \begin{bmatrix} \cos(P_\theta) & \sin(P_\theta) \\ -\sin(P_\theta) & \cos(P_\theta) \end{bmatrix} \begin{bmatrix} 1 & 0 & -P_x \\ 0 & 1 & -P_y \end{bmatrix}. \quad (9.6)$$

Using these notations, Eq. (9.1) can be written

$$\lambda \mathbf{u} = \mathbf{PU}. \quad (9.7)$$

The alternative representation for the camera state above will be called the **camera matrix**. Notice that the structure of this 2×3 matrix is

$$\mathbf{P} = \begin{bmatrix} a & b & c \\ -b & a & d \end{bmatrix}, \quad (9.8)$$

with $a^2 + b^2 = 1$. It is straightforward to obtain the elements of the camera matrix from the meter state (P_x, P_y, P_θ) and vice versa.

It is sometimes useful to consider dual image coordinates

$$\alpha \longleftrightarrow \mathbf{v} = \begin{bmatrix} -\sin(\alpha) & \cos(\alpha) \end{bmatrix}, \quad (9.9)$$

so that $\mathbf{vu} = 0$. This is particularly useful since it simplifies the camera constraint given by Eq. (9.7) to

$$\lambda \mathbf{vu} = 0 = \mathbf{vPU}. \quad (9.10)$$

We will often use capital I to denote image number and capital J to denote point number. Thus $\mathbf{u}_{i,j}$ denotes the image direction for point J in image I, $\mathbf{P}_I$ denotes camera matrix for image I and $\mathbf{U}_J$ denotes object point number J.

9.3 Problem Formulation

Motivated by the previous sections the structure and motion problem will now be defined.

Problem 1. Given n bearings from m different positions

$$\mathbf{u}_{I,J}, \quad \forall (I, J) \in \mathbb{I} \quad (9.11)$$

the **surveying problem** is to find the depths $\lambda_{I,J} > 0$, the reconstructed points

$$\mathbf{U}_J = \begin{pmatrix} X_J \\ Y_J \\ 1 \end{pmatrix} \quad (9.12)$$

and the camera matrices

$$\mathbf{P}_I = \begin{pmatrix} a_I & b_I & c_I \\ -b_I & a_I & d_I \end{pmatrix}, \quad (9.13)$$

such that

$$\lambda_{I,J} \mathbf{u}_{I,J} = \mathbf{P}_I \mathbf{U}_J, \quad (I, J) \in \mathbb{I}. \quad (9.14)$$

Here $\mathbb{I}$ is a subset of $\{1,\ldots,n\} \times \{1,\ldots,m\}$ that indicates which points are visible in which images. If all points are visible in all images, i.e. if $\mathbb{I} = \{1,\ldots,n\} \times \{1,\ldots,m\}$, we say that there is no missing data.

It is often convenient to consider things to be equal if they are equal up to scale. The notation $\sim$ will be used to denote equality up to scale. As an example two camera matrices $\mathbf{P}$ and $\widetilde{\mathbf{P}}$ are considered equal if $\mathbf{P} \sim \widetilde{\mathbf{P}}$. The reason for this is that $\mathbf{P}$ and $\widetilde{\mathbf{P}}$ give the same projections. Only the scale factor λ is different.

Definition 1. *The group of similarity transformations is defined as*

$$\mathcal{S} = \left\{ \mathbf{S} \sim \begin{pmatrix} e\cos(\theta) & -e\sin(\theta) & f \\ e\sin(\theta) & e\cos(\theta) & g \\ 0 & 0 & 1 \end{pmatrix} \right\}, \qquad (9.15)$$

where θ denotes rotation, e change of scale and (f,g) translation.

We consider two solutions $(\lambda_{I,J}, \mathbf{U}_J, \mathbf{P}_I)$ and $(\widetilde{\lambda}_{I,J}, \widetilde{\mathbf{U}}_J, \widetilde{\mathbf{P}}_I)$ to the surveying problem to be the same if they are related by a similarity transformation. If there exists a transformation matrix $\mathbf{S}$ such that

$$\widetilde{\mathbf{U}}_J = \mathbf{S}\mathbf{U}_J,$$

$$\widetilde{\mathbf{P}}_I = \mu \mathbf{P}_I \mathbf{S}^{-1},$$

$$\widetilde{\lambda}_{I,J} = \mu \lambda_{I,J},$$

then both $(\lambda_{I,J}, \mathbf{U}_J, \mathbf{P}_I)$ and $(\widetilde{\lambda}_{I,J}, \widetilde{\mathbf{U}}_J, \widetilde{\mathbf{P}}_I)$ give the same projections $\mathbf{u}_{I,J}$, since

$$\lambda_{I,J}\mathbf{u}_{I,J} = \mathbf{P}_I\mathbf{U}_J, \qquad \forall \quad (I,J) \in \mathbb{I}.$$

$$\widetilde{\lambda_{I,J}}\mathbf{u}_{I,J} = \widetilde{\mathbf{P}}_I\widetilde{\mathbf{U}}_J, \qquad \forall \quad (I,J) \in \mathbb{I}.$$

In order to understand how much information is needed in order to solve the structure and motion problem, it is useful to calculate the number of degrees of freedom of the problem and the number of constraints given by the projection equation. Each object point has two degrees of freedom, and each camera state has three. The solution is only defined up to a similarity transformation, cf. Eq (9.15). This manifold $\mathcal{S}$ has dimension 4. Using n points and m cameras, we thus have $2n+3m-4$ degrees of freedom in the parameters. Each measured bearing gives one constraint on the estimated parameters. Assuming that each point is visible in every camera, we get mn constraints. The number of excess constraints $mn - (2n + 3m - 4)$ is given in Table 9.1. Disregarding the case of 1 point in 1 image, there are two interesting cases where the number of constraints is exactly equal to the number of degrees of freedom in the estimated parameters. These two cases

1. three images of five points ($m = 3, n = 5$)
2. four images of four points ($m = 4, n = 4$)

will be called the **minimal cases of the structure and motion problem**.

274   Kalle Åström

Table 9.1. The number of excess constraints $mn - (2n + 3m - 4)$ for the structure and motion problem with m images of n points.

	\multicolumn{7}{c}{n}
m	1	2	3	4	5	6	7
1	0	−1	−2	−3	−4	−5	−6
2	−2	−2	−2	−2	−2	−2	−2
3	−4	−3	−2	−1	0	1	2
4	−6	−4	−2	0	2	4	6
5	−8	−5	−2	1	4	7	10
6	−10	−6	−2	2	6	10	14

9.4 Structure and Motion Problems Without Missing Data

9.4.1 Intersection and the Discrete Trilinear Constraint

In this section the simpler problem of determining the position of an object point using bearings from several known locations is studied. This problem is usually referred to as **intersection** or **reconstruction** in the literature [23].

The intersection problem for three bearings is connected to what is called the trilinear constraint. These trilinear constraints were originally developed for understanding of multiple view problems in ordinary vision [22, 24]. These constraints are interesting for several reasons.

First, they can be used to solve the more difficult surveying problem for three images. The relative motion of the cameras can be calculated from bearing measurements alone without calculating the structure of the object points explicitly. This gives a way to calculate an initial estimate of motion and of structure.

Second, the multilinear constraints can be used to eliminate faulty image correspondences, and to find new correct ones. This is essential for a robust, automatic structure and motion algorithm.

Only the calibrated case will be studied here because it is the natural situation for laser guided vehicles. Other camera models can be dealt with in a similar manner.

Problem 2. Given m bearing directions

$$\mathbf{u}_I, \quad I = 1, \ldots, m \qquad (9.16)$$

from m known meters states

$$\mathbf{P}_I, \quad I = 1, \ldots, m \qquad (9.17)$$

to one object point $\mathbf{U}$ in unknown position the **intersection problem** is to find the depths $\lambda_I > 0$ and the object point $\mathbf{U}$ such that

$$\lambda_I \mathbf{u}_I = \mathbf{P}_I \mathbf{U}, \qquad \forall I = 1, \ldots, m. \tag{9.18}$$

Each measured bearing from known position constrains the location of the object point to the line of sight. The equation for this line is easy to derive using dual image coordinates $\mathbf{v}$. Recall that

$$\underbrace{\mathbf{v}_I \mathbf{P}_I}_{\mathbf{l}_I} \mathbf{U} = 0, \tag{9.19}$$

thus $\mathbf{l}_I = \mathbf{v}_I \mathbf{P}_I$ is the line of sight. The geometric interpretation of the intersection problem is to intersect these m lines $(\mathbf{l}_1, \ldots, \mathbf{l}_m)$ at a point. The problem has no solution using only one measurement, but using two bearings the solution is, in general, unique.

9.4.2 The Calibrated Trilinear Tensor

The case of three cameras is of particular importance. Using three measured bearings from three different known locations, the object point is found by intersecting three lines. This is only possible if the three lines actually do intersect. This gives an additional constraint, which can be formulated in the following way

Theorem 1. *Let $\mathbf{u}_{1,J}$, $\mathbf{u}_{2,J}$ and $\mathbf{u}_{3,J}$ be the bearing directions to the same object point from three different camera states. Then the trilinear constraint*

$$\sum_{i,j,k} T_{i,j,k} \mathbf{u}_{1,J}^i \mathbf{u}_{2,J}^j \mathbf{u}_{3,J}^k = 0, \tag{9.20}$$

is fulfilled for some $2 \times 2 \times 2$ tensor T.

Proof. By lining up the camera equations

$$\underbrace{\begin{pmatrix} \mathbf{P}_1 & \mathbf{u}_{1,J} & 0 & 0 \\ \mathbf{P}_2 & 0 & \mathbf{u}_{2,J} & 0 \\ \mathbf{P}_3 & 0 & 0 & \mathbf{u}_{3,J} \end{pmatrix}}_{M} \begin{pmatrix} \mathbf{U}_J \\ -\lambda_{1,J} \\ -\lambda_{2,J} \\ -\lambda_{3,J} \end{pmatrix} = \mathbf{0} \tag{9.21}$$

we see that the 6×6 matrix M has a nontrivial right nullspace. Therefore its determinant is zero. Since the determinant is linear in each column, it follows that it can be written as

$$\det M = \sum_{i,j,k} T_{i,j,k} \mathbf{u}_{1,J}^i \mathbf{u}_{2,J}^j \mathbf{u}_{3,J}^k = 0, \tag{9.22}$$

for some $2 \times 2 \times 2$ tensor T.

The calibrated trilinear tensor $T = T_{i,j,k}$ in Eq. (9.20) will now be analyzed in more detail. Note that the constraint above only involves the *motion* parameters and the bearing directions. It does not involve the *structure* parameters U. The tensor components can be calculated from the *motion* parameters. If we denote the rows of camera matrix $\mathbf{P}_I$ by $\mathbf{P}_I^1\ \mathbf{P}_I^2$ it is straightforward to see that the tensor components are subdeterminants of the first three columns of the matrix M. In fact, the components can be obtained as

$$T_{ijk} = \wedge_{ii'}\wedge_{jj'}\wedge_{kk'} \det \begin{bmatrix} \mathbf{P}_1^{i'} \\ \mathbf{P}_2^{j'} \\ \mathbf{P}_3^{k'} \end{bmatrix}, \qquad (9.23)$$

where the tensor $\wedge$ is defined as

$$\wedge_{11} = 0, \qquad \wedge_{12} = -1, \qquad \wedge_{21} = 1, \qquad \wedge_{22} = 0 \ . \qquad (9.24)$$

If the object coordinate system is changed

$$\mathbf{P}_1 \mapsto \widetilde{\mathbf{P}}_1 = \mathbf{P}_1 \mathbf{S}, \quad \mathbf{P}_2 \mapsto \widetilde{\mathbf{P}}_2 = \mathbf{P}_2 \mathbf{S}, \quad \mathbf{P}_3 \mapsto \widetilde{\mathbf{P}}_3 = \mathbf{P}_3 \mathbf{S}, \qquad (9.25)$$

where $\mathbf{S} \in \mathcal{S}$ denotes a 3×3 transformation matrix, the tensor components change according to

$$\begin{aligned}\widetilde{T}_{i,j,k} &= \wedge_{ii'}\wedge_{jj'}\wedge_{kk'} \det \begin{bmatrix} \widetilde{\mathbf{P}}_1^{i'} \mathbf{S} \\ \widetilde{\mathbf{P}}_2^{j'} \mathbf{S} \\ \widetilde{\mathbf{P}}_3^{k'} \mathbf{S} \end{bmatrix} \\ &= \wedge_{ii'}\wedge_{jj'}\wedge_{kk'} \det \begin{bmatrix} \mathbf{P}_1^{i'} \\ \mathbf{P}_2^{j'} \\ \mathbf{P}_3^{k'} \end{bmatrix} \det \mathbf{S} = (\det \mathbf{S}) T_{i,j,k}\ . \end{aligned} \qquad (9.26)$$

A change of coordinate system only changes a common scale of the tensor.

It is natural to think of the tensor as being defined only up to scale. Two tensors T and $\widetilde{T}$ are considered equal if they differ only by a scale factor

$$T \sim \widetilde{T}. \qquad (9.27)$$

Let $\mathcal{T}_u$ denote the set of equivalence classes of trilinear tensors.

As discussed in Sect. 2 only the relative motion of the camera is important.

Definition 2. *Let the manifold of* **relative orientation** *of three cameras be defined as the set of equivalence classes of three ordered camera matrices:*

$$\mathcal{P} = \left\{ (\mathbf{P}_1, \mathbf{P}_2, \mathbf{P}_3) \,|\, \mathbf{P}_I = \begin{pmatrix} a_I & b_I & c_I \\ -b_I & a_I & d_I \end{pmatrix} \right\} \Big/ \simeq \qquad (9.28)$$

where the equivalence is defined as

$$(\mathbf{P}_1, \mathbf{P}_2, \mathbf{P}_3) \simeq (\widetilde{\mathbf{P}}_1, \widetilde{\mathbf{P}}_2, \widetilde{\mathbf{P}}_3), \quad \exists \mathbf{S} \in \mathcal{S}, \widetilde{\mathbf{P}}_I \sim \mathbf{P}_I \mathbf{S}, I = 1, 2, 3\ . \qquad (9.29)$$

Thus the above discussion states that the map $(\mathbf{P}_1, \mathbf{P}_2, \mathbf{P}_3) \mapsto T$ is in fact, a well-defined map from the manifold of equivalence classes $\mathcal{P}$ to $\mathcal{T}_u$.

It turns out that this mapping is in essence a two-to-one mapping. In fact, the following properties can be shown, [6].

Theorem 2. *A tensor $T_{i,j,k} \in \mathcal{T}_u$ is a calibrated trilinear tensor if and only if*

$$\begin{aligned} -T_{111} + T_{122} + T_{212} + T_{221} &= 0, \\ T_{112} + T_{121} + T_{211} - T_{222} &= 0. \end{aligned} \quad (9.30)$$

When these constraints are fulfilled it is possible to solve for the camera matrices in Eq. (9.23). There are, in general, two solutions, possibly nonreal.

Corollary 1. *Let $\mathcal{T} \subset \mathcal{T}_u$ denote the submanifold fulfilling the constraint Eq. (9.30) then the map*

$$T: \mathcal{P} \longrightarrow \mathcal{T}$$

$$T(\mathbf{P}_1, \mathbf{P}_2, \mathbf{P}_3)_{ijk} = \wedge_{ii'} \wedge_{jj'} \wedge_{kk'} \det \begin{bmatrix} \mathbf{P}_1^{i'} \\ \mathbf{P}_2^{j'} \\ \hat{\mathbf{P}}_3^{k'} \end{bmatrix} \quad (9.31)$$

is a well-defined two-to-one mapping.

9.4.3 The Surveying Problem for Three Images

The previous section on the calibrated trilinear tensor provided us with the tool for solving the structure and motion problem for three cameras of at least five points.

Algorithm 1 Structure and motion from three images.

1. Given three images of at least five points,

   $$\mathbf{u}_{I,J}, \quad I = 1, \ldots, 3, \, J = 1, \ldots, n, \, for \, n \geq 5 \, .$$

2. Calculate the trilinear tensor T that fulfills the linear constraints given in Eq. (9.30) and $\sum_{i,j,k} T_{i,j,k} \mathbf{u}_{1,J}^i \mathbf{u}_{2,J}^j \mathbf{u}_{3,J}^k = 0, \, \forall \, J = 1, \ldots n$.
3. Calculate the two possible solutions to the relative orientation $(\mathbf{P}_1, \mathbf{P}_2, \mathbf{P}_3)$ from T according to the proof of Theorem 2.
4. For each solution to the motion calculate structure using intersection.

Note that five point correspondences give five linear constraints in Eq. (9.20). The fact that the camera is calibrated gives two additional constraints in Eq. (9.30). These seven constraints determine the eight components of T

uniquely up to scale. Additional point correspondences will, in the ideal noise-free case, not give any additional constraints on T. There is thus a twofold ambiguity in the solution of the structure and motion problem, irrespective of the number of corresponding points. The calculations above do, however, not take the sign of the directions into account. Thus some of the reconstructed points sometimes have negative depth. This does not, however, guarantee uniqueness.

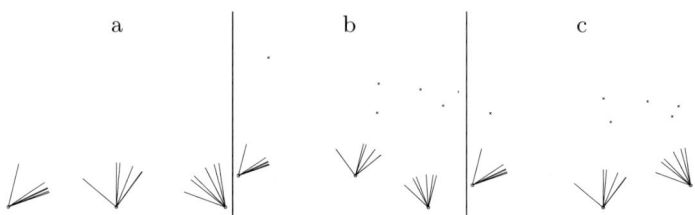

Fig. 9.2. **a**. The figure illustrates three images used as input in example 1. **b**. The first solution obtained from the analysis of the multilinear constraints. **c**. The second solution of structure and motion as obtained from the analysis of the multilinear constraints

Example 1. We illustrate the discussion above with a simple example. Fig. 9.2a shows three images of the same object points. Algorithm 1 is used to find the two possible solutions to the structure and motion problem T Fig. 9.2b,c. Note that in this example all points in both reconstructions have the correct orientation (positive depth). □

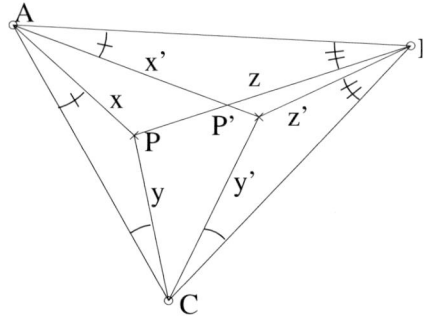

Fig. 9.3. A point P and its isogonal conjugate point P' with respect to triangle ABC

9.4.4 Understanding the Ambiguity

Looking at the solutions of the numerical examples (Fig. 9.2), one would like a geometric interpretation of the two possible solutions. It turns out that the ambiguity is a consequence of the following theorem about **isogonal conjugacy**, which is illustrated in Fig. 9.3 [7].

Theorem 3. *Let ABC be a triangle. Let x, y and z be lines through A, B and C respectively, that intersect in one point, say P. Let the line x' be the reflection of x in the bisector of the angle ABC, and similarly for y' and z'. Then the three lines x' y' and z' intersect at one point P'.*

A proof of the theorem is given in [11] and [1].

The points P and P' are called **isogonal conjugate points** with respect to the triangle ABC. An interesting property of such points is that they are focal points for a conic inscribed in the triangle, i.e. a conic that is tangent to all three sides of the triangle.

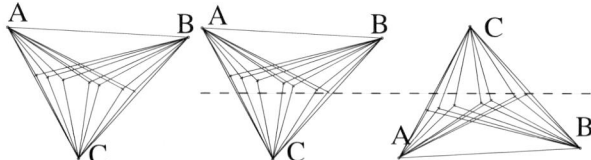

Fig. 9.4. The *leftmost picture* illustrates a triangle ABC, with drawn lines from the vertices to a number of points. The *centre picture* shows the same triangle, but using instead the corresponding isogonal conjugate points. The set of angles as seen from the three corners are the same but have different orientation. The *rightmost picture* is obtained from the centre picture by mirroring in the broken line. By turning the centre picture upside down the same angles are observed from the three corners, although the shape of points in the interior is different

Making the same constructions as in Fig. 9.3 for another pair of points Q and Q', we see that seen from each of the positions A, B and C, the absolute values of the angles between P, Q and P', Q' are equal. However, the orientation is wrong. To get the same orientation the construction needs to be turned up-side-down (or mirrored in an arbitrary line). This leads to the following corollary:

Corollary 2. *For every solution to the structure and motion problem for three cameras, another solution can be constructed by first changing the object positions to their isogonal conjugates with respect to the three camera positions, and then mirroring the camera and object positions in some line.*

An illustration is given in Fig. 9.4.

The other minimal case of four points in four images can be solved in a similar technique, by introducing a dual quadrilinear tensor. For more details on this, see [6]. This case is, however, *dual* to the case of three views of five points solved previously. This duality is described in the following section.

Example 2. Using the following bearing measurements

$\alpha_{I,J}$	J			
I	1	2	3	4
1	-2.3562	2.3562	1.1844	-0.7602
2	-2.1588	2.6779	0.7833	-0.9956
3	-2.5536	2.5536	1.7567	-1.1651
4	-2.3562	2.8198	2.0054	-1.3120

we obtain two possible solutions on the meter states and the object positions, which are illustrated in Fig. 9.5.

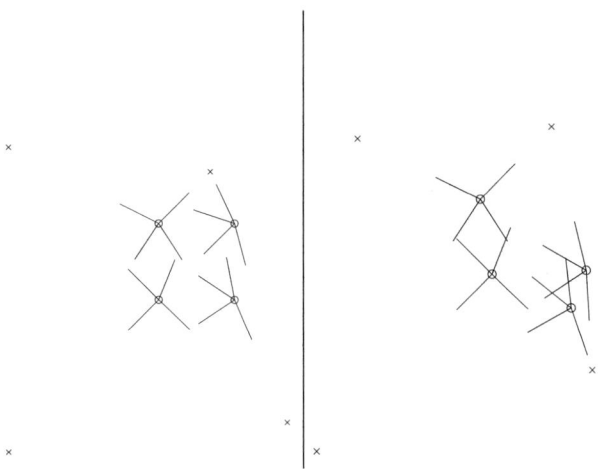

Fig. 9.5. Two different solutions to the structure and motion problem with 4 images of 4 points. The image directions are shown as unit vectors from the center of the camera. The object points are shown as small stars

9.4.5 Duality Between Number of Points and Number of Images

Table 9.2 give the number of solutions in general to the surveying problem of m images of n points.

Table 9.2 has a symmetric appearance. This can be shown using a technique that Carlsson developed in [8, 9]. The proof that he did for uncalibrated projections from 3D to 2D can be used even in the case of uncalibrated projections from 2D to 1D:

Table 9.2. The number of solutions to the surveying problem with m images of n points. *Superscript stars* indicate overdetermined situations

m \ n	3	4	5	6	7
2	∞	∞	∞	∞	∞
3	∞	∞	2	$2^\star$	$2^\star$
4	∞	2	$1^\star$	$1^\star$	$1^\star$
5	∞	$2^\star$	$1^\star$	$1^\star$	$1^\star$
6	∞	$2^\star$	$1^\star$	$1^\star$	$1^\star$

Theorem 4 (Carlsson Duality). *The uncalibrated surveying problem with n points and m images is equivalent to the uncalibrated surveying problem with $m + 3$ points and $n - 3$ images.*

Proof. This is simplest seen by choosing the image coordinates of the first three points according to

$$\mathbf{u}_1 = \begin{pmatrix} 1 \\ 0 \end{pmatrix}, \quad \mathbf{u}_2 = \begin{pmatrix} 0 \\ 1 \end{pmatrix}, \quad \mathbf{u}_3 = \begin{pmatrix} 1 \\ 1 \end{pmatrix}, \tag{9.32}$$

and the object coordinates of the first three points according to

$$\mathbf{U}_1 = \begin{pmatrix} 1 \\ 0 \\ 0 \end{pmatrix}, \quad \mathbf{U}_2 = \begin{pmatrix} 0 \\ 1 \\ 0 \end{pmatrix}, \quad \mathbf{U}_3 = \begin{pmatrix} 0 \\ 0 \\ 1 \end{pmatrix}. \tag{9.33}$$

Since $\lambda_i \mathbf{u}_i = \mathbf{P}_i \mathbf{U}_i$: it follows that the camera matrix has the following form:

$$\mathbf{P} = \begin{pmatrix} V_1 & 0 & V_3 \\ 0 & V_2 & V_3 \end{pmatrix}.$$

The camera equation for the remaining points,

$$\mathbf{u}_{i,j} = \begin{pmatrix} U_1 V_1 + U_3 V_3 \\ U_2 V_2 + U_3 V_3 \end{pmatrix}, \tag{9.34}$$

is symmetric in camera parameters (V_1, V_2, V_3) and structure parameters (U_1, U_2, U_3). Thus any algorithm for solving n points in m images can be used to solve the $m + 3$ points in $n - 3$ images.

Theorem 5. *The calibrated surveying problem with n points and m images is equivalent to the calibrated surveying problem with $m + 1$ points and $n - 1$ images.*

Proof. This follows immediately from Theorems 6 and 4.

According to Theorem 5 the 4 points in 4 images problem is equivalent to the 5 points in 3 images-problem. This explains the symmetry in Table 9.2.

9.4.6 Connection to Uncalibrated Cameras

In Sect. 3.2 it was shown that the problem of 5 points in 3 images has in general two solutions. It was also shown that if there is a solution to the problem of more than 5 points in 3 images, then there are 2 solutions. Similarly in Sect. 3.4 it was shown that the problem of 4 points in 4 images has two solutions. If there is a solution to the problem of 4 points in more than 4 images then there are two solutions. The problem of at least 5 points in at least 4 images is overdetermined, and if there is a solution it is in general unique. The situation is illustrated in Table 9.2.

In this paper it has been assumed that bearings are measured and therefore that the camera matrix has the special form given by Eq. (9.8). In many situations it can be of interest to study the structure and motion problem for so-called uncalibrated cameras. This is identical to the surveying problem, except that the camera matrix is allowed to be a general 2×3 matrix. The difference in the study of minimal cases is, however, slight, due to the following theorem.

Theorem 6. *Knowing that the camera is corrected for internal calibration is equivalent to seeing two extra points (the circular points) in each image.*

Thus it follows that for the uncalibrated surveying problem, in the two minimal cases are three images of seven points and four images of six points. In both of these situations there is a twofold ambiguity in the solution. The ambiguity is not resolved by adding points in the 3 image case or adding images in the 6 point case.

9.4.7 Solution to all Cases with Nonmissing Data

If it is possible to solve a case with a subset of cameras and beacons, then it is often relatively easy to extend that solution to other cameras and points by resection and intersection. If all points are visible in all images then any well-defined case above can be solved by first solving for one of the two minimal cases with nonmissing data, i.e. 4 views of 4 beacons and 3 views of 5 beacons.

9.5 Structure and Motion Problems With Missing Data

The aim of this section is to solve all structure and motion problems for the case of missing data. Depending on the index set $\mathbb{I}$, which describes which points are visible in which images, a structure and motion problem can be either

- ill-defined, if there is not generally enough data to constrain all unknown variables

- well-defined and minimal, if there is exactly enough data to constrain the unknown variables (up to a discrete number of solutions)
- well-defined but overconstrained, if there is more than enough data to constrain the unknown variables

The goal of this section is to classify the possible index sets $\mathbb{I}$ into these three categories.

Some of the minimal cases contain a minimal case as a subproblem. An example of this is the case with 4 points seen in 5 images, but where the fourth point is missing from the fifth image. It is minimal, but contains a subproblem (the problem with the first 4 views only), which is well defined and minimal. We will use the notation **prime problem** for a minimal problem which does not contain a well defined minimal problem as a subproblem. A minimal but not prime problem may in some cases be solved by first solving the contained prime problem and then extending the solution using resection and intersection. In other cases the prime problem may be embedded in the minimal problem in a more complicated manner. We first observe that similar to the case of nonmissing data a well-defined but overconstrained problem contains as a subset a problem that is well-defined but minimal. Thus by finding the minimal cases and solving them, we should be able to solve all well-defined problems by the following algorithm:

1. Find whether a problem contains a well-defined minimal problem as a subset.
2. Solve the structure and motion problem for this subset.
3. Extend the solution to the original problem.

As the classification is based on the index set $\mathbb{I}$ alone, it is interesting to study these sets. In this paper we consider these sets as binary matrices, that is visibility matrices A of size $m \times n$, where black denotes missing data and white denotes a measurement that is present. Another way of viewing these index sets is as bipartite graphs with $m + n$ nodes. There is an edge between node I in the first set and node J in the second set if the point J is visible in image I. Thus a well-defined minimal case can be considered to be a subgraph of a well-defined but overconstrained problem. Here we will use the notation $|\mathbb{I}|$ to denote the number of elements in the set $\mathbb{I}$.

In the following two sections the problem of classifying structure and motion problems for 1d retina will be addressed. In Sect. 9.5.9 we will consider the classification of 2D retina problems.

9.5.1 Classification of Structure and Motion Problems

The goal of this section is to give some conditions on what constitutes a well-defined minimal problem. From these minimal problems the prime problems can be determined.

9.5.2 Equivalence Classes of Index Sets

The labeling of the cameras and of the beacons are of no consequence to the structure of the problem under study. Two index sets are considered equivalent if one results from the other by suitable relabelings. This means that there are many structure and motion problems that have different $\mathbb{I}$ but that correspond in principle to the same problem.

Definition 3. *An index set $\mathbb{I}$ is said to be of type (m, n, l) if it represents a situation with m images and n points, in which exactly l points are not visible in all of the images, that is, if $|\mathbb{I}| = mn - l$.*

From this definition it is clear that an index set $\mathbb{I}$ of type (m, n, l) can be represented by a binary $m \times n$ matrix $A = (a_{IJ})$ with $a_{IJ} = 1$ if $(I, J) \in \mathbb{I}$, and $a_{IJ} = 0$ otherwise, and such that $\sum_{IJ} a_{IJ} = mn - l$. The possible index sets of type (m, n, l) are thus in one-to-one correspondence with the set

$$M(m, n, l) = \{A \in \mathrm{Mat}_{m \times n}(\mathbb{Z}_2) : \sum_{IJ} a_{IJ} = mn - l\}.$$

Let S_k denote the group of permutations on k symbols. With each permutation $\sigma \in S_k$ is associated a $k \times k$ permutation matrix, which will also be denoted by σ.

Definition 4. *Two $m \times n$ matrices A and B are said to be permutation equivalent if there exist permutations $\sigma \in S_m$ and $\tau \in S_n$ such that $B = \sigma^T A \tau$. If A and B are permutation equivalent then we write $A \sim B$.*

The notion of equivalence of index sets can now be given a formal definition

Definition 5. *Two index sets $\mathbb{I}$ and $\mathbb{I}'$ are called* equivalent *and we write $\mathbb{I} \sim \mathbb{I}'$ if their corresponding matrix representations are permutation equivalent.*

The relation $\sim$ is easily seen to be an equivalence relation. It follows that $M(m, n, l)$ (or the corresponding index sets) can be partitioned into equivalence classes $M_1, \ldots, M_\omega$ of matrices (or index sets). The number of essentially different index sets is thus seen to be exactly the same as the number $\omega = \omega(m, n, l)$ of equivalence classes. This is the number of principally different problems of type (m, n, l). The number ω also represents the number of different bipartite graphs with l edges from m to n nodes.

9.5.3 The Germs

A first characterization of a well defined minimal structure and motion problem is that it contains exactly the same number of equations as unknowns. If we concentrate on the case of calibrated cameras with 1D retina, then each object point has two degrees of freedom and each camera state has three. The solution is only defined up to a similarity transformation. This manifold has dimension 4. Using n points and m cameras we thus have $2n+3m-4$ degrees of freedom in the parameters. Each measured bearing gives one constraint on the estimated parameters. Thus for a problem with visibility index set $\mathbb{I}$ we have $|\mathbb{I}|$ equations. This means that minimal problems have $|\mathbb{I}| = 2n+3m-4$. Since the maximum number of equations with m views of n points is mn, it is easy to see how many measurements l that have to be occluded to obtain minimal problems, $l = mn - (2n+3m-4)$. This number is shown in Table 9.1.

In order to find the minimal problems we concentrate our efforts on problems of type $[m, n, mn - (2n + 3m - 4)]$.

Definition 6. *A structure and motion problem of type $[m, n, mn-(2n+3m-4)]$ is said to be a* germ *of a minimal problem.*

For a structure and motion problem to be minimal and/or prime the condition of being a germ is only a necessary condition.

9.5.4 The Prime Condition

For a given germ the corresponding structure and motion problem can be minimal or ill-posed. If it is minimal it may or may not be prime. The question of which class a germ belongs to can be categorized in terms of the graph of the index set. We will use the following intuitive assumption.

Conjecture 1. For a given germ with index set $\mathbb{I}$, the corresponding structure and motion problem is minimal iff no subgraph of $\mathbb{I}$ is overdetermined.

An empirical method for determining whether a problem is minimal and well defined is to calculate the Jacobian of the bundle adjustment problem and study its singular values. We have used this technique to empirically check our conjecture.

It is clear that if a subgraph of a germ with index set $\mathbb{I}$ is overdetermined then there has to be a part of the problem that is underdetermined, and hence the whole problem is ill-posed.

Theorem 7. *Given a germ with index set $\mathbb{I}$, at least one subgraph of $\mathbb{I}$ is overdetermined $\Rightarrow$ the corresponding structure and motion problem is ill-posed*

We will henceforth identify the class of minimal problems with those that fulfill Conjecture 1. Under this assumption the notion of being a prime problem can be given the following formal definition:

Definition 7. *A* prime problem *is a germ with index set $\mathbb{I}$ such that all strict subgraphs of $\mathbb{I}$ are underdetermined.*

A minimal problem that is not prime is an extension of a prime problem. The extended minimal problem can in many cases be solved by a succession of resections and intersections based on the solution to the prime case. In other cases the extension can be more complicated.

Definition 8. *An* extension of type (m, n) *is an extension with m extra cameras and n extra points of a prime problem. That is, the problem is extended by m additional cameras and n additional points, see e.g. Fig. 9.9.*

9.5.5 The Germ Investigation

We now concentrate our efforts on finding out how many germs there are for different numbers of cameras and points. From these germs we then determine which are minimal and which are prime.

9.5.6 Equivalence Classes of Germs

Let the type (m, n, l) be fixed throughout the remainder of the discussion. To compute $\omega = \omega(m, n, l)$ notions and results from group theory will be used. Our reference here is to Sect. 3.6 of Fraleigh's text [13].

First, denote the product group $S_m \times S_n$ by G. Second, if $g = (\sigma, \tau) \in G$ and $A \in M = M(m, n, l)$, then a group action of G on M is defined by the formula

$$g \cdot A = \sigma^T A \tau. \tag{9.35}$$

Thus two matrices $A, B \in M$ satisfy $A \sim B$ if and only if there exists $g \in G$ such that $g \cdot A = B$. The equivalence classes $M_1, \ldots, M_\omega$ of $\sim$ correspond to **the orbits in M under the action of G**. Therefore ω can be computed by the following well-known formula of Burnside: for any $g \in G$ let $M_g = \{A \in M : g \cdot A = A\}$ denote the set of matrices that are fix-points under action by g. Then

$$\omega = \frac{1}{|G|} \sum_{g \in G} |M_g|. \tag{9.36}$$

While Eq. (9.36) solves our problem in theory, there are still some practical problems to overcome. First, given $g \in G$, how do we compute $|M_g|$? Second, the sum $\sum_{g \in G} |M_g|$ must be evaluated, but as $|G| = m!n!$ becomes very large very quickly, the sheer size of G may become an obstacle, unless the evaluation is performed cleverly.

A permutation $g = (\sigma, \tau) \in G$ may be regarded as an element of S_{mn}, as $A \mapsto \sigma^T A \tau$ permutes the mn entries of A. Let $g = g_1 g_2 \cdots g_s$ be the factorization in S_{mn} of g into a product of commuting (or disjoint) cyclic permutations. It is now easy to see that $A \in M_g$ if and only if the entries in A, which equal zero, are arranged in such a manner that any cycle g_i is either completely occupied by entries equal to zero, or contains no such entry at all. It follows that $|M_g|$ equals the number of ways in which l zeros can be allocated to $m \times n$ entries, such that the condition just described is satisfied. It is clear from this discussion that $|M_g|$ only depends on g's cycle structure (the number of cycles and their lengths).

Definition 9. *If $\sigma \in S_k$ is a permutation in k symbols, let $n_i(\sigma), i = 1, \ldots, k$ denote the number of i-cycles in the factorization of σ into commuting cycles. The cycle index of σ is the polynomial*

$$P_\sigma(x_1, x_2, \ldots, x_k) = x_1^{n_1(\sigma)} x_2^{n_2(\sigma)} \cdots x_k^{n_k(\sigma)}. \tag{9.37}$$

If $H < S_k$ is a (sub)group of permutations, then the cycle index of H is the polynomial

$$P_H(x_1, x_2, \ldots, x_k) = |H|^{-1} \sum_{h \in H} P_h(x_1, x_2, \ldots, x_k).$$

It follows from the theory developed in [26] that

$$|M_g| = (l!)^{-1} (\mathrm{d}/\mathrm{d}x)^l P_g(1+x, 1+x^2, \ldots, 1+x^{mn})|_{x=0},$$

for any $g \in G$. This formula solves the first of our two problems. Furthermore, it follows from Burnside's formula Eq. (9.36) that

$$\omega = \frac{1}{l!} \left(\frac{\mathrm{d}}{\mathrm{d}x}\right)^l P_G(1+x, 1+x^2, \ldots, 1+x^{mn}) \bigg|_{x=0}. \tag{9.38}$$

It turns out that the cycle index P_H is reasonably easy to compute when H is all of S_k. Now, $G = S_m \times S_n$ is a proper subgroup of S_{mn}, so in view of Eq. (9.38) our second problem above becomes: How do we compute P_G when the cycle indices of S_m and S_n are known? Again the authors of [26] provide the answer: they introduce a new operation beside the usual addition and multiplication, denoted $*$, on the ring of polynomials in infinitely many variables $x_1, x_2, x_3, \ldots$, and with rational coefficients. The "product" is associative, commutative and distributive over both $+$ and $\cdot$, so it suffices to describe $*$ on monomial factors x_i^m and x_j^n, in which case

$$x_i^m * x_j^n = x_{[i,j]}^{imjn/[i,j]}, \tag{9.39}$$

where $[i, j]$ is the least common multiple of i and j. The authors of [26] then proceed to prove the following beautiful result, which we have used to compute P_G:

Theorem 8 (Wei and Xu). *If $H < S_m$ and $K < S_n$ are (sub)groups, then $H \times K < S_{mn}$, and $P_{H \times K} = P_H * P_K$.*

Example The cycle index of S_3 is $\frac{1}{6}(x_1^3 + 3x_1x_2 + 2x_3)$ so if $G = S_3 \times S_3$ then

$$P_G = \frac{1}{6}(x_1^3 + 3x_1x_2 + 2x_3) * \frac{1}{6}(x_1^3 + 3x_1x_2 + 2x_3)$$
$$= \frac{1}{36}(x_1^9 + 6x_1^3 x_2^3 + 9x_1 x_2^4 + 12 x_3 x_6 + 8 x_3^3),$$

so

$$P_G(1+x, \ldots, 1+x^9) = 1 + x + 3x^2 + 6x^3 + 6x^6 + 7x^4 + 7x^5 + 3x^7 + x^8 + x^9.$$

It then follows from Eq. (9.38) that

$$\omega(3,3,3) = \frac{1}{3!}\left(\frac{\mathrm{d}}{\mathrm{d}x}\right)^3 P_G(1+x, \ldots, 1+x^9)\bigg|_{x=0} = 6.$$

The procedure for calculating ω, described above, was implemented in Maple (Maplesoft, Canada). Using this program we are able to compute ω for any given (m, n, l) and in particular for the germs. Table 9.3 contains ω for the first few types (m, n, l), with l given by Table 9.1.

9.5.7 Finding and Classifying Germs

We calculated the equivalence classes for some of the first germs using algorithms described in [20]. In table 9.3 the number of distinct germs for these cases are given. These germs where then classified as being minimal and pos-

Table 9.3. The number ω of different germs for different m and n

ω m	|	4	5	6	n 7	8	9
3	|	–	1	1	3	6	11
4	|	1	3	16	62	225	
5	|	1	16	155	1402		
6	|	3	79	1799			
7	|	6	361				
8	|	16					

sibly also prime. The numbers of such problems are shown in Tables 9.4 and 9.5.

Table 9.4. The number of minimal configurations for different m and n

m	|	4	5	6	n 7	8	9
3	|	–	1	1	2	3	4
4	|	1	3	12	41	118	
5	|	1	12	110	876		
6	|	2	48	1050			
7	|	3	159				
8	|	5					

Table 9.5. The number of prime configurations for different m and n

m	|	4	5	6	n 7	8	9
3	|	–	1	0	0	0	0
4	|	1	1	3	5	8	
5	|	0	3	22	145		
6	|	0	6	136			
7	|	0	0				
8	|	0					

In Fig. 9.6 and Fig. 9.7 the prime problems for the configurations of type $(5,5,4)$ and $(4,6,4)$ are given. The configurations inf Fig. 9.6a-c seem to be connected to configurations in Fig. 9.7a-c. The similarity can be explained by the Carlsson duality.

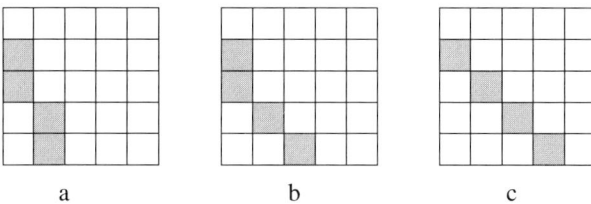

a   b   c

Fig. 9.6. The three distinct configurations **a-c** for prime cases of type $(5,5,4)$

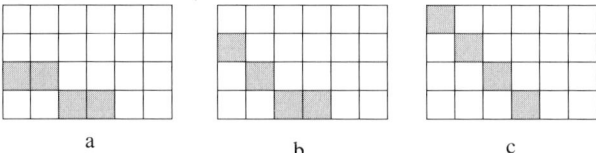

a   b   c

Fig. 9.7. The three distinct configurations **a-c** for prime cases of type $(4,6,4)$

If one looks at Table 9.5, the number of prime configurations seem to increase quickly as both m and n increase. This leads to the question whether this is true or if the number of prime cases after some time stops growing. One can at least give the result in Theorem 9.

Theorem 9. *There are infinitely many prime configurations.*

Proof. Given a germ of type $(m, m, m^2 - 5m + 4)$, one can construct the following prime configuration: the first point is seen in all images. The remaining $m - 1$ points are seen in exactly 4 images each. Of these $m - 1$ points, the first 4 cameras see exactly 3 points, and the remaining $m - 4$ cameras see exactly 4 points. The construction is illustrated in Fig. 9.8. For $\widetilde{m} \leq m - 2$: in order to use as much information as possible one should choose the $\widetilde{m}$ cameras close together. This gives in the best case $3\widetilde{m} + 2\widetilde{m} - 4 = 5\widetilde{m} - 4$ unknowns and $\widetilde{m} + 4(\widetilde{m} - 3) + 3 \cdot 2 = 5\widetilde{m} - 6$ constraints, so in this case it is always underdetermined. For $\widetilde{m} = m - 1$ the same reasoning gives at best $3\widetilde{m} + 2(\widetilde{m} + 1) - 4 = 5\widetilde{m} - 2$ unknowns and $\widetilde{m} + 4(\widetilde{m} - 3) + 3 \cdot 3 = 5\widetilde{m} - 3$ constraints. So also in this case it is always underdetermined. Finally, for $\widetilde{m} = m$ we have $3\widetilde{m} + 2\widetilde{m} - 4 = 5\widetilde{m} - 4$ unknowns matching the $\widetilde{m} + 4(\widetilde{m} - 1) = 5\widetilde{m} - 4$ constraints.

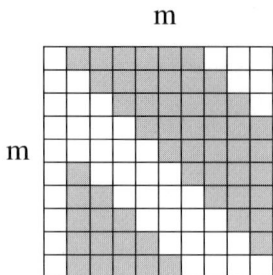

Fig. 9.8. A prime configuration of type (10, 10, 54)

9.5.8 Extensions

Comparing Table 9.4 with Table 9.5, we see that there is only one prime case for four points. Similarly, there is only one prime case for three cameras. The extensions in these cases are of type $(m, 0)$ and $(0, n)$. These types of extensions can always be solved using resection and intersection, respectively. Extensions of type $(1, n)$ and $(m, 1)$ can always also be solved using only combinations of resection and intersection. The first more complicated extension occurs for the type $(2, 2)$. In order for the extension to be unsolvable with intersection and resection all cameras and points must be underdetermined with respect to the prime configuration. And all cameras and points should be exactly determined with the information contained in the remaining four measurements. For an extension of type $(2, 2)$ this can essentially only be done in one way. This extension is shown in Fig. 9.9. It would be interesting to classify

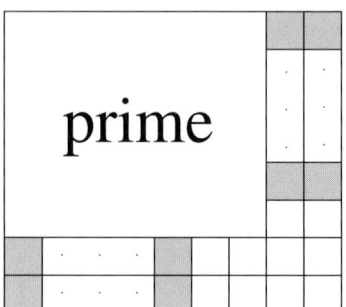

Fig. 9.9. The extension of type (2,2)

prime extensions. This would make it possible to express any minimal case as a prime problem extended by a number of prime extensions.

9.5.9 Classification of 2D Retina Problems

The tools of Sects. 9.5.1 and 9.5.5 were developed for calibrated cameras with 1D retina viewing point features. These methods work equally well for other types of cameras and other types of features. In this section we will look at the classification of minimal problems for uncalibrated cameras with 2D retina, where the features are points.

Table 9.6. The number of excess constraints for m images of n points

m	6	7	8	9	10	11
2	−1	0	1	2	3	4
3	0	3	6	9	12	
4	1	6	11	16		
5	2	9	16			
6	3	12				
7	4					

A projective camera has 11 degrees of freedom, and a point has 3 degrees of freedom. Each point gives two constraints on the camera. In addition, we have the freedom to choose a projective coordinate system that has 15 degrees of freedom. This means that for m cameras viewing n points the number of excess constraints l is:

$$l = 2mn - 11m - 3n + 15. \qquad (9.40)$$

This number is shown in Table 9.6. Since an occlusion will reduce the number of constraints by two, it follows that there are minimal cases only when l is even. This means that we only have minimal cases for m and n points when $m + n$ is odd, see Table 9.6 and Eq. (9.40).

We can now use the same procedures as we did in Sects. 9.5.1 and 9.5.5 for 1D retina for classifying the minimal cases. The number of prime config-

Table 9.7. The number of prime configurations for m 2D cameras and n points

m	6	7	8	9	10	11
2	−	1	−	0	−	0
3	1	−	1	−	1	
4	−	1	−	14		
5	0	−	26			
6	−	4				
7	0					

urations for 2D retina is shown in Table 9.7. There is only one prime case for two cameras, the unoccluded case. The same is true for six points.

For the case of three uncalibrated views of 8 points there are 6 distinct germs. Of these 6 germs, 4 are ill-defined, and of the two remaining minimal problems one is prime. This prime problem was solved in [17]. For three views of 9 points there are no minimal configurations. For three views of 10 points there are 33 distinct germs (30 are ill-defined, two are minimal but not prime and one is prime, Fig. 9.10.) Just like the case of 1D retina, there are infinitely

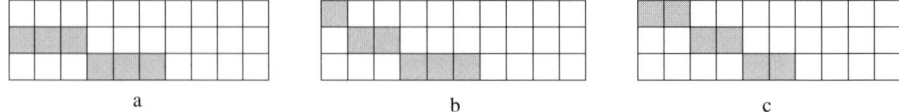

Fig. 9.10. The three minimal cases of type $(3, 10, 6)$ for 2D retina. Of these (a) and (b) are minimal but not prime, and (c) is prime

many prime configurations for 2D retina. This can be shown by a similar construction as was done for 1D retina in the proof of Theorem 9. Given m cameras, $m + 3$ points and $(2m^2 - 7m + 15)/2$ occlusions, where the first four points are visible in each view, and every subsequent point is visible in three views (Fig. 9.11) then this is a prime configuration. The proof is completely analogous to the proof of Theorem 9.

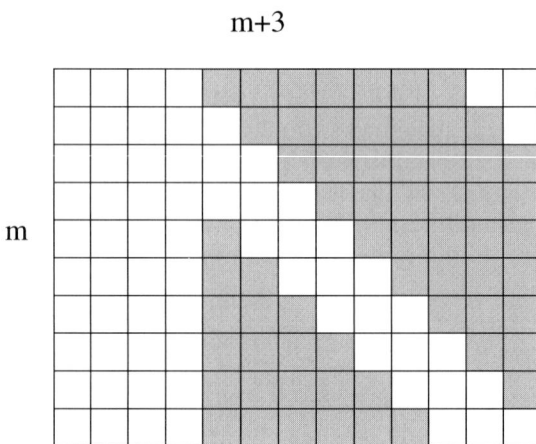

Fig. 9.11. A prime configuration with m cameras and $m + 3$ points. This configuration can be constructed for any m and is always prime

9.5.10 Solution of Some Prime Problems

We now turn our attention to the task of solving some of the prime problems for 1D retina. As such, we will consider the problem with both beacons and cameras in general positions. As in ordinary vision there exist so-called critical configurations where there is an inherent ambiguity of the solutions to the structure and motion problem irrespective of the number cameras and points. In this chapter we assume noncritical configurations. For nonmissing data and 1D retina the issue of critical configurations was completely resolved in [16]. For missing data it is not known what the critical configurations are, but in order to understand which they are, an understanding of the minimal cases for missing data is desirable.

In the previous section it was also shown that some minimal problems can be solved by extending a prime problem. One such case of a prime extension is also considered in this chapter.

9.5.11 The Case of Five Points in Four Images

There is only one prime configuration for the case of 5 points in 4 images. This is the case where one sees all 5 points in 2 images. In image 3, one point is occluded and in image 4 another point is occluded. We will start by finding the solutions to this case.

Theorem 10. *The structure and motion problem with 4 views of 5 points*

$$\lambda_{IJ}\mathbf{u}_{IJ} = \mathbf{P}_I\mathbf{U}_J, \qquad \forall (I,J) \in \mathbb{I},$$

with $\mathbb{I}$ such that point 1 is missing in view 3 and point 2 is missing in view 4 (see Table 9.8) has in general three solutions.

9.5.12 The Case of Five Points in Five Images

There are three prime problems for the case of 5 points in 5 images. We will now solve these three prime problems and their three dual cases.

Theorem 11. *The structure and motion problem for 5 images of 5 points,*

$$\lambda_{IJ}\mathbf{u}_{IJ} = \mathbf{P}_I\mathbf{U}_J, \qquad \forall (I,J) \in \mathbb{I}.$$

with $\mathbb{I}$ given by Fig. 9.6a has in general three solutions.

The dual to this case of 5 points in 5 images is the case of 6 points in 4 images given by Fig. 9.7a. This means that there are three solutions to this case of 6 points in 4 images.

Corollary 3. *The structure and motion problem for 4 images of 6 points,*

$$\lambda_{IJ}\mathbf{u}_{IJ} = \mathbf{P}_I\mathbf{U}_J, \qquad \forall (I,J) \in \mathbb{I},$$

with $\mathbb{I}$ given by Fig. 9.7a has in general three solutions.

Using the same kind of parameterization as in the previous cases, we can solve the prime problem given by figure 9.6b.

Theorem 12. *The structure and motion problem for five images of five points,*
$$\lambda_{IJ}\mathbf{u}_{IJ} = \mathbf{P}_I \mathbf{U}_J, \qquad \forall (I,J) \in \mathbb{I},$$
with $\mathbb{I}$ given by Fig. 9.6b has in general four solutions.

The dual to this case of 5 points in 5 images is the case of 6 points in 4 images given by Fig. 9.7b. This means that there are three solutions to this case of 6 points in 4 images.

Corollary 4. *The structure and motion problem for 4 images of 6 points,*
$$\lambda_{IJ}\mathbf{u}_{IJ} = \mathbf{P}_I \mathbf{U}_J, \qquad \forall (I,J) \in \mathbb{I},$$
with $\mathbb{I}$ given by Fig. 9.7b has in general four solutions.

Finally, the last prime case of 5 images of 5 points can be shown to have five solutions.

Theorem 13. *The structure and motion problem for 5 images of 5 points,*
$$\lambda_{IJ}\mathbf{u}_{IJ} = \mathbf{P}_I \mathbf{U}_J, \qquad \forall (I,J) \in \mathbb{I},$$
with $\mathbb{I}$ given by Fig. 9.6c has in general five solutions.

The dual to this case of 5 points in 5 images is the case of 6 points in 4 images given by Fig. 9.7c. This means that there are five solutions to this case of 6 points in 4 images.

Corollary 5. *The structure and motion problem for 4 images of 6 points,*
$$\lambda_{IJ}\mathbf{u}_{IJ} = \mathbf{P}_I \mathbf{U}_J, \qquad \forall (I,J) \in \mathbb{I}.$$
with $\mathbb{I}$ given by Fig. 9.7c has in general five solutions.

9.5.13 The Two-by-Two Extension

Apart from simple extensions based on intersection and resection, the first extension of a prime problem is the extension by two cameras and two points. This type of extension is shown in Fig. 9.9. We assume that we have a solution to a prime problem with m cameras and n points. The task is then to extend this solution to the solution of the extended problem with $m+2$ cameras and $n+2$ points. Two extra cameras and two extra points means that we have $2 \cdot 3 + 2 \cdot 2 = 10$ unknowns to solve for. Using intersection, the known cameras give two linear constraints on the unknown points. And using resection the known points give four linear constraints on the unknown cameras. This leaves four parameters, one for each camera (A_I, $I = 1, 2$) and one for each point

(U_J, $J = 1, 2$), to solve for. The two new points are seen in both the two new views. This gives four quadratic constraints on the four parameters,

$$a_{IJ}U_J A_I + b_{IJ}U_J + c_{IJ}A_I + d_{IJ} = 0; \quad I = 1, 2, \ J = 1, 2;$$

with the coefficients $(a_{IJ}, b_{IJ}, c_{IJ}, d_{IJ})$ only depending on the images. This means that there could be up to $2^4 = 16$ solutions according to Bezout's theorem, cf. [10]. But because of the sparseness of the polynomials this is not the case. Taking resultants pairwise we can eliminate U_J. This leaves two polynomial equations in A_I:

$$a'_J A_1 A_2 + b'_J A_1 + c'_J A_2 + d'_J = 0, \quad J = 1, 2.$$

Taking the resultant of the two polynomials with respect to A_2 leaves the following quadratic equation in A_1:

$$a'' A_1^2 + b'' A_1 + c'' = 0.$$

This discussion leads to Theorem 14:

Theorem 14. *Given an extension of type $(2, 2)$ to a structure and motion problem with m cameras and n points (as depicted in Fig. 9.9), the number of solutions are in general $2 \times N$, where N is the number of solutions of the original problem with m cameras and n points.*

9.5.14 Some Experimental Results

The methods described in the proof of Theorem 10 can easily be implemented. In Table 9.8 bearings for an example of the minimal case described in Theorem 10 is shown. The resulting solutions are illustrated in Fig. 9.12. In this case there were three real solutions with all depths positive.

Table 9.8. Some bearing measurements

0.6929	−0.7825	−1.9347	0.3263	−0.6421
0.3206	−0.9479	−1.8732	−0.0041	−0.8289
−	−2.5202	2.4474	−0.9746	−2.3323
2.3024	−	−1.0540	1.8991	0.6499

In Table 9.9 a setting with 5 cameras and 7 points is shown. This is a minimal problem but not a prime one. The subproblem of the first 3 cameras and the first 5 points is a prime problem which can be solved using algorithm for three views of five points. This gives two solutions. The solution of the whole problem can then be found by the $(2, 2)$ extension described in Sect. 9.5.13. This gives two solutions for each of the original solutions, in total four solutions. In this particular case the solutions all had positive depths. The resulting solutions are shown in Fig. 9.13.

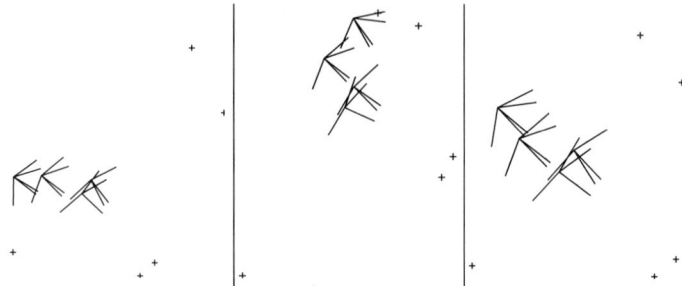

Fig. 9.12. Three solutions to the minimal case of 5 points in 4 images. Beacons are indicated by '+'

Table 9.9. Some bearing measurements

−1.9786	−0.5736	0.8046	−1.0507	0.5931	−	−
−2.5202	−1.1710	1.5796	−1.8684	1.2136	−	−
−0.8188	0.6134	3.0339	−0.0885	2.6759	−0.1730	−2.4311
−2.2663	−0.6898	−	−	−	−1.3617	1.5151
−1.8792	−0.3047	−	−	−	−1.1115	2.5170

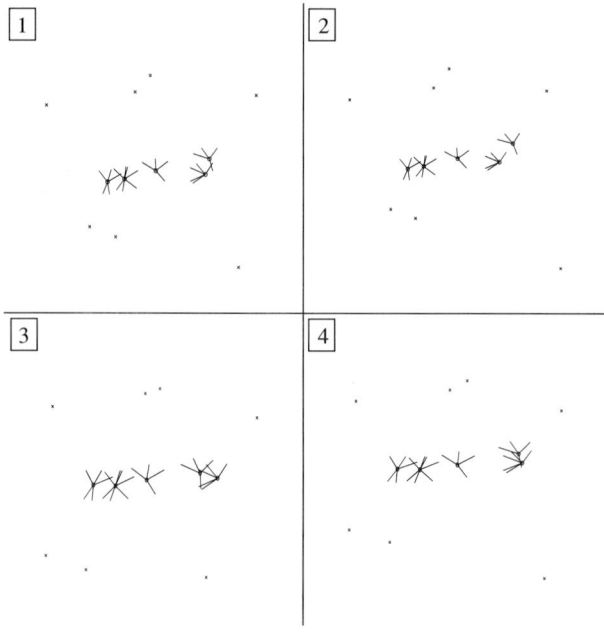

Fig. 9.13. Four solutions to one minimal case of 7 points in 5 images. Beacons are indicated by '+'

9.6 Ambiguous Cases of Structure and Motion Problems

We have seen previously that a solution to the structure and motion problem is only determined up to an unknown projective transformation. Also, for three cameras and any number of points, there is accordingly a twofold ambiguity. Additionally, there are two basic ambiguities that will be discussed here.

For the intersection problem there is one critical configuration for which there is not a unique solution.

Theorem 15. *Consider the case of several views of one point with known camera matrices. The intersection problem is ambiguous if and only if all camera centres and the point lie on a line.*

Resection is the problem of calculating camera positions using image measurements and known object points. In this case, the critical configurations are not obvious.

Theorem 16. *Consider $m > 4$ object points with known positions and one unknown camera. The resection problem is ambiguous if and only if all points and the camera centre lie on a conic curve.*

Proof. Consider first the critical configuration for the intersection problem, i.e. 1 point and $n > 1$ camera centres lying on a single line. The configuration is still ambiguous if more object points are added. Four points (where no three are on a line) and several $n > 1$ cameras are ambiguous if and only if 1 of the points and all camera centres are on a single line. If the other 3 points are taken as base points, the dual statement is: $m > 4$ object points and one camera centre are ambiguous if and only if all points and the camera centre lie on a conic curve.

The intersection and resection ambiguities are illustrated in Fig. 9.14. The calibrated version follows.

Corollary 6. *Consider $m > 2$ object points with known positions and one unknown camera. The calibrated resection problem is ambiguous if and only if all points and the camera centre lie on a circle.*

9.6.1 Three View Ambiguities

A structure and motion problem with three views can be ambiguous in three ways

1. The alternative reconstructions have the same relative camera motion.
2. The alternative reconstruction have different relative camera motion, but the corresponding trilinear tensor is the same.
3. The alternative reconstructions have different relative camera motion, and the corresponding trilinear tensor is different.

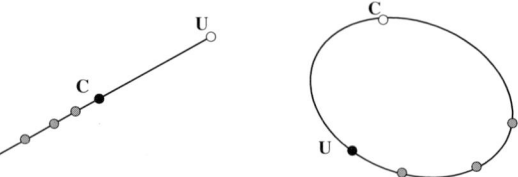

Fig. 9.14. (**left**) The intersection ambiguity where a point **U** and several camera centres **C** lie on a line. (**right**) The dual resection ambiguity where a camera centre **C** and several points **U** lie on a conic curve

For case 1 there is a unique relative motion, so one can without loss of generality assume that the camera positions are known. The alternative reconstruction differs in at least one of the object points. This can only happen if the camera centres and that point are collinear, see Theorem 15. For case 2, Theorem 2 shows that for each trilinear tensor there are two possible relative orientations. Thus any three-view problem is critical in the sense that there are at least two possible solutions. For the third case, we ask if there are cases where there might be more than two solutions to the structure and motion problem, i.e. when the tensor is not uniquely defined. We will call this case a *three-view ambiguity*.

We are now ready to state the theorem describing exactly when there are three-view ambiguities. For an example, see Fig. 9.15.

Theorem 17. *The structure and motion problem for three views and arbitrary number of points is ambiguous if and only if the three camera centres and all the object points lie on a cubic curve.*

There is an interesting special case when all the points and at least one of the camera centres lie on a conic. It fits into the theorem since there is a cubic consisting of the conic through the points and one camera centre and a line through the remaining camera centres. The cubic thus covers all points and camera centres. The problem is then critical in the sense that the resection problem for the first camera is critical, cf. Theorem 16.

Proof. Consider a situation where there is an ambiguity. Consider one of the solutions to the problem. For this solution there is a placement of cameras, **A**, **B** and **C**. The condition that there is an ambiguous solution is equivalent to saying that there is an alternative tensor T_{ijk} such that

$$\sum T_{ijk} \mathbf{a}^i \mathbf{b}^j \mathbf{c}^k = 0,$$

where **a**, **b** and **c** are image points in the three images, respectively. Since

$$a^i = \mathbf{A}^i \mathbf{X}, \quad b^i = \mathbf{B}^i \mathbf{X}, \quad c^i = \mathbf{C}^i \mathbf{X},$$

the constraint on the object point is a third-degree polynomial in $\mathbf{X} \in \mathbb{P}^2$:

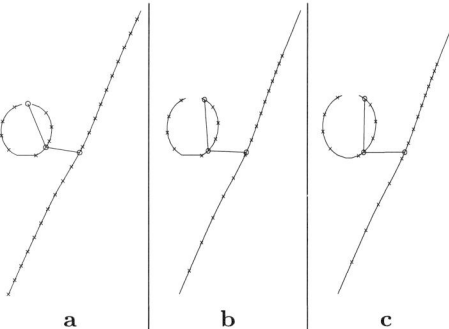

Fig. 9.15. Three cameras (*circles*) are viewing 22 points (*crosses*). All three configurations (**a-c**)(out of a one-parameter family) are consistent with the 1D image points. The 25 plane points lie on a cubic

$$p(\mathbf{X}) = \sum T_{ijk} \mathbf{A}^i \mathbf{X} \mathbf{B}^i \mathbf{X} \mathbf{C}^i \mathbf{X} = 0 \,.$$

This shows that all object points pass through this cubic curve. To see that the camera centres lie on the same curve it is sufficient to observe that $\mathbf{AF} = 0$, when $\mathbf{F}$ is the camera centre for camera 1. This gives directly that $p(\mathbf{F}) = 0$. Notice that the structure and motion problem is in general well defined for 7 points in 3 views. With 6 object points and 3 camera centres, there is in general a unique cubic curve passing through these points. That the 7th object point also lie on this curve is exceptional. To show the only if part, we consider an object where all camera centres and object points lie on an arbitrary third-degree polynomial. Without loss of generality we may change both object coordinate system and image coordinate system so that

$$\mathbf{A} = \begin{pmatrix} 1 & 0 & 0 \\ 0 & 1 & 0 \end{pmatrix}, \quad \mathbf{B} = \begin{pmatrix} 1 & 0 & 0 \\ 0 & 0 & 1 \end{pmatrix}, \quad \mathbf{C} = \begin{pmatrix} 0 & 1 & 0 \\ 0 & 0 & 1 \end{pmatrix},$$

as long as the three cameras are not on a line. The mapping from ambiguous tensors to cubic curves is a linear mapping. Each ambiguous tensor that has eight parameters $T = (T_{111}, T_{112}, T_{121}, T_{122}, T_{211}, T_{212}, T_{221}, T_{222})$ corresponds to a cubic curve

$$p(\mathbf{X}) = \sum T_{ijk} \mathbf{A}^i \mathbf{X} \mathbf{B}^i \mathbf{X} \mathbf{C}^i \mathbf{X} = 0 \,,$$

where the coefficients $c = (c_{x^3}, c_{x^2y}, c_{x^2z}, c_{xy^2}, c_{xyz}, c_{xz^2}, c_{y^3}, c_{y^2z}, c_{yz^2}, c_{z^3})$ of the polynomial $p(\mathbf{X})$ depend linearly on the tensor coefficients

$$c = MT \,. \tag{9.41}$$

For this particular choice of coordinates the matrix M becomes

$$M = \begin{bmatrix} 0 & 0 & 0 & 0 & 0 & 0 & 0 & 0 \\ 1 & 0 & 0 & 0 & 0 & 0 & 0 & 0 \\ 0 & 1 & 0 & 0 & 0 & 0 & 0 & 0 \\ 0 & 0 & 0 & 0 & 1 & 0 & 0 & 0 \\ 0 & 0 & 1 & 0 & 0 & 1 & 0 & 0 \\ 0 & 0 & 0 & 1 & 0 & 0 & 0 & 0 \\ 0 & 0 & 0 & 0 & 0 & 0 & 0 & 0 \\ 0 & 0 & 0 & 0 & 0 & 0 & 1 & 0 \\ 0 & 0 & 0 & 0 & 0 & 0 & 0 & 1 \\ 0 & 0 & 0 & 0 & 0 & 0 & 0 & 0 \end{bmatrix}.$$

It is straightforward to see that the matrix M has rank 7. Notice that the true tensor is a null vector to the matrix, so M must have rank ≤ 7. If the three camera centres happen to be on a line, it is easy to check that the corresponding mapping is also linear with rank 7. The mapping given by Eq. (9.41) is in fact a bijective mapping from the star of tensors through the true tensor (which can be identified with $\mathbb{P}^6$) to the manifold of cubic curves that pass through the three camera centres (also $\mathbb{P}^6$).

Since the mapping is bijective, our arbitrary third-degree curve on which the object points lie corresponds to an ambiguous tensor. Thus the structure and motion problem for that case is critical. This concludes the proof.

>From the principle of duality, the following theorem is obtained.

Theorem 18. *The structure and motion problem for any number of views of 6 points is ambiguous if and only if the camera centres and the object points lie on a cubic curve.*

Proof. The image under the Carlsson map of a cubic curve through the base points is again a cubic curve through the base point. The dual of 3 cameras and n points is $m = n - 3$ cameras and 6 points. So by the principle of duality and Theorem 17, the statement is proved.

9.6.2 General n Points in m Views Ambiguity

Up to now, we have limited either the number of cameras or the number of points considered. Based on the previous results, the general problem will now be solved. A natural generalization of the three-view case for the word "ambiguous" is that the alternative reconstructions have different relative camera motion, and (at least) one triplet of cameras has a different trilinear tensor.

Theorem 19. *A 1D structure and motion problem is ambiguous regardless of the number of cameras and points if and only if all the camera centres and the object points lie on a common third-degree curve.*

Proof. We begin by showing that a problem is ambiguous if all points lie on a third-degree curve. Assume that camera centres and object points lie on a third-degree curve c. By first restricting the problem to only 6 points, we know from Theorem 18 that the configuration is ambiguous and there is (at least) a one-parameter family of solutions. Now, consider a seventh point on the curve c. We need to show that the constraints generated by the projection equation for this extra point do not break the ambiguity. However, all these constraints

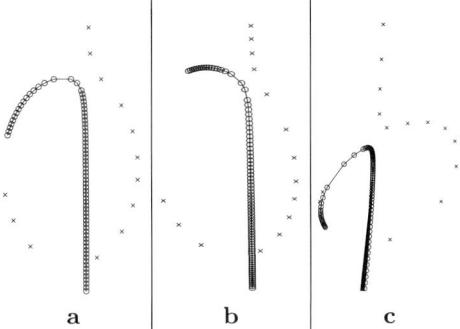

Fig. 9.16. The figure illustrates three solutions (**a-c**) (out of a one-parameter family) to the same structure and motion problem. There are 82 cameras (*circles*) viewing 15 points (*crosses*). It is critical because all 97 points lie on a cubic

reduce to trilinear constraints, as there are no higher-order constraints for 1D camera motion [6]. Thus, it suffices to consider three arbitrary cameras $\mathbf{P}_i$, $\mathbf{P}_j$ and $\mathbf{P}_k$. In the proof of Theorem 17, we showed that the map from stars of tensors to cubic curves (through the camera centres) is bijective. So, from c and the three camera centres, a star of tensors $\lambda T_1 + \mu T_2$, where $(\lambda, \mu) \in \mathbb{P}^1$, is obtained. But according to the proof of Theorem 17, as long as the seventh point is on c, all tensors in $\lambda T_1 + \mu T_2$ are still valid solutions.

To show that each ambiguous problem has the property that all points lie on a third-degree curve we use a proof by contradiction. Thus, assume that there exist ambiguous problems with m views of n points such that the $m+n$ points do not lie on a common third-degree curve. Such problems must have $m > 3$ and $n > 6$ because of Theorems 17 and 18, respectively. Study such a problem where $m+n$ is minimal. If we remove one point or one camera, we obtain an ambiguous problem with one point less. By the assumption, these $m+n-1$ points must lie on a third-degree curve. In particular, this means that all 10 point subconfigurations must lie on a third-degree curve. But then all $m+n$ points lie on a cubic curve. Thus all ambiguous configurations have the property that all $m+n$ points lie on a third-degree curve.

In Fig. 9.16 an example of a critical configuration is illustrated. Even though there are 82 views of 15 points, the 1D images alone cannot disambiguate between a one-parameter family of solutions. Another example is illustrated in Fig. 9.17, where a camera moves along a corridor, which frequently occurs in practical situations.

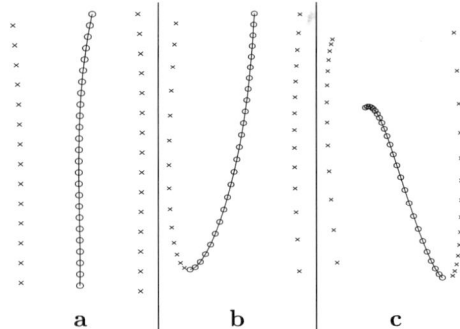

Fig. 9.17. Three solutions (out of a one-parameter family) to the same structure and motion problem. In the example, the camera moves in a corridor with scene points on both walls, which is quite common in robot navigation. There are 25 cameras (*circles*) viewing 29 points (*crosses*). It is critical because all 54 points lie on a cubic

9.7 Conclusions

In this paper we introduced the minimal conditions for solving the structure and motion problem for cameras with one-dimensional retinae. The emphasis was on calibrated cameras.

For the minimal case of 3 images with 5 points it was shown how to solve the problem using the calibrated trilinear tensor. It was shown that there is a two-to-one map from the relative orientation of three cameras to the calibrated trilinear tensor. This explains why there are two solutions to the structure and motion problem for three cameras. A geometric interpretation of this ambiguity is also given.

For the minimal case of 4 images with 4 points it was shown how to solve the problem using the dual calibrated quadrilinear tensor. It was shown that there is a two-to-one map from the shape of four planar points to this tensor. This explains why there are two solutions to the structure and motion problem for four points.

Notice that the trilinear tensor encodes the relative motion of the camera at three instants and that the dual quadrilinear tensor encodes the structure of four scene points. Thuss there are no relations at all between these two tensors.

The connection between the calibrated and the uncalibrated cameras is given. From this it follows that similar results hold for 3 images with 7 points and for 4 images with 6 points. Using the Carlsson duality it was then shown that the above two types of ambiguities are in fact dual to each other.

Furthermore, a categorisation of minimal cases for structure and motion is given. Some of the simpler minimal cases are solved. It is shown that there are infinitely many such minimal cases.

Finally a complete categorisation of ll ambiguous configurations for the structure and motion problem in 1D retina vision is presented. The main ambiguity is when all object points (regardless of how many) and all camera centres (again, regardless of the number of cameras) lie on a cubic curve.

Acknowledgements

This work was supported by Netzler Dahlgren Co., the ESPRIT Reactive LTR project no. 21914, CUMULI and the Swedish Research Council for Engineering Sciences (TFR), project no. 95-64-222.

References

1. N. Altshiller-Court. *College Geometry*. Barnes and Noble, New York, 1952.
2. M. Armstrong, A. Zisserman, and R. Hartley. Self-calibration from image triplets. In: *Proc. 4th European Conf. on Computer Vision, Cambridge, UK*, pages 3–16. Springer-Verlag, 1996.
3. K. Åström. Where am I and what am I seeing? Algorithms for a laser guided vehicle. Master's thesis, Dept. of Mathematics, Lund Institute of Technology, Sweden, 1991.
4. K. Åström. Automatic mapmaking. In: D. Charnley (ed.) *Selected Papers from the 1st IFAC International Workshop on Intelligent Autonomous Vehicles, Southampton, UK, 18-21 April 1993*, pages 181–186. Pergamon Press, Great Britain, 1993.
5. K. Åström. *Invariancy Methods for Points, Curves and Surfaces in Computational Vision*. PhD thesis, Dept of Mathematics, Lund University, Sweden, 1996.
6. K. Åström and M. Oskarsson. Solutions and ambiguities of the structure and motion problem for 1D retinal vision. *Journal of Mathematical Imaging and Vision*, 12:121–135, 2000.
7. H. F. Baker. *An Introduction to Plane Geometry*. Cambridge University Press, Cambridge, 1943.
8. S. Carlsson. Duality of reconstruction and positioning from projective views. In *IEEE Workshop on Representation of Visual Scenes*, pages 85–92. IEEE, 1995.
9. S. Carlsson and D. Weinshall. Dual computation of projective shape and camera positions from multiple images. *Int. Journal of Computer Vision*, 27(3):227–241, 1998.
10. D. Cox, J. Little, and D. O'Shea. *Using Algebraic Geometry*. Springer, Berlin Heidelberg New York, 1998.
11. H. S. M. Coxeter. *The Real Projective Plane*. Springer, Berlin Heidelberg New York, 3rd edition, 1993.
12. O. D. Faugeras, L. Quan, and P. Sturm. Self-calibration of a 1d projective camera and its application to the self-calibration of a 2D projective camera. In *Proc. 5th European Conf. on Computer Vision, Freiburg, Germany*, pages 36–52. Springer-Verlag, 1998.

13. J.B. Fraleigh. *A first course in abstract algebra*, 5th edn. Addison-Wesley Boston, 1994.
14. R Gupta and R. I. Hartley. Linear pushbroom cameras. *Pattern Analysis and Machine Intelligence*, 19(9):963–975, 1997.
15. K. Hyyppä. Optical navigation system using passive identical beacons. In Louis O. Hertzberger and Frans C. A. Groen, editors, *Intelligent Autonomous Systems, An International Conference, Amsterdam, The Netherlands, 8-11 December 1986*, pages 737–741. North-Holland, 1987.
16. F. Kahl and K. Åström. Ambiguous configurations for the 1d structure and motion problem. In *Proc. 8th Int. Conf. on Computer Vision, Vancouver, Canada*, pages 184–189, 2001.
17. F. Kahl, A. Heyden, and L. Quan. Projective reconstruction from minimal missing data. *IEEE Trans. Pattern Analysis and Machine Intelligence*, 23(4):418–424, 2001.
18. C. B. Madsen, C. S. Andersen, and J. S. Sørensen. A robustness analysis of triangulation-based robot self-positioning. In *The 5th Symposium for Intelligent Robotics Systems, Stockholm, Sweden*, 1997.
19. J. Neira, I. Ribeiro, and J. D. Tardos. Mobile robot localization and map building using monocular vision. In *The 5th Symposium for Intelligent Robotics Systems, Stockholm, Sweden*, pages 275–284, 1997.
20. M. Oskarsson. *Solutions and their Ambiguities for Structure and Motion Problems*. PhD thesis, Dept of Mathematics, Lund University, Sweden, 2002.
21. L. Quan and T. Kanade. Affine structure from line correspondences with uncalibrated affine cameras. *IEEE Trans. Pattern Analysis and Machine Intelligence*, 19(8):834–845, August 1997.
22. A. Shahsua. Algebraic functions for recognition. *IEEE Trans. Pattern Analysis and Machine Intelligence*, 17(8):779–789, 1995.
23. C.C. Slama (ed.) *Manual of Photogrammetry*, 4th edn. American Society of Photogrammetry, Falls Church, VA, 1984.
24. M. E. Spetsakis and J. Aloimonos. A unified theory of structure from motion. In *Proc. DARPA IU Workshop, Pittsburgh, PA*, pages 271–283, 1990.
25. B. Triggs. Matching constraints and the joint image. In *Proc. 5th Int. Conf. on Computer Vision, MIT, Boston, MA*, pages 338–343, IEEE Computer Society Press, Los Alamitos, 1995.
26. W. Wei and J. Xu. Cycle index of direct product of permutation groups and number of equivalence classes of subsets of z_ν. *Discrete Mathematics*, 123:179–188, 1993.

10
Three-Dimensional Geometric Computer Vision

Anders Heyden

Applied Mathematics Group, School of Technology and Society, Malmo University, Malmo, Sweden
heyden@ts.mah.se

10.1 Introduction

One of the central problems in computer vision is to calculate the three-dimensional *structure* of an unknown object from an image sequence (e.g., taken by a camcorder) or from several still images. Usually, the positions of the cameras where the images were taken are unknown, i.e., the *motion* of the camera is unknown. Therefore, this problem is often called the *structure and motion problem*.

In order to solve the structure and motion problem, it is necessary to make a mathematical model of the camera. The most widely used model is the so-called *pinhole camera* and its specializations to orthographic and affine cameras. This camera model contains some calibration parameters that need to be estimated to obtain a Euclidean reconstruction of the scene.

The first step in a reconstruction system is to extract features and track them through the sequence, alternatively finding corresponding points in several still images. This is a hard problem in itself and will not be treated in this chapter. The next step is to estimate the motion of the camera, based on the multiple-view constraints. These constraints play a central role and in order to derive them some tools from projective geometry are needed. After the motion is obtained it is fairly straightforward to obtain an initial projective reconstruction, i.e., to reconstruct the object up to an unknown projective transformation. However, a nonlinear optimization, called *bundle adjustment* is needed to get a statistically optimal result, which is needed in the subsequent step. This projective reconstruction usually contains severe nonlinear distortions and in order to obtain a Euclidean reconstruction the calibration parameters need to be estimated, so-called *autocalibration*. This step is also divided into one initial linear estimation and a final nonlinear bundle adjustment to obtain the best possible reconstruction.

This chapter is organized as follows: In Sect. 10.2, an introduction to projective geometry is given, with special emphasis on the concepts and results needed for the investigation of the structure and motion problem. The process of modelling a pinhole camera along with specializations leading to different camera models is given in Sect. 10.3. In Sect. 10.4 the geometry of multiple views is investigated, leading to the multiple-view tensors. A number of reconstruction methods are presented in Sect. 10.5 and finally in Sect. 10.6 some autocalibration methods.

10.2 Projective Geometry

This section deals with the fundamentals of projective geometry, including the definitions of projective spaces, homogeneous coordinates, duality, projective transformations and affine and Euclidean imbeddings. For a traditional approach to projective geometry see [2], and for more modern treatments see [5, 6, 8].

10.2.1 Projective Spaces

In order to define projective spaces of different dimensions, the standard $\mathbb{R}^n$-spaces need to be enlarged with some extra points. Consider the set L of all lines parallel to a given line l in $\mathbb{R}^2$ and assign a point to each such set, $\mathbf{p}_{\text{ideal}}$, called an **ideal point** or **point at infinity**. The **projective plane**, $\mathbb{P}^2$, is given by
$$\mathbb{P}^2 = \mathbb{R}^2 \cup \{\text{ideal points}\} \ .$$
This means that the 2-dimensional projective plane is obtained by adding the ideal points to $\mathbb{R}^2$. The **ideal line**, l_∞, or **line at infinity** in $\mathbb{P}^2$, is defined by
$$l_\infty = \{\text{ideal points}\} \ ,$$
i.e., it consists of all ideal points. The following constructions could easily be made in $\mathbb{P}^2$:

1. Two different points define a line (called the **join** of the points).
2. Two different lines intersect at a point.

Here there are obvious interpretations for ideal points and the ideal line, e.g., the line defined by an ordinary point and an ideal point is the line incident with the ordinary point with the direction given by the ideal point.

The **projective line**, $\mathbb{P}^1$, is given by
$$\mathbb{P}^1 = \mathbb{R}^1 \cup \{\text{ideal point}\} \ .$$
This means that the 1-dimensional projective line is obtained by adding one ideal point to $\mathbb{R}^1$.

In order to define the three-dimensional projective space, $\mathbb{P}^3$, start with $\mathbb{R}^3$ and assign an ideal point to each set of parallel lines, i.e., to each direction. The **projective space**, $\mathbb{P}^3$, is thus given by

$$\mathbb{P}^3 = \mathbb{R}^3 \cup \{\text{ideal points}\} \ .$$

Observe that the ideal points in $\mathbb{P}^3$ constitute a two-dimensional manifold, called the **ideal plane** or **plane at infinity**. This plane at infinity contains lines, again called lines at infinity. Every set of parallel planes in $\mathbb{R}^3$ defines an ideal line and all ideal lines build up the ideal plane. Many geometrical constructions can be made in $\mathbb{P}^3$, e.g.,

1. Two different points define a line (called the **join** of the two points).
2. Three different points define a plane (called the **join** of the three points).
3. Two different planes intersect in a line.
4. Three different planes intersect in a point.

10.2.2 Homogeneous Coordinates

It is often advantageous to introduce coordinates in the projective spaces, so-called **analytic projective geometry**. Introduce a cartesian coordinate system, $O\mathbf{e}_x\mathbf{e}_y$ in $\mathbb{R}^2$, and define the line $l : y = 1$; see Fig. 10.1. Observe that

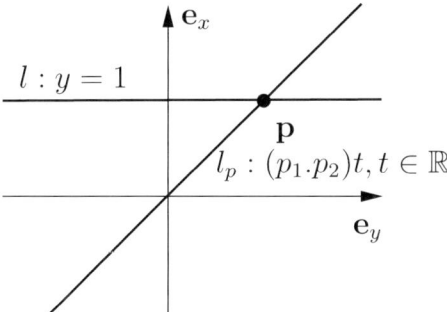

Fig. 10.1. Definition of homogeneous coordinates in $\mathbb{P}^1$

the vectors (p_1, p_2) and (q_1, q_2) determine the same line through the origin iff

$$(p_1, p_2) = \lambda(q_1, q_2), \quad \lambda \neq 0 \ .$$

Every line, $l_p = (p_1, p_2)t$, $t \in \mathbb{R}$, through the origin, except for the x-axis, intersects the line l at one point, $\mathbf{p}$. The pairs of numbers (p_1, p_2) and (q_1, q_2) are said to be **equivalent** if

$$(p_1, p_2) = \lambda(q_1, q_2), \quad \lambda \neq 0 \ ,$$

written

$$(p_1, p_2) \sim (q_1, q_2) \ .$$

There is a one-to-one correspondence between lines through the origin and points on the line l if an extra point is added on the line, corresponding to the line $x = 0$, i.e., the direction $(1,0)$. Identifying the line l augmented with this extra point, corresponding to the point at infinity, p_∞, with $\mathbb{P}^1$, the **one-dimensional projective space**, $\mathbb{P}^1$, is obtained, consisting of pairs of numbers (p_1, p_2) (under the equivalence above), where $(p_1, p_2) \neq (0, 0)$. The pair (p_1, p_2) is called the **homogeneous coordinates** for the corresponding point in $\mathbb{P}^1$. There is a natural division of $\mathbb{P}^1$ into two disjoint subsets

$$\mathbb{P}^1 = \{(p_1, 1) \in \mathbb{P}^1\} \cup \{(p_1, 0) \in \mathbb{P}^1\} ,$$

corresponding to ordinary points and the ideal point.

The introduction of homogeneous coordinates can easily be generalized to $\mathbb{P}^2$ and $\mathbb{P}^3$ using three and four homogeneous coordinates, respectively. In the case of $\mathbb{P}^2$ fix a cartesian coordinate system $Oe_x e_y e_z$ in $\mathbb{R}^3$ and define the plane $\Pi : z = 1$; see Fig. 10.2. The vectors (p_1, p_2, p_3) and (q_1, q_2, q_3)

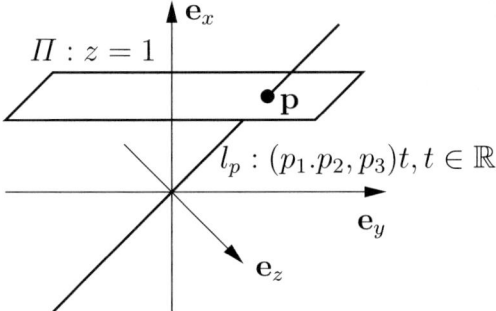

Fig. 10.2. Definition of homogeneous coordinates in $\mathbb{P}^2$

determine the same line through the origin iff

$$(p_1, p_2, p_3) = \lambda(q_1, q_2, q_3), \quad \lambda \neq 0 .$$

Every line through the origin, except for those in the x-y-plane, intersect the plane Π at one point. Again, there is a one-to-one correspondence between lines through the origin and points on the plane Π if an extra line is added, corresponding to the lines in the plane $z = 0$, i.e., the line at infinity, l_∞, built up by lines of the form $(p_1, p_2, 0)$. The plane Π augmented with this extra line, corresponding to the points at infinity, l_∞, is identified with $\mathbb{P}^2$. The pairs of numbers (p_1, p_2, p_3) and (q_1, q_2, q_3) are said to be equivalent iff

$$(p_1, p_2, p_3) = \lambda(q_1, q_2, q_3), \quad \lambda \neq 0 \quad \text{written} \quad (p_1, p_2, p_3) \sim (q_1, q_2, q_3) .$$

The two-dimensional projective space $\mathbb{P}^2$ consists of all triplets of numbers $(p_1, p_2, p_3) \neq (0, 0, 0)$. The triplet (p_1, p_2, p_3) is called the homogeneous coordinates for the corresponding point in $\mathbb{P}^2$. There is a natural division of $\mathbb{P}^2$

into two disjoint subsets

$$\mathbb{P}^2 = \{(p_1, p_2, 1) \in \mathbb{P}^2\} \cup \{(p_1, p_2, 0) \in \mathbb{P}^2\} \; ,$$

corresponding to ordinary points and ideal points (or points at infinity).

The same procedure can be carried out to construct $\mathbb{P}^3$ (and even $\mathbb{P}^n$ for any $n \in \mathbb{N}$), but it is harder to visualize. In this way, the **three-dimensional (n-dimensional) projective space** $\mathbb{P}^3$ ($\mathbb{P}^n$) is defined as the set of one-dimensional linear subspaces in a vector space, $\mathbb{V}$ (usually $\mathbb{R}^4$ ($\mathbb{R}^{n+1}$)) of dimension 4 ($n+1$). Points in $\mathbb{P}^3$ ($\mathbb{P}^n$) are represented using **homogeneous coordinates** by vectors $(p_1, p_2, p_3, p_4) \neq (0,0,0,0)$ (($p_1, \ldots, p_{n+1}) \neq (0, \ldots, 0)$), where two vectors represent the same point iff they differ by a global scale factor. There is a natural division of $\mathbb{P}^3$ ($\mathbb{P}^n$) into two disjoint subsets

$$\mathbb{P}^3 = \{(p_1, p_2, p_3, 1) \in \mathbb{P}^3\} \cup \{(p_1, p_2, p_3, 0) \in \mathbb{P}^3\}$$
$$(\mathbb{P}^n = \{(p_1, \ldots, p_n, 1) \in \mathbb{P}^n\} \cup \{(p_1, \ldots, p_n, 0) \in \mathbb{P}^n\} \; ,$$

corresponding to ordinary points and ideal points (or points at infinity). Finally, geometrical entities are defined similarly in $\mathbb{P}^3$.

10.2.3 Duality

Remember that a line in $\mathbb{P}^2$ is defined by two points $\mathbf{p}_1$ and $\mathbf{p}_2$ according to

$$l = \{\mathbf{x} = (x_1, x_2, x_3) \in \mathbb{P}^2 \mid \mathbf{x} = t_1 \mathbf{p}_1 + t_2 \mathbf{p}_2, \quad (t_1, t_2) \in \mathbb{R}^2\} \; .$$

Observe that since (x_1, x_2, x_3) and $\lambda(x_1, x_2, x_3)$ represent the same point in $\mathbb{P}^2$, the parameters (t_1, t_2) and $\lambda(t_1, t_2)$ give the same point. This gives the equivalent definition:

$$l = \{\mathbf{x} = (x_1, x_2, x_3) \in \mathbb{P}^2 \mid \mathbf{x} = t_1 \mathbf{p}_1 + t_2 \mathbf{p}_2, \quad (t_1, t_2) \in \mathbb{P}^1\} \; .$$

By eliminating the parameters t_1 and t_2 the line could also be written in the form

$$l = \{\mathbf{x} = (x_1, x_2, x_3) \in \mathbb{P}^2 \mid n_1 x_1 + n_2 x_2 + n_3 x_3 = 0\} \; , \tag{10.1}$$

where the normal vector, $\mathbf{n} = (n_1, n_2, n_3)$, could be calculated as $\mathbf{n} = \mathbf{p}_1 \times \mathbf{p}_2$. Observe that if (x_1, x_2, x_3) fulfils Eq. (10.1) then $\lambda(x_1, x_2, x_3)$ also fulfils Eq. (10.1) and that if the line, l, is defined by (n_1, n_2, n_3), then the same line is defined by $\lambda(n_1, n_2, n_3)$, which means that $\mathbf{n}$ could be considered as an element in $\mathbb{P}^2$.

The line equation in EQ. (10.1) can be interpreted in two different ways; see Fig. 10.3:

- Given $\mathbf{n} = (n_1, n_2, n_3)$, the points $\mathbf{x} = (x_1, x_2, x_3)$ that fulfil Eq. (10.1) constitute the line defined by $\mathbf{n}$.

- Given $\mathbf{x} = (x_1, x_2, x_3)$, the lines $\mathbf{n} = (n_1, n_2, n_3)$ that fulfil Eq. (10.1) constitute the lines coincident at $\mathbf{x}$.

The set of lines incident with a given point $\mathbf{x} = (x_1, x_2, x_3)$ is called a **pencil** of lines. In this way there is a one-to-one correspondence between points and lines in $\mathbb{P}^2$ given by

$$\mathbf{x} = (a, b, c) \quad \leftrightarrow \quad \mathbf{n} = (a, b, c) \ ,$$

as illustrated in Fig. 10.3.

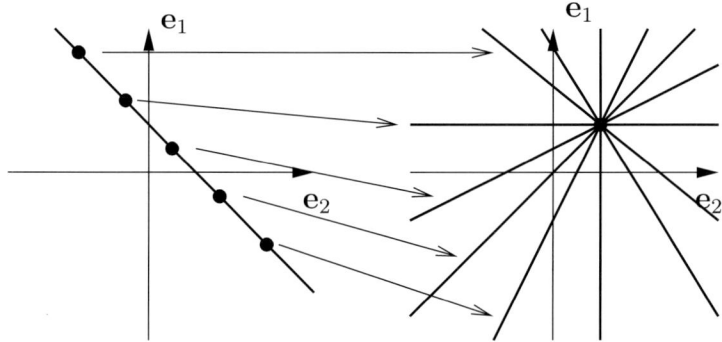

Fig. 10.3. Duality of points and lines in $\mathbb{P}^2$

Similarly, there exists a duality between points and planes in $\mathbb{P}^3$. A plane π in $\mathbb{P}^3$ consists of the points $\mathbf{x} = (x_1, x_2, x_3, x_4)$ that fulfil the equation

$$\pi = \{\mathbf{x} = (x_1, x_2, x_3, x_4) \in \mathbb{P}^3 \mid n_1 x_1 + n_2 x_2 + n_3 x_3 + n_4 x_4 = 0\} \ , \quad (10.2)$$

where $\mathbf{n} = (n_1, n_2, n_3, n_4)$ defines the plane. From Eq. (10.2) a similar argument leads to a duality between planes and points in $\mathbb{P}^3$. The following property is fundamental in projective geometry:

Given a statement valid in a projective space, then the dual to that statement is also valid, where the dual is obtained by interchanging entities with their duals, intersection with join, etc.

For instance, a line in $\mathbb{P}^3$ could be defined as the join of two points. Thus the dual to a line is the intersection of two planes, which again is a line, i.e., the dual to a line in $\mathbb{P}^3$ is a line, i.e., lines are **self-dual**. A line in $\mathbb{P}^3$ defined as the join of two points, $\mathbf{p}_1$ and $\mathbf{p}_2$, as in

$$l = \{\mathbf{x} = (x_1, x_2, x_3, x_4) \in \mathbb{P}^3 \mid \mathbf{x} = t_1 \mathbf{p}_1 + t_2 \mathbf{p}_2, \quad (t_1, t_2) \in \mathbb{P}^1\} \ ,$$

and is said to be given in **parametric form** and (t_1, t_2) can be regarded as homogeneous coordinates on the line. A line in $\mathbb{P}^3$ defined as the intersection

of two planes, π and μ, consists of the common points to the pencil of planes in
$$l : \{s\pi + t\mu \mid (s,t) \in \mathbb{P}^1\}$$
is said to be given in **intersection form**. A **conic**, c, in $\mathbb{P}^2$ is defined as
$$c = \{\mathbf{x} = (x_1, x_2, x_3) \in \mathbb{P}^2 \mid \mathbf{x}^T C \mathbf{x} = 0\}, \qquad (10.3)$$
where C denotes a 3×3-matrix. If C is nonsingular the conic is said to be **proper**, otherwise it is said to be **degenerate**. The dual to a general curve in $\mathbb{P}^2$ ($\mathbb{P}^3$) is defined as the set of tangent lines (tangent planes) to the curve. It is easy to show that the dual, c^*, to a conic $c : \mathbf{x}^T C \mathbf{x}$ is the set of lines
$$\{\mathbf{l} = (l_1, l_2, l_3) \in \mathbb{P}^2 \mid \mathbf{l}^T C' \mathbf{l} = 0\},$$
where $C' = C^{-1}$.

10.2.4 Projective Transformations

A **projective transformation** from $\mathbf{p} \in \mathbb{P}^n$ to $\mathbf{p}' \in \mathbb{P}^m$ is defined as a linear transformation in homogeneous coordinates, i.e.,
$$\mathbf{x}' \sim H\mathbf{x}, \qquad (10.4)$$
where $\mathbf{x}$ and $\mathbf{x}'$ denote homogeneous coordinates for $\mathbf{p}$ and $\mathbf{p}'$, respectively, and H denotes a $(m+1) \times (n+1)$-matrix of full rank. All projective transformations form a group, denoted $\mathcal{G}_P$. For example a projective transformation from $\mathbf{x} \in \mathbb{P}^2$ to $\mathbf{y} \in \mathbb{P}^2$ is given by
$$\begin{bmatrix} y_1 \\ y_2 \\ y_3 \end{bmatrix} \sim H \begin{bmatrix} x_1 \\ x_2 \\ x_3 \end{bmatrix},$$
where H denotes a nonsingular 3×3-matrix. Such a projective transformation from $\mathbb{P}^2$ to $\mathbb{P}^2$ is usually called a **homography**.

The subspaces
$$\mathcal{A}_i^n = \{(x_1, x_2, \ldots, x_{n+1}) \in \mathbb{P}^n \mid x_i \neq 0\}$$
of $\mathbb{P}^n$, are called **affine pieces** of $\mathbb{P}^n$. Using this construction an affine space is embedded in the projective space. The plane $H_i : x_i = 0$ is called the **plane at infinity**, corresponding to the affine piece $\mathcal{A}_i^n$. Usually, $i = n+1$ is used and called the **standard affine piece** and in this case the plane at infinity is denoted H_∞. Identify points in $\mathbb{A}^n$ with points, $\mathbf{x}$, in $\mathcal{A}_i^n \subset \mathbb{P}^n$, by
$$\mathbb{P}^n \ni (x_1, x_2, \ldots, x_n, x_{n+1}) \sim (y_1, y_2, \ldots, y_n, 1) \equiv (y_1, y_2, \ldots, y_n) \in \mathbb{A}^n.$$
The subgroup, $\mathcal{H}$, of projective transformations, $\mathcal{G}_P$, that preserves the plane at infinity consists exactly of the projective transformations of the form Eq. (10.4), with

$$H = \begin{bmatrix} A_{n \times n} & b_{n \times 1} \\ 0_{1 \times n} & 1 \end{bmatrix} ,$$

where the indices denote the sizes of the matrices. Identify the affine transformations in $\mathbb{A}$ with the subgroup $\mathcal{H}$:

$$\mathbb{A} \ni \mathbf{x} \mapsto A\mathbf{x} + b \in \mathbb{A} ,$$

which gives the affine structure in $\mathcal{A}_i^n \subset \mathbb{P}^n$. When a plane at infinity has been chosen, two lines are said to be **parallel** if they intersect at a point in the plane at infinity.

The (singular, complex) conic, Ω, in $\mathbb{P}^n$ defined by

$$x_1^2 + x_1^2 + \ldots + x_n^2 = 0 \quad \text{and} \quad x_{n+1} = 0$$

is called the **absolute conic**. Observe that the absolute conic is located in the plane at infinity, it contains only complex points, and it is singular. The dual to the absolute conic, denoted Ω', is given by the set of planes

$$\Omega' = \{\Pi = (\Pi_1, \Pi_2, \ldots \Pi_{n+1}) \mid \Pi_1^2 + \ldots + \Pi_n^2 = 0\} .$$

In matrix form Ω' can be written as $\Pi^T C' \Pi = 0$ with

$$C' = \begin{bmatrix} I_{n \times n} & 0_{n \times 1} \\ 0_{1 \times n} & 0 \end{bmatrix} .$$

The subgroup, $\mathcal{K}$, of projective transformations, $\mathcal{G}_P$, that preserves the absolute conic consists exactly of the projective transformations of the form Eq. (10.4), with

$$H = \begin{bmatrix} cR_{n \times n} & t_{n \times 1} \\ 0_{1 \times n} & 1 \end{bmatrix} ,$$

where $0 \neq c \in \mathbb{R}$ and R denotes an orthogonal matrix, i.e., $RR^T = R^T R = I$. Observe that the corresponding transformation in the affine space $\mathbb{A} = \mathcal{A}_n^{n+1}$ can be written as

$$\mathbb{A} \ni \mathbf{x} \mapsto cR\mathbf{x} + t \in \mathbb{A} ,$$

which is a similarity transformation. This imposes a Euclidean structure (to be precise a similarity structure) in $\mathbb{P}^n$ given by the absolute conic.

10.3 Camera Modelling

This chapter deals with the task of building a mathematical model of a camera. It starts with a brief review of optics and continues with a mathematical model of the standard pinhole camera. Intrinsic and extrinsic parameters are defined and the six-point algorithm for camera calibration is given. The affine camera and the scaled orthographic camera are also defined. For more detailed treatments see [8, 6], and for a different approach see [10].

10.3.1 Review of Optics

A **thin lens optical system** consists of a single spherical lens. In a first-order approximation the lens focuses parallel light rays in the direction of the **optical axis** onto a single point, P_f, called the **focal point**. The distance from the lens to the focal point is called the **focal length** of the lens. The focal length can be calculated from the so-called **lens-makers' formula**:

$$\frac{1}{f} = \frac{n_2 - n_1}{n_1}\left(\frac{1}{R_1} - \frac{1}{R_2}\right), \tag{10.5}$$

where n_1 and n_2 denote the refraction indices of the surrounding media ($n_2 = 1$ in the case of air) and the lens material, respectively, and R_1 and R_2 denote the radii of the two lens surfaces.

Given an object at finite distance from the lens, the image can be constructed using the following simple principles:

- A light ray that passes the centre of the lens will pass the lens system unaltered.
- A light ray that enters the lens system parallel to the optical axis will pass through the focal point.

These light rays meet in the **focal plane**, Π_f, of the optical system, where a sharp image of the object can be found; see Fig. 10.4. The distance, d_i, from

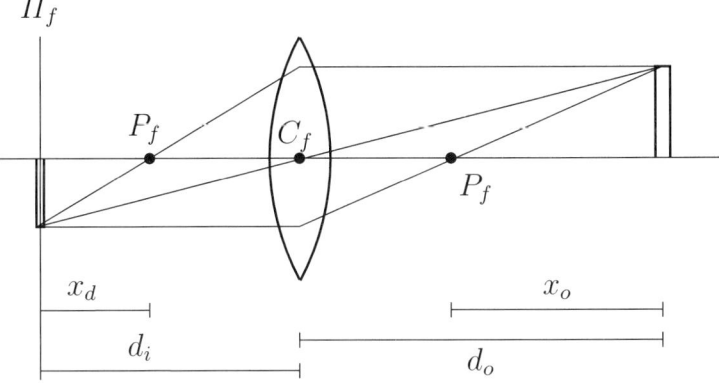

Fig. 10.4. Construction of the image in a single lens system

the lens to the focal plane can be calculated using the **lens formula**

$$\frac{1}{f} = \frac{1}{d_i} + \frac{1}{d_o}, \tag{10.6}$$

where d_o denotes the distance from the object to the lens. d_i is sometimes called the **camera constant**, and in computer vision often the **focal length**, although these coincide only when the object is infinitely far away. The centre of projection, C_f, is in computer vision often referred to as the **camera centre** or as the **focal point**.

Another useful formula for calculating the distance to the object and the image is the so-called **Newton's formula**

$$x_o x_d = f^2 \,, \tag{10.7}$$

where x_o and x_d denote the distances from the focal point to the object and focal plane, respectively; see Fig. 10.4.

10.3.2 The Pinhole Camera

The simplest optical system used for modelling cameras is the so-called **pinhole camera**. The camera is modelled as a box with a small hole in one of the sides and a photographic plate at the opposite side; see Fig. 10.5. Introduce a coordinate system as in Fig. 10.5. Observe that the origin of the coordinate system is located at the centre of projection, the so-called **focal point**, and that the z-axis coincides with the optical axis. The distance from the focal point to the image, f, is called the **focal length**. Similar triangles give

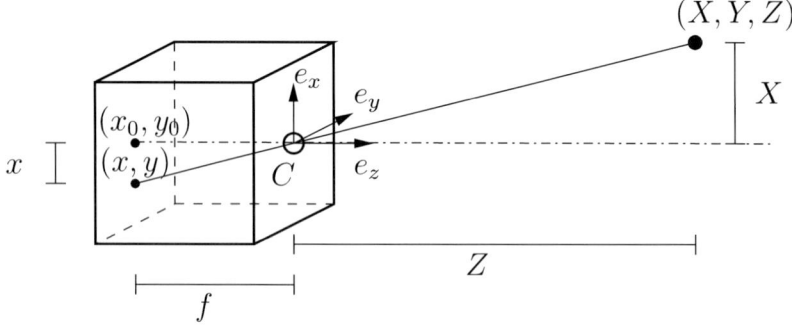

Fig. 10.5. The pinhole camera with a coordinate system

$$\frac{x}{f} = \frac{X}{Z} \quad \text{and} \quad \frac{y}{f} = \frac{Y}{Z} \,. \tag{10.8}$$

This equation can be written in matrix form, using homogeneous coordinates, as

$$\lambda \begin{bmatrix} x \\ y \\ 1 \end{bmatrix} = \begin{bmatrix} f & 0 & 0 & 0 \\ 0 & f & 0 & 0 \\ 0 & 0 & 1 & 0 \end{bmatrix} \begin{bmatrix} X \\ Y \\ Z \\ 1 \end{bmatrix} \,, \tag{10.9}$$

where the **depth**, λ, is equal to Z. Introducing the notation

$$K = \begin{bmatrix} f & 0 & 0 \\ 0 & f & 0 \\ 0 & 0 & 1 \end{bmatrix} \;,\quad \mathbf{x} = \begin{bmatrix} x \\ y \\ 1 \end{bmatrix} \;,\quad \mathbf{X} = \begin{bmatrix} X \\ Y \\ Z \\ 1 \end{bmatrix} \;, \qquad (10.10)$$

in (10.9) gives

$$\lambda \mathbf{x} = K [\, I_{3\times 3} \mid \mathbf{0}_{3\times 1} \,] \mathbf{X} = P\mathbf{X} \;, \qquad (10.11)$$

where $P = K [\, I_{3\times 3} \mid \mathbf{0}_{3\times 1} \,]$. A 3×4-matrix P relating extended image coordinates $\mathbf{x} = (x, y, 1)$ to extended object coordinates $\mathbf{X} = (X, Y, Z, 1)$ via the equation

$$\lambda \mathbf{x} = P\mathbf{X} \qquad (10.12)$$

is called a **camera matrix**, and Eq. (10.12) is called the **camera equation**. Observe that the focal point is given as the right nullspace to the camera matrix, since $P\mathbf{C} = 0$, where $\mathbf{C}$ denotes homogeneous coordinates for the focal point, C.

10.3.3 The Intrinsic Parameters

In a refined camera model, the matrix K in Eq. (10.10) is replaced by

$$K = \begin{bmatrix} \gamma f & sf & x_0 \\ 0 & f & y_0 \\ 0 & 0 & 1 \end{bmatrix} \;, \qquad (10.13)$$

where the parameters have the following interpretations:
- f : **focal length** – also called camera constant;
- γ : **aspect ratio** – modelling nonquadratic light-sensitive elements;
- s : **skew** – modelling nonrectangular light-sensitive elements;
- (x_0, y_0) : **principal point** – orthogonal projection of the focal point onto the image plane; see Fig. 10.5.

These parameters are called the **intrinsic parameters**, since they model intrinsic properties of the camera. For most cameras $s \approx 0$ and $\gamma \approx 1$, and the principal point is located close to the centre of the image. A camera is said to be **calibrated** if K is known. Otherwise, it is said to be **uncalibrated**.

10.3.4 The Extrinsic Parameters

It is often advantageous to be able to express object coordinates in a different coordinate system than the camera coordinate system. This is especially the case when the relation between these coordinate systems is not known. For this purpose it is necessary to model the relation between two different coordinate systems in 3D. The natural way to do this is to model the relation as a

Euclidean transformation. Denote the camera coordinate system with e_c and points expressed in this coordinate system with index c, e.g., (X_c, Y_c, Z_c), and similarly denote the object coordinate system with e_o and points expressed in this coordinate system with index o. A Euclidean transformation from the object coordinate system to the camera coordinate system can be written in homogeneous coordinates as

$$\begin{bmatrix} X_c \\ Y_c \\ Z_c \\ 1 \end{bmatrix} = \begin{bmatrix} R & 0 \\ 0 & 1 \end{bmatrix} \begin{bmatrix} I & -t \\ 0 & 1 \end{bmatrix} \begin{bmatrix} X_o \\ Y_o \\ Z_o \\ 1 \end{bmatrix} \implies \mathbf{X}_c = \begin{bmatrix} R & -Rt \\ 0 & 1 \end{bmatrix} \mathbf{X}_o , \quad (10.14)$$

where R denotes an orthogonal matrix and t a vector, encoding the rotation and translation in the rigid transformation. These parameters, R and t, are called the **extrinsic parameters**. Observe that the focal point $(0,0,0)$ in the c-system corresponds to the point t in the o-system. Inserting Eq. (10.14) into Eq. (10.11), and taking into account that $\mathbf{X}$ in Eq. (10.11) is the same as $\mathbf{X}_c$ in Eq. (10.14), gives

$$\lambda \mathbf{x} = KR[\,I\,|\,-t\,]\mathbf{X}_o = P\mathbf{X}_o , \quad (10.15)$$

with $P = KR[\,I\,|\,-t\,]$. Usually, it is assumed that object coordinates are expressed in the object coordinate system and the index o in $\mathbf{X}_o$ is omitted. Observe that the focal point, $C_f = t = (t_x, t_y, t_z)$, is given from the right nullspace to P according to

$$P \begin{bmatrix} t_x \\ t_y \\ t_z \\ 1 \end{bmatrix} = KR[\,I\,|\,-t\,] \begin{bmatrix} t_x \\ t_y \\ t_z \\ 1 \end{bmatrix} = \mathbf{0} .$$

Given a camera, described by the camera matrix P, this camera could also be described by the camera matrix μP, $0 \neq \mu \in \mathbb{R}$, since these give the same image point for each object point. This means that the camera matrix is only defined up to an unknown scale factor. Moreover, the camera matrix P can be regarded as a projective transformation from $\mathbb{P}^3$ to $\mathbb{P}^2$; cf. Eq. (10.11) and Eq. (10.4).

Observe also that replacing t by μt and (X, Y, Z) with $(\mu X, \mu Y, \mu Z)$, $0 \neq \mu \in \mathbb{R}$, gives the same image coordinates since

$$KR[\,I\,|\,-\mu t\,] \begin{bmatrix} \mu X \\ \mu Y \\ \mu Z \\ 1 \end{bmatrix} = \mu KR[\,I\,|\,-t\,] \begin{bmatrix} X \\ Y \\ Z \\ 1 \end{bmatrix} .$$

This ambiguity is referred to as the **scale ambiguity**.

It is now possible to calculate the number of parameters in the camera matrix, P:

- K: 5 parameters (f, γ, s, x_0, y_0)
- R: 3 parameters
- t: 3 parameters

Summing up gives a total of 11 parameters, which is the same as in a general 3×4-matrix defined up to scale. This means that for an uncalibrated camera, the factorization $P = KR[\,I\,|\,-t\,]$ has no meaning and instead P can be considered as a general 3×4-matrix.

Given a calibrated camera with camera matrix $P = KR[\,I\,|\,-t\,]$ and the corresponding camera equation

$$\lambda \mathbf{x} = KR[\,I\,|\,-t\,]\mathbf{X}\ ,$$

it is often advantageous to make a change of coordinates from $\mathbf{x}$ to $\hat{\mathbf{x}}$ in the image according to $\mathbf{x} = K\hat{\mathbf{x}}$, which gives

$$\lambda K\hat{\mathbf{x}} = KR[\,I\,|\,-t\,]\mathbf{X} \quad \Rightarrow \quad \lambda \hat{\mathbf{x}} = R[\,I\,|\,-t\,]\mathbf{X} = \hat{P}\mathbf{X}\ .$$

Now the camera matrix becomes $\hat{P} = R[\,I\,|\,-t]$. A camera represented with a camera matrix of this form is called a **normalized camera**.

10.3.5 Properties of the Pinhole Camera

This section will be concluded with some properties of the pinhole camera.

1. The set of 3D-points that projects to an image point, $\mathbf{x}$, is given by

$$\mathbf{X} = \mathbf{C} + \mu P^+ \mathbf{x}, \quad 0 \neq \mu \in \mathbb{R}\ ,$$

   where $\mathbf{C}$ denotes the focal point in homogeneous coordinates and P^+ denotes the pseudoinverse of P. This set of points is called the **optical ray** corresponding to $\mathbf{x}$.

2. The set of 3D-points that projects to a line, $\mathbf{l}$, is the points lying on the plane $\Pi = P^T \mathbf{l}$. (Proof: It is obvious that the set of points lie on the plane defined by the focal point and the line $\mathbf{l}$. A point $\mathbf{x}$ on $\mathbf{l}$ fulfils $\mathbf{x}^T \mathbf{l} = 0$ and a point $\mathbf{X}$ on the plane Π fulfils $\Pi^T \mathbf{X} = 0$. Finally, $\mathbf{x} \sim P\mathbf{X}$ implies $(P\mathbf{X})^T \mathbf{l} = \mathbf{X}^T P^T \mathbf{l} = 0$ and identification with $\Pi^T \mathbf{X} = 0$ gives $\Pi = P^T \mathbf{l}$.)

3. The projection of a quadric, $\mathbf{X}^T C \mathbf{X} = 0$ (dually $\Pi^T C' \Pi = 0$, $C' = C^{-1}$), is an image conic, $\mathbf{x}^T c \mathbf{x} = 0$ (dually $\mathbf{l}^T c' \mathbf{l} = 0$, $c' = c^{-1}$), with $c' = PC'P^T$. (Proof: Use the previous property.)

4. The image of the absolute conic is given by the conic $\mathbf{x}^T \omega \mathbf{x} = 0$ (dually $\mathbf{1}^T \omega' \mathbf{l} = 0$), where $\omega' = KK^T$. (Proof: The result follows from the previous property and

$$\omega' \sim P\Omega' P^T \sim KR\begin{bmatrix} I & -t \end{bmatrix} \begin{bmatrix} I & 0 \\ 0 & 0 \end{bmatrix} \begin{bmatrix} I \\ -t^T \end{bmatrix} R^T K^T = KRR^T K^T = KK^T\ .)$$

10.3.6 Specializations of the Perspective Camera

There exist several important special cases of the previously derived perspective camera. The most common are the orthographic, the scaled orthographic and the affine camera.

The Orthographic Camera

Consider a parallel projection along the optical axis (considered as the z-axis). The image coordinates can in this case be calculated simply as

$$\begin{bmatrix} x \\ y \end{bmatrix} = \begin{bmatrix} X \\ Y \end{bmatrix} \implies \begin{bmatrix} x \\ y \\ 1 \end{bmatrix} = \begin{bmatrix} 1 & 0 & 0 & 0 \\ 0 & 1 & 0 & 0 \\ 0 & 0 & 0 & 1 \end{bmatrix} \begin{bmatrix} X \\ Y \\ Z \\ 1 \end{bmatrix}. \tag{10.16}$$

A Euclidean change of coordinate system for the object gives

$$\begin{bmatrix} x \\ y \\ 1 \end{bmatrix} = \begin{bmatrix} 1 & 0 & 0 & 0 \\ 0 & 1 & 0 & 0 \\ 0 & 0 & 0 & 1 \end{bmatrix} \begin{bmatrix} R & t \\ 0 & 1 \end{bmatrix} \begin{bmatrix} X_o \\ Y_o \\ Z_o \\ 1 \end{bmatrix} = \begin{bmatrix} \mathbf{r}_1 & t_1 \\ \mathbf{r}_2 & t_2 \\ 0 & 1 \end{bmatrix} \begin{bmatrix} X_o \\ Y_o \\ Z_o \\ 1 \end{bmatrix}, \tag{10.17}$$

where $\mathbf{r}_1$ and $\mathbf{r}_2$ denote the first two rows in the orthogonal matrix R and $t = (t_1, t_2, t_3)$. A camera matrix as in Eq. (10.17) is called an **orthographic camera**.

The Scaled Orthographic Camera

Consider the camera model, where points first are projected using a parallel projection onto the plane $\Pi_o : Z = Z_0$ followed by a perspective projection onto the image plane. This camera model can obtained from Eq. (10.8) with Z replaced by Z_0, i.e.,

$$\begin{cases} \dfrac{x}{f} = \dfrac{X}{Z_0} \\ \dfrac{y}{f} = \dfrac{Y}{Z_0} \end{cases} \implies \begin{bmatrix} x \\ y \end{bmatrix} = \frac{f}{Z_0} \begin{bmatrix} X \\ Y \end{bmatrix}. \tag{10.18}$$

After a Euclidean change of coordinate system for the object the camera matrix becomes

$$P = \begin{bmatrix} \lambda \mathbf{r}_1 & \lambda t_1 \\ \lambda \mathbf{r}_2 & \lambda t_2 \\ 0 & 1 \end{bmatrix} \begin{bmatrix} X_o \\ Y_o \\ Z_o \\ 1 \end{bmatrix} = \begin{bmatrix} \lambda \mathbf{r}_1 & \hat{t}_1 \\ \lambda \mathbf{r}_2 & \hat{t}_2 \\ 0 & 1 \end{bmatrix} \begin{bmatrix} X_o \\ Y_o \\ Z_o \\ 1 \end{bmatrix}, \tag{10.19}$$

where $\lambda = f/Z_0$, $\mathbf{r}_1$ and $\mathbf{r}_2$ denote the first two rows of the orthogonal matrix R and $\hat{t}_i = \lambda t_i$. A camera matrix as in Eq. (10.19) is called a **scaled orthographic camera**.

The scaled orthographic camera can be seen as an approximation of the calibrated pinhole camera in the case where the distance from the camera to the object is much larger than the relative distance for the individual points in the object. This can be seen as follows: Consider a normalized perspective camera with the focal point at the origin and the optical axis coinciding with the z-axis:
$$x = \frac{X}{Z}, \qquad y = \frac{Y}{Z}.$$
Assume further that all object points are located close to the plane $Z = Z_0$ and write the object points as $(X, Y, Z_0 + \Delta Z)$, giving
$$x = \frac{X}{Z_0 + \Delta Z} = \frac{X}{Z_0} \frac{1}{1 + \Delta Z/Z_0}$$
$$= \frac{X}{Z_0}\left(1 - \frac{\Delta Z}{Z_0} + \cdots\right) \approx \frac{X}{Z_0},$$
using the Taylor series expansion of the quotient, which is a good approximation if $\Delta Z/Z_0$ is small.

The Affine Camera

Start with the scaled orthographic camera and introduce intrinsic parameters similar to the perspective camera case
$$P = \begin{bmatrix} \gamma & s \\ 0 & 1 \end{bmatrix} \begin{bmatrix} \lambda \mathbf{r}^1 & \hat{t}_1 \\ \lambda \mathbf{r}^2 & \hat{t}_2 \\ 0 & 1 \end{bmatrix} = \begin{bmatrix} \mathbf{a}_1 & b_1 \\ \mathbf{a}_2 & b_2 \\ \mathbf{0} & 1 \end{bmatrix}, \qquad (10.20)$$
where γ denotes the aspect ratio, s the skew and $\mathbf{a}_1$ and $\mathbf{a}_2$ are arbitrary vectors. A camera matrix as in Eq. (10.20) is called an **affine camera**.

In the same way as for the scaled orthographic it can be argued that the affine camera is an approximation of the uncalibrated perspective camera, valid when the distance to the object is much larger than the relative depths between the object points.

Finally, observe that a camera matrix of the form
$$P = \begin{bmatrix} \mathbf{a}_1 & b_1 \\ \mathbf{a}_2 & b_2 \\ \mathbf{0} & 1 \end{bmatrix}$$
with $\mathbf{a}_1$ and $\mathbf{a}_2$ arbitrary vectors is an affine camera, with $|\mathbf{a}_1| = |\mathbf{a}_2|$ and $\mathbf{a}_1 \cdot \mathbf{a}_2 = 0$ is a scaled orthographic camera, and with $|\mathbf{a}_1| = |\mathbf{a}_2| = 1$ and $\mathbf{a}_1 \cdot \mathbf{a}_2 = 0$ is an orthographic camera.

10.3.7 Camera Calibration

Usually, the intrinsic parameters of the camera are not provided by the manufacturer, and for a zooming camera these will change during zooming. However, there is a very simple way to estimate the intrinsic parameters and calibrate the camera. Prepare an object with at least six easily identifiable points, e.g., the corners of a box, measure the coordinates of these points in some arbitrary coordinate system and take an image of the object. After identifying the points in the image, the image consists of six points from a known object. Write down the camera equation for these points

$$\lambda_j \mathbf{x}_j = P \mathbf{X}_j, \quad j = 1, \ldots 6 , \qquad (10.21)$$

where λ_j and P are unknown and $\mathbf{x}_j$ and $\mathbf{X}_j$ are known.

Observe that the constraints in Eq. (10.21) are linear in the unknown parameters. Every point gives 3 equations, implying in total 18 constraints and there are 12 parameters in P and 6 different depths, λ_j, implying 18 unknown parameters. Write Eq. (10.21) for these six points as a linear system of equations

$$M\mathbf{u} = \mathbf{0} , \qquad (10.22)$$

where M is a 12×12-matrix built up from image coordinates and $\mathbf{u}$ contains the unknown intrinsic parameters. In the case of noisy data Eq. (10.22) may be solved in the least squares sense using the so-called singular value decomposition (SVD). This method is called **DLT** (Direct Linear Transformation).

Now, it remains to calculate the intrinsic parameters from P obtained from the DLT algorithm. Let Q denote the first 3×3-block in P, i.e., $Q = KR$, and observe that

$$QQ^T = KRR^T K^T = KK^T , \qquad (10.23)$$

from which K can be solved by Cholesky factorization of $(QQ^T)^{-1}$.

For numerical reasons it is often advantageous to rescale the coordinates by making suitable changes of coordinates in the images as well as in the object.

This linear method works well in situations with a small noise level in the images. However for noisy date it doesn't take the noise into account in an optimal way. An optimal estimate of the camera parameters can be obtained by solving the optimization problem

$$\min_P \sum_{j=1}^n |x_j(P) - x_j^m|^2 + |y_j(P) - y_j^m|^2 ,$$

where (x_i^m, y_i^m) denotes measured image coordinates and $(x_i(P), y_i(P))$ image coordinates calculated from P. This optimization problem can be solved by using the Gauss-Newton method for nonlinear least squares problems, and the result of the DLT can be used as a starting point.

Recently, several more practical methods for camera calibration have been proposed. For instance, it is possible to use several images of a planar calibration grid, by estimating the homographies between the calibration plane and the images; see [28].

10.4 Multiple-View Geometry

Multiple-view geometry is the subject where relations between coordinates of feature points in different views are studied. It is an important tool for understanding the image formation process for several cameras and for designing reconstruction algorithms. For a more detailed treatment see [12] or [8], and for a different approach see [10].

10.4.1 The Structure and Motion Problem

The investigation of multiple-view geometry is motivated by the structure and motion problem:

Problem 1 (structure and motion). Given a sequence of images with corresponding feature points $\mathbf{x}_{ij}$, taken by a perspective camera, i.e.,

$$\lambda_{ij}\mathbf{x}_{ij} = P_i\mathbf{X}_j, \quad i=1,\ldots,m, \quad j=1,\ldots,n ,$$

determine the camera matrices, P_i, i.e., the **motion**, and the 3D-points, $\mathbf{X}_j$, i.e., the **structure**, under different assumptions on the intrinsic and/or extrinsic parameters. This is called the **structure and motion problem**.

It turns out that there is a fundamental limitation on the solutions to the structure and motion problem, when the intrinsic parameters are unknown and possibly varying, a so-called **uncalibrated image sequence**. Assume that $\mathbf{X}_j$ is a reconstruction of n points in m images, with camera matrices P_i according to
$$\mathbf{x}_{ij} \sim P_i\mathbf{X}_j, \; i=1,\ldots,m, \; j=1,\ldots,n .$$
Then $H\mathbf{X}_j$ is also a reconstruction, with camera matrices $P_i H^{-1}$, for every nonsingular 4×4-matrix H, since

$$\mathbf{x}_{ij} \sim P_i\mathbf{X}_j \sim P_i H^{-1} H \mathbf{X}_j \sim (P_i H^{-1})(H\mathbf{X}_j) .$$

The transformation
$$\mathbf{X} \mapsto H\mathbf{X}$$
corresponds to all projective transformations of the object. Thus, given an uncalibrated image sequence with corresponding points, it is only possible to reconstruct the object up to an unknown projective transformation.

10.4.2 The Two-View Case

The investigation of multiple-view geometry will start with a detail treatment of the two-view case.

The Epipoles

Consider two images of the same point $\mathbf{X}$ as in Fig. 10.6. The **epipole**, $\mathbf{e}_{i,j}$,

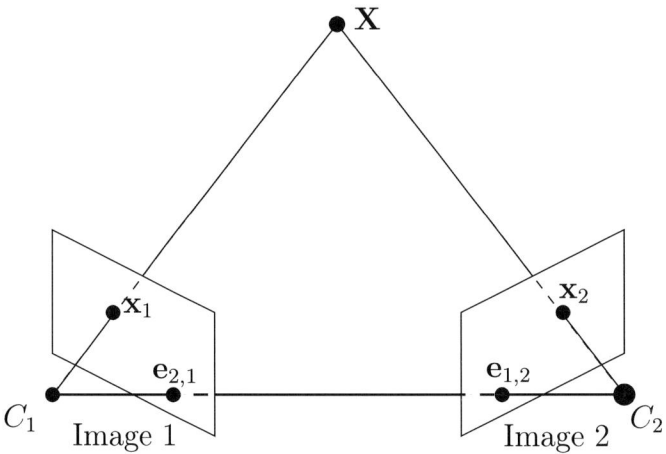

Fig. 10.6. Two images of the same point and the epipoles

is defined as the projection of the focal point of camera i in image j. Let

$$P_1 = [\, A_1 \mid b_1 \,] \quad \text{and} \quad P_2 = [\, A_2 \mid b_2 \,] \ .$$

The focal point of camera 1, C_1, is given by

$$P_1 \begin{bmatrix} C_1 \\ 1 \end{bmatrix} = [\, A_1 \mid b_1 \,] \begin{bmatrix} C_1 \\ 1 \end{bmatrix} = A_1 C_1 + b_1 = 0 \ ,$$

i.e., $C_1 = -A_1^{-1} b_1$ and then the epipole is obtained from

$$\mathbf{e}_{1,2} = P_2 \begin{bmatrix} C_1 \\ 1 \end{bmatrix} = [\, A_2 \mid b_2 \,] \begin{bmatrix} C_1 \\ 1 \end{bmatrix} = A_2 C_1 + b_2 = -A_2 A_1^{-1} b_1 + b_2 \ .$$

It is convenient to use the notation $A_{12} = A_2 A_1^{-1}$. Assume that the two camera matrices, representing the two-view geometry, have been calculated:

$$P_1 = [\, A_1 \mid b_1 \,] \quad \text{and} \quad P_2 = [\, A_2 \mid b_2 \,] \ .$$

According to the projective ambiguity these camera matrices can be multiplied with
$$H = \begin{bmatrix} A_1^{-1} & -A_1^{-1}b_1 \\ 0 & 1 \end{bmatrix}$$
from the right, obtaining
$$\bar{P}_1 = P_1 H = [\, I \mid 0 \,], \quad \bar{P}_2 = P_2 H = [\, A_2 A_1^{-1} \mid b_2 - A_2 A_1^{-1} b_1 \,] \;.$$

Thus, it may always be assumed that the first camera matrix is $[\, I \mid 0 \,]$. Observe that $\bar{P}_2 = [\, A_{12} \mid \mathbf{e} \,]$, where $\mathbf{e}$ denotes the epipole in the second image. Observe also that it is possible to multiply again with
$$\bar{H} = \begin{bmatrix} I & 0 \\ \mathbf{v}^T & 1 \end{bmatrix}$$
without changing $\bar{P}_1$, but
$$\bar{P}_2 \bar{H} = [\, A_{12} + \mathbf{e}\mathbf{v}^T \mid \mathbf{e} \,] \;,$$
i.e., the last column of the second camera matrix still represents the epipole. A pair of camera matrices is said to be in **canonical form** if $P_1 = [\, I \mid 0 \,]$ and $P_2 = [\, A_{12} + \mathbf{e}\mathbf{v}^T \mid \mathbf{e} \,]$, where $\mathbf{v}$ denote a three-parameter ambiguity.

The Fundamental Matrix

The fundamental matrix was originally discovered in the calibrated case in [15] and in the uncalibrated case in [4]. Consider a fixed point, $\mathbf{X}$, in 2 views:
$$\lambda_1 \mathbf{x}_1 = P_1 \mathbf{X} = [\, A_1 \mid b_1 \,]\mathbf{X}, \quad \lambda_2 \mathbf{x}_2 = P_2 \mathbf{X} = [\, A_2 \mid b_2 \,]\mathbf{X} \;.$$
Use the first camera equation to solve for X, Y, Z
$$\lambda_1 \mathbf{x}_1 = P_1 \mathbf{X} = [\, A_1 \mid b_1 \,]\mathbf{X} = A_1 \begin{bmatrix} X \\ Y \\ Z \end{bmatrix} + b_1 \quad \Rightarrow \quad \begin{bmatrix} X \\ Y \\ Z \end{bmatrix} = A_1^{-1}(\lambda_1 \mathbf{x}_1 - b_1)$$
and insert into the second one
$$\lambda_2 \mathbf{x}_2 = A_2 A_1^{-1}(\lambda_1 \mathbf{x}_1 - b_1) + b_2 = \lambda_1 A_{12} \mathbf{x}_1 + (-A_{12} b_1 - b_2) \;,$$
i.e., $\mathbf{x}_2$, $A_{12}\mathbf{x}_1$ and $\mathbf{t} = -A_{12}b_1 + b_2 = \mathbf{e}_{1,2}$ are linearly dependant. Observe that $\mathbf{t} = \mathbf{e}_{1,2}$, i.e., the epipole in the second image. This condition can be written as $\mathbf{x}_1^T A_{12}^T T_\mathbf{e} \mathbf{x}_2 = \mathbf{x}_1^T F \mathbf{x}_2 = 0$, with $F = A_{12}^T T_\mathbf{e}$, where $T_\mathbf{x}$ denotes the skew-symmetric matrix corresponding to the vector $\mathbf{x}$, i.e., $T_\mathbf{x}(y) = x \times y$. The bilinear constraint
$$\mathbf{x}_1^T F \mathbf{x}_2 = 0$$
is called the **epipolar constraint** and

$$F = A_{12}^T T_\mathbf{e}$$

is called the **fundamental matrix**.

Observe that the epipole in the second image is obtained as the right nullspace to the fundamental matrix and the epipole in the left image is obtained as the left nullspace to the fundamental matrix and that the fundamental matrix is singular, i.e., $\det F = 0$.

Given a point, $\mathbf{x}_1$, in the first image, the coordinates of the corresponding point in the second image fulfil

$$0 = \mathbf{x}_1^T F \mathbf{x}_2 = (\mathbf{x}_1^T F)\mathbf{x}_2 = \mathbf{l}(\mathbf{x}_1)^T \mathbf{x}_2 = 0 \ ,$$

where $\mathbf{l}(\mathbf{x}_1)$ denotes the line represented by $\mathbf{x}_1^T F$. The line $\mathbf{l} = F^T \mathbf{x}_1$ is called the **epipolar line** corresponding to $\mathbf{x}_1$.

The geometrical interpretation of the epipolar line is the following geometric construction. The points $\mathbf{x}_1$, C_1 and C_2 define a plane, Π, intersecting the second image plane in the line $\mathbf{l}$, containing the corresponding point.

From the previous considerations consider the following pair

$$F = A_{12}^T T_\mathbf{e} \quad \Leftrightarrow \quad P_1 = [\, I \,|\, 0 \,], \quad P_2 = [\, A_{12} \,|\, \mathbf{e} \,] \ . \qquad (10.24)$$

Observe that

$$F = A_{12}^T T_\mathbf{e} = (A_{12} + \mathbf{e}\mathbf{v}^T)^T T_\mathbf{e}$$

for every vector $\mathbf{v}$, since

$$(A_{12} + \mathbf{e}\mathbf{v})^T T_\mathbf{e}(\mathbf{x}) = A_{12}^T(\mathbf{e}\times\mathbf{x}) + \mathbf{v}\mathbf{e}^T(\mathbf{e}\times\mathbf{x}) = A_{12}^T T_\mathbf{e}\mathbf{x} \ ,$$

since $\mathbf{e}^T(\mathbf{e}\times\mathbf{x}) = \mathbf{e}\cdot(\mathbf{e}\times\mathbf{x}) = 0$. This ambiguity corresponds to the transformation

$$\bar{H}\bar{P}_2 = [\, A_{12} + \mathbf{e}\mathbf{v}^T \,|\, \mathbf{e} \,] \ .$$

Thus there are three free parameters in the choice of the second camera matrix when the first is fixed to $P_1 = [\, I \,|\, 0 \,]$.

The Infinity Homography

Consider a plane in the three-dimensional object space, Π, defined by a vector $\mathbf{V}$: $\mathbf{V}^T \mathbf{X} = 0$ and the following construction. Given a point in the first image, construct the intersection with the optical ray corresponding to this point and the plane Π and project to the second image. This procedure gives a homography between points in the first and second image, which depends on the chosen plane Π.

Assume that

$$P_1 = [\, I \,|\, 0 \,], \quad P_2 = [\, A_{12} \,|\, \mathbf{e} \,] \ .$$

Write $\mathbf{V} = [\, v_1 \, v_2 \, v_3 \, 1 \,]^T = [\, \mathbf{v} \, 1 \,]^T$ (assuming $v_4 \neq 0$, i.e., the plane is not incident with the origin, i.e., the focal point of the first camera) and $\mathbf{X} = [\, X \, Y \, Z \, W \,]^T = [\, \mathbf{w} \, W \,]^T$, which gives

$$\mathbf{V}^T\mathbf{X} = \mathbf{v}^T\mathbf{w} + W , \qquad (10.25)$$

which implies that $\mathbf{v}^T\mathbf{w} = -W$ for points in the plane Π. The first camera equation gives

$$\mathbf{x}_1 \sim [\,I \mid 0\,]\mathbf{X} = \mathbf{w} ,$$

and using Eq. (10.25) gives $\mathbf{v}^T\mathbf{x}_1 = -W$. Finally, the second camera matrix gives

$$\mathbf{x}_2 \sim [\,A_{12} \mid \mathbf{e}\,] \begin{bmatrix} \mathbf{x}_1 \\ -\mathbf{v}^T\mathbf{x}_1 \end{bmatrix} = A_{12}\mathbf{x}_1 - \mathbf{e}\mathbf{v}^T\mathbf{x}_1 = (A_{12} - \mathbf{e}\mathbf{v}^T)\mathbf{x}_1 ,$$

which shows that the homography corresponding to the plane $\Pi : \mathbf{V}^T\mathbf{X} = 0$ is given by the matrix

$$H_\Pi = A_{12} - \mathbf{e}\mathbf{v}^T ,$$

where $\mathbf{e}$ denotes the epipole and $\mathbf{V} = [\,\mathbf{v}\, 1\,]^T$.

Observe that when $\mathbf{V} = (0,0,0,1)$, i.e., $\mathbf{v} = (0,0,0)$ the plane Π is the plane at infinity. The homography

$$H_\infty = H_{\Pi_\infty} = A_{12}$$

is called the **homography corresponding to the plane at infinity** or the **infinity homography**.

Note that the epipolar line through the point $\mathbf{x}_2$ in the second image can be written as $\mathbf{x}_2 \times \mathbf{e}$, implying

$$(\mathbf{x}_2 \times \mathbf{e})^T H\mathbf{x}_1 = \mathbf{x}_1^T H^T T_\mathbf{e} \mathbf{x}_2 = 0 ,$$

i.e., the epipolar constraint, implying

$$F = H^T T_e .$$

This shows that there is a one-to-one correspondence between planes in 3D, homographies between two views and factorization of the fundamental matrix as $F = H^T T_e$.

Finally, note that the matrix $H_\Pi^T T_e H_\Pi = F H_\Pi$ is skew-symmetric (since $\mathbf{x}^T H_\Pi^T T_e H_\Pi \mathbf{x} = \mathbf{y}^T T_e \mathbf{y} = \mathbf{y} \cdot (\mathbf{e} \times \mathbf{y}) = 0$ for all x, $\mathbf{y} = H_\Pi \mathbf{x}$), implying that

$$FH_\Pi + H_\Pi^T F^T = 0 . \qquad (10.26)$$

10.4.3 Multiple View Constraints and Tensors

This section deals with the general case of multiple-view geometry, i.e., any number of cameras. A short introduction to tensor calculus is also included.

Tensor Calculus

Tensor calculus is a natural tool to use when the objects under study are expressed in a specific coordinate system but have physical properties that are independent of the chosen coordinate system. Another advantage is that it gives a simple and compact notation and the rules of tensor algebra make it easy to remember even quite complex formulas. For a more detailed treatment see [22] and for an engineering approach see [16].

An **affine tensor** is an object in a linear space, $\mathcal{V}$, that consists of a collection of numbers that are related to a specific choice of coordinate system in $\mathcal{V}$, indexed by one or several indices;

$$A^{i_1,i_2,\cdots,i_n}_{j_1,j_2,\cdots,j_m} \ .$$

Furthermore, this collection of numbers transforms in a predefined way when a change of coordinate system in $\mathcal{V}$ is made. The number of indices $(n+m)$ is called the **degree** of the tensor. The indices may take any value from 1 to the dimension of $\mathcal{V}$. The upper indices are called **contravariant** indices and the lower indices are called **covariant** indices.

There are some simple conventions that have to be remembered:

- *The index rule:* When an index appears in a formula, the formula is valid for every value of the index, e.g., $a_i = 0 \Rightarrow a_1 = 0, a_2 = 0, \ldots$.
- *The summation convention:* When an index appears twice in a formula, it is implicitly assumed that a summation takes place over that index, e.g., $a_i b^i = \sum_{i=1,\dim \mathcal{V}} a_i b^i$.
- *The compatibility rule:* A repeated index must appear once as a subindex and once as a superindex.
- *The maximum rule:* An index can not be used more than twice in a term.

When the coordinate system in $\mathcal{V}$ is changed from $\mathbf{e}$ to $\hat{\mathbf{e}}$ and the points with coordinates $\mathbf{x}$ are changed to $\hat{\mathbf{x}}$, according to

$$\hat{\mathbf{e}}_j = S^i_j \mathbf{e}_i \quad \Leftrightarrow \quad \mathbf{x}_i = S^i_j \hat{\mathbf{x}}_j \ ,$$

then the affine tensor components change according to

$$\hat{\mathbf{u}}_k = S^j_k \mathbf{u}_j \quad \text{and} \quad \mathbf{v}^j = S^j_k \hat{\mathbf{v}}^k \ ,$$

for lower and upper indices, respectively.

From this definition the terminology for indices can be motivated, since the covariant indices covary with the basis vectors and the contravariant indices contravary with the basis vectors. It turns out that a vector (e.g., the coordinates of a point) is a contravariant tensor of degree one and that a one-form (e.g., the coordinate of a vector defining a line in $\mathbb{R}^2$ or a hyperplane in $\mathbb{R}^n$) is a covariant tensor of degree one.

Finally, the second-order tensor

$$\delta_{ij} = \begin{cases} 1, & i = j; \\ 0, & i \neq j \end{cases}$$

is called the **Kronecker delta**. When $\dim \mathcal{V} = 3$, the third-order tensor

$$\epsilon_{ijk} = \begin{cases} 1, & (i,j,k) \text{ an even permutation}; \\ -1, & (i,j,k) \text{ an odd permutation}; \\ 0, & (i,j,k) \text{ has a repeated value} \end{cases}$$

is called the **Levi-Cevita epsilon**.

Matrix Formulation of Multiple-View Constraints

Consider one object point, $\mathbf{X}$, and its m images, $\mathbf{x}_i$, according to the camera equations $\lambda_i \mathbf{x}_i = P_i \mathbf{X}$, $i = 1 \ldots m$. These equations can be written as

$$\underbrace{\begin{bmatrix} P_1 & \mathbf{x}_1 & 0 & 0 & \cdots & 0 \\ P_2 & 0 & \mathbf{x}_2 & 0 & \cdots & 0 \\ P_3 & 0 & 0 & \mathbf{x}_3 & \cdots & 0 \\ \vdots & \vdots & \vdots & \vdots & \ddots & \vdots \\ P_m & 0 & 0 & 0 & \cdots & \mathbf{x}_m \end{bmatrix}}_{M} \begin{bmatrix} \mathbf{X} \\ -\lambda_1 \\ -\lambda_2 \\ -\lambda_3 \\ \vdots \\ -\lambda_m \end{bmatrix} = \begin{bmatrix} 0 \\ 0 \\ 0 \\ \vdots \\ 0 \end{bmatrix} . \quad (10.27)$$

From Eq. (10.27) it follows that the matrix, M, is rank deficient, i.e.,

$$\operatorname{rank} M < m + 4 \; ,$$

which is referred to as the **rank condition**. The rank condition implies that all $(m+4) \times (m+4)$ minors of M are equal to 0. These can be written using Laplace expansions as sums of products of determinants of *four rows* taken from the first four columns of M and of image coordinates. There are 3 different categories of such minors depending on the number of rows taken from each image, since one row has to be taken from each image and then the remaining 4 rows can be distributed freely. The three different types are:

1. Take the 2 remaining rows from one camera matrix and the 2 remaining rows from another camera matrix, giving 2-view constraints.
2. Take the 2 remaining rows from one camera matrix, 1 row from another and 1 row from a third camera matrix, giving 3-view constraints.
3. Take 1 row from each of four different camera matrices, giving 4-view constraints.

Observe that the minors of M can be factorized as products of the 2-, 3- or 4-view constraints and image coordinates in the other images. It is convenient to use (x^1, x^2, x^3) instead of (x, y, z) for homogeneous image coordinates and also to denote row number i of a camera matrix P by P^i.

The Monofocal Tensor

Before proceeding to the multiple-view tensors observe that the epipole in image 2 from camera 1, $\mathbf{e} = (e^1, e^2, e^3)$ in homogeneous coordinates, can be written as

$$e^j = \det \begin{bmatrix} P_1^1 \\ P_1^2 \\ P_1^3 \\ P_2^j \end{bmatrix} . \tag{10.28}$$

The numbers e^j constitute a first-order contravariant tensor, called the **monofocal tensor**, where the transformations of the tensor components are related to projective transformations of the image coordinates.

The Bifocal Tensor

Consider minors obtained by taking 3 rows from one image, and 3 rows from another image:

$$\det \begin{bmatrix} P_1 & \mathbf{x}_1 & 0 \\ P_2 & 0 & \mathbf{x}_2 \end{bmatrix} = \det \begin{bmatrix} P_1^1 & x_1^1 & 0 \\ P_1^2 & x_1^2 & 0 \\ P_1^3 & x_1^3 & 0 \\ P_2^1 & 0 & x_2^1 \\ P_2^2 & 0 & x_2^2 \\ P_2^3 & 0 & x_2^3 \end{bmatrix} = 0 ,$$

which gives a **bilinear constraint**, called the **bifocal constraint**:

$$\sum_{i,j=1}^{3} F_{ij} \mathbf{x}_1^i \mathbf{x}_2^j = F_{ij} \mathbf{x}_1^i \mathbf{x}_2^j = 0 , \tag{10.29}$$

where

$$F_{ij} = \sum_{i',i'',j',j''=1}^{3} \epsilon_{ii'i''} \epsilon_{jj'j''} \det \begin{bmatrix} P_1^{i'} \\ P_1^{i''} \\ P_2^{j'} \\ P_2^{j''} \end{bmatrix} .$$

The numbers F_{ij} constitute a second-order covariant tensor, called the **bifocal tensor**, which is the same as the fundamental matrix encountered earlier. Here the transformations of the tensor components are related to projective transformations of the image coordinates.

Observe that the indices tell us which row to *exclude* from the corresponding camera matrix when forming the determinant. The geometric interpretation of the bifocal constraint is that corresponding view-lines in two images intersect in 3D; see Fig. 10.6.

The bifocal tensor can also be used to transfer a point to the corresponding epipolar line, according to $l_j^2 = F_{ij} x_1^i$. This transfer can be extended to a homography between epipolar lines in the first view and epipolar lines in the second view according to

$$l^1_i = F_{ij}\epsilon^{jj'}_{j''}l^2_{j'}\mathbf{e}^{j''} \;,$$

since $\epsilon^{jj'}_{j''}l^2_{j'}\mathbf{e}^{j''}$ gives the vector product between the epipole $\mathbf{e}$ and the line l^2, which gives a point on the epipolar line.

The Trifocal Tensor

The trifocal tensor was originally discovered in the calibrated case in [24] and in the uncalibrated case in [20]. Considering minors obtained by taking 3 rows from one image, 2 rows from another image, and 2 rows from a third image, e.g.,

$$\det \begin{bmatrix} P^1_1 & x^1_1 & 0 & 0 \\ P^2_1 & x^2_1 & 0 & 0 \\ P^3_1 & x^3_1 & 0 & 0 \\ P^1_2 & 0 & x^1_2 & 0 \\ P^2_2 & 0 & x^2_2 & 0 \\ P^1_3 & 0 & 0 & x^1_3 \\ P^3_3 & 0 & 0 & x^3_3 \end{bmatrix} = 0 \;,$$

gives a **trilinear constraint**, called the **trifocal constraint**:

$$\sum_{i,j,j',k,k'=1}^{3} T^{jk}_i \mathbf{x}^i_1 \epsilon_{jj'j''} \mathbf{x}^{j'}_2 \epsilon_{kk'k''} \mathbf{x}^{k'}_3 = 0 \;, \tag{10.30}$$

where

$$T^{jk}_i = \sum_{i',i''=1}^{3} \epsilon_{ii'i''} \det \begin{bmatrix} P^{i'}_1 \\ P^{i''}_1 \\ P^j_2 \\ P^k_3 \end{bmatrix} \;. \tag{10.31}$$

Note that there are in total 9 constraints indexed by j'' and k'' in Eq. (10.30). The numbers T^{jk}_i constitute a third-order mixed tensor, called the **trifocal tensor**, that is covariant in i and contravariant in j and k.

Again the lower index tells us which row to <u>exclude</u> from the first camera matrix and the upper indices tell us which rows to <u>include</u> from the second and third camera matrices, respectively, and these indices become covariant and contravariant, respectively. Observe that the orders of the images are important, since the first image is treated differently. If the images are permuted another set of coefficients is obtained. The geometric interpretation of the trifocal constraint is that the view-line in the first image and the planes corresponding to arbitrary lines coincident with the corresponding points in the second and third images (together with the focal points), respectively, intersect in 3D; see Fig. 10.7.

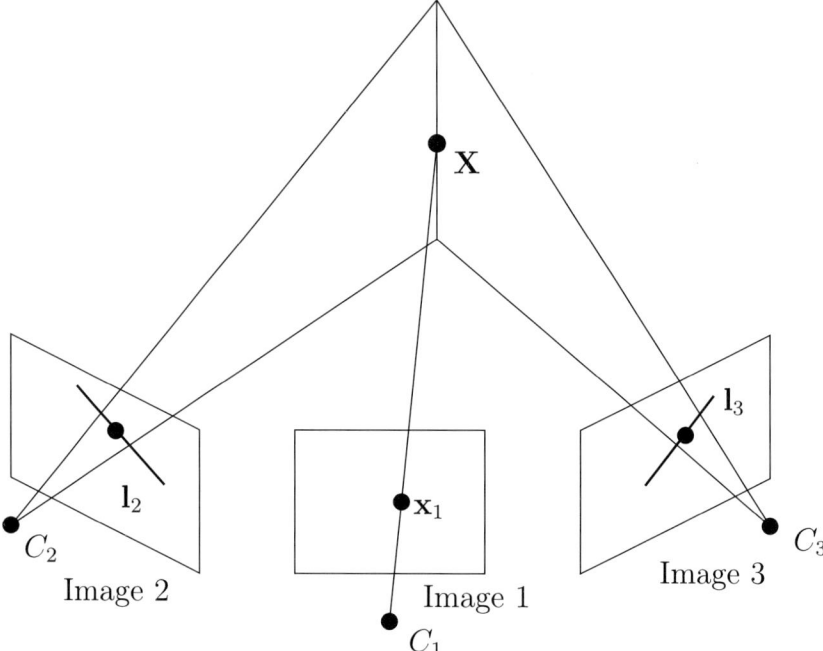

Fig. 10.7. Geometrical interpretation of the trifocal constraint.

It is easy to prove that given three corresponding lines, $\mathbf{l}^1$, $\mathbf{l}^2$ and $\mathbf{l}^3$ in three images, represented by the vectors $(\mathbf{l}_1^1, \mathbf{l}_2^1, \mathbf{l}_3^1)$, etc., then

$$\mathbf{l}_k^3 = T_k^{ij} \mathbf{l}_i^1 \mathbf{l}_j^2 \ . \tag{10.32}$$

From this result it is possible to transfer the images of a line seen in two images to a third image, a so-called **tensorial transfer**. The geometrical interpretation is that two corresponding lines define two planes in 3D (one plane from each line together with the corresponding focal point) that intersect in a line that can be projected onto the third image. There are also other transfer equations, such as

$$\mathbf{x}_2^j = T_i^{jk} \mathbf{x}_1^i \mathbf{l}_k^3 \quad \text{and} \quad \mathbf{x}_3^k = T_i^{jk} \mathbf{x}_2^j \mathbf{l}_k^3 \ ,$$

with obvious geometrical interpretations.

The Quadrifocal Tensor

The quadrifocal tensor was independently discovered in several papers, e.g., [27, 10]. Considering minors obtained by taking 2 rows from each one of 4 different images gives a *quadrilinear constraint,* called the **quadrifocal constraint**:

$$\sum_{i,i',j,j',k,k',l,l'=1}^{3} Q^{ijkl} \epsilon_{ii'i''} \mathbf{x}^1_{i'} \epsilon_{jj'j''} \mathbf{x}^2_{j'} \epsilon_{kk'k''} \mathbf{x}^3_{k'} \epsilon_{ll'l''} \mathbf{x}^4_{l'} = 0 \ , \qquad (10.33)$$

where

$$Q^{ijkl} = \det \begin{bmatrix} P^i_1 \\ P^j_2 \\ P^k_3 \\ P^l_4 \end{bmatrix} \ .$$

Note that there are in total 81 constraints indexed by i'', j'', k'' and l'' in Eq. (10.33). The numbers Q^{ijkl} constitute a fourth-order contravariant tensor, called the **quadrifocal tensor**.

Note that there are in total 81 constraints indexed by i'', j'', k'' and l''. Again, the upper indices tell us which rows to include from each camera matrix and they become contravariant indices. The geometric interpretation of the quadrifocal constraint is that the four planes corresponding to arbitrary lines coincident with the corresponding points in the images intersect in 3D.

Finally, observe that the multiview tensors have special properties for cameras other than the uncalibrated perspective one. However, it will lead to far too investigate all these cases.

10.5 Structure and Motion from Image Sequences

The structure and motion problem will now be studied in detail. Firstly, the problem when the structure is known will be solved, so-called resection, then when the motion is known, so-called intersection. Then a linear algorithm to solve for both structure and motion using the multifocal tensors will be presented, and finally a factorization algorithm will be presented. Again the reader is referred to [8] for a more detailed treatment.

10.5.1 Resection

Problem 2 (Resection). Assume that the structure is given, i.e., the object points, $\mathbf{X}_j$, $j = 1, \ldots, n$ are given in some coordinate system. Calculate the camera matrices P_i, $i = 1, \ldots, m$ from the images, i.e., from $\mathbf{x}_{i,j}$.

The most simple solution to this problem is the classical DLT algorithm based on the fact that the camera equations

$$\lambda_j \mathbf{x}_j = P \mathbf{X}_j, \quad j = 1, \ldots, n$$

are linear in the unknown parameters, λ_j and P.

10.5.2 Intersection

Problem 3 (Intersection). Assume that the motion is given, i.e., the camera matrices, P_i, $i = 1, \ldots, m$ are given in some coordinate system. Calculate the structure $\mathbf{X}_j$, $j = 1, \ldots, n$ from the images, i.e., from $\mathbf{x}_{i,j}$.

Consider the image of $\mathbf{X}$ in cameras 1 and 2

$$\begin{cases} \lambda_1 \mathbf{x}_1 = P_1 \mathbf{X}, \\ \lambda_2 \mathbf{x}_2 = P_2 \mathbf{X}, \end{cases} \tag{10.34}$$

which can be written in matrix form as (cf. Eq. (10.27))

$$\begin{bmatrix} P_1 & \mathbf{x}_1 & \mathbf{0} \\ P_2 & \mathbf{0} & \mathbf{x}_2 \end{bmatrix} \begin{bmatrix} \mathbf{X} \\ -\lambda_1 \\ -\lambda_2 \end{bmatrix} = \mathbf{0} \ , \tag{10.35}$$

which again is linear in the unknowns, λ_i and $\mathbf{X}$. This linear method can of course be extended to an arbitrary number of images.

10.5.3 Linear Estimation of Tensors

The general scheme for solving the structure and motion problem, Problem 1, is as follows

1. Estimate the components of a multiview tensor linearly from image correspondences.
2. Extract the camera matrices from the tensor components.
3. Reconstruct the object using intersection, i.e., Eq. (10.35).

The Eight-Point Algorithm

Each point correspondence gives one linear constraint on the components of the bifocal tensor according to the bifocal constraint:

$$F_{ij} \mathbf{x}_1^i \mathbf{x}_2^j = 0 \ . \tag{10.36}$$

Thus given at least eight corresponding points the tensor components can be extracted linearly (e.g., by SVD). After the bifocal tensor (fundamental matrix) has been calculated it has to be factorized as $F = A_{12} T_{\mathbf{e}}^T$, which can be done by first solving for $\mathbf{e}$ using $F\mathbf{e} = 0$ (i.e., finding the right nullspace to F) and then for A_{12}, by solving a linear system of equations. One solution is

$$A_{12} = \begin{bmatrix} 0 & 0 & 0 \\ F_{13} & F_{23} & F_{33} \\ -F_{12} & -F_{22} & -F_{32} \end{bmatrix} \ ,$$

which can be seen from the definition of the tensor components. In the case of noisy data it might happen that $\det F \neq 0$ and the right nullspace does not exist. One solution is to solve $F\mathbf{e} = 0$ in the least-squares sense using SVD. Another possibility is to project F to the closest rank-2 matrix, again using SVD. Then the camera matrices can be calculated from Eq. (10.35) and finally using intersection Eq. (10.34) the structure can be obtained. Similar to the case of camera calibration, it is numerically advantageous to rescale the image coordinates to the interval $[-1, 1]$.

The Seven-Point Algorithm

A similar algorithm can be constructed for the case of corresponding points in three images. The trifocal constraint in Eq. (10.30) contains 4 linearly independent constraints in the tensor components T_i^{jk}. Thus, at least 7 corresponding points in three views are needed in order to estimate the 27 homogeneous components of the trifocal tensor. The main difference to the eight-point algorithm is that it is not obvious how to extract the camera matrices from the trifocal tensor components. Start with the transfer equation

$$\mathbf{x}_2^j = T_i^{jk} \mathbf{x}_1^i \mathbf{l}_k^3 ,$$

which can be seen as a homography between the first two images, by fixing a line in the third image. The homography is that corresponding to the plane Π defined by the focal point of the third camera and the fixed line in the third camera. Thus, from Eq. (10.26), the fundamental matrix between image 1 and image 2 obeys

$$FT^{\cdot J} + (T^{\cdot J})^T F^T = 0 ,$$

where $T^{\cdot J}$ denotes the matrix obtained by fixing the index J. Since this is a linear constraint on the components of the fundamental matrix, it can easily be extracted from the trifocal tensor. Then the camera matrices P_1 and P_2 could be calculated and, finally, the entries in the third camera matrix P_3 can be recovered linearly from the definition of the tensor components in Eq. (10.31); see [12].

An advantage of using three views is that lines could be used to constrain the geometry, using Eq. (10.32), giving two linearly independent constraints for each corresponding line.

The Six-Point Algorithm

Again a similar algorithm can be constructed for the case of corresponding points in four images. The quadrifocal constraint in Eq. (10.33) contains 16 linearly independent constraints in the tensor components Q^{ijkl}. It seems as though 5 corresponding points would be sufficient to calculate the 81 homogeneous components of the quadrifocal tensor. However, the quadrifocal constraint in Eq. (10.33) for 2 corresponding points contains 31 linearly independent constraints in the tensor components Q^{ijkl}, because of the following relation

$$\sum_{i,i',i'',j,j',j'',k,k',k'',l,l',l''=1}^{3} Q^{ijkl} \epsilon_{ii'i''} \mathbf{x}^1_{i'} \hat{\mathbf{x}}^1_{i''} \epsilon_{jj'j''} \mathbf{x}^2_{j'} \hat{\mathbf{x}}^2_{j''} \epsilon_{kk'k''} \mathbf{x}^3_{k'} \hat{\mathbf{x}}^3_{k''} \epsilon_{ll'l''} \mathbf{x}^4_{l'} \hat{\mathbf{x}}^4_{l''} = 0 \ ,$$

where $\hat{\mathbf{x}}$ denotes coordinates of another corresponding point quadruple. Thus, at least 6 corresponding points in three views are needed in order to estimate the 81 homogeneous components of the quadrifocal tensor. Since one independent constraint is lost for each pair of corresponding points in four images, there is $6 \cdot 16 - \binom{6}{2} = 81$ linearly independent constraints.

Again, it is not obvious how to extract the camera matrices from the trifocal tensor components. First, a trifocal tensor has to be extracted and then a fundamental matrix and finally the camera matrices. It is outside the scope of this work to give the details for this, instead the reader is referred to [12]. Also in this case corresponding lines can be used by looking at transfer equations for the quadrifocal tensor.

10.5.4 Other Cameras

The structure and motion problem in the case of calibrated and affine cameras will be briefly treated.

Calibrated Cameras

Consider the case of two calibrated cameras, assumed to be normalized, i.e.,

$$\lambda_1 \mathbf{x}_1 = P_1 \mathbf{X} = [\ R_1 \mid t_1\ ]\mathbf{X}, \quad \lambda_2 \mathbf{x}_2 = P_2 \mathbf{X} = [\ R_2 \mid t_2\ ]\mathbf{X} \ .$$

Suppose that $\mathbf{X}_j$ is one reconstruction of n points in m images, with camera matrices P_i, then

$$\mathbf{x}_{ij} = P_i \mathbf{X}_j, \ i = 1, \ldots, m, \ j = 1, \ldots, n \ ,$$

where $P_i = [\ R_i \mid t_i\ ]$ represent normalized cameras. Then, $H \mathbf{X}_j$ is also a feasible reconstruction, with camera matrices $P_i H^{-1}$, for every nonsingular 4×4-matrix H, of the form

$$\begin{bmatrix} \lambda R_{3 \times 3} & b_{3 \times 1} \\ \mathbf{0}_{1 \times 3} & 1 \end{bmatrix}$$

where R denotes an orthogonal matrix, as H leaves the normalized form of the camera matrices unchanged. The transformation

$$\mathbf{X} \mapsto H\mathbf{X}$$

gives all similarity transformations of the object. Thus it is only possible to reconstruct an object up to an unknown similarity transformation for calibrated cameras.

From this ambiguity it may be assumed that the camera matrices have the form
$$P_1 = [\,I \mid 0\,], \quad P_2 \mathbf{X} = [\,R_{12} \mid \mathbf{e}\,]\mathbf{X} \ ,$$
with $R_{12} = R_2 R_1^{-1}$. The same algebra that led to the fundamental matrix can be applied in this case, leading to exactly the same result, by replacing A_{12} with $R_{12} = R_2 r_1^{-1}$, i.e., an orthogonal matrix. This gives the following form of the epipolar constraint
$$\mathbf{x}_1^T E \mathbf{x}_2 = 0 \ ,$$
with $E = R_{12}^T T_\mathbf{e}$. The matrix
$$E = R_{12}^T T_\mathbf{e}$$
is called the **essential matrix**. It can be shown that a 3×3-matrix represents an essential matrix if and only if it has one zero and two equal singular values.

One way to estimate the structure of the scene from two calibrated cameras is to just ignore the special form of the essential matrix and use the eight-point algorithm as above to estimate F. Then a simple SVD can be used to project F to an essential matrix and from this extract R and t, building up the camera matrix and via intersection obtain the structure.

Affine Cameras

Consider two affine cameras
$$P_1 = \begin{bmatrix} A_1 & b_1 \\ \mathbf{0} & 1 \end{bmatrix} \quad \text{and} \quad P_2 = \begin{bmatrix} A_2 & b_2 \\ \mathbf{0} & 1 \end{bmatrix} \ .$$

The fundamental matrix for two affine cameras has the form:
$$F_A = \begin{bmatrix} 0 & 0 & f_{1,3} \\ 0 & 0 & f_{2,3} \\ f_{3,1} & f_{3,2} & f_{3,3} \end{bmatrix} \ .$$

Notice that the epipoles have coordinates
$$\mathbf{e}_{1,2} = (f_{2,3}, -f_{1,3}, 0), \quad \mathbf{e}_{2,1} = (f_{3,2}, -f_{3,1}, 0) \ ,$$
which are points at infinity.

Suppose $\mathbf{X}_j$ is one reconstruction of n points in m images with camera matrices P_i:
$$\mathbf{x}_{ij} = P_i \mathbf{X}_j, \ i = 1, \ldots, m, \ j = 1, \ldots, n \ ,$$
where P_i represent affine cameras. Then, $H \mathbf{X}_j$ is another feasible reconstruction with camera matrices $P_i H^{-1}$, for every nonsingular 4×4-matrix H, of the form

$$\begin{bmatrix} A_{3\times 3} & b_{3\times 1} \\ \mathbf{0}_{1\times 3} & 1 \end{bmatrix}$$

as H leaves the affine form of the camera matrices unchanged. The transformation

$$\mathbf{X} \mapsto H\mathbf{X}$$

gives all affine transformations of the object. Thus, it is only possible to reconstruct an object up to an unknown affine transformation with affine cameras from given point correspondences.

As F has only 5 homogeneous parameters it is sufficient with 4 corresponding points to estimate F. Again it is fairly straightforward to extract the camera matrices and obtain the structure from intersection.

10.5.5 Minimal Cases

There are two interesting minimal cases for uncalibrated cameras: 7 points in 2 images and 6 points in 3 images. Consider the first case, which gives 7 linear constraints as in Eq. (10.36) on the 8 homogeneous components of the fundamental matrix F. This gives a two-parameter homogeneous solution for F:

$$F = sF_1 + tF_2 \ . \tag{10.37}$$

Inserting Eq. (10.37) in the constraint $\det F = 0$ gives a third-order homogeneous polynomial in s and t, resulting in 3 different solutions for the fundamental matrix.

The second case, 6 points in 3 images, also has 3 different solutions, but this is not so easy to prove; see [19, 11].

In the case of calibrated cameras, the only interesting minimal case is 5 points in 2 images. This case is extremely difficult and it has been shown that there are in general 10 different solutions; see [3, 13].

10.5.6 Factorization

A disadvantage with using multiview tensors to solve the structure and motion problem is that when many images ($\gg 4$) are available, the information in all images can not be used with equal weight. An alternative is to use a so-called factorization method; see [23] and [25].

Write the camera equations

$$\lambda_{i,j}\mathbf{x}_{i,j} = P_i\mathbf{X}_j, \quad i = 1,\ldots,m, \quad j = 1,\ldots,n$$

for a fixed image i in matrix form as

$$\mathfrak{X}_i \Lambda_i = P_i \mathbb{X} \ , \tag{10.38}$$

where

$$\mathfrak{X}_i = \begin{bmatrix} \mathbf{x}_{i,1}^T & \mathbf{x}_{i,2}^T & \ldots & \mathbf{x}_{i,n}^T \end{bmatrix}, \quad \mathbb{X} = \begin{bmatrix} \mathbf{X}_1^T & \mathbf{X}_2^T & \ldots & \mathbf{X}_n^T \end{bmatrix},$$
$$\Lambda_i = \text{diag}(\lambda_{i,1}, \lambda_{i,2}, \ldots, \lambda_{i,n}) \ .$$

The camera matrix equations for all images can now be written as

$$\hat{\mathfrak{X}} = \mathbb{P}\mathbb{X} \ , \tag{10.39}$$

where

$$\hat{\mathfrak{X}} = \begin{bmatrix} \mathfrak{X}_1 \Lambda_1 \\ \mathfrak{X}_2 \Lambda_2 \\ \vdots \\ \mathfrak{X}_m \Lambda_m \end{bmatrix}, \quad \mathbb{P} = \begin{bmatrix} P_1 \\ P_2 \\ \vdots \\ P_3 \end{bmatrix} \ .$$

Observe that $\hat{\mathfrak{X}}$ only contains image measurements apart from the unknown depths. It follows from Eq. (10.39) that

$$\text{rank}\,\hat{\mathfrak{X}} \leq 4$$

since $\hat{\mathfrak{X}}$ is a product of a $3m \times 4$- and a $4 \times n$-matrix. Assume that the depths, i.e., Λ_i are known, corresponding to affine cameras, the following simple **factorization** algorithm may be used:

1. Build up the matrix $\hat{\mathfrak{X}}$ from image measurements.
2. Factorize $\hat{\mathfrak{X}} = U\Sigma V^T$ using SVD.
3. Extract $\mathbb{P} = $ the first four columns of $U\Sigma$ and $\mathbb{X} = $ the first four rows of V^T.

In the perspective case this algorithm can be extended to the so-called **iterative factorization** algorithm:

1. Set $\lambda_{i,j} = 1$.
2. Build up the matrix $\hat{\mathfrak{X}}$ from image measurements and the current estimate of $\lambda_{i,j}$.
3. Factorize $\hat{\mathfrak{X}} = U\Sigma V^T$ using SVD.
4. Extract $\mathbb{P} = $ the first four columns of $U\Sigma$ and $\mathbb{X} = $ the first four rows of V^T.
5. Use the current estimate of $\mathbb{P}$ and $\mathbb{X}$ to improve the estimate of the depths from the camera equations $\mathfrak{X}_i \Lambda_i = P_i \mathbb{X}$.
6. If the error (reprojection errors or σ_5) is too large goto 2.

The fifth singular value, σ_5 in the SVD above, is called the **proximity measure** and is a measure of the accuracy of the reconstruction. Fig. 10.8 shows an example of a reconstruction using the iterative factorization method applied on four images of a toy block scene. Observe that the proximity measure decreases quite fast and the algorithm converges in about 20 steps.

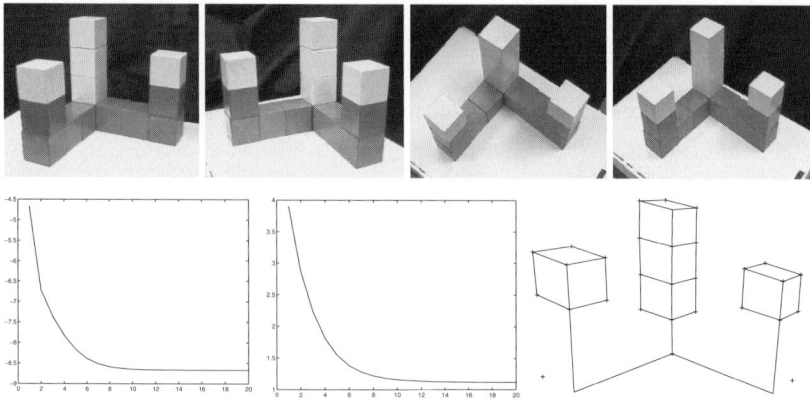

Fig. 10.8. Above: Four images. Below: The proximity measure for each iteration, the standard deviation of reprojection errors and the reconstruction

10.5.7 Long Image Sequences – Missing Data

In the case of long image sequences it often happens that features disappear and new features appear during the sequence. This implies that all features are not visible in all images, so-called **missing data**. There are several different approaches to solving this problem:

- The factorization method can be modified to handle missing data by investigating suitable submatrices.
- Use iteratively resection and intersection as soon as an initial structure and motion is obtained for the first number of frames.

10.5.8 Bundle Adjustment

The previously described methods for structure and motion recovery either ignore the nonlinear constraints on the multiview tensors or use an iterative approach. Thus, the solution obtained from these algorithms doesn't give an optimal result. Assuming that the measured coordinates of the image points are corrupted by Gaussian noise of zero mean and equal standard deviation, statistically optimal results (maximum likelihood) can be obtained by solving the nonlinear optimization problem

$$\min_{P_i, \mathbf{X}_j} \sum_{i=1}^{m} \sum_{j=1}^{n} (x_{i,j}^m - \hat{x}(P_i, \mathbf{X}_j))^2 + (y_{i,j}^m - \hat{y}(P_i, \mathbf{X}_j))^2 \ , \qquad (10.40)$$

where $x_{i,j}^m$ and $y_{i,j}^m$ denote the measured x- and y-coordinates of point i in image j, respectively, and $\hat{x}(P_i, \mathbf{X}_j)$ and $\hat{y}(P_i, \mathbf{X}_j)$ denote the x- and y-coordinates of the reprojected point from P_i and $\mathbf{X}_j$.

In detail, introduce parameters for all 3D-points, $\mathbf{X}_j$, all unknown intrinsic parameters in K_i, all rotation matrices R_i and all translation vectors t_i, as in Eq. (10.15). Given these parameters, calculate the coordinates of the resulting image points $\hat{\mathbf{x}}_{i,j}$,

$$\hat{\mathbf{x}}_{i,j} = f(K_i, R_i, t_i, \mathbf{X}^j) \ . \tag{10.41}$$

The goal of the bundle adjustment algorithm is to minimize the deviation of these reprojected coordinates to the actual measured coordinates in the 2-norm, i.e.,

$$\min_{K_i, R_i, t_i, \mathbf{X}_j} \sum_{i,j} ||\mathbf{x}_{i,j}^m - \hat{\mathbf{x}}_{i,j}||^2 \ . \tag{10.42}$$

In general, the Gauss-Newton method is used to find the minimum; see [21, 1]. Other variants of this method can also be found, e.g., Levenberg-Marquardt; see [9].

Let m denote the number of images and n the number of points. Denote by $\mathfrak{m}$ the bundle of all unknown parameters, $\mathfrak{m} = \{P_1, \ldots, P_m, \mathbf{X}_1, \ldots, \mathbf{X}_n\}$. Each such element belongs to a nonlinear manifold, $\mathcal{M}$. Introduce a **local parameterization** $\mathfrak{m}(\Delta \mathbf{x})$, around $\mathfrak{m}_0 \in \mathcal{M}$, according to

$$\mathcal{M} \times R^N \ni (\mathfrak{m}_0, \Delta \mathbf{x}) \mapsto \mathfrak{m}(\mathfrak{m}_0, \Delta \mathbf{x}) \in \mathcal{M} \ , \tag{10.43}$$

where $N = 11m + 3n$. (11 parameters in each camera matrix and 3 parameters for the coordinates of each reconstructed 3D-point.) For convenience the local parameter $\Delta \mathbf{x}$ is divided into two parts according to

$$\Delta \mathbf{x} = [\Delta a_1, \ldots, \Delta a_m, \Delta b_1, \ldots, \Delta b_n]^T \ , \tag{10.44}$$

i.e., Δa_i parameterize changes in camera matrix P_i and Δb_j parameterize changes in reconstructed point X_j. Each camera matrix is written $P_i = K_i[R_i \mid -R_i t_i]$ and changes in K_i are parameterized as

$$K_i(\mathfrak{m}_0, \Delta \mathbf{x}) = \begin{bmatrix} (f + \Delta a_i(1))(\gamma + \Delta a_i(2)) & (f + \Delta a_i(1))(s + \Delta a_i(3)) & x_0 + \Delta a_i(4) \\ 0 & f + \Delta a_i(1) & y_0 + \Delta a_i(5) \\ 0 & 0 & 1 \end{bmatrix} \ , \tag{10.45}$$

where the Δa_i are restricted differently according to the different assumptions on the intrinsic parameters. Changes in R_i are parameterized as

$$R_i(\mathfrak{m}_0, \Delta \mathbf{x}) = \exp\left(\begin{bmatrix} 0 & \Delta a_i(8) & -\Delta a_i(7) \\ -\Delta a_i(8) & 0 & \Delta a_i(6) \\ \Delta a_i(7) & -\Delta a_i(6) & 0 \end{bmatrix}\right) R_i \ , \tag{10.46}$$

changes in t_i as

$$t_i(\mathfrak{m}_0, \Delta \mathbf{x}) = \begin{bmatrix} t_x + \Delta a_i(9) \\ t_y + \Delta a_i(10) \\ t_z + \Delta a_i(11) \end{bmatrix} \ , \tag{10.47}$$

and changes in each object point, $\mathbf{X}_j$, as

$$\mathbf{X}_j(\mathfrak{m}_0, \Delta \mathbf{x}) = \begin{bmatrix} X_j + \Delta b_j(1) \\ Y_j + \Delta b_j(2) \\ Z_j + \Delta b_j(3) \\ 1 \end{bmatrix} \ . \tag{10.48}$$

Introduce a **residual vector Y**, formed by putting all deviations between measured and reprojected image coordinates in a column vector. These residuals depend on the measured image positions $\mathbf{x}_{i,j}^m$ as well as on the estimated parameters $\mathfrak{m}$. The residual vector $\mathbf{Y}(\Delta \mathbf{x})$ is a nonlinear function of the local parameterization vector $\Delta \mathbf{x}$. The sum of squared residuals $f = \mathbf{Y}^T \mathbf{Y}$ is minimized with respect to the unknown parameters $\Delta \mathbf{x}$, using the Gauss-Newton method as follows. A linearization of $\mathbf{Y}(\Delta \mathbf{x})$ gives

$$\mathbf{Y}(\Delta \mathbf{x}) \approx \mathbf{Y}(0) + \frac{\partial \mathbf{Y}}{\partial \Delta \mathbf{x}}(0) \Delta \mathbf{x} \ . \tag{10.49}$$

To find $\Delta \mathbf{x}$ so that $\mathbf{Y}(\Delta \mathbf{x}) = 0$, solve

$$\mathbf{Y}(0) + \frac{\partial \mathbf{Y}}{\partial \Delta \mathbf{x}}(0) \Delta \mathbf{x} = 0 \ . \tag{10.50}$$

The above linear equation is solved in the least-squares sense giving the update

$$\Delta \mathbf{x} = - \left(\frac{\partial \mathbf{Y}}{\partial \Delta \mathbf{x}}(0) \right)^{\dagger} \mathbf{Y}(0) \ , \tag{10.51}$$

where $A^{\dagger}$ denotes the More-Penrose pseudoinverse of A. In practice it is useful to use the singular value decomposition or the Levenberg-Marquardt method. Let

$$A = \frac{\partial \mathbf{Y}}{\partial \Delta \mathbf{x}}(0) \quad \text{and} \quad b = \mathbf{Y}(0) \ .$$

Instead of solving

$$\Delta \mathbf{x} = -(A^T A)^{-1} A^T b \ , \tag{10.52}$$

which might be numerically sensitive if $(A^T A)$ has small singular values, the singular value decomposition of A, $A = U \Sigma V^T$, may be used. Truncate the smallest singular values so that $A \approx U_0 \Sigma_0 V_0^T$, where U_0 consists of the first N rows of U, V_0 consists of the first N rows of V, and Σ_0 consists of the $N \times N$-matrix with the N largest singular values on the diagonal. Now update the parameters using

$$\Delta \mathbf{x} = V_0 \Sigma_0^{-1} U_0^T b \ . \tag{10.53}$$

An alternative is to use the Levenberg-Marquardt method. The update of the parameters is computed as

$$\Delta \mathbf{x} = -(A^T A + \epsilon I)^{-1} A^T b \ , \tag{10.54}$$

where ϵ is a small positive number. ϵ has to be chosen carefully. A common strategy is to start with a fairly large value and decrease it gradually when approaching the minima.

10.5.9 Robust Methods

When trying to find corresponding features in a sequence of images, a fairly large number of false matches, called **outliers**, is usually obtained. This implies that robust methods have to be used to obtain a meaningful result. The most widely used method is **RANSAC** (Random Sampling Consensus); see [7] and [26]. The main idea behind this method is to randomly select a minimal number of corresponding points needed to estimate the unknown parameters and then check if the other points agree with these parameters or not. This process is repeated a fixed number of times and the parameters that most points agree on are selected. In the case of estimating the fundamental matrix, the following algorithm can be used:

1. Select randomly 8 corresponding points.
2. Calculate the fundamental matrix, F.
3. Record the number of corresponding points $(\mathbf{x}_1, \mathbf{x}_2)$ that fulfil $\mathbf{x}_1^T F \mathbf{x}_2 \leq \epsilon$.
4. Go to 1.

Here ϵ is a fixed accuracy and the process is repeated a fixed number of times. Finally, the F with the highest number of agreeing points is selected. Often, those points that agree on this F are used to reestimate F once again.

10.6 Autocalibration

In Sect. 10.4 it was shown that it is only possible to reconstruct the scene up to an unknown projective transformation from a sequence taken by uncalibrated cameras. This projective reconstruction is in most cases useless, since the unknown projective transformation from the true scene can introduce severe nonlinear distortions. However, it was shown in Sect. 10.5 that if the cameras are calibrated, it is possible to obtain a Euclidean reconstruction (more precisely, a reconstruction up to an unknown similarity transformation), which is more useful. In Sect. 10.3 it was shown that a camera could be calibrated from an image of a calibration grid consisting of at least 6 points. However, this procedure has several drawbacks:

- It is time consuming and cumbersome to precalibrate using a special calibration object.
- The intrinsic parameters might change during the sequence, e.g., when zooming.
- Several sequences and/or images from cameras with different intrinsic parameters might be available.
- Sequences might be available from unknown cameras without the possibility to calibrate.

Therefore, it is desirable to be able to calibrate the camera during the reconstruction process, so-called **autocalibration**.

10.6.1 Problem Formulation

Starting with a projective reconstruction in the form of a sequence of camera matrices, P_i, defined up to an unknown projective transformation, H, auto-calibration implies finding this projective transformation, H, such that $P_i H$ can be factorized as

$$P_i H \sim K_i R_i [\, I \,|\, -t_i \,] \;, \tag{10.55}$$

where R_i denote orthogonal matrices and K_i contains intrinsic parameters obeying the constraints at hand. For instance, in the case of constant intrinsic parameters $K_i = K =$ constant.

10.6.2 Constant Intrinsic Parameters

Assume constant intrinsic parameters and let Ω denote the dual to the absolute conic, represented by a singular 4×4-matrix. Then the image of Ω is given by

$$\omega \sim P_i \Omega P_i^T \;, \tag{10.56}$$

where ω denotes the dual to the image of the absolute conic and is related to the intrinsic parameters via the equation $\omega = KK^T$. Actually Eq. (10.56) is valid for other constraints than constant intrinsic parameters by replacing ω with ω_i. By fixing the gauge freedoms in the Euclidean coordinate system, i.e., fixing the origin and a reference orientation, by putting $P_1 = [\, I \,|\, 0 \,]$,

$$H = \begin{bmatrix} K & 0 \\ \bar{n} & 1 \end{bmatrix} \;, \tag{10.57}$$

and Ω can be parameterized as

$$\Omega = TT^T = \begin{bmatrix} K \\ \bar{n} \end{bmatrix} \begin{bmatrix} K \\ \bar{n} \end{bmatrix}^T \;, \tag{10.58}$$

where $(\bar{n}, 1)$ denotes the normal to the plane at infinity. In the case of other constraints than constant intrinsic parameters K has to be replaced by K_1, i.e., the intrinsic parameters for the first camera. To sum up there are from Eq. (10.56) 5 equations for each image (apart from the first one) and 8 unknowns (5 intrinsic parameters plus 3 parameters for the normal to the plane at infinity). Again, by counting the number of equations and unknowns, three images are sufficient.

Equation (10.58) consists of a system of nonlinear equations that has to be solved; see [14]. Observe that an initial guess of the plane at infinity can easily be obtained linearly from Eq. (10.56) when the guess for the intrinsic parameters is inserted.

10.6.3 The Modulus Constraint

Let Q_i denote the first 3×3 submatrix of P_i, i.e., $P_i = Q_i[\,I\,|\,q_i\,]$. Then

$$P_i H = Q_i[\,I\,|\,q_i\,]H = [\,Q_i K - Q_i t_i \bar{n}\,|\,-Q_i q_i\,] \sim [\,KR_i\,|\,-KR_i t_i\,]$$
$$\Rightarrow \quad Q_i K - Q_i q_i \bar{n} \sim KR_i \quad \Rightarrow \quad Q_i - Q_i q_i \bar{n} K^{-1} \sim KR_i K^{-1} \;, \tag{10.59}$$

saying that $Q_i - Q_i q_i \bar{n} K^{-1}$ is similar to a scaled orthogonal matrix. In fact, $KR_i K^{-1} = H_\infty^i$, the homography of the plane at infinity induced by the fundamental matrix between image 1 and image i. Thus all eigenvalues of $Q_i - Q_i q_i \bar{n} K^{-1}$ have equal moduli, hence the name **modulus constraint**. The characteristic polynomial of $Q_i - q_i \bar{n} K^{-1}$,

$$p(\lambda) = \det(\lambda I - Q_i + q_i \bar{n} K^{-1}) = \lambda^3 + l_2 \lambda^2 + l_1 \lambda + l_0 \;,$$

obeys this modulus constraint,

$$l_1^3 = l_2^3 l_0 \;,$$

which is a polynomial equation of degree 4 in the unknown parameters, $\bar{n}K^{-1}$. Hence, a unique solution can in general be obtained from at least four images; see [17]. When the three parameters in $\bar{n}K^{-1}$ are known it is easy to solve Eq. (10.56) for the intrinsic parameters. In fact a linear method can be used to solve for the dual to the image of the absolute conic and then a Cholesky factorization gives the intrinsic parameters in the following way: Set $H_\infty^i = KR_i K^{-1}$, i.e., the homography of the plane at infinity, obtained from Eq. (10.59) when $\bar{n}K^{-1}$ is known – observe that the scale of H_∞^i is known since $\det(H_\infty^i) = 1$. Thus $R_i = K^{-1} H_\infty^i K$, which gives

$$H_\infty^i K K^T (H_\infty^i)^T = K K^T \quad \Rightarrow \quad H_\infty^i \omega (H_\infty^i)^T = \omega \;, \tag{10.60}$$

which is a linear system of equations in the five intrinsic parameters. However, only four equations are linearly independent, requiring three views to calculate K.

10.6.4 A Linear Method

Assume that only the focal length is unknown (and possibly varying), which is a reasonable assumption in order to obtain an initial estimate of the intrinsic parameters. The only unknown parameters are then the focal lengths and the plane at infinity. Inserting Eq. (10.58) into Eq. (10.56) gives

$$K_i K_i^T \sim P_i^T \begin{bmatrix} K_i K_i^T & K_1 \bar{n}^T \\ \bar{n} K_1^T & \bar{n}\bar{n}^T \end{bmatrix} P_i \;. \tag{10.61}$$

Assuming zero skew, unit aspect ratio and the principal point situated in the centre of the image, Eq. (10.61) simplifies to

$$\lambda_i \begin{bmatrix} f_i^2 & 0 & 0 \\ 0 & f_i^2 & 0 \\ 0 & 0 & 1 \end{bmatrix} = P_i^T \begin{bmatrix} f_1^2 & 0 & 0 & a \\ 0 & f_1^2 & 0 & b \\ 0 & 0 & 1 & c \\ a & b & c & a^2+b^2+c^2 \end{bmatrix} P_i \ , \qquad (10.62)$$

where $\bar{n} = (a, b, c)$. Observe that Eq. (10.62) contains 6 linear equations in the 7 unknowns λ_i, $\lambda_i f_i^2$, f_1^2, a, b, c and $a^2 + b^2 + c^2$. For $i = 1$ a pure identity due to the specific coordinate change is obtained and thus no constraints. For $i = 2$, 7 unknowns and 6 equations are obtained. For each further image 2 new unknowns and 6 equations are added. Thus $\bar{n}$ and f_i can be solved for using a quasilinear method, when at least three images are available; see [18].

10.6.5 Nonlinear Refinement

When an initial estimate of the dual to the absolute conic has been found this estimate needs to be refined by solving the nonlinear optimization problem

$$\min_{(\Omega, \omega_i)} \sum_{i=1}^{m} ||\omega_i - P_i \Omega P_i^T||^2 \ , \qquad (10.63)$$

using e.g., nonlinear least squares. Observe that a suitable normalization of ω_i and $P_i \Omega P_i^T$ is needed to obtain a non-trivial solution. Finally, bundle adjustment, with the proper choice of constraints on the intrinsic parameters, gives the optimal solution.

Figure 10.9 shows one of 42 images of a scene containing point markers and some curves and silhouettes. The images were taken by the same camera without zooming or focusing. Firstly a projective reconstruction was made using iterative factorization followed by projective bundle adjustment. Secondly, the linear autocalibration method was used to calculate the initial Euclidean structure and motion assuming known skew, aspect ratio and principal point. This result was then used as an initial value to the bundle adjustment routine using only the constraint that the skew is zero. In Fig. 10.9 is also shown a histogram of errors between estimated point positions and reprojected point positions. In the same figure are shown the two focal lengths (f_x and f_y) and the coordinates of the principal point (x_0 and y_0) for the image sequence.

10.7 Conclusions

This chapter has described how to model a camera, using the perspective pinhole camera model and its specializations. The multiview constraints and multiview tensors have been studied using tools from projective geometry. The structure and motion problem has been solved using these constraints as well as using factorization methods and bundle adjustment. Finally, autocalibration methods have been derived to obtain a final Euclidean reconstruction of the scene.

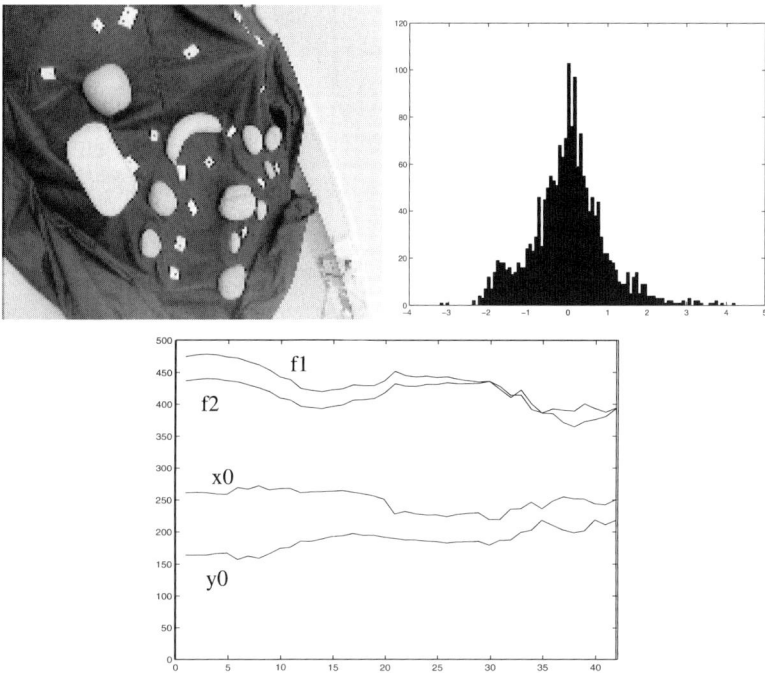

Fig. 10.9. One of 42 images used in the experiment. Histogram of reprojected residuals. Estimated focal lengths and principal point coordinates for the 42 images

Acknowledgements

This work has been supported by a Junior Individual Grant from the Swedish Research Council and the Swedish Strategic Foundation.

References

1. Atkinson K. B. (1996) *Close Range Photogrammetry and Machine Vision*. Whittles.
2. Coxeter H.S.M. (1964) *Projective Geometry*. Blaisdell.
3. Demazure M. (1988) Sur deux problemes de reconstruction. Technical Report 882, INRIA, Rocquencourt, France.
4. Faugeras O. (1992) What can be seen in three dimensions with an uncalibrated stereo rig? In: *Proc. European Conf. on Computer Vision*, pages 563–578.
5. Faugeras O. (1993) *Three-Dimensional Computer Vision*. MIT Press, Cambridge, Mass. .
6. Faugeras O. and Luong Q.-T. (2001) *The Geometry of Multiple Images*. MIT Press, Cambridge, Mass. .
7. Fischler M.A. and Bolles R.C. (1981) Random sample consensus, a paradigm for model fitting with application to image analysis and automated cartography. *Commun. Assoc. Comp. Mach.*, 24:381–395.
8. Hartley R. and Zisserman A. (2000) *Multiple View Geometry in Computer Vision*. Cambridge University Press.
9. Hartley R.I. (1994) Euclidean reconstruction from uncalibrated views. In: *Applications of Invariance in Computer Vision*, volume 825 of *Lecture notes in Computer Science*, pages 237–256. Springer-Verlag.
10. Heyden A. (1995) *Geometry and Algebra of Multipe Projective Transformations*. PhD thesis, Lund University, Sweden.
11. Heyden A. (1995) Reconstruction and prediction from three images of uncalibrated cameras. In: *Proc. Scandinavian Conf. on Computer Vision*, pages 57–66.
12. Heyden A. (2000) Tensorial properties of multilinear constraints. *Mathematical Methods in the Applied Sciences*, 23:169–202.
13. Heyden A. and Sparr G. (1999) Reconstruction from calibrated cameras: A new proof of the Kruppa-Demazure Theorem. *Journal of Mathematical Imaging and Vision*, 10(2):123–142.
14. Heyden A. and Åström K. (1996) Euclidean reconstruction from constant intrinsic parameters. In: *Proc. Int. Conf. on Pattern Recognition*, volume 1, pages 339–343.
15. Longuet-Higgins H.C. (1981) A computer algorithm for reconstructing a scene from two projections. *Nature*, 293:133–135.
16. Myklestad N.O. (1967) *Cartesian Tensors – the Mathematical Language of Engineering*. Van Nostrand, Princeton, 1967.
17. Pollefeys M., Van Gool L. and Oosterlinck M. (1996) The modulus constraint: A new constraint for self-calibration. In: *Proc. Int. Conf. on Pattern Recognition*, pages 349–353.
18. Pollefeys M., Koch R. and Van Gool L. (1998) Self-calibration and metric reconstruction in spite of varying and unknown internal camera parameters. In: *Proc. Int. Conf. on Computer Vision*.
19. Quan L. (1994) Invariants of 6 points from 3 uncalibrated images. In: *Proc. European Conf. on Computer Vision*, pages B:459–470.
20. Shashua A. (1994) Trilinearity in visual recognition by alignment. In: *Proc. European Conf. on Computer Vision*.
21. Slama C.C., editor. (1984) *Manual of Photogrammetry*. American Society of Photogrammetry, Falls Church, VA, 4th edition.

22. Spain B. (1953) *Tensor Calculus*. University Mathematical Texts, Oliver and Boyd, Edinburgh.
23. Sparr G. (1996) Simultaneous reconstruction of scene structure and camera locations from uncalibrated image sequences. In: *Proc. Int. Conf. on Pattern Recognition*.
24. Spetsakis M.E. and Aloimonos J. (1990) A unified theory of structure from motion. In: *Proc. DARPA Image Understanding Workshop*.
25. Sturm P. and Triggs B. (1996) A factorization based algorithm for multi-image projective structure and motion. In: *Proc. European Conf. on Computer Vision*, pages 709–720.
26. Torr P. (1995) *Outlier Detection and Motion Segmentation*. PhD thesis, University of Oxford.
27. Triggs B. (1995) Matching constraints and the joint image. In: *Proc. Int. Conf. on Computer Vision*, 1995.
28. Zhang Z. (2000) A flexible new technique for camera calibration. *IEEE Trans. Pattern Analysis and Machine Intelligence*, 22(11):1330–1334.

11

Dynamic $\mathcal{P}^n$ to $\mathcal{P}^n$ Alignment

Amnon Shashua[1] and Lior Wolf[2]

[1] School of Engineering and Computer Science,
  The Hebrew University of Jerusalem
  Jerusalem, 91904, Israel
  shashua@cs.huji.ac.il
[2] M.I.T. Center for Computational and Biological Learning (CBCL), Cambridge,
  MA, USA, 02139.
  lwolf@cs.huji.ac.il

We introduce in this chapter a generalization of the classical collineation of $\mathcal{P}^n$. The generalization allows for a certain degree of freedom in the localization of points in $\mathcal{P}^n$ at the expense of using multiple $m > 2$ views. The degree of freedom per point is governed by an additional parameter $1 \leq k < n$, which stands for the dimension of the subspace in which the indvidual points are allowed to move while the projective change of coordinates take place. In other words, the point set is not necessarily stationary, allowing the configuration to change while the entire coordinate system undergoes a projective change of coordinates. If we denote a change of coordinates as a "view" of the physical set of points, then in this chapter we discuss the multiview relations that can be determined from observations (views) of a dynamically changing point configuration – the underlying transformations and how they can be recovered.

For example, for a point set in $\mathcal{P}^2$ (planar configuration) undergoing linear motion, the multiple views of the point set generate a multilinear constraint across three views governed by a $3 \times 3 \times 3$ contravariant tensor $\mathcal{H}$. The tensor, referred to as *homography tensor*, can be recovered linearly from 26 observations (matching points across three views), and once recovered can be unfolded to yield the global coordinate change (the individual pair of homography matrices). A point set in $\mathcal{P}^3$ (3D configuration) can undergo motion in a plane or along a line (each point independently). For the line motion, the multi-view constraints are governed by a $4 \times 4 \times 4$ family of contravariant tensors $\mathcal{J}$ that capture the dynamic 3D-to-3D alignment problem. More generally, the family of homography tensors is captured by three parameters: the dimension n of the observation space, the dimension $k < n$ of the subspace along which each point of the point set is allowed to move, and the number of "views" m. Formally, the homography tensors form a $GL(V)$ module, denoted by $V(n, m, k)$, defined by the set of all tensors $v_1 \otimes \cdots \otimes v_m \in V^{\otimes m}$, where v_i are n-dimensional vectors and $\dim \text{Span}\{v_1, ..., v_m\} \leq k$. We will be interested in the structure and dimension of $V(n, m, k)$.

The notion of using multiview analysis for nonrigid scenes is interesting and useful on its own right. In a way, this work extends the notion of "stereo triangulation" (a stationary point observed by two or more views) to the notion of "what can be recovered from line of sight measurements only?". The chapter includes a detailed exposition of these tensors for $\mathcal{P}^2$ and $\mathcal{P}^3$, their properties and applications, and derives the dimension of $V(n, m, k)$ in the general case.

11.1 Introduction

Consider the classic problem of "3D to 3D" alignment of point sets. We are given a set of 3D points $P_1, ..., P_n$ measured by some device such as a structured light range sensor [14] or a stereo rig of cameras. When the sensor changes its position in space while the 3D points remain stationary, the 3D positions of the measured points $P'_1, ..., P'_n$ have undergone a coordinate transformation. In a projective setting, five of these matching pairs in general position are sufficient to recover the 4×4 collineation A such that $AP_i \cong P'_i$, $i = 1, ..., n$. In a rigid motion setting the coordinate transformation consists of translation and rotation, which can be recovered using four matching points; elegant techniques using Singular Value Decomposition (SVD) have been developed for this purpose [4].

In the same vein, consider another popular group of transformations that includes planar collineations between two sets of points on the projective plane $\mathcal{P}^2$ undergoing a projective mapping. The planar collineations (homographies) are the 3×3 nonsingular matrices that map between point sets undergoing a general projectivity. The planar homographies form a fundamental building block in multiple-view geometry in computer vision. The object stands on its own as a point-transfer vehicle for planar scenes (aerial photographs, for example) and in applications of mosaicing, camera stabilization and tracking [8]. A homography matrix is a standard building block in handling 3D scenes from multiple 2D projections: the "plane+parallax" framework [3, 6, 7, 11] uses a homography matrix for setting up a parallax residual field relative to a planar reference surface, and the trifocal tensor of three views is represented by a "homography-epipole" structure whose slices are homography matrices as well [5, 12].

The two examples above, general collineations of $\mathcal{P}^2$ and $\mathcal{P}^3$, readily extend to n-dimensional projective spaces $\mathcal{P}^n$. A change of coordinates in an n-dimensional projective space $\mathcal{P}^n$ is determined by an $(n+1) \times (n+1)$ matrix. Converseley, given two sets of points in $\mathcal{P}^n$ that result by having one set undergo some collineation, the alignment of the two sets can be achieved by a homography matrix which can be determined uniquely from $n+1$ matching points.

In this chapter we introduce a "dynamic" version of the $\mathcal{P}^n$ to $\mathcal{P}^n$ alignment problem by allowing the individual points of the point set to undergo independent motion within k-dimensional subspaces while the entire point set undergoes a general collineation successively. For example, in the dynamic $\mathcal{P}^2 \to \mathcal{P}^2$ version, we allow for the possibility that any number of the points may move along straight-line paths during the change of view. A change of view results in a global change of coordinates (a collineation), but while doing so the individual points of the point set have changed relative position to one another. Points that remain in place are called *stationary*, and points that move are called *dynamic*. There can be any number of dynamic points – including the possibility that *all* points are dynamic – and the system need not know in advance which of the points are stationary and which are dynamic (an unsegmented configuration). Under these conditions we wish to find the multiple projective coordinate changes from the point-match observations of the point set under successive coordinate changes. We will show that this type of transformation is governed by a $3 \times 3 \times 3$ tensor that captures the multiview relation of the changing planar point set. The tensor is formed by a bilinear product of the global pair of homography matrices, which are responsible for the changes of coordinates between the first view and the other two views. For every triplet of matching points across the three views p, p', p'' the following contravariant relation $p^i p'^j p''^k H_{ijk} = 0$ vanishes. The vanishing constraint provides a linear equation on the elements of the tensor, and the global coordinate changes can later be recovered (also linearly) from the tensor.

The dynamic $\mathcal{P}^3 \to \mathcal{P}^3$ alignment problem can be viewed as a 3D sensor that changes position in 3D space (thus creating global coordinate changes) while the physical points in space undergo independent motion either along straight-line paths or along planar subspaces (or they can stay put). We will show that this type of transformation is governed by a $4 \times 4 \times 4$ family of tensors that vanishes on each of the matching triplets induced by a physical point under three coordinate systems.

More generally, the family of tensors governing the $\mathcal{P}^n \to \mathcal{P}^n$ alignment problem is captured by three parameters: the dimension n of the observation space, the dimension $k < n$ of the subspace along which each point of the point set is allowed to move, and the number of "views" m. Formally, these tensors form a $GL(V)$ module, denoted by $V(n, m, k)$, defined by the set of all tensors $v_1 \otimes \cdots \otimes v_m \in V^{\otimes m}$, where v_i are n-dimensional vectors and $\dim \mathrm{Span}\{v_1, \ldots, v_m\} \leq k$. We will be interested in the structure and dimension of $V(n, m, k)$.

We will describe in detail the tensor families that are associated with $\mathcal{P}^2$ and $\mathcal{P}^3$, their definition, the way they can be recovered from observations, their properties and their applications. The general case will be discussed at a reduced scope, where we will address only the dimension of $V(n, m, k)$ (the number of independent linear constraints possible for a given value of n, m, k). Other issues that are addressed in derivations for $\mathcal{P}^2$ and $\mathcal{P}^3$, such as mixed stationary and dynamic motions, are left open in the general case. Part of the material described in Sects. 11.2 and 11.3 appeared in the proceedings of [13, 16] and the material of Sect. 11.4 appeared in technical report [10].

11.1.1 Background and Notation

We will be working with the projective space $\mathcal{P}^n$. A point in $\mathcal{P}^n$ is defined by $n+1$ numbers, not all zero, that form a coordinate vector defined up to a scale factor. The dual projective space represents the space of hyperplanes, which are also defined by a $(n+1)$-tuple of numbers. For example, a point p in the projective plane $\mathcal{P}^2$ coincides with a line s if and only if $p^\top s = 0$, i.e., the scalar product vanishes. In other words, the set of lines coincident with the point p is represented by the coordinate vectors s that satisfy $p^\top s = 0$, and vice versa: a point represented by the coordinate vector p can be thought of as the set of lines through it (also known as the pencil of lines through p). A line s going through two points p_1, p_2 is represented by the cross product $s \cong p_1 \times p_2$, where $\cong$ denotes equality up to scale. Likewise, the point of intersection p of the lines s_1, s_2 is represented by $p \cong s_1 \times s_2$. In projective 3D space $\mathcal{P}^3$, a point p lies on a plane π if and only if $p^\top \pi = 0$. In other words, in $\mathcal{P}^2$ points and lines are dual to each other, and in $\mathcal{P}^3$ points and planes are dual to each other - generally, points and hyperplanes are duals.

In projective space any $n+1$ points in general position (i.e., no subset of n points lie on a hyperplane) can be uniquely mapped into any other $n+1$ general point configuration. Such a mapping is called a *collineation* and is defined by an invertible $(n+1) \times (n+1)$ matrix (also known as the homography matrix) defined up to scale. In particular, the change of coordinates of a planar configuration induced by taking a photograph by a pinhole camera moving freely in the 3D world is represented by a 3×3 homography matrix, and the change of coordinates of a 3D point configuration caused by the motion of the sensor is represented by a 4×4 homography matrix. If H is a homography matrix (defined by $n+1$ matching pairs of points), then H^{-T} (inverse transpose) is the dual homography that maps hyperplanes onto hyperplanes.

The projective plane is useful to model the image plane in a pinhole camera model. Consider a collection of planar points $P_1, ..., P_k$ in space living on a plane π viewed from two views. The projections of P_i are p_i, p'_i in views 1 and 2, respectively. Because the collineations form a group, there exists a unique homography matrix H_π that satisfies the relation $H_\pi p_i \cong p'_i$, $i = 1, ..., k$, and where H_π is uniquely determined by four matching pairs from the set of k matching pairs. Moreover, $H_\pi^{-T} s \cong s'$ maps between matching lines s, s' arising from 3D lines living in the plane π. Likewise, $H_\pi^\top s' \cong s$ maps between matching lines from view 2 to view 1.

It is most convenient to use tensor notations from now on because the material we are using in this chapter involves coupling together pairs of collineations into a "joint" object. The distinction of when coordinate vectors stand for points or hyperplanes matters when using tensor notation. A point is an object whose coordinates are specified with superscripts, i.e., $p = (p^0, p^1, ..., p^n)$, thus p^i stands for the ith entry of the vector. These are called contravariant vectors. A hyperplane in $\mathcal{P}^n$ is called a covariant vector and is represented by subscripts, i.e., $s = (s_0, s_1, ..., s_n)$. Indices repeated in covariant and contravariant forms are summed over, i.e., $p^i s_i = p^0 s_0 + p^1 s_1 + ... + p^n s_n$. This is known as a contraction. For example, if p is a point incident to a line s in $\mathcal{P}^2$, then $p^i s_i = 0$.

Vectors are also called 1-valence tensors; 2-valence tensors (matrices) have two indices and the transformation they represent depends on the covariant-contravariant positioning of the indices. For example, a_i^j is a mapping from points to points (a collineation, for example) and hyperplanes to hyperplanes, because $a_i^j p^i = q^j$ and $a_i^j s_j = r_i$ (in matrix form: $Ap = q$ and $A^\top s = r$); a_{ij} maps points to hyperplanes; and a^{ij} maps hyperplanes to points. When viewed as a matrix the row and column positions are determined accordingly: in a_i^j and a_{ji} the index i runs over the columns and j runs over the rows, thus $b_j^k a_i^j = c_i^k$ is $BA = C$ in matrix form.

An outer product of two 1-valence tensors (vectors) $a_i b^j$ is a 2-valence tensor c_i^j whose i, j entries are $a_i b^j$. Note that in matrix form $C = ba^\top$. A 3-valence tensor has three indices, say H_i^{jk}. The positioning of the indices reveals the geometric nature of the mapping: for example, $p^i s_j H_i^{jk}$ must be a point because the i, j indices drop out in the contraction process and we are left with a contravariant vector (the index k is a superscript). Thus, H_i^{jk} maps a point in the first coordinate frame and a hyperplane in the second coordinate frame into a point in the third coordinate frame. A single contraction, say $p^i H_i^{jk}$, of a 3-valence tensor leaves us with a matrix. Note that when p is $(1, 0, 0)$ or $(0, 1, 0)$ or $(0, 0, 1)$ the result is a "slice" of the tensor.

In the projective plane $\mathcal{P}^2$ we will make use of the "cross product tensor" ϵ defined next. The cross product (vector product) operation $c = a \times b$ is defined for vectors in $\mathcal{P}^2$. The product operation can also be represented as the product $c = [a]_\times b$ where $[a]_\times$ is called the "skew-symmetric matrix of a" and has the form:

$$[a]_\times = \begin{pmatrix} 0 & -a_2 & a_1 \\ a_2 & 0 & -a_0 \\ -a_1 & a_0 & 0 \end{pmatrix}.$$

In tensor form we have $\epsilon_{ijk} a^i b^j = c_k$ representing the cross product of two points (contravariant vectors) resulting in the line (covariant vector) c_k. Similarly, $\epsilon^{ijk} a_i b_j = c^k$ represents the point intersection of the two lines a_i and b_j. The tensor ϵ is defined such that $\epsilon_{ijk} a^i$ produces the matrix $[a]_\times$ (i.e., ϵ contains $0, -1, 1$ in its entries such that its operation on a single vector produces the skew-symmetric matrix of that vector).

11.2 Homography Tensor $\mathcal{H}$ of the Projective Plane

Consider some plane π whose features (points or lines) are projected onto three views and let A be the collineation from view 2 to view 1, and B the collineation from view 3 to 1 (we omit the reference to π in our notation). Let P be some point on the plane π, and its projections are p, p', p'' in views 1, 2 and 3, respectively. We consider two possibilities: first, the point P on the plane π is *stationary*, i.e., the three optical rays from the camera centers to the image points p, p', p'' meet at P, and second, the point P *moves* along a straight line (in the plane) path, therefore the three optical rays meet at a line in π instead of a point (see Fig. 11.1). We summarize these two possibilities in the following definition:

Definition 1. *A triplet of points p, p', p'' is said to be* **matching with respect to a stationary point** *if the points are matching in the usual sense of the term, i.e., the corresponding optical rays meet at a single point. The triplet is said to be* **matching with respect to a moving point** *if the three optical rays meet at a line on a plane.*

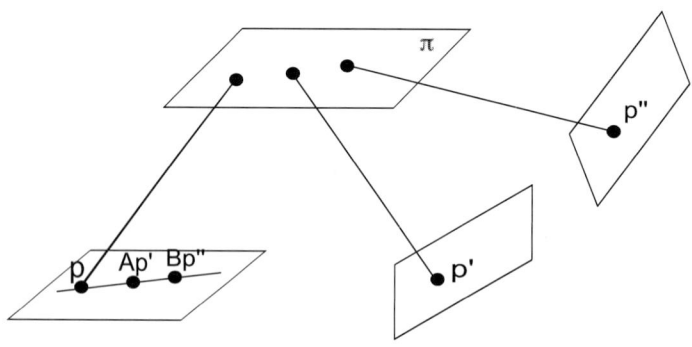

Fig. 11.1. The homography tensor of $\mathcal{P}^2$ and moving points. The collineations A, B are from views 2 to 1 and 3 to 1, respectively. If the triplet p, p', p'' is a projection of a moving point along a line on π, then p, Ap', Bp'' are collinear in view 1. Thus, $p^\top (Ap' \times Bp'') = 0$, or $p^i p'^j p''^k \mathcal{H}_{ijk} = 0$, where $\mathcal{H}_{ijk} = \epsilon_{inu} a_j^n b_k^u$.

The constraint that satisfies both the moving and stationary possibilities is:
$$\det(p, Ap', Bp'') = p^\top (Ap' \times Bp'') = 0.$$

In other words, $\det(p, Ap', Bp'') = 0$ when the rank of the 3×3 matrix $[p, Ap', Bp'']$ is either 1 or 2. The rank is 1 when the point P is stationary (three optical rays meet at a point) and is 2 when the point P moves along a straight line path (three optical rays meet at a line on π). The constraint $p^\top (Ap' \times Bp'') = 0$ is bilinear in the entries of the unknown collineations A, B and is trilinear in the observations p, p', p''. Using tensorial notation we can combine the pair of collineations into a single object, a $3 \times 3 \times 3$ tensor, as follows. We define indices i, j, k such that index i runs over view 1, index j runs over view 2 and index k runs over view 3. For example, the operation Ap' is translated to $a^i_j p'^j$ producing a point in view 1. The cross product $Ap' \times Bp''$ is translated to $\epsilon_{inu}(a^n_j p'^j)(b^u_k p''^k)$, where parenthesis are added for clarity only (that is, the positions of symbols are not important, only the positions of the indices). Taken together we have:

$$p^i \epsilon_{inu}(a^n_j p'^j)(b^u_k p''^k) = 0.$$

After rearranging the symbol positions we obtain:

$$p^i p'^j p''^k (\epsilon_{inu} a^n_j b^u_k) = 0,$$

where the object in parenthesis is the *homography tensor* of $\mathcal{P}^2$, referred to as Htensor:

$$\mathcal{H}_{ijk} = \epsilon_{inu} a^n_j b^u_k, \tag{11.1}$$

whose triple contraction $p^i p'^j p''^k \mathcal{H}_{ijk}$ vanishes on observations p, p', p'' arising from stationary or moving points on the plane π. Each such triplet of matching points provides a linear constraint on the 27 entries of $\mathcal{H}$; thus 26 matching triplets are necessary to solve for $\mathcal{H}$ uniquely (up to scale).

We see from the above that the tensor $\mathcal{H}$ applies to both stationary and moving points coming from the planar surface π. The possibility of working with stationary and moving elements was first introduced in [1, 2], where it was shown that if a moving point along a general (in 3D) straight path is observed in five views, and the camera projection matrices are *known*, then it is possible to set up a linear system for estimating the 3D line. With the Htensor $\mathcal{H}$, on the other hand, we have no knowledge of the camera projection matrices, but conversely we require that the straight paths the points are taking should all be coplanar. This makes it possible to work with three views instead of five and not require prior information on camera positions. We will address the following issues:

- What are the minimal point configurations that allow a unique solution for $\mathcal{H}$? If all points are moving, then 26 of them are needed, and we will address the issue of the necessary point-set configuration. If some of the points are *known* to be stationary, how many constraints (i.e., moving points) are minimally necessary for a unique solution? If some of the points are stationary (without the system being told about it), what would be the minimal number of moving points required for a unique solution?

- Contraction properties of $\mathcal{H}$ and the manners in which $\mathcal{H}$ acts as a mapping.
- How to recover the component collineations A, B from $\mathcal{H}$.

11.2.1 Recovering Htensor from Image Measurements

The measurements available for recovering $\mathcal{H}$ are triplets of matching points p, p', p'' across the three views and *prior* information whether a triplet arises from a moving or stationary point. Assuming first that all measurements are induced by moving points, a triplet of matching points contributes one linear constraint $p^i p'^j p''^k \mathcal{H}_{ijk} = 0$ on the elements of $\mathcal{H}$. Therefore 26 triplets are necessary for a unique[3] solution. The 26 points should be distributed on the plane π in such a way that they cover at least 4 lines in general position, such that no more than 8 points are on the first line, no more than 7 points on the second line, no more than 6 on the third line and no more than 5 on the fourth line. This distribution guarantees a unique solution for $\mathcal{H}$. Of course, more than 26 points are allowed, where in that case a least-squares approximation is recovered.

Theorem 1. *A minimal configuration of 26 matching triplets arising from moving points on π is necessary for a unique recovery of $\mathcal{H}$ provided that the distribution of the points is such that the motion trajectories cover at least 4 lines in general position on π and that no more than 8 of the points lie on the first trajectory, no more than 7 on the second trajectory, no more than 6 on the third and no more than 5 points lie on the fourth line trajectory.*

Proof. Consider a line L_1 on the plane π. Let the projections of L onto the three views be denoted by q_1, s_1, r_1 (Fig. 11.2). Since each line is determined by two points, we can have at most $2^3 = 8$ linearly independent constraints of the form $p^i p'^j p''^k \mathcal{H}_{ijk} = 0$, where the points p, p', p'' are coincident with the lines q_1, s_1, r_1, respectively. Consider a second line $L_2 \in \pi$ projecting onto lines q_2, s_2, r_2. Since each of the image lines is spanned by two points, choose one of those points to be the projection of $L_1 \cap L_2$ denoted by p, p', p''. Among the eight choices of choosing three points from the three pairs of points, the choice p, p', p'' is already covered by the span of the eight constraints induced by L_1 thus we are left with seven linearly independent constraints in $\mathcal{H}$. This argument continues by induction over additional lines L_i each inducing one less constraint than the one before it. The process ends with 4 lines inducing $8 + 7 + 6 + 5 = 26$ linearly independent constraints.□

Next, we consider the contribution of stationary points to the system of linear equations for $\mathcal{H}$. A stationary point, *known* as such (referred to as *labeled*), contributes nine linear constraints of rank seven, as follows: let p, p', p'' be a triplet of matching points arising from a known stationary point on π.

---

[3] The dimension of the $GL(V)$ module $\text{Span}\{p \otimes p' \otimes p'' \in V^{\otimes 3} : \dim \text{Span}\{p, p', p''\} = 2\}$ is 26 (see Sect. 11.4).

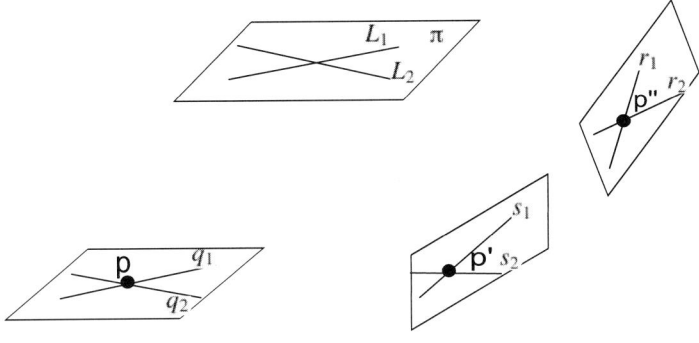

Fig. 11.2. A straight line path L_1 can induce at most eight independent linear constraints, as the projections q_1, s_1, r_1 in the three views are determined by two points each. A second straight-line path L_2 can contribute at most seven independent constraints since the constraint $p^i p'^j p''^k \mathcal{H}_{ijk} = 0$ induced by the projection of the intersection $L_1 \cap L_2$ onto p, p', p'' is spanned by the eight constraints from L_1

The rank of the matrix $[p, Ap', Bp'']$ is 1, which in turn translates to the three sets of constraints: $p \times Ap' = 0, p \times Bp'' = 0$ and $Ap' \times Bp'' = 0$. In tensor form, the contractions $p^i p'^j \mathcal{H}_{ijk}, p^i p''^k \mathcal{H}_{ijk}$ and $p'^j p''^k \mathcal{H}_{ijk}$ are null vectors. The nine constraints are explicitly written below (allow the vector e to vary over the standard basis $(1,0,0), (0,1,0)$ and $(0,0,1)$):

$$p^i p'^j e^k \mathcal{H}_{ijk} = 0 \quad \forall e, \qquad (11.2)$$
$$p^i e^j p''^k \mathcal{H}_{ijk} = 0 \quad \forall e,$$
$$e^i p'^j p''^k \mathcal{H}_{ijk} = 0 \quad \forall e.$$

Note that the constraint $p^i p'^j p''^k \mathcal{H}_{ijk} = 0$ is in the span of the three sets of constraints thus making a total of 7 linearly independent constraints (a system of nine linear equations of rank 7). Therefore we arrive at the conclusion:

Proposition 1. *The matching triplets induced by four labeled stationary points in general position on π provide a unique solution for $\mathcal{H}$.*

We consider next the contribution of *unlabeled* stationary points. A stationary point can provide nine constraints (of rank 7) provided it is known to be stationary otherwise it provides only a single constraint. Consider the case where all the measurements arise from unlabeled stationary points. It is easy to see that the rank of the estimation matrix for $\mathcal{H}$ is at most 10 (compared to 26 when moving points are used). Each row of the estimation matrix for $\mathcal{H}$ is some "constraint tensor" G^{ijk} such that $G^{ijk} \mathcal{H}_{ijk} = 0$. It is sufficient to prove this statement for the case where $A = B = I$ (the identity matrix) because all other cases are transformed into this one by local change of coordinates. In the case $A = B = I$, G^{ijk} is a symmetric tensor, i.e., remains the same under permutation of indices. Hence G^{ijk} contains only ten different groups of indices

$$111, 222, 333, 112, 113, 221, 223, 331, 332, 123$$

up to permutations. Generally speaking, the m-fold symmetric powers $\text{Sym}^m V$ of an n-dimensional vector space V is a vector space of dimension $\binom{n+m-1}{m}$ (substitute $n = 3, m = 3$ to get 10). We arrive at the following conclusion:

Proposition 2. *In a collection of unlabeled matching triplets, there could be at most ten of which that are induced by stationary points. In other words, there should be at least 16 moving points in an input collection of unlabeled points for a unique linear solution for $\mathcal{H}$.*

Finally, we consider the situation of *mixed* labeled and unlabeled triplets. Consider the case where $x \leq 4$ of the triplets are labeled as arising from stationary points. We saw above that a labeled stationary point is equivalent to seven constraints; however, some of those constraints may be already included in the span of the unlabeled stationary points. The theorem below addresses the question of how many matching triplets arising from moving points are necessary, given that $x \leq 4$ matching triplets are labeled as stationary. Clearly, when $x = 4$ there is no need for further measurements, but when $x < 4$ we obtain the following result:

Theorem 2. *In a situation of matching triplets arising from a mixture of stationary and moving points, let $x \leq 4$ be the number of matching triplets that are known a priori to arise from stationary points. To obtain a unique linear solution for $\mathcal{H}$, the minimal number of matching triplets arising from moving points is $16 - 4x$, and at most $10 - 3x$ can be (unlabeled) stationary points.*

Proof: Each row of the estimation matrix for $\mathcal{H}$ is some "constraint tensor" G^{ijk} such that $G^{ijk} H_{ijk} = 0$. It is sufficient to prove this statement for the case where $A = B = I$ (the identity matrix) because all other cases are transformed into this one by local change of coordinates. Therefore, a stationary point induces a symmetric tensor $G^{ijk} = p^i p^j p^k$. The case $x = 0$ was discussed above with the conclusion that a minimum of 16 moving points is required.

Consider the case $x = 1$, i.e., one of the matching triplets contributed nine constraints of rank 7:

$$p^i p'^j e_1^k H_{ijk} = 0, \; p^i e_1^j p''^k H_{ijk} = 0, \; e_1^i p'^j p''^k H_{ijk} = 0,$$
$$p^i p'^j e_2^k H_{ijk} = 0, \; p^i e_2^j p''^k H_{ijk} = 0, \; e_2^i p'^j p''^k H_{ijk} = 0,$$
$$p^i p'^j e_3^k H_{ijk} = 0, \; p^i e_3^j p''^k H_{ijk} = 0, \; e_3^i p'^j p''^k H_{ijk} = 0,$$

where e_1, e_2, e_3 are the standard basis $(1, 0, 0), (0, 1, 0), (0, 0, 1)$. Add the three constraints in the first row:

$$E^{ijk} = p^i p^j e_1^k + p^i e_1^j p^k + e_1^k p^j p^k.$$

Then, E^{ijk} is a symmetric tensor and thus is spanned by the 10-dimensional subspace of the unlabeled stationary points. Likewise, the constraint tensors resulting from adding the constraint of the second and third row above are also symmetric. Taken together, three out of the seven constraints contributed by a labeled stationary point are already accounted for by the space of unlabeled stationary points. Therefore, each labeled stationary point adds only 4 linearly independent constraints.□

11.2.2 Contraction Properties of $\mathcal{H}$ and Recovery of A, B

We turn our attention next to single and double contractions of the Htensor: what can be extracted from them and what is their geometric significance? Those contractions hold the key for decoupling the collineations A, B from $\mathcal{H}$.

The double contractions perform mapping operations. Consider, for example, $p^i p'^j H_{ijk}$, which by the index arrangements must be a contravariant vector (a line in $\mathcal{P}^2$) denoted by l''. Since the remaining index is k, l'' is a line in view 3. Consider the line $L \in \pi$ defined by the projection p, p' in views 1 and 2. Since $p^i p'^j p''^k H_{ijk} = 0$ for all points p'' in view 3 that are the projections from L, we conclude that l'' is the projection of L onto view 3.

The single contractions produce matrices that form the key for decoupling the collineations A, B from $\mathcal{H}$. Consider, for example, $\delta^k H_{ijk}$ for some contravariant vector (a point in view 3) to δ. The result is a matrix E with index structure suggesting it maps points to lines (a correlation matrix) and between views 1 and 2. By substitution in the definition of $\mathcal{H}$ we obtain:

$$\delta^k H_{ijk} = \epsilon_{inu} a_j^n (b_k^u \delta^k) = [B\delta]_\times A.$$

Let $E = [B\delta]_\times A$, and note that the point $\mu = B\delta$ is the matching point to δ in view 1, i.e., it is the projection onto view 1 of the point defined by the intersection of the plane π with the optical ray associated with δ (Fig. 11.3). The matching points to δ, the points $\mu = B\delta$ and $\eta = A^{-1}B\delta$, can be recovered directly from E since:

$$E^\top \mu = -A^\top [\mu]_\times \mu = 0,$$
$$E\eta = [\mu]_\times A\eta \cong [\mu]_\times \mu = 0.$$

The matrix E forms a point-to-line mapping from view 2 to view 1 as follows. Consider any point p' in view 2, then $Ep' = p'^j \delta^k H_{ijk}$ is the projection of the line in π, defined by the optical rays associated with δ and p', onto view 1. Therefore, any point p coincident with the projected line satisfies $p^\top Ep' = 0$. We conclude that the bilinear form $p^\top Ep' = 0$ is satisfied for all pairs of p, p' that are on matching lines through the fixed points μ, η (Fig. 11.3).

Finally, the collineation A can be recovered from single contractions by the fact that $A^\top E$ is a skew-symmetric matrix:

$$A^\top E + E^\top A = A^\top [\mu]_\times A - A^\top [\mu]_\times A = 0,$$

which provides six linearly independent equations on the entries of A. By taking δ to range over the standard basis $(1,0,0),(0,1,0),(1,0,0)$ we obtain three slices of $\mathcal{H}$ denoted by E_1, E_2, E_3, each producing six linear equations on A. Taken together A can be recovered linearly from the slices of $\mathcal{H}$. Likewise, B can be recovered from the slices $\delta^j \mathcal{H}_{ijk}$ in the same manner, as can the collineation $A^{-1}B$ (between views 2 and 3) from the slices $\delta^i H_{ijk}$. These findings are summarized in the theorem below:

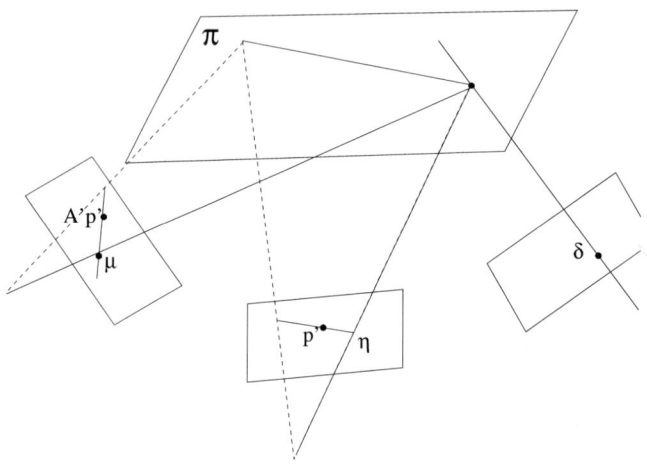

Fig. 11.3. A single contraction, say $\delta^k H_{ijk}$, is a mapping E between views 1 and 2 from points to concurrent lines. The null spaces of E and $E^\top$ are the matching points μ, η of δ in views 1 and 2. The image points p' are mapped by E to the lines $Ap' \times \mu$, and the image points p are mapped by $E^\top$ to the lines $A^{-1}p \times \eta$ in view 2. The bilinear relation $p^\top E p' = 0$ is satisfied for all pairs of p, p' on matching lines through the fixed points μ, η

Theorem 3. *Each of the contractions*

$$\delta^k \mathcal{H}_{ijk} \qquad (11.3)$$
$$\delta^j \mathcal{H}_{ijk} \qquad (11.4)$$
$$\delta^i \mathcal{H}_{ijk} \qquad (11.5)$$

represents a point-to-line (correlation) mapping between views $(1,2), (1,3)$ *and* $(2,3)$, *respectively. By setting δ to be $(1,0,0),(0,1,0)$ or $(0,0,1)$ we obtain three different slicings of the tensor: denote the slices of $\delta^i \mathcal{H}_{ijk}$ by the matrices G_1, G_2, G_3, the slices of $\delta^j \mathcal{H}_{ijk}$ by the matrices W_1, W_2, W_3, and the slices of $\delta^k \mathcal{H}_{ijk}$ by the matrices E_1, E_2, E_3. Then these slices provide sufficient (and overdetermined) linear constraints for the constituent homography matrices A, B and for $C = A^{-1}B$:*

$$CG_i^\top + G_i C^\top = 0, \tag{11.6}$$
$$BW_i^\top + W_i B^\top = 0, \tag{11.7}$$
$$AE_i^\top + E_i A^\top = 0, \tag{11.8}$$

for $i = 1, 2, 3$.

In summary, the homography tensor in $\mathcal{P}^2$ applies to both cases: optical rays meet at a single point (matching points with respect to a stationary point) and optical rays meet at a line on π (matching points with respect to a moving point). In the case where no distinction can be made to the source of a matching triplet p, p', p'' (stationary or moving) then we saw that in a set of at least 26 such matching triplets, 16 of them *must* arise from moving points. In case a number $x \leq 4$ of these triplets are *known a priori* to arise from stationary points, then $16 - 4x$ must arise from moving points. Once $\mathcal{H}$ is recovered from image measurements, it forms a mapping of both moving and stationary points and in particular can be used to *distinguish* between moving and stationary points (a triplet p, p', p'' arising from a stationary point is mapped to null vectors $p^i p'^j H_{ijk}$, $p^i p''^k H_{ijk}$ and $p'^j p''^k H_{ijk}$). The Htensor can be useful in practice to handle situations rich in dynamic motion seen from a monocular sequence; some experiments are shown in Sect. 11.6.

We will next describe the homography tensors of $\mathcal{P}^3$ where points lie in the 3D projective space, the collineations which are responsible for the coordinate changes are 4×4 matrices and the points are allowed to move along straight lines or planar subspaces while coordinate changes take place.

11.3 Homography Tensors of $\mathcal{P}^3$

We consider stepping up one dimension. Namely, the point configuration lies in $\mathcal{P}^3$, the collineations are 4×4 matrices and the dimension in which the points are allowed to move while the global collineations take place are $k = 1, 2, 3$, where $k = 1$ stands for stationary points, $k = 2$ stands for motion along a straight line path and $k = 3$ stands for motion along a planar subspace. We focus on the constraint of straight line motion and stationary points $k = 2, 1$, which induce a $4 \times 4 \times 4$ homography tensor. The situation of planar dynamic motion $k = 3$ induces a 4^4 tensor, which we will not consider in detail here and leave for the discussion on general dynamic alignment in Sect. 11.4.

Let X be some stationary point in 3D space with coordinate vector P. Let P' be the coordinate representation of the point X at some other time instant (i.e., the measurement sensor has changed its viewing position) and let P'' be the coordinate representation of X at a third time instant. Let A, B be the collineations mapping the second and third coordinate representations back to the first representation, i.e., $P \cong AP'$ and $P \cong BP''$.

If the point X happens to *move* along some straight-line path during the change of coordinate systems, then P, AP', BP'' do not coincide, but they form a rank-2 matrix (Fig. 11.4):

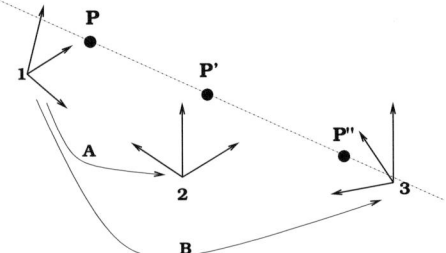

Fig. 11.4. The points P, P' and P'' are measured at three time instants from different viewing positions of the sensor, i.e., each point is given in a different coordinate system. While the measuring device changes position, the physical point in space moves along a straight-line path. In other words, the rank of the 4×3 matrix $[P, AP', BP'']$ is 2 for a moving point and 1 for a stationary point. The 4×4 matrices A, B are responsible for the change of coordinate system back to the starting position

$$\operatorname{rank} \begin{pmatrix} | & | & | \\ P & AP' & BP'' \\ | & | & | \end{pmatrix} = 2.$$

And for every column vector V we have

$$\det \begin{pmatrix} | & | & | & | \\ P & AP' & BP'' & V \\ | & | & | & | \end{pmatrix} = 0. \tag{11.9}$$

Note that because V is spanned by a basis of size 4, we can obtain at most four linearly independent constraints on some object consisting of A, B from a triplet of matching points P, P', P''. Note also that the null vector of a 4×3 matrix can be represented by the 3×3 determinant expansion. For example, let X, Y, Z be three column vectors in a 4×3 matrix, then the vector $W = (w_1, ..., w_4)$ representing the plane defined by the points X, Y, Z is

$$w_1 = \det \begin{pmatrix} x_2 & y_2 & z_2 \\ x_3 & y_3 & z_3 \\ x_4 & y_4 & z_4 \end{pmatrix}, \quad w_2 = -\det \begin{pmatrix} x_1 & y_1 & z_1 \\ x_3 & y_3 & z_3 \\ x_4 & y_4 & z_4 \end{pmatrix},$$

$$w_3 = \det \begin{pmatrix} x_1 & y_1 & z_1 \\ x_2 & y_2 & z_2 \\ x_4 & y_4 & z_4 \end{pmatrix}, \quad w_4 = -\det \begin{pmatrix} x_1 & y_1 & z_1 \\ x_2 & y_2 & z_2 \\ x_3 & y_3 & z_3 \end{pmatrix}.$$

We can write the relationship between W and X, Y, Z as a tensor operation

$$w_i = \epsilon_{ijkl} x^j y^k z^l,$$

where the entries of ϵ consist of $+1, -1, 0$ in the appropriate places. We will refer to ϵ as the "cross product" tensor. Note that the determinant of a 4×4 matrix whose columns consist of $[X, Y, Z, T]$ can be compactly written as

$$t^i x^j y^k z^l \epsilon_{ijkl}.$$

Using the cross product tensor we can write the constraint (11.9)

$$\begin{aligned} 0 &= \det \begin{pmatrix} | & | & | & | \\ P & AP' & BP'' & V \\ | & | & | & | \end{pmatrix}, \\ &= P^i(\epsilon_{ilmu}(a_j^l P'^j)(b_k^m P''^k)v^u), \\ &= P^i P'^j P''^k (\epsilon_{ilmu} a_j^l b_k^m v^u). \end{aligned}$$

Note that the tensor form allows us to separate the measurements P, P', P'' from the unknowns A, B (and vector V), and we denote the expression in parentheses as

$$\mathcal{J}_{ijk} = \epsilon_{ilmu} a_j^l b_k^m v^u \tag{11.10}$$

as the the homography tensor of $\mathcal{P}^3$. Note that for every choice of the vector V we get an Htensor. As previously mentioned, since V is spanned by a basis of dimension 4, there are at most four such tensors; each tensor is defined by the constraints

$$P^i P'^j P''^k \mathcal{J}_{ijk} = 0.$$

These are linear constraints on the 64 elements of the Htensor. Since there are four Htensors compatible with the observations, the linear system of equations for solving for $\mathcal{J}$ from the matching triplets P, P', P'' has a four-dimensional null space. The vectors of the null space are spanned by the Htensors. In practical terms, given $N \geq 60$ matching triplets P, P', P'', each triplet contributes one linear equation $P^i P'^j P''^k \mathcal{J}_{ijk} = 0$ for the 64 entries of $\mathcal{J}$. The eigenvectors associated with the four smallest eigenvalues of the estimation matrix are *the Htensors of the dynamic 3D-to-3D alignment problem*. We summarize this in the following theorem:

Theorem 4 (Htensors in $\mathcal{P}^2$). *Each matching triplet P, P', P'' arising from a dynamic point contributes one linear equation $P^i P'^j P''^k \mathcal{J}_{ijk} = 0$ to a $4 \times 4 \times 4$ tensor $\mathcal{J}$. Any $N \geq 60$ matching triplets in general position provide an estimation matrix for $\mathcal{J}_{ijk}$ with a four-dimensional null space. The 60 points should be distributed along at least 10 lines, 5 of which can hold up to 8 dynamic points, and the remaining 5 up to 4 dynamic points.*

In the remainder of this section we discuss tensor slices and the extraction of the constituent collineations A, B from the four Htensors. We also discuss the use of Htensors for direct mapping between coordinate systems (without extracting A, B along the way) and the use of Htensors to distinguish between dynamic and stationary points. Finally, we discuss the relationship between the number of stationary and dynamic points for estimating the Htensors in unsegmented and segmented configurations.

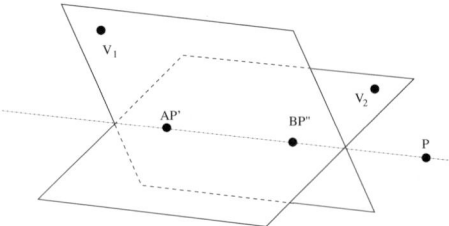

Fig. 11.5. The points AP', BP'' and V define a plane. AP', BP'' and V' define another plane. The line of intersection of these planes contains P

11.3.1 Tensor Slices and the Extraction of the Collineations A, B

The role of $\mathcal{J}$ is symmetric with respect to the position of the points P, P', P'', which is true for every purely covariant or contravariant tensor, unlike the mixed covariant-contravariant tensor. It is therefore sufficient to investigate $P'^j P''^k \mathcal{J}_{ijk}$ as one of the tensor double-contractions; the others, $P^i P''^k \mathcal{J}_{ijk}$ and $P^i P'^j \mathcal{J}_{ijk}$, follow by symmetry.

Consider any Htensor with its associated vector V. Recall that from observations we can recover four Htensors that span the null space of the measurement matrix each Htensor has a different vector V associated with it. We describe next how to recover the vector V, referred to as the "principal point" of the tensor, from the Htensor.

Consider the plane π defined by $\pi_i = P'^j P''^k \mathcal{J}_{ijk}$ that contains the three points V, AP' and BP'':

$$\pi_i = P'^j P''^k \mathcal{J}_{ijk} = \epsilon_{ilmu}(a_j^l P'^j)(b_k^m P''^k)v^u,$$

which by definition of the cross-product tensor provides the plane associated with the three points acted upon by ϵ. By varying P' and P'' we obtain a star of planes all coincident with the point V. As a result, the principal point V of the tensor can be recovered by taking three double slices of the tensor and finding their intersection.

We next recover the line in space coincident with the points AP' and BP''. Consider two Htensors denoted by $\mathcal{J}^1$ and $\mathcal{J}^2$ (recall that we have four Htensors at our disposal). The intersection of the planes $P'^j P''^k \mathcal{J}^1_{ijk}$ and $P'^j P''^k \mathcal{J}^2_{ijk}$ is the line passing through AP' and BP'' (Fig. 11.5).

The collineations A, B can be recovered (linearly) from the matrices resulting from single contractions of the Htensors. A single contraction $H_{ij} = P''^k \mathcal{J}_{ijk}$ is a 4×4 matrix H that maps points to planes. As mentioned above, $P'^j H_{ij} = P'^j P''^k \mathcal{J}_{ijk}$ is the plane passing through V, AP', BP''; thus by varying P' one obtains a pencil of planes coincident with the line through V and BP''. Hence the rank of the matrix H must be 2.

Because HP' is the plane through V, AP', BP'', we have $P'^\top A^\top H P' = 0$ for every choice of P'. Therefore $A^\top H$ is a skew-symmetric matrix and thus provides ten linear constraints for A. By varying P'' and thereby obtaining other H-matrices $P''^k \mathcal{J}_{ijk}$ we can obtain more constraints on A, but this is not sufficient to obtain a unique solution for A. A unique solution requires the H-matrix of at least another Htensor because the principal point must vary as well. Likewise, one can recover B from the contractions $P'^j \mathcal{J}_{ijk}$ by varying P' and taking at least two Htensors.

11.3.2 Direct Mapping

We can use the Htensor to map points between the coordinate frames without the need to extract the collineations A and B. Consider, for example, the direct mapping $P \cong BP''$ between the third and the first coordinate frames. The contraction $\gamma^j P''^k \mathcal{J}_{ijk}$ for some arbitrary vector γ is a plane in 3D containing the points BP'', $A\gamma$ and V (the principal point of $\mathcal{J}$), all represented in the first coordinate frame. By varying γ over the standard basis and taking the four different Htensors (so that V also varies), we get a collection of 16 planes. These planes intersect in the point $P \cong BP''$. It is sufficient to use a subset of these planes (at least three) as long as not all of them are generated using the same Htensor or the same γ.

As a result, the Htensor can play the same role as a collineation (i.e., direct) mapping between coordinate frames. The direct mapping can be used, for example, to distinguish between *stationary* and *dynamic* points. If P is equal to the direct mapping BP'', then the corresponding physical point X is stationary; otherwise (ignoring noise considerations) X is dynamic.

The segmentation of stationary and dynamic points can be achieved in other ways as well. For example, from Eq. (11.9) we know that for a stationary point X with coordinate vectors P, P' and P'' in the three frames, any double contraction vanishes:

$$P^i P'^j \mathcal{J}_{ijk} = P'^j P''^k \mathcal{J}_{ijk} = P^i P''^k \mathcal{J}_{ijk} = 0.$$

Hence a vanishing double contraction (under all three possibilities) indicates a stationary point. In practice, since the double contraction provides only an algebraic (rather than geometric) measure of error, better segmentation results are achieved by measuring the distance between the point P and the direct mapping BP''.

11.3.3 Constraints from Stationary Points

We have seen that a matching triplet P, P' and P'' satisfies the Htensor constraint

$$P^i P'^j P''^k \mathcal{J}_{ijk} = 0,$$

regardless of whether the corresponding physical point X is moving along a straight-line path (dynamic) or is stationary. For a dynamic point, the rank of the 4×3 matrix $[P, AP', BP'']$ is 2, and for a stationary point the rank is

1. In other words, admissible measurements for recovering the Htensors come from dynamic and stationary points alike. The natural question is, how much alike? That is, can all the measurements arise (unknowingly) from stationary points? If not, what is the maximal number number of stationary points after which the contributions of additional stationary points become redundant? These questions are exactly the same as those addressed for $\mathcal{H}$ in the context of $\mathcal{P}^2$.

The contribution of unlabeled stationary points, i.e., recovering $\mathcal{J}$ from constraints $P^i P'^j P''^k \mathcal{J}_{ijk} = 0$ where the triplet P, P', P'' are induced by stationary points only, can fill up a 20-dimensional subspace only (out of 60). Without loss of generality we can assume that $A = B = I$, which in turn makes each constraint $G^{ijk}\mathcal{J}_{ijk} = 0$, where $G^{ijk} = P^i P^j P^k$ is a symmetric tensor (remains the same under permutation of indices). The three-fold symmetric powers $\text{Sym}^3 V$ of a four-dimensional vector space V is $\binom{4+3-1}{3} = 20$. In other words, there are only 20 different groups of indices:

$$111, 222, 333, 444, 112, 113, 114, 221, 223, 224,$$
$$331, 332, 334, 441, 442, 444, 123, 124, 134, 234.$$

This analysis is summarized in the theorem below:

Theorem 5. *The constraints $P^i P'^j P''^k \mathcal{J}_{ijk} = 0$ made solely from stationary points span at most a 20-dimensional space.*

Consequently, in the unsegmented situation when stationary and dynamic points are treated alike, it is not possible to obtain a unique solution from stationary points alone; one needs at least 40 dynamic points in the collection of $N \geq 60$ matching triplets. We consider next the contribution arising from *labeled* stationary points, i.e., how many constraints would a triplet P, P', P'' contribute if it were known that the corresponding physical point X is stationary? In this case, for every $\delta_{4\times 1}$ and for every V, the determinant

$$\det \begin{pmatrix} | & | & | & | \\ P & AP' & B\delta & V \\ | & | & | & | \end{pmatrix}$$

vanishes. Since this is true for every pair of the three points, then for each of the four Htensors we get:

$$\begin{aligned}
P^i P'^j e_1^k \mathcal{J}_{ijk} = 0, \quad & P^i e_1^j P''^k \mathcal{J}_{ijk} = 0, \quad & e_1^i P'^j P''^k \mathcal{J}_{ijk} = 0, \\
P^i P'^j e_2^k \mathcal{J}_{ijk} = 0, \quad & P^i e_2^j P''^k \mathcal{J}_{ijk} = 0, \quad & e_2^i P'^j P''^k \mathcal{J}_{ijk} = 0, \\
P^i P'^j e_3^k \mathcal{J}_{ijk} = 0, \quad & P^i e_3^j P''^k \mathcal{J}_{ijk} = 0, \quad & e_3^i P'^j P''^k \mathcal{J}_{ijk} = 0, \\
P^i P'^j e_4^k \mathcal{J}_{ijk} = 0, \quad & P^i e_4^j P''^k \mathcal{J}_{ijk} = 0, \quad & e_4^i P'^j P''^k \mathcal{J}_{ijk} = 0,
\end{aligned} \quad (11.11)$$

where e_1, e_2, e_3, e_4 are the standard basis $(1, 0, 0, 0), (0, 1, 0, 0), (0, 0, 1, 0), (0, 0, 0, 1)$. Note that the constraint

$$P^i P'^j P''^k \mathcal{J}_{ijk} = 0$$

can be spanned by each row separately, hence the rank of the above system is at most 10. We thus arrive at the conclusion:

Theorem 6. *A labeled stationary point can provide at most ten linearly independent constraints for the solution of $\mathcal{J}$.*

These constraints came from one stationary point, but how many of them are spanned by the subspace of constraints obtained from unlabeled stationary points? This question is answered next:

Theorem 7. *Out of the ten linearly independent constraints arising from a labeled stationary point, four lie in the rank-20 subspace spanned by unlabeled stationary points, and six lie in the subspace spanned only by dynamic points.*

Proof. Again, it is sufficient to prove this theorem for the case where $A = B = I$. In this case a stationary point satisfies $P \cong P' \cong P''$.

We look at the 12 constraints of rank 10 described in Eq. (11.11). Adding the three constraints in the first row gives

$$G^{ijk} = P^i P^j e_1^k + P^i e_1^j P^k + e_1^k P^j P^k,$$

which is a symmetric tensor and thus is spanned by the 20-dimensional subspace of the unlabeled stationary points. Similarly, the constraint tensors resulting from adding the other three rows are also symmetric. One can verify that except for those four constraints (and the ones they span) there are no other symmetric constraints.

Taken together, four out of the ten constraints contributed by a labeled stationary point lie in the subspace of unlabeled stationary points, and six constraints lie in the subspace of dimension 40 spanned by dynamic points. □

As a corollary, we can deduce that seven labeled stationary points are necessary to fill up the 60-dimensional subspace necessary for a solution for $\mathcal{J}$. Since the ten constraints contributed by a labeled stationary point include four that are spanned by the subspace of unlabeled stationary points, then five labeled stationary points will fill up the 20-dimensional subspace of unlabeled stationary point. Each additional labeled stationary point can contribute at most six linearly independent constraints.

Corollary 1. *A minimum of seven labeled stationary points are necessary for a unique (up to a 4-dimensional solution space) solution for $\mathcal{J}$.*

Note that we used the term "unique" for the solution of $\mathcal{J}$ (despite the fact that $\mathcal{J}$ can be recovered only up to a four-fold linear subspace) because the collineations A, B can be recovered uniquely from the four-dimensional $\mathcal{J}$ tensor space.

Finally, we consider the situation of a *mixed* labeled and unlabeled triplets. Consider the case where $x \leq 7$ of the triplets are labeled as arising from stationary points. The corollary below addresses the question of how many matching triplets arising from moving points are necessary given that $x \leq 7$ matching triplets are labeled as stationary. Clearly, when $x = 7$ there is no need for further measurements, but when $x < 7$ we obtain the following result:

Corollary 2. *In a situation of matching triplets arising from a mixture of stationary and moving points, let $x \leq 7$ be the number of matching triplets that are known a priori to arise from stationary points. To obtain a unique linear solution for $\mathcal{J}$ (up to a four-dimensional solution space), the minimal number of unlabeled matching triplets required is:*

$$\begin{cases} 60 - 10x & x \leq 5 \\ 4 & x = 6 \\ 0 & x = 7 \end{cases},$$

out of which $40 - 6x$, $x < 7$, should be dynamic and at most $20 - 4x$, $x \leq 5$, could be unlabeled stationary points.

11.4 Homography Tensors for $\mathcal{P}^n$

The tensors $\mathcal{H}$ and $\mathcal{J}$ we have encountered so far belong to the general class of tensors defined as follows. Let $V(n, m, k)$, where $n > k$, be a $GL(V)$ module defined by the set of all tensors $v_1 \otimes \cdots \otimes v_m \in V^{\otimes m}$, where $v_i \in V$ are n-dimensional vectors and $\dim \mathrm{Span}\{v_1, \ldots, v_m\} \leq k$. What is the structure and dimension of $V(n, m, k)$? In the terminology of the previous sections, we considered the space $\mathcal{P}^{n-1}$, the number of views to be m and the motion of the dynamic points are limited to a k-dimensional subspace. Thus we have encountered $V(3, 3, 2)$ and $V(3, 3, 1)$, which stand for dynamic and stationary points in $\mathcal{P}^2$, and encountered $V(4, 3, 2)$ and $V(4, 3, 1)$, which stand for dynamic motion along straight lines and stationary points in $\mathcal{P}^3$. To generalize the construction of homography tensors to $\mathcal{P}^n$ we need to find out:

1. The dimension of $V(n, m, k)$. Namely, given linear constraints generated by a multilinear form over the m-fold Htensor from known observations of m points moving inside k-dimensional subspaces, what would be the maximal space those measurements could fill? For example, for $V(3, 3, 2)$ the maximal space is 26, which means we can obtain a unique solution for the $3 \times 3 \times 3$ Htensor, but for $V(4, 3, 2)$ the maximal dimension is 60, which means we can pinpoint the $4 \times 4 \times 4$ tensor up to a four-fold linear space. This will be the focus of this section.
2. Is the dimension of $V(n, m, k)$ sufficient for uniquely recovering the $m - 1$ individual collineations? How can we recover those collineations using the tensor slices? For example, we saw that the two collineations A, B can be recovered uniquely from $\mathcal{H}$ and also uniquely from $\mathcal{J}$, even though $\mathcal{J}$ cannot be uniquely recovered from the measurements. In other words, we

recovered A, B from the four-dimensional linear space of solutions for $\mathcal{J}$. This generalization is an open area for future research.
3. What are the constraints contributed from a labeled $k' < k$-dimensional point? For example, we saw that the stationary points $k' = 1$ for $V(3, 3, 2)$ contribute seven independent constraints, and ten independent constraints for $V(4, 3, 2)$. This is left open for future research.
4. What would be the dimension of the space covered by *mixed* observations, i.e., from labeled $k' < k$, and unlabeled points from k and $k' < k$? For example, we saw that the labeled stationary $k' = 1$ points provide only four new constraints, as three of the seven provided by labeled stationary points constraints are included in the space of dimension $V(3, 3, 1)$ covered by unlabeled stationary points. This topic is left for future research.

We focus below on item 1 which is the dimension of $V(n, m, k)$. The simple cases are $\dim V(n, m, 1) = \binom{n+m-1}{m}$, because $V(n, m, 1) = \mathrm{Sym}^m V$, and $\dim V(n, m, m-1) = n^m - \binom{n}{m}$, which arises by naive introspection. For example, $\dim V(3, 3, 2) = 26$, which means that the Htensor requires 26 matching triplets across three views of a dynamic planar configuration for a unique solution $(27 - 26 = 1)$, whereas if all the measurements arise from "stationary" points then $\dim V(3, 3, 1) = 10$. Likewise, $\dim V(4, 3, 2) = 60$, which means the Jtensors are spanned by 4 tensors $(64 - 60 = 4)$, and 60 matching triplets of 3D points across changes of coordinate systems of a dynamic 3D configuration are required for a solution, and if all the measurements arise from stationary points then $\dim V(4, 3, 1) = 20$.

We show next that the question of structure and dimension of the $GL(V)$ module $V(n, m, k)$ can be generally solved by counting irreducibles using the tools of *representation theory* [15]. The notation and a brief primer on representation theory can be found in Sect. 11.8 The central result of this section is proving that

$$V(n, m, k) = \bigoplus_{\lambda_{k+1}=0} S_\lambda(V)^{\oplus f_\lambda} ,$$

and, in particular,

$$\dim V(n, m, k) = \sum_{\lambda_{k+1}=0} f_\lambda \dim \mathcal{S}_\lambda(V) ,$$

where λ is a partition of m, the direct sum is over all partitions with at most k parts, f_λ is the number of standard tableaux on λ and $S_\lambda(V)$ is Schur's module.

The mathematics of representation theory may be somewhat unfamiliar as it is not yet in use in computer vision literature, yet it uncovers some beautiful connections between the recent new efforts of extending the envelope of structure from motion (SFM) theory and applications to nonrigid scenes and the representations of finite groups and of $GL(V)$ on the m-fold tensor product.

11.5 The Structure of $V(n,m,k)$

We would like to prove the following claim:

Claim.
$$V(n,m,k) = \bigoplus_{\lambda_{k+1}=0} S_\lambda(V)^{\oplus f_\lambda} .$$

In particular,
$$\dim V(n,m,k) = \sum_{\lambda_{k+1}=0} f_\lambda s_\lambda .$$

Proof. Suppose $\lambda \vdash m$ and $\lambda_{k+1} = 0$. Let t be the tableau given by $t(i,j) = \sum_{l=1}^{i-1} \lambda_l + j$. Noting that $V(n,r,1) = \operatorname{Sym}^r V$, it follows that
$$V^{\otimes m} \cdot a_t = \operatorname{Sym}^{\lambda_1} V \otimes \cdots \otimes \operatorname{Sym}^{\lambda_k} V,$$
$$= V(n,\lambda_1,1) \otimes \cdots \otimes V(n,\lambda_k,1) \subset V(n,m,k) .$$

Therefore,
$$S_t(V) = V^{\otimes m} \cdot a_T \cdot b_T \subset V(n,m,k) \cdot b_T \subset V(n,m,k) ,$$

hence,
$$\bigoplus_{\lambda_{k+1}=0} S_\lambda(V)^{\oplus f_\lambda} \subset V(n,m,k).$$

To show the other direction let $(\cdot,\cdot)$ be a Hermitian form on V, and let the induced form on $V^{\otimes m}$ be given by
$$(u_1 \otimes \cdots \otimes u_m, v_1 \otimes \cdots \otimes v_m) = \prod_{i=1}^m (u_i, v_i) .$$

Note that
$$(u_1 \wedge \cdots \wedge u_m, v_1 \otimes \cdots \otimes v_m) = \frac{1}{m!}(u_1 \wedge \cdots \wedge u_m, v_1 \wedge \cdots \wedge v_m),$$
$$= \frac{1}{m!} \det[(u_i, v_j)]_{i,j=1}^m .$$

Let $\lambda \vdash m$ with $\lambda_{k+1} \neq 0$, then the conjugate partition $\mu = (\mu_1 \geq \mu_2 \geq \ldots \geq \mu_t)$ satisfies $\mu_1 \geq k+1$. Let $l_j = \sum_{r=1}^j \mu_r$, and let t be the tableau given by $t(i,j) = l_{j-1} + i$. Then
$$S_t(V) = V^{\otimes m} \cdot a_t \cdot b_t \subset V^{\otimes m} \cdot b_t,$$
$$= \wedge^{\mu_1} V \otimes \cdots \otimes \wedge^{\mu_l} V .$$

Suppose now that $v_1, \ldots, v_m \in V^{\otimes m}$ satisfy $\dim \operatorname{Span}\{v_1, \ldots, v_m\} \leq k$. Then $v_1 \wedge \cdots \wedge v_{\mu_1} = 0$; therefore for any $u_1, \ldots, u_m \in V$

$$((u_1 \otimes \cdots \otimes u_m) \cdot b_T, v_1 \otimes \cdots \otimes v_m) = \prod_{r=1}^{l} \frac{1}{\mu_r!} \left(\bigwedge_{i=l_{r-1}+1}^{l_r} u_i, \bigwedge_{i=l_{r-1}+1}^{l_r} v_i \right) = 0.$$

It follows that $V(n,m,k)$ is orthogonal to
$$\bigoplus_{\lambda_{k+1} \neq 0} S_\lambda(V)^{\oplus f_\lambda}.$$

Hence,
$$\dim V(n,m,k) \leq \dim \bigoplus_{\lambda_{k+1}=0} S_\lambda(V)^{\oplus f_\lambda}.$$

☐

This claim can be used to give explicit formulas for $\dim V(n,m,k)$ when either k or $m-k$ are small. In the latter case we write
$$\dim V(n,m,k) = n^m - \sum_{\lambda_{k+1} \neq 0} f_\lambda d_\lambda(n),$$

and note that the partitions of m with $\lambda_{k+1} \neq 0$ correspond to all partitions of all numbers up to $m-k-1$.

Examples. To calculate $\dim V(n,m,m-1)$ note that only $\lambda = (1^m)$ must be excluded; thus
$$f_{(1^m)} = 1, \quad d_{(1^m)}(n) = \binom{n}{m},$$

hence,
$$\dim V(n,m,m-1) = n^m - \binom{n}{m}.$$

To calculate $\dim V(n,m,m-2)$ we must exclude, in addition to the above, the partition $(2, 1^{m-2})$; thus
$$f_{(2,1^{m-2})} = m-1, \quad d_{(2,1^{m-2})}(n) = (m-1)\binom{n+1}{m},$$

hence,
$$\dim V(n,m,m-2) = n^m - \left[\binom{n}{m} + (m-1)^2 \binom{n+1}{m} \right].$$

To calculate $\dim V(n,m,m-3)$ we must exclude, in addition to the above, the partitions $(3, 1^{m-3})$ and $(2^2, 1^{m-4})$; thus
$$f_{(3,1^{m-3})} = \binom{m-1}{2}, \quad d_{(3,1^{m-3})}(n) = \binom{m-1}{2}\binom{n+2}{m}$$

$$f_{(2^2,1^{m-4})} = \frac{m(m-3)}{2},$$

$$d_{(2^2,1^{m-4})}(n) = \frac{(m-3)n}{2}\binom{n+1}{m-1}.$$

Hence,

$$\dim V(n,m,m-3) = n^m - \left[\binom{n}{m} + (m-1)^2\binom{n+1}{m}\right.$$
$$\left. + \binom{m-1}{2}^2\binom{n+2}{m} + \frac{m(m-3)^2 n}{4}\binom{n+1}{m-1}\right].$$

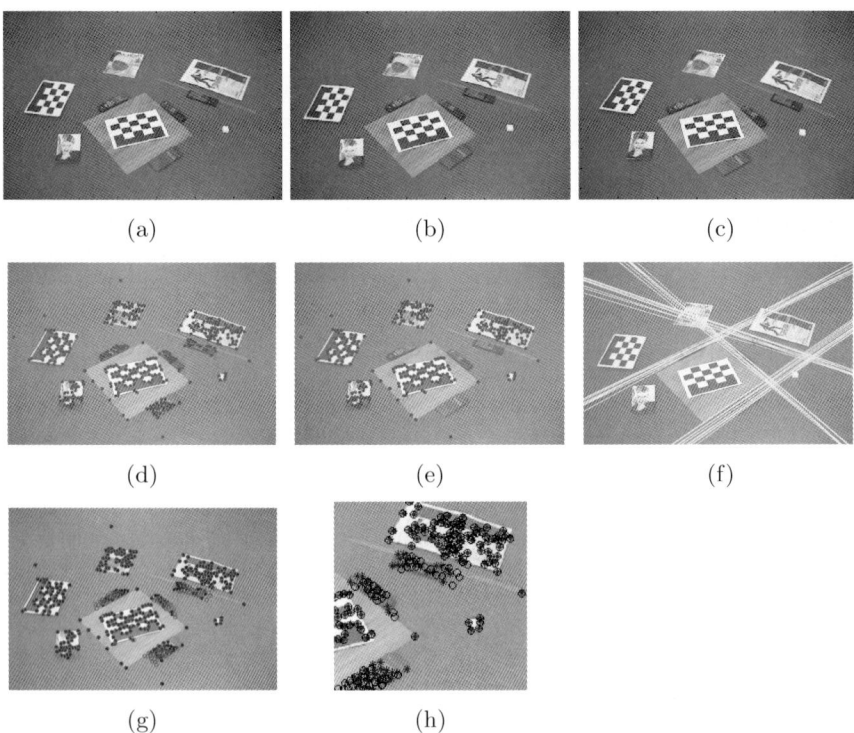

Fig. 11.6. a-c Three views of a planar scene with four remotes moving on straight lines. d The first view with the points that were tracked across the sequence. These points were used for computing the homography tensor $\mathcal{H}$ in a least-squares manner. e Segmentation: the homography tensor was used to choose the stationary points. Only the stationary points are shown. f Trajectory lines: the homography tensor was used to calculate the trajectory lines. In this figure we see the trajectory lines in the *third image*. g Reprojection: Using the homography tensor we reprojected the points in *view 1* to *view 3*. The reprojected points are shown as *circles* and the tracked points as *stars*. h A zoom of the previous image

11.6 Experiments and Applications

We start with an experiment for separating dynamic from stationary points from a planar configuration. The projections of a planar configuration are governed by collineations. The conventional way to separate the moving from the stationary points is to treat the dynamic points as outliers and use robust estimation to recover the collineations [9]. Using homography tensors we can treat the dynamic and stationary points alike and recover the governing Htensor $\mathcal{H}$ instead.

The point configuration is illustrated in Fig. 11.6. The moving points were part of four remote controls that were in motion while the camera changed position from one view to the next. The points were tracked along the three views, without knowledge of what was stationary and what was moving. The triplet of matching points was fed into a least-square estimation for $\mathcal{H}$. We then checked the error of reprojection on the stationary points these were at the subpixel level, as can be seen in Fig. 11.6h and the accuracy of the line trajectory of the moving points. Because the moving points were clustered on only four objects (the remote controls), then the accuracy was measured by "eyeballing" the parallelism of the trajectories of all points within a moving object. The lines are closely parallel, as can be seen in Fig. 11.6f. The Htensor can also be used to segment the scene into stationary and moving points, as shown in Fig. 11.6e.

To illustrate the use of the homography tensor $\mathcal{J}$, consider the problem of 3D reconstruction of an object that extends beyond the field of view of the sensor. For this purpose we can use a stereo rig that contains a texture pattern projector for obtaining matching points on textureless areas of the object. Because the field of view of the cameras does not cover the entire object, the stereo rig must acquire images from multiple viewing positions. Each image provides a 3D patch of the object, and the goal is to "stitch" these patches together by aligning their coordinate systems. In other words, we must recover the relative 3D motion of the rig. The problem is conventional if the texture projection is stationary (i.e., remains in place while the rig changes position), but here the projector moves with the rig. In this domain, the dynamic points are the points arising from the projected texture, and the stationary points arise from texture markings on the object's surface. Hence, if the rig moves in a piecewise straight-line path and the object is polyhedral, Htensor theory is an appropriate tool for aligning the coordinate systems of the 3D patches.

Once the Htensor $\mathcal{J}$ are recovered, one can align the reconstructed patches using two different approaches. The first approach is to align all the patches to one coordinate frame using direct mapping (Sect. 11.3.2) or by recovering the transformations A and B (Sect. 11.3.1). The second approach is to first segment the tracked points into stationary and dynamic points. Then, using only the stationary points, we can recover the collineation between the coordinate frames A and B.

(a) Left view, time 1   (b) Right view, time 1

(c) Left view, time 2   (d) Right view, time 2

(e) Left view, time 3   (f) Right view, time 3

Fig. 11.7. a-f A pair of views from a stereo rig taken at three time instants. The rig is moving with the texture pattern. The scene therefore contains both stationary and dynamic points

We apply the Htensor to the scene with multiple objects shown in Fig. 11.7. Most of the objects are textureless, but there are stationary features throughout the scene. A texture was projected, and 236 features were tracked between the images in each stereo pair and across the three stereo pairs. The feature set contains both stationary and dynamic points, see Fig. 11.8. It can be seen from the last row of Fig. 11.9 that the correct motion was captured because the stationary points were stabilized, whereas the dynamic points are moving on straight-line paths. The last image shows the segmented stationary points. Note that in our framework we use only projective reconstruction, and we do not use any calibration. If Euclidean reconstruction is desired, a 4×4 projective-to-Euclidean transformation can be applied later on.

(a) Left-hand image of first pair

(b) Right-hand image of first pair

(c) Tracked points, shown on (a)

(d) Zoomed part of (c)

Fig. 11.8. a-d Application of the Htensor $\mathcal{J}$ to 3D reconstruction. *Row 1* displays two images from one stereo pair. The images show the projected texture. The stereo rig and the projector are moved together at subsequent time instants (not shown). *Row 2* displays the tracked points. Some of the points are stationary features (physical objects) and some are from the projected texture

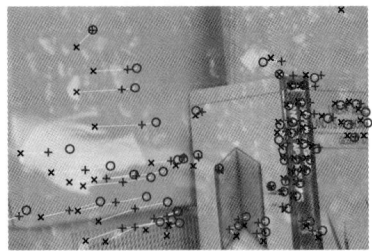
(a) Stabilized points, shown on Fig. 11.8.a

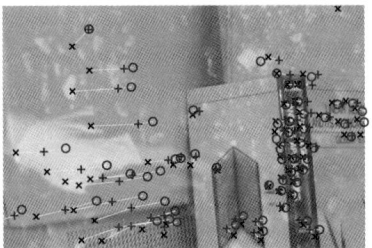
(b) Zoomed part of (a)

(c) Segmentation of moving/static points

Fig. 11.9. a-c Application of the Htensor $\mathcal{J}$ to 3D reconstruction. *Row 1* displays the points after the motion was canceled with the Htensor. Notice that points that are stationary were stabilized, meaning that the Htensor captured the correct 3D motion. *Row 2* shows the stationary points, which were identified by the Htensor

11.7 Summary

In this chapter we have introduced the m-view analogue of the classical collineation (homography matrix). The extension from 2 to m views introduces an additional parameter $k < n$, which endows the individual points of the point configuration that is being transformed projectively from view to view with the ability to become "dynamic". The value of k stands for the dimension of the subspace in which the individual points are allowed to move while the projective change of coordinates take place. For example, when $k = 1$ the points are not allowed to move (are stationary) just like with conventional collineations, and when $k = 2$ the individual points are allowed to move along straight-line paths, and so forth.

The m-view tensors for $\mathcal{P}^n$ and for $k < n$, referred to as homography tensors, were developed in detail for the case $n = 3, 4$ and the case $k = 2, 1$, which are instances of practical value for applications. In the derivation of the homography tensor the following issues need to be addressed: the maximal space contributed by dynamic points of subdimension k, number of constraints contributed by *mixed* points where some are labeled to move in k-subspace and some are unlabeled, and the use of the homography tensor as a mapping and the recovery of the individual projective mappings between views from the elements of the tensor.

Those issues were covered in detail for $n = 3, k = 2, 1$ (the $\mathcal{H}$ tensor for planar configurations) and for $n = 4, k = 2, 1$ (the $\mathcal{J}$ tensor for 3D configurations). For general n, m, k we have covered only the first issue above, that of dimension of the $GL(V)$ module $V(n, m, k)$ associated with the question of how many independent linear constraints are possible for a given value of n, m, k.

As for applications, we presented two instances, in 2D and 3D, of the problem of recovering the global alignment under dynamic motion. Without homography tensors, a recovery of alignment requires the use of statistical methods of sampling, where the points undergoing dynamic motion are considered as outliers. Whereas with the homography tensors both stationary and moving points can be considered alike, and part of a global transformation can be recovered analytically from observations (matching points across m views).

Generally, the homography tensors can be used to recover linear models under linear uncertainty. This generalization is quite straightforward, although the size of the resulting tensors grows exponentially. The use of such tensors in dimensions larger than $\mathcal{P}^3$ ($n > 4$) is not straightforward and is left for further research.

11.8 Representation Theory Digest

In this section we briefly recall some relevant facts concerning the representation theory of the general linear group. For a thorough introduction see [15].

Let V be a finite n-dimensional vector space over the complex numbers. The collection of invertible $n \times n$ matrices is denoted by $GL(n)$, which is the group of automorphisms of V denoted by $GL(V)$. The vector space $V^{\otimes m}$ (m-fold tensor product) is spanned by decomposable tensors of the form $v_1 \otimes \cdots \otimes v_m$, where the vectors v_i are in V. Hence the dimension of $V^{\otimes m}$ is n^m. The vector space $V^{\oplus m}$ is the m-fold direct sum of V, and thus is of dimension nm.

The *exterior* powers $\wedge^m V$ of V, $n \geq m$, is the vector space spanned by the $m \times m$ minors of the $n \times m$ matrix $[v_1, ..., v_m]$, where the vectors v_i are in V. Hence the dimension of $\wedge^m V$ is $\binom{n}{m}$. The exterior powers are the images of the map $V^{\times m} \to V^{\otimes m}$ given by

$$(v_1, \cdots, v_m) \to \sum_{\sigma \in S_m} \text{sgn}(\sigma) v_{\sigma(1)} \otimes, \cdots, v_{\sigma(m)},$$

where S_m denotes the symmetric group (of *permutations* of m letters).

The *symmetric powers* $\text{Sym}^m V$ are the images of the map $V^{\times m} \to V^{\otimes m}$ given by

$$(v_1, \cdots, v_m) \to \sum_{\sigma \in S_m} v_{\sigma(1)} \otimes, \cdots, v_{\sigma(m)}.$$

Hence the vector space $\text{Sym}^m V$ is of dimension $\binom{n+m-1}{m}$. Note that

$$V \otimes V = \text{Sym}^2 V \oplus \wedge^2 V,$$

with the appropriate dimension: $n^2 = \binom{n+1}{2} + \binom{n}{2}$. This decomposition into irreducibles (see later) is not true for $V^{\otimes m}$, $m > 2$. The remainder of this section is devoted to the necessary notation for representing $V^{\otimes m}$ as a decomposition of irreducibles.

A *representation* of a group G on a complex finite-dimensional space U is a homomorphism G to $GL(U)$, the group of linear automorphisms of U. The action of $g \in G$ on $u \in U$ is denoted by $g \cdot u$. The G-module U is *irreducible* if it contains no nontrivial G-invariant subspaces. Any finite-dimensional representation of a compact group G can be decomposed as a direct sum of irreducible representations. This basic property, called *complete reducibility*, also holds for all holomorphic representations of the general linear group $GL(V)$.

The main focus of Sect. 11.4 is the space

$$V(n, m, k) = \text{Span}\{v_1 \otimes \cdots \otimes v_m \in V^{\otimes m} :$$
$$\dim \text{Span}\{v_1, \ldots, v_m\} \leq k \}.$$

Since $V(n, m, k)$ is invariant under the $GL(V)$ action given by $g \cdot v_1 \otimes \cdots \otimes v_m = g(v_1) \otimes \cdots \otimes g(v_m)$, it is natural to study its structure by decomposing it into irreducible $GL(V)$-modules.

The description of the finite-dimensional irreducible representations (irreps) of $GL(V)$ depends on the combinatorics of partitions and Young diagrams, which we now describe. A *partition* of m is an ordered set $\lambda = (\lambda_1, ..., \lambda_k)$ such that $\lambda_1 \geq ... \geq \lambda_k \geq 1$ and $\sum \lambda_i = m$. A partition is represented by its *Young diagram* (also called *shape*), which consists of k left-aligned rows of boxes with λ_i boxes in row i. The *conjugate partition* $\mu = (\mu_1, ..., \mu_r)$ to a partition λ is defined by interchanging rows and columns in the Young diagram. Or, without reference to the diagram, μ_i is the number of terms in λ that are greater than or equal to i.

An assignment of the numbers $\{1, ..., m\}$ to each of the boxes of the diagram of λ, one number to each box, is called a *tableau*. A tableau in which all the rows and columns of the diagram are increasing is called a *standard tableau*. We denote by f_λ the number of standard tableaux on λ, i.e., the number of ways to fill the Young diagram of λ with the numbers from 1 to m, such that all rows and columns are increasing. Let (i,j) denote the coordinates of the boxes of the diagram, where $i = 1,..,k$ denotes the row number and j denotes the column number, i.e., $j = 1,..., \lambda_i$ in the ith row. The *hook length* h_{ij} of a box at position (i,j) in the diagram is the number of boxes directly below plus the number of boxes to the right plus 1 (without reference to the diagram, $h_{ij} = \lambda_i + \mu_j - i - j + 1$). Then,

$$f_\lambda = \frac{m!}{\prod_{(i,j)} h_{ij}},$$

where the product of the hook lengths is over all boxes of the diagram. We denote by $d_\lambda(n)$ the number of *semistandard tablaeux*, which is the number of ways to fill the diagram with the numbers from 1 to n, such that all rows are nondecreasing and all columns are increasing. We have:

$$d_\lambda(n) = \prod_{(i,j)} \frac{n - i + j}{h_{ij}}.$$

Let S_m denote the symmetric group on $\{1, \ldots, m\}$. The *group algebra* $\mathsf{C}S_m$ is the algebra spanned by the elements of S_m

$$\mathsf{C}G = \{ \sum_{\sigma \in S_m} \alpha_\sigma \sigma \mid \alpha_\sigma \in \mathsf{C} \},$$

where addition and multiplication are defined as follows:

$$\alpha(\sum_{\sigma \in S_m} \alpha_\sigma \sigma) + \beta(\sum_{\sigma \in S_m} \beta_\sigma \sigma) = \sum_{\sigma \in S_m} (\alpha \alpha_\sigma + \beta \beta_\sigma)\sigma,$$

and

$$(\sum_{\sigma \in S_m} \alpha_\sigma \sigma)(\sum_{\tau \in S_m} \beta_\tau \tau) = \sum_{g \in S_m} (\sum_{g = \sigma \tau} \alpha_\sigma \beta_\tau) g,$$

for $\alpha, \beta, \alpha_\sigma, \beta_\sigma \in \mathsf{C}$.

Let t be a tableau on λ (a numbering of the boxes of the diagram), and let $P(t)$ denote the group of all permutations $\sigma \in S_m$ that permute only the rows of t. Similarly, let $Q(t)$ denote the group of permutations that preserve the columns of t. Let a_t, b_t be two elements in the group algebra $\mathsf{C}S_m$ defined as

$$a_t = \sum_{g \in P(t)} g, \quad b_t = \sum_{g \in Q(t)} \text{sgn}(g) g.$$

The group algebra $\mathsf{C}S_m$ acts on $V^{\otimes m}$ on the right by permuting factors, i.e., $(v_1 \otimes \cdots \otimes v_m) \cdot \sigma = v_{\sigma(1)} \otimes \cdots \otimes v_{\sigma(m)}$. For a general shape λ and a tableau t on λ the image of a_t, $V^{\otimes m} \cdot a_t$, is the subspace:

$$V^{\otimes m} \cdot a_t = \text{Sym}^{\lambda_1} V \otimes \cdots \otimes \text{Sym}^{\lambda_k} V \subset V^{\otimes m},$$

and the image of b_t is

$$V^{\otimes m} \cdot b_t = \wedge^{\mu_1} V \otimes \cdots \otimes \wedge^{\mu_r} V \subset V^{\otimes m},$$

where μ is the conjugate partition to λ. The *Young symmetrizer* is defined by $c_t = a_t \cdot b_t \in \mathbb{C}S_m$. The image of the Young symmetrizer

$$S_t(V) = V^{\otimes m} \cdot c_t,$$

is the *Schur module* associated with t and is an irreducible $GL(V)$ module. The isomorphism type of $S_t(V)$ depends only on the shape λ, so we may write $S_t(V) = S_\lambda(V)$. It turns out that all the polynomial irreps of $GL(V)$ are of the form $S_\lambda(V)$ for some m and a partition $\lambda \vdash m$.

Let $\mathcal{T}_\lambda$ denote the set of standard tableaux on λ. Then the direct sum decomposition of $V^{\otimes m}$ into irreducible $GL(V)$ modules is given by

$$V^{\otimes m} = \bigoplus_{\lambda \vdash m} \bigoplus_{t \in \mathcal{T}_\lambda} S_t(V) \cong \bigoplus_{\lambda \vdash m} \mathcal{S}_\lambda(V)^{\oplus f_\lambda}.$$

Since $d_\lambda(n) = \dim \mathcal{S}_\lambda(V)$, it follows that

$$\dim V^{\otimes m} = n^m = \sum_{\lambda \vdash m} d_\lambda(n) f_\lambda.$$

For example, consider $n = m = 3$, i.e., $V \otimes V \otimes V$, where $\dim V = 3$. There are three possible partitions λ of 3: these are $(3), (1,1,1)$ and $(2,1)$. From the above, $S_{(3)}(V) = \text{Sym}^3 V$ and $S_{(1,1,1)}V = \wedge^3 V$. There are two, $f_{(2,1)} = 2$, standard tableaux for $\lambda = (2,1)$, and these are 123 and 132 (numbering of boxes left to right and top to bottom). There are eight, $d_{(2,1)}(3) = 8$, semistandard tableaux, which are: $112, 113, 122, 123, 132, 133, 223$ and 233. We have the decomposition:

$$V \otimes V \otimes V = \text{Sym}^3 V \oplus \wedge^3 V \oplus (S_{(2,1)}V)^{\oplus 2},$$

with the appropriate dimensions: $27 = 10 + 1 + (8 + 8)$.

References

1. S. Avidan and A. Shashua. Trajectory triangulation of lines: Reconstruction of a 3d point moving along a line from a monocular image sequence. In *Proceedings of the IEEE Conference on Computer Vision and Pattern Recognition*, June 1999.
2. S. Avidan and A. Shashua. Trajectory triangulation: 3D reconstruction of moving points from a monocular image sequence. *IEEE Transactions on Pattern Analysis and Machine Intelligence*, 22(4):348–357, 2000.
3. A. Criminisi, I. Reid, and A. Zisserman. Duality, rigidity and planar parallax. In *Proceedings of the European Conference on Computer Vision*, Frieburg, Germany, 1998. Springer, LNCS 1407.
4. G.H. Golub and C.F. Van Loan. *Matrix computations*. Johns Hopkins University Press, 1989.
5. R.I. Hartley. Lines and points in three views and the trifocal tensor. *International Journal of Computer Vision*, 22(2):125–140, 1997.
6. M. Irani and P. Anandan. Parallax geometry of pairs of points for 3D scene analysis. In *Proceedings of the European Conference on Computer Vision*, LNCS 1064, pages 17–30, Cambridge, UK, April 1996. Springer-Verlag.
7. M. Irani, P. Anandan, and D. Weinshall. From reference frames to reference planes: Multiview parallax geometry and applications. In *Proceedings of the European Conference on Computer Vision*, Frieburg, Germany, 1998. Springer, LNCS 1407.
8. M. Irani, B. Rousso, and S. Peleg. Recovery of ego-motion using image stabilization. In *Proceedings of the IEEE Conference on Computer Vision and Pattern Recognition*, pages 454–460, Seattle, Washington, June 1994.
9. P. Meer, D. Mintz, D. Kim, and A. Rosenfeld. Robust regression methods for computer vision: A review. *International Journal of Computer Vision*, 6(1):59–70, 1991.
10. A. Shashua, R. Meshulam, L. Wolf, A. Levin, and G. Kalai. On representation theory in computer vision problems. Technical report, School of Computer Science and Eng., The Hebrew University of Jerusalem, July 2002.
11. A. Shashua and N. Navab. Relative affine structure: Canonical model for 3D from 2D geometry and applications. *IEEE Transactions on Pattern Analysis and Machine Intelligence*, 18(9):873–883, 1996.
12. A. Shashua and M. Werman. Trilinearity of three perspective views and its associated tensor. In *Proceedings of the International Conference on Computer Vision*, June 1995.
13. A. Shashua and Lior Wolf. Homography tensors: On algebraic entities that represent three views of static or moving planar points. In *Proceedings of the European Conference on Computer Vision*, Dublin, Ireland, June 2000.
14. C.C. Slama. *Manual of Photogrammetry*. American Society of Photogrammetry and Remote Sensing, 1980.
15. W.Fulton and J.Harris. *Representation Theory: a First Course*. Springer-Verlag, 1991.
16. Lior Wolf, A. Shashua, and Y. Wexler. Join tensors: on 3d-to-3d alignment of dynamic sets. In *Proceedings of the International Conference on Pattern Recognition*, Barcelona, Spain, September 2000.

12

Detecting Independent 3D Movement

Abhijit S. Ogale, Cornelia Fermüller, Yiannis Aloimonos

Center for Automation Research, University of Maryland, College Park,
MD 20742, USA
`ogale,fer,yiannis@cfar.umd.edu`

12.1 Introduction

Motion segmentation is the problem of finding parts of the scene which possess independent 3D motion (such as people, animals or other objects like vehicles). This process is conceptually straightforward if the camera is stationary, and is often referred to as background subtraction. However, if the camera itself is also moving, then the problem becomes more complicated, since the image motion is generated by the combined effects of camera motion, structure and the motion of the independently moving objects. Isolating the contribution of each of these three factors is needed to solve the more general independent motion problem, which involves motion segmentation (finding the moving objects) and also finding their 3D motion. In this chapter, we shall restrict ourselves to the problem of finding moving objects only and not worry about finding their 3D motion. In the beginning, we present our philosophy that visual problems such as motion segmentation are inextricably linked with other problems in vision, and must be approached with a compositional outlook which attempts to solve multiple problems simultaneously. This is followed by a brief review of existing algorithms which detect independently moving objects. The main body of this chapter presents our approach to motion segmentation[1] which classifies moving objects and demonstrates that motion segmentation is compositional and is not about motion alone, but can also utilize information from sources such as occlusions to detect a wider array of moving objects.

---

[1] This chapter is based on our paper which is due to appear in the IEEE Transactions on Pattern Analysis and Machine Intelligence [1]; portions from [1] have been reprinted with permission (© 2004 IEEE). The support of the National Science Foundation is gratefully acknowledged.

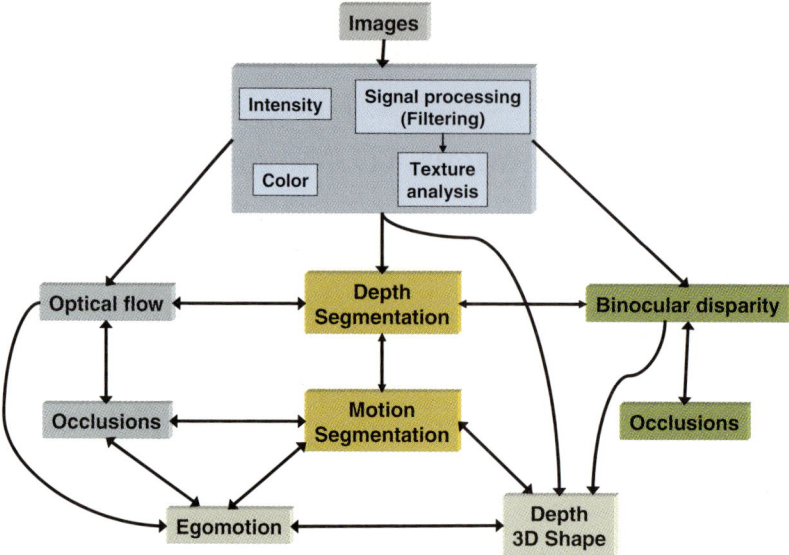

Fig. 12.1. Compositional problems

12.2 A Compositional Viewpoint

Finding the independently moving objects in an image sequence involves solving a host of related problems. In Fig. 12.1, we attempt to give our viewpoint about the relationships between the motion segmentation problem and other problems. At the beginning, the image stream is described in terms of intensity and/or color, and may be subjected to signal processing operations using various localized filters which describe features such as edges or the components of textures using spatial frequency channels. These local measurements are aggregated within a global framework to compute quantities such as optical flow (2D motion) and occlusions. Occlusions are parts of an image frame which disappear in the next frame. If binocular input is present, then disparity measurements and binocular occlusion information can also be computed. Detection of depth edges is affected by evidence from optical flow, binocular disparity and also monocular image measurements such as intensity and texture; in turn, these edges influence the estimation of each of these quantities.

Optical flow is used to compute egomotion (the motion of the camera), detect independently moving objects, and recover the background depth map concurrently. As we shall see, this process is often performed by finding clusters with consistent 3D motion. Later, we also show that occlusions provide information about ordinal scene structure which can be used to find new types of moving objects. The depth map of the scene is influenced by structure estimates from motion, binocular disparity measurements (if present), and influences from monocular image measurements such as intensity (i.e., via shape from shading) and texture (i.e., via shape from texture). Overall, the problem of motion segmentation requires a compositional solution which utilizes the relationships between different modules to obtain better solutions.

12.3 Existing Approaches

Prior research can mostly be classified into two groups: (a) The approaches relying, prior to 3D motion estimation, on 2D motion field measurements only [2, 3, 4, 5]. The limitations of these techniques are well understood. Depth discontinuities and independently moving objects both cause discontinuities in the 2D optical flow, and it is not possible to separate these factors without involving 3D motion estimation. (b) Approaches which assume that partial or full information about egomotion is available or can be recovered. Adiv [6] first segments on the basis of optical flow, and then groups the segments by searching for agreeable 3D motion parameters. Zhang et al. [7] utilize rigidity constraints on a sequence of stereo images to find egomotion and moving objects. Thompson and Pong's [8] first method finds inconsistencies between the egomotion and the flow field by using the motion epipolar constraint, while the second method relies on external depth information. Nelson [9] discusses two approaches, the first of which is similar to Thompson and Pong, while the second relies on acceleration detection. Sinclair [10] uses the angular velocity field and the premise that independently moving objects violate the epipolar constraint. Torr and Murray [11] find a set of fundamental matrices to describe the observed correspondences by hypothesizing clusters using robust statistics. Costeira and Kanade [12] use the factorization method along with a feature grouping step (block diagonalization of the shape interaction matrix).

Some techniques, such as [13], which address both 3D motion estimation and moving object detection, are based on alternate models of image formation, such as weak perspective. Such additional constraints can be justified for domains such as aerial imagery. In this case, the planarity of the scene allows a registration process [14, 15, 16, 17], and uncompensated regions correspond to independent movement. This idea has been extended to cope with general scenes by selecting models depending on the scene complexity [18], or by fitting multiple planes using the plane plus parallax constraint [19, 20]. The former [19] uses the best of 2D and 3D techniques, progressively increasing the complexity based on the situation. The latter [20] also develops constraints on the structure using three frames. Clearly, improvement in motion detection can be gained using temporal integration. Yet questions related to the integration of 3D motion and scene structure are not yet well understood, as the extension of the rigidity constraint to multiple frames is nontrivial. We therefore restrict ourselves to detecting moving objects using two or three frames only.

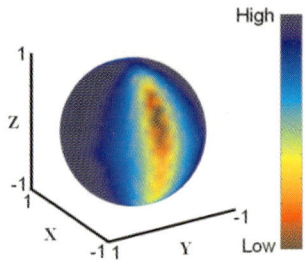

Fig. 12.2. Motion valley (red) visualized as an error surface in the 2D space of directions of translation, represented by the surface of a sphere. The error is found after finding the optimal rotation and structure for each translation direction. (Reproduced from [1] with permission © 2004 IEEE)

12.4 Ambiguity in 3D Motion Estimation

Many techniques detect independently moving objects based on 3D motion estimates, either explicitly or implicitly. Some utilize inconsistencies between egomotion estimates and the observed flow field, while some utilize additional information such as depth from stereo, or partial egomotion from other sensors. Nevertheless, the central problem faced by all motion-based techniques is that, in general, it is extremely difficult to uniquely estimate 3D motion from flow. Several studies have addressed the issue of noise sensitivity in structure from motion. In particular, it is known that for a moving camera with a small field of view observing a scene with insufficient depth variation, translation and rotation are easily confused [21, 22]. This can be intuitively understood by examining the differential flow equation:

$$\begin{aligned} u &= \tfrac{-t_x f + x t_z}{Z} + \alpha \tfrac{xy}{f} - \beta \left(\tfrac{x^2}{f} + f \right) + \gamma y, \\ v &= \tfrac{-t_y f + y t_z}{Z} + \alpha \left(\tfrac{y^2}{f} + f \right) - \beta \tfrac{xy}{f} - \gamma x \end{aligned} \qquad (12.1)$$

In the above equation, (u, v) is the optical flow, (t_x, t_y, t_z) is the translation, (α, β, γ) is the rotation and $Z(x, y)$ is the depth map. Notice that for a planar scene, up to zeroth order, we have $u \approx -t_x f/Z - \beta f$ and $v \approx -t_y f/Z + \alpha f$. Intuitively, we see how translation along the x-axis t_x can be confused with rotation β along the y-axis, and t_y with α for a small field of view.

Maybank [23] and Heeger and Jepson [24] have also shown that if the scene is sufficiently nonplanar, then the minima of the cost function resulting from the epipolar constraint lie along a line in the space of translation directions, which passes through the true translation direction and the viewing direction. In [25], an algorithm-independent stability analysis of the structure from motion problem has been carried out.

Thus, given a noisy flow field, any motion estimation technique will yield a region of solutions in the space of translations instead of a unique solution;

we refer to this region as the *motion valley*. Each translation direction in the motion valley, along with its best corresponding rotation and structure estimate, will agree with the observed noisy flow field. Fig. 12.2 shows a typical error function obtained using the motion estimation technique of Brodsky et al. [26] plotted on the 2D spherical surface of translational directions. Motion-based clustering can only succeed if a scene entity has a motion which does not lie in the background motion valley. In the following sections, we go beyond motion-based clustering and present a classification of moving objects with algorithms for detecting each class, laying particular emphasis on the role of occlusions.

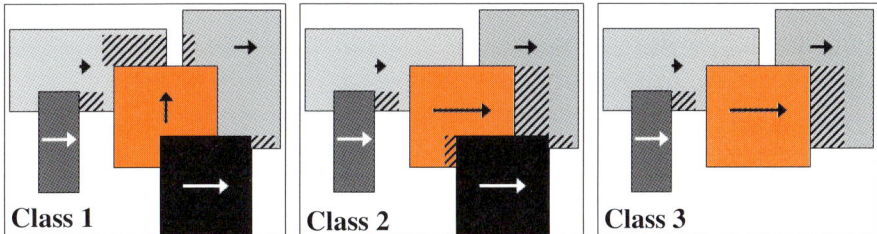

Fig. 12.3. Toy examples of three classes of moving objects. In each case, the independently moving object is red colored. Portions of objects which disappear in the next frame (i.e., occlusions) are shown in a dashed texture. (Reproduced from [1] with permission © 2004 IEEE)

12.5 Types of Independently Moving Objects

We now discuss three distinct classes of independently moving objects; the moving objects belonging to *Class 1* can be detected using motion-based clustering, the objects in *Class 2* are detected by detecting conflicts between depth from motion and ordinal depth from occlusions, and objects in *Class 3* are detected by finding conflicts between depth from motion and depth from another source (such as stereo). Any specific case will consist of a combination of objects from these three classes. Fig. 12.3 shows illustrative examples of the three classes.

12.5.1 Class 1: 3D Motion-Based Clustering

The first column of Fig. 12.3 shows a situation in which the background objects (non-independently moving) are translating horizontally, while the red object is moving vertically. In this scenario, motion-based clustering approaches will be successful, since the motion of the red object is not contained in the motion valley of the background. Thus, *Class 1* objects can be detected using motion alone. Our strategy for quickly performing motion-based clustering and detecting *Class 1* objects is discussed in Sect. 12.7.

12.5.2 Class 2: Ordinal Depth Conflict between Occlusions and Structure from Motion

The second column of Fig. 12.3 shows a situation in which the background objects are translating horizontally to the right, and the red object also moves towards the right. In this scenario, motion estimation will not be sufficient to detect the independently moving object, since motion estimation yields a single valley of solutions. An additional constraint, which may be termed the *ordinal depth conflict* or the *occlusion-structure from motion (SFM) conflict* needs to be used to detect the moving object.

Notice the occluded areas in the figure: we can use our knowledge of these occlusions to develop ordinal depth (i.e., *front/back*) relationships between regions of the scene. In this example, the occlusions tell us that the red object is *behind* the black object. However, if we compute structure from motion, since the motion is predominantly a translation, the result would indicate that the red object is in front of the black object (since the red object moves faster). This conflict between ordinal depth from occlusions and structure from motion permits the detection of *Class 2* moving objects. In Sect. 12.8, we present a novel algorithm for finding ordinal depth.

12.5.3 Class 3: Cardinal Depth Conflict

The third column of Fig. 12.3 shows a situation similar to the second column, except that the black object which was in front of the red object has been removed. Due to this situation, the ordinal depth conflict which helped us detect the red object in the earlier scenario is no longer present. In order to detect the moving object in this case, we must employ cardinal comparisons between structure from motion and structure from another source (such as stereo) to identify deviant regions as *Class 3* moving objects. In our experiments, we have used a calibrated stereo pair of cameras to detect objects of *Class 3*. The calibration allows us to compare the depth from motion directly with the depth from stereo up to a scale. We use k-means clustering (with $k = 3$) on the depth ratios to detect the background (the largest cluster). The reason for using $k = 3$ is to allow us to find three groups: the background, pixels with depth ratio greater than the background, and pixels with depth ratio less than the background. Pixels not belonging to the background cluster are the *Class 3* moving objects. At this point, it may be noted that alternative methods exist in the literature (e.g., [7]) for performing motion segmentation on stereo images, which can also be used to detect *Class 3* moving objects.

12.6 Phase Correlation

Before we move on to our approach for motion-based clustering, let us explain a simple technique which allows us to recover a four-parameter transformation between a pair of images using phase correlation (see [27]). We use this technique for initializing background motion estimation, and it may also be used for stabilizing a jittery video.

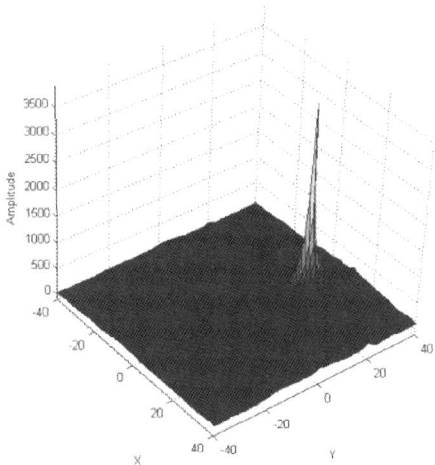

Fig. 12.4. An example of a peak generated by the phase correlation method

12.6.1 Basic Process

Consider an image which is moving in its own plane, i.e., every point on the image has the same flow. Thus, if the image $I_2(x, y)$ is such a translated version of the original image $I_1(x, y)$, then we can use phase correlation to recover the translation in the following manner.

If $I_2(x, y) = I_1(x + t_x, y + t_y)$, then their Fourier transforms are related by:

$$f_2(\omega_x, \omega_y) = f_1(\omega_x, \omega_y) e^{-i(\omega_x t_x + \omega_y t_y)} \quad (12.2)$$

The phase correlation $PC(x, y)$ of the two images is then given by:

$$PC(x, y) = \mathcal{F}^{-1}\left[\frac{f_1^* \cdot f_2}{|f_1^* \cdot f_2|}\right] = \mathcal{F}^{-1}\left[e^{-i(\omega_x t_x + \omega_y t_y)}\right] \quad (12.3)$$

$$PC(x, y) = \delta(x - t_x, y - t_y) \quad (12.4)$$

where $\mathcal{F}^{-1}$ is the inverse Fourier transform, * denotes the complex conjugate and δ is the delta function. Thus, if we use phase correlation, we can recover this global image translation (t_x, t_y) since we get a peak at this position (see example in Fig. 12.4).

12.6.2 Log-Polar Coordinates

The log-polar coordinate system (ρ, α) is related to the cartesian coordinates (x, y) by the following transformation:

$$x = e^\rho cos(\alpha) \tag{12.5}$$
$$y = e^\rho sin(\alpha) \tag{12.6}$$

If the image is transformed into the log-polar coordinate system (see Fig. 12.5), then changes in scale and rotation about the image center in the cartesian coordinates are transformed into translations in the log-polar coordinates. Hence, if we perform the phase correlation procedure mentioned above in the log-polar domain, we can also recover the scale change s, and a rotation about the center γ, between two images.

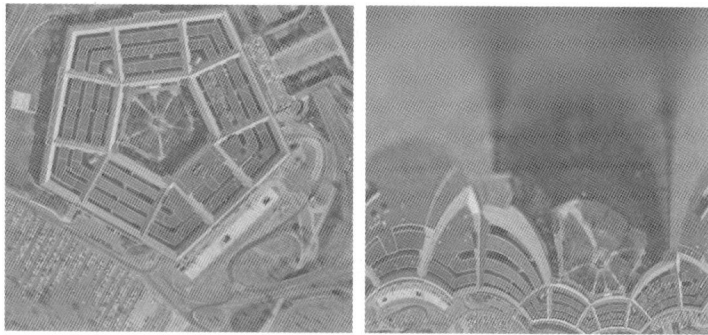

Fig. 12.5. An image in cartesian coordinates (left) and its log-polar representation (right)

12.6.3 Four-Parameter Estimation

Given two images which are related by 2D translation, rotation about the center and scaling, we can perform phase correlation in both the cartesian and log-polar domains to compute a four-parameter transformation T between the two images:

$$T = \begin{bmatrix} s \cdot cos(\gamma) & s \cdot sin(\gamma) & t_x \\ -s \cdot sin(\gamma) & s \cdot cos(\gamma) & t_y \\ 0 & 0 & 1 \end{bmatrix} \tag{12.7}$$

In practice, initializing this process is tricky, since dominant 2D translation will cause problems in the log-polar phase correlation by introducing many large additional peaks, and, similarly, dominant scaling and rotation will cause problems in the cartesian phase correlation.

To address this, we first perform phase correlation in both the cartesian and log-polar representations on the original images. Then, for each of the results, we find the ratio of the magnitude of the tallest peak to the overall median peak amplitude. If this ratio is greater for the cartesian computation, it means that translation is dominant over scaling and rotation, and must be removed first. Then we can estimate scaling and rotation again on the corrected images. Similarly, if the ratio is greater for the log-polar computation, we perform the correction the other way around. This process can be iterated a few times until the transformations converge.

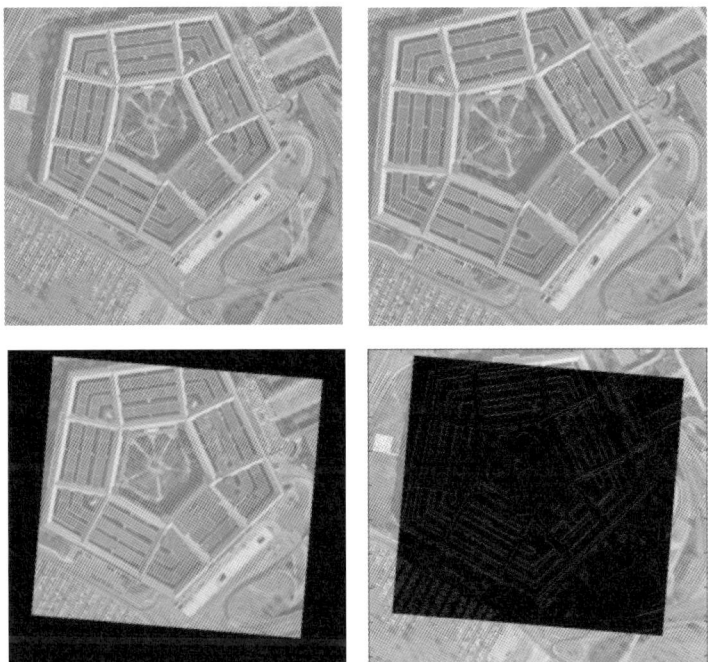

Fig. 12.6. Top row shows two input images I_1 and I_2. Image I_2 was created from I_1 by rotating by 5 degrees, scaling by a factor of 1.2, and translating by $(-10, 20)$ pixels. Bottom row: The left image shows image I'_2 obtained by unwarping I_2 using the results of the phase correlation. The right-hand side shows the absolute intensity difference between I_1 and the unwarped image I'_2 to reveal the accuracy of the registration. Notice that the difference is nearly zero in the area of overlap

12.6.4 Results

Fig. 12.6 shows the results on a pair of test images which are related by significantly large values of translation, rotation and scaling. These results can be improved to subpixel accuracy by using the method of Foroosh et al. [28]. We have applied the method mentioned above to video sequences and have achieved good stabilization over long durations, even in the presence of independently moving objects.

12.7 Motion-Based Clustering

Motion-based clustering is in itself a difficult problem, since the process of finding the background motion and finding the independently moving clusters has to be performed concurrently. The problem is a chicken-and-egg problem: if we knew the background pixels, we could find the background motion, and vice versa. In Sect. 12.3, we have cited several novel approaches which find motion clusters by concurrently performing segmentation and motion estimation. Here, we present a fast and simple method which consists of two steps.

The first step consists of using phase correlation on two frames in the cartesian representation (to find 2D translation t_x, t_y), and in the log-polar representation (to find scale S and z-rotation γ); we obtain a four-parameter transformation between frames (see the previous section). Phase correlation can be thought of as a voting approach [29], and hence we find empirically that these four parameters depend primarily on the background motion even in the presence of moving objects. This assumption is true as long as the background edges dominate the edges on the moving objects. This four-parameter transform predicts a flow direction at every point in the image. We select a set of points S in the image whose true flow direction lies within an angle of η_1 degrees about the direction predicted by phase correlation or its exact opposite direction (we use $\eta_1 = 45°$) .

In the second step, optical flow values at the points in set S are used to estimate the background motion valley using the 3D motion estimation technique of Brodsky et al. [26]. Since all points in the valley predict similar flows on the image (which is why the valley exists in the first place), we can pick any solution in the valley and compare the reprojected flow with the true flow. Regions where the two flows are not within η_2 degrees of each other are considered to be *Class 1* independently moving objects (we use $\eta_2 = 45°$) .

This procedure allows us to find the background and *Class 1* moving objects without iterative processes. The voting nature of phase correlation helps us to get around the chicken-and-egg aspect of the problem. To find optical flow, we can use any algorithm which finds dense flow (e.g., [30, 31]; we use the former). Although we have not used occlusions here, it is worthwhile to note that occlusions can be used to reduce the size of the motion valley.

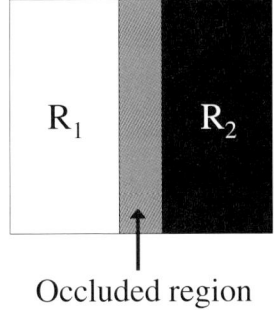

Occluded region

Fig. 12.7. If the occluded region belongs to R_1, then R_1 is behind R_2, and vice versa. (Reproduced from [1] with permission © 2004 IEEE)

12.8 Ordinal Depth from Occlusion Filling Using Three Frames

12.8.1 Why Occlusions Must Be *Filled*?

Given two frames from a video, occlusions are points in one frame which have no corresponding point in the other frame. However, merely knowing the occluded regions is not sufficient to deduce ordinal depth. In Fig. 12.7, we show a situation where an occluded region O is surrounded by two regions R_1 and R_2 which are visible in both frames.

If the occluded region O belongs to region R_1, then we know that R_1 must be behind R_2, and vice versa.

This statement is extremely significant, since it holds true even when the camera undergoes general motion, and even when we have independently moving objects in the scene! Thus, we need to know *'who occluded what'* as opposed to merely knowing *'what was occluded'*. Since optical flow estimation provides us with a segmentation of the scene (regions of continuous flow), we now have to *assign flows to the occluded regions*, and *merge* them with existing segments to find ordinal depth.

12.8.2 Occlusion Filling (Rigid Scene, No Independently Moving Objects)

In the absence of independently moving objects, knowledge of the focus of expansion (FOE) or contraction (FOC) can be used to fill occluded regions. Since camera rotation does not cause occlusions [32], knowing the FOE is enough. In the simplest case, shown in Fig. 12.8a, where the camera translates to the right, if object A is in front of object B then object A moves more to the left than B, causing a part of B on the left of A to become occluded. Thus, if the camera translates to the right, occluded parts in the first frame always belong to segments on their left. For general egomotion: First, draw a line L

from the FOE/FOC to an occluded pixel O. Then: (A) If we have an *FOE* (see Fig. 12.8b), the flow at O is obtained using the flow at the nearest visible pixel P on this line L, such that O lies between P and the FOE. (B) If we have an *FOC* (see Fig. 12.8c), then fill in with the nearest pixel Q on line L, such that Q lies between O and the FOC.

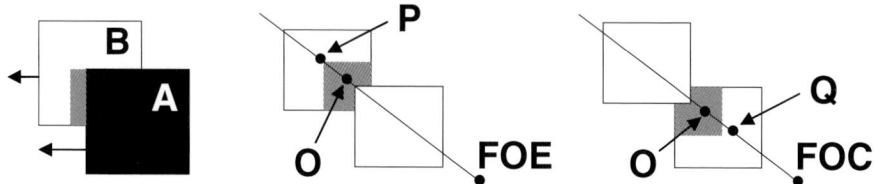

Fig. 12.8. Occlusion filling: from left (a) to (c). Gray regions indicate occlusions (portions which disappear in the next frame) (Reproduced from [1] with permission © 2004 IEEE)

12.8.3 Generalized Occlusion Filling (in the Presence of Moving Objects)

In the presence of moving objects, even the knowledge of the FOE provides little assistance for filling occlusions, since the occlusions no longer obey the criteria presented above; a more general strategy must be devised. The simplest idea which comes to mind is the following: if an occluded region O lies between regions R_1 and R_2, then we can decide how to fill O based on its *similarity* with R_1 and R_2. However, *similarity* is an ill-defined notion in general, since it may mean similarity of gray value, color, texture or some other feature. Using such measures of image similarity creates many failure modes. We present below a novel and robust strategy utilizing optical flow alone (see Fig. 12.9) for filling occlusions in the general case using three frames instead of two.

Given three consecutive frames F_1, F_2, F_3 of a video sequence, we use an optical flow algorithm which finds dense flow and occlusions (e.g., [30, 31]; we use the former) to compute the following:

1. Using F_1 and F_2, we find flow $\mathbf{u}_{12}$ from frame F_1 to F_2, and the reverse flow $\mathbf{u}_{21}$ from frame F_2 to F_1. The algorithm also gives us occlusions O_{12} which are regions of frame F_1 which are not visible in frame F_2. Similarly, we also have O_{21}.
2. Using frames F_2 and F_3, we find $\mathbf{u}_{23}$ and $\mathbf{u}_{32}$, and O_{23} and O_{32}.

Our objective is to fill the occlusions O_{21} and O_{23} in frame F_2 to deduce the ordinal depth. The idea is simple: O_{23} *denotes areas of F_2 which have*

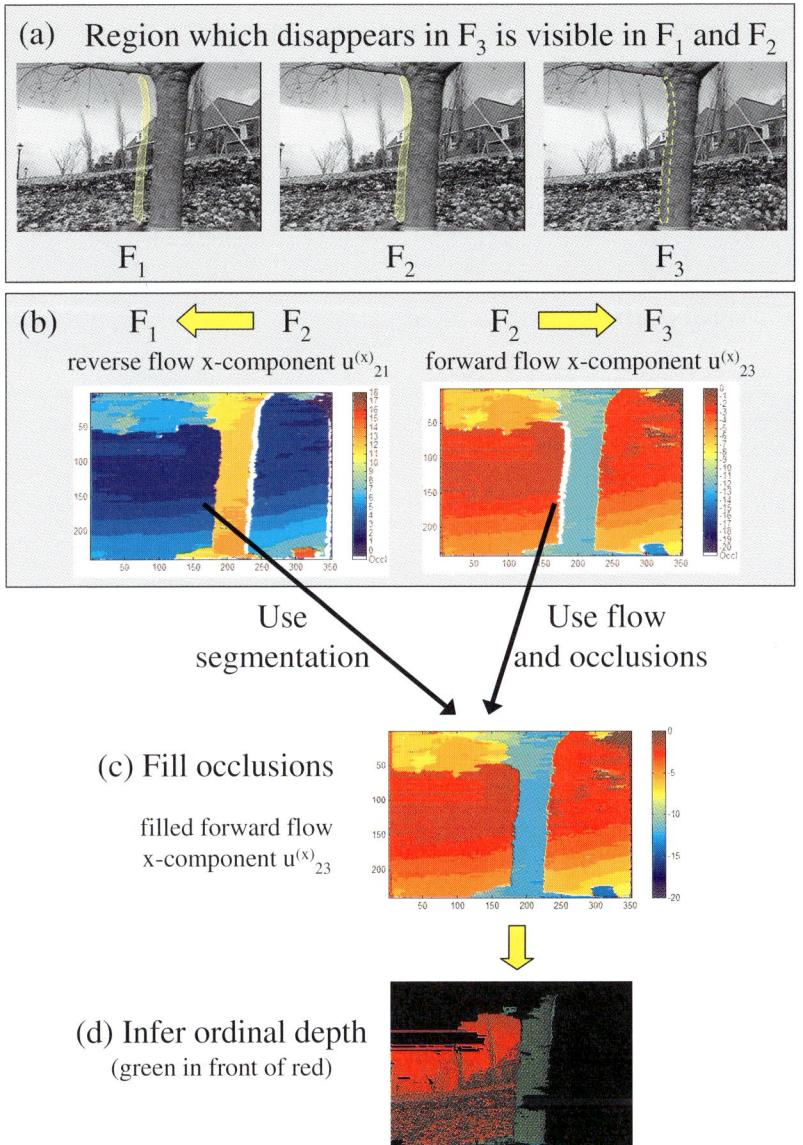

Fig. 12.9. Generalized occlusion filling and ordinal depth estimation. (a) Three frames of a video sequence. The yellow region which is visible in F_1 and F_2 disappears behind the tree in F_3. (b) Forward and reverse flow (only the x-components are shown). Occlusions are colored white. (c) Occlusions in $\mathbf{u}_{23}$ are filled using the segmentation of $\mathbf{u}_{21}$. Note that the white areas have disappeared. (d) Deduce ordinal depth relation. In a similar manner, we can also fill occlusions in $\mathbf{u}_{21}$ using the segmentation of $\mathbf{u}_{23}$ to deduce ordinal depth relations for the right side of the tree. (Reproduced from [1] with permission © 2004 IEEE)

no correspondence in F_3. However, these areas were visible in both F_1 and F_2; hence in $\mathbf{u}_{21}$ these areas have already been grouped with their neighboring regions. Therefore, we can use the segmentation of flow $\mathbf{u}_{21}$ to fill the occluded areas O_{23} in the flow field $\mathbf{u}_{23}$. Similarly, we can use the segmentation of $\mathbf{u}_{23}$ to fill the occluded areas O_{21} in the flow field $\mathbf{u}_{21}$. After filling, deducing ordinal depth is straightforward: if an occlusion is bounded by R_1 and R_2, and if R_1 was used to fill it, then R_1 is below R_2. This method is able to fill the occlusions and the find ordinal depth in a robust fashion.

12.9 Algorithm summary

1. Input video sequence: $V = (F_1, F_2, ..., F_n)$
2. For each $F_i \in V$ do
    a) find forward $\mathbf{u}_{i,i+1}$ and reverse $\mathbf{u}_{i,i-1}$ flows with occlusions $O_{i,i+1}$ and $O_{i,i-1}$
    b) select a set S of pixels using phase correlation between F_i and F_{i+1}
    c) find background motion valley using the flows for pixels in S
    d) detect *Class 1* moving objects and background B_1
    e) find ordinal depth relations using results of step (a)
    f) for pixels in B_1, detect *Class 2* moving objects, and new background B_2
    g) if depth from stereo is available, detect *Class 3* objects present in B_2

12.10 Experiments

Fig. 12.10 shows a situation in which the background is translating horizontally, while a *teabox* is moved vertically. In this scenario, since the *teabox* is not contained in the motion valley of the background, it is detected as a *Class 1* moving object.

Fig. 12.11 shows three frames of a video in which the camera translates horizontally, while a *coffee mug* is moved vertically upward, and a *red Santa Claus toy* is moved horizontally parallel to the background motion. The *coffee mug* is detected as a *Class 1* moving object, while the *red toy* is detected as a *Class 2* moving object using the conflict between ordinal depth from occlusions and structure from motion. A handwaving analysis indicates that since the *red toy* is moving faster than the foreground boxes, structure from motion (since the motion is predominantly a translation) naturally suggests that the *red toy* is in front of the two boxes. But the occlusions clearly indicate that the *toy* is behind the two boxes, thereby generating a conflict.

Finally, Fig. 12.12 shows a situation in which the background is translating horizontally to the right, and the leopard is dragged horizontally towards the right. In this case, a single motion valley is found, the depth estimates are all positive, and no ordinal depth conflicts are present. (Although this case shows the simplest situation, we can also imagine the same situation as in Fig. 12.11, with the exception that the *red toy* does not move fast enough so as to appear in front of the two boxes and generate an occlusion-motion conflict.) In this case, depth information from stereo (obtained using a calibrated stereo pair of cameras) was compared with depth information from motion. k-means clustering (with $k = 3$) of the depth ratios was used to detect the background (the largest cluster). The pixels which did not belong to the background cluster are labeled as *Class 3* moving objects.

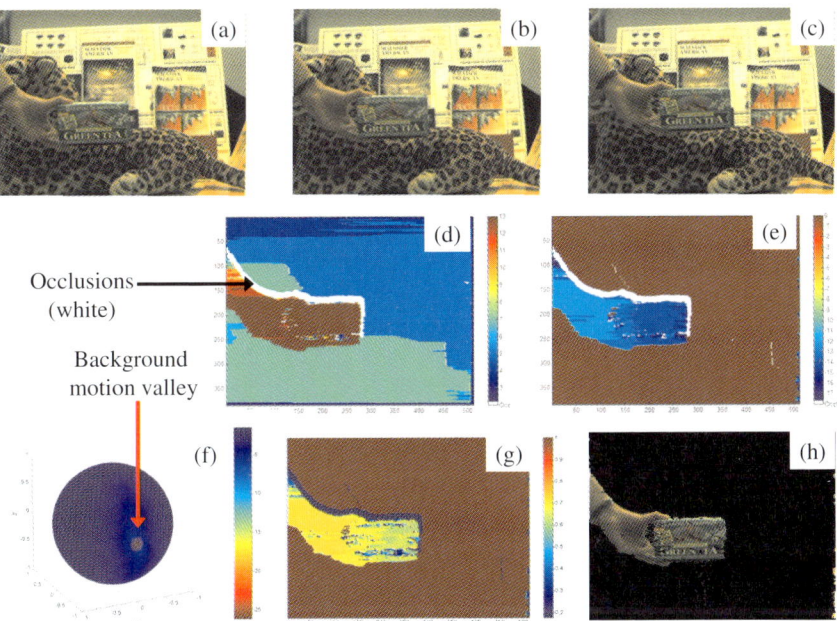

Fig. 12.10. Class 1: (a,b,c) show three frames of the *teabox* sequence. (d,e) show X and Y components of the optical flow using frames (b) and (c). Occlusions are colored white. (f) shows the computed motion valley for the background. (g) shows the cosine of the angular error between the reprojected flow (using the background motion) and the true flow. (h) shows the detected *Class 1* moving object (Reproduced from [1] with permission © 2004 IEEE)

12.11 Conclusion

In this chapter, we have discussed the motion segmentation problem within a compositional framework. Moving objects were classified into three classes,

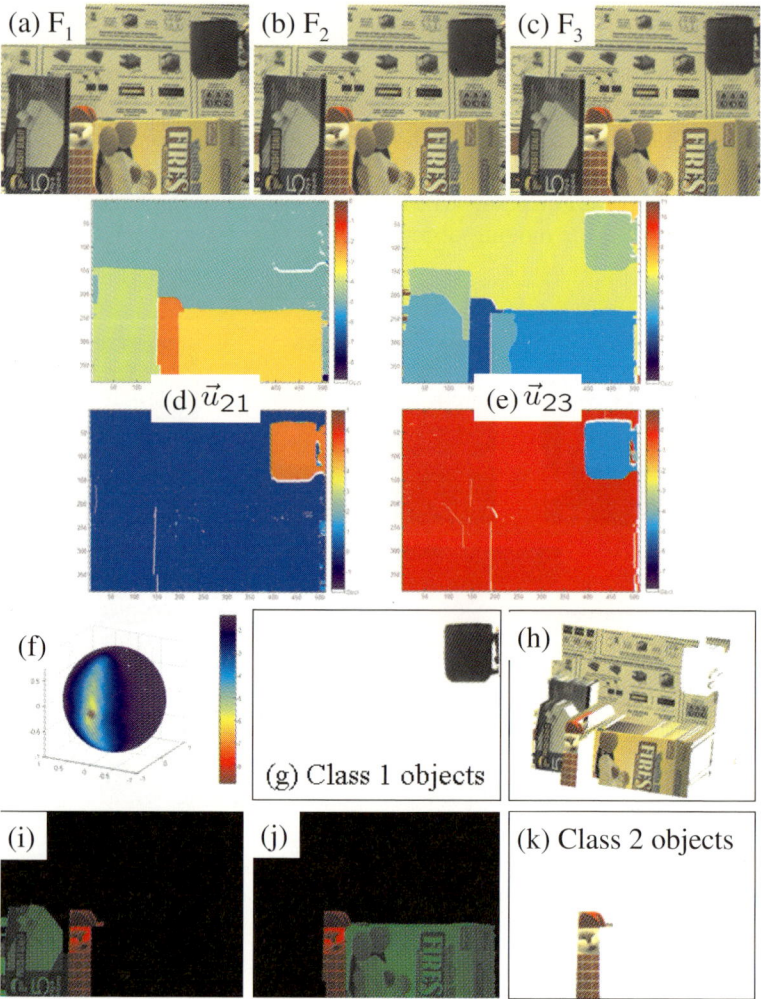

Fig. 12.11. (a,b,c) show three frames F_1, F_2, F_3 of the *santa-coffee* sequence. The camera translates horizontally to the left, hence the scene moves to the right. The *coffee mug* is lifted up, and the red toy *santa* is pulled by a person (not seen) to the right. (d) and (e) show optical flow $\mathbf{u}_{21}$ from frame F_2 to F_1, and $\mathbf{u}_{23}$ from frame F_2 to F_3 respectively. Note that each flow is shown as two images, with the X-component image above the Y-component image. Occlusions are colored white. (f) shows the estimated background motion. (g) shows the *coffee mug* detected as a *Class 1* object. (h) shows the computed structure from motion (SFM) for the background. Note that the toy *santa* appears *in front* of the two boxes. (i) and (j) show two ordinal depth relations obtained from occlusions which tell us that the santa (marked in red) *is behind* the boxes (marked in green). (k) shows the toy *santa* detected as a *Class 2* moving object using the ordinal depth conflict (Reproduced from [1] with permission © 2004 IEEE)

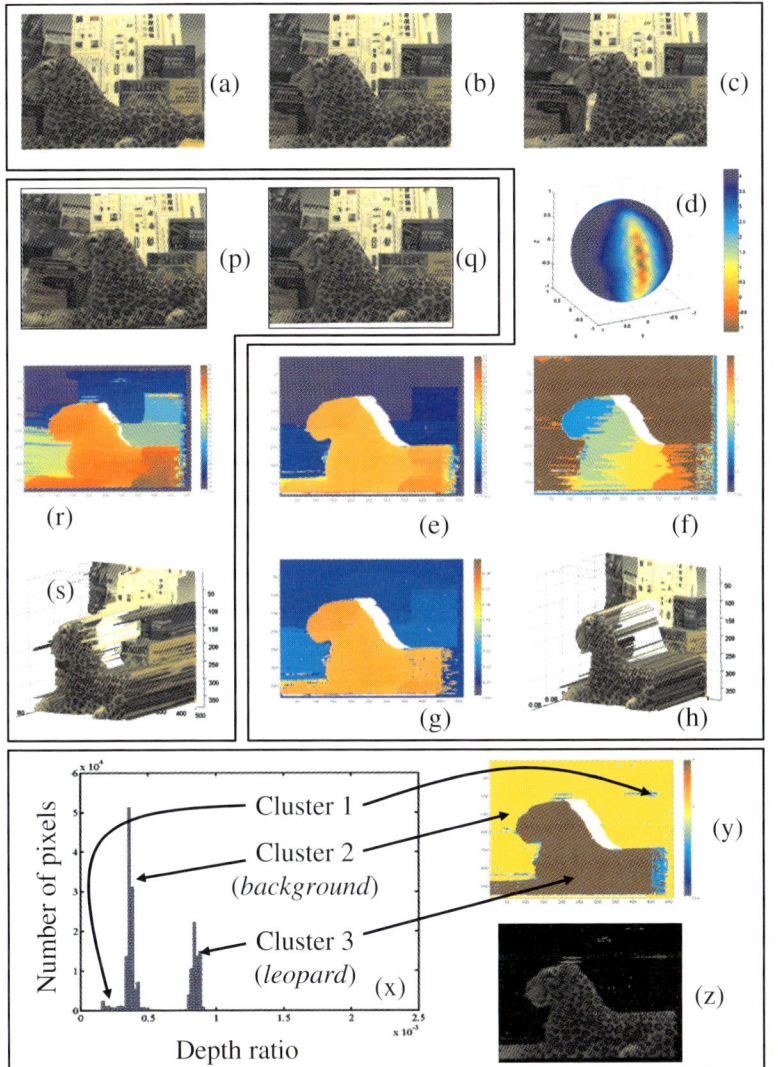

Fig. 12.12. Class 3: (a,b,c) show three frames F_1, F_2, F_3 of the *leopardB* sequence. (d) shows the computed motion valley. (e,f) show X and Y components of the flow $\mathbf{u}_{23}$ between F_2 and F_3. White regions denote occlusions. (g) shows inverse depth from motion. (h) shows 3D structure from motion. (p,q) show rectified stereo pair of images. (q) is the same as (b). (r) shows inverse depth from stereo. (s) shows 3D structure from stereo. Compare (s) with (h) to see how the background objects appear closer to the leopard in (s) than in (h). (x) shows the histogram of depth ratios and clusters detected by k-means ($k = 3$). (y) shows cluster labels: cluster 2 (yellow) is the background, cluster 3 (red) is the leopard, cluster 1 (light blue) is mostly due to errors in the disparity and flow. (z) shows the moving objects of *Class 3* (clusters other than 2) (Reproduced from [1] with permission © 2004 IEEE)

and constraints for detecting each class of objects were presented: *Class 1* was detected using motion alone, *Class 2* was detected using conflicts between ordinal depth from occlusions and depth from motion, while *Class 3* required cardinal comparisons between depth from motion and depth from another source.

References

1. A.S. Ogale, C. Fermüller and Y. Aloimonos, Motion segmentation using occlusions, *IEEE Transactions on Pattern Analysis and Machine Intelligence*, in press.
2. M. Bober and J. Kittler, Robust motion analysis, in *Proc. IEEE Conference on Computer Vision and Pattern Recognition*, 947–952, 1994.
3. P.J. Burt, J.R. Bergen, R. Hingorani, R. Kolczynski, W.A. Lee, A. Leung, J. Lubin, and H. Shvaytser, Object tracking with a moving camera, in *Proc. IEEE Workshop on Visual Motion*, 2–12, 1989.
4. J.-M. Odobez and P. Bouthemy, MRF-based motion segmentation exploiting a 2D motion model and robust estimation, in *Proc. International Conference on Image Processing*, III:628–631, 1995.
5. Y. Weiss, Smoothness in layers: Motion segmentation using nonparametric mixture estimation, in *Proc. IEEE Conference on Computer Vision and Pattern Recognition*, 520–526, 1997.
6. G. Adiv, Determining 3D motion and structure from optical flow generated by several moving objects, *IEEE Transactions on Pattern Analysis and Machine Intelligence*, 7:384–401, 1985.
7. Z. Zhang, O.D. Faugeras, and N. Ayache, Analysis of a sequence of stereo scenes containing multiple moving objects using rigidity constraints, in *Proc. Second International Conference on Computer Vision*, 177–186, 1988.
8. W.B. Thompson and T.-C. Pong, Detecting moving objects, *International Journal of Computer Vision*, 4:39–57, 1990.
9. R.C. Nelson, Qualitative detection of motion by a moving observer, *International Journal of Computer Vision*, 7:33–46, 1991.
10. D. Sinclair, Motion segmentation and local structure, in *Proc. Fourth International Conference on Computer Vision*, 366–373, 1993.
11. P.H.S. Torr and D.W. Murray, Stochastic motion clustering, in *Proc. Third European Conference on Computer Vision*. Springer, 328–337, 1994.
12. J. Costeira and T. Kanade, A multi-body factorization method for motion analysis, in *Proc. International Conference on Computer Vision*, 1071–1076, 1995.
13. J. Weber and J. Malik, Rigid body segmentation and shape description from dense optical flow under weak perspective, *IEEE Transactions on Pattern Analysis and Machine Intelligence*, 19(2):139–143, 1997.
14. Q.F. Zheng and R. Chellappa, Motion detection in image sequences acquired from a moving platform, in *Proc. IEEE International Conference on Acoustics, Speech, and Signal Processing*, 201–204, 1993.
15. S. Ayer, P. Schroeter, and J. Bigün, Segmentation of moving objects by robust motion parameter estimation over multiple frames, in *Proc. Third European Conference on Computer Vision*. Springer, 316–327, 1994

16. C.S. Wiles and M. Brady, Closing the loop on multiple motions, in *Proc. Fifth International Conference on Computer Vision*, 308–313, 1995.
17. B. Triggs, P. McLauchlan, R. Hartley, and A. Fitzgibbon, Bundle adjustment – a modern synthesis, in *Vision Algorithms: Theory and Practice*, B. Triggs, A. Zisserman, and R. Szeliski, Eds. Springer, 2000.
18. P.H.S. Torr, Geometric motion segmentation and model selection, in *Philosophical Transactions of the Royal Society A*, J. Lasenby, A. Zisserman, R. Cipolla, and H. Longuet-Higgins, Eds., 1321–1340, 1998.
19. M. Irani and P. Anandan, A unified approach to moving object detection in 2D and 3D scenes, *IEEE Transactions on Pattern Analysis and Machine Intelligence*, 20:577–589, 1998.
20. H. Sawhney, Y. Guo, and R. Kumar, "Independent motion detection in 3D scenes," *IEEE Transactions on Pattern Analysis and Machine Intelligence*, 22:1191–1199, 2000.
21. G. Adiv, Inherent ambiguities in recovering 3D motion and structure from a noisy flow field, *IEEE Transactions on Pattern Analysis and Machine Intelligence*, 11:477–489, 1989.
22. K. Daniilidis and M.E. Spetsakis, Understanding noise sensitivity in structure from motion, in *Visual Navigation: from Biological Systems to Unmanned Ground Vehicles*, Series on Advances in Computer Vision, Y. Aloimonos, Ed., Lawrence Erlbaum Associates, ch. 4, 1997.
23. S.J. Maybank, A theoretical study of optical flow, Ph.D. dissertation, University of London, 1987.
24. D.J. Heeger and A.D. Jepson, Subspace methods for recovering rigid motion I: Algorithm and implementation, *International Journal of Computer Vision*, 7:95–117, 1992.
25. C. Fermüller and Y. Aloimonos, Observability of 3D motion, *International Journal of Computer Vision*, 37:43–63, 2000.
26. T. Brodský, C. Fermüller, and Y. Aloimonos, Structure from motion: beyond the epipolar constraint, *International Journal of Computer Vision*, 37:231–258, 2000.
27. B.S. Reddy and B. Chatterji, "An FFT-based technique for translation, rotation and scale-invariant image registration," *IEEE Transactions on Image Processing*, 5(8):1266–1271, August 1996.
28. H. Foroosh, J. Zerubia, and M. Berthod, Extension of phase correlation to subpixel registration, *IEEE Transactions on Image Processing*, 11(3):188–200, March 2002.
29. D. Fleet, Disparity from local weighted phase-correlation, *IEEE International Conference on SMC*, 48–56, October 1994.
30. A.S. Ogale, The compositional character of visual correspondence, Ph.D. dissertation, University of Maryland, College Park, USA, www.cfar.umd.edu/users/ogale/thesis/thesis.html, August 2004.
31. V. Kolmogorov and R. Zabih, Computing visual correspondence with occlusions using graph cuts, in *Proc. International Conference on Computer Vision*, 2:508–515, 2001.
32. C. Silva and J. Santos-Victor, Motion from occlusions, *Robotics and Autonomous Systems*, 35(3–4):153–162, June 2001.

Part V

Perception and Action

13

Robot Perception and Action Using Conformal Geometric Algebra

Eduardo Bayro-Corrochano

CINVESTAV, Centro de Investigación y de Estudios Avanzados,
Unidad Guadalajara, Computer Science Department, GEOVIS Laboratory,
P.O. Box 31-438, Plaza La Luna, Guadalajara, Jalisco 44550, Mexico,
edb@gdl.cinvestav.mx, http://www.gdl.cinvestav.mx/~edb

13.1 Introduction

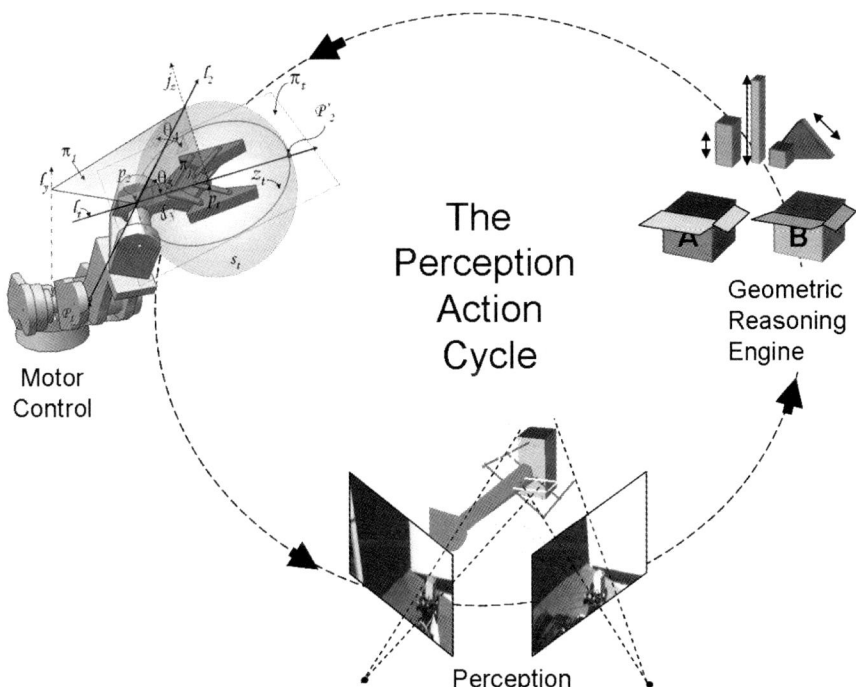

Fig. 13.1. Abstract rendering of the perception action cycle

In this chapter we make use of conformal Clifford geometric algebra for dealing with the algebra of incidence and directed distance involving Euclidean transformations that are represented as spinors. Our mathematical approach appears promising for the development of perception–action cycle systems (see Fig. 13.1). This work represents an improvement over previous research [1, 2, 3, 13] because by using conformal Clifford geometric algebra we are now able to incorporate group transformations into our computations. Other researchers have used Grassmann–Cayley algebra in computer vision [6] and robotics [14]. While the Grassmann–Cayley approach can express the ideas of projective geometry, such as meet and join, mathematically, it lacks an inner (regressive) product (and, indeed, practitioners often go to great lengths to "create" an inner product) as well as the ability to define some other key concepts such as duality and projective split; thus, it cannot easily include group transformations in its computations. Because of these limitations, we believe a better approach is to use conformal geometric algebra for robotic vision applications. In this chapter we present: 3-D motion estimation, body-eye calibration, 3-D reconstruction, omnidirectional vision, navigation, reaching and 3-D object manipulation. This work introduces a suitable computational framework for geometric computations, which, although illustrated for the cases of stereo- and omnidirectional-guided robotics, can be also of great value for applications using range data-, laser- and odometry-based systems.

The chapter is organized as follows: Sect. 2 presents a brief introduction to geometric algebra; Sect. 3 explains conformal geometric algebra, and Sect. 4 explains the n-dimensional affine plane; Sect. 5 presents an estimation of 3-D motion using points or lines as observations; Sect. 6 describes our method for body-eye calibration; in Sect. 7 we present real applications of reconstruction, navigation, reaching, and grasping; Sect. 8 illustrates strategies for object manipulation; and, finally, Sect. 9 presents our conclusions.

In the text, vectors in 3-D Euclidean space will be represented in boldface type using lowercase letters. Vectors of any geometric algebra (except for basis multivectors) will be represented using lowercase letters, italics, and boldface type, and multivectors in general will be denoted using italics and uppercase letters.

13.2 Geometric Algebra: An Outline

The algebras of Clifford and Grassmann are well known to pure mathematicians, but they were long ago abandoned by physicists in favor of the vector algebra of Gibbs, which is still in common use today for most areas of physics. The approach to Clifford algebra we adopt here was pioneered in the 1960s by David Hestenes [8], who has since further developed his version of Clifford algebra into a unifying language for both mathematics and physics. Here, we will refer to Clifford algebra as *geometric algebra* [9, 1]. This mathematical system includes also the antisymmetrical Grassmann–Cayley algebra.

13.2.1 Basic Definitions

Let $\mathcal{G}_n$ denote a geometric algebra of n dimensions–that is, a graded linear space. In addition to vector addition and scalar multiplication there is a noncommutative product that is associative and distributive over addition: this is the *geometric* or *Clifford product*. A further distinguishing feature of this type of algebra is that the square of any vector is a scalar. The geometric product of two vectors $\boldsymbol{a}$ and $\boldsymbol{b}$ is written $\boldsymbol{ab}$ and can be expressed as the sum of the vectors' symmetric and antisymmetric parts,

$$\boldsymbol{ab} = \boldsymbol{a}\cdot\boldsymbol{b} + \boldsymbol{a}\wedge\boldsymbol{b}. \tag{13.1}$$

Using this definition, we can express the inner product $\boldsymbol{a}\cdot\boldsymbol{b}$ and the outer product $\boldsymbol{a}\wedge\boldsymbol{b}$ in terms of the noncommutative geometric products,

$$\boldsymbol{a}\cdot\boldsymbol{b} = \frac{1}{2}(\boldsymbol{a}\cdot\boldsymbol{b} + \boldsymbol{a}\wedge\boldsymbol{b} + \boldsymbol{a}\cdot\boldsymbol{b} - \boldsymbol{a}\wedge\boldsymbol{b}) = \frac{1}{2}(\boldsymbol{ab} + \boldsymbol{ba}),$$
$$\boldsymbol{a}\wedge\boldsymbol{b} = \frac{1}{2}(\boldsymbol{a}\cdot\boldsymbol{b} + \boldsymbol{a}\wedge\boldsymbol{b} - \boldsymbol{a}\cdot\boldsymbol{b} + \boldsymbol{a}\wedge\boldsymbol{b}) = \frac{1}{2}(\boldsymbol{ab} - \boldsymbol{ba}). \tag{13.2}$$

The inner product of two vectors is the standard *scalar* or *dot* product and produces a scalar. The outer or wedge product of two vectors is a new quantity, which we call a *bivector*. We think of a bivector as an oriented area in the plane containing $\boldsymbol{a}$ and $\boldsymbol{b}$ formed by sweeping $\boldsymbol{a}$ along $\boldsymbol{b}$ (see Fig. 13.2a). Thus, $\boldsymbol{b}\wedge\boldsymbol{a}$ will have an opposite orientation, making the wedge product anticommutative, as given in Eq. (13.2). The outer product is immediately generalizable to

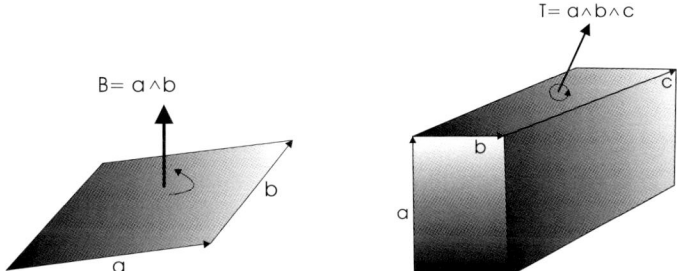

Fig. 13.2. (a) *(left)* The directed area, or bivector, $\boldsymbol{a}\wedge\boldsymbol{b}$; (b) *(right)* the oriented volume, or trivector, $\boldsymbol{a}\wedge\boldsymbol{b}\wedge\boldsymbol{c}$

higher dimensions: for example, $(\boldsymbol{a}\wedge\boldsymbol{b})\wedge\boldsymbol{c}$, a *trivector*, is interpreted as the oriented volume formed by sweeping the area $\boldsymbol{a}\wedge\boldsymbol{b}$ along vector $\boldsymbol{c}$. In Fig. 13.2b it can be seen that the outer product of k vectors is a k-vector, or k-blade, and that such a quantity is said to have *grade* k. A *multivector* (a linear combination of objects of different types) is *homogeneous* if it only contains terms of the same grade. Geometric algebra provides a means for manipulating multivectors which allows us to keep track of different grade objects simultaneously–much as one does with complex number operations.

13.2.2 Geometric Algebra of n-D Space

In an n-dimensional space we can introduce an orthonormal basis of vectors $\{\sigma_i\}$, $i = 1, ..., n$, such that $\sigma_i \cdot \sigma_j = \delta_{ij}$. This leads to a basis for the entire algebra:

$$1, \quad \{\sigma_i\}, \quad \{\sigma_i \wedge \sigma_j\}, \quad \{\sigma_i \wedge \sigma_j \wedge \sigma_k\}, \ldots \quad , I = \sigma_1 \wedge \sigma_2 \wedge \ldots \wedge \sigma_n. \quad (13.3)$$

Note that the basis vectors are not represented using boldface type. Any multivector can be expressed in terms of this basis. In this chapter, we specify a geometric algebra $\mathcal{G}_n$ of the n-dimensional space by the expression $\mathcal{G}_{p,q,r}$, where p, q, and r stand for the number of basis vectors which square to 1, -1 and 0, respectively, and fulfill the condition $n = p + q + r$. If a subalgebra is generated only by a multivector basis of even grade, it is called an even subalgebra and is denoted by $\mathcal{G}^+_{p,q,r}$.

In the n-D space there are multivectors of grade 0 (scalars), grade 1 (vectors), grade 2 (bivectors), grade 3 (trivectors), etc., ... up to grade n. The blade of grade n $\sigma_1 \wedge \sigma_2 \wedge \ldots \wedge \sigma_n$ is called the unit *pseudoscalar*, which will denoted by I_n. Its magnitude is 1 and its square defines the signature of the geometric algebra $\mathcal{G}_n$. It is *non-degenerated* if its pseudoscalar has a non-zero magnitude.

A fundamental concept algebraically related to the unit pseudoscalar I is that of *duality*. In a geometric algebra $\mathcal{G}_n$ we find dual multivectors and dual operations. The dual of a multivector $\boldsymbol{A} \in \mathcal{G}_n$ is defined as follows:

$$\boldsymbol{A}^* = \boldsymbol{A} I_n^{-1}, \quad (13.4)$$

where I_n^{-1} differs from I_n at most by a sign. Note that, in general, I^{-1} might not necessarily commute with $\boldsymbol{A}$.

The multivector bases of a geometric algebra $\mathcal{G}_n$ have 2^n basis elements. It can be shown that the second half is the dual of the first half. For example, in $\mathcal{G}_3$ the dual of the scalar is the pseudoscalar, and the dual of a vector is a bivector $\sigma_{23} = I\sigma_1$. In general, the dual of an r-blade is an $(n-r)$-blade.

The operations related directly to the Clifford product are the inner and outer products, which are dual to one another. This can be written as follows:

$$(\boldsymbol{x} \cdot \boldsymbol{A}) I_n = \boldsymbol{x} \wedge (\boldsymbol{A} I_n), \quad (13.5)$$
$$(\boldsymbol{x} \wedge \boldsymbol{A}) I_n = \boldsymbol{x} \cdot (\boldsymbol{A} I_n). \quad (13.6)$$

where $\boldsymbol{x}$ is any vector and $\boldsymbol{A}$ any multivector.

By using the ideas of duality, we are then able to relate the inner product to incidence operators in the following manner. In an n-D space, suppose we have an r-vector $\boldsymbol{A}$ and an s-vector $\boldsymbol{B}$ where the dual of $\boldsymbol{B}$ is given by $\boldsymbol{B}^* = \boldsymbol{B}I^{-1} \equiv \boldsymbol{B} \cdot I^{-1}$. Since $\boldsymbol{B}I^{-1} = \boldsymbol{B} \cdot I^{-1} + \boldsymbol{B} \wedge I^{-1}$, we can replace the geometric product by the inner product alone (in this case, the outer product equals zero, and there can be no $(n+1)$-D vector). Now, using the identity

$$\boldsymbol{A}_r \cdot (\boldsymbol{B}_s \cdot \boldsymbol{C}_t) = (\boldsymbol{A}_r \wedge \boldsymbol{B}_s) \cdot \boldsymbol{C}_t \quad \text{for} \quad r + s \leq t, \tag{13.7}$$

we can write

$$\boldsymbol{A} \cdot (\boldsymbol{B}I^{-1}) = \boldsymbol{A} \cdot (\boldsymbol{B} \cdot I^{-1}) = (\boldsymbol{A} \wedge \boldsymbol{B}) \cdot I^{-1} = (\boldsymbol{A} \wedge \boldsymbol{B})I^{-1}. \tag{13.8}$$

This expression can be rewritten using the definition of the dual as follows:

$$\boldsymbol{A} \cdot \boldsymbol{B}^* = (\boldsymbol{A} \wedge \boldsymbol{B})^*. \tag{13.9}$$

The equation shows the relationship between the inner and outer products in terms of the duality operator. Now, if $r + s = n$, then $\boldsymbol{A} \wedge \boldsymbol{B}$ is of grade n and is therefore a pseudoscalar. Using Eq. (13.8), we can employ the involved pseudoscalar in order to get an expression in terms of a bracket:

$$\begin{aligned} \boldsymbol{A} \cdot \boldsymbol{B}^* &= (\boldsymbol{A} \wedge \boldsymbol{B})^* = (\boldsymbol{A} \wedge \boldsymbol{B})I^{-1} = ([\boldsymbol{A} \wedge \boldsymbol{B}]I)I^{-1} \\ &= [\boldsymbol{A} \wedge \boldsymbol{B}]. \end{aligned} \tag{13.10}$$

We see, therefore, that the bracket relates the inner and outer products to nonmetric quantities.

When we work with lines, planes, and spheres, however, it will clearly be necessary to employ operations for computing the *meets* (intersections) or *joins* (expansions) of geometric objects. For this, we will need a geometric means of performing the set-theory operations of intersection, ∩, and union, ∪. Herewith, ∪ and ∩ will stand for the algebra of incidence operations of join and meet, respectively.

If in an n-dimensional geometric algebra the r-vector A and the s-vector B do not have a common subspace (null intersection), one can define the *join* of both vectors as

$$\boldsymbol{C} = \boldsymbol{A} \cup \boldsymbol{B} = \boldsymbol{A} \wedge \boldsymbol{B}, \tag{13.11}$$

so that the join is simply the outer product (an $r+s$ vector) of the two vectors. However, if $\boldsymbol{A}$ and $\boldsymbol{B}$ have common blades, the join would not simply be given by the wedge but by the subspace the two vectors span. The operation join ∪ can be interpreted as a *common dividend of lowest grade* and is defined up to a scale factor. The join gives the pseudoscalar if $(r+s) \geq n$. We will use ∪ to represent the join only when the blades A and B have a common subspace; otherwise, we will use the ordinary exterior product, ∧, to represent the join.

If there exists a k-vector $\boldsymbol{D}$ such that for $\boldsymbol{A}$ and $\boldsymbol{B}$ we can write $\boldsymbol{A} = \boldsymbol{A}'\boldsymbol{D}$ and $\boldsymbol{B} = \boldsymbol{B}'\boldsymbol{D}$ for some $\boldsymbol{A}'$ and $\boldsymbol{B}'$, then we can define the *intersection* or *meet* using the duality principle as follows

$$\boldsymbol{D} \cdot \boldsymbol{I}^{-1} = \boldsymbol{D}^* = (\boldsymbol{A} \cap \boldsymbol{B})^* = \boldsymbol{A}^* \cup \boldsymbol{B}^*. \tag{13.12}$$

This is a beautiful result, telling us that the dual of the meet is given by the join of the duals. Since the dual of $\boldsymbol{A} \cap \boldsymbol{B}$ will be taken with respect to the *join* of $\boldsymbol{A}$ and $\boldsymbol{B}$, we must be careful to specify which space we will use for the dual in Eq. (13.12). However, in most cases of practical interest, this join will indeed cover the entire space, and therefore we will be able to obtain a more useful expression for the meet using Eq. (13.9). Thus,

$$\boldsymbol{A} \cap \boldsymbol{B} = ((\boldsymbol{A} \cap \boldsymbol{B})^*)^* = (\boldsymbol{A}^* \cup \boldsymbol{B}^*)\boldsymbol{I} = (\boldsymbol{A}^* \wedge \boldsymbol{B}^*)(\boldsymbol{I}^{-1}\boldsymbol{I})\boldsymbol{I} = (\boldsymbol{A}^* \cdot \boldsymbol{B}). \tag{13.13}$$

The above concepts are discussed further in [11].

Let us now analyze a geometric product involving any of two kinds of multivectors. Consider two multivectors $\boldsymbol{A}_r$ and $\boldsymbol{B}_s$ of grades r and s, respectively. The geometric product of $\boldsymbol{A}_r$ and $\boldsymbol{B}_s$ can be written as

$$\boldsymbol{A}_r \boldsymbol{B}_s = \langle \mathbf{AB} \rangle_{r+s} + \langle \mathbf{AB} \rangle_{r+s-2} + \ldots + \langle \mathbf{AB} \rangle_{|r-s|}, \tag{13.14}$$

where $\langle \boldsymbol{M} \rangle_t$ denotes the t-grade part of multivector $\boldsymbol{M}$. An example is the geometric product of two vectors: $\boldsymbol{ab} = \langle \boldsymbol{ab} \rangle_0 + \langle \boldsymbol{ab} \rangle_2 = \boldsymbol{a} \cdot \boldsymbol{b} + \boldsymbol{a} \wedge \boldsymbol{b}$.

Another simple illustration is the geometric product of $\boldsymbol{A} = 5\sigma_3 + 3\sigma_1\sigma_2$ and $\boldsymbol{b} = 9\sigma_2 + 7\sigma3$:

$$\begin{aligned}\boldsymbol{Ab} &= 35(\sigma_3)^2 + 27\sigma_1(\sigma_2)^2 + 45\sigma_3\sigma_2 + 21\sigma_1\sigma_2\sigma_3 \\ &= 35 + 27\sigma_1 - 45\sigma_2\sigma_3 + 21I.\end{aligned} \tag{13.15}$$

Note here that for $\sigma_i\sigma_i = (\sigma_i)^2 = \sigma_i \cdot \sigma_i = 1$ and $\sigma_i\sigma_j = \sigma_i \wedge \sigma_j$ the geometric product of equal unit basis vectors equals 1 and the product of different unit bases is equal to their wedge, which for simple notation can be omitted.

Let us now focus on the general definition of the inner product between the r blades $\boldsymbol{A}_r = \boldsymbol{a}_1 \wedge \boldsymbol{a}_2 \wedge \ldots \boldsymbol{a}_r$ and $\boldsymbol{B}_s = \boldsymbol{b}_1 \wedge \boldsymbol{b}_2 \wedge \ldots \boldsymbol{b}_s$. The inner product can be defined recursively as follows:

$$\boldsymbol{A}_r \cdot \boldsymbol{B}_s = \begin{cases} ((\boldsymbol{a}_1 \wedge \boldsymbol{a}_2 \wedge \ldots \boldsymbol{a}_r) \cdot \boldsymbol{b}_1) \cdot (\boldsymbol{b}_2 \wedge \boldsymbol{b}_3 \wedge \ldots \boldsymbol{b}_s) & \text{if } r \geq s \\ (\boldsymbol{a}_1 \wedge \boldsymbol{a}_2 \wedge \ldots \boldsymbol{a}_{r-1}) \cdot (\boldsymbol{a}_r(\boldsymbol{b}_1 \wedge \boldsymbol{b}_2 \wedge \ldots \boldsymbol{b}_s)) & \text{if } r < s, \end{cases} \tag{13.16}$$

where

$$(\boldsymbol{a}_1 \wedge \boldsymbol{a}_2 \wedge \ldots \boldsymbol{a}_r) \cdot \boldsymbol{b}_1 = \sum_{i=1}^{r} (-1)^{r-i} \boldsymbol{a}_1 \wedge \boldsymbol{a}_2 \wedge \ldots \boldsymbol{a}_{i-1} \wedge (\boldsymbol{a}_i \cdot \boldsymbol{b}_1) \wedge \boldsymbol{a}_{i+1} \wedge \ldots \boldsymbol{a}_r, \tag{13.17}$$

and

$$\boldsymbol{a}_r \cdot (\boldsymbol{b}_1 \wedge \boldsymbol{b}_2 \wedge ... \boldsymbol{b}_s) = \sum_{i=1}^{s} (-1)^{i-1} \boldsymbol{b}_1 \wedge \boldsymbol{b}_2 \wedge ... \boldsymbol{b}_{i-1} \wedge (\boldsymbol{a}_r \cdot \boldsymbol{b}_i) \wedge \boldsymbol{b}_{i+1} \wedge ... \wedge \boldsymbol{b}_s. \tag{13.18}$$

From Eq. (13.14) and Eq. (13.16), we can express the interior and exterior products for the multivectors as

$$\boldsymbol{A}_r \cdot \boldsymbol{B}_s = \langle \boldsymbol{A}_r \boldsymbol{B}_s \rangle_{|r-s|}, \tag{13.19}$$

and

$$\boldsymbol{A}_r \wedge \boldsymbol{B}_s = \langle \boldsymbol{A}_r \boldsymbol{B}_s \rangle_{r+s}. \tag{13.20}$$

We illustrate the use of these equations with the following examples

$$\boldsymbol{a} \cdot (\boldsymbol{b} \wedge \boldsymbol{c}) = (\boldsymbol{a} \cdot \boldsymbol{b})\boldsymbol{c} - (\boldsymbol{a} \cdot \boldsymbol{c})\boldsymbol{b}. \tag{13.21}$$

Given $\boldsymbol{a}_r = \boldsymbol{a}_1$ and $\boldsymbol{B}_s = \boldsymbol{b}_1 \wedge \boldsymbol{b}_2 \wedge \boldsymbol{b}_3 \wedge \boldsymbol{b}_4$

$$\begin{aligned}
\boldsymbol{a}_r \cdot \boldsymbol{B}_s &= \boldsymbol{a}_1 \cdot (\boldsymbol{b}_1 \wedge \boldsymbol{b}_2 \wedge \boldsymbol{b}_3 \wedge \boldsymbol{b}_4) = \sum_{k=1}^{n} (-1)^{k+1} (\boldsymbol{b}_1 \wedge ... (\boldsymbol{a}_1 \cdot \boldsymbol{b}_k) ... \boldsymbol{b}_4) \\
&= (\boldsymbol{a}_1 \cdot \boldsymbol{b}_1)(\boldsymbol{b}_2 \wedge \boldsymbol{b}_3 \wedge \boldsymbol{b}_4) - (\boldsymbol{a}_1 \cdot \boldsymbol{b}_2)(\boldsymbol{b}_1 \wedge \boldsymbol{b}_3 \wedge \boldsymbol{b}_4) \\
&\quad + (\boldsymbol{a}_1 \cdot \boldsymbol{b}_3)(\boldsymbol{b}_1 \wedge \boldsymbol{b}_2 \wedge \boldsymbol{b}_4) - (\boldsymbol{a}_1 \cdot \boldsymbol{b}_4)(\boldsymbol{b}_1 \wedge \boldsymbol{b}_2 \wedge \boldsymbol{b}_3),
\end{aligned} \tag{13.22}$$

and, given $\boldsymbol{A}_r = \boldsymbol{a}_1 \wedge \boldsymbol{a}_2 \wedge \boldsymbol{a}_3$ and $\boldsymbol{B}_s = \boldsymbol{b}_1 \wedge \boldsymbol{b}_2$,

$$\begin{aligned}
\boldsymbol{A}_r \cdot \boldsymbol{B}_s &= (\boldsymbol{a}_1 \wedge \boldsymbol{a}_2 \wedge \boldsymbol{a}_3) \cdot (\boldsymbol{b}_1 \wedge \boldsymbol{b}_2) \\
&= [(\boldsymbol{a}_1 \wedge \boldsymbol{a}_2 \wedge \boldsymbol{a}_3) \cdot \boldsymbol{b}_1] \cdot \boldsymbol{b}_2 \\
&= [\boldsymbol{a}_1 \wedge \boldsymbol{a}_2 (\boldsymbol{a}_3 \cdot \boldsymbol{b}_1) - \boldsymbol{a}_1 \wedge \boldsymbol{a}_3 (\boldsymbol{a}_2 \cdot \boldsymbol{b}_1) + \boldsymbol{a}_2 \wedge \boldsymbol{a}_3 (\boldsymbol{a}_1 \cdot \boldsymbol{b}_1)] \cdot \boldsymbol{b}_2 \\
&= (\boldsymbol{a}_3 \cdot \boldsymbol{b}_1)[\boldsymbol{a}_1 \wedge \boldsymbol{a}_2 \cdot \boldsymbol{b}_2] - (\boldsymbol{a}_2 \cdot \boldsymbol{b}_1)[\boldsymbol{a}_1 \wedge \boldsymbol{a}_3 \cdot \boldsymbol{b}_2] + (\boldsymbol{a}_1 \cdot \boldsymbol{b}_1)[\boldsymbol{a}_2 \wedge \boldsymbol{a}_3 \cdot \boldsymbol{b}_2] \\
&= (\boldsymbol{a}_3 \cdot \boldsymbol{b}_1)[\boldsymbol{a}_1 (\boldsymbol{a}_2 \cdot \boldsymbol{b}_2) - \boldsymbol{a}_2 (\boldsymbol{a}_1 \cdot \boldsymbol{b}_2)] - (\boldsymbol{a}_2 \cdot \boldsymbol{b}_1)[\boldsymbol{a}_1 (\boldsymbol{a}_3 \cdot \boldsymbol{b}_2) \\
&\quad - \boldsymbol{a}_3 (\boldsymbol{a}_1 \cdot \boldsymbol{b}_2)] + (\boldsymbol{a}_1 \cdot \boldsymbol{b}_1)[\boldsymbol{a}_2 (\boldsymbol{a}_3 \cdot \boldsymbol{b}_2) - \boldsymbol{a}_3 (\boldsymbol{a}_2 \cdot \boldsymbol{b}_2)] \\
&= [(\boldsymbol{a}_3 \cdot \boldsymbol{b}_1)(\boldsymbol{a}_2 \cdot \boldsymbol{b}_2) - (\boldsymbol{a}_2 \cdot \boldsymbol{b}_1)(\boldsymbol{a}_3 \cdot \boldsymbol{b}_2)]\boldsymbol{a}_1 \\
&\quad + [(\boldsymbol{a}_1 \cdot \boldsymbol{b}_1)(\boldsymbol{a}_3 \cdot \boldsymbol{b}_2) - (\boldsymbol{a}_3 \cdot \boldsymbol{b}_1)(\boldsymbol{a}_1 \cdot \boldsymbol{b}_2)]\boldsymbol{a}_2 \\
&\quad + [(\boldsymbol{a}_2 \cdot \boldsymbol{b}_1)(\boldsymbol{a}_1 \cdot \boldsymbol{b}_2) - (\boldsymbol{a}_1 \cdot \boldsymbol{b}_1)(\boldsymbol{a}_2 \cdot \boldsymbol{b}_2)]\boldsymbol{a}_3 \\
&= \alpha_1 \boldsymbol{a}_1 + \alpha_2 \boldsymbol{a}_2 + \alpha_3 \boldsymbol{a}_3,
\end{aligned} \tag{13.23}$$

we get a new point which represents the intersection of the plane $\boldsymbol{A}_r = \boldsymbol{a}_1 \wedge \boldsymbol{a}_2 \wedge \boldsymbol{a}_3$ and the line $\boldsymbol{B}_s = \boldsymbol{b}_1 \wedge \boldsymbol{b}_2$. It is expressed as a linear combination of $\boldsymbol{a}_1, \boldsymbol{a}_2, \boldsymbol{a}_3$ or the points expanding the plane $\boldsymbol{A}_r$.

Finally, for an r-grade multivector $\boldsymbol{A}_r = \sum_{i=0}^{r} \langle \boldsymbol{A}_r \rangle_i$, the following operations are defined:

$$\text{Grade Involution: } \widehat{\boldsymbol{A}}_r = \sum_{i=0}^{r}(-1)^i\langle\boldsymbol{A}_r\rangle_i, \tag{13.24}$$

$$\text{Reversion: } \widetilde{\boldsymbol{A}}_r = \sum_{i=0}^{r}(-1)^{\frac{i(i-1)}{2}}\langle\boldsymbol{A}_r\rangle_i, \tag{13.25}$$

$$\text{Clifford Conjugation: } \overline{\boldsymbol{A}}_r = \widetilde{\widehat{\boldsymbol{A}}}_r = \sum_{i=0}^{r}(-1)^{\frac{i(i+1)}{2}}\langle\boldsymbol{A}_r\rangle_i. \tag{13.26}$$

The grade involution simply negates the odd-grade blades of a multivector. The reversion can also be obtained by reversing the order of basis vectors making up the blades in a multivector and then rearranging them to their original order using the anticommutativity of the Clifford product. The Clifford conjugation can be used to compute the inverse of a vector $\boldsymbol{a}$ as

$$\boldsymbol{a}^{-1} = \frac{\bar{\boldsymbol{a}}}{\boldsymbol{a}\bar{\boldsymbol{a}}}. \tag{13.27}$$

This formula for the inverse can also be applied for homogeneous multivectors but cannot be used for all multivectors in general.

13.2.3 Geometric Algebra of 3-D Space

The basis for the geometric algebra $\mathcal{G}_{3,0,0}$ of 3-D space has $2^3 = 8$ elements and is given by

$$\underbrace{1}_{scalar}, \underbrace{\{\sigma_1, \sigma_2, \sigma_3\}}_{vectors}, \underbrace{\{\sigma_1\sigma_2, \sigma_2\sigma_3, \sigma_3\sigma_1\}}_{bivectors}, \underbrace{\{\sigma_1\sigma_2\sigma_3\}}_{trivector} \equiv I. \tag{13.28}$$

It can easily be verified that the trivector or pseudoscalar $\sigma_1\sigma_2\sigma_3$ squares to -1 and commutes with all multivectors in the 3-D space. We therefore give it the symbol I, noting that this is not the uninterpreted commutative scalar imaginary j used in quantum mechanics and engineering.

Multiplication of the three basis vectors σ_1, σ_2, and σ_3 by I results in the three basis bivectors $\sigma_1\sigma_2 = I\sigma_3$, $\sigma_2\sigma_3 = I\sigma_1$, and $\sigma_3\sigma_1 = I\sigma_2$. These simple bivectors rotate vectors in their own plane by 90°–e.g., $(\sigma_1\sigma_2)\sigma_2 = \sigma_1$, $(\sigma_2\sigma_3)\sigma_2 = -\sigma_3$, etc. By identifying the $\boldsymbol{i}, \boldsymbol{j}, \boldsymbol{k}$ of the quaternion algebra with $I\sigma_1, -I\sigma_2, I\sigma_3$, the famous Hamilton relations $\boldsymbol{i}^2 = \boldsymbol{j}^2 = \boldsymbol{k}^2 = \boldsymbol{ijk} = -1$ can be recovered. Since the $\boldsymbol{i}, \boldsymbol{j}, \boldsymbol{k}$ are bivectors, it comes as no surprise that they represent 90° rotations in orthogonal directions and that they provide a well-suited system for the representation of general 3-D rotations.

In geometric algebra a rotor (short name for rotator), $\boldsymbol{R}$, is an even-grade element of the algebra which satisfies the expression $\boldsymbol{R}\widetilde{\boldsymbol{R}} = 1$, where $\widetilde{\boldsymbol{R}}$ stands for the conjugate of $\boldsymbol{R}$.

If $\mathcal{A} = \{a_0, a_1, a_2, a_3\} \in \mathcal{G}_{3,0,0}$ represents a unit quaternion, then the rotor which performs the same rotation is simply given by

$$\boldsymbol{R} = \underbrace{a_0}_{scalar} + \underbrace{a_1(I\sigma_1) - a_2(I\sigma_2) + a_3(I\sigma_3)}_{bivectors}$$

$$= a_0 + a_1\sigma_2\sigma_3 - a_2\sigma_3\sigma_1 + a_3\sigma_3\sigma_1. \tag{13.29}$$

The quaternion algebra is therefore seen to be a subset of the geometric algebra of 3-D space. The conjugate of the rotor is computed as follows:

$$\widetilde{\boldsymbol{R}} = a_0 - a_1\sigma_2\sigma_3 + a_2\sigma_3\sigma_1 - a_3\sigma_3\sigma_1. \tag{13.30}$$

The transformation in terms of a rotor $\boldsymbol{a} \mapsto \boldsymbol{R}\boldsymbol{a}\widetilde{\boldsymbol{R}} = \boldsymbol{b}$ is a very general way of handling rotations; this works for multivectors of any grade and in spaces of any dimension, in contrast to quaternion calculus. Rotors combine in a straightforward manner, i.e., a rotor $\boldsymbol{R}_1$ followed by a rotor $\boldsymbol{R}_2$ is equivalent to a total rotor $\boldsymbol{R}$, where $\boldsymbol{R} = \boldsymbol{R}_2\boldsymbol{R}_1$.

13.3 Conformal Geometry

The geometric algebra of 3D Euclidean space $\mathcal{G}_{3,0,0}$ has a point basis and the motor algebra $\mathcal{G}_{3,0,1}$ a line basis. In the latter the lines expressed in terms of Plücker coordinates can be used to represent points and planes as well [4, 5]. The reader can find a comparison of representations of points, lines and planes using $\mathcal{G}_{3,0,1}$ and $\mathcal{G}_{3,0,1}$ in [4].

Interesting enough in the case of the conformal geometric algebra we find that the unit element is the sphere which allows us to represent the other geometric primitives in its terms. To see how this is possible, we follow the same formulation presented in [10] and show how the Euclidean vector space $\mathbb{R}^n$ is represented in $\mathbb{R}^{n+1,1}$. This space has an orthonormal vector basis given by $\{e_1, ..., e_n, e_+, e_-\}$, with the properties

$$e_i^2 = 1, \quad i = 1..., n; \tag{13.31}$$
$$e_\pm^2 = \pm 1; \tag{13.32}$$
$$e_i \cdot e_+ = e_i \cdot e_- = e_+ \cdot e_- = 0, \quad i = 1, ..., n. \tag{13.33}$$

Note that this basis is not written in bold.

A *null basis* $\{e_0, e_\infty\}$ can be introduced by

$$e_0 = \frac{(e_- - e_+)}{2}, \tag{13.34}$$
$$e_\infty = e_- + e_+, \tag{13.35}$$

with the properties

$$e_0^2 = e_\infty^2 = 0, \ e_\infty \cdot e_0 = -1. \tag{13.36}$$

e_0 and e_∞ are the origin and the point at infinity. A unit pseudoscalar $E \in \mathbb{R}^{1,1}$, which represents the Minkowski plane, is defined by

$$\boldsymbol{E} = e_\infty \wedge e_0 = e_+ \wedge e_- = e_+ e_-, \tag{13.37}$$

having the properties

$$\boldsymbol{E}^2 = 1, \ \widetilde{\boldsymbol{E}} = -\boldsymbol{E}, \tag{13.38}$$
$$\boldsymbol{E} e_\pm = e_\mp = -e_\pm, \tag{13.39}$$
$$\boldsymbol{E} e_\infty = -e_\infty \boldsymbol{E} = -e_\infty, \ \boldsymbol{E} e_0 = -e_0 \boldsymbol{E} = e_0 \ \text{(absorption)}, \tag{13.40}$$
$$1 - \boldsymbol{E} = -e_\infty e_0, \ 1 + \boldsymbol{E} = -e_0 e_\infty. \tag{13.41}$$

The dual of $\boldsymbol{E}$ is given by

$$\boldsymbol{E}^* = \boldsymbol{E} I^{-1} = -\widetilde{\boldsymbol{E} I}, \tag{13.42}$$

where I is the pseudoscalar for $\mathbb{R}^{n+1,1}$.

Euclidean points $\mathbf{x}_e \in \mathbb{R}^n$ can be represented in $\mathbb{R}^{n+1,1}$ in a general way, as

$$\boldsymbol{x}_c = \mathbf{x}_e + \alpha e_0 + \beta e_\infty, \tag{13.43}$$

where α and β are arbitrary scalars. A conformal point $\boldsymbol{x}_c \in \mathbb{R}^{n+1,1}$ can be divided into its Euclidean and conformal parts by an operation called the *conformal split*. This split is defined by the projection operators P_E (projection) and $P_E^\perp$ (rejection) as follows:

$$P_E(\boldsymbol{x}_c) = (\boldsymbol{x}_c \cdot \boldsymbol{E})\boldsymbol{E} = \alpha e_0 + \beta e_\infty \in \mathbb{R}^{1,1}, \tag{13.44}$$
$$P_E^\perp(\boldsymbol{x}_c) = (\boldsymbol{x}_c \cdot \boldsymbol{E}^*)\widetilde{\boldsymbol{E}^*} = (\boldsymbol{x}_c \wedge \boldsymbol{E})\boldsymbol{E} = \mathbf{x}_e \in \mathbb{R}^n. \tag{13.45}$$
$$\boldsymbol{x}_c = P_E(\boldsymbol{x}_c) + P_E^\perp(\boldsymbol{x}_c). \tag{13.46}$$

The names "projection" and "rejection" stem from the geometrical meaning of these operators. The first returns the component of $\boldsymbol{x}_c$ which is parallel to E by a projection (dot product). The latter produces the component of $\boldsymbol{x}_c$ which is orthogonal to $\boldsymbol{E}$, hence the name (see Fig. 13.3).

To improve our model, we would like to use *homogeneous coordinates*, as in the case of projective geometry. In homogeneous coordinates, all points are equal up to a scale factor. Therefore, we need to define some way to fix the scale of the points. A point $\boldsymbol{x}_c \in \mathbb{R}^{n+1,1}$ is normalized or expressed in standard form when

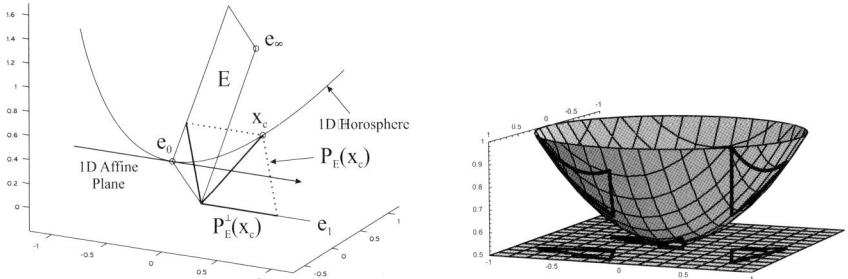

Fig. 13.3. (a) The projection and rejection of vector $x_c \in \mathbb{R}^{2,1}$ from the $\boldsymbol{E}$ plane. The operators are illustrated for the case of 1-D. (b) Horosphere of R^2 with triangles of the 2-D affine plane projected into the horosphere

$$\boldsymbol{x}_c \cdot e_\infty = -1. \tag{13.47}$$

Now, recall that a hyperplane $\mathbb{P}(\boldsymbol{n}, \boldsymbol{a}) \in \mathbb{R}^{n+1,1}$ with normal $\boldsymbol{n}$ and passing through the point $\boldsymbol{a}$ is the solution to the equation

$$\boldsymbol{n} \cdot (\boldsymbol{x} - \boldsymbol{a}) = 0, \ \boldsymbol{x} \in \mathbb{R}^{n+1,1}. \tag{13.48}$$

The normalization condition $\boldsymbol{x}_c \cdot e_\infty = e_\infty \cdot e_0 = -1$ is equivalent to the equation

$$e_\infty \cdot (\boldsymbol{x}_c - e_0) = 0, \tag{13.49}$$

which is the equation of a hyperplane $\mathbb{P}(e_\infty, e_0)$. Thus, the normalization condition of Eq. (13.47) constrains the points $\boldsymbol{x}_c$ to lie in a hyperplane passing though e_0 with normal e_∞. Equation (13.47) fixes scale; however, for the conformal model, another constraint is needed to fix $\boldsymbol{x}_c$ as a unique representation of $\mathbf{x}_e \in \mathbb{R}^n$.

To complete the definition of *generalized homogeneous coordinates* for points in $\mathbb{R}^{n+1,1}$, the last constraint is that $\boldsymbol{x}_c^2 = 0$. The set $\mathbb{N}^{n+1}$ of vectors that square to zero is called the *null cone*. Therefore, conformal points are required to lie in the intersection of the null cone $\mathbb{N}^{n+1}$ with the hyperplane $\mathbb{P}(e_\infty, e_0)$. The resulting surface $\mathbb{N}_e^n$ is called the *horosphere*:

$$\mathbb{N}_e^n = \mathbb{N}^{n+1} \cap \mathbb{P}(e_\infty, e_0) = \{\boldsymbol{x}_c \in \mathbb{R}^{n+1,1} | \boldsymbol{x}_c^2 = 0, \boldsymbol{x}_c \cdot e_\infty = -1\}. \tag{13.50}$$

These two constraints finally define the mapping from Euclidean space to conformal space. To see how this mapping is obtained, first we see that any point $\boldsymbol{x}_c = \mathbf{x}_e + \alpha e_0 + \beta e_\infty \in \mathbb{N}_e^n$ can be expressed as $\boldsymbol{x}_c = \mathbf{x}_e + k_1 e_+ + k_2 e_-$, for some scalars k_1, k_2, since e_0 and e_∞ are linear combinations of the basis vectors e_+ and e_-. Then, by applying the conformal split to x_c we get

$$x_c = x_c E^2 = (x_c \wedge E + x_c \cdot E)E = (x_c \wedge E)E + (x_c \cdot E)E, \qquad (13.51)$$

since $E^2 = 1$. Now, recall that $(x_c \wedge E)E = \mathbf{x}_e$ is the rejection (see Eq. (13.46)). The expression $(x_c \cdot E)E$ can be expanded as

$$(x_c \cdot E)E = (x_c \cdot (e_\infty \wedge e_0))E = e_0 + (k_1 + k_2)e_\infty. \qquad (13.52)$$

Now, applying the condition that $x_c^2 = 0$, we find from Eq. (13.51) that

$$\begin{aligned} x_c^2 &= ((x_c \wedge E)E + (x_c \cdot E)E)^2, \\ 0 &= (\mathbf{x}_e + e_0 + (k_1 + k_2)e_\infty)^2 = \mathbf{x}_e^2 - (k_1 + k_2), \\ \mathbf{x}_e^2 &= (k_1 + k_2). \end{aligned} \qquad (13.53)$$

Finally, using Eq. (13.51), and substituting Eq. (13.53) into Eq. (13.52), we get

$$x_c = (x_c \wedge E)E + (x_c \cdot E)E = \mathbf{x}_e + e_0 + \frac{1}{2}(k_1 + k_2)e_\infty = \mathbf{x}_e + \frac{1}{2}\mathbf{x}_e^2 e_\infty + e_0. \qquad (13.54)$$

An illustration of the null cone, the hyperplane, and the horosphere can be seen in Fig. 13.4a.

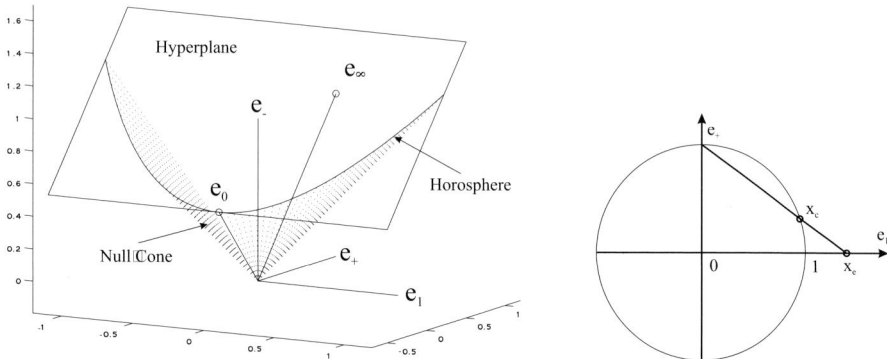

Fig. 13.4. (a) (*left*) The null cone (dotted lines), the hyperplane $\mathbb{P} + (e_\infty, e_0)$, and the horosphere for 1-D. Note that even though the normal of the hyperplane is e_∞, the plane is actually *geometrically parallel* to this vector. (b) (*right*) Stereographic projection for 1-D

We can gain further insight into the geometrical meaning of the null vectors by analyzing Eq. (13.54). For instance, by setting $\mathbf{x}_e = 0$, we find that e_0

represents the origin of $\mathbb{R}^n$ (hence the name). Similarly, dividing this equation by $\boldsymbol{x}_c \cdot e_0 = -\frac{1}{2}\mathbf{x}_e^2$ gives

$$\frac{\boldsymbol{x}_c}{\boldsymbol{x}_c \cdot e_0} = -\frac{2}{\mathbf{x}_e^2}(\mathbf{x}_e + \frac{1}{2}\mathbf{x}_e^2 e_\infty + e_0) = -\frac{2\mathbf{x}_e^2}{\mathbf{x}_e^2}(\frac{1}{\mathbf{x}_e} + \frac{1}{2}e_\infty + \frac{e_0}{\mathbf{x}_e^2})$$

$$= -2(\frac{1}{\mathbf{x}_e} + \frac{1}{2}e_\infty + \frac{e_0}{\mathbf{x}_e^2}) \xrightarrow[\mathbf{x}_e \to \infty]{} e_\infty. \quad (13.55)$$

Thus, we conclude that e_∞ represents the point at infinity.

13.3.1 Stereographic Projection

Conformal geometry is equivalent to stereographic projection in Euclidean space. Generally speaking, a stereographic projection is a mapping, taking points lying on a hypersphere to points lying on a hyperplane and following a simple geometric construction. It is well known that this projection is used in cartography to make maps of the earth. In that case, the projection plane passes through the equator and the sphere is centered at the origin. To make a projection, a line is drawn from the North Pole to each point on the sphere, and the intersection of this line with the projection plane constitutes the stereographic projection.

Next, we will illustrate the equivalence between stereographic projection and conformal geometric algebra in $\mathbb{R}^1$. We will be working in $\mathbb{R}^{2,1}$, with the basis vectors $\{e_1, e_+, e_-\}$ having the usual properties. The projection plane will be the x-axis, and the sphere will be a circle centered at the origin with unitary radius.

Given a scalar $\mathbf{x}_e$ representing a point on the x-axis, we wish to find the point $\boldsymbol{x}_c$ lying on the circle that projects to it (see Fig. 13.4b). The equation of the line passing through the north pole and $\mathbf{x}_e$ is given by $f(\mathbf{x}) = -\frac{1}{\mathbf{x}_e}\mathbf{x}+1$. The equation of the circle is $\mathbf{x}^2 + f(\mathbf{x})^2 = 1$. Substituting the equation of the line on the circle, we get $\mathbf{x}^2 - 2\mathbf{x}\mathbf{x}_e + \mathbf{x}^2\mathbf{x}_e^2 = 0$, which has the two solutions $\mathbf{x} = 0$, $\mathbf{x} = 2\frac{\mathbf{x}_e}{\mathbf{x}_e^2+1}$. Only the latter solution is meaningful. Substituting in the equation of the line, we get $f(\mathbf{x}) = \frac{\mathbf{x}_e^2-1}{\mathbf{x}_e^2+1}$. Hence, $\boldsymbol{x}_c$ has coordinates $\boldsymbol{x}_c = \left(2\frac{\mathbf{x}_e}{\mathbf{x}_e^2+1}, \frac{\mathbf{x}_e^2-1}{\mathbf{x}_e^2+1}\right)$, which can be represented in homogeneous coordinates as the vector

$$\boldsymbol{x}_c = 2\frac{\mathbf{x}_e}{\mathbf{x}_e^2+1}e_1 + \frac{\mathbf{x}_e^2-1}{\mathbf{x}_e^2+1}e_+ + e_-. \quad (13.56)$$

If we take the limits to Eq. (13.56), we get the expected points at infinity and the origin of $\mathbb{R}^n$

$$\lim_{\mathbf{x}_e \to \infty} \boldsymbol{x}_c = e_+ + e_- = e_\infty, \quad (13.57)$$

$$\lim_{\mathbf{x}_e \to 0} \frac{\boldsymbol{x}_c}{2} = \frac{e_- - e_+}{2} = e_0.$$

This result is a first confirmation that stereographic projection is equivalent to a conformal mapping given by Eq. (13.54). For a second proof we note that Eq. (13.56) can be rewritten as

$$\boldsymbol{x}_c = 2\frac{\mathbf{x}_e}{\mathbf{x}_e^2+1}e_1 + \frac{\mathbf{x}_e^2-1}{\mathbf{x}_e^2+1}e_+ + e_-$$
$$= \frac{2}{\mathbf{x}_e^2+1}\left(\mathbf{x}_e e_1 + \frac{1}{2}(\mathbf{x}_e^2-1)e_+ + \frac{1}{2}(\mathbf{x}_e^2+1)e_-\right). \quad (13.58)$$

Dividing by the scale factor $\frac{2}{\mathbf{x}_e^2+1}$ in order to achieve the constraint imposed by Eq. (13.47), $\boldsymbol{x}_c \cdot e_\infty = -1$, we arrive at

$$\boldsymbol{x}_c = \mathbf{x}_e e_1 + (\mathbf{x}_e^2-1)e_+ + (\mathbf{x}_e^2+1)e_- = \mathbf{x}_e e_1 + \frac{1}{2}\mathbf{x}_e^2 e_\infty + e_0,$$
$$= \mathbf{x}_e + \frac{1}{2}\mathbf{x}_e^2 e_\infty + e_0, \quad (13.59)$$

where $\mathbf{x}_e = x_e e_1$, which is precisely Eq. (13.54). Hence, we have demonstrated that conformal geometric algebra is projectively equivalent to a stereographic projection (i.e., *up to a scale factor*).

13.3.2 Spheres and Planes

The equation of a sphere of radius ρ centered at point $\mathbf{p}_e \in \mathbb{R}^n$ can be written as

$$(\mathbf{x}_e - \mathbf{p}_e)^2 = \rho^2. \quad (13.60)$$

Since $\boldsymbol{x}_c \cdot \boldsymbol{y}_c = -\frac{1}{2}(\mathbf{x}_e - \mathbf{y}_e)^2$, we can rewrite the formula above in terms of homogeneous coordinates as

$$\boldsymbol{x}_c \cdot \boldsymbol{p}_c = -\frac{1}{2}\rho^2. \quad (13.61)$$

Since $\boldsymbol{x}_c \cdot e_\infty = -1$, we can factor the expression above to

$$\boldsymbol{x}_c \cdot (\boldsymbol{p}_c - \frac{1}{2}\rho^2 e_\infty) = 0. \quad (13.62)$$

This equation then yields the simplified equation for the sphere as

$$\boldsymbol{x}_c \cdot \boldsymbol{s} = 0, \quad (13.63)$$

where

$$\boldsymbol{s} = \boldsymbol{p}_c - \frac{1}{2}\rho^2 e_\infty = \boldsymbol{p}_e + e_0 + \frac{\mathbf{p}_e^2 - \rho^2}{2}e_\infty \quad (13.64)$$

is the equation of the sphere (note from this equation that a point is just a sphere with zero radius). The vector s has the properties

$$s^2 = \rho^2 > 0, \tag{13.65}$$
$$e_\infty \cdot s = -1. \tag{13.66}$$

From these properties, we conclude that the sphere s is a point lying on the hyperplane but *outside* the null cone. In particular, all points on the hyperplane outside the horosphere determine spheres with positive radius, points lying on the horosphere define spheres of zero radius (i.e., points), and points lying inside the horosphere have imaginary radius. Finally, note that spheres of the same radius form a surface which is parallel to the horosphere.

The radius and center of a sphere can be recovered from s using Eq. (13.65) and Eq. (13.66), as

$$\rho^2 = \frac{s^2}{(s \cdot e_\infty)^2}, \quad \text{and} \tag{13.67}$$

$$\boldsymbol{p}_c = \frac{s}{-(s \cdot e_\infty)} + \frac{1}{2}\rho^2 e_\infty. \tag{13.68}$$

With the normalization $s \cdot e_\infty = -1$, each sphere is represented by a unique vector and the set $\{\boldsymbol{x}_c \in \mathbb{R}^{n+1,1} | \boldsymbol{x}_c \cdot s > 0\}$ represents the interior of the sphere.

Alternatively, spheres can be dualized and represented as $(n+1)$-vectors $s^* = sI^{-1}$. Since

$$\widetilde{I} = (-1)^{\frac{1}{2}(n+2)(n+1)} I = -I^{-1}, \tag{13.69}$$

we can express the constraints of Eqs. (13.65) and (13.66) as

$$s^2 = -\widetilde{s^*}s^* = \rho^2,$$
$$e_\infty \cdot s = e_\infty \cdot (s^*I) = (e_\infty \wedge s^*)I = -1. \tag{13.70}$$

The equation for the sphere now becomes

$$\boldsymbol{x}_c \wedge s^* = 0. \tag{13.71}$$

The advantage of the dual form is that the sphere can be directly computed from four points (in 3-D) as

$$s^* = \boldsymbol{x}_{c_1} \wedge \boldsymbol{x}_{c_2} \wedge \boldsymbol{x}_{c_3} \wedge \boldsymbol{x}_{c_4}. \tag{13.72}$$

If we replace one of these points for the point at infinity, we get

$$\pi^* = \boldsymbol{x}_{c_1} \wedge \boldsymbol{x}_{c_2} \wedge \boldsymbol{x}_{c_3} \wedge e_\infty. \tag{13.73}$$

Developing the products, we get

$$\boldsymbol{x}_{c_1} \wedge \boldsymbol{x}_{c_2} = \left(\mathbf{x}_{e_1} + \frac{1}{2}\mathbf{x}_{e_1}^2 e_\infty + e_0\right) \wedge \left(\mathbf{x}_{e_2} + \frac{1}{2}\mathbf{x}_{e_2}^2 e_\infty + e_0\right),$$

$$= \mathbf{x}_{e_1} \wedge \mathbf{x}_{e_2} + \frac{1}{2}(\mathbf{x}_{e_2}^2 \mathbf{x}_{e_1} - \mathbf{x}_{e_1}^2 \mathbf{x}_{e_2}) \wedge e_\infty$$

$$+ (\mathbf{x}_{e_1} - \mathbf{x}_{e_2}) \wedge e_0 + \frac{1}{2}(\mathbf{x}_{e_1}^2 - \mathbf{x}_{e_2}^2)\boldsymbol{E}, \quad (13.74)$$

$$\boldsymbol{x}_{c_1} \wedge \boldsymbol{x}_{c_2} \wedge e_\infty = \mathbf{x}_{e_1} \wedge \mathbf{x}_{e_2} \wedge e_\infty - (\mathbf{x}_{e_1} - \mathbf{x}_{e_2}) \wedge \boldsymbol{E}, \quad (13.75)$$

$$\boldsymbol{x}_{c_3} \wedge \boldsymbol{x}_{c_1} \wedge \boldsymbol{x}_{c_2} \wedge e_\infty = \mathbf{x}_{e_3} \wedge \mathbf{x}_{e_1} \wedge \mathbf{x}_{e_2} \wedge e_\infty - \mathbf{x}_{e_1} \wedge \mathbf{x}_{e_2} \wedge \boldsymbol{E}$$

$$+ \mathbf{x}_{e_3} \wedge (\mathbf{x}_{e_2} - \mathbf{x}_{e_1}) \wedge \boldsymbol{E}. \quad (13.76)$$

Since $\mathbf{x}_{e_1} \wedge \mathbf{x}_{e_2} = \mathbf{x}_{e_1} \wedge (\mathbf{x}_{e_2} - \mathbf{x}_{e_1})$, we get

$$\boldsymbol{x}_{c_3} \wedge \boldsymbol{x}_{c_1} \wedge \boldsymbol{x}_{c_2} \wedge e_\infty = \mathbf{x}_{e_3} \wedge \mathbf{x}_{e_1} \wedge \mathbf{x}_{e_2} \wedge e_\infty$$

$$- \mathbf{x}_{e_1} \wedge (\mathbf{x}_{e_2} - \mathbf{x}_{e_1}) \wedge \boldsymbol{E} \boldsymbol{x}_{e_3} \wedge (\mathbf{x}_{e_2} - \mathbf{x}_{e_1}) \wedge \boldsymbol{E},$$

$$= \mathbf{x}_{e_3} \wedge \mathbf{x}_{e_1} \wedge \mathbf{x}_{e_2} \wedge e_\infty + (\mathbf{x}_{e_3} \wedge (\mathbf{x}_{e_2} - \mathbf{x}_{e_1})$$

$$- \mathbf{x}_{e_1} \wedge (\mathbf{x}_{e_2} - \mathbf{x}_{e_1})) \wedge \boldsymbol{E}, \quad (13.77)$$

$$= \mathbf{x}_{e_3} \wedge \mathbf{x}_{e_1} \wedge \mathbf{x}_{e_2} \wedge e_\infty + ((\boldsymbol{x}_{e_3} - \boldsymbol{x}_{e_1}) \wedge (\boldsymbol{x}_{e_2} - \boldsymbol{x}_{e_1})) \wedge \boldsymbol{E}.$$

But since $\boldsymbol{x}_e \cdot \boldsymbol{E} = 0$, we can rewrite this as

$$\pi^* = \boldsymbol{x}_{c_3} \wedge \boldsymbol{x}_{c_1} \wedge \boldsymbol{x}_{c_2} \wedge e_\infty$$

$$= \mathbf{x}_{e_3} \wedge \mathbf{x}_{e_1} \wedge \mathbf{x}_{e_2} \wedge e_\infty + ((\mathbf{x}_{e_3} - \mathbf{x}_{e_1}) \wedge (\mathbf{x}_{e_2} - \mathbf{x}_{e_1}))\boldsymbol{E}, \quad (13.78)$$

which is the equation of the plane passing through the points $\mathbf{x}_{e_1}$, $\mathbf{x}_{e_2}$, and $\mathbf{x}_{e_3}$. We can easily see that $\mathbf{x}_{e_1} \wedge \mathbf{x}_{e_2} \wedge \mathbf{x}_{e_3}$ is a scalar representing the volume of the parallelepiped with sides $\mathbf{x}_{e_1}$, $\mathbf{x}_{e_2}$, and $\mathbf{x}_{e_3}$. Also, since $(\mathbf{x}_{e_1} - \mathbf{x}_{e_2})$ and $(\mathbf{x}_{e_3} - \mathbf{x}_{e_2})$ are two vectors on the plane, the expression $((\mathbf{x}_{e_1} - \mathbf{x}_{e_2}) \wedge (\mathbf{x}_{e_3} - \mathbf{x}_{e_2}))$ is the normal to the plane. Therefore, planes are spheres passing through the point at infinity.

13.3.3 Geometric Identities, Duals, and Incidence Algebra Operations

A circle z can be regarded as the intersection of two spheres s_1 and s_2. This means that each point on the circle, $\boldsymbol{x}_c \in z$, will be on both spheres: $\boldsymbol{x}_c \in s_1$ and $\boldsymbol{x}_c \in s_2$. So, $\boldsymbol{x}_c \wedge s_i^* = 0$ and for duality $\boldsymbol{x}_c \cdot s_i = 0$, $i = 1, 2$. So, assuming that s_1 and s_2 are linearly independent, and $\boldsymbol{x}_c \in z$ we can write, using Eq. (13.16) backwards

$$0 = (\boldsymbol{x}_c \cdot s_1)s_2 - (\boldsymbol{x}_c \cdot s_2)s_1 = x_c \cdot (s_1 \wedge s_2), \quad (13.79)$$

so $\boldsymbol{x}_c \in \boldsymbol{s}_1 \wedge \boldsymbol{s}_2$, and hence $\boldsymbol{z} = \boldsymbol{s}_1 \wedge \boldsymbol{s}_2$. This means that a circle can be expressed as the intersection of two spheres. It is easy to see that the intersection with a third sphere leads to a point pair. We have derived algebraically that the wedge of two linearly independent spheres yields their intersecting circle (see Fig. 13.5), this toplogical relation between two spheres can also be conveniently described using the dual of the meet operation, namely

$$\boldsymbol{z} = (\boldsymbol{z}^*)^* = (\boldsymbol{s}_1^* \vee \boldsymbol{s}_2^*)^* = \boldsymbol{s}_1 \wedge \boldsymbol{s}_2. \tag{13.80}$$

This new equation says that the dual of a circle can be computed via the meet of two spheres in their dual form: $\boldsymbol{z}^* = \boldsymbol{s}_1^* \vee \boldsymbol{s}_2^*$. This equation confirms geometrically our previous algebraic computation from Eq. (13.79).

The dual form of the circle (in 3D) can be expressed by three points lying on it as

$$\boldsymbol{z}^* = \boldsymbol{x}_{c_1} \wedge \boldsymbol{x}_{c_2} \wedge \boldsymbol{x}_{c_3}, \tag{13.81}$$

see Fig. 13.5a.

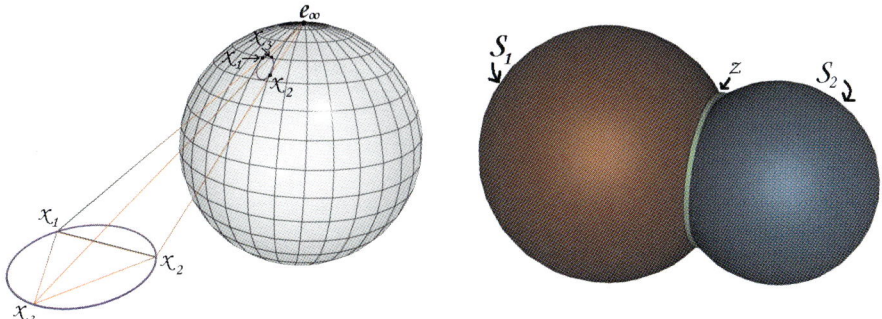

Fig. 13.5. (a) (*left*) Circle computed using three points, note its stereographic projection. (b) (*right*) Circle computed using the meet of two spheres

Similar to the case of planes shown in Eq. (13.73), lines can be defined by circles passing through the point at infinity as

$$\boldsymbol{L}^* = \boldsymbol{x}_{c_1} \wedge \boldsymbol{x}_{c_2} \wedge e_\infty. \tag{13.82}$$

This can be demonstrated by developing the wedge products as in the case of the planes to yield

$$\boldsymbol{x}_{c_1} \wedge \boldsymbol{x}_{c_2} \wedge e_\infty = \boldsymbol{x}_{e_1} \wedge \boldsymbol{x}_{e_2} \wedge e_\infty + (\boldsymbol{x}_{e_2} - \boldsymbol{x}_{e_1}) \wedge \boldsymbol{E}, \tag{13.83}$$

from where it is evident that the expression $\boldsymbol{x}_{e_1} \wedge \boldsymbol{x}_{e_2}$ is a bivector representing the plane where the line is contained and $(\boldsymbol{x}_{e_2} - \boldsymbol{x}_{e_1})$ is the direction of the line.

The dual of a point $\boldsymbol{p}$ is a sphere $\boldsymbol{s}$. The intersection of four spheres yields a point, see Fig. 13.6b. The dual relationships between a point and its dual, the sphere, are:

$$\boldsymbol{s}^* = \boldsymbol{p}_1 \wedge \boldsymbol{p}_2 \wedge \boldsymbol{p}_3 \wedge \boldsymbol{p}_4 \leftrightarrow \boldsymbol{p}^* = \boldsymbol{s}_1 \wedge \boldsymbol{s}_2 \wedge \boldsymbol{s}_3 \wedge \boldsymbol{s}_4, \tag{13.84}$$

where the points are denoted as $\boldsymbol{p}_i$ and the spheres $\boldsymbol{s}_i$ for $i = 1, 2, 3, 4$.

A summary of the basic geometric entities and their duals is presented in Table 13.1.

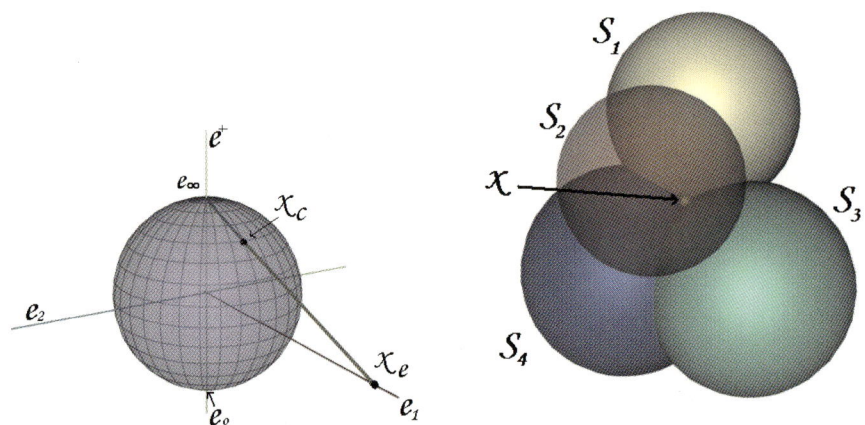

Fig. 13.6. (a) *(left)* Conformal point generated by projecting a point of the affine plane to the unit sphere. (b) *(right)* Point generated by the meet of four spheres

There is another very useful relationship between a $r - 2$-dimensional sphere $\boldsymbol{A}_r$ and the sphere $\boldsymbol{s}^*$ (computed as the dual of a point $\boldsymbol{s}$). If from the sphere $\boldsymbol{A}_r$ we can compute the hyperplane $\boldsymbol{A}_{r+1} \equiv \boldsymbol{e}_\infty \wedge \boldsymbol{A}_r \neq 0$, we can express the meet between the dual of the point $\boldsymbol{s}$ (a sphere) and the hyperplane $\boldsymbol{A}_{r+1}$, which gives us the sphere $\boldsymbol{A}_r$ one dimension lower:

$$(-1)^\epsilon \boldsymbol{s}^* \cap \boldsymbol{A}_{r+1} = (\boldsymbol{s}^* I) \cdot \boldsymbol{A}_{r+1} = \boldsymbol{s} \boldsymbol{A}_{r+1} = \boldsymbol{A}_r. \tag{13.85}$$

This result demonstrates an interesting relationship: that the sphere $\boldsymbol{A}_r$ and the hyperplane $\boldsymbol{A}_{r+1}$ are related via the point $\boldsymbol{s}$ (dual of the sphere $\boldsymbol{s}^*$). Thus, we then rewrite the Eq. (13.85) as follows:

$$\boldsymbol{s} = \boldsymbol{A}_r \boldsymbol{A}_{r+1}^{-1}. \tag{13.86}$$

Let us see some particular cases of the relation Eq. (13.86). Given the circle $\boldsymbol{z}^*$, $(\boldsymbol{A}_r)$, we can generate the plane $\pi^* = \boldsymbol{z}^* \wedge \boldsymbol{e}_\infty$, $(\boldsymbol{A}_{r+1})$, and compute the sphere $\boldsymbol{s}$

13 Robot Perception and Action Using Conformal Geometric Algebra

Table 13.1. Entities in conformal geometric algebra

Entity	Representation	Grade	Dual Representation	Grade
Sphere	$s = \mathbf{p} + \frac{1}{2}(\mathbf{p}^2 - \rho^2)e_\infty + e_0$	1	$s^* = a \wedge b \wedge c \wedge d$	4
Point	$x = \mathbf{x}_c + \frac{1}{2}\mathbf{x}_c^2 e_\infty + e_0$	1	$x^* = s_1 \wedge s_2 \wedge s_3 \wedge s_4$ $= (-Ex - \frac{1}{2}x^2 e_\infty + e_0)I_E$	4
Plane	$\pi = nI_E - de_\infty$ $n = (a-b) \wedge (a-c)$ $d = (a \wedge b \wedge c)I_E$	1	$\pi^* = e_\infty \wedge a \wedge b \wedge c$	4
Line	$L = \pi_1 \wedge \pi_2$ $L = nI_E - e_\infty m I_E$ $n = (a-b)$ $d = (a \wedge b)$	2	$L^* = e_\infty \wedge a \wedge b$	3
Circle	$z = s_1 \wedge s_2$	2	$z^* = a \wedge b \wedge c$	3
Point Pair	$P_p = s_1 \wedge s_2 \wedge s_3$ $P_p = s \wedge L$	3 3	$P_p^* = a \wedge b, \ X^* = e_\infty \wedge x$	2

$$s = z^*(\pi^*)^{-1} = z(\pi)^{-1}, \quad (13.87)$$

see Fig. 13.7a . Observe that z^* is the great circle in s. Note that we can apply either $s = z^*(\pi^*)^{-1}$ or $s = z(\pi)^{-1}$, because the pseudoscalar of z^* is canceled with the inverse pseudoscalar from of $(\pi^*)^{-1}$.

From Eq. (13.87) we can obtain the center of a circle. Let $z^* = x_1 \wedge x_2 \wedge x_3$ be a circle and the plane $\pi^* = e_\infty \wedge x_1 \wedge x_2 \wedge x_3$. From Eq. (13.68), $p_c = \frac{s}{-(s \cdot e_\infty)} + \frac{1}{2}\rho^2 e_\infty$ is the center of the sphere s, and then the same center for the great circle $z^* = s\pi^*$, where the sphere can be obtained as

$$s = z^*(\pi^*)^{-1} = \frac{z^*}{\pi^*} = \frac{x_1 \wedge x_2 \wedge x_3}{e_\infty \wedge x_1 \wedge x_2 \wedge x_3}, \quad (13.88)$$

see Fig. 13.7c.

Similarly we can compute another important geometric relationship called the *pair of points*. Let P_p or (A_r) two points and L or (A_{r+1}), the line $L = P_p \wedge e_\infty$. Then there is a sphere s whose intersection with the line L is the *pair of points* P_p,

$$s = P_p L^{-1}. \quad (13.89)$$

Observe that the sphere s is such that the two points P_p are the extremes of one of its diameters. Now using this result, given the line L and the sphere s we can compute the pair of points P_p, see Fig. 13.7b.

$$P_p = sL = s \wedge L. \quad (13.90)$$

Finally it can be shown that the pair of points can also be computed by the meet of three spheres as follows

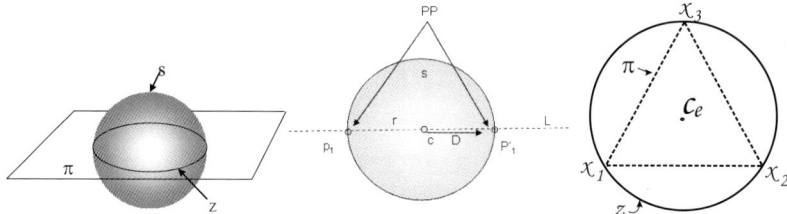

Fig. 13.7. (a) (*left*) The meet of a sphere and a plane. (b) (*middle*) Pair of points resulting from the meet between a line and a circle. (c) (*right*) The center of a circumscribed triangle

$$P_p = s_1 \wedge s_2 \wedge s_3 = s_1 \wedge (s_2 \wedge s_3) = s_1 \wedge z. \tag{13.91}$$

Note that the intersection of two of these spheres yields a circle and its meet with a third sphere will produce the pair of points. If we consider that one of the three points that makes up the circle z is at infinity, the circle turns out to be a line. As a result the meet of the sphere with this line will yield again a pair of points, as in Eq. (13.90).

13.3.4 Conformal Transformations

A transformation of geometric figures is said to be *conformal* if it preserves the *shape* of the figures, that is, if it preserves the angles and hence the shapes of straight lines and circles. Any conformal transformation in $\mathbb{R}^n$,

$$x \longmapsto x \frac{1}{1 + ax}, \tag{13.92}$$

can be expressed as a composite of inversions and a translation

$$x \xrightarrow{inversion} \frac{x}{x^2}, \tag{13.93}$$

$$\xrightarrow{translation} \frac{x}{x^2} + a, \tag{13.94}$$

$$\xrightarrow{inversion} \frac{\frac{x}{x^2} + a}{(\frac{x}{x^2} + a)(\frac{x}{x^2} + a)} = x \frac{1}{1 + ax}. \tag{13.95}$$

The conformal transformation in geometric algebra uses a versor representation,

$$g(x_c) = Gx_c(G^*)^{-1} = \sigma x'_c, \tag{13.96}$$

where $x_c \in \mathbb{R}^{n+1,1}$, G is a versor, and σ is a scalar. G can be expressed in geometric algebra as a composite of versors for *inversions* in spheres and *reflections* in hyperplanes. These individual versors are explained next.

Inversions

The general form of a reflection about a vector is

$$s(\bm{x}_c) = -\bm{s}\bm{x}_c\bm{s}^{-1} = \bm{x}_c - 2(\bm{s}\cdot\bm{x}_c)\bm{s}^{-1} = \sigma\bm{x}'_c, \quad (13.97)$$

where $\bm{s}\bm{x} + \bm{x}\bm{s} = 2(\bm{s}\cdot\bm{x})$, from the definition of the Clifford product between two vectors. We will now analyze what happens when s represents a sphere. Recall that the equation of a sphere of radius ρ centered at point $\bm{c}_c$ is the vector

$$s = \bm{c}_c - \frac{1}{2}\rho^2 e_\infty. \quad (13.98)$$

If $\bm{s}$ represents the unit sphere centered at the origin, then $\bm{s}$ and $\bm{s}^{-1}$ reduce to $e_0 - \frac{1}{2}e_\infty$. Hence, $-2(\bm{s}\cdot\bm{x}_c) = \mathbf{x}_e^2 - 1$, and Eq. (13.97) becomes

$$\sigma\bm{x}'_c = (\mathbf{x}_e + \frac{1}{2}\mathbf{x}_e^2 + e_\infty) + (\mathbf{x}_e^2 - 1)(e_0 - e_\infty) = \mathbf{x}_e^2(\mathbf{x}_e^{-1} + \mathbf{x}_e^{-2}e_\infty + e_0), \quad (13.99)$$

which is the conformal mapping for $\mathbf{x}_c^{-1}$.

To see how a general sphere inverts a point, we return to Eq. (13.98) to get

$$\bm{s}\cdot\bm{x}_c = \bm{c}_c\cdot\bm{x}_c - \rho^2 e_\infty \cdot \bm{x}_c = -[(\mathbf{x}_e - \mathbf{c}_e)^2 - \rho^2]. \quad (13.100)$$

Insertion into Eq. (13.97) and a little algebra gives

$$\sigma\bm{x}'_c = \left(\frac{\mathbf{x}_e - \mathbf{c}_e}{\rho}\right)^2 [g(\bm{x}_c) + \frac{1}{2}g^2(\bm{x}_c)e_\infty + e_0], \quad (13.101)$$

where

$$g(\mathbf{x}_e) = \frac{\rho^2}{\mathbf{x}_e - \mathbf{c}_e} + \mathbf{c}_e = \frac{\rho^2(\mathbf{x}_e - \mathbf{c}_e)}{(\mathbf{x}_e - \mathbf{c}_e)^2} + \mathbf{c}_e \quad (13.102)$$

is the inversion in $\mathbb{R}^n$.

Reflections

A hyperplane with unit normal $\mathbf{n}$ and signed distance δ from the origin in $\mathbb{R}^n$ can be represented by the vector

$$s = \mathbf{n} + \delta e_\infty. \quad (13.103)$$

Inserting $\bm{s}\cdot\bm{x}_c = \mathbf{n}\cdot\mathbf{x}_e$ into Eq. (13.97), we find that

$$g(\mathbf{x}_e) = \mathbf{n}\mathbf{x}_e\mathbf{n}^\dagger + 2\delta\mathbf{n} = \mathbf{n}(\mathbf{x}_e - \delta\mathbf{n})\mathbf{n}^\dagger + \delta\mathbf{n}, \quad (13.104)$$

where $\mathbf{n}^\dagger = -s^{-1}$. This expression is equivalent to a reflection $\mathbf{n}\mathbf{x}_e\mathbf{n}^\dagger$ at the origin, translated by δ along the direction of $\mathbf{n}$. A point $\mathbf{c}_e$ is on the hyperplane when $\delta = \mathbf{n} \cdot \mathbf{c}_e$, in which case Eq. (13.103) can be written as

$$s = \mathbf{n} + e_\infty \mathbf{n} \cdot \mathbf{c}_e. \tag{13.105}$$

Via Eq. (13.104), this vector represents the reflection in a hyperplane through point $\mathbf{c}_e$.

Translations

Translations can be modeled as two reflections about two parallel hyperplanes. Without loss of generality, we can assume that both planes have been normalized so that the magnitude of their normals is 1 and one of them passes through the origin. Then, from Eq. (13.103), we can represent the operator for translation (called a translator) as

$$\begin{aligned}\boldsymbol{T_a} = \pi_1\pi_2 &= (\mathbf{n} + \delta e_\infty)(\mathbf{n} + 0e_\infty), \\ &= 1 + \mathbf{a}e_\infty,\end{aligned} \tag{13.106}$$

where $\mathbf{a} = 2\delta\mathbf{n}$ and $\|\mathbf{n}\| = 1$. The translation distance is twice the separation between the hyperplanes.

Transversions

The transversion can be generated from two inversions and a translation. The transversor has the form

$$\boldsymbol{K_b} = e_+ \boldsymbol{T_b} e_+ = (e_\infty - e_0)(1 + \mathbf{b}e_\infty)(e_\infty - e_0) = 1 + \mathbf{b}e_0. \tag{13.107}$$

The transversion generated by $\boldsymbol{K_b}$ can be expressed in various forms:

$$g(\mathbf{x}_e) = \frac{\mathbf{x}_e - \mathbf{x}_e^2 \mathbf{b}}{1 - 2\mathbf{b} \cdot \mathbf{x}_e + \mathbf{x}_e^2 \mathbf{b}^2} = \mathbf{x}_e(1 - \mathbf{b}\mathbf{x}_e)^{-1} = (\mathbf{x}_e^{-1} - \mathbf{b})^{-1}. \tag{13.108}$$

The last form can be written down directly as an inversion followed by a translation and another inversion.

Rotations

Rotations can be modeled by the composition of two reflections about two hyperplanes intersecting in a common point $\mathbf{c}_e$, as in

$$\boldsymbol{R} = (\mathbf{a} + e_\infty \mathbf{a} \cdot \mathbf{c}_e)(\mathbf{b} + e_\infty \mathbf{b} \cdot \mathbf{c}) = \mathbf{ab} + e_\infty \mathbf{c}_e \cdot (\mathbf{a} \wedge \mathbf{b}), \tag{13.109}$$

where $\mathbf{a}$ and $\mathbf{b}$ are unit normals. Rotations about the origin can also be written in exponential form, as in

$$\boldsymbol{R} = e^{\frac{1}{2}\alpha \boldsymbol{B}}, \tag{13.110}$$

where $\boldsymbol{B}$ is a unit bivector representing the axis of rotation and α is the magnitude of the angle of rotation.

Dilations

Dilations are the composite of two inversions centered at the origin. Using the unit sphere $s_1 = e_0 - \frac{1}{2}e_\infty$, another sphere of arbitrary radius ρ, and $s_2 = e_0 - \frac{1}{2}\rho^2 e_\infty$ as inversors, we get

$$(e_0 - e_\infty)(e_0 - \rho^2 e_\infty) = (1 - \boldsymbol{E}) + (1 + \boldsymbol{E})\rho^2. \tag{13.111}$$

Normalizing to unity, we have

$$\boldsymbol{D}_\rho = (1 + \boldsymbol{E})\rho + (1 - \boldsymbol{E})\rho^{-1} = e^{\boldsymbol{E}\phi}, \tag{13.112}$$

where $\phi = \ln \rho$. To prove that this is indeed a dilation, we note that

$$\boldsymbol{D}_\rho e_\infty \boldsymbol{D}_\rho^{-1} = \rho^{-2} e_\infty, \quad \text{and similarly,} \tag{13.113}$$

$$\boldsymbol{D}_\rho e_0 \boldsymbol{D}_\rho^{-1} = \rho^2 e_0. \tag{13.114}$$

Therefore,

$$\boldsymbol{D}_\rho(\mathbf{x}_e + \mathbf{x}_e^2 e_\infty + e_0)\boldsymbol{D}_\rho^{-1} = \rho^2[\rho^{-2}\mathbf{x}_e + (\rho^{-2}\mathbf{x}_e)^2 e_\infty + e_0], \tag{13.115}$$

which is the conformal mapping $g(\boldsymbol{x}_e) = \sigma \boldsymbol{x}_c'$ with $\boldsymbol{x}_c' = \boldsymbol{x}_e' + \boldsymbol{x}'^2 e_\infty + e_0$, where $\boldsymbol{x}_e' = \rho^{-2}\mathbf{x}_e$.

Involutions

The bivector E (the Minkowski plane of $\mathbb{R}^{1,1}$) represents an operation which corresponds to the main involution, but for an r-blade $\boldsymbol{A}_r$, $\bar{\boldsymbol{A}}_r = (-1)^r \boldsymbol{A}_r$. In particular, for vectors $\bar{\mathbf{x}}_e = -\boldsymbol{x}_e$, which can be easily obtained by applying the versor E,

$$E(\mathbf{x}_e + \mathbf{x}_e^2 e_\infty + e_0)E = -(-\mathbf{x}_e + \frac{1}{2}\mathbf{x}_e^2 e_\infty + e_0). \tag{13.116}$$

This expression corresponds to the conformal mapping of $-\boldsymbol{x}_e$, thus confirming that the versor E represents the main involution for $\mathbb{R}^n$. This means that the main involution is a reflection via the Minkowski plane $\mathbb{R}^{1,1}$.

Finally, using previous results, we can write down a canonical decomposition in terms of individual versors, which reveals the structure of the 3-parameter group $\{G \in \mathbb{R}_{1,1} | G^* G^\dagger = 1\} \simeq GL_2(\mathbb{R})$ as follows:

$$G = K_\mathbf{b} T_\mathbf{a} R_\alpha. \tag{13.117}$$

13.4 The 3-D Affine Plane

We have described the general conformal framework and its transformations. However, many of these operators were not employed in the present work. Indeed, since in this case only rigid transformations are needed, we will limit ourselves to the use of the *affine plane*, which is an $n+1$-dimensional subspace of the hyperplane of reference $\mathbb{P}(e_\infty, e_0)$.

We have chosen to work in $\mathcal{G}_{4,1}$ algebra. Since we deal with homogeneous points, the particular choice of null vectors does not affect the properties of the conformal geometry. Thus, for this work we choose to define these vectors as

$$e = \frac{1}{2}(e_4 + e_5), \tag{13.118}$$

$$\bar{e} = e_4 - e_5, \tag{13.119}$$

with the following properties:

$$e_i^2 = 1, \quad \text{for } i = 1, ..., 4, \tag{13.120}$$

$$e_5^2 = -1, \tag{13.121}$$

$$e^2 = \bar{e}^2 = 0, \tag{13.122}$$

$$e \cdot \bar{e} = 1. \tag{13.123}$$

Points in the affine plane $\boldsymbol{x} \in \mathbb{R}^{4,1}$ are formed with

$$\boldsymbol{x}^a = \mathbf{x}_e + e, \tag{13.124}$$

where $\mathbf{x}_e \in \mathbb{R}^3$. From this equation, we note that e represents the origin (by setting $\mathbf{x}_e = 0$), and that, similarly, $\bar{e}$ represents the point at infinity. The equation allows the normalization equation (13.47) to be expressed as

$$\bar{e} \cdot \boldsymbol{x}^a = 1. \tag{13.125}$$

In this framework, the conformal mapping equation is expressed by

$$\boldsymbol{x}_c = \mathbf{x}_e - \mathbf{x}_e^2 \bar{e} + e = \boldsymbol{x}^a - \mathbf{x}_e^2 \bar{e}. \tag{13.126}$$

For working on the affine plane exclusively, we will be mainly concerned with a simplified version of *rejection*. By noting that $\boldsymbol{E} = e_\infty \wedge e_0 = \bar{e} \wedge e$, then Eq. (13.46) becomes

$$P_E^\perp(\boldsymbol{x}_c) = (\boldsymbol{x}_c \wedge \boldsymbol{E})\boldsymbol{E} = (\boldsymbol{x}_c \wedge \boldsymbol{E}) \cdot \boldsymbol{E} = (\bar{e} \wedge e) \cdot e + (\boldsymbol{x}_c \wedge \bar{e}) \cdot e,$$
$$\boldsymbol{x}_e = -e + (\boldsymbol{x}_c \wedge \bar{e}) \cdot e. \tag{13.127}$$

Now, since the points in the affine plane have the form $\boldsymbol{x}_a = \boldsymbol{x}_e + e$, we conclude that

$$\boldsymbol{x}^a = (\boldsymbol{x}_c \wedge \bar{e}) \cdot e \tag{13.128}$$

is the mapping from the horosphere to the affine plane.

13.4.1 Lines and Planes

The lines and planes in the affine plane are expressed in a similar fashion to their conformal counterparts, as the *join* of 2 and 3 points, respectively:

$$\boldsymbol{L}^a = \boldsymbol{x}_1^a \wedge \boldsymbol{x}_2^a, \tag{13.129}$$
$$\Pi^a = \boldsymbol{x}_1^a \wedge \boldsymbol{x}_2^a \wedge \boldsymbol{x}_3^a. \tag{13.130}$$

Note that unlike their conformal counterparts, the line is a *bivector* and the plane is a *trivector*. As seen earlier, these equations produce the expected moment-direction representation. Thus,

$$\boldsymbol{L}^a = e\mathbf{d} + \boldsymbol{B}, \tag{13.131}$$

where $\mathbf{d}$ is a vector representing the direction of the line and B is a bivector representing the (orthogonal) moment of the line. Similarly, we have

$$\Pi^a = e\mathbf{n} + \delta e_{123}, \tag{13.132}$$

where $\mathbf{n}$ is the normal vector to the plane and δ is a scalar representing the distance from the plane to the origin. Note that in any case, the direction and normal can be retrieved with $\mathbf{d} = \bar{e} \cdot \boldsymbol{L}^a$ and $\mathbf{n} = \bar{e} \cdot \Pi^a$, respectively.

In this framework, the intersection or *meet* has a simple expression, too. If $\boldsymbol{A}^a = a_1^a \wedge ... \wedge a_r^a$ and $\boldsymbol{B}^a = b_1^a \wedge ... \wedge b_s^a$, then the meet is defined as

$$\boldsymbol{A}^a \cap \boldsymbol{B}^a = \boldsymbol{A}^a \cdot (\boldsymbol{B}^a \cdot \bar{I}_{\boldsymbol{A}^a \cup \boldsymbol{B}^a}), \tag{13.133}$$

where $\bar{I}_{\boldsymbol{A}^a \cup \boldsymbol{B}^a}$ is either $e_{12}\bar{e}$, $e_{23}\bar{e}$, $e_{31}\bar{e}$, or $e_{123}\bar{e}$, according to which basis vectors span the largest common space of $\boldsymbol{A}^a$ and $\boldsymbol{B}^a$ (see Sect. 13.2.2).

13.4.2 Rigid Transformations in the 3-D Affine Plane

Rotations and translations have the same form as previously stated. In our new definition, the rotor becomes

$$\boldsymbol{R} = e^{\frac{\theta}{2}\boldsymbol{B}} = \cos(\frac{\theta}{2}) + \boldsymbol{B}\sin(\frac{\theta}{2}), \tag{13.134}$$

where θ is the angle of rotation and B is a unit bivector representing the axis of rotation.

The translator is defined by

$$\boldsymbol{T} = e^{\frac{1}{2}\mathbf{t}\bar{e}} = 1 + \mathbf{t}\bar{e}, \tag{13.135}$$

where $\mathbf{t}$ is a vector representing the translation. For the particular case of rotors and translators, the inverse is equal to the Clifford conjugate, which in turn is equal to the reversion. Thus, the transformation rule of a rotor and translator may be written as

$$\boldsymbol{X}' = \widetilde{\boldsymbol{R}}\boldsymbol{X}\boldsymbol{R}, \text{ and} \tag{13.136}$$

$$\boldsymbol{X}' = \widetilde{\boldsymbol{T}}\boldsymbol{X}\boldsymbol{T}. \tag{13.137}$$

Note that when a translator is applied to a geometric entity the result will lie in the horosphere. Therefore, it is necessary to perform *partial rejection*, as defined by Eq. (13.128), followed by the normalization $\bar{e}\cdot xa = 1$, if applicable.

Rotors and translators can be combined multiplicatively to produce *motors*. The motor $\boldsymbol{M}$ is defined, in general, as

$$\boldsymbol{M} = \boldsymbol{T}\boldsymbol{R}, \tag{13.138}$$
$$= (1 + \mathbf{t}\bar{e})\,\boldsymbol{R}, \tag{13.139}$$
$$= \boldsymbol{R} + \mathbf{t}\bar{e}\boldsymbol{R} = \boldsymbol{R} + \bar{e}\boldsymbol{R}'. \tag{13.140}$$

Therefore, it is rather simple to find the rotational part of a motor by simple inspection. The translational part can then be computed by right-multiplying the remainder of $\boldsymbol{M}$ by $\boldsymbol{R}^{-1}$.

Note, in particular, that the motor defined by

$$\boldsymbol{M} = \widetilde{\boldsymbol{T}}\boldsymbol{R}\boldsymbol{T} \tag{13.141}$$

represents a rotation about an arbitrary axis. There is a simple relationship between the axis of rotation and the form of a motor that produces the rotation about it. Let $\boldsymbol{L} = \boldsymbol{m} + e \wedge \mathbf{n}$ be a line such that $\|\mathbf{n}\| = 1$. Then, the motor M that rotates by α radians about the axis defined by L is given by

$$\boldsymbol{M} = \cos(\frac{\alpha}{2}) + \sin(\frac{\alpha}{2})[\bar{e} \wedge (e_{123}\boldsymbol{m}) - (e_{123}\mathbf{n})]. \tag{13.142}$$

13 Robot Perception and Action Using Conformal Geometric Algebra

The converse formula, to extract the axis of rotation from a given motor, is also simple. First, notice that from the previous equation M can be written as

$$M = k_1 + k_2 \bar{e} \wedge (e_{123} m) - k_2 e_{123} n. \tag{13.143}$$

From this equation, it is easy to see that

$$m' = e_{123}(e \cdot (M - k_1)), \text{ and} \tag{13.144}$$
$$n' = -e_{123} \cdot (M - k_1), \tag{13.145}$$

where m' and n' differ from the original m and n by a scale factor. Thus, the line representing the axis of rotation is given by $L = m' + e \wedge n'$.

13.4.3 Directed Distance

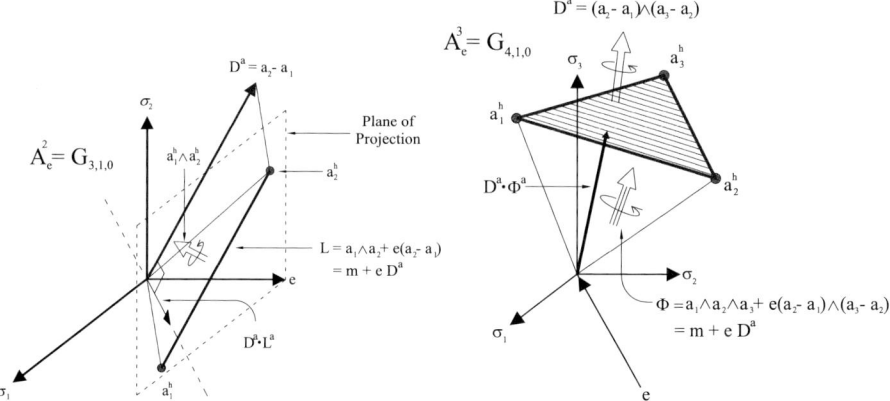

Fig. 13.8. (a) (*left*) Line in 2-D affine space. (b) (*right*) Plane in the 3-D affine space (note that the 3-D space is "lifted" by a null vector e)

To derive our equations we can use the line and the plane depicted in Fig. 13.8. We see that any k-plane A^a consists of a momentum of k degree and a direction of $k-1$ degree. Thus, if we take the inner product between the unit direction and the moment, we will gain a directed distance as the vector p^a. If we dot the equation of A^a with $\bar{e}$ and divide this result by its norm, we get the unit direction D_u^a,

$$D^a = \bar{e} \cdot A_a \longrightarrow D_u^a = \frac{D_a^a}{|D_a^a|}, \tag{13.146}$$

which we can further use to compute a directed distance as follows:

$$p^a = D_u^a \cdot A^a. \tag{13.147}$$

Here, the dot operation basically takes place between D_u^a and the moment part of A^a. The point p^a is attached to the origin and it touches orthogonally the k-plane A^a. Interestingly enough, the norm of p^a is equal to the Hesse distance. For the sake of simplicity, in Fig. 13.8a and in Fig. 13.8b only $D^a \cdot L^a$ and $D^a \cdot \varPhi^a$ are respectively shown.

Now, having calculated this point on the first object, we can use it to compute the directed distance from the k-plane A^a parallel to the object B^a as follows:

$$d[A^a, B^a] = d[D^a \cdot A^a, B^a] = d[(\bar{e} \cdot A^a) \cdot A^a, B^a]. \tag{13.148}$$

13.4.4 Incidence Relations in the Affine 3-D Plane

The distance from a point b^a to a line $L^a = a_1^a \wedge a_2^a$ is the *magnitude* or *norm* of their directed distance,

$$|d| = \left| \left[\{\bar{e} \cdot (a_1^a \wedge a_2^a)\} \wedge \{\bar{e} \cdot (b^a)\} \right]^{-1} \left[\bar{e} \cdot (a_1^a \wedge a_2^a \wedge b^a) \right] \right|. \tag{13.149}$$

The distance of a point b_a to the plane $A^a = a_1^a \wedge a_2^a \wedge a_3^a$ is

$$|d| = \left| \left[\{\bar{e} \cdot (a_1^a \wedge a_2^a \wedge a_3^a)\} \wedge \{\bar{e} \cdot (b^a)\} \right]^{-1} \left[\bar{e} \cdot (a_1^a \wedge a_2^a \wedge a_3^a \wedge b_a^a) \right] \right| \tag{13.150}$$

The incidence relation between the lines $L_1^a = a_1^a \wedge a_2^a$ and $L_2^a = b_1^a \wedge b_2^a$ is completely determined by their join $I_{L_1 \cup L_2} = L_1 \cup L_2^h$.

If $I_{L_1^a \cup L_2^a}$ is a bivector, the lines coincide and $L_1^a = tL_2^a$ for some $t \in \mathbb{R}$. If $I_{L_1^a \cup L_2^a}$ is a trivector, the lines are either parallel or intersect at a common point; in this case, the meet

$$p^a = L_1^a \cap L_2^a = L_1^a \cdot [(L_2^a \cdot \bar{I}_{L_1^a \cup L_2^a})]. \tag{13.151}$$

If $\bar{e} \cdot p^a = 0$, the lines are parallel; otherwise, they intersect at the point $p^a = \frac{p^a}{\bar{e} \cdot p^a}$ in the affine 3-space $\mathcal{A}_e^3$. Finally, if $I_{L_1^a \cup L_2^a}$ is a 4-vector, the lines are skew.

The incidence relation between a line $L^a = a_1^a \wedge a_2^a$ and a plane $B^a = b_1^a \wedge b_2^a \wedge b_3^a$ is also determined by their join $L^a \cup B^a$. Clearly, if the join is a trivector, the line L^a lies in the plane B^a. The only other possibility is that their join is the pseudoscalar $I = \sigma_{123}e$. In this case, we calculate the meet

$$p^a = L^a \cap B^a = L^a \cdot [(B^a \cdot \bar{I})]. \tag{13.152}$$

If $\bar{e} \cdot p^a = 0$, the line is parallel to the plane, with the directed distance determined by Eq. (13.146). Otherwise, $p^a = \frac{p^a}{\bar{e} \cdot p^a}$ is their point of intersection in the affine plane.

Two planes $\boldsymbol{A}^a = \boldsymbol{a}_1^a \wedge \boldsymbol{a}_2^a \wedge \boldsymbol{a}_3^a$ and $\boldsymbol{B}^a = \boldsymbol{b}_1^a \wedge \boldsymbol{b}_2^a \wedge \boldsymbol{b}_3^a$ in the affine plane $\mathcal{A}_e^3$ are either parallel, intersect in a line, or coincide. If their join is a trivector, i.e., if $\boldsymbol{A}^a = t\boldsymbol{B}^a$ for some $t \in \mathbb{R}$, they obviously coincide. If they do not coincide, then their join is the pseudoscalar $I = \sigma_{123}e$. In this case, we calculate the meet,

$$\boldsymbol{L}^a = \boldsymbol{A}^a \cap \boldsymbol{B}^a = [(\overline{I}) \cdot \boldsymbol{A}^a] \cdot \boldsymbol{B}^a. \tag{13.153}$$

If $\overline{e} \cdot \boldsymbol{L} = 0$, the planes are parallel, with the directed distance determined by Eq. (13.146). Otherwise, $\boldsymbol{L}$ represents the line of intersection in the affine plane having the direction $\overline{e} \cdot \boldsymbol{L}$.

13.5 3-D Rigid-Motion Estimation

In the following sections we describe a series of algorithms and techniques that help us to calibrate a mobile robot for building a map for navigation purposes. For that we have to solve first the problem of coordinate-system calibration between multiple sensors. In this work we will assume that the cameras have been calibrated and thus metric reconstruction can be easily achieved.

The problem of coordinate-system calibration is basically a problem of motion estimation. We proceed first to describe two methods for motion estimation which we have employed in this work.

13.5.1 Point-Based Rigid-Motion Estimation

The most basic feature that can be detected is the point. The algorithm we describe here to perform rigid-motion estimation based on point observations was originally presented in [12]. The algorithm proceeds thus. Given the 3-D coordinates of two sets of n corresponding points $\boldsymbol{x}_i$ and $\boldsymbol{x}_i'$ in $\mathcal{G}_{3,0,0}$, we wish to find the rotation $\boldsymbol{R}$ and translation $\boldsymbol{t}$ that minimizes

$$S = \sum_{i=1}^{n} \left[\boldsymbol{x}_i' - \boldsymbol{R}(\boldsymbol{x}_i - \boldsymbol{t})\widetilde{\boldsymbol{R}} \right]^2. \tag{13.154}$$

This therefore involves minimizing S with respect to $\boldsymbol{R}$ and $\boldsymbol{t}$. The differentiation with respect to $\boldsymbol{t}$ is straightforward:

$$\partial_{\boldsymbol{t}} S = \sum_{i=1}^{n} \left[\boldsymbol{x}_i' - R(\boldsymbol{x}_i - \boldsymbol{t})\widetilde{R} \right] \partial_{\boldsymbol{t}}(R\boldsymbol{t}\widetilde{R}) = 0. \tag{13.155}$$

Equation (13.155) can be easily solved for $\boldsymbol{t}$ to give

$$\boldsymbol{t} = \frac{1}{n} \sum_{i=1}^{n} \left[\boldsymbol{x}_i - \widetilde{R}\boldsymbol{x}_i' R \right], \tag{13.156}$$

which can be rewritten as

$$t = \bar{x} - \widetilde{R}\bar{x}'R, \qquad (13.157)$$

where $\bar{x}$ and $\bar{x}'$ are the centroids of the data points in the two views.

The differentiation with respect to the rotor R is much more involved. We direct the interested reader to [12] for details. The result of this differentiation produces

$$\sum_{i=1}^{n} x'_i \wedge R(x_i - t)\widetilde{R} = 0. \qquad (13.158)$$

If we substitute the value of t into the previous equation, we get

$$\sum_{i=1}^{n} x'_i \wedge R(x_i - \bar{x} - \widetilde{R}\bar{x}'R)\widetilde{R} = 0. \qquad (13.159)$$

Noting that the term $\sum_{i=1}^{n} x'_i \wedge \bar{x}'$ vanishes, the previous equation is thus reduced to

$$\sum_{i=1}^{n} w_i \wedge Ru_i\widetilde{R} = 0, \qquad (13.160)$$

where $u_i = x_i - \bar{x}$ and $w_i = x'_i$. With these new values, we obtain a matrix $\mathrm{F}_{\alpha\beta}$:

$$\mathrm{F}_{\alpha\beta} = \sum_{i=1}^{n} (e_\alpha \cdot \mathbf{u}_i)(e_\beta \cdot \mathbf{w}_i). \qquad (13.161)$$

Applying the SVD to matrix $\mathrm{F}_{\alpha\beta} = \mathrm{USV}^\mathrm{T}$, we can find the rotation matrix R:

$$\mathrm{R} = \mathrm{VU}^\mathrm{T}. \qquad (13.162)$$

Once the rotor is R is obtained from R, the translation t can be computed from Eq. (13.157).

13.5.2 Line-based rigid-motion estimation

Since lines are more robust to noise than points, it is important to describe how rigid-motion estimation can be performed based on them. The algorithm we describe here was originally presented in [4, 1] using motors or dual quaternions from motor algebra $\mathcal{G}_{3,0,1}$, but we have extended it here for the 3-D affine plane. In this framework, the motion of the line can be expressed as

$$\boldsymbol{L}_B^a = M\boldsymbol{L}_A^a \widetilde{M}, \qquad (13.163)$$

where $\widetilde{\boldsymbol{x}}$ is the Clifford conjugate of $\boldsymbol{x}$ and, in the case of motors $\boldsymbol{M}\widetilde{\boldsymbol{M}} = 1$,

$$\boldsymbol{L}_B^a = \boldsymbol{b} + e\boldsymbol{b}',$$
$$\boldsymbol{L}_A^a = \boldsymbol{a} + e\boldsymbol{a}' = \boldsymbol{R}\boldsymbol{b}\widetilde{\boldsymbol{R}} + e(\boldsymbol{R}\boldsymbol{b}\widetilde{\boldsymbol{R}}' + \boldsymbol{R}\boldsymbol{b}'\widetilde{\boldsymbol{R}} + \boldsymbol{R}'\boldsymbol{b}\widetilde{\boldsymbol{R}}). \quad (13.164)$$

By separating the real and dual parts,

$$\boldsymbol{a} = \boldsymbol{R}\boldsymbol{b}\widetilde{\boldsymbol{R}}, \quad (13.165)$$
$$\boldsymbol{a}' = \boldsymbol{R}\boldsymbol{b}\widetilde{\boldsymbol{R}}' + \boldsymbol{R}\boldsymbol{b}'\widetilde{\boldsymbol{R}} + \boldsymbol{R}'\boldsymbol{b}\widetilde{\boldsymbol{R}}, \quad (13.166)$$

and then multiplying on the right by $\boldsymbol{R}$, knowing that $\widetilde{\boldsymbol{R}}\boldsymbol{R}' + \widetilde{\boldsymbol{R}}'\boldsymbol{R} = 0$, we get

$$\boldsymbol{a}\boldsymbol{R} - \boldsymbol{R}\boldsymbol{b} = 0,$$
$$(\boldsymbol{a}'\boldsymbol{R} - \boldsymbol{R}\boldsymbol{b}') + (\boldsymbol{a}\boldsymbol{R}' - \boldsymbol{R}'\boldsymbol{b}) = 0. \quad (13.167)$$

This can be rewritten in matrix form for computational purposes as

$$\begin{bmatrix} \boldsymbol{a} - \boldsymbol{b} & [\boldsymbol{a} + \boldsymbol{b}]_\times & 0_{3\times 1} & 0_{3\times 3}, \\ \boldsymbol{a}' - \boldsymbol{b}' & [\boldsymbol{a}' + \boldsymbol{b}']_\times & \boldsymbol{a} - \boldsymbol{b} & [\boldsymbol{a} + \boldsymbol{b}]_\times \end{bmatrix} \begin{bmatrix} \boldsymbol{R} \\ \boldsymbol{R}' \end{bmatrix}. \quad (13.168)$$

Taking $n \geq 2$ observations, we stack the n line parameters in the left matrix, which is decomposed, using SVD, as $\mathtt{C} = \mathtt{U}\mathtt{\Sigma}\mathtt{V}^T$. Since the rank of the left matrix is 6, we use two vectors which span the null space for computing R and R'. For more details of the method, refer to [4].

13.6 Body-Eye Calibration

The so called hand-eye calibration problem involves the computation of the transformation between a coordinate system attached to a robotic hand and the camera on top of it. This works on the premise that the camera is always fixed at the same position with respect to the robotic arm, which is not the case for pan-tilt units, where the camera is constantly changing its position and orientation with respect to the robot's reference frame. To solve this problem, we proposed a novel algorithm in the affine 3-space $\mathcal{A}_e^3$ which we describe next.

The robot-to-sensor relation can be seen as a series of joints $J_1, J_2, ..., J_n$ (where a rotation about joint J_i affects all joints $J_{i+1}, ..., J_n$) and a measurement system U which is rigidly attached to the last joint J_n. The problem can be stated as the computation of the transformations $\boldsymbol{M}_1, \boldsymbol{M}_2, ..., \boldsymbol{M}_{n-1}$

between the robot frame and the last joint and the transformation M_n between the last joint and the measurement device U, using only data gathered with U.

Note that this formulation is independent of the type of sensor U used; however, we will discuss how this calibration can be implemented to solve the robot-to-camera relationship and, later on, how it was slightly modified to calibrate a laser sensor against the robot coordinate system.

Furthermore, we would like to solve this problem in a way that enables a real-time response when the spatial location of the joints varies. Therefore, we have divided our algorithm into two stages. The first stage computes the *screw axes* of the joints, and the second stage uses these axes to compute the final transformation between the coordinate systems.

13.6.1 Screw Axes Computation

To compute the axes of rotation, we use a motion estimator such as the one described in Sect. 13.5.2. Each joint J_i is moved in turn while leaving the others at their home position (see Fig. 13.9a). From the resulting motor M_i, the axis of rotation S_i can be extracted using Eqs. (13.144) and (13.145). For our particular robot, the sequence of motions is presented in Fig. 13.9b. The general procedure is presented as Algorithm I in Table 13.2.

Table 13.2. Algorithm I–Computation of the axes of rotation

For each joint J_i, $i = 1, ...n$:
1. Set all joints to their home position.
2. Rotate joint J_i by $-\alpha_1$ degrees.
3. Measure a set of 3-D points x_j (or lines L_k^a) using the stereo camera.
4. Return joint J_i to its home position and rotate it by α_2 degrees.
5. Compute the corresponding set of points x'_j (or lines $L_j'^a$).
6. Compute the motor M_i such that $x'_j = M_i x_j \widetilde{M}$ (or $L_j'^a = M_i L_j \widetilde{M}$).
7. Compute the axis of rotation S_i using Eq. (13.144) and Eq. (13.145) as

$$S_i = e_{123}(e \cdot (M_i - k_1)) + e \wedge [-e_{123} \cdot (M_i - k_1)],$$

where k_1 is the scalar part of M_i.

13.6.2 Calibration

Note that Algorithm I will produce a set of lines S_i in the camera's coordinate system. Once these axes are known, the transformation taking one point x_k measured in the camera's framework to the robot's coordinate system is easy to derive, provided that we know the angles α_i applied to each joint J_i.

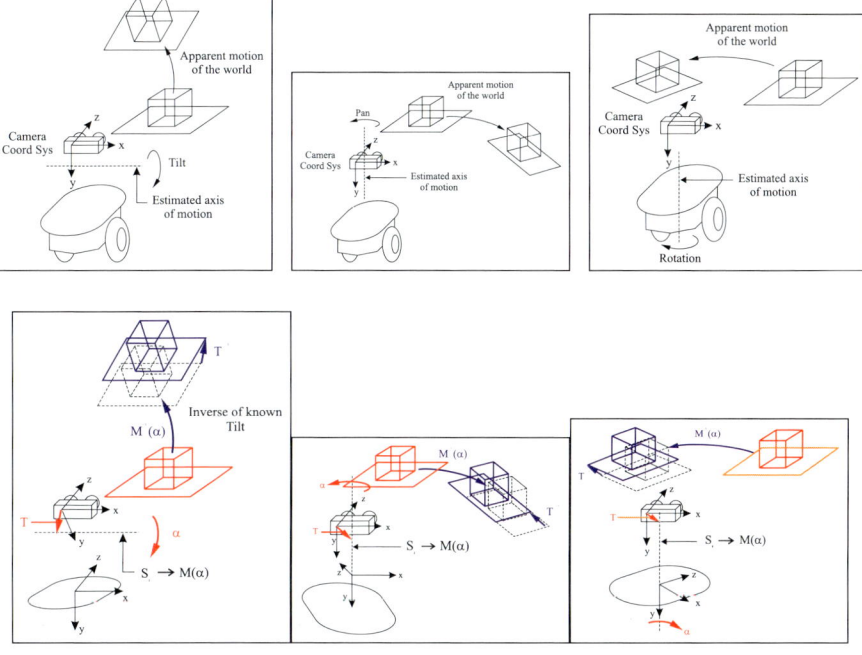

Fig. 13.9. (a) (*upper row*) Estimation of the screw axes. (b) (*lower row*) Correction of the rotation and relocation of the screw axes

Basically, the algorithm undoes the implicit transformations applied on the camera's framework by first rotating about joint J_k and then translating the joint (and the framework, along with the other joints) to the origin (see Fig. 13.9b). The full procedure is described as Algorithm II in Table 13.3.

The functions used in Algorithm II are defined as follows:

$$\texttt{nearest}(\boldsymbol{x}) = \frac{(\bar{e} \cdot \boldsymbol{x}) \cdot \boldsymbol{x}}{\bar{e} \cdot [(\bar{e} \cdot \boldsymbol{x}) \cdot \boldsymbol{x}]}, \tag{13.169}$$

$$\texttt{makeTranslator}(\mathbf{t}) = 1 + \frac{\boldsymbol{t}}{2}\bar{e}, \tag{13.170}$$

$$\texttt{lineToMotor}(\boldsymbol{L}^a, \alpha) = \cos(\frac{\alpha}{2}) + \sin(\frac{\alpha}{2})[\bar{e} \wedge (e_{123}\boldsymbol{m}) - (e_{123}\boldsymbol{n})], \tag{13.171}$$

where $\boldsymbol{L}^a = \boldsymbol{m} + e \wedge \boldsymbol{n}$, and $\boldsymbol{n} = 1$. The function $\texttt{nearest}(\boldsymbol{x})$ returns the point on $\boldsymbol{x}$ which is nearest to the origin, $\texttt{makeTranslator}(\mathbf{t})$ returns a translator displacing by an amount $\boldsymbol{t}$, and $\texttt{lineToMotor}(\boldsymbol{L}^a, \alpha)$, previously explained in Eq. (13.142), simply returns a motor that rotates α radians about the axis $\boldsymbol{L}^a$.

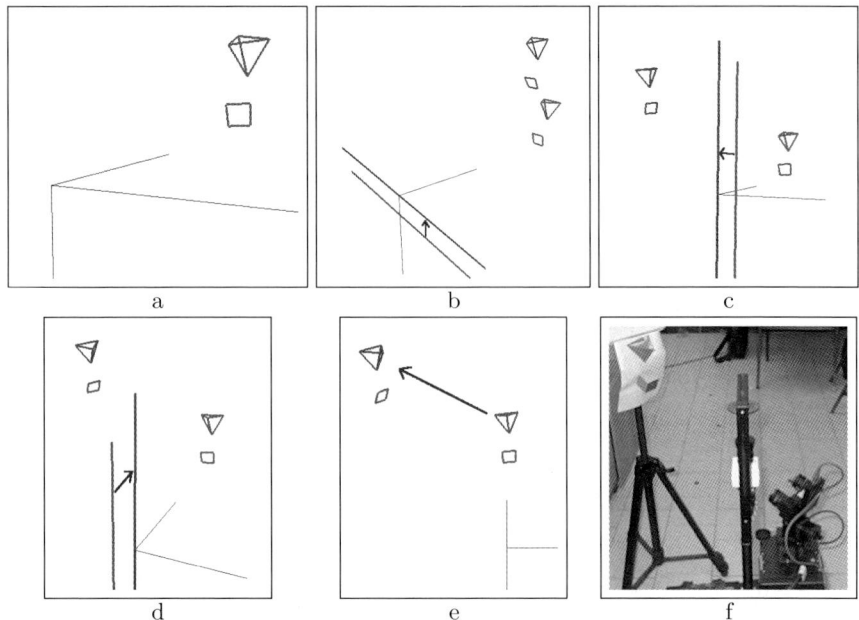

Fig. 13.10. (a) Reconstruction without calibration. (b-d) Relocation of the screws according to Algorithm II in Table 13.3. (e) Comparison of the final reconstruction with the real view

13.6.3 Reconstruction and Navigation

This subsection presents a series of tasks related to the perception and action of a mobile robot equipped with a 5-degrees-of-freedom manipulator. First, we present the reconstruction, which confirms that our body-eye calibration method works. Thereafter, the accuracy of the binocular head and laser calibration is shown by a test: approaching and grasping a target object.

In this experiment, we applied a certain pan and tilt to our robot and reconstructed the scene shown in Fig. 13.10a. The reconstruction was initially aligned with the camera coordinate system. Applying Algorithm II from Table 13.3, the reconstruction was transformed to match the robot's coordinate system (Figs. 13.10b-d). The comparison between the final reconstruction and the initial location of the robot is shown in Fig. 13.10e.

The second test consisted making a 3-D map of the surroundings of the robot by superimposing multiple reconstructions obtained at different pan and tilt angles. For each pan and tilt position, a reconstruction was obtained and transformed to the robot's reference-coordinate system. The robot was not moved throughout the process. Some of the images used for the reconstruction can be seen in the top row of Fig. 13.11, and the resulting reconstruction can be seen in the bottom row.

13 Robot Perception and Action Using Conformal Geometric Algebra   439

Table 13.3. Algorithm II–Computation of the transformation $X_i \mapsto X'_i$ for a system of n joints and m points where X_i is a point measured in the camera's framework and X'_i is the same point in the robot's coordinate system

Input: the number of joints n, a set of m points x_i, the screw axes S_i, and their angles of rotation α_i.
1. (Initialization) set: $k \leftarrow n$, $x'_i \leftarrow x_i$, for $i = 1, ..., m$, $S'_i \leftarrow S'_i$, for $i = 1, ..., n$. 2. $P \leftarrow \texttt{nearest}(S'_k)$. 3. $T \leftarrow \texttt{makeTranslator}(-(P-e))$. 4. $M \leftarrow T\,\texttt{lineToMotor}(S'_k, -\alpha_k)$. 5. $x'_i \leftarrow M x'_i \widetilde{M}$, for $i = 1, ..., m$. 6. $S'_i \leftarrow T S'_i \widetilde{T}$, for $i = 1, ..., k-1$. 7. $k \leftarrow k - 1$. 8. Repeat steps 2–7 until $k = 0$. The final corrected points are x'_i, $i = 1, ..., m$.

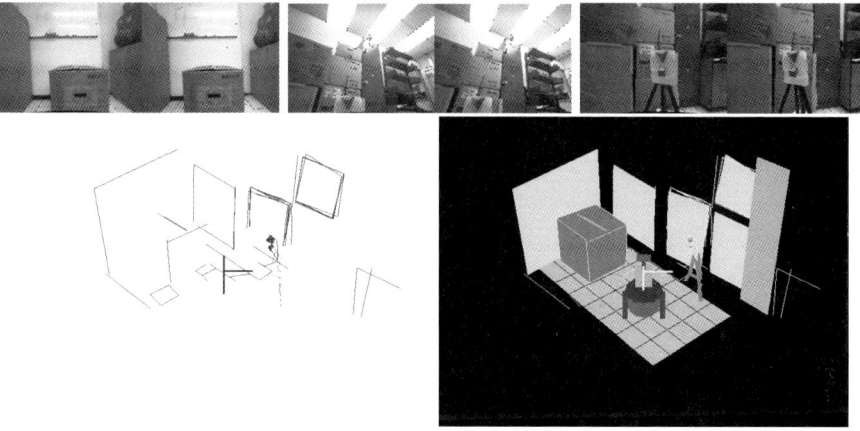

Fig. 13.11. (a) (*top row*) Some of the stereo pairs used in the reconstruction. (b) (*bottom left*) Extracted 3-D data. (c) (*bottom right*) Texture-mapped reconstruction

As an extension of this experiment which demanded more accuracy, an approaching and grasping task was tested. First, the laser system of the robot was calibrated according to Algorithm I in Table 13.2 (note that, in this case, the only axis of rotation is the robot itself). Then, an object was placed on a box and the robot's pan-tilt unit was rotated to look at it. The object's coordinates were transformed into the robot's reference frame according to Algorithm II in Table 13.3 and then further correlated with the laser system. Once the coordinates were located in the laser's frame, the robot automatically navigated to place itself in position to grab the object. In order to guarantee a robust navigation, a very simple algorithm for the control of the robot's position was implemented. This algorithm was based on the line measurements obtained with the laser and on the motion estimation algorithm presented in Section 13.5.2. Some of the images in the video sequence can be seen in Fig. 13.12.

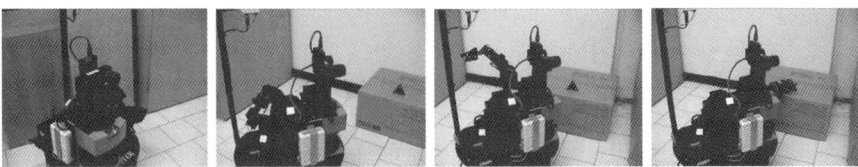

Fig. 13.12. Full calibration: after the laser and the binocular systems are calibrated, the robot can navigate and reach objects in 3-D accurately

13.7 Strategies for Object Manipulation

In this section we show how to perform certain object manipulation tasks in the context of conformal geometric algebra. First, we solve the inverse kinematics for a pan-tilt unit so that the binocular head will be able to follow the end-effector and to position the gripper of the arm in a certain position in space. Then, we illustrate how the robotic arm can follow linear and spherical paths. Finally, we show how to grasp an object in space.

13.7.1 Inverse Kinematics for a Pan-Tilt Unit

In this task we apply a language of spheres for solving the inverse kinematics; this can be seen as an extension of an early approach [5], when a language of points, lines and planes was used instead.

In the inverse kinematics for a pan-tilt unit problem we aim to determine the angles θ_{tilt} and θ_{pan} of the stereo-head, so that the cameras fix at the point p_t. We will now show how we find the values of θ_{pan} and θ_{tilt} using the conformal approach. The problem will be divided into three steps to be solved.

Step 1: Determine the point p_2.

When the θ_{tilt} rotates and the bases rotate (θ_{pan}) around l_y (see Fig. 13.13), the point p_2 describes a sphere s_1. This sphere has center at the point p_1 and radius d_2.

$$S_1 = p_1 - \frac{d_2^2}{2}e_\infty. \tag{13.172}$$

Also the point p_t can be locked from every point around it. that is the point p_2 is in the sphere:

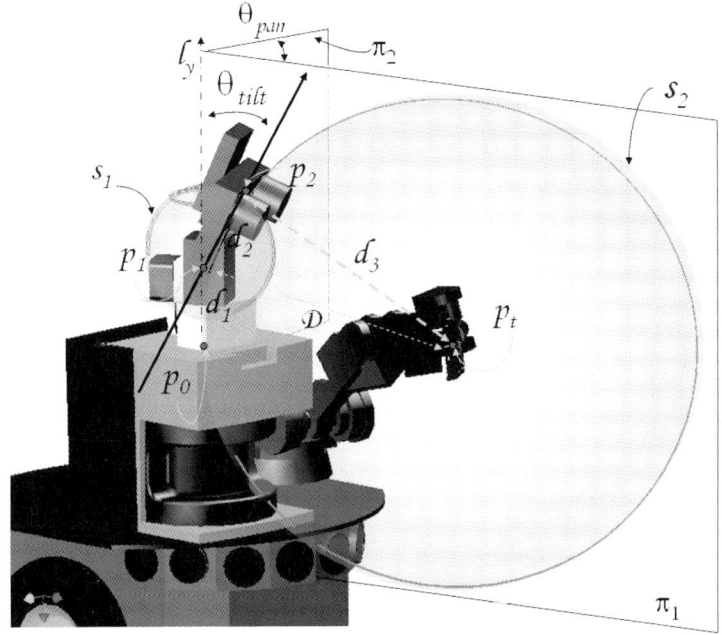

Fig. 13.13. Point p_2 given by intersection of the plane π_1 and the spheres s_1 and s_2

$$S_2 = p_t - \frac{d_3^2}{2} e_\infty, \tag{13.173}$$

where d_3 is the distance between point p_t and the cameras, and we can calculate d_3 using a Pythagorean theorem $d_3^2 = D^2 - d_2^2$, where D is the direct distance between p_t and p_1. We have restricted the position of the point p_2, but there is another restriction: the vector going from p_2 to the point p_t must exist on the plane π_1 generated by the l_y axis ($l_y^* = p_0 \wedge p_1 \wedge e_\infty$) and the point p_t, as we can see in Fig. 13.13. So p_2 can be determined by intersecting the plane π_1 with the spheres s_1 and s_2 as follows

$$\pi_1^* = l_y^* \wedge p_t, \qquad P_{p_2} = s_1 \wedge \pi_1 \wedge s_2. \tag{13.174}$$
$$\tag{13.175}$$

Step 2: Determine the lines and planes.

Once p_2 has been determined, the line l_2 and the plane π_2 can be defined. This line and plane will be useful for calculating the angles θ_{tilt} and θ_{pan}.

$$l_2^* = p_1 \wedge p_2 \wedge e_\infty, \qquad \pi_2^* = l_y^* \wedge e_3. \tag{13.176}$$

Step 3: Find the angles θ_{tilt} and θ_{pan}.

Once we have all the geometric entities, the computation of the angles is a trivial step:

$$\cos(\theta_{pan}) = \frac{\pi_1^* \cdot \pi_2^*}{|\pi_1^*||\pi_2^*|}, \qquad \cos(\theta_{tilt}) = \frac{l_1^* \cdot l_y^*}{|l_1^*||l_y^*|}. \qquad (13.177)$$

13.7.2 Touching a Point

In order to reconstruct the point of interest, we back-project two rays extending from two views of a given scene (see Fig. 13.14). These rays will not intersect, in general, due to noise. Hence, we compute the directed distance between these lines and use the middle point as target. Once the 3-D point p_t is computed with respect to the cameras' framework, we transform it to the arm's coordinate system.

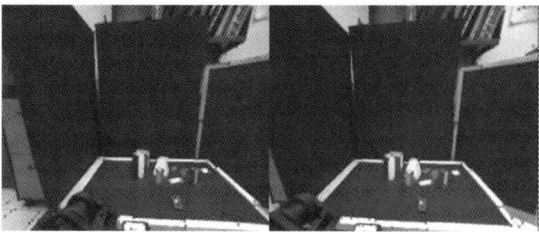

Fig. 13.14. Point of interest in both cameras (p_t)

Once we have a target point with respect to the arm's framework, there are three possibilities to consider: there might be several solutions (see Figs. 13.15a and 13.16a); there might be only a single solution (see Fig. 13.15b); or the point may be impossible to reach (see Fig. 13.16b).

In order to distinguish between these cases, we create a sphere $S_t = p_t - \frac{1}{2}d_3^2 e_\infty$ centered at the point p_t and intersect it with the bounding sphere $S_e = p_0 - \frac{1}{2}(d_1 + d_2)^2 e_\infty$ of the other joints (see Figs. 13.15a and 13.15b), producing the circle $z_s = S_e \wedge S_t$.

If the spheres S_t and S_e intersect, then we have a solution circle z_s which represents all the possible positions that the point p_2 (see Fig. 13.15) may have in order to reach the target. If the spheres are tangential, then there is only one point of intersection and a single solution to the problem, as shown in Fig. 13.15b.

If the spheres do not intersect, then there are two possibilities. The first is that S_t is outside the sphere S_e. In this case, there is no solution, since the arm cannot reach the point p_t, as shown in Fig. 13.16b. On the other hand, if the sphere S_t is inside S_e, then we have a sphere of solutions. In other words, we can place the point p_2 anywhere inside S_t, as shown in Fig. 13.16a. For this case, we arbitrarily choose the upper point of the sphere S_t.

13 Robot Perception and Action Using Conformal Geometric Algebra

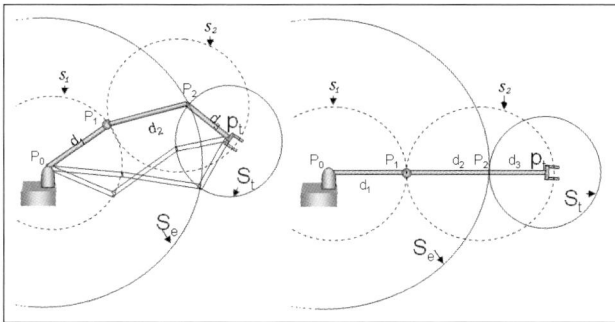

Fig. 13.15. (a) (*left*) S_e and S_t meet (infinitely solutions). (b) (*right*) S_e and S_t are tangential (single solution)

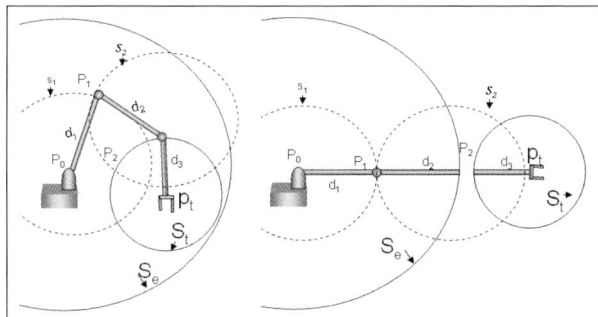

Fig. 13.16. (a) (*left*) S_t inside S_e produces infinitely solutions. (b) (*right*) S_t outside S_e, no possible solution

In the experiment shown in Fig. 13.17a, the sphere S_t is placed inside the bounding sphere S_e and therefore the point selected by the algorithm is the upper limit of the sphere, as shown in Figs. 13.17a and 13.17b. The last joint is completely vertical.

13.7.3 Line of Intersection of Two Planes

In industry, mainly in the industrial sector dedicated to car assembly, it is often necessary to weld together various assembly components. However, for a variety of reasons, the components are not always in the same position, thus complicating the task and making it almost impossible to automate. In many cases, the task requires the welding of components in a straight line when no points on the line are available. This is the problem we attempt to solve in the following experiment.

If one does not have data for points along the line of interest, then the line can still be deduced via the intersection of the two planes (the welding planes). In order to determine each plane, we need three points. The 3-D

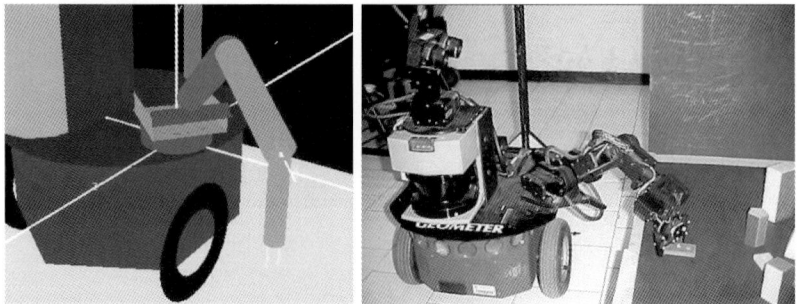

Fig. 13.17. (a) Simulation of the robotic arm touching a point. (b) Robot "Geometer" touching a point with its arm

coordinates of the points are triangulated using the stereo vision system of the robot, which yields a configuration like the one shown in Fig. 13.18.

Fig. 13.18. Images acquired by the binocular system of the robot "Geometer" showing the points on each plane

Once the 3-D coordinates of the points in space are computed, we are able to find each plane with $\pi^* = \boldsymbol{x}_1 \wedge \boldsymbol{x}_2 \wedge \boldsymbol{x}_3 \wedge e_\infty$ and $\pi'^* = \boldsymbol{x}'_1 \wedge \boldsymbol{x}'_2 \wedge \boldsymbol{x}'_3 \wedge e'_\infty$. The line of intersection is computed via the meet operator $l = \pi' \cap \pi$. Figure 13.19a shows a simulation of the arm as it follows the line produced by the intersection of these two planes.

Once the line of intersection l is determined, it is sufficient to translate it on the plane $\psi = l^* \wedge e_2$ (see Fig. 13.19b) by using the translator $\boldsymbol{T}_1 = 1 + \gamma e_2 e_\infty$ in the direction of e_2 (the y-axis) for the distance γ. Furthermore, we are able to build the translator $\boldsymbol{T}_2 = 1 + d_3 e_2 e_\infty$, with the same direction (e_2) but with a separation d_3, which corresponds to the size of the gripper. Once the translators have been computed, we find the lines l' and l'' by translating the line l with $l' = \boldsymbol{T}_1 l \boldsymbol{T}_1^{-1}$, and $l'' = \boldsymbol{T}_2 l' \boldsymbol{T}_2^{-1}$.

The next step, after computing the lines, is to find the points $\boldsymbol{p}_t$ and $\boldsymbol{p}_2$, which represent the points at which the arm will start and finish its motion,

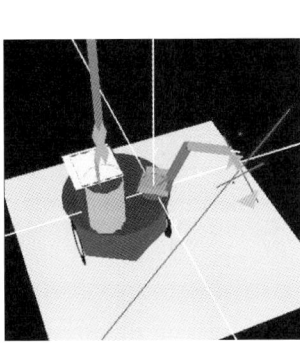 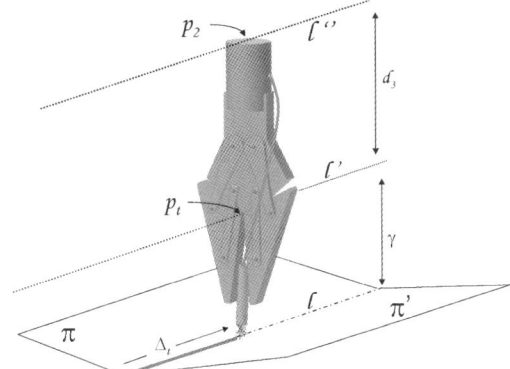

Fig. 13.19. (a) (*left*) Simulation of the arm as it follows the path of a line produced by the intersection of two planes. (b) (*right*) Guiding lines for the robotic arm produced by the intersection (meet) of the planes and vertical translation

respectively. These points were given manually, but they may be computed by using the intersection of lines l' and l'' with the plane that defines the desired depth. In order to make the motion over the line, we build a translator $T_L = 1 - \Delta_L l e_\infty$ with the same direction as l, as shown in Fig. 13.19b. Then, this translator is applied to the points $\boldsymbol{p}_2 = \boldsymbol{T}_L \boldsymbol{p}_2 \boldsymbol{T}_L^{-1}$ and $\boldsymbol{p}_t = \boldsymbol{T}_L \boldsymbol{p}_t \boldsymbol{T}_L^{-1}$ in an iterative fashion to yield a displacement Δ_L on the robotic arm.

By placing the end point over the lines and p_2 over the translated line, and by following the path with a translator in the direction of l, we get a motion over l, as seen in the image sequence of Fig. 13.20.

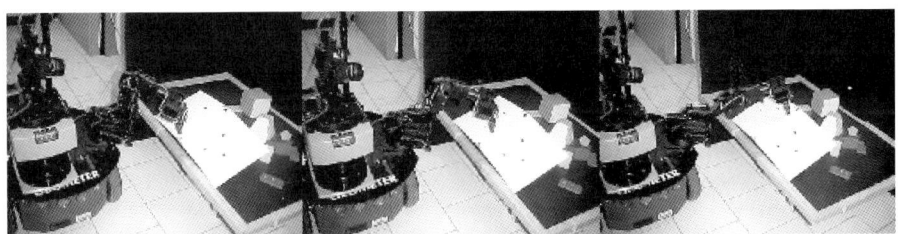

Fig. 13.20. Image sequence of a linear-path motion

13.7.4 Following a Spherical Path

This experiment consists of following the path of a spherical object at a certain fixed distance from it. For this experiment, only four points on the object are available (see Fig. 13.21a).

After acquiring the four 3-D points, we compute the sphere $\boldsymbol{S}^* = \boldsymbol{x}_1 \wedge \boldsymbol{x}_2 \wedge \boldsymbol{x}_3 \wedge \boldsymbol{x}_4$. In order to place the point $\boldsymbol{p}_2$ in such a way that the arm points

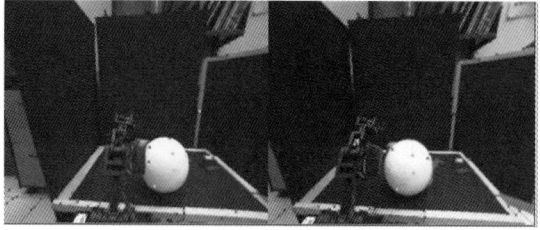 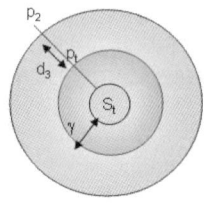

Fig. 13.21. (a) (*left*) Points over the sphere as seen by the robot "Geometer". (b) (*right*) Guiding spheres for the arm's motion

toward the sphere, the sphere was expanded using two different dilators. This produces a sphere that contains S^* and ensures that a fixed distance between the arm and S^* is preserved, as shown in Fig. 13.21b.

The dilators are computed as follows:

$$D_\gamma = e^{-\frac{1}{2}\ln(\frac{\gamma+\rho}{\rho})}E, \qquad (13.178)$$

$$D_d = e^{-\frac{1}{2}\ln(\frac{d_3+\gamma+\rho}{\rho})}E. \qquad (13.179)$$

The spheres S_1 and S_2 are computed by dilating S_t:

$$S_1 = D_\gamma S_t D_\gamma^{-1}, \qquad (13.180)$$

$$S_2 = D_d S_t D_d^{-1}. \qquad (13.181)$$

We decompose each sphere in their parametric form as

$$p_t = M_1(\varphi)M_1(\phi)p_{s_1}M_1^{-1}(\phi)M_1^{-1}(\varphi), \qquad (13.182)$$

$$p_2 = M_2(\varphi)M_2(\phi)p_{s_2}M_2^{-1}(\phi)M_2^{-1}(\varphi), \qquad (13.183)$$

where p_s is any point on the sphere. In order to simplify the problem, we select the upper point on the sphere. To perform the motion on the sphere, we vary the parameters φ and ϕ and compute the corresponding p_t and p_2 using Eq. (13.182) and Eq. (13.183). The results of the simulation are shown in Fig. 13.22a, whereas the results of the real experiment can be seen in Figs. 13.22b and 13.22c.

13.7.5 Grasping an Object

Other interesting experiments involve tasks of grasping objects. First, we consider only approximately cubic objects (i.e., objects with nearly the same width, length, and height). We begin with four non-coplanar points belonging to the corners of the object and use them to build a sphere. With this sphere,

13 Robot Perception and Action Using Conformal Geometric Algebra 447

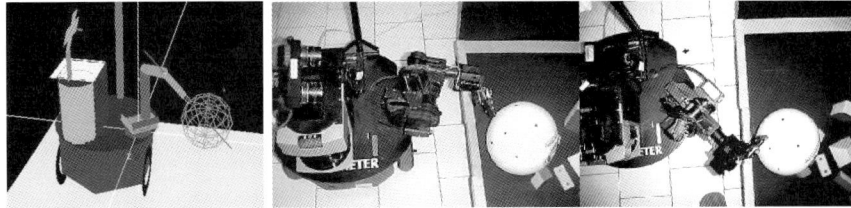

Fig. 13.22. (a) (*left*) Simulation of the motion over a sphere. (b) and (c) (*right*) Two of the images in the sequence of the real experiment

we can make either a horizontal or transversal section, so as to grasp the object from above, below, or in a horizontal fashion. Figure 13.23a shows the sphere obtained using our simulator; the corners of the cube are shown in Fig. 13.23b; and Fig. 13.23c shows the robot arm moving its gripper toward the object after computing its inverse kinematics.

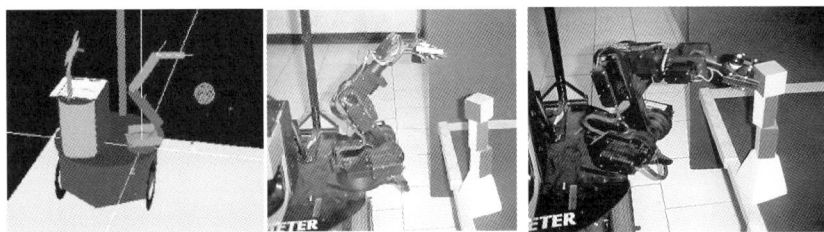

Fig. 13.23. (a) Algorithm simulation showing the sphere containing the cube. (b) Image of the object we wish to grasp with the robotic arm. (c) The robot "Geometer" grasping a wooden cube

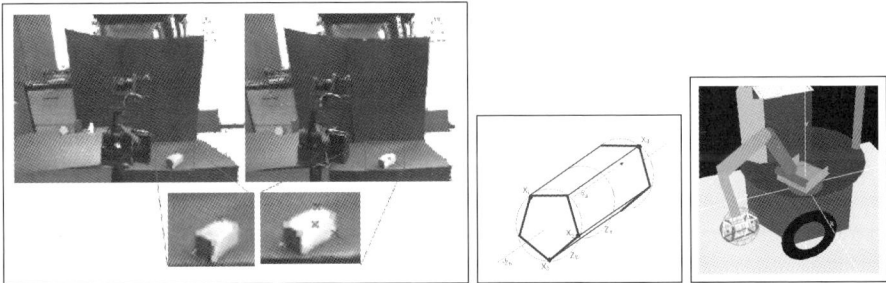

Fig. 13.24. (a) Points on the object as seen by the robot. (b) Regular prism with height d_4 and main axis j_{z_b}. (c) Simulation of the algorithm showing the robotic arm grasping the object

The disadvantage of this algorithm is that not all the objects are cubes; therefore, we need to design a more general algorithm that allows the grasping of objects that are in the shape of regular prisms. That procedure is described next:

1. Take a calibrated stereo pair of images of the object.
2. Extract four non-coplanar points from these images (see, for example Fig. 13.24a).
3. Compute the corresponding 3-D points x_i, $i = 1, ..., 4$ using the stereo vision system and triangulation.
4. Compute the directed distances (see (13.147)):

$$d_1 = Dist(\boldsymbol{x}_1, \boldsymbol{x}_2 \wedge \boldsymbol{x}_3 \wedge \boldsymbol{x}_4 \wedge e_\infty),$$
$$d_2 = Dist(\boldsymbol{x}_2, \boldsymbol{x}_1 \wedge \boldsymbol{x}_3 \wedge \boldsymbol{x}_4 \wedge e_\infty),$$
$$d_3 = Dist(\boldsymbol{x}_3, \boldsymbol{x}_2 \wedge \boldsymbol{x}_1 \wedge \boldsymbol{x}_4 \wedge e_\infty),$$
$$d_4 = Dist(\boldsymbol{x}_4, \boldsymbol{x}_2 \wedge \boldsymbol{x}_3 \wedge \boldsymbol{x}_1 \wedge e_\infty).$$

5. Select the point with the greatest distance as the apex x_a and label the others $\boldsymbol{x}_{b_1}, \boldsymbol{x}_{b_2}, \boldsymbol{x}_{b_3}$ as belonging to the base of the object.
6. Compute the circle $\boldsymbol{z}_b = \boldsymbol{x}_{b_1} \wedge \boldsymbol{x}_{b_2} \wedge \boldsymbol{x}_{b_3}$.
7. Compute the directed distance $\boldsymbol{d}_a$ between $\boldsymbol{z}_b$ and $\boldsymbol{x}_a$.
8. Translate the circle z in the direction and magnitude of $\boldsymbol{d}_a$ to produce the grasping plane.

Some points of the previous algorithm can be explained in more detail. For example, for the object in Fig. 13.24b, the base circle is $\boldsymbol{z}_b^* = \boldsymbol{x}_1 \wedge \boldsymbol{x}_2 \wedge \boldsymbol{x}_3$, whereas the main axis of the object is computed by $j_{z_b} = \boldsymbol{z}_b \wedge e_\infty$. The translator that moves $\boldsymbol{z}_b$ is produced as $\boldsymbol{T} = 1 + \frac{1}{4}d_4 e_\infty$. The grasping circle can be computed with $\boldsymbol{z}_t^* = \boldsymbol{T}\boldsymbol{z}_b^*\boldsymbol{T}^{-1}$, the point of contact being the closest point from the circle to the y-axis. Finally, the grasping plane is denoted by $\pi^* = \boldsymbol{z}_t^* \wedge e_\infty$. Note that this last algorithm may grasp the object regardless of whether it is in a horizontal or vertical position. We illustrate this algorithm with the simulation shown in Fig. 13.24c.

13.8 Omnidirectional Vision Using Conformal Geometric Algebra

The model defined by Geyer and Daniilidis [7] is used in this work to find an equivalent spherical projection of a catadioptric projection. This model is very useful for simplifying the projections, but the representation is not ideal because it is defined in a projective geometry context where the basis objects are points and lines and not spheres. The computations are also complicated and difficult to follow.

Our proposal is based on conformal geometric algebra, where the basic element is the sphere. That is, all the entities (point, point pair, circle, plane) are defined in terms of the sphere (e.g., a point can be defined as a sphere of zero radius). This framework also has the advantage that the intersection operation between entities is mathematically well defined (e.g., the intersection of a sphere with a line can be defined as $L \wedge S$). In short, the unified model is more natural and concise in the context of conformal geometric algebra (spheres) than of projective algebra (points and lines).

13.8.1 Conformal Unified Model

First, in Fig. 13.25 we present the unified model in terms of conformal geometric algebra. In this regard, we assume that the optical axis of the mirror is parallel to the e_2 axis, then let $\mathbf{f}$ be a point in the Euclidean space (which represents the focus of the mirror which lies in such an optical axis), defined by

$$\mathbf{f} = \alpha_1 e_1 + \alpha_2 e_2 + \alpha_3 e_3 , \tag{13.184}$$

with conformal representation given by

$$\boldsymbol{F} = \mathbf{f} + \frac{1}{2}\mathbf{f}^2 e + e_0 . \tag{13.185}$$

Using the point F as the center, we define a unit sphere $\boldsymbol{S}$ (see Fig. 13.25) as

$$\boldsymbol{S} = \boldsymbol{F} - \frac{1}{2}e . \tag{13.186}$$

Now, if $\boldsymbol{N}$ is the point of projection (that also lies on the optical axis) at a distance l from the point $\boldsymbol{F}$, then this point can be found using a translator,

$$\boldsymbol{T} = 1 + \frac{l e_2 e}{2} , \tag{13.187}$$

and then

$$\boldsymbol{N} = \boldsymbol{T}\boldsymbol{F}\widetilde{\boldsymbol{T}} . \tag{13.188}$$

Finally, the image plane Π is perpendicular to the optical axis at a distance $-m$ from the point $\boldsymbol{F}$, and its equation is

$$\Pi = e_2 + (\mathbf{f} \cdot e_2 - m)e . \tag{13.189}$$

13.8.2 Point Projection

Let $\mathbf{p}$ be a point in the Euclidean space. Then the corresponding homogeneous point in the conformal space is

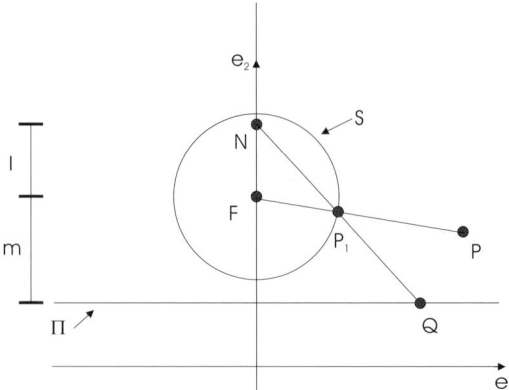

Fig. 13.25. Conformal unity model

$$P = \mathbf{p} + \frac{1}{2}\mathbf{p}^2 e + e_0 \; . \tag{13.190}$$

Now, for the projection of point P we trace a line joining the points $\mathbf{F}$ and $\mathbf{P}$. Using the definition of the line in dual form, we get

$$\mathbf{L}_1^* = \mathbf{F} \wedge \mathbf{P} \wedge e \; . \tag{13.191}$$

Then, we calculate the intersections of the line $\mathbf{L}_1$ and the sphere $\mathbf{S}$ (see Eq. (13.90)), which results in the point pair

$$\mathbf{PP}^* = (\mathbf{L}_1 \wedge \mathbf{S})^* \; . \tag{13.192}$$

From the point pair, we choose the point $\mathbf{P}_1$, which is the closest point to $\mathbf{P}$, and then we find the line passing through the points $\mathbf{P}_1$ and $\mathbf{N}$:

$$\mathbf{L}_2^* = \mathbf{P}_1 \wedge \mathbf{N} \wedge e. \tag{13.193}$$

Finally, we find the intersection of the line $\mathbf{L}_2$ with the plane Π, using

$$\mathbf{Q} = (\mathbf{L}_2 \wedge \Pi)^* \; . \tag{13.194}$$

The point $\mathbf{Q}$ is the projection in the image plane of point $\mathbf{P}$ of the space. Notice that we can project any point in the space into any type of mirror (changing l and m) using the previous procedure (see Fig. 13.26). The reader can now appreciate the simplicity and elegance of the unified model using conformal geometric algebra.

13.8.3 Inverse Point Projection

We have already seen how to project a point in space to the image plane through the sphere. Now, we want to back-project a point in the image plane

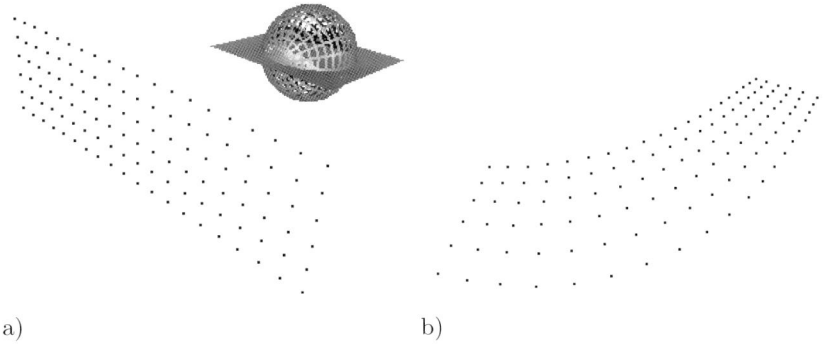

Fig. 13.26. (a) Unified model and points in the space. (b) Projection in the image plane

into 3-D space. First, if Q is a point in the image plane, then the equation of the line passing through the points Q and N is

$$L_2^* = Q \wedge N \wedge e , \qquad (13.195)$$

and the intersection of the line L_2 and the sphere S is

$$PP^* = (L_2 \wedge S)^* . \qquad (13.196)$$

From the point pair, we choose the point P_1, which is the closest point to Q, and then we find the equation of the line from point P_1 to the focus F:

$$L_1^* = P_1 \wedge F \wedge e . \qquad (13.197)$$

Point P should lie on the line L_1^*, however it cannot be calculated exactly because a coordinate was lost when the point was projected to the image plane (a single view does not allow us to know the projective depth). However, we can project this point to some plane and say that it is equivalent to the original point up to a scale factor (see Fig. 13.27).

13.8.4 Line Projection

Suppose that L is a line in the space and that we want to find its projection in the image plane (see Fig. 13.28a). First, we find the plane where L and F lie; its equation is

$$\Phi_L^* = L^* \wedge F . \qquad (13.198)$$

The intersection of the plane and the sphere is the great circle defined by

$$C^* = (\Phi_L \wedge S)^* . \qquad (13.199)$$

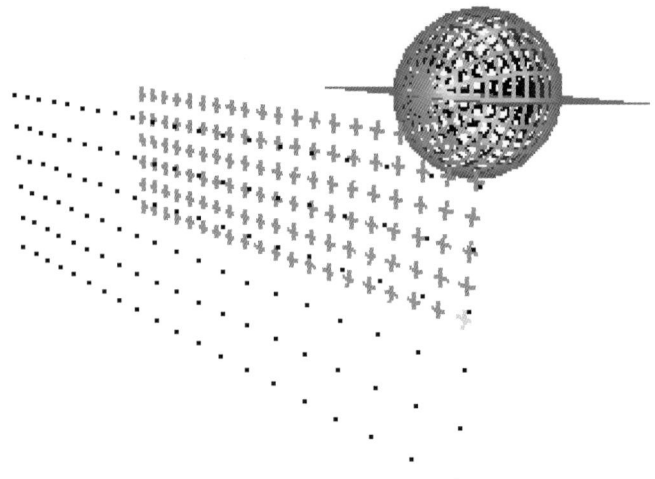

Fig. 13.27. Inverse point projection (from the image to the space). The crosses are the projected points and the dots are the original points

The line that passes through the center of the circle and is perpendicular to the plane Φ_L is

$$U^* = (C \wedge e) . \tag{13.200}$$

Using U as an axis, we make a rotor,

$$R = e^{(\frac{\theta}{2}U)}. \tag{13.201}$$

We find a point pair PP^* that lies on the circle, using

$$PP^* = (C \wedge e_2)^* . \tag{13.202}$$

We choose any point from the point pair, say P_1, and using the rotor R we can find the points in the circle,

$$P'_1 = RP_1\widetilde{R} . \tag{13.203}$$

For each point P'_1 we find the line that passes through the points P_1 and N, defined as

$$L_2^* = P'_1 \wedge N \wedge e . \tag{13.204}$$

Finally, for each line L_2 we find the intersection with the plane Φ,

$$P_2 = (L_2 \wedge \Pi)^*, \tag{13.205}$$

which is the projection of the line in the space to the image plane (see Fig. 13.28b).

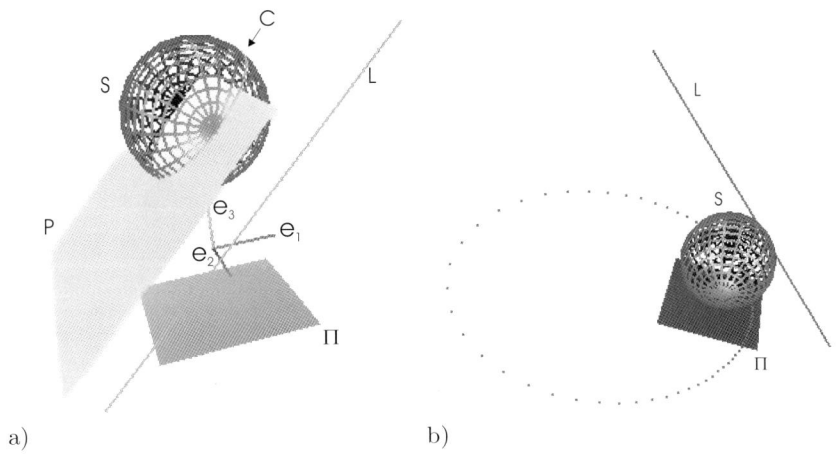

Fig. 13.28. (a) Line projection to the sphere. (b) Line projection to the image plane (note that, in this case, the result is an ellipse)

13.8.5 Image-Plane Conic Characteristics

We have shown how a line is projected to the image plane as a conic and that this conic has parameters which are dependent on the mirror and the line. Now, we shall see how to find the conic center, the major and minor axis, and their lengths $2a$ and $2b$, respectively, using the conformal unified model (see Fig. 13.29).

Using Eq. (13.200) and $\boldsymbol{N}$, we can calculate the plane where the principal axis lies as

$$\boldsymbol{P}^* = \boldsymbol{U}^* \wedge \boldsymbol{N} . \tag{13.206}$$

The principal axis of the conic is the intersection of the planes $\boldsymbol{P}$ and Π,

$$\boldsymbol{A}_1^* = (\boldsymbol{P} \wedge \Pi)^* . \tag{13.207}$$

The points at both extremes of the axis are found by the intersection of the circle $\boldsymbol{C}$ Eq. (13.81) and the plane $\boldsymbol{P}$,

$$\boldsymbol{PP}^* = (\boldsymbol{C} \wedge \boldsymbol{P})^* . \tag{13.208}$$

For each of the two points $\boldsymbol{P}_i$ ($i = 1, 2$), we trace a line,

$$\boldsymbol{L}_i^* = \boldsymbol{P}_i \wedge \boldsymbol{N} \wedge e , \tag{13.209}$$

and then we intersect both lines and the plane $\boldsymbol{M}$,

$$\boldsymbol{Q}_i = (\boldsymbol{L}_i \wedge \Pi)^*, \tag{13.210}$$

consequently the points $\boldsymbol{Q}_1$ and $\boldsymbol{Q}_2$ lie on the principal axis of the conic and the distance between them equals $2a$.

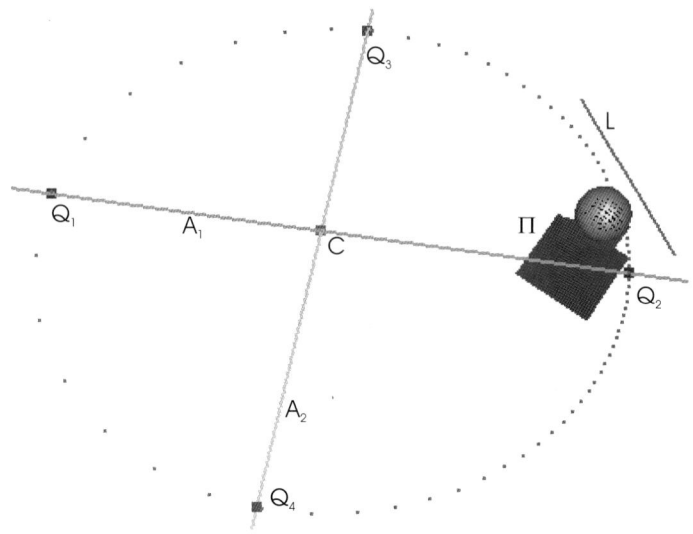

Fig. 13.29. Center and principal axis of the conic in the image plane

The center of the conic lies on the axis $\boldsymbol{A}_1^*$ at half of the distance between $\boldsymbol{Q}_1$ and $\boldsymbol{Q}_2$. We build a translator using the distance a and the axis $\boldsymbol{A}_1^*$:

$$\boldsymbol{T} = (1 + \frac{e\boldsymbol{A}_1^*}{2}a) \,. \tag{13.211}$$

The center of the conic is found as

$$\boldsymbol{C}_0 = \boldsymbol{T}\boldsymbol{Q}_1\widetilde{\boldsymbol{T}} \,. \tag{13.212}$$

The normal of the plane $\boldsymbol{P}$ is

$$\boldsymbol{n} = (\boldsymbol{P} \wedge e) \cdot e_0 \,. \tag{13.213}$$

And, now, the other axis of the conic is found easily:

$$\boldsymbol{A}_2^* = \boldsymbol{n} \wedge \boldsymbol{C}_0 \wedge e \,. \tag{13.214}$$

Then we calculate the plane $\boldsymbol{P}_2$ as

$$\boldsymbol{P}_2^* = \boldsymbol{A}_2^* \wedge \boldsymbol{N} \,. \tag{13.215}$$

At the intersection of plane $\boldsymbol{P}$ and the circle $\boldsymbol{C}$, we get a point pair,

$$\boldsymbol{R}\boldsymbol{R}^* = (\boldsymbol{C} \wedge \boldsymbol{P}_2)^* \,. \tag{13.216}$$

We trace a line $\boldsymbol{L}j$ with each of the two points $\boldsymbol{R}_i$,

$$L_j = R_i \wedge N \wedge e ,\qquad (13.217)$$

and, finally, we calculate the intersection of the lines and the image plane:

$$Q_j = (L_j \wedge \Pi)^* .\qquad (13.218)$$

The distance between Q_3 and Q_4 is equal to $2b$.

13.8.6 Navigation Using Omnidirectional Vision

In order to use the omnidirectional image for robot navigation, we first apply the inverse point projection (defined previously) to each pixel in the image so as to produce a rectified image (see Fig. 13.30a). As mentioned earlier, our system is monocular so we cannot make a 3-D reconstruction; but since we are only interested in the plane of the floor, we can project all the image points to this plane, though all the points that do not belong to such a plane will appear distorted. Once we project all the points to the floor plane, the lines in the floor will appear as lines in the image and not as curves (see Fig. 13.30b).

Robot Navigation

In this experiment, the robot was placed in the corridor of our laboratory, the objective being that the robot should navigate approximately parallel to the wall for a distance of $d = 50$ cm (see Fig. 13.31). Let L be the line of the corridor, and G the robot, which is modeled by a circle defined as

$$G^* = x_1 \wedge x_2 \wedge x_3 .\qquad (13.219)$$

Note that with the wedge of three points lying on the robot circumference we get the robot contour as an entity for further algebraic manipulations (see

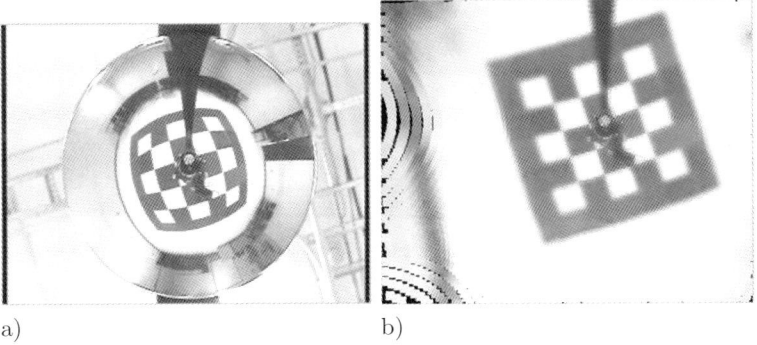

Fig. 13.30. (a) Omnidirectional image. (b) Rectified image

Ea. 13.230). Using the line of the corridor, we calculate the rotor to make the robot parallel to such a line as

$$\boldsymbol{R}_1 = e^{(\frac{\arccos(\boldsymbol{L}_1^* \cdot e_{3+-})}{2} \boldsymbol{L}_G^*)}. \tag{13.220}$$

The translator for the parallel line is

$$\boldsymbol{T}_1 = (1 + d\frac{e}{2}e_{3+-}). \tag{13.221}$$

With this translator, we can find the line $\boldsymbol{L}_2^*$ parallel to $\boldsymbol{L}_1^*$, which is defined by

$$\boldsymbol{L}_2^* = \boldsymbol{T}_1 \boldsymbol{L}_1^* \widetilde{\boldsymbol{T}}_1. \tag{13.222}$$

The point $\boldsymbol{P}_1$ is computed as

$$\boldsymbol{P}_1 = \{[[((\boldsymbol{L}_1 \wedge e) \cdot e_0)I_E] \wedge \boldsymbol{P}_c' \wedge e] \wedge e_2\}^* \wedge (\boldsymbol{L}_2^* \wedge e_2)^* \wedge \varPi, \tag{13.223}$$

and using the distance, d_2, between the points $\boldsymbol{P}_1$ and $\boldsymbol{P}_c'$ we calculate the translator,

$$\boldsymbol{T}_2 = (1 + \frac{e\boldsymbol{L}_2^*}{2}d_2). \tag{13.224}$$

To find the next point where the robot will be, we apply

$$\boldsymbol{P}_c' = \boldsymbol{T}_2 \boldsymbol{P}_1 \widetilde{\boldsymbol{T}}_2. \tag{13.225}$$

The line between points $\boldsymbol{P}_c$ and $\boldsymbol{P}_c'$ is

$$\boldsymbol{L}_3^* = \boldsymbol{P}_c \wedge \boldsymbol{P}_c' \wedge e, \tag{13.226}$$

and with this line we calculate the rotor,

$$\boldsymbol{R}_2 = e^{(\frac{\arccos(\boldsymbol{L}_3^* \cdot e_{3+-})}{2} \boldsymbol{L}_G^*)}. \tag{13.227}$$

The translator that moves the robot to its final position is

$$\boldsymbol{T}_3 = (1 + \frac{e\boldsymbol{L}_3^*}{2}d_3), \tag{13.228}$$

where d_3 is the distance between $\boldsymbol{P}_c$ and $\boldsymbol{P}_c'$. The motor is then

$$\boldsymbol{M} = \boldsymbol{T}_3 \boldsymbol{R}_2 \boldsymbol{R}_1, \tag{13.229}$$

which can be applied to the robot to obtain its next position,

$$\boldsymbol{G}_2^* = \boldsymbol{M} \boldsymbol{G}^* \widetilde{\boldsymbol{M}}. \tag{13.230}$$

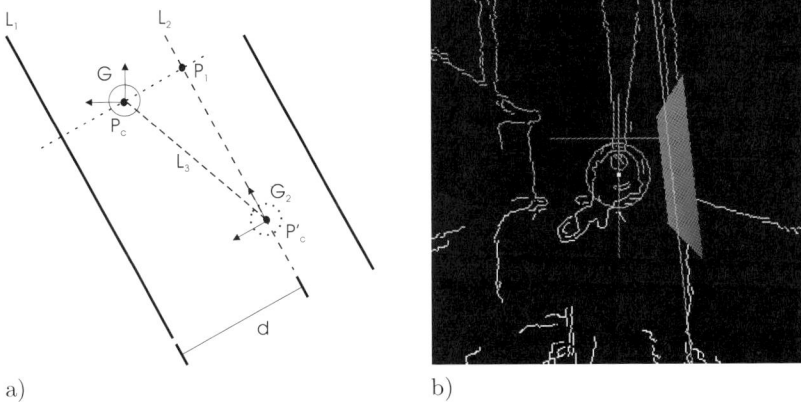

Fig. 13.31. (a) Model of the experiment. (b) Processed image and conformal objects (note that the mirror reverses the image)

13.9 Conclusion

In this chapter we have presented a series of interesting real applications of robot vision by employing non-conventional techniques for projective geometry. The key to our approach is the use of conformal geometric algebra for incidence algebra computations involving linear transformations represented efficiently as spinors. By utilizing our mathematical system, it is no longer necessary to abandon the mathematical framework in order to carry out simultaneous operations of algebra of incidence and conformal transformations. The author strongly believes that the framework of conformal geometric algebra can surely be of great advantage for representing and processing information from robots equipped with stereo-vision systems, laser sensors, omnidirectional vision, and odometry.

Acknowledgements

I am very thankful to my PhD students, Leo Reyes Lozano, Julio Zamora Esquivel and Carlos Lopez-Franco, who provided me with experimental results useful for illustrating the application of geometric algebra in robotic vision. Eduardo Bayro-Corrochano was supported by Project 49 of CONACYT Fondo Sectorial de Investigación en Salud y Seguridad Social.

References

1. Bayro-Corrochano E. (2001) *Geometric Computing for Perception Action Systems*, Springer, New York.
2. Bayro-Corrochano E. and Lasenby J. (1995) Object modelling and motion analysis using Clifford algebra. In *Proceedings of Europe-China Workshop on Geometric Modeling and Invariants for Computer Vision*, Ed. Roger Mohr and Wu Chengke, Xi'an, China, April 27–29, pp. 143–149.
3. Bayro-Corrochano E., Lasenby J. and Sommer, G. (1996) Geometric algebra: a framework for computing point and line correspondences and projective structure using n uncalibrated cameras. In: *Proceedings of the International Conference on Pattern Recognition (ICPR'96), Vienna, August 1996*, Vol. I, pp. 393–397.
4. Bayro-Corrochano E., Daniilidis K. and Sommer G. (2000) Motor algebra for 3D kinematics. The case of the hand–eye calibration. *Journal of Mathematical Imaging and Vision*, Vol. 13, pp. 79–99.
5. Bayro-Corrochano E. and Kähler D. 2000. Motor algebra approach for computing the kinematics of robot manipulators. *Journal of Robotics Systems*, Vol. 17(9), pp. 495–516.
6. Csurka G. and Faugeras O. (1998) Computing three dimensional project invariants from a pair of images using the Grassmann–Cayley algebra *Journal of Image and Vision Computing*, 16, pp. 3–12.
7. Geyer C. and Daniilidis K. (2001) Catadioptric projective geometry. *International Journal of Computer Vision*, 43, pp. 223–243.
8. Hestenes D. (1966) *Space–Time Algebra*. Gordon and Breach.
9. Hestenes D. and Sobczyk G. (1984) *Clifford Algebra to Geometric Calculus: A unified language for mathematics and physics*. D. Reidel, Dordrecht.
10. Hestenes D., Li H. and Rockwood A. (2001) New algebraic tools for classical geometry. In: *Geometric Computing with Clifford Algebra*, G. Sommer (ed.). Springer, Berlin Heidelberg, Chap. 1, pp. 3–23.
11. Hestenes D. and Ziegler R. (1991) Projective geometry with Clifford algebra. *Acta Applicandae Mathematicae*, 23, pp. 25–63.
12. Lasenby J., Lasenby A., Doran C. and Fitzgerald W. (1998) New geometric methods for computer vision – an application to structure and motion estimation. *International Journal of Computer Vision*, 26(3), 191–213.
13. Lasenby J. and Bayro–Corrochano E. (1999) Analysis and computation of projective invariants from multiple views in the geometric algebra framework. In: *Special Issue on Invariants for Pattern Recognition and Classification*, ed. M.A. Rodrigues. *Int. Journal of Pattern Recognition and Artificial Intelligence*, Vol. 13, No. 8, pp. 1105–1121.
14. White N. (1997) Geometric applications of the Grassmann–Cayley algebra. In: J.E. Goodman and J. O'Rourke (eds.) *Handbook of Discrete and Computational Geometry*, CRC Press, Florida.

Part VI

Uncertainty in Geometric Computations

14

Uncertainty Modeling and Geometric Inference

Kenichi Kanatani

Department of Computer Science, Okayama University, Okayama 700-8530
Japan kanatani@suri.it.okayama-u.ac.jp

14.1 Introduction

Statistical inference from images is one of the key components of computer vision research today. Traditionally, statistical methods have been used for recognition and classification purposes. Recently, however, there are many studies of statistical analysis for *geometric inference* based on geometric primitives such as points and lines extracted by image processing operations.

However, the term "statistical" has somewhat a different meaning for such geometric inference problems than for the traditional recognition and classification purposes. This difference has often been overlooked, causing controversies over the validity of the statistical approach to geometric problems in general. In Sect. 14.2, we take a close look at this problem, tracing back the origin of feature uncertainty to image processing operations. In Sect. 14.3, we discuss the implications of asymptotic analysis in reference to geometric fitting and geometric model selection. In Sect. 14.4, we point out that a correspondence exists between the standard statistical analysis and the geometric inference problem. We also compare the capability of the geometric AIC and the geometric MDL in detecting degeneracy. In Sect. 14.5, we review recent progress in geometric fitting techniques for linear constraints, describing the FNS method, the HEIV method, the renormalization method, and other related techniques. In Sect. 14.6, we discuss the Neyman–Scott problem and semiparametric models in relation to geometric inference. Sect. 14.7 presents our concluding remarks.

14.2 What Is Geometric Inference?

14.2.1 Ensembles for Geometric Inference

The goal of statistical methods is not to study the properties of observed data themselves but to infer the properties of the *ensemble* from which we

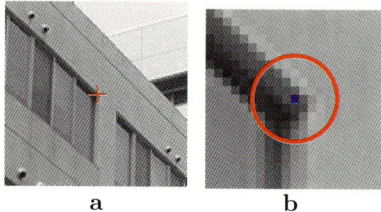

Fig. 14.1. a A feature point in an image of a building. **b** Its enlargement and the uncertainty of the feature location

regard the observed data as sampled. The ensemble may be a collection of existing entities (e.g., the entire population), but often it is a hypothetical set of conceivable possibilities. When a statistical method is employed, the underlying ensemble is often taken for granted. However, this issue is very crucial for geometric inference based on feature points.

Suppose, for example, we extract feature points, such as corners of walls and windows, from an image of a building and want to test if they are collinear. The reason why we need a statistical method is that the extracted feature positions have uncertainty. So, we have to judge the extracted feature points as collinear if they are sufficiently aligned. We can also evaluate the degree of uncertainty of the fitted line by propagating the uncertainty of the individual points. What is the ensemble that underlies this type of inference?

This question reduces to the question of why the uncertainty of the feature points occurs at all. After all, statistical methods are not necessary if the data are exact. Using a statistical method means regarding the current feature position as sampled from a set of its possible positions. But where else could it be if not in the current position?

14.2.2 Uncertainty of Feature Extraction

Many algorithms have been proposed for extracting feature points, including the Harris operator [12] and SUSAN [46], and their performance has been extensively compared [4, 41, 45]. However, if we use, for example, the Harris operator to extract a particular corner of a particular building image, the output is unique (Fig. 14.1). No matter how many times we repeat the extraction, we obtain the same point because no external disturbances exist and the internal parameters (e.g., thresholds for judgment) are unchanged. It follows that the current position is the sole possibility. How can we find it elsewhere?

If we closely examine the situation, we are compelled to conclude that other possibilities should exist because the extracted position is not necessarily correct. But if it is not correct, why did we extract it? Why didn't we extract the correct position in the first place? The answer is: *we cannot*.

14.2.3 Image Processing for Computer Vision

The reason why there exist so many feature extraction algorithms, none of them being definitive, is that they are aiming at an intrinsically impossible task. If we were to extract a point around which, say, the intensity varies to the largest degree in such and such a measure, the algorithm would be unique; variations may exist in intermediate steps, but the final output should be the same.

However, what we want is not "image properties" but "3-D properties" such as corners of a building, but the way a 3-D property is translated into an image property is intrinsically heuristic. As a result, as many algorithms can exist as the number of heuristics for its 2-D interpretation. If we specify a particular 3-D feature to extract, say a corner of a window, its appearance in the image is not unique. It is affected by many properties of the scene, including the details of its 3-D shape, the viewing orientation, the illumination condition, and the light reflectance properties of the material. A slight variation of any of them can result in a substantial difference in the image.

Theoretically, exact extraction would be possible if all the properties of the scene were exactly known, but to infer them from images is the very task of computer vision. It follows that we must make a guess in the image processing stage. For the current image, some guesses may be correct, but others may be wrong. The exact feature position could be found only by an (nonexisting) "ideal" algorithm that could guess everything correctly.

This observation allows us to interpret the "possible feature positions" to be *the positions that would be located by different (nonideal) algorithms based on different guesses*. It follows that the set of hypothetical positions should be associated with *the set of hypothetical algorithms*. The current position is regarded as produced by an algorithm sampled from it. This explains why one always obtains the same position no matter how many times one repeats extraction using that algorithm. To obtain a different position, one has to sample another algorithm.

Remark 1. We may view the statistical ensemble in the following way. If we repeat the *same* experiment, the result should always be the same. But if we declare that the experiment is the "same" if such and such are the same while other things can vary, then those variable conditions define the ensemble. The conventional view is to regard the experiment as the same if the *3-D scene* we are viewing is the same while other properties, such as the lighting condition, can vary. Then, the resulting image would be different for each (hypothetical) experiment, so one would obtain a different output each time, using the same image processing algorithm. The expected spread of the outputs measures the robustness of that algorithm. Here, however, we are viewing the experiment as the same *if the image is the same*. Then, we could obtain different results only by sampling other algorithms. The expected spread of the outputs measures the uncertainty of feature detection from *that image*. We take this view because we are analyzing the reliability of geometric inference from a particular image, while the conventional view is suitable for assessing the robustness of a *particular algorithm*.

14.2.4 Covariance Matrix of a Feature Point

The performance of feature point extraction depends on the image properties around that point. If, for example, we want to extract a point in a region with an almost homogeneous intensity, the resulting position may be ambiguous whatever algorithm is used. In other words, the positions that potential algorithms would extract should have a large spread. If, on the other hand, the intensity greatly varies around that point, any algorithm could easily locate it accurately, meaning that the positions that the hypothetical algorithms would extract should have a strong peak. It follows that we may introduce for each feature point its *covariance matrix* that measures the spread of its potential positions.

Let $V[p_\alpha]$ be the covariance matrix of the αth feature point p_α. The above argument implies that we can estimate the qualitative characteristics of uncertainty but not its absolute magnitude. So, we write the covariance matrix $V[p_\alpha]$ in the form

$$V[p_\alpha] = \varepsilon^2 V_0[p_\alpha], \tag{14.1}$$

where ε is an unknown magnitude of uncertainty, which we call the *noise level*. The matrix $V_0[p_\alpha]$, which we call the *(scale) normalized covariance matrix*, describes the relative magnitude and the dependence on orientations.

Remark 2. The decomposition of $V[p_\alpha]$ into ε^2 and $V_0[p_\alpha]$ involves scale ambiguity. We assume that the decomposition is made unique by an appropriate scale normalization such as $\mathrm{tr} V_0[p_\alpha] = 2$. However, the subsequent analysis does not depend on particular normalizations, so we do not explicitly specify it except that it should be done in such a way that ε is much smaller than the data themselves. Note that mathematically, modeling the covariance matrix by a common scale factor ε^2 and the individual matrix part $V_0[p_\alpha]$ is rather restrictive. However, this model is sufficient for most practical applications, as we describe in the following.

14.2.5 Covariance Matrix Estimation

If the intensity variations around p_α are almost the same in all directions, we can think of the probability distribution as isotropic, a typical equiprobability line, known as the *uncertainty ellipses*, being a circle (Fig. 14.1b). On the other hand, if p_α is on an object boundary, distinguishing it from nearby points should be difficult whatever algorithm is used, so its covariance matrix should have an elongated uncertainty ellipse along that boundary.

However, existing feature extraction algorithms are usually designed to output those points that have large image variations around them, so points in a region with an almost homogeneous intensity or on object boundaries are rarely chosen. As a result, the covariance matrix of a feature point extracted by such an algorithm can be regarded as nearly isotropic. This has also been confirmed by experiments [26], justifying the use of the identity as the normalized covariance matrix $V_0[p_\alpha]$.

Remark 3. The intensity variations around different feature points are usually unrelated, so their uncertainty can be regarded as statistically independent. However, if we track feature points over consecutive video frames, it has been observed that the uncertainty has strong correlations over the frames [47].

Remark 4. Many interactive applications require humans to extract feature points by manipulating a mouse. Extraction by a human is also an "algorithm", and it has been shown by experiments that humans are likely to choose "easy-to-see" points such as isolated points and intersections, avoiding points in a region with an almost homogeneous intensity or on object boundaries [26]. In this sense, the statistical characteristics of human extraction are very similar to machine extraction. This is no surprise if we recall that image processing for computer vision is essentially a heuristic that simulates human perception. It has also been reported that strong microscopic correlations exist when humans manually select corresponding feature points over multiple images [34].

14.2.6 Image Quality and Uncertainty

The uncertainty of feature points has often been identified with "image noise", giving a misleading impression as if the feature locations were perturbed by random intensity fluctuations. Of course, we may obtain better results using higher-quality images whatever algorithm is used. However, the task of computer vision is not to analyze image properties but to study the 3-D properties of the scene. As long as the image properties and the 3-D properties do not correspond one to one, any image processing inevitably entails some degree of uncertainty, however high the image quality may be, and the result must be interpreted statistically. The underlying ensemble is the set of hypothetical (inherently imperfect) algorithms of image processing. Yet, the performance of image processing algorithms has often been evaluated by adding *independent Gaussian noise* to individual pixels.

Remark 5. This also applies to *edge detection*, whose goal is to find the boundaries of 3-D objects in the scene. In reality, all existing algorithms seek *edges*, i.e., lines and curves across which the intensity changes discontinuously. Yet, this is regarded by many as an objective image processing task, and the detection performance is often evaluated by adding independent Gaussian noise to individual pixels. From the above considerations, we conclude that edge detection is also a heuristic, and hence no definitive algorithm will ever be found.

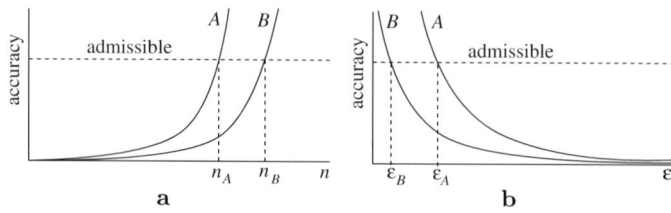

Fig. 14.2. a For the standard statistical analysis, it is desired that the accuracy increases rapidly as the number of experiments $n \to \infty$, because admissible accuracy can be reached with a smaller number of experiments. b For geometric inference, it is desired that the accuracy increases rapidly as the noise level $\varepsilon \to 0$, because larger data uncertainty can be tolerated for admissible accuracy

14.3 Asymptotic Analysis for Geometric Inference

14.3.1 What Is Asymptotic Analysis?

As stated earlier, *statistical estimation* refers to estimating the properties of an ensemble from a finite number of samples, assuming some knowledge, or a *model*, about the ensemble. If the uncertainty originates from external conditions, as in experiments in physics, the estimation accuracy can be increased by controlling the measurement devices and environments. For internal uncertainty, on the other hand, there is no way of increasing the accuracy except by repeating the experiment and doing statistical inference. However, repeating experiments usually entails costs, and in practice the number of experiments is often limited.

Taking account of this, statisticians usually evaluate the performance of estimation *asymptotically*, analyzing the growth in accuracy as the number n of experiments increases. This is justified because a method whose accuracy increases more rapidly as $n \to \infty$ can reach admissible accuracy *with fewer experiments* (Fig. 14.2a).

In contrast, the ensemble for geometric inference is, as we have seen, the set of potential feature positions that could be located if other (hypothetical) algorithms were used. As noted earlier, however, we can choose only *one* sample from the ensemble as long as we use a particular image processing algorithm. In other words, the number n of experiments is 1. Then, how can we evaluate the performance of statistical estimation?

Evidently, we want a method whose accuracy is sufficiently high *even for large data uncertainty*. This implies that we need to analyze the growth in accuracy as the noise level ε decreases, because a method whose accuracy increases more rapidly as $\varepsilon \to 0$ can tolerate larger data uncertainty for admissible accuracy (Fig. 14.2b).

14.3.2 Geometric Fitting

We now illustrate the above consideration in more specific terms. Let $\{p_\alpha\}$, $\alpha = 1, ..., N$, be the extracted feature points. Suppose each point should satisfy a parameterized constraint

$$F(p_\alpha, \mathbf{u}) = 0 \tag{14.2}$$

when no uncertainty exists. In the presence of uncertainty, Eq. (14.2) may not hold exactly. Our task is to estimate the parameter $\mathbf{u}$ from observed positions $\{p_\alpha\}$ in the presence of uncertainty.

A typical problem of this form is to fit a line or a curve to given N points in the image, but this can be straightforwardly extended to multiple images. For example, if a point (x_α, y_α) in one image corresponds to a point (x'_α, y'_α) in another, we can regard them as a single point p_α in a four-dimensional joint space with coordinates $(x_\alpha, y_\alpha, x'_\alpha, y'_\alpha)$. If the camera imaging geometry is modeled as perspective projection, constraint (14.2) corresponds to the *epipolar equation*; the parameter $\mathbf{u}$ is the *fundamental matrix* [13]. This is discussed in more detail in Sect. 14.5.1.

3.2.1 General Geometric Fitting

The above problem can be stated in abstract terms as *geometric fitting* as follows. We view a feature point in the image plane or a set of feature points in the joint space as an m-dimensional vector $\mathbf{x}$; we call it a "datum". Let $\{\mathbf{x}_\alpha\}$, $\alpha = 1, ..., N$, be observed data. Their true values $\{\bar{\mathbf{x}}_\alpha\}$ are supposed to satisfy r constraint equations

$$F^{(k)}(\bar{\mathbf{x}}_\alpha, \mathbf{u}) = 0, \quad k = 1, ..., r, \tag{14.3}$$

parameterized by a p-dimensional vector $\mathbf{u}$. We call Eq. (14.3) the *(geometric) model*. The domain $\mathcal{X}$ of the data $\{\mathbf{x}_\alpha\}$ is called the *data space*; the domain $\mathcal{U}$ of the parameter $\mathbf{u}$ is called the *parameter space*. The number r of the constraint equations is called the *rank* of the constraint. The r equations $F^{(k)}(\mathbf{x}, \mathbf{u}) = 0$, $k = 1, ..., r$, are assumed to be mutually independent, defining a manifold $\mathcal{S}$ of codimension r parameterized by $\mathbf{u}$ in the data space $\mathcal{X}$. Equation (14.3) requires that the true values $\{\bar{\mathbf{x}}_\alpha\}$ be all in the manifold $\mathcal{S}$. Our task is to estimate the parameter $\mathbf{u}$ from the noisy data $\{\mathbf{x}_\alpha\}$ (Fig. 14.3a).

3.2.2 Maximum Likelihood Estimation

Let

$$\mathsf{V}[\mathbf{x}_\alpha] = \varepsilon^2 \mathsf{V}_0[\mathbf{x}_\alpha] \tag{14.4}$$

be the covariance matrix of $\mathbf{x}_\alpha$, where ε and $\mathsf{V}_0[\mathbf{x}_\alpha]$ are the noise level and the normalized covariance matrix, respectively. If the distribution of uncertainty

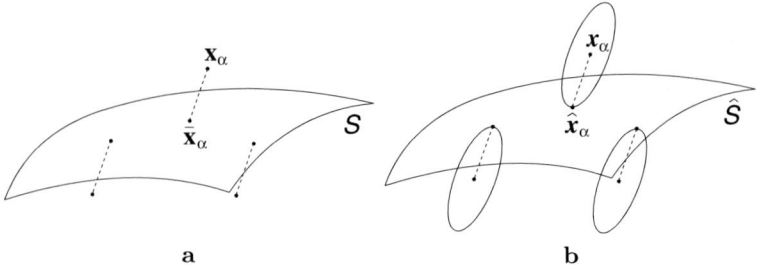

Fig. 14.3. **a** Fitting a manifold S to the data $\{\mathbf{x}_\alpha\}$. **b** Estimating $\{\bar{\mathbf{x}}_\alpha\}$ and $\mathbf{u}$ by minimizing the sum of squared Mahalanobis distance with respect to the normalized covariance matrices $V_0[\mathbf{x}_\alpha]$

is Gaussian, which we assume hereafter, the probability density of the data $\{\mathbf{x}_\alpha\}$ is given by

$$P(\{\mathbf{x}_\alpha\}) = C \prod_{\alpha=1}^{N} e^{-(\mathbf{x}_\alpha - \bar{\mathbf{x}}_\alpha, V[\mathbf{x}_\alpha]^{-1}(\mathbf{x}_\alpha - \bar{\mathbf{x}}_\alpha))/2}, \qquad (14.5)$$

where C is a normalization constant. Throughout this chapter, we denote the inner product of vectors $\mathbf{a}$ and $\mathbf{b}$ by $(\mathbf{a}, \mathbf{b})$.

Maximum likelihood estimation (MLE) is finding the values of $\{\bar{\mathbf{x}}_\alpha\}$ and $\mathbf{u}$ that maximize the *likelihood*, i.e., Eq. (14.6) into which the data $\{\mathbf{x}_\alpha\}$ are substituted, or equivalently minimize the sum of the squared *Mahalanobis distances* in the form

$$J = \sum_{\alpha=1}^{N} (\mathbf{x}_\alpha - \bar{\mathbf{x}}_\alpha, V_0[\mathbf{x}_\alpha]^{-1}(\mathbf{x}_\alpha - \bar{\mathbf{x}}_\alpha)) \qquad (14.6)$$

subject to the constraint (14.3) (Fig. 14.3b). The solution is called the *maximum likelihood (ML) estimator*. If the uncertainty is small, which we assume hereafter, constraint (14.3) can be eliminated by introducing Lagrange multipliers and applying first-order approximation. After some manipulations, we obtain the following form [14]:

$$J = \sum_{\alpha=1}^{N} \sum_{k,l=1}^{r} W_\alpha^{(kl)} F^{(k)}(\mathbf{x}_\alpha, \mathbf{u}) F^{(l)}(\mathbf{x}_\alpha, \mathbf{u}). \qquad (14.7)$$

Here, $W_\alpha^{(kl)}$ is the (kl) element of the inverse of the $r \times r$ matrix whose (kl) element is $(\nabla_\mathbf{x} F_\alpha^{(k)}, V_0[\mathbf{x}_\alpha] \nabla_\mathbf{x} F_\alpha^{(l)})$. We symbolically write

$$\left(W_\alpha^{(kl)}\right) = \left((\nabla_\mathbf{x} F_\alpha^{(k)}, V_0[\mathbf{x}_\alpha] \nabla_\mathbf{x} F_\alpha^{(l)})\right)^{-1}, \qquad (14.8)$$

where $\nabla_\mathbf{x} F^{(k)}$ is the gradient of the function $F^{(k)}$ with respect to $\mathbf{x}$. The subscript α means that $\mathbf{x} = \mathbf{x}_\alpha$ is substituted.

Remark 6. The data $\{\mathbf{x}_\alpha\}$ may be subject to some constraints. For example, each $\mathbf{x}_\alpha$ may be a unit vector. The above formulation still holds if the inverse $\mathsf{V}_0[\mathbf{x}_\alpha]^{-1}$ in Eq. (14.6) is replaced by the Moore–Penrose generalized (or pseudo) inverse $\mathsf{V}_0[\mathbf{x}_\alpha]^{-}$ [14]. Similarly, the r constraints in Eq. (14.3) may be redundant, say only r' ($< r$) of them are independent. The above formulation still holds if the inverse in Eq. (14.8) is replaced by the generalized inverse of rank r' with all but r' largest eigenvalues replaced by zero [14].

3.2.3 Accuracy of the ML Estimator

It can be shown [14] that the covariance matrix of the ML estimator $\hat{\mathbf{u}}$ has the form

$$\mathsf{V}[\hat{\mathbf{u}}] = \varepsilon^2 \mathsf{M}(\hat{\mathbf{u}})^{-1} + O(\varepsilon^4), \tag{14.9}$$

where

$$\mathsf{M}(\mathbf{u}) = \sum_{\alpha=1}^{N} \sum_{k,l=1}^{r} W_\alpha^{(kl)} \nabla_\mathbf{u} F_\alpha^{(k)} \nabla_\mathbf{u} F_\alpha^{(k)\top}. \tag{14.10}$$

Here, $\nabla_\mathbf{u} F^{(k)}$ is the gradient of the function $F^{(k)}$ with respect to $\mathbf{u}$. The subscript α means that $\mathbf{x} = \mathbf{x}_\alpha$ is substituted.

Remark 7. It can be proved that no other estimators could reduce the covariance matrix further than Eq. (14.9) except for the higher-order term $O(\varepsilon^4)$ [14, 17]. The ML estimator is optimal in this sense. Recall that we are focusing on the asymptotic analysis for $\varepsilon \to 0$. Thus, what we call the "ML estimator" should be understood to be a first approximation to the true ML estimator for small ε.

Remark 8. The p-dimensional parameter vector $\mathbf{u}$ may be constrained. For example, it may be a unit vector. If it has only p' ($< p$) degrees of freedom, the parameter space $\mathcal{U}$ is a p'-dimensional manifold in $\mathcal{R}^p$. In this case, the matrix $\mathsf{M}(\mathbf{u})$ in Eq. (14.9) is replaced by $\mathsf{P}_\mathbf{u} \mathsf{M}(\mathbf{u}) \mathsf{P}_\mathbf{u}$, where $\mathsf{P}_\mathbf{u}$ is the projection matrix onto the tangent space to the parameter space $\mathcal{U}$ at $\mathbf{u}$ [14]. The inverse $\mathsf{M}(\hat{\mathbf{u}})^{-1}$ in Eq. (14.9) is replaced by the generalized inverse $\mathsf{M}(\hat{\mathbf{u}})^{-1}$ of rank p' [14].

14.3.3 Geometric Model Selection

Geometric fitting is to estimate the parameter $\mathbf{u}$ of a given model. If we have multiple candidate models

$$\begin{aligned} F_1^{(k)}(\bar{\mathbf{x}}_\alpha, \mathbf{u}_1) &= 0, \quad F_2^{(k)}(\bar{\mathbf{x}}_\alpha, \mathbf{u}_2) = 0, \\ F_3^{(k)}(\bar{\mathbf{x}}_\alpha, \mathbf{u}_3) &= 0, \quad \ldots, \end{aligned} \tag{14.11}$$

from which we are to select an appropriate one for the observed data $\{\mathbf{x}_\alpha\}$, the problem is *(geometric) model selection* [14, 16, 18].

Suppose, for example, we want to fit a curve to given points in two dimensions. If they are almost collinear, a straight line may fit fairly well, but a quadratic curve may fit better, and a cubic curve even better. Which curve should we fit? A naive idea is to compare the *residual (sum of squares)*, i.e., the minimum value $\hat{J}$ of J in Eq. (14.6); we select the one that has the smallest residual $\hat{J}$. This does not work, however, because the ML estimator $\hat{\mathbf{u}}$ is so determined as to minimize the residual $\hat{J}$, and the residual $\hat{J}$ can be made arbitrarily smaller if the model is equipped with more parameters to adjust. So, the only conclusion would be to fit a curve of a sufficiently high degree passing through all the points.

3.3.1 Geometric AIC

The above observation leads to the idea of compensating for the negative bias of the residual caused by substituting the ML estimator. This is the principle of the *Akaike information criterion (AIC)* [1], which is derived from the asymptotic behavior of the *Kullback–Leibler information* (or *divergence*) as the number n of experiments goes to infinity. Doing a similar analysis to Akaike's and examining the asymptotic behavior as the noise level ε goes to zero, we can obtain the following *geometric AIC* [14, 15]:

$$\text{G-AIC} = \hat{J} + 2(Nd + p)\varepsilon^2 + O(\varepsilon^4). \tag{14.12}$$

Here, d is the dimension of the manifold $\mathcal{S}$ defined by the constraint (14.3) in the data space $\mathcal{X}$, and p is the dimension of $\mathbf{u}$ (i.e., the number of unknowns). The model for which Eq. (14.12) is the smallest is regarded as the best. The derivation of Eq. (14.12) is based on the following facts [14, 15]:

- The ML estimator $\hat{\mathbf{u}}$ converges to its true value as $\varepsilon \to 0$.
- The ML estimator $\hat{\mathbf{u}}$ obeys a Gaussian distribution under linear constraints, because the noise is assumed to be Gaussian. For nonlinear constraints, linear approximation can be justified in the neighborhood of the solution if ε is sufficiently small.
- A quadratic form in standardized Gaussian random variables is subject to a χ^2 distribution, whose expectation is equal to its degree of freedom.

3.3.2 Geometric MDL

Another well-known criterion for model selection is Rissanen's *minimum description length (MDL)* [42, 43, 44], which measures the goodness of a model by the minimum information theoretic code length of the data and the model. The basic idea is simple, but the following difficulties must be resolved to apply it in practice:

- Encoding a problem involving real numbers requires an infinitely long code length.
- The probability density, from which a minimum length code can be obtained, involves unknown parameters.
- The exact form of the minimum code length is very difficult to compute.

Rissanen [42, 43, 44] avoided these difficulties by quantizing the real numbers in a way that does not depend on individual models and substituting the ML estimators for the parameters. They, too, are real numbers, so they are also quantized. The quantization width is so chosen as to minimize the total description length (*two-stage encoding*). The resulting code length is evaluated asymptotically as the data length n goes to infinity. If we analyze the asymptotic behavior of encoding the geometric fitting problem as the noise level ε goes to zero, we obtain the following *geometric MDL* [20]:

$$\text{G-MDL} = \hat{J} - (Nd + p)\varepsilon^2 \log\left(\frac{\varepsilon}{L}\right)^2 + O(\varepsilon^2). \quad (14.13)$$

Here, L is a reference length chosen so that its ratio to the magnitude of data is $O(1)$, e.g., L can be taken to be the image size for feature point data. Its exact determination requires an a priori distribution that specifies where the data are likely to appear (we discuss this more in Sect. 14.4.1), but it has been observed that the model selection is not very much affected by L as long as it is within the same order of magnitude [20].

14.4 Standard Statistical Analysis vs. Geometric Inference

We now point out that a correspondence exists between the standard statistical analysis and the geometric inference problem. We also compare the capability of the geometric AIC and the geometric MDL in detecting degeneracy.

14.4.1 Standard Statistical Analysis

The asymptotic analysis in Sect. 14.3 bears a strong resemblance to the standard statistical estimation problem: after observing n data $\mathbf{x}_1, \mathbf{x}_2, ..., \mathbf{x}_n$, we want to estimate the parameter $\boldsymbol{\theta}$ of the probability density $P(\mathbf{x}|\boldsymbol{\theta})$ called the *(stochastic) model*, according to which each datum is assumed to be sampled independently.

Maximum likelihood estimation (MLE) is to find the value $\boldsymbol{\theta}$ that maximizes $\prod_{i=1}^{n} P(\mathbf{x}_i|\boldsymbol{\theta})$, or equivalently minimizes its negative logarithm $-\sum_{i=1}^{n} \log P(\mathbf{x}_i|\boldsymbol{\theta})$. It can be shown that the covariance matrix $\mathsf{V}[\hat{\boldsymbol{\theta}}]$ of the resulting ML estimator $\hat{\boldsymbol{\theta}}$ converges, under a mild condition, to O as the number n of experiments goes to infinity (*consistency*) in the form

$$V[\hat{\boldsymbol{\theta}}] = I(\boldsymbol{\theta})^{-1} + O\left(\frac{1}{n^2}\right), \tag{14.14}$$

where we define the *Fisher information matrix* $I(\boldsymbol{\theta})$ by

$$I(\boldsymbol{\theta}) = nE[(\nabla_{\boldsymbol{\theta}} \log P(\mathbf{x}|\boldsymbol{\theta}))(\nabla_{\boldsymbol{\theta}} \log P(\mathbf{x}|\boldsymbol{\theta}))^\top]. \tag{14.15}$$

The operation $E[\cdot]$ denotes expectation with respect to the density $P(\mathbf{x}|\boldsymbol{\theta})$. The first term in the right-hand side of Eq. (14.14) is called the *Cramer–Rao lower bound (CRLB)*, describing the minimum degree of fluctuations in all estimators. Thus, the ML estimator is optimal if n is sufficiently large (*asymptotic efficiency*).

If we have multiple candidate models

$$P_1(\mathbf{x}|\boldsymbol{\theta}_1), \quad P_2(\mathbf{x}|\boldsymbol{\theta}_2), \quad P_3(\mathbf{x}|\boldsymbol{\theta}_3), \quad ..., \tag{14.16}$$

from which we are to select an appropriate one for the observations $\mathbf{x}_1, \mathbf{x}_1, ..., \mathbf{x}_n$, the problem is *(stochastic) model selection*. Akaike's AIC has the following form:

$$\text{AIC} = -2\sum_{i=1}^{N} \log P(\mathbf{x}_i|\hat{\boldsymbol{\theta}}) + 2k + O\left(\frac{1}{n}\right). \tag{14.17}$$

The model for which this quantity is the smallest is regarded as the best. The derivation of Eq. (14.17) is based on the following facts [1]:

- The maximum likelihood estimator $\hat{\boldsymbol{\theta}}$ converges to its true value as $n \to \infty$ (the *law of large numbers*).
- The maximum likelihood estimator $\hat{\boldsymbol{\theta}}$ asymptotically obeys a Gaussian distribution as $n \to \infty$ (the *central limit theorem*).
- A quadratic form in standardized Gaussian random variables is subject to a χ^2 distribution, whose expectation is equal to its degree of freedom.

Rissanen's MDL has the following form [43, 44]:

$$\text{MDL} = -\sum_{i=1}^{n} \log P(\mathbf{x}_i|\hat{\boldsymbol{\theta}}) + \frac{k}{2}\log\frac{n}{2\pi} + \log\int_{\mathcal{T}} \sqrt{|I(\boldsymbol{\theta})|}d\boldsymbol{\theta} + O(1). \tag{14.18}$$

Here, $\hat{\boldsymbol{\theta}}$ is the ML estimator; the symbol $O(1)$ denotes terms of order 0 in n in the limit $n \to \infty$. In order that the integration in the right-hand side of Eq. (14.18) exists, the domain $\mathcal{T}$ of the parameter $\boldsymbol{\theta}$ must be compact. In other words, we must specify in the k-dimensional space of $\boldsymbol{\theta}$ a finite region $\mathcal{T}$ in which the true value of $\boldsymbol{\theta}$ is likely to exist. This is nothing but the *Bayesian* standpoint that requires a prior distribution for the parameter to estimate. If it is not known, we must introduce an appropriate expedient to suppress an explicit dependence on the prior. Such an expedient is also necessary for the geometric MDL, i.e., the introduction of the reference length L in Eq. (14.18).

14.4.2 Dual Interpretations of Asymptotic Analysis

Thus, we have seen that the limit $n \to \infty$ for the standard statistical analysis corresponds to the limit $\varepsilon \to 0$ for geometric inference. For example, the covariance matrix of the ML estimator agrees with the Cramer–Rao lower bound up to $O(1/n^2)$ for $n \to \infty$ (see Eq. (14.14)), while for geometric inference it agrees with the lower bound up to $O(\varepsilon^4)$ for $\varepsilon \to 0$ (see Eq. (14.9)). It follows that $1/\sqrt{n}$ for the standard statistical analysis plays the same role as ε for geometric inference.

The same correspondence exists for model selection, too. The unknowns for geometric inference are the p parameters of the constraint plus the N true positions specified by the d coordinates of the d-dimensional manifold S defined by the constraint. If Eq. (14.12) is divided by ε^2, we have $\hat{J}/\varepsilon^2 + 2(Nd+p) + O(\varepsilon^2)$, which is ($-2$ times the logarithmic likelihood)+2(the number of unknowns), the same form as Akaike's AIC given by Eq. (14.17). The same holds for Eq. (14.13), which corresponds to Rissanen's MDL given by Eq. (14.18) if ε is replaced by $1/\sqrt{n}$ [20].

This correspondence can be interpreted as follows. Since the underlying ensemble is hypothetical, we can actually observe only one sample as long as a particular algorithm is used. Suppose we hypothetically sample n different algorithms to find n different positions. The optimal estimate of the true position under the Gaussian model is their sample mean. The covariance matrix of the sample mean is $1/n$ times that of the individual samples. Hence, this hypothetical estimation is equivalent to dividing the noise level ε in Eq. (14.4) by $\sqrt{n}$.

In fact, there were attempts to generate a hypothetical *ensemble of algorithms* by randomly varying the internal parameters (e.g., the thresholds for judgments), not adding random noise to the image [5, 6]. Then, one can compute their means and covariance matrix. Such a process as a whole can be regarded as one operation that effectively achieves higher accuracy.

Thus, the asymptotic analysis for $\varepsilon \to 0$ is equivalent to the asymptotic analysis for $n \to \infty$, where n is the number of hypothetical observations. As a result, the expression $\cdots + O(1/\sqrt{n^k})$ in the standard statistical analysis turns into $\cdots + O(\varepsilon^k)$ in geometric inference.

14.4.3 Noise Level Estimation

In order to use the geometric AIC or the geometric MDL, we need to know the noise level ε. If not known, it must be estimated. Here arises a sharp contrast between the standard statistical analysis and our geometric inference.

For the standard statistical analysis, the noise magnitude is a *model parameter*, because "noise" is defined to be *the random effects that cannot be accounted for by the assumed model*. Hence, the noise magnitude should be estimated, if not known, *according to the assumed model*. For geometric inference, on the other hand, the noise level ε is *a constant that reflects the uncertainty of feature detection*. So, it should be estimated *independently of individual models*.

If we know the true model, it can be estimated from the residual $\hat{J}$ using the knowledge that $\hat{J}/\varepsilon^2$ is subject to a χ^2 distribution with $rN - p$ degrees of freedom in the first-order [14]. Specifically, we obtain an unbiased estimator of ε^2 in the form

$$\hat{\varepsilon}^2 = \frac{\hat{J}}{rN - p}. \tag{14.19}$$

The validity of this formula has been confirmed by many simulations.

One may wonder if model selection is necessary at all when the true model is known. In practice, however, a typical situation where model selection is called for is *degeneracy detection*. In 3-D analysis from images, for example, the constraint (14.3) corresponds to our knowledge about the scene such as rigidity of motion. However, the computation fails if degeneracy occurs (e.g., the motion is zero). Even if exact degeneracy does not occur, the computation may become numerically unstable in near-degeneracy conditions. In such a case, the computation can be stabilized by switching to a model that describes the degeneracy [16, 21, 24, 25, 31, 39, 53].

Degeneracy means *addition* of new constraints, such as some quantity being zero. It follows that the manifold $\mathcal{S}$ degenerates into a submanifold $\mathcal{S}'$ of it. Since the general model still holds irrespective of the degeneracy, i.e., $\mathcal{S}' \subset \mathcal{S}$, we can estimate the noise level ε from the residual $\hat{J}$ of the general model $\mathcal{S}$ using Eq. (14.19).

Remark 9. Equation (14.19) can be intuitively understood as follows. Recall that $\hat{J}$ is the sum of the square distances from $\{\mathbf{x}_\alpha\}$ to the manifold $\hat{\mathcal{S}}$ defined by the constraint $F^{(k)}(\mathbf{x}, \mathbf{u}) = 0$, $k = 1, ..., r$. Since $\hat{\mathcal{S}}$ has codimension r (the dimension of the orthogonal directions to it), the residual $\hat{J}$ should have expectation $rN\varepsilon^2$. However, $\hat{\mathcal{S}}$ is fitted by adjusting its p-dimensional parameter $\mathbf{u}$, so the expectation of $\hat{J}$ reduces to $(rN - p)\varepsilon^2$.

Remark 10. It may appear that the residual $\hat{J}$ of the general model cannot be stably computed in the presence of degeneracy. However, what is unstable is *model specification*, not the residual. For example, if we fit a planar surface to almost collinear points in 3-D, it is difficult to specify the fitted plane stably; the solution is very susceptible to noise. Yet, the residual is stably computed, since unique specification of the fit is difficult *because all the candidates have almost the same residual*.

Remark 11. Note that the noise level estimation from the general model $\mathcal{S}$ by Eq. (14.19) is still valid even if degeneracy occurs, because degeneracy means shrinkage of the model manifold $\mathcal{S}'$ *within* $\mathcal{S}$, which does not affect the data deviations in the "orthogonal" directions (in the Mahalanobis sense) to $\mathcal{S}$ that account for the residual $\hat{J}$.

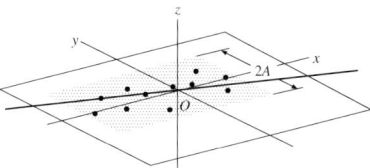

Fig. 14.4. Fitting a space line and a plane to points in space

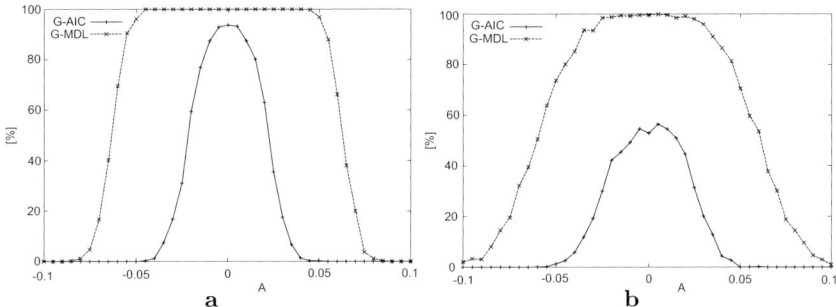

Fig. 14.5. The rate (%) of detecting a space line by the geometric AIC (*solid lines with* +) and the geometric MDL (*dotted lines with* ×) with **a** the true noise level and **b** the estimated noise level

14.4.4 Comparing the Geometric AIC and the Geometric MDL

We now illustrate the different characteristics of the geometric AIC and the geometric MDL in detecting degeneracy. Consider a rectangular region $[0, 10] \times [-1, 1]$ on the x-y plane in the x-y-z space. We randomly take 11 points in it and magnify the region A times in the y-direction. Adding Gaussian noise of mean 0 and variance ε^2 to the x, y, and z coordinates of each point independently, we fit a space line and a plane in a statistically optimal manner (Fig. 14.4). The rectangular region degenerates into a line segment as $A \to 0$.

A space line is a one-dimensional model with four degrees of freedom; a plane is a two-dimensional model with three degrees of freedom. Their geometric AIC and geometric MDL are

$$\text{G-AIC}_l = \hat{J}_l + 2(N+4)\varepsilon^2, \qquad \text{G-AIC}_p = \hat{J}_p + 2(2N+3)\varepsilon^2,$$
$$\text{G-MDL}_l = \hat{J}_l - (N+4)\varepsilon^2 \log\left(\frac{\varepsilon}{L}\right)^2, \quad \text{G-MDL}_p = \hat{J}_p - (2N+3)\varepsilon^2 \log\left(\frac{\varepsilon}{L}\right)^2,$$
(14.20)

where the subscripts l and p refer to lines and planes, respectively. For each A, we compare the geometric AIC and the geometric MDL of the fitted line and plane and choose the one that has the smaller value. We used the reference length $L = 1$.

Figure 14.5a shows the percentage of choosing a line for $\varepsilon = 0.01$ after 1000 independent trials for each A. If there were no noise, it should be 0% for

$A \neq 0$ and 100% for $A = 0$. In the presence of noise, the geometric AIC has a high capability of distinguishing a line from a plane, but it judges a line to be a plane with some probability. In contrast, the geometric MDL judges a line to be a line almost 100%, but it judges a plane to be a line over a wide range of A.

In Fig. 14.5a, we used the true value of ε^2. Figure 14.5b shows the corresponding result using its estimate obtained from the general plane model by Eq. (14.19). We observe somewhat degraded but similar performance characteristics.

Thus, we can observe that the geometric AIC has a higher capability for detecting degeneracy than the geometric MDL, but the general model is chosen with some probability when the true model is degenerate. In contrast, the percentage for the geometric MDL to detect degeneracy when the true model is really degenerate approaches 100% as the noise decreases. This is exactly the dual statement to the well-known fact, called the *consistency of the MDL*, that the percentage for Rissanen's MDL to identify the true model converges to 100% in the limit of an infinite number of observations. Rissanen's MDL is regarded by many as superior to Akaike's AIC because the latter lacks this property.

At the cost of this consistency, however, the geometric MDL regards a wide range of nondegenerate models as degenerate. This is no surprise, since the penalty $-(Nd + p)\varepsilon^2 \log(\varepsilon/L)^2$ for the geometric MDL in Eq. (14.13) is heavier than the penalty $2(Nd + p)\varepsilon^2$ for the geometric AIC in Eq. (14.12). As a result, the geometric AIC is more faithful to the data than the geometric MDL, which is more likely to choose a degenerate model. This contrast has also been observed in many applications [23, 31].

Remark 12. Despite the fundamental difference of geometric model selection from the standard (stochastic) model selection, many attempts have been made in the past to apply Akaike's AIC and their variants to computer vision problems based on the asymptotic analysis of $n \to \infty$, where the interpretation of n is different from problem to problem [48, 49, 50, 51, 52]. Rissanen's MDL is also used in computer vision applications. Its use may be justified if the problem has the standard form of linear/nonlinear regression [3, 32]. Often, however, the solution having a shorter description length was chosen with a rather arbitrary definition of the complexity [11, 27, 33].

Remark 13. Note that one cannot compare different model selection criteria in general terms, because each is based on its own logic. Not only that, one cannot prove that a particular criterion works at all. In fact, although Akaike's AIC and Rissanen's MDL are based on rigorous mathematics, there is no guarantee that they work well in practice. The mathematical rigor is in their *reduction* from their starting principles (the Kullback–Leibler information and the minimum description length principle), which are beyond proof. What one can tell is which criterion is more suitable for a particular application when used in a particular manner. The geometric AIC and the geometric MDL have shown to be effective in many computer vision applications [19, 22, 23, 24, 25, 31, 39, 53], but other criteria may be better in other applications. The important thing is, however, to understand the underlying logic of each criterion.

14.5 Linear Geometric Fitting

Now, we consider a special type of geometric fitting problem that most frequently arises in computer vision applications: the constraint is linear in both data and unknowns. We systematically review existing methods.

14.5.1 Linear Constraints

In many geometric inference problems of computer vision, the constraint (14.3) has the form

$$(\boldsymbol{\xi}(\bar{\mathbf{x}}_\alpha), \mathbf{u}) = 0, \tag{14.21}$$

where $\boldsymbol{\xi}(\,\cdot\,)$ is generally a nonlinear mapping from an m-dimensional vector to a p-dimensional vector. Evidently, the magnitude of $\mathbf{u}$ is unconstrained, so we normalize it to a unit vector: $\|\mathbf{u}\| = 1$.

Example 1. Suppose we are given N points $\{(x_\alpha, y_\alpha)\}$, $\alpha = 1, ..., N$, in two dimensions. Their true positions $\{(\bar{x}_\alpha, \bar{y}_\alpha)\}$ are assumed to be on a *conic* (a circle, an ellipse, a parabola, a hyperbola, or their degeneracy). Our task is to estimate the curve from the noisy data $\{(x_\alpha, y_\alpha)\}$. The constraint on $\{(\bar{x}_\alpha, \bar{y}_\alpha)\}$ is

$$A\bar{x}_\alpha^2 + 2B\bar{x}_\alpha \bar{y}_\alpha + C\bar{y}_\alpha^2 + 2(D\bar{x}_\alpha + E\bar{y}_\alpha) + F = 0 \tag{14.22}$$

for some coefficients A, B, ..., D, not all being zero. This constraint reduces to Eq. (14.21) if we put

$$\boldsymbol{\xi}(x, y) = \begin{pmatrix} x^2 & 2xy & y^2 & 2x & 2y & 1 \end{pmatrix}^\top, \quad \mathbf{u} = \begin{pmatrix} A & B & C & D & E & F \end{pmatrix}^\top. \tag{14.23}$$

The data space $\mathcal{X}$ is a two-dimensional manifold in the six-dimensional space $\mathcal{R}^6$; the parameter space $\mathcal{U}$ is the five-dimensional unit sphere S^6 centered on the origin of $\mathcal{R}^6$.

Example 2. Suppose N points in a 3-D scene are projected to (x_α, y_α) in the first image and (x'_α, y'_α) in the second, $\alpha = 1, ..., N$. If the camera imaging geometry is perspective projection, there exists a matrix F of determinant 0 such that

$$\left(\begin{pmatrix} \bar{x}_\alpha \\ \bar{y}_\alpha \\ 1 \end{pmatrix}, \mathsf{F} \begin{pmatrix} \bar{x}'_\alpha \\ \bar{y}'_\alpha \\ 1 \end{pmatrix} \right) = 0, \tag{14.24}$$

which is called the *epipolar equation* [13]. The matrix F is known as the *fundamental matrix*. For 3-D reconstruction from the images, we need to estimate the fundamental matrix F from the noisy data $\{(x_\alpha, y_\alpha)\}$ and $\{(x'_\alpha, y'_\alpha)\}$. Equation (14.24) reduces to Eq. (14.21) if we put

$$\boldsymbol{\xi}(x, y, x', y') = \begin{pmatrix} xx' & xy' & x & yx' & yy' & y & x' & y' & 1 \end{pmatrix}^\top,$$
$$\mathbf{u} = \begin{pmatrix} F_{11} & F_{12} & F_{13} & F_{21} & F_{22} & F_{23} & F_{31} & F_{32} & F_{33} \end{pmatrix}^\top. \quad (14.25)$$

The data space $\mathcal{X}$ is a four-dimensional manifold in the nine-dimensional space $\mathcal{R}^9$; the parameter space $\mathcal{U}$ is a seven-dimensional manifold defined by $\det \mathsf{F} = 0$ and $\|\mathsf{F}\| = 1$, where the matrix norm is define by $\|\mathsf{F}\| = \sqrt{\sum_{i,j=1}^3 F_{ij}^2}$.

For the linear constraint (14.21), the function J in Eq. (14.7) reduces to

$$J = \sum_{\alpha=1}^N \frac{(\boldsymbol{\xi}_\alpha, \mathbf{u})^2}{(\mathbf{u}, \mathsf{V}_0[\boldsymbol{\xi}_\alpha]\mathbf{u})}, \quad (14.26)$$

where $\mathsf{V}_0[\boldsymbol{\xi}_\alpha]$ is the normalized covariance matrix of $\boldsymbol{\xi}_\alpha$; we use the abbreviation $\boldsymbol{\xi}_\alpha = \boldsymbol{\xi}(\mathbf{x}_\alpha)$. The matrix $\mathsf{V}_0[\boldsymbol{\xi}_\alpha]$ can be expressed to a first approximation in the form

$$\mathsf{V}_0[\boldsymbol{\xi}_\alpha] = \nabla_\mathbf{x}\boldsymbol{\xi}|_{\mathbf{x}=\mathbf{x}_\alpha}^\top \mathsf{V}_0[\mathbf{x}_\alpha] \nabla_\mathbf{x}\boldsymbol{\xi}|_{\mathbf{x}=\mathbf{x}_\alpha}, \quad (14.27)$$

where $\nabla_\mathbf{x}\boldsymbol{\xi}$ is the $m \times p$ Jacobian matrix of $\boldsymbol{\xi}(\mathbf{x})$:

$$\nabla_\mathbf{x}\boldsymbol{\xi} = \begin{pmatrix} \partial\xi_1/\partial x_1 & \cdots & \partial\xi_p/\partial x_1 \\ \vdots & & \vdots \\ \partial\xi_1/\partial x_m & \cdots & \partial\xi_p/\partial x_m \end{pmatrix}. \quad (14.28)$$

The covariance matrix $\mathsf{V}[\hat{\mathbf{u}}]$ of the ML estimator $\hat{\mathbf{u}}$ given by Eq. (14.9) now reads

$$\mathsf{V}[\hat{\mathbf{u}}] = \varepsilon^2 \Big(\sum_{\alpha=1}^N \frac{\mathsf{P}_\mathbf{u} \boldsymbol{\xi}_\alpha \boldsymbol{\xi}_\alpha^\top \mathsf{P}_\mathbf{u}}{(\mathbf{u}, \mathsf{V}_0[\boldsymbol{\xi}_\alpha]\mathbf{u})}\Big)^- + O(\varepsilon^4), \quad (14.29)$$

where the superscript $-$ denotes the Moore–Penrose generalized inverse. The matrix $\mathsf{P}_\mathbf{u}$ denotes projection onto the tangent space to the parameter space $\mathcal{U}$ at $\mathbf{u}$ (cf. Remark 8). Since the leading term is the lower bound on the covariance matrix of any estimation (Remark 7), the ML estimator is optimal up to higher-order terms in ε.

Remark 14. Since we are focusing on the asymptotic analysis for $\varepsilon \to 0$, what we call the "ML estimator" is a first approximation to the true ML estimator for small ε (Remark 7). Note that if the parameter $\mathbf{u}$ is not constrained, the generalized inverse in Eq. (14.29) can be replaced by the usual inverse, and the projection matrix $\mathsf{P_u}$ is not necessary. However, $\mathbf{u}$ is at least constrained to be a unit vector, and often additional constraints exist, e.g., $\det \mathsf{F} = 0$ on the fundamental matrix F. If no constraints exist other than $\|\mathbf{u}\| = 1$, the covariance matrix $V[\hat{\mathbf{u}}]$ has rank $p-1$, and its null space is in the direction of $\mathbf{u}$. The projection matrix $\mathsf{P_n}$ in this case is

$$\mathsf{P_u} = \mathsf{I} - \mathbf{u}\mathbf{u}^\top. \tag{14.30}$$

14.5.2 Least-Squares Method

If $\mathbf{u}$ is constrained, the minimization of Eq. (14.26) should be carried out subject to the constraint, but this is very difficult in many cases. A practical approach to this is to ignore all the constraints except the normalization $\|\mathbf{u}\| = 1$ and do minimization over the $(p-1)$-dimensional sphere S^{p-1} in $\mathcal{R}^p$. This expedient is motivated by the fact that if the data $\{\mathbf{x}_\alpha\}$ are exact, the solution should automatically satisfy the remaining constraints. It follows that if the data uncertainty is very small, which we always assume, the resulting solution $\hat{\mathbf{u}}$ should satisfy all the constraints up to higher-order terms in ε.

However, the minimization of Eq. (14.26) is still nonlinear even if all constraints other than $\|\mathbf{u}\| = 1$ are ignored. The simplest approach is to solve Eq. (14.21) directly by (total) least squares, minimizing

$$J_{\mathrm{LS}} = \sum_{\alpha=1}^{N} (\boldsymbol{\xi}_\alpha, \mathbf{u})^2. \tag{14.31}$$

If we define the second-order moment matrix

$$\mathsf{M} = \sum_{\alpha=1}^{N} \boldsymbol{\xi}_\alpha \boldsymbol{\xi}_\alpha^\top, \tag{14.32}$$

Eq. (14.31) is rewritten as

$$J_{\mathrm{LS}} = (\mathbf{u}, \mathsf{M}\mathbf{u}). \tag{14.33}$$

The unit vector $\mathbf{u}$ that minimizes this is the unit eigenvector of M for the smallest eigenvalue. The resulting *least-squares (LS) solution* $\hat{\mathbf{u}}_{\mathrm{LS}}$ is a very crude approximation to the ML estimator $\hat{\mathbf{u}}$. However, because of the ease of the computation, it is often used as an initial guess for computing the ML estimator $\hat{\mathbf{u}}$ by iterations.

14.5.3 Naive Method

If we define

$$\mathsf{M}(\mathbf{u}) = \sum_{\alpha=1}^{N} \frac{\boldsymbol{\xi}_\alpha \boldsymbol{\xi}_\alpha^\top}{(\mathbf{u}, V_0[\boldsymbol{\xi}_\alpha]\mathbf{u})}, \tag{14.34}$$

Eq. (14.26) is written as

$$J = (\mathbf{u}, \mathsf{M}(\mathbf{u})\mathbf{u}). \qquad (14.35)$$

This inspires the following iterations for computing the ML estimator:

1. Guess an appropriate initial value $\mathbf{u}_0$, say the LS solution $\hat{\mathbf{u}}_{\mathrm{LS}}$.
2. Assuming that $\mathbf{u}_{i-1}$ is obtained (initially $i = 1$), let $\mathbf{u}_i$ be the unit eigenvector of $\mathsf{M}(\mathbf{u}_{i-1})$ for the smallest eigenvalue.
3. Return $\mathbf{u}_i$ if $\mathbf{u}_i$ is sufficiently close to $\mathbf{u}_{i-1}$ except for the sign. Otherwise, let $\mathbf{u}_{i-1} \leftarrow \mathbf{u}_i$, and go back to step 2.

This scheme does not work, however, because the resulting solution $\hat{\mathbf{u}}$ is the value $\mathbf{u}$ that minimizes $(\mathbf{u}, \mathsf{M}(\hat{\mathbf{u}})\mathbf{u})$, not $(\mathbf{u}, \mathsf{M}(\mathbf{u})\mathbf{u})$. In other words,

$$(\hat{\mathbf{u}}, \mathsf{M}(\hat{\mathbf{u}})\hat{\mathbf{u}}) < (\hat{\mathbf{u}} + \Delta\mathbf{u}, \mathsf{M}(\hat{\mathbf{u}})(\hat{\mathbf{u}} + \Delta\mathbf{u})) \qquad (14.36)$$

for any nonzero perturbation $\Delta\mathbf{u}$, but not

$$(\hat{\mathbf{u}}, \mathsf{M}(\hat{\mathbf{u}})\hat{\mathbf{u}}) < (\hat{\mathbf{u}} + \Delta\mathbf{u}, \mathsf{M}(\hat{\mathbf{u}} + \Delta\mathbf{u})(\hat{\mathbf{u}} + \Delta\mathbf{u})). \qquad (14.37)$$

A detailed analysis shows that $\hat{\mathbf{u}}$ is *biased* by $O(\varepsilon^2)$ [14]. Namely, if the fluctuations of the data $\{\mathbf{x}_\alpha\}$ are centered on their true values $\{\bar{\mathbf{x}}_\alpha\}$, the corresponding fluctuations of $\hat{\mathbf{u}}$ are around a value different from its true value by $O(\varepsilon^2)$. This causes inadmissible errors in many practical applications.

14.5.4 FNS Method

If the constraint on $\mathbf{u}$ is ignored, the solution that minimizes Eq. (14.26) is obtained by solving $\nabla_\mathbf{u} J = \mathbf{0}$. Since

$$\nabla_\mathbf{u} J = \sum_{\alpha=1}^{N} \frac{2(\boldsymbol{\xi}_\alpha, \mathbf{u})\boldsymbol{\xi}_\alpha}{(\mathbf{u}, \mathsf{V}_0[\boldsymbol{\xi}_\alpha]\mathbf{u})} - \sum_{\alpha=1}^{N} \frac{2(\boldsymbol{\xi}_\alpha, \mathbf{u})^2 \mathsf{V}_0[\boldsymbol{\xi}_\alpha]\mathbf{u}}{(\mathbf{u}, \mathsf{V}_0[\boldsymbol{\xi}_\alpha]\mathbf{u})^2}, \qquad (14.38)$$

the equation $\nabla_\mathbf{u} J = \mathbf{0}$ is written in the form

$$\mathsf{X}(\mathbf{u})\mathbf{u} = \mathbf{0}, \qquad (14.39)$$

where

$$\mathsf{X}(\mathbf{u}) = \sum_{\alpha=1}^{N} \frac{\boldsymbol{\xi}_\alpha \boldsymbol{\xi}_\alpha^\top}{(\mathbf{u}, \mathsf{V}_0[\boldsymbol{\xi}_\alpha]\mathbf{u})} - \sum_{\alpha=1}^{N} \frac{(\boldsymbol{\xi}_\alpha, \mathbf{u})^2 \mathsf{V}_0[\boldsymbol{\xi}_\alpha]}{(\mathbf{u}, \mathsf{V}_0[\boldsymbol{\xi}_\alpha]\mathbf{u})^2}. \qquad (14.40)$$

From this, we have the following scheme for solving Eq. (14.39):

1. Guess an appropriate initial value $\mathbf{u}_0$, say the LS solution $\hat{\mathbf{u}}_{\mathrm{LS}}$.

2. Assuming that $\mathbf{u}_{i-1}$ is obtained (initially $i = 1$), solve the eigenvalue problem

$$\mathsf{X}(\mathbf{u}_{i-1})\mathbf{u} = \lambda\mathbf{u}. \tag{14.41}$$

Let $\mathbf{u}_i$ be the unit eigenvector for the eigenvalue λ closest to 0.

3. Return $\mathbf{u}_i$ if $\mathbf{u}_i$ is sufficiently close to $\mathbf{u}_{i-1}$ except for the sign. Otherwise, let $\mathbf{u}_{i-1} \leftarrow \mathbf{u}_i$, and go back to step 2.

The resulting solution $\hat{\mathbf{u}}$ satisfies Eq. (14.39). In fact, the value $\hat{\mathbf{u}}$ produced by the above iterations should satisfy

$$\mathsf{X}(\hat{\mathbf{u}})\hat{\mathbf{u}} = \lambda\hat{\mathbf{u}} \tag{14.42}$$

for some λ. Taking the inner product of $\hat{\mathbf{u}}$ and both sides, we have

$$(\hat{\mathbf{u}}, \mathsf{X}(\hat{\mathbf{u}})\hat{\mathbf{u}}) = \lambda. \tag{14.43}$$

Equation (14.40) implies that

$$(\hat{\mathbf{u}}, \mathsf{X}(\hat{\mathbf{u}})\hat{\mathbf{u}}) = \sum_{\alpha=1}^{N} \frac{(\hat{\mathbf{u}}, \boldsymbol{\xi}_\alpha)^2}{(\hat{\mathbf{u}}, V_0[\boldsymbol{\xi}_\alpha]\hat{\mathbf{u}})} - \sum_{\alpha=1}^{N} \frac{(\boldsymbol{\xi}_\alpha, \hat{\mathbf{u}})^2(\hat{\mathbf{u}}, V_0[\boldsymbol{\xi}_\alpha]\hat{\mathbf{u}})}{(\hat{\mathbf{u}}, V_0[\boldsymbol{\xi}_\alpha]\hat{\mathbf{u}})^2} = 0, \tag{14.44}$$

meaning that $\lambda = 0$. Thus, $\hat{\mathbf{u}}$ is indeed the solution of Eq. (14.39). This method was proposed by Chojnacki et al. [7] and is called the *fundamental numerical scheme (FNS) method*. Usually, the iterations converge very quickly.

Remark 15. Equation (14.44) is a consequence of the fact that the right-hand side of Eq. (14.26) is a *homogeneous function of degree 0* in $\mathbf{u}$. Since multiplying $\mathbf{u}$ by any nonzero constant does not change the value of J, the gradient $\nabla_\mathbf{u} J$ is necessarily orthogonal to $\mathbf{u}$. Thus, $(\mathbf{u}, \nabla_\mathbf{u} J) = 2(\mathbf{u}, \mathsf{X}(\mathbf{u})\mathbf{u})$ is identically 0.

14.5.5 HEIV Method

Equation (14.39) can also be written as

$$\mathsf{M}(\mathbf{u})\mathbf{u} = \mathsf{L}(\mathbf{u})\mathbf{u}, \tag{14.45}$$

where

$$\mathsf{M}(\mathbf{u}) = \sum_{\alpha=1}^{N} \frac{\boldsymbol{\xi}_\alpha \boldsymbol{\xi}_\alpha^\top}{(\mathbf{u}, V_0[\boldsymbol{\xi}_\alpha]\mathbf{u})},$$

$$\mathsf{L}(\mathbf{u}) = \sum_{\alpha=1}^{N} \frac{(\boldsymbol{\xi}_\alpha, \mathbf{u})^2 V_0[\boldsymbol{\xi}_\alpha]}{(\mathbf{u}, V_0[\boldsymbol{\xi}_\alpha]\mathbf{u})^2}. \tag{14.46}$$

This implies the following scheme:

1. Guess an appropriate initial value $\mathbf{u}_0$, say the LS solution $\hat{\mathbf{u}}_{LS}$.
2. Assuming that $\mathbf{u}_{i-1}$ is obtained (initially $i = 1$), solve the generalized eigenvalue problem

$$\mathsf{M}(\mathbf{u}_{i-1})\mathbf{u} = \lambda \mathsf{L}(\mathbf{u}_{i-1})\mathbf{u}. \qquad (14.47)$$

Let $\mathbf{u}_i$ be the generalized eigenvector for the generalized eigenvalue closest to 1. The norm of $\mathbf{u}_i$ is normalized to be

$$(\mathbf{u}_i, \mathsf{L}(\mathbf{u}_{i-1})\mathbf{u}_i) = 1. \qquad (14.48)$$

3. Return $\mathbf{u}_i$ if $\mathbf{u}_i$ is sufficiently close to $\mathbf{u}_{i-1}$ except for the sign. Otherwise, let $\mathbf{u}_{i-1} \leftarrow \mathbf{u}_i$, and go back to step 2.

The resulting solution $\hat{\mathbf{u}}$ should satisfy

$$\mathsf{M}(\hat{\mathbf{u}})\hat{\mathbf{u}} = \lambda \mathsf{L}(\hat{\mathbf{u}})\hat{\mathbf{u}}, \qquad (14.49)$$

for some λ. Taking the inner product of $\hat{\mathbf{u}}$ and both sides, we have

$$(\hat{\mathbf{u}}, \mathsf{M}(\hat{\mathbf{u}})\hat{\mathbf{u}}) = \lambda, \qquad (14.50)$$

because of the normalization convention given in Eq. (14.48), which implies from the second of Eqs. (14.46) that

$$1 = (\hat{\mathbf{u}}, \mathsf{L}(\hat{\mathbf{u}})\hat{\mathbf{u}}) = \sum_{\alpha=1}^{N} \frac{(\boldsymbol{\xi}_\alpha, \mathbf{u})^2 (\hat{\mathbf{u}}, V_0[\boldsymbol{\xi}_\alpha]\hat{\mathbf{u}})}{(\mathbf{u}, V_0[\boldsymbol{\xi}_\alpha]\mathbf{u})^2}$$

$$= \sum_{\alpha=1}^{N} \frac{(\boldsymbol{\xi}_\alpha, \mathbf{u})^2}{(\mathbf{u}, V_0[\boldsymbol{\xi}_\alpha]\mathbf{u})}. \qquad (14.51)$$

From the first of Eqs. (14.46), we see that

$$(\hat{\mathbf{u}}, \mathsf{M}(\hat{\mathbf{u}})\hat{\mathbf{u}}) = \sum_{\alpha=1}^{N} \frac{(\hat{\mathbf{u}}, \boldsymbol{\xi}_\alpha)^2}{(\mathbf{u}, V_0[\boldsymbol{\xi}_\alpha]\mathbf{u})} = 1, \qquad (14.52)$$

meaning that $\lambda = 1$. Thus, $\hat{\mathbf{u}}$ is indeed the solution of Eq. (14.45). However, the matrix $\mathsf{L}(\mathbf{u})$ is usually singular, because the matrix $V_0[\mathbf{x}_\alpha]$ in the second of Eqs. (14.46) is likely to degenerate. This is easily seen from Eq. (14.27): the dimension p of $\boldsymbol{\xi}_\alpha$ is generally larger than the dimension m of $\mathbf{x}_\alpha$. Hence, the generalized eigenvalue problem in Eq. (14.47) needs to be reduced to subproblems of smaller dimensions. The reduced form (we omit the details, see [9]) was proposed by Leedan and Meer [28] and Matei and Meer [30] and called the *heteroscedastic errors-in-variables* (HEIV) method.

14.5.6 Renormalization Method

The reason why the solution of the naive method of Sect. 14.5.3 is biased is that the matrix $M(\mathbf{u})$ in Eq. (14.34) is biased. If we decompose the datum $\boldsymbol{\xi}_\alpha$ into its true value $\bar{\boldsymbol{\xi}}_\alpha$ and the noise term $\Delta\boldsymbol{\xi}_\alpha$, the expectation of Eq. (14.34) is

$$\begin{aligned} E[\boldsymbol{\xi}_\alpha \boldsymbol{\xi}_\alpha^\top] &= E[(\bar{\boldsymbol{\xi}}_\alpha + \Delta\boldsymbol{\xi}_\alpha)(\bar{\boldsymbol{\xi}}_\alpha + \Delta\boldsymbol{\xi}_\alpha)^\top] \\ &= E[\bar{\boldsymbol{\xi}}_\alpha \bar{\boldsymbol{\xi}}_\alpha^\top] + E[\bar{\boldsymbol{\xi}}_\alpha \Delta\boldsymbol{\xi}_\alpha^\top] + E[\Delta\boldsymbol{\xi}_\alpha \bar{\boldsymbol{\xi}}_\alpha^\top] + E[\Delta\boldsymbol{\xi}_\alpha \Delta\boldsymbol{\xi}_\alpha^\top] \\ &= \bar{\boldsymbol{\xi}}_\alpha \bar{\boldsymbol{\xi}}_\alpha^\top + V_0[\boldsymbol{\xi}_\alpha]. \end{aligned} \tag{14.53}$$

Thus,

$$E[M(\mathbf{u})] = \bar{M}(\mathbf{u}) + \varepsilon^2 N(\mathbf{u}) + O(\varepsilon^4), \tag{14.54}$$

where $\bar{M}(\mathbf{u})$ is the value of $M(\mathbf{u})$ evaluated using the true values $\{\bar{\boldsymbol{\xi}}_\alpha\}$ and

$$N(\mathbf{u}) = \sum_{\beta=1}^{N} \frac{V_0[\boldsymbol{\xi}_\beta]}{(\mathbf{u}, V_0[\boldsymbol{\xi}_\beta]\mathbf{u})}. \tag{14.55}$$

Equation (14.54) implies that an unbiased solution can be obtained if the matrix $M(\mathbf{u})$ in Eq. (14.35) is replaced by

$$\hat{M}(\mathbf{u}) = M(\mathbf{u}) - \varepsilon^2 N(\mathbf{u}). \tag{14.56}$$

The square noise level ε^2 is unknown, but if we note that the smallest eigenvalue of $\bar{M}(\mathbf{u})$ is 0, we can estimate ε^2 so that the smallest eigenvalue of $\hat{M}(\mathbf{u})$ is 0. Thus, we obtain the following scheme:

1. Guess an appropriate initial value $\mathbf{u}_0$, say the LS solution $\hat{\mathbf{u}}_{\text{LS}}$, and let $c_0 = 0$.
2. Assuming that $\mathbf{u}_{i-1}$ and c_{i-1} are obtained (initially $i = 1$), solve the eigenvalue problem

$$(M(\mathbf{u}_{i-1}) - c_{i-1} N(\mathbf{u}_{i-1}))\mathbf{u} = \lambda \mathbf{u}. \tag{14.57}$$

   Let $\mathbf{u}_i$ be the unit eigenvector for the smallest eigenvalue λ.
3. Return $\mathbf{u}_i$ if λ is sufficiently close to 0. Otherwise, let

$$c_i = c_{i-1} + \frac{\lambda}{(\mathbf{u}_{i-1}, N(\mathbf{u}_{i-1})\mathbf{u}_{i-1})}. \tag{14.58}$$

4. Let $\mathbf{u}_{i-1} \leftarrow \mathbf{u}_i$, and go back to step 2.

Equations (14.57) and (14.58) imply that if c_i is close to 0 we have

$$(M(\mathbf{u}_{i-1}) - c_i N(\mathbf{u}_{i-1}))\mathbf{u}_{i-1} = \mathbf{0}. \tag{14.59}$$

In fact, the inner product of $\mathbf{u}_{i-1}$ and the left-hand side is

$$(\mathbf{u}_{i-1}, (\mathsf{M}(\mathbf{u}_{i-1}) - c_i \mathsf{N}(\mathbf{u}_{i-1}))\mathbf{u}_{i-1}) = (\mathbf{u}_{i-1}, (\mathsf{M}(\mathbf{u}_{i-1}) - c_{i-1}\mathsf{N}(\mathbf{u}_{i-1}))\mathbf{u}_i) - \frac{\lambda(\mathbf{u}_{i-1}, \mathsf{N}(\mathbf{u}_{i-1})\mathbf{u}_{i-1})}{(\mathbf{u}_{i-1}, \mathsf{N}(\mathbf{u}_{i-1})\mathbf{u}_{i-1})}$$
$$= \lambda - \lambda = 0. \qquad (14.60)$$

If c_i is close to 0, the matrix $\mathsf{M}(\mathbf{u}_{i-1}) - c_i \mathsf{N}(\mathbf{u}_{i-1})$ is positive semidefinite, so Eq. (14.60) implies that $\mathbf{u}_{i-1}$ is included in the null space of $\mathsf{M}(\mathbf{u}_{i-1}) - c_i \mathsf{N}(\mathbf{u}_{i-1})$, proving Eq. (14.59). Hence, the solution satisfies

$$(\mathsf{M}(\hat{\mathbf{u}}) - c\mathsf{N}(\hat{\mathbf{u}}))\hat{\mathbf{u}} = \mathbf{0}, \qquad (14.61)$$

and c gives an estimate of ε^2. This scheme was proposed by Kanatani [14] and called *renormalization*.

Remark 16. Historically, this method was proposed first; the HEIV and FNS methods were proposed as refinements to it. However, the renormalization solution and the HEIV/FNS solution (FNS and HEIV produce the same value) are both optimal in the sense that their covariance matrices differ only in the term $O(\varepsilon^4)$ in Eq. (14.29) [14]. This is confirmed by numerical simulations [7, 8, 9].

Remark 17. Renormalization tries to eliminate the bias term in Eq. (14.54) by "subtraction" in the form of Eq. (14.56). An alternative strategy would be to remove the bias by "division". In fact, if we let $\tilde{\mathsf{M}}(\mathbf{u}) = \mathsf{N}(\mathbf{u})^{-1/2}\mathsf{M}(\mathbf{u})\mathsf{N}(\mathbf{u})^{-1/2}$ (the negative square root is defined by replacing all its eigenvalues λ by $1/\sqrt{\lambda}$ in the canonical form), $E[\tilde{\mathsf{M}}(\mathbf{u})]$ and $\tilde{\mathsf{M}}(\mathbf{u})$ share the same eigenvectors up to $O(\varepsilon^4)$. If $\tilde{\mathbf{u}}$ is an eigenvector of $\tilde{\mathsf{M}}(\mathbf{u})$, the corresponding eigenvector of $\mathsf{M}(\mathbf{u})$ is $\mathsf{N}(\mathbf{u})^{-1/2}\tilde{\mathbf{u}}$. This implies that an unbiased solution is obtained by applying the naive method of Sect. 14.5.3 to $\tilde{\mathsf{M}}(\mathbf{u})$. This strategy is known as *equilibration* or *whitening*. However, the matrix $\mathsf{N}(\mathbf{u})$ is often singular due to the degeneracy of $V_0[\boldsymbol{\xi}_\alpha]$ (cf. Sect. 14.5.5), so $\mathsf{N}(\mathbf{u})^{-1/2}$ cannot be computed. Still, it has been applied to a few problems for which $\mathsf{N}(\mathbf{u})$ does not degenerate [29, 35, 36].

14.5.7 Optimal Correction

In deriving the FNS, HEIV, and renormalization methods, we ignored all constraints on $\mathbf{u}$ except $\|\mathbf{u}\| = 1$. Let the remaining constraints be

$$\phi^{(k)}(\mathbf{u}) = 0, \quad k = 1, ..., r. \qquad (14.62)$$

From Eq. (14.29), the normalized covariance of the ML estimator $\hat{\mathbf{u}}$ is given by

$$V_0[\hat{\mathbf{u}}] = \left(\mathsf{P}_{\hat{\mathbf{u}}}\mathsf{M}(\hat{\mathbf{u}})\mathsf{P}_{\hat{\mathbf{u}}}\right)^{-}, \qquad (14.63)$$

where $\mathsf{M}(\mathbf{u})$ is defined in Eq. (14.34) (or in Eqs. (14.46)). The maximum likelihood solution of $\mathbf{u}$ that satisfies the constraint (14.62) is obtained to a first approximation by minimizing

$$J = (\hat{\mathbf{u}} - \mathbf{u}, \mathsf{V}_0[\hat{\mathbf{u}}]^-(\hat{\mathbf{u}} - \mathbf{u})) \tag{14.64}$$

subject to Eq. (14.62). Introducing Lagrange multipliers and first-order approximation, we obtain the following solution [14]:

$$\mathbf{u}^* = \hat{\mathbf{u}} - \mathsf{V}_0[\hat{\mathbf{u}}] \sum_{k,l=1}^{r} w^{(kl)} \hat{\phi}^{(k)} \nabla_\mathbf{u} \hat{\phi}^{(l)}. \tag{14.65}$$

Here, $w^{(kl)}$ is the (kl) element of the inverse of the $r \times r$ matrix whose (kl) element is $(\nabla_\mathbf{u} \hat{\phi}^{(k)}, \mathsf{V}_0[\hat{\mathbf{u}}] \nabla_\mathbf{u} \hat{\phi}^{(l)})$, i.e.,

$$\left(w^{(kl)}\right) = \left((\nabla_\mathbf{u} \hat{\phi}^{(k)}, \mathsf{V}_0[\hat{\mathbf{u}}] \nabla_\mathbf{u} \hat{\phi}^{(l)})\right)^{-1}. \tag{14.66}$$

The hat means that the ML estimator $\hat{\mathbf{u}}$ is substituted for $\mathbf{u}$. The normalized covariance matrix of the corrected value $\mathbf{u}^*$ of Eq. (14.65) is

$$\mathsf{V}_0[\mathbf{u}^*] = \mathsf{V}_0[\hat{\mathbf{u}}] - \sum_{k,l=1}^{r} w^{(kl)} (\mathsf{V}_0[\hat{\mathbf{u}}] \nabla_\mathbf{u} \hat{\phi}^{(k)}) (\mathsf{V}_0[\hat{\mathbf{u}}] \nabla_\mathbf{u} \hat{\phi}^{(k)})^\top \tag{14.67}$$

up to $O(\varepsilon^2)$ [14]. For a single constraint, Eqs. (14.65) and (14.67) reduce to

$$\mathbf{u}^* = \hat{\mathbf{u}} - \frac{\hat{\phi} \mathsf{V}_0[\hat{\mathbf{u}}] \nabla_\mathbf{u} \hat{\phi}}{(\nabla_\mathbf{u} \hat{\phi}, \mathsf{V}_0[\hat{\mathbf{u}}] \nabla_\mathbf{u} \hat{\phi})}, \tag{14.68}$$

$$\mathsf{V}_0[\mathbf{u}^*] = \mathsf{V}_0[\hat{\mathbf{u}}] - \frac{(\mathsf{V}_0[\hat{\mathbf{u}}] \nabla_\mathbf{u} \hat{\phi})(\mathsf{V}_0[\hat{\mathbf{u}}] \nabla_\mathbf{u} \hat{\phi})^\top}{(\nabla_\mathbf{u} \hat{\phi}, \mathsf{V}_0[\hat{\mathbf{u}}] \nabla_\mathbf{u} \hat{\phi})}. \tag{14.69}$$

Remark 18. If the r constraints in Eq. (14.62) are redundant, say only r' ($< r$) of them are independent, the inverse in Eq. (14.66) is replaced by the generalized inverse of rank r' (cf. Remark 6).

Remark 19. If all the r constraints in Eq. (14.62) are independent, the rank of the matrix $\mathsf{V}_0[\mathbf{u}^*]$ given by Eq. (14.65) is smaller than $\mathsf{V}_0[\hat{\mathbf{u}}]$ by r. Intuitively, the ellipsoid that represents the uncertainty of $\mathbf{u}$ in $\mathcal{R}^p$ "collapses" in the r directions in which the constraint (14.62) is violated, while it keeps its shape in the directions orthogonal to them. Hence, the optimality of the ML estimator is not affected by doing this type of posterior correction [14].

Remark 20. Equation (14.65) enforces all the constraints only to a first approximation, so $\phi^{(k)}(\mathbf{u}^*)$, $k = 1, ..., r$, may not exactly be 0, and $\mathbf{u}^*$ may not exactly be a unit vector. Such higher-order discrepancies can be eliminated by iterating Eqs. (14.68) and (14.69) in the form

$$\mathbf{u}^* \leftarrow N[\hat{\mathbf{u}} - \frac{\hat{\phi} V_0[\hat{\mathbf{u}}] \nabla_{\mathbf{u}} \hat{\phi}}{(\nabla_{\mathbf{u}} \hat{\phi}, V_0[\hat{\mathbf{u}}] \nabla_{\mathbf{u}} \hat{\phi})}], \tag{14.70}$$

$$V_0[\mathbf{u}^*] \leftarrow \mathsf{P}_{\mathbf{u}^*} \left(V_0[\hat{\mathbf{u}}] - \frac{(V_0[\hat{\mathbf{u}}] \nabla_{\mathbf{u}} \hat{\phi})(V_0[\hat{\mathbf{u}}] \nabla_{\mathbf{u}} \hat{\phi})^\top}{(\nabla_{\mathbf{u}} \hat{\phi}, V_0[\hat{\mathbf{u}}] \nabla_{\mathbf{u}} \hat{\phi})}, \right) \mathsf{P}_{\mathbf{u}^*}, \tag{14.71}$$

where $N[\,\cdot\,]$ denotes normalization to a unit vector ($N[\mathbf{v}] = \mathbf{v}/\|\mathbf{v}\|$), and $\mathsf{P}_{\mathbf{u}^*}$ is the projection matrix defined by Eq. (14.30). Equation (14.71) makes the null space of the $V_0[\mathbf{u}^*]$ exactly compatible with $\mathbf{u}^*$.

14.6 Nuisance Parameters and Semiparametric Model

Finally, we discuss some new topics related to the use of statistical methods for geometric inference.

14.6.1 Asymptotic Parameters

The number n that appears in the standard statistical analysis is the *number of experiments*. It is also called the *number of trials*, the *number of observations*, and the *number of samples*. Evidently, the properties of the ensemble are revealed more precisely as more data are sampled from it.

However, the number n is often called the *number of data*, which has caused considerable confusion. For example, if we observe a 100-dimensional vector datum in one experiment, one may think that the "number of data" is 100, but this is wrong: the number n of experiments is 1. We are observing 1 sample from an ensemble of 100-dimensional vectors.

For character recognition, the underlying ensemble is the set of possible character images, and the learning process concerns the number n of training steps necessary to establish satisfactory responses. This is independent of the dimension N of the vector that represents each character. The learning performance is evaluated asymptotically as $n \to \infty$, not $N \to \infty$.

For geometric inference, however, many researchers have taken the dimension of the data as the "number of data" perhaps because the ensemble is hypothetical and one cannot sample more than one datum from it. However, if we extract, for example, 50 feature points, they constitute a 100-dimensional vector consisting of their x and y coordinates. If no other information, such as the image intensity, is used, the image is completely characterized by that vector. Applying a statistical method means regarding it as a sample from a hypothetical ensemble of 100-dimensional vectors.

14.6.2 Neyman–Scott Problem

In the past, many computer vision researchers have analyzed the asymptotic behavior as $N \to \infty$ without explicitly mentioning what the underlying ensemble is. This is perhaps motivated by a similar formulation in the statistical literature. Suppose, for example, a rodlike structure lies on the ground in the distance. We emit a laser beam toward it and estimate its position and orientation by observing the reflection of the beam, which is contaminated by noise. We assume that the laser beam can be emitted in any orientation any number of times, but the emission orientation is measured with noise. The task is to estimate the position and orientation of the structure as accurately as possible by emitting as small a number of beams as possible. Naturally, the estimation performance should be evaluated in the asymptotic limit $n \to \infty$ with respect to the number n of emissions.

The underlying ensemble is the set of all response times for all possible directions of emission. Usually, we are interested in the position and orientation of the structure but not the exact orientation of each emission, so the variables for the former are called the *structural parameters*, which are fixed in number, while the latter are called the *nuisance parameters*, which increase indefinitely as the number n of experiments increases [2]. Such a formulation is called the *Neyman–Scott problem* [37]. Since the constraint is an implicit function in the form of Eq. (14.3), we are considering an *errors-in-variables model* [10]. If we linearize the constraint by changing variables, the noise characteristics differs for each data component, so the problem is *heteroscedastic* [28].

To solve this problem, one can introduce a parametric model for the distribution of possible laser emission orientations, regarding the actual emissions as random samples from it. This formulation is called a *semiparametric model* [2]. An optimal solution can be obtained by finding a good *estimating function* [2, 40].

14.6.3 Semiparametric Model for Geometric Inference

Since the semiparametric model has something different from the geometric inference problem described in Sect. 14.3.2, a detailed analysis is required for examining if application of a semiparametric model to geometric inference will yield a desirable result [38, 40]. In any event, one should explicitly state what kind of ensemble (or ensemble of ensembles) is assumed before doing statistical analysis.

This is not merely a conceptual issue. It also affects the performance evaluation of simulation experiments. In doing a simulation, one can freely change the number N of feature points and the noise level ε. If the accuracy of method A is higher than method B for particular values of N and ε, one cannot conclude that method A is superior to method B, because opposite results may come out for other values of N and ε. Here, we have two alternatives for performance evaluation: fixing ε and varying N to see if admissible accuracy is attained for a smaller number of feature point; fixing N and varying ε to see if larger data uncertainty can be tolerated for admissible accuracy. These two types of evaluation have different meanings. Our conclusion is that the results of one type of evaluation cannot directly be compared with the results of the other.

14.7 Conclusions

We have investigated the meaning of "statistical methods" for geometric inference based on image feature points. Tracing back the origin of feature uncertainty to image processing operations, we discussed the implications of asymptotic analysis in reference to geometric fitting and geometric model selection. We pointed out that a correspondence exists between the standard statistical analysis and the geometric inference problem. We also compared the capability of the geometric AIC and the geometric MDL in detecting degeneracy. Next, we reviewed recent progress in geometric fitting techniques for linear constraints, describing the FNS method, the HEIV method, the renormalization method, and other related techniques. Finally, we discussed the Neyman–Scott problem and semiparametric models in relation to geometric inference.

From these discussions, we conclude that applications of statistical methods require careful consideration about the nature of the problem in question and that different statistical theories are necessary for different classes of problems. In this sense, there is much room for new statistical theories to emerge as the scope of computer vision research expands. The important thing is, however, to always make clear the underlying hypotheses and assumptions, and to not simply use the methods in the statistical literature.

References

1. Akaike H. (1977) A new look at the statistical model identification. *IEEE Trans. Autom. Control* **16**:716–723
2. Amari S., Kawanabe M. (1997) Information geometry of estimating functions in semiparametric statistical models. *Bernoulli* **3**:29–54
3. Bubna K., Stewart C.V. (2000) Model selection techniques and merging rules for range data segmentation algorithms. *Comput. Vision Image Understand.* **80**:215–245
4. Chabat F., Yang G.Z., Hansell D.M. (1999) A corner orientation detector. *Image Vision Comput.* **17**:761–769
5. Cho K., Meer P. (1997) Image segmentation form consensus information. *Comput. Vision Image Understand.* **68**:72–89
6. Cho K., Meer P., Cabrera J. (1997) Performance assessment through bootstrap. *IEEE Trans. Patt. Anal. Mach. Intell.* **19**:1185–1198

7. Chojnacki W., Brooks M.J., van den Hengel A., Gawley D. (2000) On the fitting of surfaces to data with covariances. *IEEE Trans. Patt. Anal. Mach. Intell.* **22**:1294–1303
8. Chojnacki W., Brooks M.J., van den Hengel A. (2001) Rationalising the renormalisation method of Kanatani. *J. Math. Imaging Vision* **14**:21–38
9. Chojnacki W., Brooks M.J., van den Hengel A., Gawley D. (2004) From FNS to HEIV: A link between two vision parameter estimation methods. *IEEE Trans. Patt. Anal. Mach. Intell.* **26**:264–268
10. Fuller W.A. (1987) *Measurement Error Models*, Wiley, New York
11. Gu H., Shirai Y., Asada M. (1996) MDL-based segmentation and motion modeling in a long sequence of scene with multiple independently moving objects *IEEE Trans. Patt. Anal. Mach. Intell.* **18**:58–64
12. Harris C., Stephens M. (1988) A combined corner and edge detector. In: *Proceedings of Fourth Alvey Vision Conference*. Manchester, UK, pp. 147–151
13. Hartley R., Zisserman A. (2000) *Multiple View Geometry in Computer Vision*. Cambridge University Press, Cambridge, UK
14. Kanatani K. (1996) *Statistical Optimization for Geometric Computation: Theory and Practice*. Elsevier, Amsterdam
15. Kanatani K. (1998) Geometric information criterion for model selection. *Int. J. Comput. Vision* **26**:171–189
16. Kanatani K. (1998) Statistical optimization and geometric inference in computer vision. *Phil. Trans. Roy. Soc. Lond. A***356**:1303–1320
17. Kanatani K. (1988) Cramer–Rao lower bounds for curve fitting. *Graphical Models Image Process.* **60**:93–99
18. Kanatani K. (2000) Model selection criteria for geometric inference. In: Bab-Hadiashar A., Suter D. (eds.) *Data Segmentation and Model Selection for Computer Vision: A Statistical Approach*, Springer, Berlin Heidelberg New York, pp. 91–115
19. Kanatani K. (2001) Motion segmentation by subspace separation and model selection. In: *Proceedings of Eighth International Conference on Computer Vision, vol. 2*. Vancouver, Canada, pp. 301–306
20. Kanatani K. (2002) Model selection for geometric inference, plenary talk. In: *Proceedings of Fifth Asian Conference on Computer Vision, vol. 1*. Melbourne, Australia, pp. xxi–xxxii
21. Kanatani K. (2002) Motion segmentation by subspace separation: Model selection and reliability evaluation. *Int. J. Image Graphics* **2**:179–197
22. Kanatani K. (2002) Evaluation and selection of models for motion segmentation. In: *Proceedings of Seventh European Conference on Computer Vision, vol. 3*. Copenhagen, Denmark, pp. 335–349
23. Kanatani K., Matsunaga C. (2002) Estimating the number of independent motions for multibody motion segmentation. In: *Proceedings of Fifth Asian Conference on Computer Vision, vol. 1*. Melbourne, Australia, pp. 7–12
24. Kanazawa Y., Kanatani K. (1997) Infinity and planarity test for stereo vision. *IEICE Trans. Inf. & Syst.* **E80-D**:774–779
25. Kanazawa Y., Kanatani K. Stabilizing image mosaicing by model selection. In: Pollefeys M., Van Gool L., Zisserman A., Fitzgibbon A. (eds.) *3D Structure from Images–SMILE 2000. LNCS, vol. 2018*. Springer, Berlin Heidelberg New York, 2001, pp. 35–51
26. Kanazawa Y., Kanatani K. (2001) Do we really have to consider covariance matrices for image features? In: *Proceedings of Eighth International Conference on Computer Vision, vol. 2*. Vancouver, Canada, pp. 586–591

27. Leclerc Y.G. (1989) Constructing simple stable descriptions for image partitioning. *Int. J. Comput. Vision* **3**:73–102
28. Leedan Y., Meer P. (2000) Heteroscedastic regression in computer vision: Problems with bilinear constraint. *Int. J. Comput. Vision.* **37**:127–150
29. MacLean W.J. (1999) Removal of translation bias when using subspace methods. In: *Proceedings of Seventh International Conference on Computer Vision, vol. 2*. Kerkyra, Greece, pp. 753–758
30. Matei B., Meer P. (2000) A generalized method for errors-in-variables problem in computer vision. In: *Proceedings of Fifteenth International Conference on Pattern Recognition, vol. 2*. Barcelona, Spain, pp. 18–25
31. Matsunaga C., Kanatani K. (2000) Calibration of a moving camera using a planar pattern: Optimal computation, reliability evaluation and stabilization by model selection. In: *Proceedings of Sixth European Conference on Computer Vision, vol. 2*. Dublin, Ireland, pp. 595–609
32. Maxwell B.A. (2000) Segmentation and interpretation of multicolored objects with highlights. *Comput. Vision Image Understand.* **77**:1–24
33. Maybank S.J., Sturm P.F. (1999) MDL, collineations and the fundamental matrix. In: *Proceedings of Tenth British Machine Vision Conference*. Nottingham, UK, pp. 53–62
34. Morris D.D., Kanatani K., Kanade T. (2001) Gauge fixing for accurate 3D estimation. In: *Proceedings of IEEE Conference on Computer Vision Pattern Recognition, vol. 2*. Kauai, Hawaii, pp. 343–350
35. Mühlich M., Mester R. (1998) The role of total least squares in motion analysis. In: *Proceedings of Fifth European Conference on Computer Vision, vol. 2*. Freiburg, Germany, pp. 305–321
36. Mühlich M., Mester R. (2001) A considerable improvement in pure parameter estimation using TLS and equilibration. *Patt. Recog. Lett.* **22**:1181–1189
37. Neyman J., Scott E.L. (1948) Consistent estimates based on partially consistent observations. *Econometrica* **16**:1–32
38. Ohta N. (2003) Motion parameter estimation from optical flow without nuisance parameters. In: *Third International Workshop on Statistical and Computational Theory of Vision*. Nice, France: http://www.stat.ucla.edu/~yuille/meetings/2003_workshop.php. Cited 12 October 2003
39. Ohta N., Kanatani K. (1998) Moving object detection from optical flow without empirical thresholds. *IEICE Trans. Inf. & Syst.* **E81-D**:243–245
40. Okatani T., Deguchi K. (2003) Toward a statistically optimal method for estimating geometric relations from noisy data: Cases of linear relations. In: *Proceedings of IEEE Conference on Computer Vision Pattern Recognition, vol. 1*. Madison, WI, pp. 432–439
41. Reisfeld D., Wolfson H., Yeshurun Y. (1995) Context-free attentional operators: The generalized symmetry transform. *Int. J. Comput. Vision* **14**:119–130
42. Rissanen J. (1984) Universal coding, information, prediction and estimation. *IEEE Trans. Inform. Theory* **30**:629–636
43. Rissanen J. (1989) *Stochastic Complexity in Statistical Inquiry*. World Scientific, Singapore
44. Rissanen J. (1996) Fisher information and stochastic complexity, *IEEE Trans. Inform. Theory* **42**:40–47
45. Schmid C., Mohr R., Bauckhage C. (2000) Evaluation of interest point detectors. *Int J. Comput. Vision* **37**:151–172

46. Smith S.M., Brady J.M. (1997) SUSAN—A new approach to low level image processing. *Int. J. Comput. Vision* **23**:45–78
47. Sugaya Y., Kanatani K. (2003) Outlier removal for feature tracking by subspace separation. *IEICE Trans. Inf. & Syst.* **E86-D**:1095–1102
48. Torr P.H.S. (1997) An assessment of information criteria for motion model selection. In: *Proceedings of IEEE Conference on Computer Vision Pattern Recognition*. Puerto Rico, pp. 47–53
49. Torr P.H.S. (1998) Geometric motion segmentation and model selection. *Phil. Trans. Roy. Soc. Lond. A***356**:1321–1340
50. Torr P.H.S. (2002) Bayesian model estimation and selection for epipolar geometry and generic manifold fitting. *Int. J. Comput. Vision* **50**:35–61
51. Torr P.H.S., FitzGibbon A., Zisserman A. (1998) Maintaining multiple motion model hypotheses through many views to recover matching and structure. In: *Proceedings of Sixth International Conference on Computer Vision*. Bombay, India, pp. 485–492
52. Torr P.H.S., Zisserman A. (2000) Concerning Bayesian motion segmentation, model averaging, matching and the trifocal tensor. In: *Proceedings of Sixth European Conference on Computer Vision, vol. 1*. Dublin, Ireland, pp. 511–528
53. Triono I., Ohta N., Kanatani K. (1998) Automatic recognition of regular figures by geometric AIC, *IEICE Trans. Inf. & Syst.* **E81-D**:246–248

15

Uncertainty and Projective Geometry

Wolfgang Förstner

Institut für Photogrammetrie, Universität Bonn
Nussallee 15, D-53121 Bonn, wf@ipb.uni-bonn.de

15.1 Introduction

Uncertainty is present in computer vision in all analysis steps: in image processing, in feature extraction, pose estimation, grouping and also in recognition and interpretation. Problems are, among others, the adequate representation of uncertainty, propagation of uncertainty, estimation under uncertainty and decision making under uncertainty. Recently, statistical inference has become a major thread of research at all levels of image analysis. This certainly is due to the rich arsenal of tools, which allows us to precisely model uncertainty, to check the validity of the assumptions made and to reason under uncertainty.

This paper is about uncertainty in geometric reasoning, specifically using projective geometry. Algebraic projective geometry has become the basic tool for representing geometry of multiple views, cf. the two classical text books [10, 18]. The two examples in Figs. 15.1 and 15.2 show two applications where algebraic projective geometry can be used to advantage.

In both cases the geometric relations can be expressed as multi-linear forms of the entities involved, which would not have been possible when not using projective geometry.

On the other hand, rigorous estimation techniques, e.g. used in bundle adjustment for image orientation to minimize the reprojection errors, have been accepted as reference for suboptimal techniques and as a final step in order to obtain statistically optimal results. The need to exploit the full information about the statistics is demonstrated in the example of Fig. 15.3: all geometric entities with a certain probability lie within a certain region, whose shape and size vary individually. Therefore pure geometric measures are not useful for reasoning under uncertainty.

We only want to mention two prominent representative publications where projective geometry and statistics have been integrated to a larger extent. Kanatani [25] apparently was the first to integrate geometry in 2D and 3D and statistics in a rigorous manner. He aimed at completeness in uncertain

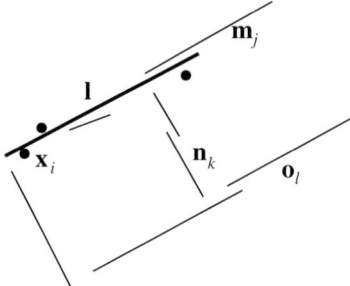

Fig. 15.1. 2D grouping of points and lines, e.g. resulting from a image preprocessing step, consists of two steps: (1) testing of hypothesized mutual relations, taking uncertainty of image features into account, and (2) joint estimation of geometric features. All points and lines in the figure may be grouped. The result may be an optimal estimate, e.g. of the line l using incident points x_i, collinear lines m_j, orthogonal n_k and parallel lines o_l. In algebraic projective geometry all these relations are linear, thereby easing statistical testing and estimation

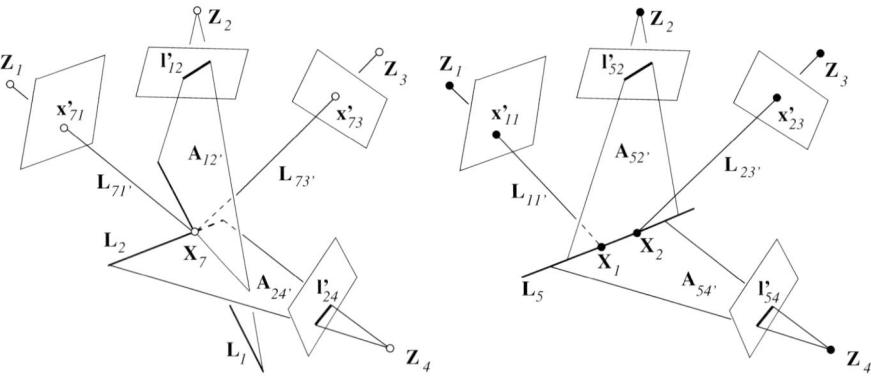

Fig. 15.2. Triangulation of points and lines: estimation of 3D point X_7 (*left*) and 3D line L_5 (*right*) from image points x' and image lines l'. The spatial relations between image and space points and lines can easily be expressed as a function of the projection matrices and for points and lines, respectively. Also in this case, using algebraic projective geometry eases statistical testing of these relations and the joint optimal estimation of the 3D line

geometric reasoning, and discussed motion estimation and optical flow. He proposed rigorous tests and optimal, i.e. maximum likelihood, estimates. However, though he used homogeneous vectors for representing geometric entities, he required 2D and 3D points to be Euclidian normalized. This was motivated by the otherwise indefinite scaling of the vectors but does not allow for handling of points at infinity. The partitioning of the vectors into homogeneous and Euclidean parts, which as such is reasonable for interpretation

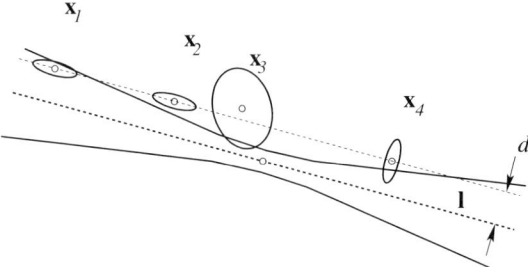

Fig. 15.3. Testing spatial relations, especially the necessity of taking the uncertainty into account rigorously, e.g. when testing a point-line incidence: in the presence of uncertainty the geometric distance d of the points $\mathbf{x}_i$ from the line $\mathbf{l}$ is not useful for testing. The uncertainty of a point in a first approximation can be represented by a confidence ellipse, while the uncertainty of a line can be represented by a confidence hyperbola, which is the collection of confidence regions across the line of all points. Though the situation is much more complex in 3D, it can easily be handled using the covariance matrices of the entities in concern

[3], leads to cumbersome expressions in the covariance matrices, especially as Kanatani aimed at giving explicit expressions, including both the error propagation and the normalization to Euclidean homogenous vectors.

In his thesis, Criminisi [9] integrated uncertainty reasoning into all steps of single- and multiple-view analysis. For a great number of geometric reasoning tasks, also including the determination of transformations, he gave explicit expressions for covariance matrices. He also analyzed the degree of approximation introduced by linearization. Unfortunately, in both cases the beauty of projective geometry got lost on the way to the integration of statistical reasoning.

The ease of handling multilinear, and especially bilinear, forms was questioned by Haddon and Forsyth [16] They demonstrated that significant bias and deviations from a Gaussian distribution might occur when partitioning bilinear forms, i.e. when solving for structure and motion from image observations. Their examples, however, seem to be caused by a very low signal-to-noise ratio, resulting from comparably short base lines in image sequences.

Altogether there appears to be a clear agreement to use both statistics and projective geometry for spatial reasoning under uncertainty. The main problem left is to find an adequate representation and adequate procedures for geometric reasoning. The approach described in this paper tries to avoid the disadvantages mentioned above.

In statistics uncertainty of measurements usually is represented by covariance matrices. Generally speaking, instead of working with the probability densities one uses the first two moments of the distribution. This appears to be widely accepted and is adequate as long as the signal-to-noise ratio is high enough, say much better than 10:1, which nearly always is the case in the first steps of image analysis. Moreover, propagation of uncertainty can be performed within the calculus of linear algebra to a sufficient accuracy.

Therefore, among the many *representations*, homogeneous vectors and matrices appear to be the right choice for geometric entities and transformations. An embedding into more fundamental concepts, such as the double algebra or the Grassman–Cayley algebra [4, 11] or even the geometric algebra [21, 22] does not seem to be possible, however – and this was the motivation for the approach presented here, cf. [14, 15, 23] – the beautiful structures of these algebras should be kept, as far as possible.

Together with the nearly always approximate representation of uncertainty using covariance matrices, there exists a rich and powerful arsenal of tools for statistical reasoning. This especially holds for estimation techniques [31]. The versatility of these tools is the basis for the broad experience in the field of geodesy, where estimation of geometric quantities is a standard task [20, 27, 29]. The key to an easy to use concept of estimation procedures lies in the generic representation of the given functional relations between all observed quantities and all unknowns, not so much in the optimization function or in the optimization procedures e.g. the trust region method to increase convergence [19], which appears to be a necessary second step.

Therefore, among the many *procedures* for estimation, the maximum likelihood estimation based on the so-called Gauss–Helmert model appears to be the right choice. It is generic, as it allows us to represent all estimation problems with nonlinear constraints. Similarly to all other alternative models, the estimation processes iteratively improve approximate values for all parameters. The approximate values have to be reasonably good, which in our context can be achieved based on the rich work of the past decade.

The paper is organized as follows: Sect. 2 discusses issues of uncertainty: the representation, especially in the context of homogeneous entities; the propagation, especially the effect of linearization; the estimation under generic constraints; and basic elements from testing. Sect. 3 discusses issues of projective geometry: the representation of geometric entities; their construction from given ones; and their relations, including homogeneous transformations as a basis for uncertainty propagation, statistical testing and estimation. Finally, Sect. 4 discusses when conditioning of the geometric entities and normalization of the covariance matrices is necessary to overcome the proposed approximations.

The goal of this chapter is to provide simple-to-use tools for uncertain geometric reasoning. We do not discuss the sources of uncertainty (cf. Kanatani's valuable discussion in this volume) and do not address refinements concerning computational efficiency.

Notation: Vectors are boldface letters, such as $\boldsymbol{x}$ or $\boldsymbol{X}$, matrices are boldface sans serif letters, such as $\mathsf{A} = [a_{ij}]$ or H. Homogeneous vectors and matrices are upright letters, such as x or H, Euclidean vectors are italic letters, such as x or X. Vectors representing geometric 2D entities are lowercase letters, such as $\boldsymbol{x}$ or l, and vectors representing geometric 3D entities are uppercase letters, such as **A** and **X**. Planes are denoted with letters **A**, **B**, and so on from the beginning of the alphabet, lines are denoted with letters **l**, **L**, **m**, and so on from the middle of the alphabet, and points are denoted with letters **X**, **Y**, and so on from the end of the alphabet. The $n \times n$ unit matrix is denoted with I_n. The ith n unit vector is denoted with $\boldsymbol{e}_i^{(n)}$. Stochastic variables are underscored, such as $\underline{x}$. The density function of the stochastic variable x, possibly being a vector $\boldsymbol{x}$, is denoted with $p_x(\cdot)$. The expectation, the variance and the covariance operators are E($\cdot$), V($\cdot$) and Cov($\cdot,\cdot$), respectively. Covariance matrices are indexed with two indices, e.g. V($\underline{\boldsymbol{x}}$) = $\boldsymbol{\Sigma}_{xx} = [\sigma_{x_i x_j}]$, allowing us to densely write the covariance of two different vectors Cov($\underline{\boldsymbol{x}},\underline{\boldsymbol{y}}$) = $\boldsymbol{\Sigma}_{xy} = [\sigma_{x_i y_j}]$. The determinant of a matrix is $|A|$.

We will use the vec-operator, columnwise stacking the columns of an $n \times m$ matrix A into a nm vector vec$A$, thus vec(A^T) contains the nm elements of A rowwise. We will use the Kronecker product $A \otimes B = [A_{ij} B]$. With the vec-operator we use the two relations vec(ABC) = ($C^\mathsf{T} \otimes A$)vec$B$ and as vec($AB\boldsymbol{c}$) = vec($\boldsymbol{c}^\mathsf{T} B^\mathsf{T} A^\mathsf{T}$) and vec($AB\boldsymbol{c}$) = ($\boldsymbol{c}^\mathsf{T} \otimes A$)vec$B$ = ($A \otimes \boldsymbol{c}^\mathsf{T}$)vec($B^\mathsf{T}$). Rowwise concatenation of two matrices A and B leads to the matrix $[A|B]$.

15.2 Uncertainty

15.2.1 Representation and Propagation of Uncertainty

Basics

Probability theory is a classical tool for representing uncertainty. In our context we are concerned with representing the uncertainty of coordinate vectors $\underline{\boldsymbol{x}}$, which is usually done via the probability density function (PDF) $p_x(\boldsymbol{x})$, or the cumulative probability density function (CPDF) $P_x(\boldsymbol{x})$, the function in contrast to the independent variable in brackets. In many cases one can reasonably well use the Gaussian or normal distribution with density

$$g_x(\boldsymbol{x}; \boldsymbol{\mu}_x, \boldsymbol{\Sigma}_{xx}) = \frac{1}{\sqrt{(2\pi)^n |\boldsymbol{\Sigma}_{xx}|}} e^{-\frac{1}{2}(\boldsymbol{x}-\boldsymbol{\mu}_x)^\mathsf{T} \boldsymbol{\Sigma}_{xx}^{-1}(\boldsymbol{x}-\boldsymbol{\mu}_x)} \,, \qquad (15.1)$$

which depends on two parameters, the n vector $\boldsymbol{\mu}_x$ and the symmetric $n \times n$ matrix $\boldsymbol{\Sigma}_{xx}$. Observe that the density function of the normal distribution only is defined for regular $\boldsymbol{\Sigma}_{xx}$. For practical reasons one uses the short notation $\underline{\boldsymbol{x}} \sim N(\boldsymbol{\mu}_x, \boldsymbol{\Sigma}_{xx})$ to indicate the stochastic variable $\underline{\boldsymbol{x}}$ to be normally distributed with the parameters $\boldsymbol{\mu}_x$ and $\boldsymbol{\Sigma}_{xx}$.

Often reasoning can be restricted to the so-called moments of the distribution. With the expectation operator E($f(\underline{\boldsymbol{x}})$) = $\int f(\boldsymbol{x}) p_x(\boldsymbol{x}) \mathrm{d}\boldsymbol{x}$ we will regularly use the mean, which is the first moment of the distribution, the variance, which is the central second moment, and the kurtosis, which is the central fourth moment μ_{4x} normalized with $3\sigma_x^4$.

$$\mu_x = \mathrm{E}(\underline{x}) = \int x p_x(x)\,\mathrm{d}x\,, \quad \sigma_x^2 = \mathrm{V}(\underline{x}) = \mathrm{E}((\underline{x}-\mu_x)^2),$$

$$\kappa = \frac{\mu_{4x}}{3\sigma_x^4} = \frac{\mathrm{E}((\underline{x}-\mu_x)^4)}{3\sigma_x^4},$$

which leads to $\kappa = 1$ for Gaussian variables.

For vector-valued stochastic variables we have the covariance matrix defined as $\boldsymbol{\Sigma}_{xx} = [\sigma_{x_i x_j}] = \mathrm{V}(\underline{\boldsymbol{x}}) = \mathrm{E}((\underline{\boldsymbol{x}}-\boldsymbol{\mu}_x)(\underline{\boldsymbol{x}}-\boldsymbol{\mu}_x)^\mathsf{T})$. In case two stochastic variables are statistically independent, their joint distribution $p_{xy}(x,y)$ is separable; thus

$$p_{xy}(x,y) = p_x(x)\,p_y(y), \quad \text{and} \quad P_{xy}(x,y) = P_x(x)\,P_y(y)\,. \qquad (15.2)$$

For normally distributed variables $\underline{\boldsymbol{x}} \sim N(\boldsymbol{\mu}_x, \boldsymbol{\Sigma}_{xx})$ the two parameters $\boldsymbol{\mu}_x$ and $\boldsymbol{\Sigma}_{xx}$ are the mean and the covariance matrix.

In general, propagating uncertainty through chains of nonlinear functions is intractable. For very specific distributions and simple functions this can be done explicitely, using various techniques, depending on the situation, cf. the discussion and the many examples given by Papoulos [30]. We, however, may restrict ourselves to the propagation of the first two moments.

If the two first moments of a stochastic vector are used to describe its distribution $\underline{\boldsymbol{x}} \sim M_x(\boldsymbol{\mu}_x, \boldsymbol{\Sigma}_{xx})$, then the vector valued nonlinear function $\boldsymbol{y} = \boldsymbol{f}(\boldsymbol{x})$ has a distribution $\underline{\boldsymbol{y}} \sim M_y(\boldsymbol{\mu}_y, \boldsymbol{\Sigma}_{yy})$ with the first two moments (cf. [27])

$$\boxed{\boldsymbol{\mu}_y = \boldsymbol{f}(\boldsymbol{\mu}_x), \quad \boldsymbol{\Sigma}_{yy} = J_{yx}\boldsymbol{\Sigma}_{xx}J_{yx}^\mathsf{T}, \quad \text{with} \quad J_{yx} = \left.\frac{\partial \boldsymbol{f}(\boldsymbol{x})}{\partial \boldsymbol{x}}\right|_{x=\mu_x}.} \qquad (15.3)$$

This error propagation law holds rigorously for any distributions with finite first- and second-order moments in case the relation $\boldsymbol{f}(\boldsymbol{x})$ is linear. In the case of nonlinear functions it is an approximation, which we discuss below in the context of geometric reasoning. The basic idea is to attach a covariance matrix to each uncertain entity during geometric reasoning.

Representing Uncertain Homogeneous Vectors

Attaching a covariance matrix to homogeneous vectors can be done straightforwardly and has been extensively done by Kanatani and Criminisi. For example, in case the Euclidean coordinates $\underline{\boldsymbol{x}} = (\underline{x}, \underline{y})^\mathsf{T} \sim M(\boldsymbol{\mu}_x, \boldsymbol{\Sigma}_{xx})$ is given, the corresponding covariance matrix of the homogeneous 3-vector $\mathbf{x} = (\underline{x}, \underline{y}, 1)^\mathsf{T}$ is given by

$$\boldsymbol{\Sigma}_{\mathbf{xx}} = \begin{bmatrix} \sigma_x^2 & \sigma_{xy} & 0 \\ \sigma_{xy} & \sigma_y^2 & 0 \\ 0 & 0 & 0 \end{bmatrix}. \qquad (15.4)$$

This approach is correct and circumvents the problem of discussing projective entities (cf. Fig. 15.4). That is, the transition from uncertain Euclidean entities, which are primary observations, to uncertain homogeneous entities is simple and can be done statistically rigorously for points. One goal of this chapter is to show, that the transition to other uncertain homogeneous entities, e.g. by construction, is simple, and a good approximation. The same holds for the derivation of Euclidean entities from homogeneous ones. Thus instead of working with a nonredundant representation in Euclidean space, e.g. in $\mathbb{R}^2$ for 2D points, one uses a redundant representation in a higher dimensional Euclidean space, e.g. in $\mathbb{R}^3$ for 2D points.

The beauty of algebraic projective geometry for geometric reasoning and multiple-view analysis shown in the work of Faugeras and Papadopoulo [11] was the key motivation to use homogeneous coordinates for representing uncertain geometric entities but to stay as close as possible to the concepts of the Grassman–Cayley algebra in order to preserve the transparency of the geometric relations. However, the redundancy k in the representation with homogeneous entities, which is $k = 1$ for all entities, except for 3D lines, where it is $k = 2$, leads to some difficulties:

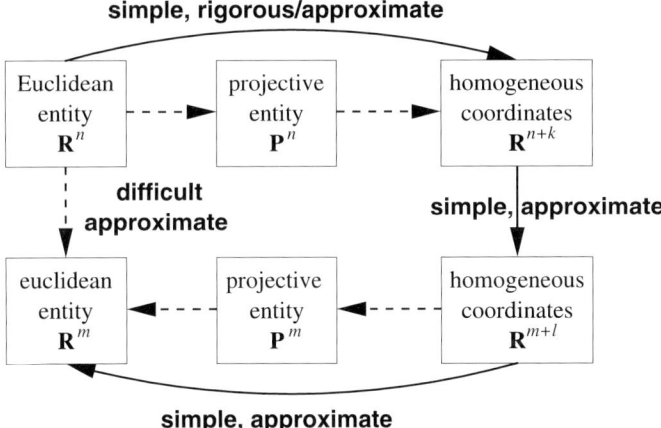

Fig. 15.4. Reasoning under uncertainty in projective geometry. Instead of performing calculations with Euclidean entities one works with homogeneous entities, only implicitly using them as representations of projective entities

1. The redundancy in the representation immediately leads to *singular covariance matrices* as e.g. in Eq. (15.4). Thus there is no proper PDF for homogeneous entities.[1]

---

[1] This was a criticism of a reviewer of an earlier conference paper using this approach.

2. Construction of new entities from given ones, using the classical tools of statistical error propagation, generally leads to regular covariance matrices. This is in contrast to the singularity of the covariance matrices derived from Euclidean entities, cf. Eq. (15.4), and prevents a simple comparison of uncertain stochastic vectors having covariance matrices of different rank. Kanatani [25] proposed to use the Euclidean normalized homogeneous coordinates for points. This imposes a constraint, e.g. the component $\underline{w}$ of $\mathbf{x} = [\underline{u}, \underline{v}, \underline{w}]^\mathsf{T}$ to be normalized to 1, and results in covariance matrices with the desired rank. Interestingly, he normalizes 2D lines, 3D lines and planes spherically. This normalization, however, will be shown to be generally unnecessary. Therefore also homogeneous vectors with a full rank covariance matrix will be allowed. The *equivalence relation* for homogeneous vectors needs to be redefined for this reason. Only in case one uses the proposed test statistics for correctly sorting multiple hypotheses in search problems does one need to condition and normalize the geometric entities. However, spherical normalization can always be applied, which enables the inclusion of points at infinity.
3. The representation and propagation of uncertainty with the second moments is an approximation. The degree of approximation needs to be known to safely apply the proposed approach.

On Singular Covariance Matrices

We first discuss the PDF of random vectors containing fixed entities. When representing fixed values, such as the third component in $(\underline{x}, \underline{y}, 1)^\mathsf{T}$, we might track this property through the reasoning chain or just treat the value 1 as a stochastic variable with mean 1 and variance 0. The second alternative has implicitly been chosen by Kanatani [25] and Criminisi [9].

One can easily construct a 2-vector with a singular 2×2 covariance matrix. Assume $\underline{x} \sim N(\mu_x, 1)$ and $\underline{y} \sim N(\mu_y, 0)$ are independent stochastic variables; thus

$$\begin{bmatrix} \underline{x} \\ \underline{y} \end{bmatrix} \sim N\left(\begin{bmatrix} \mu_x \\ \mu_y \end{bmatrix}, \begin{bmatrix} 1 & 0 \\ 0 & 0 \end{bmatrix}\right).$$

The distribution of $\underline{y}$ can be defined as a limiting process:

$$p_y(y) = \lim_{\sigma_y \to 0} g(y; \mu_y, \sigma_y^2) = \delta(y - \mu_y).$$

The resulting δ-function is a so-called *generalized function*, which is only definable via a limiting process. As $\underline{x}$ and $\underline{y}$ are stochasticly independent, their *joint generalized PDF* is, cf. Eq. (15.2)

$$g_{xy} = g_x(x; \mu_x, 1)\, \delta(y - \mu_y).$$

Obviously, working with a mixture of Gaussians and δ-functions will be cumbersome in cases when stochastic variables are not independent.

Again, in most cases reasoning can done using the moments, therefore the complicated distribution is not of primary concern. The propagation of uncertainty using the second moments only relies on the covariance matrices, not on their inverses, and can be derived using the so-called moment generating function [30], which is also defined for generalized PDFs. Thus uncertainty propagation can also be performed in mixed cases.

Equivalence Relation for Uncertain Homogeneous Entities

A more critical problem is the equivalence relation of uncertain vectors. The equivalence of two fixed, i.e. statistically certain, homogeneous n-vectors $\mathbf{x}$ and $\mathbf{y}$ usually is represented as

$$\mathbf{x} \cong \mathbf{y} \iff \mathbf{x} = \lambda \mathbf{y} \tag{15.5}$$

for some factor $\lambda \in \mathbb{R} \setminus 0$. In case two stochastic n-vectors $\underline{\mathbf{x}}$ and $\underline{\mathbf{y}}$ are given with their PDF $p_x(\mathbf{x})$ and $p_y(\mathbf{y})$, the equivalence relation would transfer to

$$\underline{\mathbf{x}} \cong \underline{\mathbf{y}} \iff p_x(\mathbf{x}) = \frac{1}{\lambda^n} p_y\left(\frac{\mathbf{y}}{\lambda}\right) \tag{15.6}$$

for some factor $\lambda \in \mathbb{R} \setminus 0$.

This equivalence relation does not allow us to use regular covariances for homogeneous entities as they may occur. As an example, assume

$$\underline{\mathbf{p}} = \begin{bmatrix} x_1 \\ y_1 \\ 1 \end{bmatrix} \sim N(\boldsymbol{\mu}_p, \boldsymbol{\Sigma}_{pp}), \qquad \underline{\mathbf{q}} = \begin{bmatrix} x_2 \\ y_2 \\ 1 \end{bmatrix} \sim N(\boldsymbol{\mu}_q, \boldsymbol{\Sigma}_{qq}),$$

with

$$\boldsymbol{\Sigma}_{pp} = \boldsymbol{\Sigma}_{qq} = \sigma^2 \begin{bmatrix} 1 & 0 & 0 \\ 0 & 1 & 0 \\ 0 & 0 & 0 \end{bmatrix}.$$

Using Eq. (15.3), the covariance matrix of the joining line $\underline{\mathbf{l}} = \underline{\mathbf{p}} \times \underline{\mathbf{q}}$ is (cf. the discussion before Table 15.1)

$$\boldsymbol{\Sigma}_{ll} = \sigma^2 \begin{bmatrix} 2 & 0 & -(x_1 + x_2) \\ 0 & 2 & -(y_1 + y_2) \\ -(x_1 + x_2) & -(y_1 + y_2) & x_1^2 + x_2^2 + y_1^2 + y_2^2 \end{bmatrix},$$

and has determinant

$$|\boldsymbol{\Sigma}_{ll}| = 2\sigma^6 ((x_2 - x_1)^2 + (y_2 - y_1)^2).$$

Thus the covariance matrix of $\underline{\mathbf{l}}$ always has full rank. This is in contrast to the fact that, given the two parameters of a line, the resulting covariance matrix of its homogeneous vector has rank 2.

This conflict arises because the equivalence relation given by Eq. (15.6) does not allow the factor λ to be uncertain.

For example, if two 2-vectors $\underline{\mathbf{x}} = \underline{\mathbf{y}} \in \mathbb{P}^1$ follow a Gaussian distribution $p_x(\mathbf{x}) = p_y(\mathbf{y})$ with covariance matrix $\boldsymbol{\Sigma}_{xx} = \boldsymbol{\Sigma}_{yy}$, they certainly are equivalent. If now $\underline{\lambda}$ is 2 and 3 with probability $1/2$, then $\underline{\mathbf{z}} = \underline{\lambda}\,\underline{\mathbf{y}}$ follows a mixture of two equally probable Gaussians with $4\boldsymbol{\Sigma}_{yy}$ and $9\boldsymbol{\Sigma}_{yy}$; thus

$$p_z(\mathbf{y}) = \frac{1}{2}\cdot\frac{1}{4}\cdot p_y\left(\frac{\mathbf{y}}{2}\right) + \frac{1}{2}\cdot\frac{1}{9}\cdot p_y\left(\frac{\mathbf{y}}{3}\right) = \frac{1}{8}\,p_y\left(\frac{\mathbf{y}}{2}\right) + \frac{1}{18}\,p_y\left(\frac{\mathbf{y}}{3}\right),$$

shown in Fig. 15.5. This density function is not equivalent to $p_x(\mathbf{x})$ when using the equivalence relation given by Eq. (15.6). However, any realization

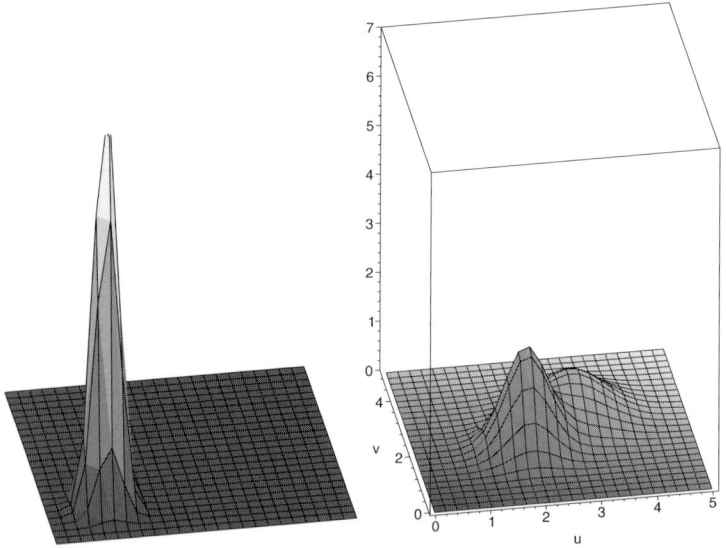

Fig. 15.5. Two PDF's of a homogeneous 2-vector. *Left:* original; *right:* modified. The two distributions represent the same 1D point $\underline{x} = \underline{u}/\underline{v}$ with $\boldsymbol{\mu}_x = (\mu_u, \mu_v)^\mathsf{T} = (1,1)^\mathsf{T}$. However, the distributions cannot be easily related; in particular, without further knowledge, the right distribution cannot be derived from the left one

comes either from $1/4\,p_y(\mathbf{y}/2)$ or from $1/9\,p_y(\mathbf{y}/3)$ and thus is equivalent to $\underline{\mathbf{x}}$. Therefore the equivalence relation in Eq. (15.6) is too restrictive.

We therefore propose to use the following equivalence relation for fixed vectors:

$$\boxed{\mathbf{x} \cong \mathbf{y} \iff \mathbf{x}^s = \mathbf{y}^s} \qquad (15.7)$$

with the spherically normalized vectors

$$\mathbf{x}^s = \frac{\mathbf{x}}{|\mathbf{x}|}, \qquad \mathbf{y}^s = \frac{\mathbf{y}}{|\mathbf{y}|}.$$

This equivalence relation can be directly transferred to uncertain vectors using the PDF of the normalized vectors

$$\boxed{\underline{\mathbf{x}} \cong \underline{\mathbf{y}} \iff p_{\underline{x}^s}(\mathbf{x}^s) = p_{\underline{y}^s}(\mathbf{y}^s)} \tag{15.8}$$

In case one wants to be sure to not deal with generalized functions, one also can use the equivalence relation based on the cumulative distributions

$$\underline{\mathbf{x}} \cong \underline{\mathbf{y}} \iff P_{\underline{x}^s}(\mathbf{x}^s) = P_{\underline{y}^s}(\mathbf{y}^s). \tag{15.9}$$

In the case of a normally distributed homogeneous vector $\underline{\mathbf{x}} \sim N(\boldsymbol{\mu}_x, \boldsymbol{\Sigma}_{xx})$ with a covariance matrix of arbitrary rank, one directly can determine the covariance matrix of the normalized vector from

$$\boldsymbol{\Sigma}_{x^s x^s} = J_s \boldsymbol{\Sigma}_{xx} J_s^\top$$

with the Jacobian

$$J_s = \frac{\partial \mathbf{x}^s}{\partial \mathbf{x}} = \frac{1}{|\mathbf{x}|}\left(I - \frac{\mathbf{x}\mathbf{x}^\top}{\mathbf{x}^\top \mathbf{x}}\right).$$

Obviously, the Jacobian has rank deficiency 1 and null space $\mathcal{N}(\boldsymbol{\Sigma}_{xx}) = \mathbf{x}$, therefore the covariance matrix at least has rank deficiency 1 and $\mathbf{x}$ is in its null space.

Thus the equivalence relations given by Eqs. (15.8) and (15.9) explicitely state that *only the direction* of a homogeneous vector is of concern, and if the PDF or CPDF of the direction is the same for two homogenous vectors, they are equivalent. This allows us to use covariance matrices of any rank to represent uncertain homogeneous vectors, cf. Fig. 15.6

Comment: There is a close relation of this equivalence relation to the equivalence relation for covariance matrices referring to different gauges (cf. [26]): In both cases only estimable quantities, i.e. quantities that are estimable, are of concern. The invariance here refers to arbitrary distributions not only to Gaussians, though the transformations discussed later all refer to the second moments. The notion of *gauge* is known as *datum* in the geodetic literature (cf. [2] and [12]). Gauge transformations, due to the special application, are called S-transformations (*S* from similarity, as in the application in [26]). The problem was identified 1987 in robotics [34].

Bias in Uncertainty Propagation

From Fig. 15.4 we observe three essential steps where we use an approximation: (1) generating a homogenous vector from Euclidean coordinates, (2) generating a homogenous vector from given ones, and (3) deriving Euclidean parameters for the geometric entity in concern.

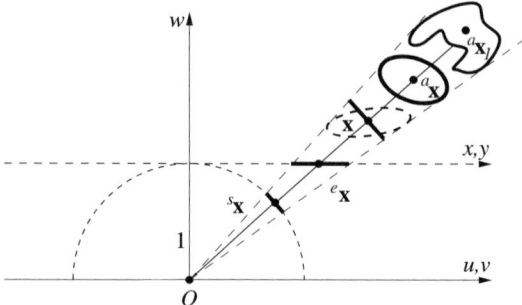

Fig. 15.6. Equivalent uncertain homogeneous vectors. The Euclidean plane $\mathbb{R}^2$ with Euclidean vectors (x, y) is embedded in $\mathbb{R}^3$ with coordinates $(u, v, w)^\mathsf{T}$. From *upper-right to center*: homogeneous vector $^a\mathbf{x}$ with general and regular covariance matrix; homogeneous vector $\mathbf{x}$ with equivalent covariance matrices, *dashed*: regular, *solid*: singular, normalized such that null space of Σ_{xx} is $\mathbf{x}$; Euclidean normalized homogeneous vector $^e\mathbf{x}$ with $w = 1$ and original covariance matrix from observations, null space $\Sigma_{e_x e_x}$ is $(0, 0, 1)^\mathsf{T}$, cf. Eq. (15.4) and [25]; spherically normalized homogeneous vector $^s\mathbf{x}$, null space of $\Sigma_{s_x s_x}$ is $\mathbf{x}$. In principle the distribution of a homogeneous vector may have any form (cf. point $\mathbf{x}_1^a$): it still is equivalent to another one, in the case after spherical normalization, they have the same distribution, as only the direction is of concern

We want to show that in most practical cases the induced bias in mean and variance is negligible and only in unlikely cases does the distribution of the resulting entities significantly deviate from a Gaussian. Though this has been discussed in literature (cf. e.g. [8], [9] or [30]) we present it here for completeness, adapted to the problem in concern.

Bias in mean and variance

The above-mentioned rule for propagating uncertainty given in Eq. (15.3) results from a Taylor expansion of the nonlinear function $\boldsymbol{y} = \boldsymbol{f}(\boldsymbol{x})$. Including higher-order terms yields bias terms.

For a scalar function in one variable we obtain the following result: if the PDF of a stochastic variable $\underline{x}$ is symmetrical, the mean and the variance for $y = f(x)$ can be shown to be

$$E(\underline{y}) = \mu_y = f(\mu_x) + \frac{1}{2}f''(\mu_x)\sigma_x^2 + \frac{1}{24}f^{(4)}\mu_{4_x} + O(f^{(n)}, m_n), \ n > 4 \quad (15.10)$$

for normally distributed variables, with the central fourth moment $\mu_{4_x} = 3\sigma_x^4$

$$V(\underline{y}) = \sigma_y^2$$
$$= f'^2(\mu_x)\sigma_x^2 + \left(f'(\mu_x)f'''(\mu_x) + \frac{1}{2}f''(\mu_x)\right)\sigma_x^4 + O(f^{(n)}, m_n), \ n > 4.$$

Obviously the bias, i.e. the second term, depends on the variance and the higher-order derivatives: the larger the variance and the higher the curvature or the third derivative, the higher the bias. Higher-order terms depend on derivatives and moments of order higher than 4.

If the PDF of a stochastic vector $\boldsymbol{x}$ is symmetrical, the mean of the scalar function $y = f(\boldsymbol{x})$ can be shown to be

$$E(\underline{y}) = \mu_y = f(\boldsymbol{\mu}_x) + \frac{1}{2}\text{trace}(H|_{x=\mu_x} \cdot \boldsymbol{\Sigma}_{xx}) + O(f^{(n)}, m_n), \quad n \geq 3, \quad (15.11)$$

with the Hessian matrix $H = (\partial f^2/\partial x_i \partial x_j)$ of the function $f(x)$. This is a generalization of Eq. (15.10).

We now want to discuss two cases: first, the product $\underline{z} = \underline{xy}$ of two random variables, which is the most simple case of a bilinear form, occurring when constructing new geometric elements with homogeneous coordinates, and second, normalizing a vector to unity $^s\mathbf{x} = \mathbf{x}/|\mathbf{x}|$.

Bias and distribution of the bilinear form $z = xy$

The Taylor series at the mean in this case is finite. Therefore we can derive rigorous expressions for the mean for arbitrary distribution M

$$\mu_z = E(\underline{z}) = \mu_x \mu_y + \sigma_{xy}. \quad (15.12)$$

As fourth moments are involved in the determination of the variance, we assume M to be normal. Then we obtain the rigorous expression for the variance of $\underline{z}$

$$\sigma_z^2 = V(\underline{z}) = \mu_y^2 \sigma_x^2 + \mu_x^2 \sigma_y^2 + 2\mu_x \mu_y \sigma_{xy} + \sigma_x^2 \sigma_y^2 + \sigma_{xy}^2. \quad (15.13)$$

Obviously, the linear approximation of the mean and the variance are

$$\mu_z^{(1)} = \mu_x \mu_y, \qquad \sigma_z^{2(1)} = \mu_x^2 \mu_y^2 + \mu_y^2 \sigma_x^2 + 2\mu_x \mu_y \sigma_{xy}. \quad (15.14)$$

The bias in mean is

$$b_{\mu_z} = \mu_z^{(1)} - \mu_z = -\sigma_{xy}. \quad (15.15)$$

It is zero if the two variables are uncorrelated. The bias in variance is

$$b_{\sigma_z^2} = \sigma_z^{2(1)} - \sigma_z^2 = -\sigma_x^2 \sigma_y^2 - \sigma_{xy}^2 = -\sigma_x^2 \sigma_y^2 (1 + \rho_{xy}^2). \quad (15.16)$$

It is *not* zero for uncorrelated variables. Actually, the variance is *underestimated* if one relies on classical error propagation, as $\sigma_z^{2(1)} < \sigma_z^2$ for uncorrelated variables [16].

In order to get an impression of the size we assume $\sigma_x = \sigma_y = \sigma$ and $\sigma_{xy} = 0$, and obtain the relative bias in variance

$$r_{\sigma_z^2} = \frac{b_{\sigma_z^2}}{\sigma_z^2} = -\frac{\sigma^2}{\mu_x^2 + \mu_y^2 + \sigma^2}. \quad (15.17)$$

Thus only in the case $\mu_x^2 + \mu_y^2 < \sigma^2$ is the relative bias in variance larger than 50% of the variance. This is very unlikely to occur, as the relative precision σ_x/μ_x of homogeneous vectors in computer vision applications is usually better than $1/100$.

We finally want to show the type distribution of the product for an extreme case, especially for $\mu_x = \mu_y = 0$. In the case of independent zero-mean Gaussian variables

$$\underline{x} \sim N(0, \sigma^2), \qquad \underline{y} \sim N(0, \sigma^2),$$

we find the probability density function of $\underline{z}$ from

$$p_z(z) = \int_0^\infty \frac{2}{u} p_x(u) p_y\left(\frac{z}{u}\right) du,$$

yielding

$$p_z(z) = \frac{\mathrm{BesselK}\left(0, \dfrac{|z|}{\sigma^2}\right)}{\pi \sigma^2},$$

with the Bessel function $\mathrm{BesselK}(v, x)$ of the second kind. It definitely is not normally distributed (cf. Fig. 15.7), but has the variance

$$V(\underline{z}) = 2 \int_{z=0}^\infty z^2 p_z(z) \mathrm{d}z = \sigma^4,$$

in accordance with Eq. (15.13).

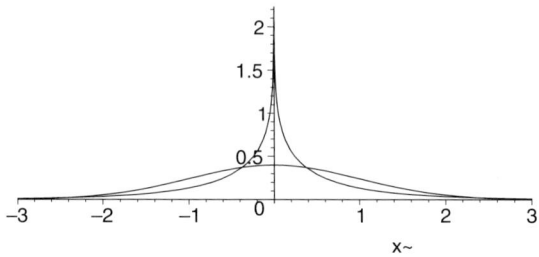

Fig. 15.7. Bessel function of the second kind and Gaussian with the same variance. The PDF of the product of two zero–mean Gaussian random variables is a Bessel function, thus it significantly deviates from a Gaussian PDF.

Bias of spherical normalization

The findings are confirmed when analyzing the bias of normalization of a vector. Let the vector $\underline{\mathbf{w}} \sim N(\boldsymbol{\mu}_w, \sigma_w^2 I)$ be normally distributed with independent components with the same variance $\sigma_{w_i} = \sigma_w$. The normalized vector is determined from

$$\underline{z} = \frac{\boldsymbol{w}}{|\boldsymbol{w}|}, \quad \text{or} \quad \underline{z}_i = \frac{w_i}{|\boldsymbol{w}|} .$$

With Eq. (15.11) this leads to the second-order approximation of the mean

$$E(\underline{z}) = \frac{\boldsymbol{\mu}_w}{|\boldsymbol{\mu}_w|} - \frac{1}{2} \frac{\boldsymbol{\mu}_w}{|\boldsymbol{\mu}_w|^3} \sigma_w^2 = \frac{\boldsymbol{\mu}_w}{|\boldsymbol{\mu}_w|} \left(1 - \frac{1}{2} \frac{\sigma_w^2}{|\boldsymbol{\mu}_w|^2}\right) .$$

Thus the relative bias, i.e. the bias related to the standard deviation is

$$\frac{|b_{\mu_z}|}{\sigma_{\mu_z}} = 0.7 \frac{\sigma_w}{|\boldsymbol{\mu}_w|},$$

and is approximately identical to the directional error $\sigma_w/|\boldsymbol{\mu}_w|$, which usually is below $1/100$. One can also show that the relative bias in the standard deviation is approximately identical to the directional error

$$\frac{|b_{\sigma_z}|}{\sigma_{\sigma_z}} \sim \frac{\sigma_w}{|\boldsymbol{\mu}_w|} .$$

These findings are useful for the constructions, the tests and the estimation procedures discussed below. The strong bias, found by Haddon and Forsyth [16], refers to partitioning bilinear forms with mean values close to zero. There the relative errors are large; thus the bias cannot be neglected.

15.2.2 Estimation

Estimation of unknown parameters is a classical task in statistical inference. Maximum-likelihood estimation is one of the standard tools.

In our context, many relations are multilinear, but they are linear in the unknown parameters. Therefore direct solutions minimizing the so-called algebraic error are frequently used. Though they usually lead to sufficiently good approximate values, they are suboptimal. As a result, techniques for improving these approximations have been proposed, such as renormalization [25] and [5], total least squares, the heteroscedastic errors in variables (HEIV) method, meaning variables with errors of different weight [28] and their improvements [6], [7]. The motivation for these techniques stems from deficiencies in historically older methods: renormalization compensates for bias, total least squares takes the stochastic properties of the coefficients in a regression model into account, and HEIV takes the correct weights in the algebraic minimization scheme. All of them are iterative.

We propose to directly use the given constraints and minimize the weighted sum of residuals of the observations, the weights being inversely proportional to the covariance matrix. The resulting estimates are local maximum likelihood estimates in case the given observations are Gaussian. We will show that the HEIV method, in a modified version, is a special case, and furthermore, the method of minimizing the algebraic error is a further simplification.

The proposed scheme has two definite advantages:

- It can handle *any number of constraints at the same time*. In our context, this allows us to simultaneously take into account the fixed length of a homogeneous vector or matrix and additional constraints, such as the singularity of the fundamental matrix or the Plücker constraint for space lines. Also, problems such as space curve fitting for observed space points $(x, y, z)_l, l = 1, ...L$, can easily be handled, e.g. using a space conic, being the intersection of a vertical conical cylinder with the six parameters $a_{ij}, 0 \leq i+j \leq 2$ and a plane with four parameters $b_{ijk}, 0 \leq i+j+k \leq 1$ via

$$\sum_{0 \leq i+j \leq 2} a_{ij} x_l^i y_l^j = 0, \qquad \sum_{0 \leq i+j+k \leq 1} b_{ijk} x_l^i y_l^j z_l^k = 0,$$

$$\sum_{i+j \leq 2} a_{ij}^2 = 1, \qquad \sum_{i+j+k \leq 1} b_{ijk}^2 = 1,$$

including two constraints, guaranteeing the space conic to be parameterized by eight parameters. The proposed estimation method covers the technique in [36] as special case.

- It can *handle groups of mutually correlated observations*. In many cases this might not be an issue. However, assume space points are derived from a pair of images by correspondence analysis, including relative orientation and triangulation: then the coordinates of the space points are mutually correlated due to the common relative orientation, e.g. represented by an estimated fundamental matrix $\widehat{\mathsf{F}}$. In case these points are used for further processing, e.g. surface or curve fitting, one may take the mutual correlations of the space coordinates of *all* points into account.

Gauss–Helmert Model

The used *mathematical model* for the estimation may be partitioned into a *functional model* and a *stochastic model*. The *functional model*, the so-called Gauss–Helmert model, proposed by Helmert in 1872 [20], with constraints between the unknown parameters, starts from G constraints $\boldsymbol{g} = (g_g)$ among N observations $\boldsymbol{l} = (l_n)$ and U unknown parameters $\boldsymbol{x} = (x_u)$ with additional H constraints $\boldsymbol{h} = (h_h)$ among the unknowns. The constraints should hold for the fitted observations $\widehat{\boldsymbol{l}} = \boldsymbol{l} + \widehat{\boldsymbol{v}}$, including the estimated corrections $\widehat{\boldsymbol{v}}$ and the estimated parameters $\widehat{\boldsymbol{x}}$:

$$\boldsymbol{g}(\boldsymbol{l} + \widehat{\boldsymbol{v}}, \widehat{\boldsymbol{x}}) = \boldsymbol{0}, \quad \text{or} \quad \boldsymbol{g}(\widehat{\boldsymbol{l}}, \widehat{\boldsymbol{x}}) = \boldsymbol{0}, \tag{15.18}$$
$$\boldsymbol{h}(\widehat{\boldsymbol{x}}) = \boldsymbol{0}. \tag{15.19}$$

Starting from approximate values $\widehat{\boldsymbol{x}}^{(0)}$ and $\widehat{\boldsymbol{l}}^{(0)}$ by Taylor expansion and neglection of terms of order higher than 2, one obtains the linear Gauss–Helmert model with constraints between the unknown parameters

$$A\widehat{\Delta x} + B^\mathsf{T}\widehat{v} = w_g\,, \qquad H^\mathsf{T}\widehat{\Delta x} = w_h\,, \tag{15.20}$$

with the residuals of the constraints and the corrections of the unknown parameters

$$w_g = -g(\widehat{l}^{(0)}, \widehat{x}^{(0)}) - B^\mathsf{T}(l - \widehat{l}^{(0)})\,, \quad w_h = -h(\widehat{x}^{(0)})\,, \tag{15.21}$$
$$\widehat{\Delta x} = x - \widehat{x}^{(0)}\,,$$

and the Jacobians evaluated at the approximate values

$$\underset{G\times U}{A} = \left.\frac{\partial g(l,x)}{\partial x}\right|_{\substack{l = l^{(0)} \\ x = x^{(0)}}}, \qquad \underset{N\times G}{B} = \left.\left(\frac{\partial g(l,x)}{\partial l}\right)^\mathsf{T}\right|_{\substack{l = l^{(0)} \\ x = x^{(0)}}},$$

$$\underset{U\times H}{H} = \left.\left(\frac{\partial h(x)}{\partial x}\right)^\mathsf{T}\right|_{x=x^{(0)}}.$$

The matrices B and H are defined as transposed of the Jacobians, to ensure they have more rows than columns.

The *stochastic model* is assumed to be simple: we assume an initial covariance matrix $\Sigma_{ll}^{(0)}$ of the observations l to be known. It may be singular. We assume the true covariance matrix to be

$$\Sigma_{ll} = \sigma_0^2 \Sigma_{ll}^{(0)}\,.$$

We start the estimation with the initial value $\sigma_0^{2(0)} = 1$, thus start with the approximation $\Sigma_{ll} = \Sigma_{ll}^{(0)}$. Generalizing the stochastic model is straight forward, e.g. by assuming variance components [13, 27]

$$\Sigma_{ll} = \sum_k \sigma_{0k}^2 \Sigma_{ll,k}^{(0)}\,.$$

Then, the variance factors σ_{0k}^2 need to be estimated simultaneously with the unknown parameters x.

Maximum Likelihood Estimation

We now give explicit expressions for the estimated parameters, the covariance matrices of the parameters, the corrections and the fitted observations and the estimated variance factor. We derive the locally best linear estimators, i.e. estimators having the smallest variance in the linearized models. Moreover, in case the observations are samples from a normal distribution with $\underline{l} \sim N(\widetilde{l}, \Sigma_{ll})$, the estimates are local maximum likelihood estimators, local, as they depend on the approximate values, and far-off the global optimum might exist.

Minimizing the quadratic form

$$\Omega = (\widehat{l} - l)^\mathsf{T} \Sigma_{ll}^+ (\widehat{l} - l), \tag{15.22}$$

under the constraints $A\widehat{\Delta x} + B^\mathsf{T}(\widehat{l} - l) = w_g$ and $H^\mathsf{T}\widehat{\Delta x} = w_h$, we have to minimize the form

$$\Phi = (\widehat{l} - l)^\mathsf{T} \Sigma_{ll}^+ (\widehat{l} - l) + 2\lambda^\mathsf{T}(A\widehat{\Delta x} + B^\mathsf{T}(\widehat{l} - l) - w_g) + 2\mu^\mathsf{T}(H^\mathsf{T}\widehat{\Delta x} - w_h),$$

where λ and μ are Lagrangian multipliers. In case the covariance matrix Σ_{ll} of the observations is singular, one needs to take its pseudoinverse. As Σ_{ll}^+ is positive semidefinite and the constraints are linear we obtain a unique minimum.

Setting the partials of Φ zero, we obtain with $\widehat{v} = \widehat{l} - l$

$$\frac{1}{2}\frac{\partial \Phi}{\partial \widehat{l}^\mathsf{T}} = \Sigma_{ll}^+ \widehat{v} + B\lambda = 0, \tag{15.23}$$

$$\frac{1}{2}\frac{\partial \Phi}{\partial \widehat{x}^\mathsf{T}} = A^\mathsf{T}\lambda + H\mu = 0, \tag{15.24}$$

$$\frac{1}{2}\frac{\partial \Phi}{\partial \lambda^\mathsf{T}} = -w_g + A\widehat{\Delta x} + B^\mathsf{T}\widehat{v} = 0, \tag{15.25}$$

$$\frac{1}{2}\frac{\partial \Phi}{\partial \mu^\mathsf{T}} = -w_h + H^\mathsf{T}\widehat{\Delta x} = 0. \tag{15.26}$$

From Eq. (15.25) follows the relation

$$\widehat{v} = -\Sigma_{ll} B\lambda. \tag{15.27}$$

When substituting Eq. (15.27) into Eq. (15.25), solving for λ yields

$$\lambda = (B^\mathsf{T} \Sigma_{ll} B)^{-1}(A\widehat{\Delta x} - w_g). \tag{15.28}$$

Substitution in Eq. (15.26) yields the symmetric normal equation system

$$\begin{bmatrix} A^\mathsf{T}(B^\mathsf{T}\Sigma_{ll}B)^{-1}A & H \\ H^\mathsf{T} & 0 \end{bmatrix} \begin{bmatrix} \widehat{\Delta x} \\ \mu \end{bmatrix} = \begin{bmatrix} A^\mathsf{T}(B^\mathsf{T}\Sigma_{ll}B)^{-1}w_g \\ w_h \end{bmatrix}. \tag{15.29}$$

The Lagrangian multipliers can be obtained from Eq. (15.28) which then yields the estimated residuals in Eq. (15.27).

The estimated variance factor is given by

$$\widehat{\sigma}_0^2 = \frac{\widehat{v}^\mathsf{T} \Sigma_{ll}^+ \widehat{v}}{R} \sim F_{R,\infty}, \tag{15.30}$$

with the redundancy

$$R = G + H - U \,. \tag{15.31}$$

The redundancy R is the difference of the number $G + H$ of constraints and the number U of unknown parameters. In case of normally distributed observations $\boldsymbol{l}$ and in case the model holds, the estimated variance factor is Fisher distributed with (R, ∞) degrees of freedom. Thus $\widehat{\sigma_0}^2$ may be used to check the validity of the model.

We finally obtain the *estimated* covariance matrix

$$\widehat{\boldsymbol{\Sigma}}_{\widehat{x}\widehat{x}} = \widehat{\sigma}_0^2 \boldsymbol{\Sigma}_{\widehat{x}\widehat{x}} \tag{15.32}$$

of the estimated parameters, where $\boldsymbol{\Sigma}_{\widehat{x}\widehat{x}}$ results from the inverted reduced normal equation matrix

$$\begin{bmatrix} \boldsymbol{\Sigma}_{\widehat{x}\widehat{x}} & \boldsymbol{S} \\ \boldsymbol{S}^{\mathsf{T}} & \boldsymbol{T} \end{bmatrix} = \begin{bmatrix} \overline{\boldsymbol{N}} & \boldsymbol{H} \\ \boldsymbol{H}^{\mathsf{T}} & \boldsymbol{0} \end{bmatrix}^{-1} \tag{15.33}$$

using

$$\overline{\boldsymbol{N}} = \boldsymbol{A}^{\mathsf{T}} (\boldsymbol{B}^{\mathsf{T}} \boldsymbol{\Sigma}_{ll} \boldsymbol{B})^{-1} \boldsymbol{A} \,.$$

Equation (15.33) can be used even if $\overline{\boldsymbol{N}}$ is singular. The covariance matrix $\boldsymbol{\Sigma}_{\widehat{x}\widehat{x}}$ has null space $\boldsymbol{H}$.

The estimation needs to be iterated using improved approximate values in the next, say the $(\nu+1)$th, iteration

$$\widehat{\boldsymbol{x}}^{(\nu+1)} = \boldsymbol{x}^{(\nu)} + \widehat{\boldsymbol{\Delta x}}^{(\nu)}, \qquad \widehat{\boldsymbol{l}}^{(\nu+1)} = \widehat{\boldsymbol{l}}^{(\nu)} + \widehat{\boldsymbol{v}}^{(\nu)},$$

from Eqs. (15.21) and (15.27). This requires recomputation of the Jacobians $\boldsymbol{A}$, $\boldsymbol{B}$ and $\boldsymbol{H}$.

We now discuss various specializations of these relations leading to well-known statistical tools.

The Special Case of Implicit Error Propagation

Observe, in case the number G of independent constraints $\boldsymbol{g}(\boldsymbol{l}, \boldsymbol{x}) = \boldsymbol{0}$ and the number U of unknowns is identical, and there are no other constraints, the redundancy R is zero, the matrices $\boldsymbol{A}$ and $\boldsymbol{B}$ are of size $U \times U$, and the covariance matrix of the unknown parameters is

$$\boldsymbol{\Sigma}_{\widehat{x}\widehat{x}} = \boldsymbol{A}^{-1} \boldsymbol{B}^{\mathsf{T}} \boldsymbol{\Sigma}_{ll} \boldsymbol{B} \boldsymbol{A}^{-\mathsf{T}},$$

which is the implicit error propagation law. Observe the definition of $\boldsymbol{B}$ as the transpose of the Jacobian of $\boldsymbol{g}$ w.r.t. the observations $\boldsymbol{l}$.

Relation to the HEIV Method

Up to this point all given derivations are well known. We now want to assume the constraints g to be linear in the unknown parameters, and only the constraint h on the length of the unknown parameter vector should hold. Thus we have the model

$$g(\widehat{l}, \widehat{x}) = A(\widehat{l})\widehat{x} = 0, \qquad h = \frac{1}{2}(\widehat{x}^\mathsf{T}\widehat{x} - 1).$$

This leads to $H = \widehat{x}$. In the case of convergence we have $\widehat{\Delta x} = 0$ and $w_g = A(l)\widehat{x}$, and therefore the first equation of (15.29) leads to the iteration sequence [14]

$$\mu \cdot \widehat{x}^{(\nu)} = A^\mathsf{T}\left(\widehat{l}^{(\nu-1)}\right) \left(B\left(\widehat{x}^{(\nu-1)}\right) \Sigma_{ll} B^\mathsf{T}\left(\widehat{x}^{(\nu-1)}\right)\right)^{-1} A(l) \cdot \widehat{x}^{(\nu)}. \quad (15.34)$$

This shows the unknown parameter vector to be an eigenvector of an asymmetric matrix. The Jacobian B is to be evaluated at the fitted values $\widehat{x}^{(\nu-1)}$, causing the iteration process. This method is equivalent to Matei and Meer's HEIV method [28]. In case of additional constraints, such as the singularity of the fundamental matrix, the authors propose to impose these constraints in a second step.

Imposing Constraints Onto a Stochastic Vector

Imposing a set of constraints onto a stochastic vector, already tackled by Helmert in 1872 [20], is a special case of the Gauss–Helmert model mentioned above and is used for normalizing a vector or imposing additional constraints onto the vector, such as the Plücker constraint for space lines or the singularity constraint for the fundamental matrix.

The stochastic vector is treated as observational vector $l \sim M(\widetilde{l}, \Sigma_{ll})$. Then we only have the G constraints

$$g(\widehat{l}) = 0,$$

as no unknowns are involved. The resulting fitted observations $\widehat{l}$ can be derived iteratively from Eqs. (15.27), (15.28) and (15.21)

$$\widehat{l}^{(\nu+1)} = l + \widehat{v}^{(\nu)} = l - \Sigma_{ll} B (B^\mathsf{T} \Sigma_{ll} B)^{-1} (g(\widehat{l}^{(\nu)}) + B^\mathsf{T}(l - \widehat{l}^{(\nu)})),$$

or in the case of linear constraints from

$$\widehat{l} = l + \widehat{v} = l - \Sigma_{ll} B (B^\mathsf{T} \Sigma_{ll} B)^{-1} g(l).$$

The covariance matrix of the fitted observations $\widehat{l}$ is

$$\boldsymbol{\Sigma}_{\widehat{ll}} = \boldsymbol{P}_B^{\mathsf{T}} \boldsymbol{\Sigma}_{ll} \boldsymbol{P}_B = \boldsymbol{\Sigma}_{ll} - \boldsymbol{\Sigma}_{ll} \boldsymbol{B} (\boldsymbol{B}^{\mathsf{T}} \boldsymbol{\Sigma}_{ll} \boldsymbol{B})^{-1} \boldsymbol{B}^{\mathsf{T}} \boldsymbol{\Sigma}_{ll} = \boldsymbol{\Sigma}_{ll} \boldsymbol{P}_B, \quad (15.35)$$

with the $N-G$ rank projection matrix

$$\boldsymbol{P}_B = \boldsymbol{I} - \boldsymbol{B}(\boldsymbol{B}^{\mathsf{T}} \boldsymbol{\Sigma}_{ll} \boldsymbol{B})^{-1} \boldsymbol{B}^{\mathsf{T}} \boldsymbol{\Sigma}_{ll}$$

fulfilling $\boldsymbol{P}_B \boldsymbol{B} = \boldsymbol{0}$, leading to a singular covariance matrix $\boldsymbol{\Sigma}_{\widehat{ll}}$ with nullspace $\boldsymbol{B}$. This is an example – demonstrating the generality of the Gauss–Helmert model.

Minimizing the Algebraic Error

Minimizing the algebraic error $\boldsymbol{w}_g = \boldsymbol{A}(\boldsymbol{l})\widehat{\boldsymbol{x}}$ under the constraint $|\widehat{\boldsymbol{x}}| = 1$ leads to the simple eigenvalue problem

$$\mu \cdot \widehat{\boldsymbol{x}} = \boldsymbol{A}^{\mathsf{T}}(\boldsymbol{l}) \boldsymbol{A}(\boldsymbol{l}) \cdot \widehat{\boldsymbol{x}},$$

demonstrating the neglection of the weighting matrix $(\boldsymbol{B} \boldsymbol{\Sigma}_{ll} \boldsymbol{B}^{\mathsf{T}})^{-1}$, when compared to the rigorous solution.

15.2.3 Testing

We only need very little from testing theory, namely testing a vector to be zero. Let the observed n-vector be $\boldsymbol{c}$. We want to test the null hypothesis $H_0 : \boldsymbol{\mu}_c = \boldsymbol{0}$, that the mean $\boldsymbol{\mu}_c$ of the vector $\underline{\boldsymbol{c}}$ is zero against the alternative hypothesis $H_a : \boldsymbol{\mu}_c \neq \boldsymbol{0}$ that the mean is not zero. For the test we need the distribution of $\underline{\boldsymbol{c}}|H_0$ of the vector, provided the hypothesis H_0 holds. In case we can assume $\underline{\boldsymbol{c}}|H_0 \sim N(\boldsymbol{0}, \boldsymbol{\Sigma}_{cc})$ with a full-rank covariance matrix, we obtain the test optimal statistic [27]

$$T = \boldsymbol{c}^{\mathsf{T}} \boldsymbol{\Sigma}_{cc}^{-1} \boldsymbol{c} \sim \chi_d^2.$$

If $\text{rank}(\boldsymbol{\Sigma}_{cc}) = r \leq d$ then we obtain the test statistic

$$T = \boldsymbol{c}^{\mathsf{T}} \boldsymbol{\Sigma}_{cc}^{+} \boldsymbol{c} \sim \chi_r^2.$$

The test compares the test statistic T with a critical value. Specifying a (smaller) significance number α or a (large) significance level $S = 1 - \alpha$ one uses the $1 - \alpha$ percentile of the distribution of the test statistic as critical value. If

$$T > \chi_{r,1-\alpha}^2,$$

then we may reject H_0. Thus there is reason to assume that the difference of $\boldsymbol{c}$ from $\boldsymbol{0}$ cannot be explained by random errors, leading to deviations of $\boldsymbol{c}$ from $\boldsymbol{0}$. Otherwise, H_0 cannot be rejected, i.e. there is no reason to assume H_0 to be incorrect. This does not say H_0 is accepted, as other hypotheses H_{0i} might be valid, which are not tested for.

15.3 Geometric Relations

15.3.1 Representations

We represent all geometric entities and transformations with homogeneous vectors or matrices.

Points x and lines l in 2D are 3-vectors, given respectively by

$$\mathbf{x} = \begin{bmatrix} u \\ v \\ w \end{bmatrix} = \begin{bmatrix} x_1 \\ x_2 \\ x_3 \end{bmatrix} = \begin{bmatrix} \boldsymbol{x}_0 \\ x_h \end{bmatrix}, \quad \text{and} \quad \mathbf{l} = \begin{bmatrix} a \\ b \\ c \end{bmatrix} = \begin{bmatrix} l_1 \\ l_2 \\ l_3 \end{bmatrix} = \begin{bmatrix} \boldsymbol{l}_h \\ l_0 \end{bmatrix}.$$

Sometimes it is useful to distinguish the Euclidean part, indexed 0, and the homogeneous part, indexed h, of a vector or a matrix [3]: in case the homogeneous part is normalized to 1, the Euclidean part can be interpreted metrically. In case the homogeneous part of a geometric entity is zero, the entity is at infinity.

Analogously, in 3D points $\mathbf{X}$ and planes $\mathbf{A}$ are represented with 4-vectors

$$\mathbf{X} = \begin{bmatrix} U \\ V \\ W \\ T \end{bmatrix} = \begin{bmatrix} X_1 \\ X_2 \\ X_3 \\ X_4 \end{bmatrix} = \begin{bmatrix} \boldsymbol{X}_0 \\ X_h \end{bmatrix}, \quad \text{and} \quad \mathbf{A} = \begin{bmatrix} A \\ B \\ C \\ D \end{bmatrix} = \begin{bmatrix} A_1 \\ A_2 \\ A_3 \\ A_4 \end{bmatrix} = \begin{bmatrix} \boldsymbol{A}_h \\ A_0 \end{bmatrix}.$$

Furthermore, 3D lines $\mathbf{L}$ are represented with so-called Plücker coordinates

$$\mathbf{L} = [L_i] = \begin{bmatrix} \boldsymbol{L}_h \\ \boldsymbol{L}_0 \end{bmatrix},$$

where the first 3-vector $\boldsymbol{L}_h$ is the direction of the line, and the second 3-vector $\boldsymbol{L}_0$ is the normal on the plane through the line and the origin, such that the three vectors $\boldsymbol{L}_h$, $\boldsymbol{L}_0$ and the normal from the origin onto the 3D line span a right-handed orthogonal coordinate system. Any 6-vector fulfilling the so-called Plücker constraint

$$\boldsymbol{L}_h^\mathsf{T} \boldsymbol{L}_0 = 0$$

represents a 3D line.

Each geometric element $\mathbf{g}$ has a dual, denoted by $\overline{\mathbf{g}}$. Vice versa, $\mathbf{g}$ is the dual of $\overline{\mathbf{g}}$; thus $\mathbf{g} = \overline{\overline{\mathbf{g}}}$.

In 2D the dual $\overline{\mathbf{x}}$ of a point $\mathbf{x}$ is the line $[u, v, w]^\mathsf{T}$ with the same coordinates, and vice versa, the dual $\overline{\mathbf{l}}$ of a line $\mathbf{l}$ is the point $[a, b, c]^\mathsf{T}$ with the same coordinates:

$$\overline{\mathbf{x}} = I_3\, \mathbf{x}, \quad \overline{\mathbf{l}} = I_3\, \mathbf{l}.$$

In 3D the dual $\overline{\mathbf{X}}$ of a point $\mathbf{X}$ is the plane $[U, V, W, T]^\mathsf{T}$ with the same coordinates, and vice versa, the dual $\overline{\mathbf{A}}$ of a plane $\mathbf{A}$ is the point $[A, B, C, D]^\mathsf{T}$ with the same coordinates:

$$\overline{\mathbf{X}} = I_4\,\mathbf{X}, \qquad \overline{\mathbf{A}} = I_4\,\mathbf{A}.$$

The dual $\overline{\mathbf{L}}$ of a 3D line $\mathbf{L} = (\boldsymbol{L}_h^\mathsf{T}, \boldsymbol{L}_0)^\mathsf{T}$ is the 3D line $(\boldsymbol{L}_0^\mathsf{T}, \boldsymbol{L}_h^\mathsf{T})^\mathsf{T}$ with the homogeneous and the Euclidean parts exchanged:

$$\overline{\mathbf{L}} = D_6\,\mathbf{L} = \begin{bmatrix} \boldsymbol{L}_0 \\ \boldsymbol{L}_h \end{bmatrix},$$

with the dualizing matrix

$$D_6 = \begin{bmatrix} \mathbf{0} & I_3 \\ I_3 & \mathbf{0} \end{bmatrix}.$$

This representation of 3D entities is consistent with the geometric algebra G_4 [22] when using the bases (e_1, e_2, e_3, e_4) for 3D points, $(e_{41}, e_{42}, e_{43}, e_{23}, e_{31}, e_{12})$ for 3D lines and $(e_{234}, e_{314}, e_{124}, e_{321})$ for planes, which can be easily verified with the geometric algebra package of Ashdown [1].

Representation and Visualization of the Uncertainty of Geometric Entities

All homogenous vectors and matrices involved will get covariance matrices attached to them. Thus we obtain the pairs

$$(\mathbf{x}, \boldsymbol{\Sigma}_{xx}), \quad (\mathbf{l}, \boldsymbol{\Sigma}_{ll}), \quad (\mathbf{X}, \boldsymbol{\Sigma}_{XX}), \quad (\mathbf{L}, \boldsymbol{\Sigma}_{LL}), \quad (\mathbf{A}, \boldsymbol{\Sigma}_{XX}), \qquad (15.36)$$

for points and lines in 2D and for points, lines and planes in 3D. We will later also transfer this representation to transformation matrices.

The uncertainty of the geometric entities can be visualized by the confidence regions. Fig. 15.3 shows confidence regions for 2D points, which are ellipses, and for 2D lines, which are hyperbolae, namely the set of all one-dimensional confidence regions of points sitting on the line. They directly transfer to 3D points, being ellipsoids, and planes, being hypeboloids of two sheets, cf. Fig. 15.8. The situation is more complicated for 3D lines. The set of confidence ellipses of the 3D points sitting on the 3D line, measured across the line, yields a shape as in Fig. 15.9 left. It has different minima in different planes through the line, and thus in general is not a hyperboloid of one sheet, but is closely related to the ray configuration of an astigmatism.

15.3.2 Constructions

Constructions in 2D

Geometric entities easily can be constructed from given ones.
(1) A 2D line $\mathbf{l}$ joining two points $\mathbf{x}$ and $\mathbf{y}$ is given by

$$\boxed{\mathbf{l} = \mathbf{x} \wedge \mathbf{y} = -\mathbf{y} \wedge \mathbf{x}: \quad \mathbf{l} = \mathbf{x} \times \mathbf{y} = \mathsf{S}(\mathbf{x})\mathbf{y} = -\mathsf{S}(\mathbf{y})\mathbf{x}.} \qquad (15.37)$$

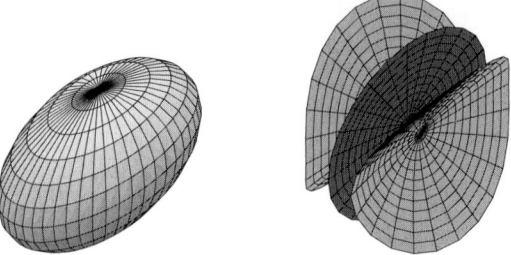

Fig. 15.8. Confidence regions for a 3D point and a plane. The hyperboloid of two sheets is the set of the 1D confidence regions of all points in the plane measured across the plane

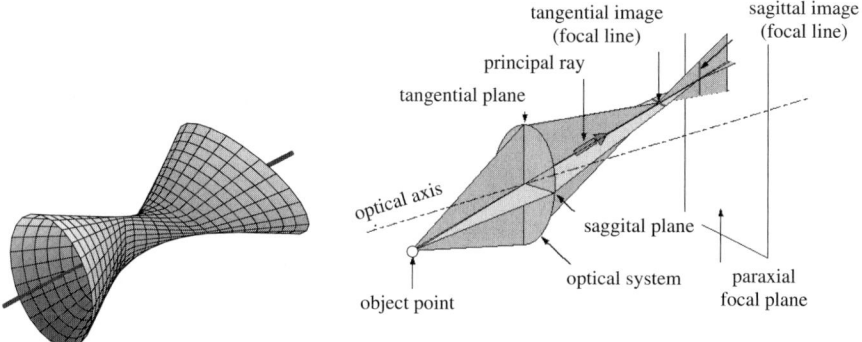

Fig. 15.9. Confidence region for a 3D line (*left*): it is the set of the 2D confidence regions of all points on the 3D line measured across the line. Compare the structure of the isosurface with the structure of the bundle of rays for astigmatism (*right*, after http://www.mellesgriot.com/): in cases where the 3D line only is uncertain in direction, in general there are two points, where the elliptic confidence region degenerates to a straight line segment, corresponding to the two focal lines of an astigmatism. However, the straight line segments need not be perpendicular, whereas the two focal lines are

Herein, we used the Jacobians

$$\mathsf{S}(\mathbf{x}) = \frac{\partial(\mathbf{x} \times \mathbf{y})}{\partial \mathbf{y}}, \quad \text{and} \quad \mathsf{S}(\mathbf{y}) = \frac{\partial(\mathbf{y} \times \mathbf{x})}{\partial \mathbf{x}},$$

and the skew symmetric matrices of a 3-vector, denoted by

$$\mathsf{S}(\mathbf{x}) = \mathsf{S}_x = [\mathbf{x}]_\times = \begin{bmatrix} 0 & -x_3 & x_2 \\ x_3 & 0 & -x_1 \\ -x_2 & x_1 & 0 \end{bmatrix}. \tag{15.38}$$

This representation of constructions as matrix-vector products is given explicitely for the most important constructions in 2D and 3D and is very useful for statistical error propagation.

(2) The 2D point **x** being the intersection of two lines **l** and **m** is given by

$$\mathbf{x} = \mathbf{l} \cap \mathbf{m} = -\mathbf{m} \cap \mathbf{l} : \quad \mathbf{x} = \mathbf{l} \times \mathbf{m} = \mathsf{S}(\mathbf{l})\mathbf{m} = -\mathsf{S}(\mathbf{m})\mathbf{l}. \tag{15.39}$$

Here the Jacobians are

$$\mathsf{S}(\mathbf{l}) = \frac{\partial(\mathbf{l} \times \mathbf{m})}{\partial \mathbf{m}}, \quad \text{and} \quad \mathsf{S}(\mathbf{m}) = \frac{\partial(\mathbf{m} \times \mathbf{l})}{\partial \mathbf{l}}.$$

The constructions are collected in Table 15.1

As the expressions for constructions are bilinear, we easily can find the covariance matrix of the generated elements. Therefore, we always give the two expressions

$$\mathbf{c} = \mathsf{U}(\mathbf{a})\mathbf{b} = \mathsf{V}(\mathbf{b})\mathbf{a} \tag{15.40}$$

for the bilinear form, where the matrices

$$\mathsf{U}(\mathbf{a}) = \frac{\partial \mathbf{c}}{\partial \mathbf{b}}, \quad \text{and} \quad \mathsf{V}(\mathbf{b}) = \frac{\partial \mathbf{c}}{\partial \mathbf{a}}$$

are the Jacobians of **c** with respect to **b** and **a**, resp. They are used to determine a first-order approximation of the covariance matrix Σ_{cc} of $\underline{\mathbf{c}}$, in case $\underline{\mathbf{a}}$ and $\underline{\mathbf{b}}$ are given together with their covariance matrices Σ_{aa} and Σ_{bb}:

$$\Sigma_{cc} = [\mathsf{V}(\boldsymbol{b}) \ \mathsf{U}(\boldsymbol{a})] \begin{bmatrix} \Sigma_{aa} & \Sigma_{ab} \\ \Sigma_{ba} & \Sigma_{bb} \end{bmatrix} \begin{bmatrix} \mathsf{V}^\mathsf{T}(\boldsymbol{b}) \\ \mathsf{U}^\mathsf{T}(\boldsymbol{a}) \end{bmatrix}, \tag{15.41}$$

or

$$\Sigma_{cc} = \mathsf{U}(\mathbf{a})\Sigma_{bb}\mathsf{U}(\mathbf{a})^\mathsf{T} + \mathsf{V}(\mathbf{b})\Sigma_{aa}\mathsf{V}(\mathbf{b})^\mathsf{T} + \mathsf{U}(\mathbf{a})\Sigma_{ba}\mathsf{V}(\mathbf{b})^\mathsf{T} + \mathsf{V}(\mathbf{b})\Sigma_{ab}\mathsf{U}(\mathbf{a})^\mathsf{T},$$

allowing the given entities to be statistically dependent.

For example, from column 2 in Table 15.1 one can easily read out the Jacobians that are the matrices in the bilinear forms, e.g. for the intersection point **x** we have $\partial \mathbf{x}/\partial \mathbf{m} = \mathsf{S}(\mathbf{l})$ and $\partial \mathbf{x}/\partial \mathbf{l} = -\mathsf{S}(\mathbf{m})$, cf. after Eq. (15.39).

Constructions in 3D

Here we mention six different cases, which we give without proof.

(1) A 3D line **L** joining two 3D points **X** and **Y** is given by

$$\mathbf{L} = \mathbf{X} \wedge \mathbf{Y} = -\mathbf{Y} \wedge \mathbf{X} : \mathbf{L} = \begin{bmatrix} X_h \mathbf{Y}_0 - Y_h \mathbf{X}_0 \\ \mathbf{X}_0 \times \mathbf{Y}_0 \end{bmatrix} = \mathsf{\Pi}(\mathbf{X})\mathbf{Y} = -\mathsf{\Pi}(\mathbf{Y})\mathbf{X}.$$

$$\tag{15.42}$$

Here the Jacobians are

Table 15.1. Construction of new 2D geometric entities. The structure of the matrix S is given in Eq. (15.38). All forms are linear in the coordinates of the given entities, allowing simple error propagation

New entity	Algebraic construction	Equation
$l = x \wedge y$	$l = S(x)y = -S(y)x$	(15.37)
$x = l \cap m$	$x = S(l)m = -S(m)l$	(15.39)

$$\Pi(X) = \frac{\partial (X \wedge Y)}{\partial Y}, \quad \text{and} \quad \Pi(Y) = \frac{\partial (Y \wedge X)}{\partial X},$$

or explicitly, e.g.

$$\Pi(X)_{6\times 4} = \begin{bmatrix} X_h I_3 & -X_0 \\ S(X_0) & 0 \end{bmatrix} = \begin{bmatrix} X_4 & 0 & 0 & -X_1 \\ 0 & X_4 & 0 & -X_2 \\ 0 & 0 & X_4 & -X_3 \\ 0 & -X_3 & X_2 & 0 \\ X_3 & 0 & -X_1 & 0 \\ -X_2 & X_1 & 0 & 0 \end{bmatrix}. \quad (15.43)$$

(2) A 3-line **L** as the intersection of two planes **A** and **B** is given by

$$\boxed{L = A \cap B = -B \cap A : L = \begin{bmatrix} A_h \times B_h \\ A_0 B_h - B_0 A_h \end{bmatrix} = \overline{\Pi}(A)B = -\overline{\Pi}(B)A.} \quad (15.44)$$

Here the Jacobians are

$$\overline{\Pi}(A) = \frac{\partial (A \cap B)}{\partial B}, \quad \text{and} \quad \overline{\Pi}(B) = \frac{\partial (B \cap A)}{\partial A},$$

or explicitely

$$\overline{\Pi}(A) = D_6 \Pi(A) = \begin{bmatrix} S(A_h) & 0 \\ A_0 I_3 & -A_h \end{bmatrix}. \quad (15.45)$$

Observe, we might have obtained $L = \overline{\Pi}(A)B$ from $L = \Pi(X)Y$ using dualing, using $A = \overline{X}$ and $B = \overline{Y}$, namely $\overline{L} = \Pi(\overline{X})\overline{Y} = D_6 L = \overline{\Pi}(A)B$, and noting $D_6 = D_6^{-1}$.

Remark: The letter P in the name Pi of the Greek capital letter Π indicates this matrix referring to *p*oints and *p*lanes.

(3) 3D point $\mathbf{X}$ as intersection of the 3D line $\mathbf{L}$ and the plane $\mathbf{A}$ from

$$\mathbf{X} = \mathbf{L} \cap \mathbf{A} = \mathbf{A} \cap \mathbf{L} : \mathbf{X} = \begin{bmatrix} \mathbf{L}_0 \times \mathbf{A}_h + A_0 \mathbf{L}_h \\ -\mathbf{L}_h^\mathsf{T} \mathbf{A}_h \end{bmatrix} = \mathsf{\Gamma}^\mathsf{T}(\mathbf{L})\mathbf{X} = \mathsf{\Pi}^\mathsf{T}(\mathbf{X})\mathbf{L}. \tag{15.46}$$

The Jacobians are

$$\mathsf{\Gamma}^\mathsf{T}(\mathbf{L}) = \frac{\partial(\mathbf{A} \cap \mathbf{L})}{\partial \mathbf{L}}, \quad \text{and} \quad \mathsf{\Pi}^\mathsf{T}(\mathbf{A}) = \frac{\partial(\mathbf{L} \wedge \mathbf{A})}{\partial \mathbf{L}} = \begin{bmatrix} A_0 I_3 & -S(\mathbf{A}_h) \\ -\mathbf{A}_h^\mathsf{T} & \mathbf{0}^\mathsf{T} \end{bmatrix}.$$

Explicitely we have the Plücker matrix of the line $\mathbf{L}$

$$\mathsf{\Gamma}(\mathbf{L}) = \begin{bmatrix} -S(\mathbf{L}_0) & -\mathbf{L}_h \\ \mathbf{L}_h^\mathsf{T} & 0 \end{bmatrix} = \begin{bmatrix} 0 & L_6 & -L_5 & -L_1 \\ -L_6 & 0 & L_4 & -L_2 \\ L_5 & -L_4 & 0 & -L_3 \\ L_1 & L_2 & L_3 & 0 \end{bmatrix}. \tag{15.47}$$

One can show that the Plücker matrix of the line $\mathbf{L} = \mathbf{X} \wedge \mathbf{Y}$ is the skew-symmetric form having rank 2:

$$\mathsf{\Gamma}(\mathbf{L}) = \mathsf{\Gamma}(\mathbf{X} \wedge \mathbf{Y}) = \mathbf{X}\mathbf{Y}^\mathsf{T} - \mathbf{Y}\mathbf{X}^\mathsf{T}. \tag{15.48}$$

Remark: The Greek letter $\mathsf{\Gamma}$ is the mirror of $\mathbf{L}$ and indicates it refers to 3D lines.

(4) Dually, we obtain the plane $\mathbf{A}$ joining a 3D line $\mathbf{L}$, and a 3D point $\mathbf{X}$ is given by

$$\mathbf{A} = \mathbf{L} \wedge \mathbf{X} = \mathbf{X} \wedge \mathbf{L} : \mathbf{A} = \begin{bmatrix} \mathbf{L}_h \times \mathbf{X}_0 + X_h \mathbf{L}_0 \\ -\mathbf{L}_0^\mathsf{T} \mathbf{X}_0 \end{bmatrix} = \overline{\mathsf{\Gamma}}^\mathsf{T}(\mathbf{L})\mathbf{X} = \overline{\mathsf{\Pi}}^\mathsf{T}(\mathbf{X})\mathbf{L}, \tag{15.49}$$

with the Jacobians

$$\overline{\mathsf{\Gamma}}^\mathsf{T}(\mathbf{L}) = \frac{\partial(\mathbf{A} \cap \mathbf{L})}{\partial \mathbf{A}}, \quad \text{and} \quad \overline{\mathsf{\Pi}}^\mathsf{T}(\mathbf{A}) = \frac{\partial(\mathbf{L} \cap \mathbf{A})}{\partial \mathbf{L}} = \begin{bmatrix} -S(\mathbf{X}_0) & X_h I \\ \mathbf{0}^\mathsf{T} & -\mathbf{X}_0^\mathsf{T} \end{bmatrix},$$

and the dual Plücker matrix of the 3D line $\mathbf{L}$

$$\overline{\mathsf{\Gamma}}(\mathbf{L}) = \mathsf{\Gamma}(\overline{\mathbf{L}}) = \begin{bmatrix} -S(\mathbf{L}_h) & -\mathbf{L}_0 \\ \mathbf{L}_0^\mathsf{T} & 0 \end{bmatrix}. \tag{15.50}$$

One can show that the dual Plücker matrix of the line $\mathbf{L} = \mathbf{A} \cap \mathbf{B}$ is the skew-symmetric form of rank 2:

$$\overline{\mathsf{\Gamma}}(\mathbf{L}) = \overline{\mathsf{\Gamma}}(\mathbf{A} \cap \mathbf{B}) = \mathbf{A}\mathbf{B}^\mathsf{T} - \mathbf{B}\mathbf{A}^\mathsf{T}. \tag{15.51}$$

Finally, we obtain the plane $\mathbf{A}$ joining three points $\mathbf{X}$, $\mathbf{Y}$ and $\mathbf{Z}$ from

$$\boxed{\mathbf{A} = \mathbf{X} \wedge \mathbf{Y} \wedge \mathbf{Z} = (\mathbf{X} \wedge \mathbf{Y}) \wedge \mathbf{Z},} \tag{15.52}$$

and the point $\mathbf{X}$ as the intersection of three planes $\mathbf{A}$, $\mathbf{B}$ and $\mathbf{C}$ from

$$\boxed{\mathbf{X} = \mathbf{A} \cap \mathbf{B} \cap \mathbf{C} = (\mathbf{A} \cap \mathbf{B}) \cap \mathbf{C}} \tag{15.53}$$

with similar expressions by cyclic exchange of the geometric entities.

Table 15.2. Construction of new 3D geometric entities. The matrices Π, $\overline{\Pi}$, Γ and $\overline{\Gamma}$ are given in Eqs. (15.43), (15.45), (15.47) and (15.50), resp. All forms are linear in the coordinates of the given entities, making error propagation easy

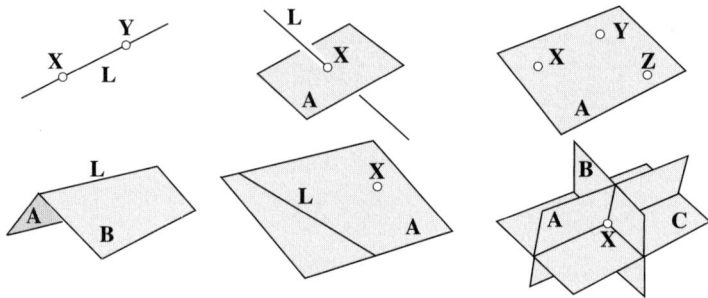

New entity	Algebraic construction	Equation
$\mathbf{L} = \mathbf{X} \wedge \mathbf{Y} = -\mathbf{Y} \wedge \mathbf{X}$	$\mathbf{L} = \Pi(\mathbf{X})\mathbf{Y} = -\Pi(\mathbf{Y})\mathbf{X}$	(15.42)
$\mathbf{L} = \mathbf{A} \cap \mathbf{B} = -\mathbf{B} \cap \mathbf{A}$	$\mathbf{L} = \overline{\Pi}(\mathbf{A})\mathbf{B} = -\overline{\Pi}(\mathbf{B})\mathbf{A}$	(15.44)
$\mathbf{X} = \mathbf{L} \cap \mathbf{A} = \mathbf{A} \cap \mathbf{L}$	$\mathbf{X} = \Gamma^\mathsf{T}(\mathbf{L})\mathbf{A} = \Pi^\mathsf{T}(\mathbf{A})\mathbf{L}$	(15.46)
$\mathbf{A} = \mathbf{L} \wedge \mathbf{X} = \mathbf{X} \wedge \mathbf{L}$	$\mathbf{A} = \overline{\Gamma}^\mathsf{T}(\mathbf{L})\mathbf{X} = \overline{\Pi}^\mathsf{T}(\mathbf{X})\mathbf{L}$	(15.49)
$\mathbf{A} = \mathbf{X} \wedge \mathbf{Y} \wedge \mathbf{Z}$	$\overline{\Gamma}^\mathsf{T}(\Pi(\mathbf{X})\mathbf{Y})\mathbf{Z} = \overline{\Gamma}^\mathsf{T}(\Pi(\mathbf{Y})\mathbf{Z})\mathbf{X} = \overline{\Gamma}^\mathsf{T}(\Pi(\mathbf{Z})\mathbf{X})\mathbf{Y}$	(15.52)
$\mathbf{X} = \mathbf{A} \cap \mathbf{B} \cap \mathbf{C}$	$\Gamma^\mathsf{T}(\overline{\Pi}(\mathbf{A})\mathbf{B})\mathbf{C} = \Gamma^\mathsf{T}(\overline{\Pi}(\mathbf{B})\mathbf{C})\mathbf{A} = \Gamma^\mathsf{T}(\overline{\Pi}(\mathbf{C})\mathbf{A})\mathbf{B}$	(15.53)

Table 15.2 collects these six cases. Observe the similarity in the representation with the operators of the Grassmann–Cayley algebra and the ease of reading out the Jacobians for statistical error propagation.

Constructions Containing Mappings

Geometric entities also can be generated by transformations. Table 15.3 collects the most important ones useful in geometric reasoning in computer vision.

15 Uncertainty and Projective Geometry

- 2D homography H for 2D points (row 1) and 2D lines (row 2)
- 3D homography H for 3D points (row 3), planes (row 4) and 3D lines (row 5)
- projection P of 3D points (row 6) and Q of 3D lines (row 7) into an image yielding 2D points and 2D lines
- back-projection of image points (row 8) and lines (row 9), yielding projection rays (3D lines) and projection planes
- 2D-2D correlation F (row 10) for mapping a point of one image to a line in another, knowing the epipolar geometry, or the relative orientation via the fundamental matrix F, used in the epipolar constraint $\mathbf{x'}^T \mathsf{F} \mathbf{x''} = 0$

All mappings are bilinear in the given geometric entity and the transformation matrix. Therefore again we give two relations, making the individual Jacobians explicit that are necessary for statistical error propagation.

Table 15.3. Mappings. 2D and 3D homographies H for points, homography H_L for space lines, projection matrices P and Q for points and lines. Projection ray $\mathbf{L}_{x'}$ and projection plane $\mathbf{A}_{l'}$. Fundamental matrix $\mathsf{F} \doteq \mathsf{F}_{12}$ for coplanarity constraint $\mathbf{x'}^T \mathsf{F} \mathbf{x''} = 0$

Number	Mapping	Relation 1	Relation 2
1	2D homography	$\mathbf{x'} = \mathsf{H}\mathbf{x}$	$\mathbf{x'} = (I_3 \otimes \mathbf{x}^T)\text{vec}(\mathsf{H}^T)$
2		$\mathbf{l'} = \mathsf{H}^{-T}\mathbf{l}$	$\mathbf{l'} = (I_3 \otimes \mathbf{l}^T)\text{vec}(\mathsf{H}^{-1})$
3	3D homography	$\mathbf{X'} = \mathsf{H}\mathbf{X}$	$\mathbf{X'} = (I_4 \otimes \mathbf{X}^T)\text{vec}(\mathsf{H}^T)$
4		$\mathbf{A'} = \mathsf{H}^{-T}\mathbf{A}$	$\mathbf{A'} = (I_3 \otimes \mathbf{A}^T)\text{vec}(\mathsf{H}^{-1})$
5		$\mathbf{L'} = \mathsf{H}_L \mathbf{L}$	$\mathbf{L'} = (I_6 \otimes \mathbf{L}^T)\text{vec}(\mathsf{H}_L^T)$
6	2D-3D projection	$\mathbf{x'} = \mathsf{P}\mathbf{X}$	$\mathbf{x'} = (I_3 \otimes \mathbf{X}^T)\text{vec}(\mathsf{P}^T)$
7		$\mathbf{l'} = \mathsf{Q}\mathbf{L}$	$\mathbf{l'} = (I_3 \otimes \mathbf{L}^T)\text{vec}(\mathsf{Q}^T)$
8	3D-2D back-projection	$\mathbf{A}_{l'} = \mathsf{P}^T \mathbf{l'}$	$\mathbf{A}_{l'} = (\mathbf{l'}^T \otimes I_4)\text{vec}(\mathsf{P}^T)$
9		$\mathbf{L}_{x'} = \overline{\mathsf{Q}}^T \mathbf{x'}$	$\mathbf{L}_{x'} = (\mathbf{x'}^T \otimes D_6)\text{vec}(\mathsf{Q}^T)$
10	2D-2D correlation	$\mathbf{l''} = \mathsf{F}^T \mathbf{x'}$	$\mathbf{l''} = (\mathbf{x'}^T \otimes I_3)\text{vec}(\mathsf{F}^T)$

Table 15.3 needs some explanation:

1. We use the vec-operator to represent uncertain homogeneous matrices as uncertain homogeneous vectors, e.g.

$$\mathbf{h} = \text{vec}(\mathsf{H}^T).$$

Thus we might want to work with the pairs

$$(\underline{\mathbf{h}}, \Sigma_{hh}), \quad (\underline{\mathbf{h}}_L, \Sigma_{h_L h_L}), \quad (\underline{\mathbf{p}}, \Sigma_{pp}), \quad (\underline{\mathbf{q}}, \Sigma_{qq}), \quad (\underline{\mathbf{f}}, \Sigma_{ff}),$$

of uncertain transformations collected in Table 15.3.

2. We use the Kronecker product, the vec-operator to express the result $\mathbf{x}'$ of the 2D-2D homography as a function of the vector $\mathbf{h} = \text{vec}(\mathsf{H}^\mathsf{T})$. With the rule $\text{vec}(\mathsf{A}\mathbf{b}) = (\mathbf{b}^\mathsf{T} \otimes \mathsf{I})\text{vec}\mathsf{A} = \text{vec}(\mathbf{b}^\mathsf{T} \mathsf{A}^\mathsf{T}) = (\mathsf{I} \otimes \mathbf{b}^\mathsf{T})\text{vec}(\mathsf{A}^\mathsf{T})$ we obtain

$$\mathbf{x}' = \mathsf{H}\mathbf{x} = (\mathsf{I}_3 \otimes \mathbf{x}^\mathsf{T})\mathbf{h}\,.$$

This is useful for deriving the covariance matrix of the transformed point $\mathbf{x}'$ in case the covariance matrices $\boldsymbol{\Sigma}_{xx}$ and $\boldsymbol{\Sigma}_{hh}$ of the point $\mathbf{x}$ and of the elements $\mathbf{h}$ of H are known

$$\boldsymbol{\Sigma}_{x'x'} = \mathsf{H}\boldsymbol{\Sigma}_{xx}\mathsf{H}^\mathsf{T} + (\mathsf{I}_3 \otimes \mathbf{x}^\mathsf{T})\boldsymbol{\Sigma}_{hh}(\mathsf{I}_3 \otimes \mathbf{x})\,,$$

in this special case, assuming statistical independence of $\underline{\mathbf{x}}$ and $\underline{\mathsf{H}}$.

3. If we want to derive the covariance matrix of transformed 2D lines, we need covariance matrix of the transposed inverse $\mathsf{M} = \mathsf{H}^{-\mathsf{T}}$. This can easily be derived: from $\mathsf{H}\mathsf{H}^{-1} = \mathsf{I}$ we have $\text{d}\mathsf{H}\,\mathsf{H}^{-1} + \mathsf{H}\,\text{d}\mathsf{H}^{-1} = \mathbf{0}$; thus $\mathsf{M}^\mathsf{T}\,\text{d}\mathsf{H}\,\mathsf{M}^\mathsf{T} + \text{d}\mathsf{M}^\mathsf{T} = \mathbf{0}$. Therefore with $\mathbf{m} = \text{vec}(\mathsf{M}^\mathsf{T}) = \text{vec}(\mathsf{H}^{-1})$ we obtain $(\mathsf{M} \otimes \mathsf{M}^\mathsf{T})\text{d}\mathbf{h} + \text{d}\mathbf{m} = \mathbf{0}$. The covariance matrix of $\mathbf{m}$ is therefore

$$\boldsymbol{\Sigma}_{mm} = (\mathsf{M} \otimes \mathsf{M}^\mathsf{T})\boldsymbol{\Sigma}_{hh}(\mathsf{M} \otimes \mathsf{M}^\mathsf{T})\,.$$

Finally, we obtain the covariance matrix of the transformed lines

$$\boldsymbol{\Sigma}_{l'l'} = \mathsf{H}^{-\mathsf{T}}\boldsymbol{\Sigma}_{ll}\mathsf{H}^{-1} + (\mathsf{I}_3 \otimes \mathbf{l}^\mathsf{T})\boldsymbol{\Sigma}_{mm}(\mathsf{I} \otimes \mathbf{l})\,,$$

or only in terms of the given values

$$\boldsymbol{\Sigma}_{l'l'} = \mathsf{H}^{-\mathsf{T}}\boldsymbol{\Sigma}_{ll}\mathsf{H}^{-1} + (\mathsf{H}^{-\mathsf{T}} \otimes \mathbf{l}'^\mathsf{T})\boldsymbol{\Sigma}_{hh}(\mathsf{H}^{-1} \otimes \mathbf{l}')\,.$$

4. The 3D line transformation is not made explicit in the table. Starting from the transformation $\mathbf{X}' = \mathsf{H}\mathbf{X}$ of 3D points $\mathbf{X}$, one obtains an expression for the transformation matrix H_L for 3D lines in terms of their Plücker coordinates

$$\boxed{\mathbf{L}' = \mathsf{H}_L\,\mathbf{L} = (\mathsf{I}_6 \otimes \mathbf{L}^\mathsf{T})\text{vec}(\mathsf{H}_L^\mathsf{T})\,,} \qquad (15.54)$$

with the transformation matrix (cf. Appendix)

$$\mathsf{H}_L = \frac{1}{2}\mathsf{J}_{\Gamma L}^\mathsf{T}(\mathsf{H} \otimes \mathsf{H})\mathsf{J}_{\Gamma L}\,,$$

using the 16×6 Jacobian

$$\mathsf{J}_{\Gamma L}_{16 \times 6} = \frac{\partial\text{vec}(\Gamma(\mathbf{L}))}{\partial \mathbf{L}}\,,$$

which via $\mathbf{L} = \frac{1}{2}\mathsf{J}_{\Gamma L}^\mathsf{T}\,\text{vec}(\Gamma(\mathbf{L}))$ maps the columns of the Plücker matrix to the Plücker coordinates. With the Jacobian of the transformed line with respect to the elements $\mathbf{h}$ of the transformation matrix (cf. Appendix)

$$J_{L'h} = \frac{\partial \mathbf{L}'}{\partial \mathbf{h}} = \frac{1}{2} J_{\Gamma L}^{\mathsf{T}}(I_4 \otimes (\Gamma(\mathbf{L})\mathsf{H}^{\mathsf{T}} - \mathsf{H}\Gamma^{\mathsf{T}}(\mathbf{L}))),$$

we obtain the covariance matrix of the transformed line from statistical error propagation:

$$\Sigma_{L'L'} = J_{L'h}\Sigma_{hh}J_{L'h}^{\mathsf{T}} + \mathsf{H}_L\Sigma_{LL}\mathsf{H}_L^{\mathsf{T}}.$$

5. Finally, we discuss how to derive the covariance matrix of the projection matrix for space lines Q from the covariance matrix of the projection matrix for space points P. The projection matrix P for points and its elements row-wise are

$$\underset{3\times 4}{\mathsf{P}} = \begin{bmatrix} \mathbf{A}^{\mathsf{T}} \\ \mathbf{B}^{\mathsf{T}} \\ \mathbf{C}^{\mathsf{T}} \end{bmatrix}, \qquad \underset{12\times 1}{\mathbf{p}} = \mathrm{vec}(\mathsf{P}^{\mathsf{T}}) = \begin{bmatrix} \mathbf{A} \\ \mathbf{B} \\ \mathbf{C} \end{bmatrix},$$

with the coordinate planes **A**, **B** and **C** of the camera coordinate system, intersecting in the projection center **Z**, cf. Table 15.10. The corresponding projection matrix Q for 3D lines and its elements row-wise are given by

$$\underset{3\times 6}{\mathsf{Q}} = \begin{bmatrix} (\mathbf{B}\wedge\mathbf{C})^{\mathsf{T}} \\ (\mathbf{C}\wedge\mathbf{A})^{\mathsf{T}} \\ (\mathbf{A}\wedge\mathbf{B})^{\mathsf{T}} \end{bmatrix}, \qquad \underset{18\times 1}{\mathbf{q}} = \mathrm{vec}(\mathsf{Q}^{\mathsf{T}}) = \begin{bmatrix} \mathbf{B}\wedge\mathbf{C} \\ \mathbf{C}\wedge\mathbf{A} \\ \mathbf{A}\wedge\mathbf{B} \end{bmatrix},$$

with the dual lines $\mathbf{B}\cap\mathbf{C}$, $\mathbf{C}\cap\mathbf{A}$ and $\mathbf{A}\cap\mathbf{B}$ of the coordinate axes of the camera system.

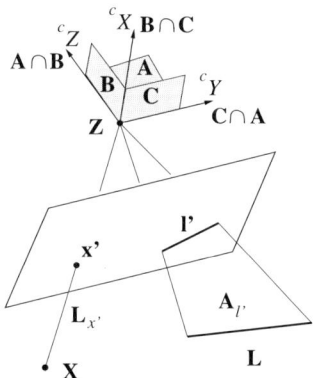

Fig. 15.10. Single-image geometry. Rows of P are coordinate planes, rows of Q are duals of coordinate axes of the camera coordinate system ($^c X$, $^c Y$, $^c Z$). The projection ray $\mathbf{L}_{x'} = \mathsf{D}_6 \mathsf{Q}^{\mathsf{T}} \mathbf{x}'$ and the projection plane $\mathbf{A}_{l'} = \mathsf{P}^{\mathsf{T}} \mathbf{l}'$ can be determined directly from the image entities using the projection matrices [11, 18].

Therefore

$$J_{qp}\atop 18\times 12 = \begin{bmatrix} 0 & -\Pi(C) & \Pi(B) \\ \Pi(C) & 0 & -\Pi(A) \\ -\Pi(B) & \Pi(A) & 0 \end{bmatrix},$$

and thus the covariance matrix of $\mathbf{q}$ of the elements of the projection matrix Q for lines reads as

$$\Sigma_{qq} = J_{qp}\Sigma_{pp}J_{qp}^\mathsf{T}. \tag{15.55}$$

In case the projection center of the image is denoted with $\mathbf{Z}$ we have

$$\mathbf{L}_{x'} = D_6 \mathbf{Q}^\mathsf{T} \mathbf{x}' = \overline{\mathbf{Q}}^\mathsf{T} \mathbf{x}' = \overline{\mathbf{Q}}^\mathsf{T} \mathbf{P}\, \mathbf{X} = \Pi(\mathbf{Z})\, \mathbf{X} = \mathbf{Z} \wedge \mathbf{X},$$

as the projection line $\mathbf{L}_{x'}$ is the join of projection center with the 3D point and is independent of the other parts of the orientation. Thus

$$\overline{\mathbf{Q}}^\mathsf{T} \mathbf{P} = \Pi(\mathbf{Z}).$$

15.3.3 Constraints

Constraints Between Geometric Entities in 2D

The incidence of a 2D point and a 2D line can be checked using

$$c = \mathbf{x}^\mathsf{T} \mathbf{l} = \mathbf{l}^\mathsf{T}\mathbf{x} \stackrel{!}{=} 0,$$

which should vanish. The identity of two points or two lines can be checked using the 3-vectors

$$\mathbf{c} = \mathsf{S}(\mathbf{x})\mathbf{y} = -\mathsf{S}(\mathbf{y})\mathbf{x} \stackrel{!}{=} \mathbf{0}, \quad \text{or} \quad \mathbf{c} = \mathsf{S}(\mathbf{l})\mathbf{m} = -\mathsf{S}(\mathbf{m})\mathbf{l} \stackrel{!}{=} \mathbf{0},$$

which should vanish in the case of identity. The reasoning behind this type of constraint is: in case two points are identical the generating line $\mathbf{x} \times \mathbf{y}$ is not defined. Similarly, in case two lines are identical, the intersection point $\mathbf{l} \times \mathbf{m}$ is not defined.

Observe, there are three constraints, but only two of them are independent, as the skew-symmetric matrices have rank 2. One may select those two constraints (m, n) where the entry S_{mn} in the skew-symmetric matrices is the largest absolute value. Then the independence of the selected constraints is guaranteed. This leads to a set of reduced constraints, e.g.:

$$\mathbf{c}^{[\mathrm{r}]} = \underbrace{\begin{bmatrix} \mathbf{e}_m^{(3)\mathsf{T}} \\ \mathbf{e}_n^{(3)\mathsf{T}} \end{bmatrix} \mathsf{S}(\mathbf{x})}_{\mathsf{S}^{[\mathrm{r}]}(\mathbf{x})} \mathbf{y}$$

$$= -\begin{bmatrix} \mathbf{e}_m^{(3)\mathsf{T}} \\ \mathbf{e}_n^{(3)\mathsf{T}} \end{bmatrix} \mathsf{S}(\mathbf{y})\mathbf{x} \stackrel{m=1,n=2}{=} \begin{bmatrix} 0 & -x_3 & x_2 \\ x_3 & 0 & -x_1 \end{bmatrix} \begin{bmatrix} y_1 \\ y_2 \\ y_3 \end{bmatrix}, \tag{15.56}$$

which are linearly independent, except for points $\mathbf{x}$ at infinity, thus in case $x_3 \neq 0$.

Table 15.4 collects the three cases mentioned.

Table 15.4. Relationships between points and lines useful for 2D grouping, together with the degree of freedom (DOF) and the essential part of the test statistic. The bullet in the last column indicates that a selection may be performed

No.	2D Entities	Relation	DOF	Test	Selection
1	Points x, y	$x \equiv y$	2	$c = S(x)y = -S(y)x$	•
2	Point x, line l	$x \in l$	1	$c = x^T l = l^T x$	−
3	Lines l, m	$l \equiv m$	2	$c = S(l)m = -S(m)l$	•

Constraints Between Geometric Entities in 3D

In a similar manner one may construct constraints for geometric relations between 3D entities collected in Table 15.5 except for the following cases (numbers refer to the rows in Table 15.5):

7. The identity of two lines **L** and **M** can be checked using the interpretation of the rows or columns of the Plücker matrix and their dual: the rows and columns of $\Gamma(\mathbf{L})$ are the intersection points of **L** with the coordinate planes $e_i^{(4)}$, and the rows and columns of $\overline{\Gamma}(\mathbf{L})$ are the planes parallel to the coordinate axes. As the intersection points of **L** with the coordinate planes lie on the planes through **M** parallel to the coordinate axes, in case $\mathbf{L} \equiv \mathbf{M}$ the product $\mathbf{C} = \overline{\Gamma}(\mathbf{L})\Gamma(\mathbf{M})$ must vanish.
8. The incidence of two lines **L** and **M** can be checked by assuming $\mathbf{L} = \mathbf{X} \wedge \mathbf{Y}$ and $\mathbf{M} = \mathbf{Z} \wedge \mathbf{T}$. Then $|\mathbf{X}, \mathbf{Y}, \mathbf{Z}, \mathbf{T}| = -\overline{\mathbf{L}}^T \mathbf{M} = 0$ only if the four points are coplanar.

Also here selection of constraints can be performed (numbers refer to rows in Table 15.5) :

4., 10. From the six constraints only three are independent. One can select those three constraints where the largest element occurs in the matrices $\overline{\Pi}(\mathbf{X})$ or $\overline{\Pi}(\mathbf{A})$.

5., 9. From the four constraints only two are independent. One can select those two constraints where the largest element occurs in the matrices $\overline{\Gamma}(\mathbf{L})$ or $\Gamma(\mathbf{L})$. The selection transfers to the corresponding matrices $\overline{\Pi}^T(\cdot)$ and $\Pi^T(\cdot)$.

7. From the 16 constraints only 4 are independent. They can be selected by taking the largest element of $\overline{\Gamma}(\mathbf{L})$ with index (m, n) and the largest element of $\Gamma(\mathbf{M})$ with index (k, l) and check the entries $\{(m, k), (m, l), (n, k), (n, l)\}$ of $\mathbf{C}$.

Table 15.5. Relationships between points, lines and planes useful for 3D grouping, together with the degree of freedom (DOF) and the essential part of the test statistic. The bullet in the last column indicates the possibility for the selection of independent constraints

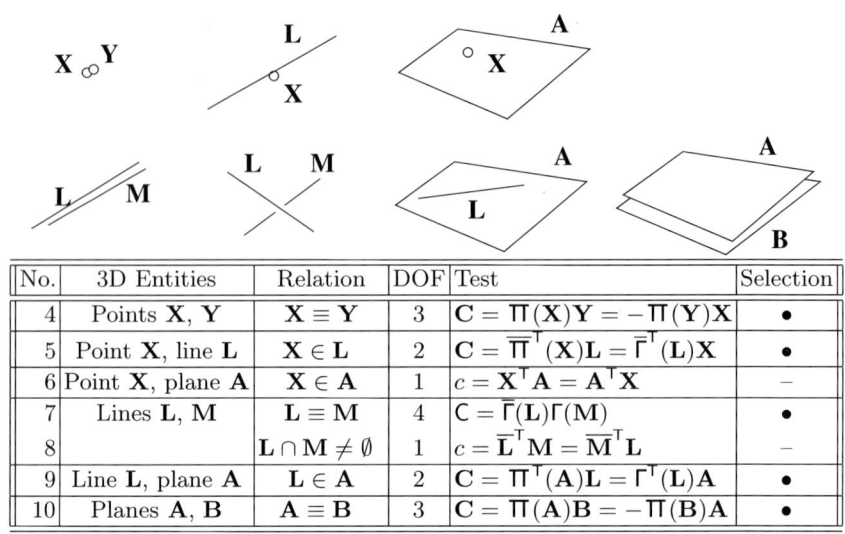

No.	3D Entities	Relation	DOF	Test	Selection
4	Points $\mathbf{X}$, $\mathbf{Y}$	$\mathbf{X} \equiv \mathbf{Y}$	3	$\mathbf{C} = \overline{\Pi}(\mathbf{X})\mathbf{Y} = -\overline{\Pi}(\mathbf{Y})\mathbf{X}$	•
5	Point $\mathbf{X}$, line $\mathbf{L}$	$\mathbf{X} \in \mathbf{L}$	2	$\mathbf{C} = \overline{\Pi}^\mathsf{T}(\mathbf{X})\mathbf{L} = \overline{\Gamma}^\mathsf{T}(\mathbf{L})\mathbf{X}$	•
6	Point $\mathbf{X}$, plane $\mathbf{A}$	$\mathbf{X} \in \mathbf{A}$	1	$c = \mathbf{X}^\mathsf{T}\mathbf{A} = \mathbf{A}^\mathsf{T}\mathbf{X}$	–
7	Lines $\mathbf{L}$, $\mathbf{M}$	$\mathbf{L} \equiv \mathbf{M}$	4	$\mathbf{C} = \overline{\Gamma}(\mathbf{L})\Gamma(\mathbf{M})$	•
8		$\mathbf{L} \cap \mathbf{M} \neq \emptyset$	1	$c = \overline{\mathbf{L}}^\mathsf{T}\mathbf{M} = \overline{\mathbf{M}}^\mathsf{T}\mathbf{L}$	–
9	Line $\mathbf{L}$, plane $\mathbf{A}$	$\mathbf{L} \in \mathbf{A}$	2	$\mathbf{C} = \overline{\Pi}^\mathsf{T}(\mathbf{A})\mathbf{L} = \Gamma^\mathsf{T}(\mathbf{L})\mathbf{A}$	•
10	Planes $\mathbf{A}$, $\mathbf{B}$	$\mathbf{A} \equiv \mathbf{B}$	3	$\mathbf{C} = \overline{\Pi}(\mathbf{A})\mathbf{B} = -\overline{\Pi}(\mathbf{B})\mathbf{A}$	•

Constraints Containing Mappings

We now easily can derive arbitrary constraints containing mappings. As an example, the constraint

$$\mathbf{c} = \mathbf{x}' \times \mathsf{H}\mathbf{x} = \mathsf{S}(\mathbf{x}')\mathsf{H}\mathbf{x} = -\mathsf{S}(\mathsf{H}\mathbf{x})\mathbf{x}' = (\mathsf{S}(\mathbf{x}') \otimes \mathbf{x}^\mathsf{T})\mathbf{h} \stackrel{!}{=} \mathbf{0}$$

would require the mapped point $\mathsf{H}\mathbf{x}$ to be identical to $\mathbf{x}'$. The fifth expression uses $\mathbf{h} = \text{vec}(\mathsf{H}^\mathsf{T})$.

Observe, the expression for $\mathbf{c}$ is trilinear, here in $\mathbf{x}$, in H and in $\mathbf{x}'$. Therefore we give three different expressions making the Jacobians of $\mathbf{c}$ with respect to the three generating entities explicit:

$$\frac{\partial \mathbf{c}}{\partial \mathbf{x}} = \mathsf{S}(\mathbf{x}')\mathsf{H}, \quad \frac{\partial \mathbf{c}}{\partial \mathbf{x}'} = -\mathsf{S}(\mathsf{H}\mathbf{x}), \quad \frac{\partial \mathbf{c}}{\partial \mathbf{h}} = \mathsf{S}(\mathbf{x}') \otimes \mathbf{x}^\mathsf{T}.$$

Other constraints easily can be found by combining the relations given in the tables. Observe, the constraints involving the mappings with H_L and Q for 3D lines are not linear in the corresponding mappings H and P for 3D points. This leads to more involved expressions.

The tests involving mappings from 3D to 2D can be read in two ways: testing incidence in the image or testing incidence in space. As the constraint is the same, both tests lead to the same result. As an example, take the incidence test of a space point $\mathbf{X}$ and image line $\mathbf{l}'$. It reads as $c = \mathbf{l}'^\mathsf{T} \mathsf{P} \mathbf{X}$. If we take the predicted point to be $\mathbf{x}'_p = \mathsf{P}\mathbf{X}$ and the projection plane to be $\mathbf{A}_{l'} = \mathsf{P}^\mathsf{T}\mathbf{l}'$ then the constraint can be written as

$$c = \mathbf{l}'^\mathsf{T} \mathbf{x}'_p \stackrel{!}{=} 0, \quad \text{or} \quad c = \mathbf{A}_{l'}^\mathsf{T} \mathbf{X} \stackrel{!}{=} 0,$$

which explicitely shows the equivalence of testing the relation in image and 3D space. Actually, the relation for the projection plane has been derived from this identity.

Comments: The essence of the constructions and constraints have been published in many places, e.g. [33]. They have also been given in [25] and can be also found in [18]. The tables in the appendix of [35] give all constructions explicitly, even for elements in $\mathbb{P}^5$. The representation chosen here, initially given in [15], allows one to easily remember the relations and explicitly have the corresponding Jacobians available. For space limitations we did not give the relations for geometric entities to be orthogonal or parallel; instead we refer to [15, 23].

15.3.4 Testing Uncertain Geometric Relations

We directly use the mentioned constraints for statistical testing. The idea is to test the c-quantities in Tables 15.4 and 15.5 having different dimensions. In case the relation holds they should be zero; thus formally, the null hypothesis $H_0 : c = 0$ is tested versus the alternative $H_a : c \neq 0$.

The generic procedure is the following:

(1) Determine the difference $\mathbf{c}$, using one of the two equations in column 5 in Table 15.4 or 15.5
(2) If necessary, select independent constraints leading to the reduced vector of differences $\mathbf{c}'$. The number of independent constraints is the degree of freedom, cf. column 4. The selection is different for the individual tests and is indicated in column 6 where necessary.
(3) Determine the covariance matrix of the difference $\mathbf{c}$ or the reduced difference $\mathbf{c}'$ using error propagation given in Eq. (15.41) and the two Jacobians from Table 4 or 5 in column 5, possibly taking the selection into account. The Jacobians can be taken from these equations all having the structure of Eq. (15.40).
(4) Determine the test statistic T

$$T = \frac{c'^2}{\sigma_c'^2} \sim \chi_1^2, \quad \text{or} \quad T = \mathbf{c}'^\mathsf{T} \mathbf{\Sigma}_{c'c'}^{-1} \mathbf{c}' \sim \chi_r^2, \qquad (15.57)$$

being χ_1^2 or χ_r^2 distributed, the degrees of freedom r given in column 4 of Tables 15.4 and 15.5.

(5) Choose a significance number α and compare T with the critical value $\chi^2_{r,1-\alpha}$. If $T > \chi^2_{r,1-\alpha}$ then the hypothesis that the spatial relation holds can be rejected.

The proposed error propagation in step 15.3.4 using Eq. (15.41) is simple, but it only holds *approximately*, in case the tested relation is not fulfilled. This is because the Jacobians of the bilinear relations are not consistent, as they depend on the observed entities, not on the true or unbiased estimated entities as required in Eq. (15.3).

A rigorous procedure would be to first impose the constraints of the relation onto the two observed entities using the estimation procedure of Sect. 15.2.2 and test the estimated variance factor $\widehat{\sigma}_0^2$. This approach was taken by Kanatani [25].

In case the test is not rejected, the covariance matrix of the differences, however, is a very good approximation. Applying this approximation of statistical testing depends on the rigor needed. As the assumed variances of the initial geometric entities will not be more precise than 10 or 20%, at best, the test statistic will also have this uncertainty. In most applications this chapter is motivated by, the tests will be used for deleting erroneous correspondences or for controlling search in grouping processes. Especially in the latter application, the monotonicity of the test statistic with respect to the rigorous one is essential. We discuss this problem below.

An Example for Testing

We want to demonstrate the test procedure for the point-line incidence. Let the 3D line $\mathbf{L}$ and the 3D point $\mathbf{X}$ be given by

$$\mathbf{L} = [3, 0, 0, 0, 3, -3]^\mathsf{T}, \qquad \mathbf{X} = [1, 1, 1, -1]^\mathsf{T}.$$

The line is parallel to the X-axis and passes through the point $[0, 1, 1]^\mathsf{T}$. The 3D point has the Euclidean coordinates $[-1, -1, -1]^\mathsf{T}$. The distance of $\mathbf{X}$ and $\mathbf{L}$ therefore is $d_{XL} = 2\sqrt{2} \approx 2.8$. For demonstration purposes we assume the covariance matrices to be a multiple of the unit matrix:

$$\mathbf{\Sigma}_{LL} = 4 \mathbf{I}_6, \qquad \mathbf{\Sigma}_{XX} = \mathbf{I}_4.$$

Thus the standard deviations of the homogeneous point coordinates are 1, and the standard deviations of the Euclidean coordinates are larger than 1, which could be verified by applying error propagation to $X = U/W$, etc. The line has a similar precision in the vicinity of the origin of the coordinate system, however, it has a very large uncertainty in direction. This could be verified by intersecting the line with the planes $X = 0$ and $X = -1$, which are parallel to the Y-Z-plane, and determining the standard deviations of the Y- and Z-coordinates of the intersection points. Thus we can expect the distance of 2.8 not to be significant.

We now follow the above-mentioned steps:

(1) The difference c is

$$c = \overline{\Gamma}^\mathsf{T}(\mathbf{L})\mathbf{X} = \begin{bmatrix} 0 & 0 & 0 & 0 \\ 0 & 0 & -3 & 3 \\ 0 & 3 & 0 & -3 \\ 0 & -3 & 3 & 0 \end{bmatrix} \begin{bmatrix} 1 \\ 1 \\ 1 \\ -1 \end{bmatrix}$$

$$= \overline{\Pi}^\mathsf{T}(\mathbf{X})\mathbf{L} = \begin{bmatrix} 0 & 1 & -1 & -1 & 0 & 0 \\ -1 & 0 & 1 & 0 & -1 & 0 \\ 1 & -1 & 0 & 0 & 0 & -1 \\ 0 & 0 & 0 & -1 & -1 & -1 \end{bmatrix} \begin{bmatrix} 3 \\ 0 \\ 0 \\ 0 \\ 3 \\ -3 \end{bmatrix} = \begin{bmatrix} 0 \\ -6 \\ 6 \\ 0 \end{bmatrix}.$$

(2) We select rows 3 and 4, as $L_6 = -3$ does not vanish (we could have taken rows 2 and 3 or rows 2 and 4 also). Thus we obtain the reduced vector

$$\mathbf{c}' = R^\mathsf{T}\mathbf{c} = \begin{bmatrix} e_3^{(4)\mathsf{T}} \\ e_4^{(4)\mathsf{T}} \end{bmatrix} \mathbf{c} = \begin{bmatrix} 0 & 0 & 1 & 0 \\ 0 & 0 & 0 & 1 \end{bmatrix} \begin{bmatrix} 0 \\ -6 \\ 6 \\ 0 \end{bmatrix} = \begin{bmatrix} 6 \\ 0 \end{bmatrix},$$

with the reduction matrix $R^\mathsf{T} = [e_3^{(4)}, e_4^{(4)}]$.

(3) The covariance matrix of $\mathbf{c}'$ is obtained from

$$\Sigma_{c'c'} = R^\mathsf{T}\left(\overline{\Gamma}^\mathsf{T}(\mathbf{L})\Sigma_{XX}\overline{\Gamma}(\mathbf{L}) +^\mathsf{T} \overline{\Pi}^\mathsf{T}(\mathbf{X})\Sigma_{LL}\overline{\Pi}(\mathbf{X})\right) R = \begin{bmatrix} 30 & -5 \\ -5 & 30 \end{bmatrix}.$$

(4) The test statistic is

$$T = \mathbf{c}'^\mathsf{T}\Sigma_{c'c'}^{-1}\mathbf{c}' = \frac{216}{175} \approx 1.23.$$

It is χ_2^2 distributed.

(5) We choose a significance number $\alpha = 0.05$. The critical value is $\chi_{2,1-\alpha} = 5.99$. As T is smaller than the critical value, we have no reason to reject the hypothesis that the point $\mathbf{X}$ sits on the line $\mathbf{L}$.

An Example for Setting up the Estimation

We want to demonstrate the use of the mentioned relations for the estimation task in Fig. 15.2 right. We assume the orientation of the images to be given. Here we only need the four projection matrices Q_k for space lines.

We have $N = 12$ observations, namely the four homogeneous 3-vectors of the image points $\mathbf{x}'$ and lines $\mathbf{l}'$ in the four images and $U = 6$ unknowns for the Plücker coordinates $\mathbf{L}_5$ of the 3D line. Thus, referring to the Gauss–Helmert model, we have the vector $\mathbf{l}$ of the observations and the vector $\mathbf{x}$ of the unknown parameters:

$$\underset{12\times 1}{l} = \begin{bmatrix} \mathbf{x}'_{11} \\ \mathbf{l}'_{52} \\ \mathbf{x}'_{23} \\ \mathbf{l}'_{54} \end{bmatrix}, \quad \underset{6\times 1}{\mathbf{x}} = \mathbf{L}_5 \; .$$

There are two constraints for the measured image points and $2 \times 2 = 4$ independent constraints for the observed image lines, using the reduced skew-symmetric matrices $\mathsf{S}^{[r]}(\cdot)$ (cf. Eq. (15.56)), altogether yielding six constraints $\mathbf{g}$ between the observations and the unknown parameters. Moreover, we have two constraints $\mathbf{h}$ on the unknown parameters, namely the length constraint and the Plücker constraint on $\mathbf{L}_5$. Thus we obtain the Gauss–Helmert model for this estimation task, where all relations should hold for the fitted, i.e. estimated values:

$$\underset{6\times 1}{\mathbf{g}(l,\mathbf{x})} = \begin{bmatrix} \mathbf{x}'^\mathsf{T}_{11} \mathbf{Q}_1 \mathbf{L}_5 \\ \mathsf{S}^{[r]}(\mathbf{l}'_{52}) \mathbf{Q}_2 \mathbf{L}_5 \\ \mathbf{x}'^\mathsf{T}_{23} \mathbf{Q}_3 \mathbf{L}_5 \\ \mathsf{S}^{[r]}(\mathbf{l}'_{54}) \mathbf{Q}_4 \mathbf{L}_5 \end{bmatrix} = \mathbf{0}, \quad \underset{2\times 1}{\mathbf{h}(\mathbf{x})} = \begin{bmatrix} \frac{1}{2}\left(\mathbf{L}_5^\mathsf{T} \mathbf{L}_5 - 1\right) \\ \frac{1}{2}\overline{\mathbf{L}}_5^\mathsf{T} \mathbf{L}_5 \end{bmatrix} = \mathbf{0} \; .$$

The three Jacobians $\mathbf{A}$, $\mathbf{B}$ and $\mathbf{H}$ for solving for the ML estimates using the Gauss–Helmert model can easily be derived using Tables 15.1 to 15.5 given above. The initial solution $\widehat{\mathbf{x}}^{(0)}$ for $\mathbf{x} = \mathbf{L}_5$ can be obtained from the right eigenvector of $\mathbf{A}$. Then the Jacobians can be evaluated at approximate values. In the first iteration one uses $\widehat{l}^{(0)} = l$ and the initial solution $\widehat{\mathbf{x}}^{(0)}$. In the following iterations the Jacobians have to be evaluated at the *fitted* observations and unknown parameters in order to avoid bias and the necessity to renormalize [24]. The redundancy of the system is $R = G + H - U = 6 + 2 - 6 = 2$. Obviously the setup can directly be transferred to the other two problems shown in Figs. 15.1 and 15.2.

15.4 Conditioning and Normalization

There are three reasons why a blind use of the approach described so far may lead to problems: (1) entries in the homogeneous vectors or matrices with highly different orders of magnitude lead to numerical instability; (2) when testing extreme deviations from the null-hypothesis leads to wrong test statistics; and (3) large deviations from the null hypothesis may lead to test statistics that do not increase monotonically with the geometric distance of the involved entities.

These problems can be cured by conditioning and normalization:

1. *Conditioning*, as the name indicates, aims at improving the condition numbers of the matrices of concern. It is achieved by centering and scaling the data such that the Euclidean coordinates of points are in the range $[-k, +k]$. Hartley [17][2] proposed $k = 1$. The monotonicity of the test statistic with the geometric distance is guaranteed when choosing $k < 1$, e.g. $k = 1/3$, cf. [23].
2. *Normalization* affects at least the covariance matrix of uncertain homogeneous entities. It imposes the constraints on the length of the complete vector or on the Euclidean part of the vector. Thus, for some uncertain homogeneous vector $\underline{\mathbf{x}} = (\underline{\boldsymbol{x}}_h, \underline{\boldsymbol{x}}_0)^\mathsf{T}$ with Euclidean part $\boldsymbol{x}_h$, and homogenous part $\boldsymbol{x}_0$, we have the constraint [23]

$$|\underline{\mathbf{x}}| = |\mathbf{x}|, \quad \text{or} \quad |\underline{\boldsymbol{x}}_h| = |\boldsymbol{x}_h|.$$

In addition, we might have the Plücker constraint $\overline{\mathbf{L}}^\mathsf{T}\mathbf{L} = 0$ for 3D lines or the singularity constraint $|\mathsf{F}| = 0$ for the fundamental matrix. When imposing these constraints the resulting covariance matrix can be determined from Eq. (15.35). Kanatani [25] applied the Euclidean normalization together with a scaling of the Euclidean part of points to 1.

15.5 Conclusions

The paper discussed an approach for uncertain geometric reasoning, which tries to keep as close to algebraic projective geometry as possible, and a generic estimation scheme for uncertain geometric entities and transformation that can handle any number of constraints. Though the various ingredients are well known, building a software system is simplified by using the presented representations, both in geometry and statistics.

The basic idea behind the approach is to exploit the multilinearity of all geometric constructions and constraints. As these multilinearities are also found in all variations of geometric algebra, e.g. conformal geometric algebra [32], it appears feasible to reformulate the geometric expressions in terms of coordinate vectors with a covariance matrix attached to them, and thus to extend the approach to a much wider field than projective geometry.

---

[2] Hartley called this procedure normalization.

15.6 Appendix: Uncertainty of Tranformed 3D Lines

Starting from the transformation $\mathbf{X}' = \mathsf{H}\mathbf{X}$ of 3D points $\mathbf{X}$, we observe

$$\Gamma(\mathbf{L}') = \mathsf{H}\Gamma(\mathbf{L})\mathsf{H}^\mathsf{T}, \qquad (15.58)$$

as $\mathsf{H}(\mathbf{X}\mathbf{Y}^\mathsf{T} - \mathbf{Y}\mathbf{X}^\mathsf{T})\mathsf{H}^\mathsf{T} = \mathbf{X}'\mathbf{Y}'^\mathsf{T} - \mathbf{Y}'\mathbf{X}'^\mathsf{T} = \Gamma(\mathbf{L}')$. This shows that the transformation of lines is quadratic in the entries of the transformation matrix H for points.

We now first derive an expression for the transformation matrix H_L for 3D lines in terms of their Plücker coordinates $\mathbf{L}' = \mathsf{H}_L\mathbf{L}$, and then derive the Jacobians that are necessary to derive the covariance matrix of the transformed line $\mathbf{L}'$.

The transformation in Eq. (15.58) may be written in terms of the elements of $\Gamma(\mathbf{L})$

$$\text{vec}(\Gamma(\mathbf{L}')) = (\mathsf{H} \otimes \mathsf{H})\text{vec}(\Gamma(\mathbf{L}))$$

containing the Plücker coordinates of the lines. We now map the 16 values of $\text{vec}(\Gamma(\mathbf{L}))$ to the 6-Plücker vector $\mathbf{L}$ using the 16×6 Jacobian

$$\underset{16\times 6}{J_{\Gamma L}} = \frac{\partial \text{vec}(\Gamma(\mathbf{L}))}{\partial \mathbf{L}}.$$

Then we have

$$\mathbf{L} = \frac{1}{2} J_{\Gamma L}^\mathsf{T} \text{vec}(\Gamma(\mathbf{L})), \qquad \text{and} \qquad \text{vec}(\Gamma(\mathbf{L})) = J_{\Gamma L} \mathbf{L}. \qquad (15.59)$$

We obtain the transformation

$$\mathbf{L}' = \mathsf{H}_L \, \mathbf{L} = \frac{1}{2} J_{\Gamma L}^\mathsf{T}(\mathsf{H} \otimes \mathsf{H}) J_{\Gamma L} \, \mathbf{L} = (I_6 \otimes \mathbf{L}^\mathsf{T}) \text{vec}(\mathsf{H}_L^\mathsf{T}), \qquad (15.60)$$

with the 6×6 transformation matrix for 3D lines

$$\mathsf{H}_L = \frac{1}{2} J_{\Gamma L}^\mathsf{T}(\mathsf{H} \otimes \mathsf{H}) J_{\Gamma L}.$$

We now want to determine the Jacobian $J_{L'h} = \partial \mathbf{L}'/\partial \mathbf{h}$. We start from the differential of $\Gamma(\mathbf{L}')$ in Eq. (15.58):

$$d\Gamma(\mathbf{L}') = d\mathsf{H}\,\Gamma(\mathbf{L})\mathsf{H}^\mathsf{T} + \mathsf{H}\,\Gamma(d\mathbf{L})\mathsf{H}^\mathsf{T} + \mathsf{H}\,\Gamma(\mathbf{L})d\mathsf{H}^\mathsf{T}.$$

With Eq. (15.59) we obtain

$$d\mathbf{L}' = \frac{1}{2} J_{\Gamma L}^\mathsf{T} \text{vec}(d\Gamma(\mathbf{L}'))$$

$$= \frac{1}{2} J_{\Gamma L}^\mathsf{T}(I_4 \otimes \Gamma(\mathbf{L})\mathsf{H}^\mathsf{T})d\mathbf{h} + \frac{1}{2} J_{\Gamma L}^\mathsf{T}(\mathsf{H} \otimes \mathsf{H}) J_{\Gamma L}\, d\mathbf{L} + \frac{1}{2} J_{\Gamma L}^\mathsf{T}(I_4 \otimes \mathsf{H}\Gamma(\mathbf{L}))d\mathbf{h}$$

$$= \underbrace{\frac{1}{2} J_{\Gamma L}^\mathsf{T}(I_4 \otimes (\Gamma(\mathbf{L})\mathsf{H}^\mathsf{T} - \mathsf{H}\Gamma^\mathsf{T}(\mathbf{L})))}_{J_{L'h}} d\mathbf{h} + \underbrace{\frac{1}{2} J_{\Gamma L}^\mathsf{T}(\mathsf{H} \otimes \mathsf{H}) J_{\Gamma L}}_{J_{L'L}=\mathsf{H}_L} d\mathbf{L},$$

which can be shortend to

$$d\mathbf{L}' = \mathsf{J}_{L'h}d\mathbf{h} + \mathsf{H}_L d\mathbf{L}, \qquad \text{with} \qquad \mathsf{J}_{L'h} = \frac{1}{2}\mathsf{J}_{\Gamma L}^\mathsf{T}(I_4 \otimes (\mathsf{\Gamma}(\mathbf{L})\mathsf{H}^\mathsf{T} - \mathsf{H}\mathsf{\Gamma}^\mathsf{T}(\mathbf{L}))) \,.$$

We can use this result for statistical error propagation:

$$\boldsymbol{\Sigma}_{L'L'} = \mathsf{J}_{L'h}\boldsymbol{\Sigma}_{hh}\mathsf{J}_{L'h}^\mathsf{T} + \mathsf{H}_L \boldsymbol{\Sigma}_{LL} \mathsf{H}_L^\mathsf{T} \,.$$

References

1. M. Ashdown. The GA package for MAPLE release V. http://www.mrao.cam.ac.uk/~clifford/software/GA/, May 2004.
2. W. Baarda (1973). *S-Transformations and Criterion Matrices*, vol. 5 of *1*. Netherlands Geodetic Commission, Delft, 1973.
3. L. Brand (1966). *Vector and Tensor Analysis*. Wiley.
4. S. Carlsson (1994). The double algebra: an effective tool for computing invariants in computer vision. In: J. Mundy, Zisserman A., and D. Forsyth (eds.) *Applications of Invariance in Computer Vision*, LNCS, vol. 825. Springer, Berlin Heidelberg New York, pp. 145–164 , 1994.
5. W. Chojnacki, M. J. Brooks, A. van den Hengel (2001). Rationalising the renormalisation method of Kanatani. *Journal of Mathematical Imaging and Vision*, 14(1):21–38.
6. W. Chojnacki, M. J. Brooks, A. van den Hengel, D. Gawley (2000). On the fitting of surfaces to data with covariances. *IEEE Trans. Pattern Analysis Machine Intelligence*, 22(11):1294–1303.
7. W. Chojnacki, M. J. Brooks, A. van den Hengel, D. Gawley (2003). From FNS to HEIV: A link between two vision parameter estimation methods. *IEEE Transactions on Pattern Analysis of Machine Intelligence*, 26(2):264–268.
8. R. Collins (1993). *Model Acquisition Using Stochastic Projective Geometry*. PhD thesis, Department of Computer Science, University of Massachusetts. Also published as UMass Computer Science Technical Report TR95-70.
9. A. Criminisi (2001). *Accurate Visual Metrology from Single and Multiple Uncalibrated Images*. Springer-Verlag London Ltd.
10. O. Faugeras, Q. Luong, with contributions by T. Papdopoulo (2001). *The geometry of multiple images*. MIT Press, Cambridge, MA.
11. O. Faugeras, T. Papadopoulo (1998). Grassmann–Cayley algebra for modeling systems of cameras and the algebraic equations of the manifold of trifocal tensors. In *Trans. of the Royal Society A*, 365:1123–1152.
12. W. Förstner (1996). 10 pros and cons against performance characterisation of vision algorithms. In: Madsen C. B. Christensen H. I., Förstner W. (eds.) *Proceedings of the ECCV Workshop on Performance Characteristics of Vision Algorithms*, pages 13–29, Cambridge, UK.
13. W. Förstner (1979). Ein Verfahren zur Schätzung von Varianz- und Kovarianzkomponenten. *Allgemeine Vermessung Nachrichten*, 11-12:446–453.
14. W. Förstner (2001). Algebraic projective geometry and direct optimal estimation of geometric entities. In Stefan Scherer (ed.) *Computer Vision, Computer Graphics and Photogrammetry – A common viewpoint.*, Proc. 25th Workshop of the Austrian Association for Pattern Recognition (ÖAGM/AAPR), Österreichische Computer Gesellschaft, pp. 67–86, 2001

15. W. Förstner, A. Brunn, S. Heuel (2000). Statistically testing uncertain geometric relations. In G. Sommer, N. Krüger, and C. Perwass (eds.) *Mustererkennung 2000*, pages 17–26. DAGM, Springer, Berlin Heidelberg New York, 2000.
16. J. Haddon, D. A. Forsyth (2001). Noise in bilinear problems. In *Proceedings of ICCV*, volume II, pages 622–627, Vancouver, IEEE Computer Society.
17. R. Hartley (1995). In defense of the 8 point algorithm. In *ICCV 95*, pages 1064–1070.
18. R. I. Hartley, A. Zisserman (2000). *Multiple View Geometry in Computer Vision.* Cambridge University Press.
19. H.-P. Helfrich, D. Zwick (1996). A trust region algorithm for parametric curve and surface fitting. *J. Comp. Appl. Math.*, 73:119–134.
20. F. R. Helmert (1872). *Die Ausgleichungsrechnung nach der Methode der Kleinsten Quadrate.* Teubner, Leipzig.
21. D. Hestenes, G. Sobczyk (1984). *Clifford algebra to geometric calculus.* D. Reidel Publishing Comp.
22. D. Hestenes, R. Ziegler (1991). Projective geometry with Clifford algebra. *Acta Applicandae Mathematicae.*
23. S. Heuel (2004). *Uncertain Projective Geometry – Statistical Reasoning for Polyhedral Object Reconstruction.* LNCS 3008. Springer, Berlin Heidelberg New York.
24. K. Kanatani (1994). Statistical bias of conic fitting and renormalization. *IEEE Trans. Pattern Analysis and Machin Intelligence*, 16(3):320–326.
25. K. Kanatani (1996). *Statistical Optimization for Geometric Computation: Theory and Practice.* Elsevier Science.
26. K. Kanatani, D.D. Morris (2001). Gauges and gauge transformations for uncertainty description of geometric structure with indeterminacy. *IEEE Transactions on Information Theory*, 47(5):1–12.
27. K.-R. Koch (1988). *Parameter estimation and hypothesis testing in linear models.* Springer, Berlin Heidelberg New York.
28. B. Matei, P. Meer (2000). A general method for errors-in-variables problems in computer vision. In *Computer Vision and Pattern Recognition Conference*, volume II, pages 18–25. IEEE.
29. E. M. Mikhail, F. Ackermann (1976). *Observations and Least Squares.* University Press of America.
30. A. Papoulis (1965). *Probability, Random Variables, and Stochastic Processes.* McGraw-Hill.
31. R. C. Rao (1973). *Linear Statistical Inference and Its Applications.* Wiley, NY.
32. B. Rosenhahn, G. Sommer (2002). Pose estimation in conformal geometric algebra. Technical Report 0206, Inst. f. Informatik u. Praktische Mathematik, Universität Kiel.
33. J. G. Semple, G. T. Kneebone (1952). *Algebraic Projective Geometry.* Oxford Science.
34. R. Smith, M. Self, P. Cheeseman (1991). A Stochastic Map for Uncertain Spatial Relationships. In: S. S. Iyengar, A. Elfes (eds.): *Autonomous Mobile Robots: Perception, Mapping, and Navigation*, vol. 1. IEEE Computer Society Press, pp. 323–330.
35. J. Stolfi (1991). *Oriented Projective Geometry: A Framework for Geometric Computations.* Academic Press, San Diego.
36. G. Taubin (1993). An improved algorithm for algebraic curve and surface fitting. In *Fourth ICCV*, Berlin, pages 658–665.

16

The Tensor Voting Framework

Gérard Medioni, Philippos Mordohai, and Mircea Nicolescu

Institute for Robotics and Intelligent Systems, University of Southern California, medioni@iris.usc.edu

16.1 Introduction

The design and implementation of a complete artificial vision system is a daunting challenge. The computer vision community has made significant progress in many areas, but the ultimate goal is still far off. A key component of a general computer vision system is a computational framework that can address a wide range of problems in a unified way. We have developed such a framework for mid-level vision over the past several years. It is based on a data representation formalism that uses *second-order symmetric non-negative definite tensors* and an information propagation mechanism termed *tensor voting*.

The term mid-level vision refers to stages of processing that aim at bridging the gap between low-level modules, which produce image primitives, such as points of interest or edges, and high-level modules, which perform the semantic analysis of what is being viewed. The tensor voting framework provides the means for organizing oriented and un-oriented primitives, such as points or curve or surface elements, into perceptual structures, which can serve as input to high-level processing modules, or be the desired output themselves in some applications. Most computer vision problems can be posed as the extraction of scene descriptions and interpretations from 2-D or 3-D (in the case of medical or range data) images. After the desired primitives have been generated, we claim that the problem of describing what is being viewed in terms of structures, such as curves, surfaces, regions, junctions and intersections, can be addressed within a perceptual organization framework guided by the gestalt principles [53]. This claim is valid as long as the input scenes consist of coherent objects that are smooth almost everywhere, according to the paradigm set by Marr [31].

The tensor voting framework offers many advantages, one of which is the unified symbolic representation of all possible types of perceptual structures in the same space, which allows different structures, for instance, curves and surfaces, to interact with each other. Since processing is not iterative and tensor voting is performed locally, the framework is efficient in terms of computational complexity, unlike other perceptual organization methodologies of exponential complexity. Performance degrades gracefully with noise as shown by our experiments, where satisfactory results were obtained under severe noise corruption [15, 50, 32]. Finally, the framework is flexible since it can operate on oriented or unoriented data, or a combination of both, and can easily be generalized to spaces of any higher dimensions.

This chapter is organized as follows: in Sect. 16.2 we briefly review related research in perceptual organization; in Sect. 16.3 we present the second order tensor voting framework and the representation, voting and structure extraction mechanisms; in Sect. 16.4 we present a first-order augmentation to the original framework for boundary inference; and in Sect. 16.5 we show results on two particular computer vision problems, namely stereo and motion analysis.

16.2 Related Work on Perceptual Organization

Perceptual organization has been an active research area. Important issues include noise robustness, detection of discontinuities and computational complexity. This section reviews related work, which can be classified according to the approach taken in the following categories:

- regularization
- relaxation labeling
- geometric techniques
- robust methods
- level set methods
- symbolic methods
- clustering
- methods based on local interactions
- psychophysiology and neuroscience inspired methods

16.2.1 Regularization

Because of the projective nature of imaging, a single image can correspond to different scene configurations. Because of the image formation ambiguity, the inverse problem, the inference of structures from images, is ill-posed. To address this ambiguity, constraints have to be imposed on the solution space. Within the regularization theory, this is achieved by selecting the appropriate objective function and optimizing it according to global constraints. Poggio et al. [39] present the application of regularization theory to computer vision problems. Terzopoulos [51] and Robert and Deriche [40] address the issue of preserving discontinuities while enforcing global smoothness in a regularization framework. A Bayesian formulation of the problem based on minimum description length is proposed by Leclerc [25]. Variational techniques are used by Horn and Schunck [20] for the estimation of optical flow, and by Morel and Solimini [34] for image segmentation. In both cases the goal is to infer functions that optimize the selected criteria, while preserving discontinuities.

16.2.2 Relaxation Labeling

A different approach to vision problems is relaxation labeling. The problems are cast as the assignment of labels, from a set of possible labels, to the elements of the scene. Haralick and Shapiro define the consistent labeling problem in [16] and [17]. Faugeras and Berthod [7] describe a gradient optimization approach to relaxation labeling. A global criterion is defined that combines the concepts of ambiguity and consistency of the labeling process. Geman and Geman discuss how stochastic relaxation can be applied to the task of image restoration in [10]. MAP estimates are obtained by a Gibbs sampler and simulated annealing. Hummel and Zucker [21] attempt to develop an underlying theory for the continuous relaxation process. One result is the development of an explicit function to maximize the relaxation process, leading to a new relaxation operator. The second result is that finding a consistent labeling is equivalent to solving a variational inequality. This work was continued by Parent and Zucker [38] for the inference of trace points in 2-D, and by Sander and Zucker [42] in 3-D.

16.2.3 Geometric Techniques

Techniques for inferring surfaces from 3-D point clouds have been reported in the computer graphics literature. Boissonnat [2] proposes a technique based on computational geometry for object representation by triangulating point clouds in 3-D. Hoppe et al. [19] infer surfaces from unorganized point clouds as the zero levels of a signed distance function from the unknown surface. The strength of their approach lies in the fact that the surface model, topology and boundaries need not be known a priori. Later, Edelsbrunner and Mücke [6] introduce the *three-dimensional alpha shapes* that are based on a 3-D Delaunay triangulation of the data. Szeliski et al. [49] describe a method for modeling surfaces of arbitrary or changing topology using a set of oriented dynamic particles that interact according to distance, coplanarity, conormality and cocircularity. The drawbacks of computational geometry based methods are their sensitivity to even a very small number of outliers and their computational complexity.

16.2.4 Robust Methods

In the presence of noise, robust techniques inspired by random sample consensus (RANSAC) [8] can be applied. Small random samples are selected from the noisy data and are used to derive model hypotheses, which are tested using the remainder of the dataset. Hypotheses that are consistent with a large number of the data points are considered valid. Variants of RANSAC include the residual consensus(RESC) [57] and the mutual inlier ratio (MIR) [24], which are mainly used for segmentation of surfaces from noisy 3-D point clouds. The extracted surfaces are limited to planar or quadratic, except for the approach in [26] which can extract high-order polynomial surfaces. In all cases an a priori parametric representation of the unknown structure is necessary, thus limiting the applicability of these methods.

16.2.5 Level Set Methods

The antipode of the explicit representation of surfaces by a set of points is the implicit representation in terms of some function. In [46], Sethian proposes a *level set* approach under which surfaces can be inferred as the zero-level isosurface of a multivariate implicit function. The technique allows for topological changes; thus it can reconstruct surfaces of any genus as well as nonmanifolds. Osher et al. [58] and Osher and Fedkiw [37] propose efficient ways of handling implicit surfaces as level sets of a function. A combination of points and elementary surfaces and curves can be provided as input to their technique, which can handle local changes locally, as well as global deformations and topological changes. All the implicit surface-based approaches are iterative and require careful selection of the implicit function and initialization. The surface in explicit form, as a set of polygons, can be extracted by a technique such as the classic Marching Cubes algorithm [29]. The simultaneous representation of surfaces, curves and junctions is impossible, and all the approaches are limited to closed surfaces.

16.2.6 Symbolic Methods

Following the paradigm set by Marr [31], many researchers developed methods for hierarchical grouping of symbolic data. Lowe [30] developed a system for 3-D object recognition based on perceptual organization of image edgels. Groupings are selected among the numerous possibilities according to the gestalt principles, viewpoint invariance and low likelihood of being accidental formations. Later, Mohan and Nevatia [33] and Dolan and Riseman [5] also proposed perceptual organization approaches based on the gestalt principles [53]. Both are symbolic and operate in a hierarchical bottom-up fashion starting from edgels and increasing the level of abstraction at each iteration. The latter approach aims at extracting curvilinear structures, while the former aims at segmentation and the extraction of 3-D scene descriptions from collations of features that have high likelihood of being projections of scene objects. Along the same lines is Jacobs' [22] technique for inferring salient convex groups among clutter since they most likely correspond to world objects. The criteria to determine the non-"accidentalness" of the potential structures are convexity, proximity and contrast of the edgels.

16.2.7 Clustering

A significant current trend in perceptual organization is clustering [23]. Data are represented as nodes of a graph, and the edges between them encode the likelihood that two nodes belong in the same partition of the graph. Clustering is achieved by cutting some of these edges in a way that optimizes global criteria. A landmark approach in the field was the introduction of *normalized cuts* by Shi and Malik [48]. They aim at maximizing the degree of dissimilarity between the partitions normalized by essentially the size of each partition, in order to remove the bias for small clusters. Boykov et al. [3] use graph cut-based algorithms to approximately optimize energy functions whose explicit optimization is NP-hard. They demonstrate the validity of their approach on a number of computer vision problems. Stochastic clustering algorithms have been developed by Cho and Meer [4] and Gdalyahu et al. [9]. A consensus of various clusterings of the data is used as a basis of the solution. Finally, Robles-Kelly and Hancock [41] present a perceptual grouping algorithm based on graph cuts and an iterative expectation maximization scheme, which improves the quality of results at the expense of increased computational complexity.

16.2.8 Methods Based on Local Interactions

We now turn our attention to perceptual organization techniques that are based on local interaction between primitives. Shashua and Ullman [47] first addressed the issue of structural saliency and how prominent curves are formed from tokens that are not salient in isolation. They define a locally connected network that assigns a saliency value to every image location according to the length and smoothness of curvature of curves going through that location. In [38], Parent and Zucker infer trace points and their curvature based on spatial integration of local information. An important aspect of this method is its robustness to noise. This work was extended to surface inference in three dimensions by Sander and Zucker [42]. Sarkar and Boyer [43] employ a voting scheme to detect a hierarchy of tokens. Voting in parameter space has to be performed separately for each type of feature, thus making the computational complexity prohibitive for generalization to 3-D. The inability of previous techniques to simultaneously handle surfaces, curves and junctions was addressed in the precursor of our research in [15]. A unified framework where all types of perceptual structures can be represented is proposed along with a preliminary version of the voting scheme presented here. The major advantages of the work of Guy and Medioni were noise robustness and computational efficiency, since it is not iterative. How this methodology evolved is presented in the remaining sections of this chapter.

16.2.9 Psychophysiology- and Neuroscience-Inspired Methods

Finally, there is an important class of perceptual organization methods that are inspired by human perception and research in psychophysiology and neuroscience. Grossberg and Mingolla [12] and Grossberg and Todorovic [13] developed the *boundary contour system* and the *feature contour system* that can group fragmented and even illusory edges to form closed boundaries and regions by feature cooperation in a neural network. Heitger and von der Heydt [18], in a classic paper on neural contour processing, claim that elementary curves are grouped into contours via convolution with a set of orientation-selective kernels, whose responses decay with distance and difference in orientation. Williams and Jacobs [55] introduce the *stochastic completion fields* for contour grouping. Their theory is probabilistic and models the contour from a source to a sink as the motion of a particle performing a random walk. Particles decay after every step, thus minimizing the likelihood of completions that are not supported by the data or between distant points. Li [28] presents a contour integration model based on excitatory and inhibitory cells and a top-down feedback loop. What is more relevant to our research, which focuses on the preattentive bottom-up process of perceptual grouping, is that connection strength decreases with distance, and that zero- or low-curvature alternatives are preferred to high-curvature ones. The model for contour extraction of Yen and Finkel [56] is based on psychophysical and physiological evidence that has many similarities to ours. It employs a voting mechanism where votes, whose strength decays as a Gaussian function of distance, are cast along the tangent of the osculating circle. Even though we do not attempt to present a biologically plausible system, the similarities between our framework and the ones presented in this paragraph are nevertheless encouraging.

16.2.10 Our Approach

Some important aspects of our approach in the context of the work presented in this section are discussed here. In case of dense, noise-free, uniformly distributed data, we are able to match the performance of surface extraction methods such as [2, 6, 52, 46, 29]. Furthermore, our results degrade much more gracefully in the presence of noise (see, for example, [14] and [32]) and the multiscale implementation allows us to overcome uneven data densities. The input can be oriented, unoriented tokens or a combination of both, while many of the techniques mentioned above require oriented inputs to proceed. In addition to the advantages this brings, we are able to extract open and closed surfaces and curves and junctions in 3-D simultaneously. Our model-free approach allows us to handle arbitrary perceptual structures that adhere to the "matter is cohesive" [31] principle. Model-based approaches cannot easily distinguish between model misfit and noise. To our knowledge, the tensor voting framework is the only methodology that can represent and infer all possible types of structures in any dimension in the same space. Our voting function has many similarities with other voting-based methods, such as the decay with distance and curvature [18, 56, 28], and the use of constant-curvature paths [38, 44, 43, 56] that result in an eight-shaped voting field (in 2-D) [18, 56].

The major difference is that in our case the votes cast are tensors and not scalars, therefore they can express much richer information.

16.3 The Original Second-Order Tensor Voting Framework

In this section we review the original second-order tensor voting framework as it was presented in [32]. The purpose of the framework is to serve as a computational mechanism for perceptual grouping of oriented and unoriented primitives. It has mainly been applied to mid-level vision problems, but it is suitable for any problem (of any dimensionality) that can be formulated as a perceptual organization problem. The novelty of our approach is that there is no objective function that is explicitly defined and optimized according to global criteria. Instead, tensor voting is performed locally, and the saliency of perceptual structures is estimated as a function of the support tokens receive from their neighbors. Tokens with compatible orientations that can form salient structures reinforce each other. The support of a token for its neighbors is expressed by *votes* that are cast according to the gestalt principles of proximity, colinearity and cocurvilinearity.

We begin by describing the representation, then illustrate the voting mechanism and introduce the concept of voting fields and how they are derived from the *2-D second-order fundamental stick voting field*. Even though the illustration is for the 3-D case for visualization reasons, all aspects of the framework can easily be generalized to any dimensions. Finally, we briefly review the way dense structures such as surfaces and curves can be extracted from sparse data.

16.3.1 Representation

The representation of a token consists of a symmetric second-order tensor that encodes *perceptual saliency*. The tensor essentially indicates the saliency of each type of perceptual structure (surface, curve or region in 3-D) the token belongs to, and its preferred normal and tangent orientations. Tensors were first used as a signal processing tool for computer vision applications by Granlund and Knutsson [11] and Westin [54]. Our use of tensors differs in that our representation is not signal based, but rather symbolic, where hypotheses for the presence of a perceptual structure at a given location are represented as tokens with associated second-order tensors that encode the likelihood for the presence of a perceptual structure at the location, the most likely type of structure and its preferred tangent or normal orientations. The power of this representation lies in that all types of saliency are encoded by the same tensor.

A representation scheme sufficient for our purposes must be able to encode both smooth perceptual structures as well as discontinuities. These occur at locations where multiple salient structures such as curves, surfaces or region boundaries meet. Curve orientation discontinuities occur at locations where multiple curve segments intersect, while surface orientation discontinuities occur where multiple surface patches intersect. In other words, whereas there is only one orientation associated with a location within a smooth curve segment, or a surface patch or a region boundary, there are multiple orientations associated with locations where a discontinuity occurs. Hence, the desirable data representation is one that can encode more than one orientation at a given location. It turns out that a second-order symmetric tensor possesses precisely this property.

An N-D, symmetric, non-negative definite, second-order tensor can be viewed as a $N \times N$ matrix or equivalently an N-D hyperellipsoid. Intuitively, its shape indicates the type of structure represented and its size the saliency of this information. In 3-D, the tensor can be decomposed as in the following equation:

$$T = \lambda_1 \hat{e}_1 \hat{e}_1^T + \lambda_2 \hat{e}_2 \hat{e}_2^T + \lambda_3 \hat{e}_3 \hat{e}_3^T$$
$$= (\lambda_1 - \lambda_2)\hat{e}_1 \hat{e}_1^T + (\lambda_1 - \lambda_2)(\hat{e}_1 \hat{e}_1^T + \hat{e}_2 \hat{e}_2^T) + \lambda_3(\hat{e}_1 \hat{e}_1^T + \hat{e}_2 \hat{e}_2^T + \hat{e}_3 \hat{e}_3^T), \tag{16.1}$$

where λ_i are the eigenvalues in decreasing order and $\hat{e}_i$ are the corresponding eigenvectors (see also Fig. 16.1). Note that the eigenvalues are non-negative since the tensor is non-negative definite and the eigenvectors are orthogonal. The first term in Eq. (16.1) corresponds to a degenerate elongated ellipsoid, termed hereafter the *stick tensor*, that indicates an elementary surface token with $\hat{e}_1$ as its surface normal. The second term corresponds to a degenerate disk-shaped ellipsoid, termed hereafter the *plate tensor*, that indicates a curve or a surface intersection with $\hat{e}_3$ as its tangent, or, equivalently normal to the subspace spanned by with $\hat{e}_1$ and $\hat{e}_2$. Finally, the third term corresponds to a sphere, termed the *ball tensor*, that corresponds to a junction that has no preference of orientation. The ball tensor can also be viewed as a measure of uncertainty.

The size of the tensor indicates the certainty of the information represented by the tensor. Therefore, the size of the stick component $(\lambda_1 - \lambda_2)$ indicates *surface saliency*, the size of the plate component $(\lambda_2 - \lambda_3)$ indicates *curve saliency* and that of the ball component (λ_3) *junction saliency*. Note that the representation is in terms of normals. Therefore, a surface patch in 3-D is represented by a stick tensor parallel to the patch's normal. A curve, which can also be viewed as a surface intersection, is represented by two salient normals, the plate tensor. Adopting this representation, as opposed to a representation by tangents, allows a structure with $N-1$ degrees of freedom in N-D (a curve in 2-D, a surface in 3-D) to be represented by a single vector, while a tangent representation would require the definition of $N-1$ vectors that form a basis for an $(N-1)$-D subspace. Our choice for the normal representation is justified since, typically, the most frequent structures in an N-D space are $(N-1)$-D

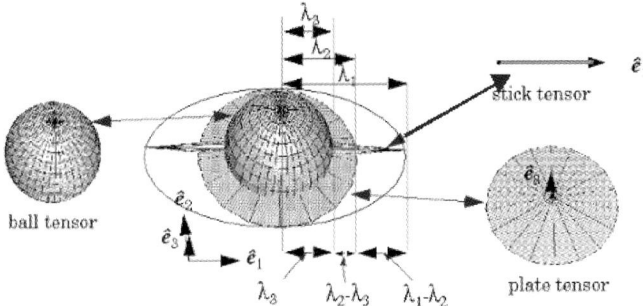

Fig. 16.1. second-order generic tensor and its decomposition into the *stick*, *plate* and *ball* components in 3-D

manifolds, which are represented by the *stick tensor*. The *stick voting field* is the basis from which all other voting fields are derived.

The tensors can be initialized as balls with no preference of orientation, or, if prior knowledge is available, with some preferred orientation. But in general, after voting, a generic tensor comprising all three components will be the representation for each token. The benefit of having this representation is that the likelihood of the token belonging to each type of structure can be encoded simultaneously and carried throughout the processing stages without having to make premature, hard decisions, or maintain separate maps for every token type.

16.3.2 Tensor Voting

The core of our framework is the way information is propagated from token to token. The question we want to answer is: assuming that a token at O with normal $\mathbf{N}$ and a token at P belong to the same smooth perceptual structure, what information should the token at O cast at P? We first answer the question for the 2-D case of a voter with a pure *stick tensor* and show how all other cases can be derived from it. We claim that, in the absence of other information, the arc of the *osculating circle* (the circle that shares the same normal as a curve at the given point) at O that goes through P is the most likely smooth path, since it maintains constant curvature. In case of straight continuation from O to P, the osculating circle degenerates to a straight line. Similar use of primitive circular arcs can also be found in [38, 44, 43].

As shown in Fig. 16.2, the second-order vote at P is also a stick tensor and has a normal lying along the radius of the osculating circle. What remains to be defined is the magnitude of the vote. According to the gestalt principles it should be a function of proximity and smooth continuation. The *saliency decay function* we have selected at [14, 32] has the following form:

$$DF(s, \kappa, \sigma) = e^{-(\frac{s^2 + c\kappa^2}{\sigma^2})}, \qquad (16.2)$$

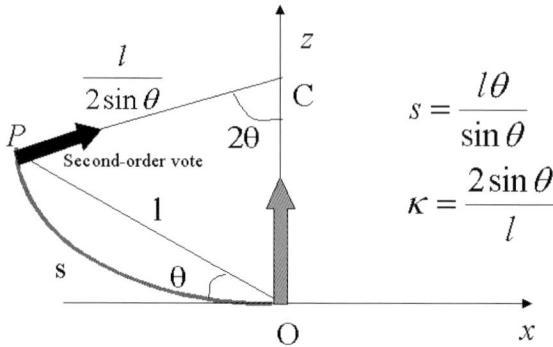

Fig. 16.2. Second-order vote cast by a stick tensor located at the origin

where s is the arc length OP, κ is the curvature, c is a constant that controls the decay with high curvature, and σ is the scale of analysis, which determines the effective neighborhood size. Since we use a Gaussian decay function for $DF(\cdot)$, the effective neighborhood size is about 3σ. The derivation is as follows: let k be the effective neighborhood size. Then, we can set $e^{-\frac{k^2}{\sigma^2}} = \epsilon$, a small number, which is the lower bound of significant magnitude. We can then derive the effective neighborhood size k given σ. Note that σ is the only free parameter in the system.

The voting process is identical whether the receiver contains a token or not, but we use the term *sparse vote* to describe a pass of voting where votes are cast to locations that contain tokens only. We use the term *dense vote* for a pass of voting from the tokens to all locations within the neighborhood regardless of the presence of tokens.

16.3.3 Voting Fields

In this section we show how all the necessary votes can be cast in the same way as described in the previous section for the 2-D stick tensor case, and how any second-order field in any dimension can be derived. Finally, we show how the votes cast by an arbitrary tensor can be computed, given the voting fields.

The second-order stick voting field is a second-order tensor field, which at every position contains a tensor that is the vote cast there by a unitary stick tensor located at the origin and aligned with the y-axis. The shape of the field in 2-D can be seen in Fig. 16.3a. Note that it is identical to a cut of the 3-D field containing the origin, since the voting stick tensor and the receiver define a plane in 3-D where the voting takes place. Depicted at every position is the eigenvector corresponding to the largest eigenvalue of the second-order tensor contained there. Its size is proportional to the magnitude of the vote. To compute a vote cast by an arbitrary stick tensor, we need to align the field

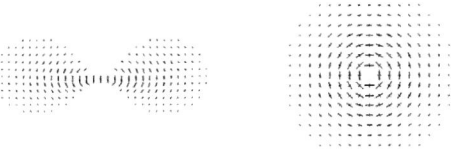

(a) 2-D stick voting field (b) 2-D ball voting field

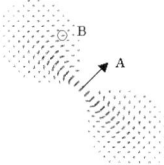

(c) The stick field is aligned with the voter A and a vote is cast at B

Fig. 16.3. a-c Voting field illustration in 2-D

with the orientation of the voter and multiply the saliency of the vote that coincides with the receiver by the saliency of the arbitrary stick tensor, as in Fig. 16.3c. Since the locations O and P and the unitary stick tensor define a plane in 3-D, the generation of stick votes is identical in 2-D, 3-D and N-D, as can be easily shown. The stick voting fields in higher dimensions, therefore, can be derived by a simple rotation of the 2-D stick field.

At the other end of the spectrum is the *ball voting field*, a cut of which can be seen in Fig. 16.3b. The ball tensor has no preference of orientation, but still it can cast meaningful information to other locations. The presence of two proximate unoriented tokens, the voter and the receiver, indicates a potential perceptual structure. In the 3-D case, this can either be a curve segment, or a pencil of planes intersecting on that segment. Even though the voters are unoriented, surfaces can be inferred, since the accumulation of votes from point to point with one degree of freedom in terms of surface orientation, from neighbors in the same surface, results in a high certainty for the correct surface normal and eliminates the degree of freedom. The case for curves is similar. The ball voting fields allow us to infer preferred orientations from unoriented tokens, thus minimizing initialization requirements.

To show the derivation of the ball voting field from the stick voting field, we can visualize the vote at P from a unitary ball tensor at the origin O as the integration of the votes of stick tensors that span the space of all possible orientations. In 2-D, this is equivalent to a rotating stick tensor that spans the unit circle at O, while in 3-D the stick tensor spans the unit sphere. The 3-D ball field can be derived from the stick field $\mathbf{S}(P)$, as follows:

$$\mathbf{B}(P) = \int_0^\pi \int_0^\pi R_{\theta\phi\gamma} \mathbf{S}(R_{\theta\phi\gamma}^{-1} P) R_{\theta\phi\gamma}^T \mathrm{d}\phi \mathrm{d}\gamma |_{\theta=0}, \qquad (16.3)$$

where $R_{\theta\phi\gamma}$ is the rotation matrix to align $\mathbf{S}$ with $\hat{e}_1$, the eigenvector corresponding to the maximum eigenvalue (the stick component), of the rotating tensor at P, and θ, ϕ, γ are rotation angles about the x-, y-, z-axes respectively.

In practice, the integration is approximated by a summation (tensor addition):

$$V = \sum_{i=0}^{\pi} \sum_{j=0}^{\pi} \mathbf{v}_{ij} \mathbf{v}_{ij}^T, \qquad (16.4)$$

where V is the accumulated vote, and $\mathbf{v}_{ij}$ are the votes from O to P cast by the stick tensor as i, j span the unit sphere. Normalization has to be performed in order to make the energy emitted by a unitary ball equal to that of a unitary stick. As a result of the integration, the second-order ball field does not contain purely stick or purely ball tensors, but arbitrary second-order symmetric tensors. The field is radially symmetric, as expected, since the voter has no preferred orientation.

The *plate voting field* completes the set of voting fields for the 3-D case. Its description illustrates how any voting field in any dimension can be generated. Since the plate tensor encodes uncertainty of orientation around one axis, it can be derived by integrating the votes of a rotating stick tensor that spans the unit circle, in other words the plate tensor. The formal derivation is analogous to that of the ball voting fields and can be written as follows:

$$\mathbf{P}(P) = \int_0^{\pi} R_{\theta\phi\gamma} \mathbf{S}(R_{\theta\phi\gamma}^{-1} P) R_{\theta\phi\gamma}^T d\gamma|_{\theta=\phi=0}, \qquad (16.5)$$

where θ, ϕ, γ, and $R_{\theta\phi\gamma}$ have the same meaning as in the previous equation.

The generalization to N dimensions is straightforward. Starting from the N-D stick field, which is a rotated version of the *2-D second-order fundamental stick field*, the remaining $N-1$ fields can be derived by integration as shown here for the 3-D case.

All voting fields are functions of the position of the receiver relative to the voter and a single parameter, the scale of the saliency decay function. After these fields have been precomputed at the desired resolution, computing the votes cast by any second-order tensor is reduced to a few look-up operations and linear interpolation. Voting takes place in a finite neighborhood within which the magnitude of the votes cast remains significant. The size of this neighborhood is obviously a function of the scale σ. As described in Sect. 16.3.1, any tensor can be decomposed into the basis components (stick, plate and ball in 3-D) according to its eigensystem. Then, the corresponding fields can be aligned with each component. Votes are retrieved by simple look-up operations, and their magnitude is multiplied by the corresponding saliency. For instance, in 3-D the saliency of the stick component is $\lambda_1 - \lambda_2$, of the plate component $\lambda_2 - \lambda_3$, and of the ball component λ_3.

The complexity of tensor voting is $O(kn)$. Each voter casts on average k votes to its neighbors, where k depends on data density and the scale of voting. In the worst case, this can lead to $O(n)$ votes per token, but that is a clear indication of incorrect setting of the size of the neighborhood. For most practical cases, k is a small fraction of the data set. Since votes are precomputed, each vote requires a fixed number of linear operations (matrix multiplication and linear interpolation) to be cast. The overhead before voting can begin comes form two sources: the computation of the voting fields, and the generation of the data structure that holds the data.

16.3.4 Vote Collection and Interpretation

Votes are cast from token to token (sparse vote), as described in the previous section, and they are accumulated by tensor addition. The resulting tensor at each token is in general an arbitrary tensor. Analysis of the second-order votes can be performed once the eigensystem of the accumulated second-order $N \times N$ tensor has been computed. In 3-D, the tensor can be decomposed into the stick, plate and ball components:

$$T = (\lambda_1 - \lambda_2)\hat{e}_1\hat{e}_1^T + (\lambda_2 - \lambda_3)(\hat{e}_1\hat{e}_1^T + \hat{e}_2\hat{e}_2^T) + \lambda_3(\hat{e}_1\hat{e}_1^T + \hat{e}_2\hat{e}_2^T + \hat{e}_3\hat{e}_3^T), \quad (16.6)$$

where $\hat{e}_1\hat{e}_1^T$ is a *stick tensor*, $\hat{e}_1\hat{e}_1^T + \hat{e}_2\hat{e}_2^T$ is a *plate tensor*, $\hat{e}_1\hat{e}_1^T + \hat{e}_2\hat{e}_2^T + \hat{e}_3\hat{e}_3^T$ is a *ball tensor*. The following cases have to be considered. If $\lambda_1 \gg \lambda_2, \lambda_3$, this indicates certainty of one normal orientation, therefore the token most likely belongs on a surface. In case of a token that belongs on a curve, or surface intersection, the uncertainty in normal orientation spans a plane perpendicular to the tangent. Hence, the inferred tensor is platelike, that is, $\lambda_1 \approx \lambda_2 \gg \lambda_3$. If the token is a point junction, $\lambda_1 \approx \lambda_2 \approx \lambda_3$, and the dominant component is the ball. An outlier receives few, inconsistent votes, so all eigenvalues are small and no preference of orientations emerges.

16.3.5 Extraction of Dense Structures

Now that the most likely type of feature at each token has been estimated, we want to compute the dense structures (curves and surfaces in 3-D) that can be inferred from the tokens. This can be achieved by casting votes to *all* locations, whether they contain a token or not (dense vote). Then, each site contains a 2-tuple $(s, \hat{v})$, indicating feature saliency and direction. Given this dense information, a modified marching algorithm [50] (based on the marching cubes algorithm [29]) is used to extract surfaces and curves that correspond to zero crossings in s along $\hat{v}$'s. Junctions are isolated and, therefore, are extracted as maxima of junction saliency. Generalization of dense structure extraction to higher dimensions is not impossible, but is definitely not an easy task, simply because of space storage complexity.

In order to reduce computational cost, the calculation of saliency tensors at locations with no prior information and structure extraction are integrated

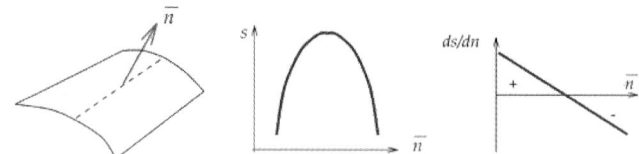

Fig. 16.4. Dense surface extraction in 3-D: (*Left*) Elementary surface patch with normal **n**. (*Center*) 3-D surface saliency along normal direction (*Right*) First derivative of surface saliency along normal direction

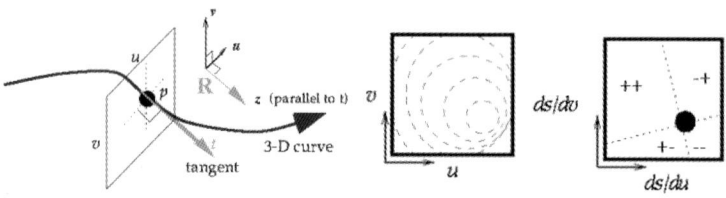

Fig. 16.5. Dense curve extraction in 3-D: (*Left*) 3-D curve with tangent **t** and the normal plane. (*Center*) Curve saliency isocontours on the normal plane. (*Right*) First derivatives of surface saliency on the normal plane

and performed as a marching process. Beginning from seeds, locations with highest saliency, we perform a dense vote only toward the directions dictated by the orientation of the features. Surfaces are extracted with subvoxel accuracy, as the zero-crossings of the first derivative of surface saliency (Fig. 16.4). Locations with high *surface* saliency are selected as seeds for surface extraction, while locations with high *curve* saliency are selected as seeds for curve extraction. The marching direction in the former case is perpendicular to the surface normal, while, in the latter case, the marching direction is along the curve's tangent (Fig. 16.5). Curve junction saliency gives rise to isolated local extrema, and therefore is not propagated in the extraction process.

16.4 First-Order Voting

In this section, we describe how the original, strictly second-order framework is augmented by including first-order properties, in order to infer discontinuities. The addition of *polarity vectors* (first-order tensors) to the representation complements the strictly second-order representation that was insufficient for encoding first-order properties, such as boundaries of perceptual structures. The new representation exploits the essential property of boundaries to have all their neighbors, at least locally, on the same side of a half-space. As described in the remainder of this section, the voting scheme is identical to that

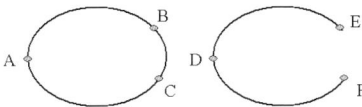

Fig. 16.6. Curve orientation still varies smoothly at endpoints of open contours. Therefore, they cannot be distinguished from interior points based on the second-order representation

of the second-order case, and the first-order vector voting fields can be easily derived from the second-order fundamental tensor voting field.

16.4.1 Motivation

To illustrate the significance of this contribution, consider the contour depicted in Fig. 16.6, keeping in mind that we represent curve elements by their normals. Consider points A and D, which are smooth inliers of the contours. The second-order tensors associated with these points are identical in terms of both saliency and orientation. The problem appears when comparing points B and C with E and F. The former are inliers of the closed contour, while the latter are the endpoints of the open contour. The second-order tensor at B has almost identical orientation as the one at E. Even though there is a difference in curve saliency since E receives less support from its neighborhood than B, the inferred description is very similar for two points that are qualitatively very different. This occurs because the second-order representation is inadequate to capture the key property of endpoints: that all their neighbors in the contour are on the same side. The first-order augmentation to the framework addresses this shortcoming by being sensitive to the direction from which votes are received.

16.4.2 Representation and Voting

The representation is augmented by the addition of the *polarity vector*, which can be viewed as a vector pointing toward the direction of maximum saliency, or, in other words, the direction from which the token receives the most salient votes. The polarity vector is used to collect the information that cannot be captured by the second-order tensor, which is insensitive to the direction from which votes are received. In the augmented framework, tokens cast second-order votes as described in the previous section and first-order votes as described here.

As shown in Fig. 16.7, the first-order vote cast by a unitary stick tensor at the origin is *tangent* to the osculating circle, the smoothest path between the voter and receiver. Its magnitude, since nothing suggests otherwise, is equal to that of the second-order vote according to Eq. (16.2). What should be noted is that polarity vectors are initialized to zero since no prior information

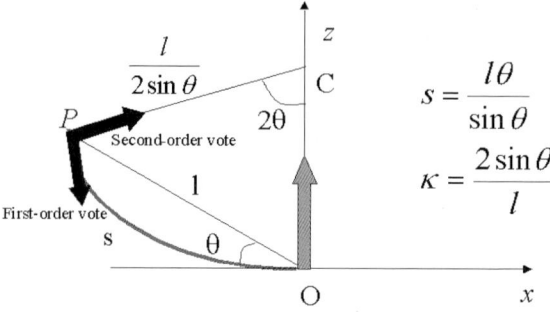

Fig. 16.7. Second- and first-order votes cast by a stick tensor located at the origin

is available on whether the token is on a boundary or not. Therefore, first-order votes are cast based on the second-order part of the representation. The *fundamental 2-D first-order stick voting field* is a vector field, which at every position holds a vector that is equal in magnitude to the stick vote that exists in the same position in the fundamental second-order stick voting field, but is tangent to the smooth path between the voter and receiver instead of normal to it. First-order stick fields in higher dimensions can be derived by rotation of the 2-D field around the unitary stick. The other fields can be derived as in the second-order case by substituting the first-order vote generation function in Eqs. (16.3) and (16.5), keeping in mind that the votes are not tensors but vectors in this case, and accumulation has to be performed by vector addition. Before voting, the second-order tensors of the voters are decomposed into the basis tensors (stick, plate and ball in 3-D) and cast both first- and second-order votes after being aligned with the appropriate fields.

Vote collection for the first-order case is performed by vector, instead of tensor, addition. The accumulated result is a vector that points to the weighted center of mass from which votes are cast, and whose magnitude encodes polarity. Since the first-order votes are weighted by the saliency of the voters and attenuate with distance and curvature, their vector sum points to the direction from which the most salient contributions were received. Low polarity indicates a token that is in the interior of a curve, surface or region, therefore surrounded by neighbors whose votes cancel each other out. On the other hand, high polarity indicates a token that is on or close to a boundary, thus receiving votes from only one side with respect to the boundary. The interpretation of polarity vectors for boundary inference is done in conjunction with second-order tensors and is described in Table 16.1.

16.4.3 Boundary Inference

In this section, we describe how the theory developed in the previous section can be used to infer boundaries of perceptual structures. We illustrate in 3-D

Table 16.1. Summary of first- and second-order tensor structure for each feature type in 3-D

3-D Feature	Saliency	Second-order tensor	Polarity	Polarity vector
Surface interior	High $\lambda_1 - \lambda_2$	Normal: $\hat{e}_1$	Low	–
Surface end-curve	High $\lambda_1 - \lambda_2$	Normal: $\hat{e}_1$	High	Orthogonal to $\hat{e}_1$ and end-curve
Curve interior	High $\lambda_2 - \lambda_3$	Tangent: $\hat{e}_3$	Low	–
Curve endpoint	High $\lambda_2 - \lambda_3$	Tangent: $\hat{e}_3$	High	Parallel to $\hat{e}_3$
Region interior	High λ_3	–	Low	–
Region boundary	High λ_3	–	High	Normal to bounding surface
Junction	Locally max λ_3	–	Low	–
Outlier	Low	–	Low	–

for visualization purposes. In this space, boundaries that can be inferred are surface boundaries, curve endpoints and region boundaries.

Surface Boundary Inference

We are interested in extracting surface end-curves, which in some applications may indicate depth discontinuities or occlusion boundaries. In the case of surfaces, both interior points and points on boundaries are characterized by a dominant stick component. The factor that differentiates between them is that the interior points have low polarity values after first-order voting, while points on the boundaries have high polarity values.

Assume we are given an open smooth surface patch in 3-D, encoded as a sparse set of tokens, possibly contained within a larger data set. The tokens are initially encoded as ball tensors since their preference of orientation is unknown. After a pass of second-order voting, the tokens that lie on the surface, both in the interior and on the boundaries, have accumulated second-order tensors with dominant stick components consistent with the normal of the surface at each location.

Then, tokens propagate first-order votes to their neighbors. As seen in Sect. 16.3.2, these votes will be along the tangent of the circular arc connecting the voter and the receiver. Therefore, the resulting polarity vector at the receiver after vote accumulation lies on a plane perpendicular to the estimated local surface normal. In case of a token in the interior of the region, the first-order votes come from all directions and cancel each other out. On the other hand, close to the surface boundaries, a large vector sum is accumulated, pointing toward the average direction (weighted by vote saliencies) from which the votes came. This direction is locally orthogonal to the boundary, at least as long as the "continuity of discontinuities" principle of Marr [31] holds. If the polarity vector is not exactly orthogonal to the estimated surface normal,

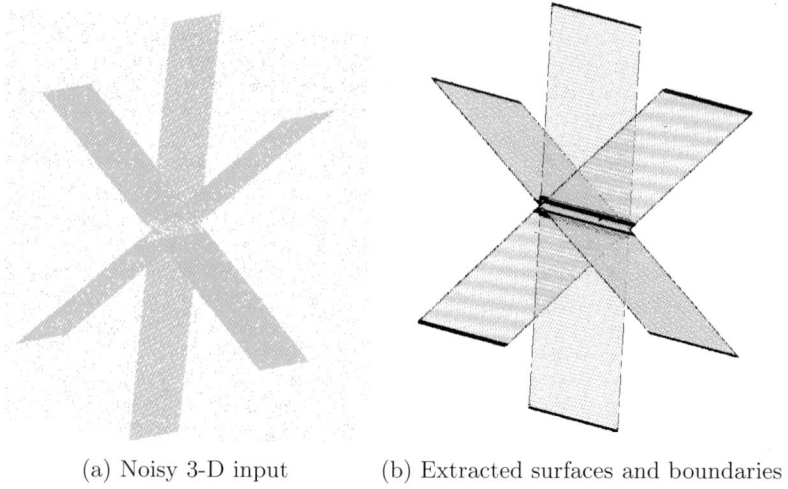

(a) Noisy 3-D input (b) Extracted surfaces and boundaries

Fig. 16.8. a, b Surface boundary inference from a synthetic dataset with additive noise

which is an indication of interference by noise, we use its projection on the plane orthogonal to the normal. The tensor is robust against this kind of interference, since it contributes to the ball component and does not affect the estimated normal, which is reliably estimated. After polarity has been inferred at all locations, we extract the curves corresponding to maxima in polarity along the tangent direction, using a modified curve marching process [50]. Results for a synthetic example that consists of three intersecting surfaces with added uniform noise can be seen in Fig. 16.8.

Curve Endpoint Inference

As in the case of surface boundaries, the second-order representation alone cannot convey whether a point is in the interior of a curve or an endpoint, since they are both characterized by a dominant plate component with curve saliency $\lambda_2 - \lambda_3$ and preferred tangent parallel to $\hat{e}_3$, the eigenvector corresponding to the minimum eigenvalue.

After second-order voting, the eigenvector corresponding to the smallest eigenvalue ($\hat{e}_3$), of the tensor inferred at each location gives the tangent orientation. We extract curve endpoints by casting first-order votes and inferring polarity. Polarity vectors are parallel to the curve tangent. This can be proven in a way analogous to the case of surface boundaries. The second-order votes collected at a curve boundary result in a dominant plate component, therefore span a plane in 3-D. Since, according to the definition of the voting fields,

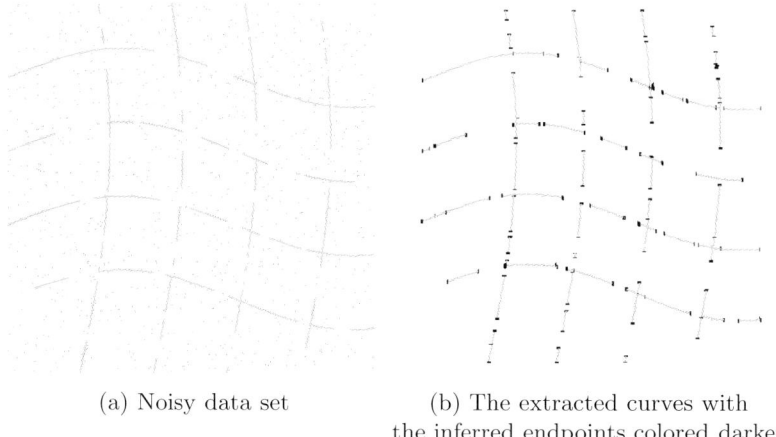

(a) Noisy data set   (b) The extracted curves with the inferred endpoints colored darker

Fig. 16.9. a, b Curve endpoint inference from noisy data

the first-order votes are orthogonal to the second-order ones, they span the remainder of the space, which has $\hat{e}_3$ as its basis.

At tokens that lie in the interior of the curve, first-order votes come from both directions and cancel each other out. At the endpoints, all first-order votes are cast from the same direction, thus combining into a large vector sum pointing toward the interior of the curve. Since curve endpoints are isolated in space, no marching process is needed for their extraction. The exterior tokens, with respect to the curve tangent, that have accumulated high polarity are selected as the endpoints. Figure 16.9 contains results from a noisy dataset.

Region Boundary Inference

Given a noisy set of points that belong to a 3-D region, we infer region boundaries in a similar way. Note that in this case, the normal refers to the vector inferred by first-order voting, since the characteristic second-order tensor of a region is a ball that has no orientation preference.

In terms of second-order tensors, regions are characterized by a dominant ball component, since they collect second-order votes from all directions in 3-D. The same holds for tokens close to the region boundaries since second-order votes are function of orientation but not direction. Once second-order information is available at each token, first-order votes are cast. The bounding surface of a 3-D region can be extracted by the modified surface marching algorithm [50] as the maximal isosurface of polarity along the normal direction, indicated by the polarity vectors. Figure 16.10 contains results on region boundary extraction for a synthetic example that consists of a peanut shape that consists of a uniform distribution of unoriented points contaminated by

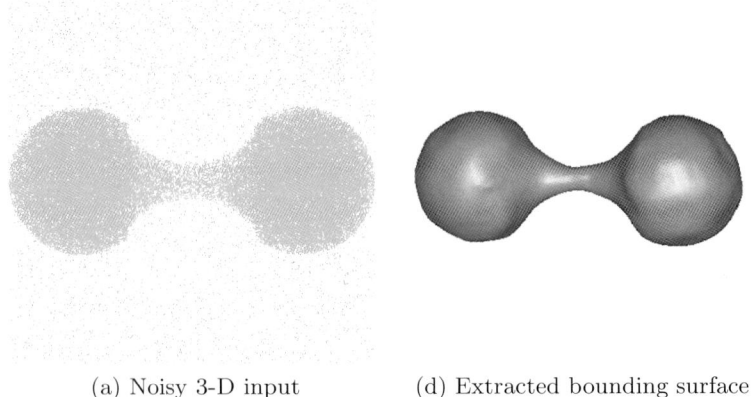

(a) Noisy 3-D input (d) Extracted bounding surface

Fig. 16.10. a, b Surface boundary inference from a synthetic dataset with additive noise

uniform noise with smaller density. The detected region boundaries are used as inputs for dense surface extraction. The extracted surfaces can be seen in Fig. 16.10b.

16.5 Applications

In this section we describe the application of the tensor voting framework to two core computer vision problems: stereo and motion analysis. Stereo vision is the process of establishing the 3-D positions of points based on their projections on two or more 2-D images. Motion analysis, on the other hand, is the process of determining the velocity of points on the image plane (optical flow) between two time instances. We treat both stereo and motion in a unified way as perceptual organization problems within the framework. We claim that tokens, generated by matching corresponding pixels in the two images, form coherent perceptual structures in the appropriate space, while erroneous matches generate outlier tokens. The tensor voting framework is suitable for these problems because it can detect perceptual structures based solely on the smoothness constraint, without using any models. This property allows us to handle arbitrary surfaces, which are unavoidable in nontrivial scenes.

A fundamental similarity between stereo and motion is that processing in both cases begins with pixel matching. It is evident from the literature and from our own experiments that there is no optimal matching method and that different methods perform better in different parts of the same image. We propose a matching scheme that combines the results of multiple methods in an attempt to the correct matches for as many pixels as possible. The novelty of the proposed method is that we do not make decisions on the correctness of matches based on their matching score, but on their perceptual saliency after tensor voting. This allows us to use a number of potential matches for every pixel through the next processing stages and make hard decisions when more information is available.

The difference between stereo and motion is the dimensionality of the space in which processing is performed. In the case of stereo with known epipolar geometry, the images can be transformed so that the epipolar lines are parallel to the horizontal axis, thus making the search space for correspondences 1-D. Therefore, since each pixel's motion is restricted on a line, the appropriate space in this case is 3-D, where each match is represented in a coordinate system that consists of the image axes and the disparity axis. In the case of motion, the search space for potential matches for each pixel is 2-D since objects can freely move anywhere within the scene. Therefore, the appropriate space for each match here is 4-D, with the axes being the two image axes and the two velocity axes. One can view stereo as a case of motion with zero vertical velocity, but that means not using the epipolar constraint, which contributes greatly to the performance of any stereo algorithm since it drastically reduces ambiguities in matching. We process stereo in 3-D to take advantage of the epipolar constraint. In both cases, potential matches are encoded as tensors and propagate their preferred orientation via tensor voting to their neighbors in 3-D and 4-D, respectively. We consider the use of 3-D and 4-D neighborhoods a critical aspect of our framework since it overcomes difficulties associated with propagating information in image neighborhoods between adjacent image pixels that are not necessarily neighboring in the scene depicted on the images. After voting, the tokens are grouped in salient perceptual structures, while the outliers of these structures correspond to false matches and are rejected.

16.5.1 Stereo

Interested readers are referred to [45] for a comprehensive review of state of the art stereo algorithms. In this section we present our approach to stereo within the tensor voting framework (an earlier version of which was published in [27]), which has the following steps:

- matching
- tensor encoding and tensor voting
- uniqueness enforcement and outlier rejection
- boundary inference
- dense surface and curve extraction

Matching

A common dilemma in the establishment of initial pixel correspondences is the need to choose a matching window large enough to contain enough intensity variations for reliable matching and small enough not to extend over pixels from different surfaces. Besides some trivial cases, there is no unique window size that is optimal for every image location. Small windows are preferable where image details exist and are necessary close to discontinuities, but they are very unreliable in textureless uniform areas. The opposite holds for large windows. Since we use the tensor voting-based stereo algorithm, which is robust to a large number of false matches, we are more interested in obtaining as many correct matches as possible, even if these are accompanied by a large number of false matches.

Given this goal, we employ a matching scheme where a number of candidate matches for every pixel are identified, in an attempt to have the correct one also included in the set, while disambiguation occurs after the tensor voting stage when uniqueness is enforced. For the results of this and the following section, we used normalized cross-correlation in rectangular windows of different sizes, starting from 3×3, and kept all peaks of cross-correlation as potential matches. The quantitative matching scores were disregarded in the subsequent stages, and all matches were initialized as ball tensors with saliency 1 in the 3-D (x, y, d) space. If two or more matches fall within the same voxel, their initial saliencies are added. This allows us to combine matches produced by totally different techniques without having to heuristically adjust different confidence measures. So far, we have experimented with normalized cross-correlation and interval matching [45], but the strategy of ignoring the matching scores and focusing on the presence or not of a likely match is a powerful tool that enables us to integrate more matching techniques into the framework without making any changes in the following processing stages.

Tensor Encoding and Tensor Voting

As mentioned in the previous paragraph, the tokens generated at the matching stage are encoded as unitary ball tensors, unless their saliency is increased, if they are confirmed by multiple matchers. First- and second-order votes are cast among neighboring tokens as described in Sects. 16.3 and 16.4. The selection of scale is not critical, in the sense that the framework's sensitivity to small variations in scale is low. Smaller values of scale are appropriate when noise levels are low and fine details exist in the data. In the presence of noise or when a considerable amount of data is missing, a larger scale would be more suitable.

Uniqueness Enforcement and Outlier Rejection

After a first pass of voting, tokens with very low saliency (surface, curve or junction) are rejected since they are not consistent with their neighbors. In addition, only the most salient among the tokens on the same line of sight is retained. The lines of sight are either defined as sets of tokens with the same (x, y) image position, in disparity space, or on the actual ray if calibration information for a metric reconstruction is available. The results after these

Table 16.2. Error rates for unoccluded pixels as a function of scale for the "sawtooth" example

Scale of voting	Error Rate (%)
10	1.32
20	1.27
50	1.09
100	0.97
200	0.92
500	0.93
1000	1.06
2000	1.10

operations for the "sawtooth" stereo pair [45] can be seen in Fig. 16.11. The displayed results are with the value of σ set at 200. Table 16.2 summarizes how the error rate for unoccluded pixels for the "sawtooth" varies with scale. Note that, even though there are differences in the results according to the desired level of smoothness expressed in the scale parameter, the overall performance of the algorithm is stable with respect to changes in scale.

Boundary Inference

A final pass of sparse first- and second-order voting is performed among the tokens that were not rejected in the previous stage. The result is a set of tokens with refined orientations. Surface boundary and curve endpoint inference are performed as described in Sect. 16.4.3. Inferred boundaries for the "sawtooth" example can be seen in Fig. 16.11e.

Dense Surface and Curve Extraction

If the goal of the stereo algorithm is the computation of depth for every pixel of the reference frame, this can be accomplished by a simple pass of voting. Disparity hypotheses are generated for pixels that have no disparity, either due to failure of the matching stage, or because all the matches for them have been rejected. Votes are collected, from the neighborhood, as before, at potential (x, y, d_i) points, where the disparities d_i range from a few disparity levels below to a few disparity levels above the range of the neighbors. The most salient point among these candidates for each optical ray is selected. Results for the "sawtooth" example can be seen in Fig. 16.11f.

Otherwise, dense surfaces and curves can be extracted as in Sect. 16.3.5. The inferred boundaries are critical in this stage since they indicate where the marching process should be terminated. Experimental results can be seen in Fig. 16.12, where we show texture-mapped surfaces obtained for an aerial stereo pair to demonstrate that our approach is not limited to planar surfaces.

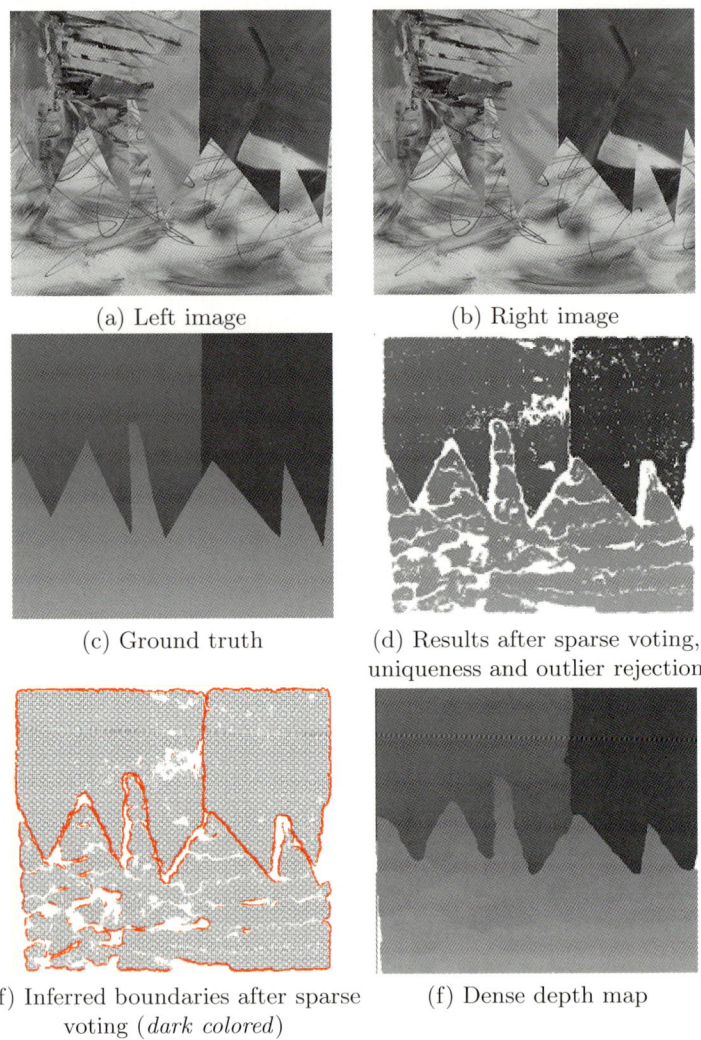

(a) Left image
(b) Right image
(c) Ground truth
(d) Results after sparse voting, uniqueness and outlier rejection
(f) Inferred boundaries after sparse voting (*dark colored*)
(f) Dense depth map

Fig. 16.11. a-f Results for the "sawtooth" stereo pair

16.5.2 Motion

Given two or more image frames, the goal of the problem of grouping from motion is to determine three types of information – a *dense velocity field*, *motion boundaries*, and *regions*. Our approach to motion analysis has also been published in [36] and [35]. From a computational point of view, the analysis can be decomposed in three processes – *matching*, *densification* and *segmentation*. The *matching* process identifies the elements (tokens) in successive views that represent the same physical entity, thus producing a possibly sparse velocity

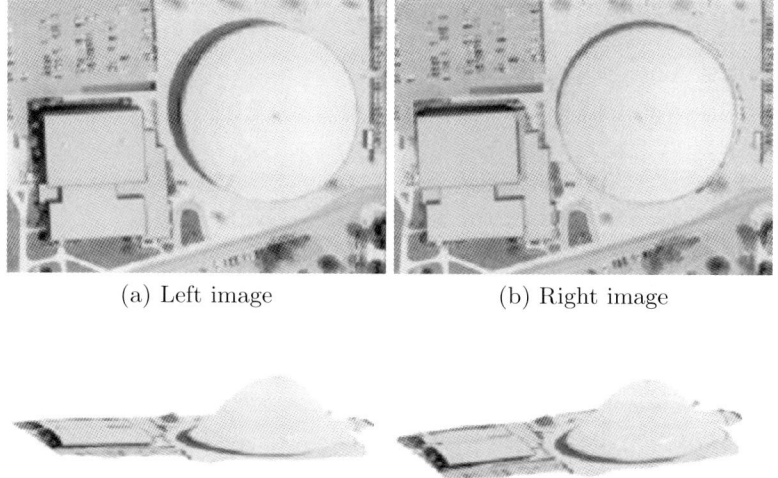

(a) Left image   (b) Right image

(c) Rotated views of the texture-mapped extracted surfaces

Fig. 16.12. a-c Results for the "arena" stereo pair where the texture has been mapped on the extracted dense surfaces

field. The *densification* process infers velocity vectors at every image location, and the *segmentation* process groups tokens into regions separated by motion boundaries.

From a computational point of view, one of the most powerful and most often used constraints is the smoothness of motion. Most approaches rely on parametric models that restrict the types of motion that can be analyzed, and also involve iterative methods that depend heavily on initial conditions and are subject to instability. Moreover, previous techniques usually encounter difficulties in image regions where motion is not smooth (i.e., around motion boundaries). This problem has led to numerous inconsistent methods, with ad hoc criteria introduced to account for motion discontinuities.

In order to address these difficulties, we developed a novel approach for motion analysis, by formulating it as a motion layers inference from a noisy and possibly sparse point set in a 4-D space within the tensor voting framework. From a possibly sparse input consisting of identical point tokens in two frames, the image position (x, y) and potential velocity (v_x, v_y) of each token are encoded into a 4-D tensor. Within this 4-D space, moving regions are conceptually represented as *smooth surface layers*, and are extracted through a voting process that enforces the smoothness constraint.

Here we focus on the problem of motion analysis from sparse sets of point tokens in two frames. Two examples of such input are shown in Fig. 16.13. If the frames in each pair are presented in a properly timed succession, a certain motion of image regions is perceived from one frame to the other.

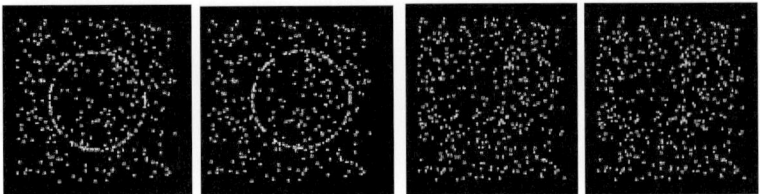

Fig. 16.13. Input frames for a rotating circle and a translating disk

However, while in one case the regions can be detected even without motion, only from monocular cues (here, different densities of points), in the other case no monocular information is available. This example shows that analysis is possible even from motion cues *only*. Another interesting aspect is the fact that the human vision system not only establishes point correspondences, but also perceives *regions* in motion, although the input consists of sparse points only.

Each token is characterized by four attributes – its image coordinates (x, y) and its velocity with the components (v_x, v_y). We encapsulate them into a (x, y, v_x, v_y)-tuple in the 4-D space, this being a natural way of expressing the spatial separation of tokens according to *both* velocities and image coordinates. In general, there may be several candidate velocities for each point (x, y), so each tuple (x, y, v_x, v_y) represents a potential match.

Both matching and densification are based on a process of communicating the affinity between tokens. In our representation, this affinity is expressed as the token preference for being incorporated into a *smooth surface layer* in the 4-D space. A necessary condition is to enforce strong support between tokens in the same layer, and weak support across layers, or at isolated tokens. In the tensor voting framework, the affinities between tokens are embedded in the concept of surface saliency exhibited by the data. By letting the tokens propagate their information through voting, wrong matches are eliminated as they receive little support, and distinct moving regions are extracted as salient smooth layers.

Voting in 4-D

The tensor voting framework is general enough to be extended to any dimension readily, except for some implementation changes, mainly for efficiency purposes. The issues to be addressed here are the *tensorial representation* of the features in the 4-D space, the generation of *voting fields* and the *data structures* used for vote collection. Table 16.3 shows all the geometric features that appear in a 4-D space and their representation as *elementary* 4-D tensors, where **n** and **t** represent normal and tangent vectors, respectively. Note that a surface in the 4-D space can be characterized by two normal vectors, or by two tangent vectors. From a *generic* 4-D tensor that results after voting, the geometric features are extracted as shown in Table 16.4. The 4-D voting fields are obtained as follows. First the 4-D stick field is generated in a similar manner to the 2-D stick field. Then the other three voting fields are built by integrating all the contributions obtained by rotating a 4-D stick field around appropriate axes.

Table 16.3. Elementary tensors in 4-D

Feature	$\lambda_1\ \lambda_2\ \lambda_3\ \lambda_4$	$e_1\ e_2\ e_3\ e_4$	Tensor
Point	1 1 1 1	Any basis	Ball
Curve	1 1 1 0	$n_1\ n_2\ n_3\ t$	C-Plate
Surface	1 1 0 0	$n_1\ n_2\ t_1\ t_2$	S-Plate
Volume	1 0 0 0	$n_1\ t_1\ t_2\ t_3$	Stick

Table 16.4. A generic tensor in 4-D

Feature	Saliency	Normals	Tangents
Point	λ_4	None	None
Curve	$\lambda_3 - \lambda_4$	$e_1\ e_2\ e_3$	e_4
Surface	$\lambda_2 - \lambda_3$	$e_1\ e_2$	$e_3\ e_4$
Volume	$\lambda_1 - \lambda_2$	e_1	$e_2\ e_3\ e_4$

Matching

We take as input two frames containing identical point tokens in a sparse configuration. For illustration purposes, we give a description of our approach by using a specific example: the point tokens represent an opaque *translating disk* (Fig. 16.13) against a static background. Candidate matches are generated as follows: in a preprocessing step, for each token in the first frame we simply create a potential match with every point in the second frame that is located within a neighborhood (whose size is given by the scale factor) of the first token. The resulting candidates appear as a cloud of (x, y, v_x, v_y) points in the 4-D space. Figure 16.14a shows a 3-D view of the candidate matches–the three dimensions shown are x and y (in the horizontal plane), and v_x (the height). The motion layers can be already perceived as their tokens are grouped in smooth surfaces surrounded by noisy matches.

Since no information is initially known, each potential match is encoded as a 4-D ball tensor. Then each token casts votes by using the ball voting field. During voting there is strong support between tokens that lie on a smooth surface (layer), while isolated tokens receive little or no support. For each

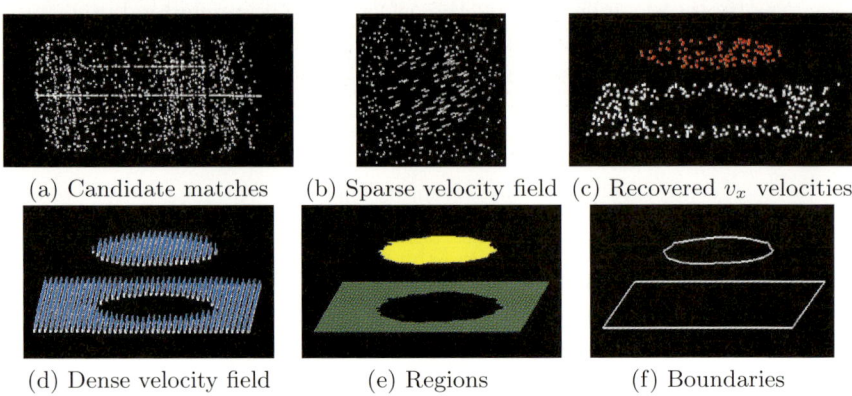

Fig. 16.14. a-f Translating disk

pixel we retain the candidate match with the highest surface saliency, and we reject the others as outliers. Figure 16.14b shows the recovered sparse velocity field, while Fig. 16.14c shows a 3-D view of the recovered matches (the height represents the v_x velocity component).

Densification

In order to recover boundaries and regions as *continuous* curves and surfaces, we need to first infer velocities and layer orientations at *every* image location. This is performed through an additional dense voting step, by generating discrete velocity candidates, collecting votes at each such location and retaining the candidate with maximal surface saliency. By following this procedure at every image location we generate a dense velocity field. Note that in this process, along with velocities we simultaneously infer layer orientations. Figure 16.14d shows a 3-D view of the dense set of tokens and their associated layer orientations (only one normal shown).

Segmentation

The next step is to group tokens into *regions* (Fig. 16.14e), by using again the smoothness constraint. We start from an arbitrary point in the image, assign a region label to it, and try to propagate this label to all its image neighbors. In order to decide whether the label must be propagated, we use the smoothness of both velocity and layer orientation as a grouping criterion. Finally, we have implemented a method to extract the *motion boundary* for each region (Fig. 16.14f), as a "partially convex hull". The process is controlled by the scale factor only, which determines the perceived level of detail (the departure from the actual convex hull).

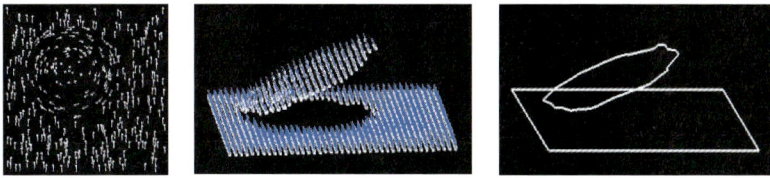

Fig. 16.15. Rotating disk – translating background

Results

1) Using motion cues only. **Rotating disk – translating background** (Fig. 16.15). The input consists of two sets of 400 points each, representing an opaque rotating disk against a translating background. After processing, only 2 matches among 400 are wrong. This is a very difficult case even for human vision, due to the fact that around the left extremity of the disk the two motions are almost identical. In that part of the image there are points on different moving objects that are not separated, even in the 4-D space. In spite of this inherent ambiguity, our method is still able to accurately recover velocities, regions and boundaries. The key fact is that we rely not only on the 4-D positions, but also on the local layer orientations that are still different and therefore provide a good affinity measure.

2) Incorporating intensity information. To further validate our approach we have also analyzed several real image sequences, where both monocular and motion cues are available. In order to incorporate monocular information into our framework, we only need to change the preprocessing step where candidate matches are generated. We run a simple intensity-based cross-correlation procedure and retain all peaks of correlation as candidate matches. The rest of our framework remains unchanged.

Yosemite sequence (Fig. 16.16). We analyzed the motion from two frames of the Yosemite sequence (without the sky) to quantitatively estimate the performance of our approach. The average angular error obtained is $3.74 \pm 4.3°$ for 100% field coverage. A result which is comparable with those in the literature [1]. Also note that our method successfully recovers nonplanar motion layers.

Barrier sequence (Fig. 16.17). For a qualitative estimation, we analyzed the motion from two frames of a sequence showing two cars moving away from the camera. The analysis is difficult due to the large ground area with very low texture, and the large amount of noise present in the set of candidates. Also note that the *image* motion is not translational – the front of each car has a lower velocity than its back. This is visible in the 3-D view of the motion layers, which appear as tilted surfaces.

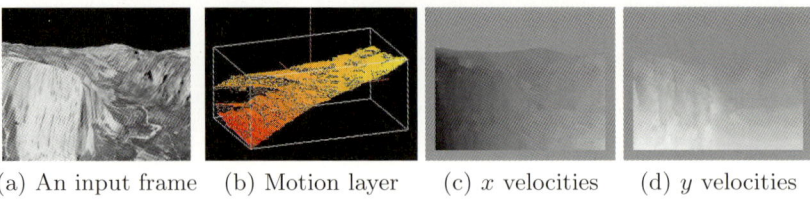

(a) An input frame (b) Motion layer (c) x velocities (d) y velocities

Fig. 16.16. a-d Yosemite

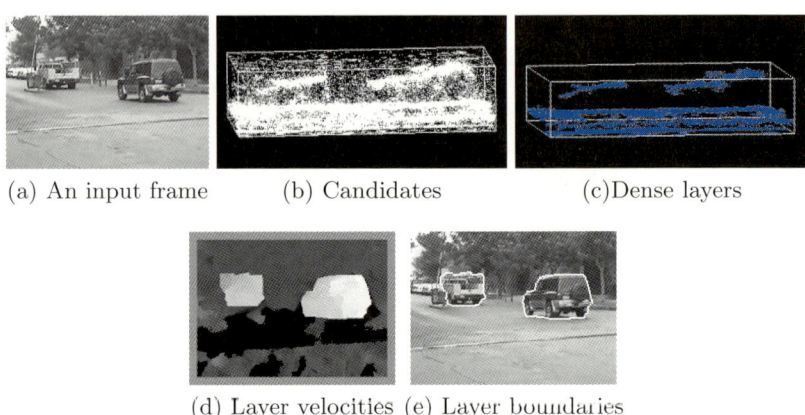

Fig. 16.17. a-e The Barrier sequence

16.6 Conclusion

We have presented the current state of the tensor voting framework, which is a product of a number of years of research, mostly at the University of Southern California. It provides a general methodology that can be applied to a large range of problems as long as they can be posed as the inference of salient structures in an N-dimensional space. The benefits from our representation and voting schemes are that no models need to be known a priori, nor do the data have to fit a parametric model. In addition, all types of perceptual structures can be represented and inferred at the same time. Processing can begin with unoriented inputs, is noniterative and there is only one free parameter, the scale of the voting field. Tensor voting facilitates the propagation of information locally and enforces smoothness while explicitly detecting and preserving discontinuities with very little initialization requirements. The local nature of the operations makes the framework efficient and applicable to very large datasets. Robustness to noise is an important asset due to the large amounts of noise that are inevitable in computer vision problems. Results, besides the organization of generic tokens, were shown in real computer vision problems such as stereo and motion analysis. Performance equivalent or superior to state-of-the-art algorithms has been achieved without the use of algorithms that are specific to each problem, but rather with general, simple and reusable modules.

References

1. Barron J., Fleet D., Beauchemin S. (1994) Performance of optical flow techniques. *Int. J. of Computer Vision*, 12(1):43–77
2. Boissonnat J. (1984) Representing 2D and 3D shapes with the Delaunay triangulation. In *Int. Conf. on Pattern Recognition*, pp. 745–748
3. Boykov Y., Veksler O., Zabih R. (2001) Fast approximate energy minimization via graph cuts. *IEEE Trans. on Pattern Analysis and Machine Intelligence*, 23(11):1222–1239
4. Cho K., Meer P. (1997) Image segmentation from consensus information. *Computer Vision and Image Understanding*, 68(1):72–89
5. Dolan J., Riseman E. (1992) Computing curvilinear structure by token-based grouping. In *Int. Conf. on Computer Vision and Pattern Recognition*, pp. 264–270
6. Edelsbrunner H., Mücke E. (1994) Three-dimensional alpha shapes. *ACM Trans. on Graphics*, 13(1):43–72
7. Faugeras O., Berthod M. (1981) Improving consistency and reducing ambiguity in stochastic labeling: An optimization approach. *IEEE Trans. on Pattern Analysis and Machine Intelligence*, 3(4):412–424
8. Fischler M., Bolles R. (1981) Random sample consensus: A paradigm for model fitting with applications to image analysis and automated cartography. *Comm. of the ACM*, 24(6):381–395
9. Gdalyahu Y., Weinshall D., Werman M. (2001) Self-organization in vision: Stochastic clustering for image segmentation, perceptual grouping, and image database organization. *IEEE Trans. on Pattern Analysis and Machine Intelligence*, 23(10):1053–1074
10. Geman S., Geman D. (1984) Stochastic relaxation, Gibbs distributions, and the Bayesian restoration of images. *IEEE Trans. on Pattern Analysis and Machine Intelligence*, 6(6):721–741
11. Granlund G., Knutsson H. (1995) *Signal Processing for Computer Vision*. Kluwer, Dordrecht
12. Grossberg S., Mingolla E. (1985) Neural dynamics of form perception: Boundary completion. *Psychological Review*, 92(2):173–211
13. Grossberg S., Todorovic D. (1988) Neural dynamics of 1-d and 2-d brightness perception: A unified model of classical and recent phenomena. *Perception and Psychophysics*, 43:723–742
14. Guy G., Medioni G. (1996) Inferring global perceptual contours from local features. *Int. J. of Computer Vision*, 20(1/2):113–133
15. Guy G., Medioni G. (1997) Inference of surfaces, 3D curves, and junctions from sparse, noisy, 3D data. *IEEE Trans. on Pattern Analysis and Machine Intelligence*, 19(11):1265–1277
16. Haralick R., Shapiro L. (1979) The consistent labeling problem. Part I. *IEEE Trans. on Pattern Analysis and Machine Intelligence*, 1(2):173–184
17. Haralick R., Shapiro L. (1980) The consistent labeling problem. Part II. *IEEE Trans. on Pattern Analysis and Machine Intelligence*, 2(3):193–203
18. Heitger F., von der Heydt R. (1993) A computational model of neural contour processing: Figure-ground segregation and illusory contours. In *Int. Conf. on Computer Vision*, pp. 32–40
19. Hoppe H., DeRose T., Duchamp T., McDonald J., Stuetzle W. (1992) Surface reconstruction from unorganized points. *Computer Graphics*, 26(2):71–78

20. Horn B., Schunck B. (1981) Determining optical flow. *AI*, 17(1-3):185–203
21. Hummel R., Zucker S. (1983) On the foundations of relaxation labeling processes. *IEEE Trans. on Pattern Analysis and Machine Intelligence*, 5(3):267–287
22. Jacobs D. (1996) Robust and efficient detection of salient convex groups. *IEEE Trans. on Pattern Analysis and Machine Intelligence*, 18(1):23–37
23. Jain A., Dubes R. (1988) *Algorithms for clustering data*. Prentice-Hall, Englewood Cliffs
24. Koster K., Spann M. (2000) MIR: An approach to robust clustering-application to range image segmentation. *IEEE Trans. on Pattern Analysis and Machine Intelligence*, 22(5):430–444
25. Leclerc Y. (1989) Constructing simple stable descriptions for image partitioning. *Int. J. of Computer Vision*, 3(1):73–102
26. Lee K., Meer P., Park R. (1998) Robust adaptive segmentation of range images. *IEEE Trans. on Pattern Analysis and Machine Intelligence*, 20(2):200–205
27. Lee M., Medioni G., Mordohai P. (2002) Inference of segmented overlapping surfaces from binocular stereo. *IEEE Trans. on Pattern Analysis and Machine Intelligence*, 24(6):824–837
28. Li Z. (1998) A neural model of contour integration in the primary visual cortex. *Neural Computation*, 10:903–940
29. Lorensen W., Cline H. (1987) Marching cubes: A high resolution 3D surface reconstruction algorithm. *Computer Graphics*, 21(4):163–169
30. Lowe D. (1985) *Perceptual Organization and Visual Recognition*. Kluwer, Dordrecht
31. Marr D. (1982) *Vision*. Freeman Press, San Francisco
32. Medioni G., Lee M., Tang C.K. (2000) *A Computational Framework for Segmentation and Grouping*. Elsevier, New York
33. Mohan R., Nevatia R. (1992) Perceptual organization for scene segmentation and description. *IEEE Trans. on Pattern Analysis and Machine Intelligence*, 14(6):616–635
34. Morel J., Solimini S. (1995) *Variational Methods in Image Segmentation*. Birkhauser, Boston
35. Nicolescu M., Medioni G. (2002) 4-D voting for matching, densification and segmentation into motion layers. In *Int. Conf. on Pattern Recognition*, III: 303–308
36. Nicolescu M., Medioni G. (2002) Perceptual grouping from motion cues using tensor voting in 4-D. In *European Conf. on Computer Vision*, III: 423–428,
37. Osher S., Fedkiw R. (2002) *The Level Set Method and Dynamic Implicit Surfaces*. Springer, Berlin Heidelberg New York
38. Parent P., Zucker S. (1989) Trace inference, curvature consistency, and curve detection. *IEEE Trans. on Pattern Analysis and Machine Intelligence*, 11(8):823–839
39. Poggio T., Torre V., Koch C. (1985) Computational vision and regularization theory. *Nature*, 317:314–319
40. Robert L., Deriche R. (1996) Dense depth map reconstruction: A minimization and regularization approach which preserves discontinuities. In *European Conf. on Computer Vision*, I:439–451
41. Robles-Kelly A., Hancock E. (2001) An expectation-maximisation framework for perceptual grouping. In *IWVF4, LNCS 2059*, pp. 594–605. Springer, Berlin Heidelberg New York

42. Sander P., Zucker S. (1990) Inferring surface trace and differential structure from 3-D images. *IEEE Trans. on Pattern Analysis and Machine Intelligence*, 12(9):833–854
43. Sarkar S., Boyer K. (1994) A computational structure for preattentive perceptual organization: Graphical enumeration and voting methods. *IEEE Trans. on Systems, Man and Cybernetics*, 24:246–267
44. Saund E. (1992) Labeling of curvilinear structure across scales by token grouping. In *Int. Conf. on Computer Vision and Pattern Recognition*, pp. 257–263
45. Scharstein D., Szeliski R. (2002) A taxonomy and evaluation of dense two-frame stereo correspondence algorithms. *Int. J. of Computer Vision*, 47(1-3):7–42
46. Sethian, J. (1996) *Level Set Methods: Evolving Interfaces in Geometry, Fluid Mechanics, Computer Vision and Materials Science.* Cambridge University Press, Cambridge
47. Shashua, A., Ullman S. (1988) Structural saliency: The detection of globally salient structures using a locally connected network. In *Int. Conf. on Computer Vision*, pp. 321–327
48. Shi J., Malik J. (2000) Normalized cuts and image segmentation. *IEEE Trans. on Pattern Analysis and Machine Intelligence*, 22(8):888–905
49. Szeliski R., Tonnesen D., Terzopoulos D. (1993) Modeling surfaces of arbitrary topology with dynamic particles. In *Int. Conf. on Computer Vision and Pattern Recognition*, pp. 82–87
50. Tang C., Medioni G. (1998) Inference of integrated surface, curve, and junction descriptions from sparse 3D data. *IEEE Trans. on Pattern Analysis and Machine Intelligence*, 20(11):1206–1223
51. Terzopoulos D. (1986) Regularization of inverse visual problems involving discontinuities. *IEEE Trans. on Pattern Analysis and Machine Intelligence*, 8(4):413–424
52. Terzopoulos D., Metaxas D. (1991) Dynamic 3D models with local and global deformations: Deformable superquadrics. *IEEE Trans. on Pattern Analysis and Machine Intelligence*, 13(7):703–714
53. Wertheimer M. (1923) Laws of organization in perceptual forms. *Psycologische Forschung, Translation by W. Ellis, A source book of Gestalt psychology (1938)*, 4:301–350
54. Westin C. (1994) *A Tensor Framework for Multidimensional Signal Processing*. Ph.D. thesis, Linkoeping University, Sweden
55. Williams L., Jacobs D. (1997) Stochastic completion fields: A neural model of illusory contour shape and salience. *Neural Computation*, 9(4):837–858
56. Yen S., Finkel L. (1998) Extraction of perceptually salient contours by striate cortical networks. *Vision Research*, 38(5):719–741
57. Yu, X. Bui T., Krzyzak A. (1994) Robust estimation for range image segmentation and reconstruction. *IEEE Trans. on Pattern Analysis and Machine Intelligence*, 16(5):530–538
58. Zhao H., Osher S., Fedkiw R. (2001) Fast surface reconstruction using the level set method. *UCLA Computational and Applied Mathematics Reports*, pp. 32–40

Part VII

Computer Graphics and Visualization

17

Methods for Nonrigid Image Registration

Lawrence H. Staib[1] and Yongmei Michelle Wang[2]

[1] Diagnostic Radiology, Biomedical Engineering and Electrical Engineering, Yale University, New Haven, CT 06520 lawrence.staib@yale.edu
[2] Diagnostic Radiology, Yale University, New Haven, CT 06520 wang@noodle.med.yale.edu

17.1 Introduction

Nonrigid image registration is the process that determines the geometric transformation of one image space with respect to another that brings into correspondence two images of the same scene, typically by optimizing some match metric. Nonrigid methods are required when the two images are not related by a simple rigid or similarity transform. Correspondence may either be known a priori, determined implicitly by the transformation or determined explicitly during the optimization process. Detailed nonrigid registration requires the determination of correspondence throughout the image. In some sense, the correspondence and the transformation are duals of each other in that a complete correspondence determines the transformation and vice versa. There are many applications that involve multiple images of an object or scene that require registration. Note, the images may be 2D, 3D or higher.

There has been tremendous growth in interest in nonrigid registration in recent years due to the increase in computing power which makes it feasible and in the emergence of high-resolution 3D images from medical imaging modalities such as magnetic resonance imaging (MRI) and associated functional images, primarily from nuclear medicine and functional magnetic resonance imaging (fMRI). Nonrigid registration of such structural images is an essential task for analysis when there is a need for detailed comparison between different individuals or when there is deformation within an individual.

Early work in image registration, initially rigid, arose from work in remote sensing [1]. One of the first methods for nonrigid image registration, by Burr [2], was applied to face images. He determined correspondences of feature points within a neighborhood and used a smoothed composite of the displacements as the transformation. The neighborhood was iteratively decreased to remove local distortions.

Most work on nonrigid registration has used medical images, and in particular brain images. Bajcsy and her collaborators [3] did some of the first work in nonrigid registration using an elastic model to warp magnetic resonance

(MR) and computed tomography (CT) brain images to an atlas. The brain is of tremendous interest because of many applications in neuroscience and neurosurgery, which present many unique challenges. Nonrigid registration of the brain is a difficult task, but it has many important applications, including comparison of shape and function between individuals or groups, development of probabilistic models and atlases, measurement of change within an individual, and determination of location with respect to a preacquired image during stereotactic surgery.

There have been some excellent reviews of image registration, including those on primarily rigid techniques (e.g. [4, 5, 6]). Others cover nonrigid techniques. Toga [7] presents a collection of methods for brain warping. Fitzpatrick *et al.* [8] present a tutorial of the current state of the art of image registration in general, and Bankman [9] includes a set of chapters on methods for image registration. Wolberg [10] introduces and presents the basic definitions of nonrigid transformations. Hajnal [11] is an excellent reference for medical image methods. Zitová [12] presents a more recent review of image registration over all application areas. In addition, there have been conferences [13, 14] and recent special issues of journals devoted to the topic [15, 16].

17.2 Categorization

There are four categorizations that can be made regarding nonrigid registration methods based on the type of transformation, geometric structure to be matched, match metric and optimization method. These choices are often interrelated, as will be seen below.

17.2.1 Transformation Constraints

The transformation may be parametrized in many ways. The simplest useful transformation is a rigid transformation typically defined by an orthonormal matrix using homogeneous coordinates [17] or quaternions [18]. The rigid transformation is defined by six parameters: three rotations and three translations. As the number of parameters increases, the flexibility increases, allowing nonrigid deformations. For a similarity transformation, a single scaling factor is added. Affine transformations are defined by 12 parameters and allow shearing and scaling in all three axes. If the above transformations are applied locally, the ability to deform is increased accordingly [19]. A regular tessellation [20] of the image volume can be used to divide a 3D image into tetrahedral volume elements [19]. Each element is defined by the locations of the four vertices of each tetrahedron. Linear [19] or higher-order models [21] can be used to interpolate. The location of the vertices determines the transformation, and these can be adjusted according to any match measure.

Spline models (e.g. cubic, b, thin plate) or other polynomials [22] can also be used to formulate flexible transformations. The flexibility of the transformation is determined by the number of control points. Such models can be formulated as free-form deformations [23], which are quite useful for computer graphics as well as image registration. The tensor b-spline formulation has been effectively applied to medical nonrigid image registration [24]. Basis function representations (Fourier, cosine, statistically based) can also be used to model the transformation [25, 26]. The most flexible representation is to define a displacement at each voxel. With no additional constraints, however, this could yield transformations that are discontinuous or folded over.

Usually, the transformation is constrained in some way in order to result in a good solution. In general, the registration problem is ill-posed (i.e. in this case, many possible solutions exist). Constraints are necessary to determine a unique solution from the enormous solution space. Constraints may be needed to ensure that the solution is, for example, smooth and one-to-one. Some transformations are constrained explicitly due to the parametrization (affine, polynomial, etc.) while others are implicitly constrained by the model used in the match metric. A penalty term in the match metric can enforce, for example, smoothness. If the constraints are too restrictive, such as from a low-order polynomial, the match will likely not be sufficiently detailed. Model constraints typically enforce continuity and smoothness properties. Physical models, including linear elastic and viscous fluid models, have been used to enforce topological properties on the deformation and constrain the solution space [27, 28, 29, 30, 31, 32, 33].

Statistical models, instead of physical models, can be powerful tools to directly capture the character of the variability of the individuals being modeled. While originally used for segmentation [34, 35, 36, 37, 38], statistical models are now being applied to registration [39, 40]. Probabilistic brain atlases can be constructed to facilitate this type of model [39, 41, 42]. Statistical models can both constrain the deformations to the most likely but also can define a parameter space in which the optimal registration can be found.

It is desirable in medical image registration that nonlinear transformations be diffeomorphisms based on the assumption that the two images are related in a smooth way such that all structures have homologous counterparts in the other image. A transformation is a homeomorphism if it is a bijection (one-to-one and onto) that is continuous and its inverse is continuous. A transformation is diffeomorphic if it is a bijection and differentiable. The Jacobian determinant needs to be greater than zero at all points to ensure that the transformation is one-to-one. In cases of abnormalities where homologous structures do not exist, such as because of surgery or congenital abnormality, more general techniques are needed. It is also desirable that the transformations form a group. While the affine transformations and diffeomorphisms form groups, some parametrizations, such as splines, do not. In principle, these transformations can be made to be subgroups of the diffeomorphisms by appropriately extending them.

17.2.2 Geometric Structure

The choice of geometric structure to be matched is very important to the performance of the nonrigid registration technique. At the lowest level, techniques can match densely at every voxel [27, 30, 33, 43, 44]. These techniques are potentially highly accurate but can be susceptible to local minima in the optimization process. The most straightforward way is to directly use the gray-level intensity at each voxel in the match metric. There may be advantages to using filtered images (such as gray-level gradients) derived from the gray levels [45]. Local geometric moment invariants can also be used for nonrigid registration [46].

Alternatively, instead of dense features, sparse point locations marking distinct features or landmarks [47, 48, 49] can be valuable and can yield strong matches at these locations, greatly constraining the warp [50, 51]. In order to take advantage of point landmarks in a match metric, knowledge of correspondence is required. Some methods explicitly determine correspondence, and this can be done in a robust way, allowing for outliers, that is, points with no match [52, 53].

Matching based on curves (in 2D) and surfaces (in 3D) can be powerful since the structures provide a strong constraint. Often structural information is needed anyway for other analysis or measurement purposes. Determining the correspondence of points between pairs of surfaces has important applications in addition to nonrigid registration, such as for comparing shape between deformable objects and developing probabilistic models and atlases.

Iterative closest point (ICP) and related methods [54, 55, 43, 56] match surfaces based on distance by minimizing the distance from points in one surface to the closest point in another surface. ICP has also been augmented with shape information for model building [57]. Other methods also match surfaces [31, 58, 59, 60, 61] using geometric or physical models. Similar methods have been developed in 2D for the related problem of nonrigid motion tracking [62, 63]. Combinations of the geometric structure can also be employed in order to take advantage of the complementary information that they provide, as will be described below [50, 64]. Fleute et al. [65] developed a surface correspondence algorithm to build a statistical shape model. Random point sets sampled from the surface are registered and matched to establish correspondence using a multiresolution octree spline approach. Minimum description length (MDL) methods have also been developed to align point sets for building statistical models [66]. Here, multiple boundaries are considered at once, and a parametrization of the boundary is optimized by minimizing the description length, yielding a compact representation.

17.2.3 Match Metric

The match metric ultimately determines the registration and should be designed in order to bring corresponding points together based on the features used. In addition, it should be formulated, if possible, so that the optimum is deep, distinct and smooth in order to help avoid local minima. The simplest metric matches image gray level using, for example, gray-level squared difference or gray-level correlation. Metrics using gray-level properties attempt to use all available information, unlike feature metrics, which attempt to be selective. Such measures are, of course, sensitive to gray level differences due to shading, inhomogeneities, calibration and individual variations that affect gray level. Features derived from the gray levels, as described above, are an alternative that can remove sensitivity to shifts in gray level and can provide strong features useful for matching. gray-level gradients and curvature features have been used for registration [45, 67, 68, 69].

In the cortex of the brain, the high anatomic variability can often result in intensity-based methods yielding inaccurate results, as was recently demonstrated [70]. In addition, with functional imaging, there is the additional problem that structure and function are not completely linked: structurally homologous regions can be registered while areas of corresponding function are not [71]. Feature-based methods have been developed to overcome such problems [58, 72, 73]. However, none of these methods is able to handle large variations in sulcal anatomy, as well as irregular sulcal branching and discontinuity.

If structure is extracted, the simplest metric to use is Euclidean distance, as was used early on for the rigid matching of surfaces [74]. Of course, a correspondence of some kind must be established in order to compute, for example, the sum of distances between corresponding points. The correspondences need to be determined efficiently when evaluating the metric, as opposed to resulting from the final registration. Closest distance (i.e. the closest distance from one structure to the nearest point on the corresponding structure) can be used as is done in ICP [54, 55]. Distance transforms are one way to efficiently compute closest distances [75, 76]. Robust distance metrics can be used to account for distances obviously too large to indicate correspondence [53, 77].

Distance, however, ignores other qualities that may be more specific to a particular match than simple proximity. Using distance in combination with more specific features, such as surface normal or curvature [62, 78, 79] or simply gray level [43] can lead to more realistic correspondences, as we will discuss below. By matching properties that are expected to be consistent, the correspondence and the resulting match are likely to be better.

Mutual Information

A more general approach using gray levels, originally applied to rigid registration, is based on the mutual information [80] of the gray-level densities [81, 82, 83, 84, 85]. These techniques are very successful for registration in many applications. Instead of directly matching gray levels between two images, the mutual information is calculated. This measure allows for a more general relationship and is insensitive to gray-level changes, such as scaling and level and even more general variation such as occurs between modalities.

A number of groups have used mutual information for nonrigid registration [86, 87, 88, 89, 90]. Estimates of correspondence are adjusted to optimize a mutual information metric using various nonrigid transformations, such as 3D thin plate splines [87, 88]. Gaens et al. [86] use a mutual information metric computing the optimal displacements of overlapping neighborhoods and interpolating between them. Rueckert et al. [90] use a global affine model with local deformations governed by spline-based free-form deformations. Mutual information metrics can also be combined with other measures, such as gray-level features [91]. A recent review covers mutual information registration of medical images [92].

The mutual information I of an image i with an atlas a transformed by T, is defined by:

$$I(i, T(a)) = H(i) + H(T(a)) - H(i, T(a)), \tag{17.1}$$

where $H(x)$ is entropy of the probability density of the gray levels in the image, and $H(x, y)$ is joint entropy, as defined below:

$$H(x) = -\int \Pr(x) \ln(\Pr(x)) \, dx, \tag{17.2}$$

$$H(x, y) = -\int \int \Pr(x, y) \ln(\Pr(x, y)) \, dx dy. \tag{17.3}$$

The probabilities can be estimated from the image histograms [81] or using Parzen window estimation [84]. Entropy is a measure of complexity or uncertainty. Here, mutual information is a measure of how well the transformed atlas explains the image. It measures how well clustered corresponding pixels are in their joint histogram, and this suits the purposes of accommodating variations in the gray levels between the atlas and the image.

A variant, normalized mutual information [93]

$$\text{NMI}(i, T(a)) = \frac{H(i) + H(T(a))}{H(i, T(a))} \tag{17.4}$$

performs better especially when there is incomplete overlap between the structures in the image.

17.2.4 Optimization

The nonrigid registration problem is typically multidimensional and multimodal. The number of dimensions is the number of degrees of freedom in the transformation, which may be large in order to accommodate detailed variation. In a 3D image for a dense displacement field, the number of degrees of freedom of the transformation are three times the number of pixels. For spline formulations, the dimensionality can be much smaller, as it is three times the number of control points.

Local optimization methods, such as conjugate gradient or Powell's method [94], are often used and rely on a starting point that is close to the optimal solution. In nonrigid registration, starting points for the parameters can be generated from the output of a prior, more constrained registration step, such as rigid or affine. Local optima are still a concern when the deformation is large.

More global methods, such as multiresolution techniques [27, 95], genetic optimization [96], simulated annealing [97] and stochastic methods [85] have all been used to help improve optimization to find a better match by avoiding local minima.

17.3 Types of Deformation

There are three types of deformation that need to be accounted for in nonrigid image registration: change within an individual due to growth, surgery, or disease, differences between individuals and warping due to image distortion, such as in echo-planar magnetic resonance imaging.

17.3.1 Within-Subject Deformation

In some applications, it is important to account for change within an individual. Some deformation is incremental.

Soft tissue structures, such as in the abdomen, constantly move and deform. The brain, however, is relatively fixed within the skull. An individual's brain may change because of growth, degenerative processes such as Alzheimer's or multiple sclerosis or malignant disease. In cases such as these, the deformation is incremental and, depending on the timing, likely to be representable in terms of a relatively small magnitude and smooth transformation. Often, the transformation is rigid over most of the space, deviating only at the area of deformation. This problem is similar to the problem of nonrigid motion tracking (e.g. [98]).

Another type of deformation within a subject is due to surgery. During surgery, the brain deforms or shifts because of changes in pressure, fluid loss and other factors [99, 100]. This shifting is also smooth and incremental. For image-guided stereotactic surgery, where the correspondence between the images and the patient must be known with high accuracy, the correction for this deformation must be done in realtime. Physical modeling of tissue has been used extensively for this purpose. In addition, there is a more severe deformation due to the removal of tissue (and subsequent remodeling). When there is a discrete change in the tissue, such as with the resection of a tumor, the deformation is much more complex, and the transformation would likely require singularities.

17.3.2 Between-Subject Deformation

Anatomic variation between individuals is usually great, especially in the brain. A detailed nonrigid transformation that brings brain images from different individuals into correspondence to account for differences is required in order to compare their anatomy. The outer portion, or cortex, of the brain is a thin folded layer. While the major folds (gyri) are fairly consistent between normal individuals, the minor folds vary greatly. There are differences in folding patterns (extra and missing folds, different branching patterns, etc.), which change the shape significantly, typically requiring transformations of high flexibility, allowing large-scale deformation and making registration quite difficult. Subcortical structures, beneath this layer, vary in shape and size between individuals in a relatively smooth manner. In order to compare a group of individuals, structural variation should be accounted for by registering an atlas image to each individual's image, in order to have a common coordinate system for comparison. In addition, this nonrigid registration facilitates atlas-based segmentation [101, 102]. By registering an individual's image to a presegmented atlas image, the segmentation can be applied to the individual. The deformations themselves can be used to characterize anatomical shape, as described below. Functional differences, as seen in corresponding functional images, can also be compared in this coordinate system. In fMRI or positron emission tomography (PET) analysis, a key step is the formation of multisubject composite activation maps where each subject is mapped to a common coordinate space.

17.3.3 Imaging Distortions

While most imaging systems yield an undistorted view of the underlying scene, in some modalities there are distortions due to the imaging process that are large enough in magnitude that they require correction. Echo planar magnetic resonance imaging (EPI) is a fast imaging technique used for fMRI that can exhibit severe geometric distortions. It is necessary to accurately map fMRI activations to corresponding structural images in order to anatomically localize them. The displacements in EPI are dependent on the configuration of tissue in the subject, the orientation of the subject within the scanner and the acquisition parameters. Such distortion is particularly bad in regions of the brain close to tissues with significantly different magnetic susceptibility such as bone or air sinuses [103]. Such spatial distortion can lead to misplacement of functional signals by many millimeters, large enough to displace a functional response into a neighboring gyrus [104]. Distortions in one part of the brain can also corrupt the global transformation estimate. The transformation can be modeled using knowledge of the physics of the image acquisition process.

17.4 Characterization of Deformation

While the primary goal of nonrigid registration is the alignment of images, the transformation that is determined can be used to characterize the deformation in terms of local structural differences. Groups can be characterized by the transformations that bring the individual images into a common space. The deformation gradient tensor allows the direct measurement of geometric properties of the deformation and can be computed at all locations from the transformation. While this representation can be used for simple measures of local expansion or contraction, it also allows a full characterization of the deformation by measures such as dilatation, principal direction of expansion/contraction and anisotropy of expansion/contraction.

In a Lagrangian representation, we describe the current configuration in terms of the initial or reference configuration: $\mathbf{x} = \mathbf{x}(\mathbf{X}, t)$, where $\mathbf{x}$ is the position at time t (current configuration), and $\mathbf{X}$ is the position at time $t = 0$ (reference configuration of the atlas). The deformation gradient tensor $\mathbf{F}$ is useful in the characterization of deformation and has components $F_{iR} = \frac{\partial x_i}{\partial X_r}$ [105]. The determinant of $\mathbf{F}$ at each point, $D(p) = \det \mathbf{F}(p)$, indicates whether there is local shrinking ($D(p) < 1$) or local expansion ($D(p) > 1$). For small deformations, $D = 1 + \Delta$ where Δ is the change of volume per unit initial volume or the dilatation. The tensor $\mathbf{F}$ can be decomposed into a rotation $\mathbf{R}$ and a deformation $\mathbf{U}$, or $\mathbf{F} = \mathbf{R}\mathbf{U}$, where $\mathbf{R}$ and $\mathbf{U}$ can be determined from the eigendecomposition of $\mathbf{F}^{\mathrm{T}}\mathbf{F}$ [105].

Local expansion and contraction from an atlas (based on the determinant) gives a good simple scalar measure of deformation [106, 107]. To more fully characterize the deformation, compute the eigendecomposition of $\mathbf{U}$ to give the three magnitudes ($\lambda_1, \lambda_2, \lambda_3$) and directions ($\epsilon_1, \epsilon_2, \epsilon_3$) of principal stretching [105]. We can also measure local anisotropy of expansion/contraction, for example, using $\sqrt{\frac{3}{2}\frac{\sum(\lambda_i - \bar{\lambda})^2}{\sum \lambda_i^2}}$, which ranges from 0 for isotropic, to 1 for extremely anisotropic. Measures such as these derived from the tensor provide a rich descriptive tool for deformation.

Measurement of deformation from the transformation is much preferred to the technique known as voxel-based morphometry [108], which looks for differences in intensity between blurred segmented images after nonrigid registration. This approach is flawed, as has been noted by Bookstein [109], in part because it relies on partial misregistration in order to reveal differences.

17.5 Specific Approaches

In this section we discuss a few selected approaches that fit into the above categories and make different choices for the components of the registration.

17.5.1 Physical Models

There is no true physical model for intersubject deformation because, for example, one individual's anatomical structure does not literally result from the deformation of another individual's. Analogous physical models are used because their properties enforce smooth and plausible deformations. Christensen et al. [30] present two physical models for nonrigid registration of the brain: elastic [33] and fluid [29].

Here we use an Eulerian reference (where a particle is tracked with respect to its final coordinates) defining the nonrigid registration by the homeomorphic mapping:

$$\mathbf{w} = (x, y, z) \rightarrow (x - u_x(\mathbf{w}), y - u_y(\mathbf{w}), z - u_z(\mathbf{w})), \quad (17.5)$$

where $\mathbf{u}(\mathbf{w}) = [u_x(\mathbf{w}), u_y(\mathbf{w}), u_z(\mathbf{w})]^T$ is the displacement at each pixel $\mathbf{w}$ with coordinate (x, y, z).

The elastic model penalizes deformation in proportion to the deformed distance, thus constraining the deformation to be small. The spatial transformation satisfies the partial differential equation (PDE):

$$\mu \nabla^2 \mathbf{u} + (\mu + \beta) \boldsymbol{\nabla}(\boldsymbol{\nabla} \cdot \mathbf{u}) = \mathbf{F}(\mathbf{u}), \quad (17.6)$$

with boundary conditions such as that $\mathbf{u}(\mathbf{w}) = 0$ for $\mathbf{w}$ on the image boundary. In this equation, μ and β are the Lamé constants, related to the rigidity and incompressibility of the material. A typical choice would be $\mu = 1.0$ and $\beta = 0.0$. However, in order to guarantee a homeomorphic transformation for large deformations, a larger value of μ is needed. The body force $\mathbf{F}(\mathbf{u})$ drives the deformation of the atlas into the subject.

For viscous fluids, the force is proportional to the time rate of change in displacement. The fluid model is much less constrained than the elastic model and allows long-distance, nonlinear deformations.

The PDE describing the fluid transformation of the atlas is given by:

$$\mu \nabla^2 \mathbf{v} + (\mu + \beta) \boldsymbol{\nabla}(\boldsymbol{\nabla} \cdot \mathbf{v}) = \mathbf{F}(\mathbf{u}), \quad (17.7)$$

where $\mathbf{v} = [v_x(\mathbf{w}, t), v_y(\mathbf{w}, t)]^T$ is the instantaneous velocity of the deformation field $\mathbf{u}$. Velocity is related to displacement $\mathbf{u}$ by:

$$\mathbf{v}(\mathbf{w}, t) = \frac{\partial \mathbf{u}(\mathbf{w}, t)}{\partial t} + \mathbf{v}(\mathbf{w}, t)^T \boldsymbol{\nabla} \mathbf{u}(\mathbf{w}, t). \quad (17.8)$$

The $\nabla^2 \mathbf{v}$ term is the viscous term of the PDE. This term constrains the velocity of neighboring particles of the displacement field to vary smoothly. The term $\boldsymbol{\nabla}(\boldsymbol{\nabla} \cdot \mathbf{v})$ is the mass source term, and it allows structures in the atlas to change in mass. The coefficients μ and β are the viscosity coefficients (typically $\mu = 1.0$, $\beta = 0.0$). The boundary conditions are typically $\mathbf{v}(\mathbf{w}) = 0$ for $\mathbf{w}$ on the image boundary. The term $\mathbf{v}(\mathbf{w}, t)^T \boldsymbol{\nabla} \mathbf{u}(\mathbf{w}, t)$ accounts for the kinematic nonlinearities of the displacement field $\mathbf{u}$ [29, 110].

The fluid PDE in Eq. (17.7) is similar in form to the elastic PDE given by Eq. (17.6) except that the displacement field **u** is replaced by the velocity field **v**. The resulting behavior of the fluid is very different due to the nonlinear relationship between **v** and **u** (Eq. (17.8)) and allows long-distance, nonlinear deformations.

The linear elastic model is derived assuming small angles of rotation and small linear deformations. Large deformations cannot be accommodated with this linear PDE. However, even though linear elasticity does not guarantee a homeomorphic transformation, in practice a homeomorphic transformation can be generated using strong elasticity (large μ). The trade-off is that only small deformations can be generated [29]. This limitation of linear elasticity is removed by using the viscous model because the restoring forces relax over time and then account for the large-distance kinematic nonlinearities, while ensuring a homeomorphic transformation (globally positive Jacobian).

Fluid models are less constraining than elastic models and allow long-distance, nonlinear deformations of small subregions. They require more computation compared to elastic models. In both of these formulations, the driving force is determined using local pixel intensity differences [30]. Lower-contrast objects deform slower than high-contrast objects, independent of their importance. Sometimes objects do not deform correctly because their gradient is relatively low and the smoothness ensured by the physical models dominates the deformation. Additional constraints can help this method bring corresponding structure together. Gee *et al.* [32] use a similar linear elastic strain model in conjunction with a Markov random field to model a displacement field for nonrigid registration.

17.5.2 Woods AIR

Woods *et al.* [111, 112] register by reducing the variance of the gray-level ratio at corresponding sites in the two images. This technique uses gray-level information, but tries to avoid dependence on the actual gray level. Reducing the variance of the pixel ratio is intended to match tissue types, because whatever their actual pixel values, the ratio is expected to be relatively constant throughout the image. The use of the variance makes the optimal point of matching less pronounced, because the variance changes gradually as the pixels become more matched. In addition, in functional images, different tissue types are usually not distinguished by their gray level, other than in a very general way. For within-modality matching, pixels can be grouped by intensity value and the sum of group variances is minimized; the method becomes dependent on these threshold values. The transformation types include rigid, affine, quadratic and cubic up to quintic polynomial (168 parameters).

17.5.3 Friston (SPM) Spatial Normalization

The methods of Friston [25, 113] are used for nonrigid registration where the sum of squares between gray levels is minimized. The relationship is modeled

by:
$$f_x(a(x)) = i(T(x)) + e(x), \tag{17.9}$$

where a is the atlas, i is the individual, f_x is the gray-level transformation (necessary primarily for intermodality matching), T is the warping transformation and e is the residual error. The relationship can be approximated by describing the spatial transformation using Fourier basis functions. The intensity transformation is also described by basis functions. Both transformations are further approximated with a first-order Taylor expansion. A least squares solution is determined. Because of the approximations involved, this method cannot account for detailed warping. The images, deformation and intensity transformation must all be smooth.

17.5.4 Cortical Features

Thompson and Toga [114, 115, 116] formulate warping using Chen surfaces (hybrid superquadrics and spherical harmonics) to extract surface models of cortical structure, including sulci. A volumetric warp defined by an octree spline grid is then computed based on the surface correspondence.

Collins et al. [101, 117] segment the brain using an elastic registration to an average brain, based on a hierarchical local correlation. The average brain provides strong prior information about the expected image data and can be used to form probabilistic brain atlases [41, 42]. They incorporate sulcal curves [118] and sulcal ribbons [72] as constraints into the intensity matching approach.

Sandor et al. [119] use nonrigid warping to elastically match a prelabeled brain atlas to a brain surface extracted using morphological operations and 3D bicubic spline surfaces. The model is attracted to the brain surface while points along major sulci (sylvian fissure, interhemispheric fissure, central sulcus) are attracted to fissure features (also extracted with morphological operations).

17.5.5 RPM

In robust point matching (RPM) [120, 121, 53] nonrigid registration is formulated in terms of point features and a corresponding match matrix:

$$M_{ij} = \frac{1}{\sqrt{2\pi t^2}} \exp\left[\frac{-|T^k(X_i) - Y_j|}{2t^2}\right] \tag{17.10}$$

where t is the temperature corresponding to the distance range of matches, and T is the transformation. The fuzziness of the match reduces as the temperature is reduced via deterministic annealing. The match matrix converges to a specification of correspondence and a corresponding nonrigid transformation, with outlier rejection, where C and R store the degree of "outlierness":

$$\forall i, \ \sum_j M_{ij} + C_i = 1; \quad \forall j, \ \sum_i M_{ij} + R_j = 1. \tag{17.11}$$

Extended RPM [121] determines the transformation (spline-based), as a compromise between the fit to the determined correspondences (weighted by outlierness) and the smoothness (bending energy) of the transformation. The process converges as the match matrix and the transformation are alternatively updated.

17.6 Surface Warping

In this section we describe an automatic method using shape-based matching and geodesic interpolation to directly identify corresponding points on pairs of surfaces [79] for nonrigid registration. While shape can provide the basis for such a correspondence, this problem remains a difficult one because of ambiguity when the surfaces are complex and variable, such as with the human cerebral cortex. In addition, the lack of ground truth for correspondence remains a problem for evaluating such methods.

An individual surface is matched to a reference (or atlas) surface. This method simultaneously triangulates the individual surface based on the generated points. An initial sparse set of points on the individual surface is determined based on proximity and shape. The interpolated points in 3D are generated by finding the shortest surface paths between the initial points and then labeling the interpolated points equally spaced along these paths.

17.6.1 Initial Point Matching and Triangulation

Triangulated surfaces (at the voxel level) are extracted from segmented (manually or automatically) images using the marching cubes algorithm [122]. A small set of points on the atlas surface is labeled manually. These points are normally either sulcal or gyral points, which would be visually identifiable from a 3D rendering. An initial triangulation of these points is constructed manually. This manual step only needs to be done once for the atlas image. For each individual brain, we use the following automatic method to determine the corresponding points of the atlas. The initial triangle connections are then inherited from the atlas. We align the individual image to the atlas image by scaling, translation and rotation. The procrustes shape distance method [123] is used to calculate the scaling parameter by using several thousand points evenly sampled on the two surfaces. Rigid registration is then performed using distance transform methods [77].

The procedure for determining the initial corresponding points of the atlas is based on matching the local surface geometry. For each initial point i on the atlas surface, the objective function to be minimized within a region for point j on the individual surface is:

$$O_{ij} = d_{ij} \cdot n_{ij} \cdot f_{ij}, \qquad (17.12)$$

where d_{ij}, n_{ij} and f_{ij} are, respectively, a Euclidean distance measure, a surface normal match measure and a feature (curvedness) match measure, formulated as follows. The Euclidean distance measure is

$$d_{ij} = 1 + \sqrt{(x_i - x_j)^2 + (y_i - y_j)^2 + (z_i - z_j)^2}, \qquad (17.13)$$

where (x, y, z) is the 3D coordinate for each surface point. Note that $1 \leq d_{ij} \leq R_w$, where R_w is the radius of the search window with center point i. The surface normal match measure is $n_{ij} = 2 - \mathbf{n}_i \cdot \mathbf{n}_j$, where $\mathbf{n}$ is the unit normal vector for each surface point. Note that $1 \leq n_{ij} \leq 3$.

We use a function of curvature as our feature match measure. We would like the feature match to bring together corresponding cortical surface geometry (sulci and gyri). Given the segmented brain image L, the Gaussian curvature K and the mean surface curvature H can be calculated from the partial derivatives of the image [124, 125]:

$$K = \frac{\sum_{(i,j,k)\in\Omega} \left[L_i^2(L_{jj}L_{kk} - L_{jk}^2) + 2L_iL_j(L_{ik}L_{jk} - L_{ij}L_{kk})\right]}{(L_i^2 + L_j^2 + L_k^2)^2},$$

$$H = \frac{\sum_{(i,j,k)\in\Omega} \left[(L_{ii} + L_{jj})L_k^2 - 2L_iL_jL_{ij}\right]}{2(L_i^2 + L_j^2 + L_k^2)^{3/2}}, \qquad (17.14)$$

where $\Omega = \{(x, y, z), (y, z, x), (z, x, y)\}$ is the set of circular shifts of (x, y, z). The two principal curvatures k_1 and k_2 are related to the Gaussian and mean curvatures [126]: $k_1 = H + \sqrt{H^2 - K}$, and $k_2 = H - \sqrt{H^2 - K}$. Shape can be characterized by two values: one describing the type of curvature and one describing the degree [127]. A shape index function, $S = \frac{2}{\pi} \arctan[(k_2 + k_1)/(k_2 - k_1)]$, can be used to classify surfaces into nine types [127]. Shape index distinguishes between sulci and gyri [125]. Curvedness measures the degree of curvature [127]: $C = \sqrt{(k_1^2 + k_2^2)/2}$.

For our feature match measure, we use a thresholded curvedness, signed according to convexity. In order to locate sulcal and gyral points t, we threshold curvedness based on the shape index (Fig. 17.1). Then,

$$t = \begin{cases} \text{gyrus,} & \text{if } C_s > K_g \text{ and } S \geq 0; \\ \text{sulcus,} & \text{if } C_s > K_s \text{ and } S < 0; \\ \text{no feature, otherwise.} \end{cases} \qquad (17.15)$$

The threshold values K_g and K_s are chosen dynamically so that the selected sulcal or gyral points are approximately a specified percentage of the total surface points. Then f_{ij} can be set to a low value when the surface points match (e.g. 1), and a high value when they do not (e.g. 3). Thus, for each labeled point on the atlas, the point on the individual surface that minimizes Eq. (17.12) within a radius R_w (normally chosen to be 15 pixels) is selected as the corresponding point.

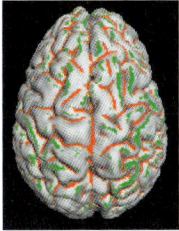

Fig. 17.1. Sulcal (red) and gyral (green) points of the individual brain computed by thresholding the curvedness

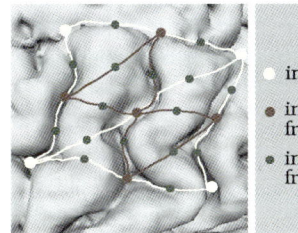

Fig. 17.2. Schematic diagram for corresponding surface points identification and surface triangulation in our hierarchical approach showing brain sulci being followed

17.6.2 Triangulation

Given the set of corresponding points and their connections, the next step is to determine the shortest paths between each pair of connected surface points (Fig. 17.2). Geodesic distance is a useful measure for understanding complex shape. One of the first uses in brain analysis was by Griffin [128], who used mean geodesic distance to characterize cortical shape. Geodesic distance and curvature have, for example, been used to follow sulci [129, 130]. The cortical surface is composed of folds (gyri) separated by sulci. The geodesic path connecting points in a sulcus will tend to follow the sulcus. Minimal paths constructed across the surface have also been used for surface patch parameterization [131].

There are many methods to determine the shortest path between points on a surface. We use an algorithm based on Kimmel's extended fast marching method [132]. The fast marching method [133] is an extremely fast numerical algorithm for solving the Eikonal equation:

$$\mid \nabla T \mid = \mathcal{F}(x, y) \qquad (17.16)$$

on a rectangular or triangulated mesh in $O(M \log M)$ steps, where M is the total number of grid points. This technique has been extended to triangulated domains with the same computational complexity [132].

For each edge in the initial triangulation, we calculate the shortest path between the surface points by the above method. Then, the midpoint on each shortest path is determined in order to decompose each triangle into four smaller ones (Fig. 17.3). In this way, a more dense geodesic triangulation is derived. Now, we simply repeat the previous two steps until we have a sufficiently dense triangulation.

Fig. 17.3. 3D synthetic surface reconstruction. *Left*: initial points and triangulation of synthetic surface (38 points and 72 triangles); *middle*: first iteration of surface triangulation; *right*: final (third) iteration of surface triangulation

The assumption of this approach is that the relative deformation of the two surfaces is approximately a uniform stretching between the initial points. Locally uniform stretching, or homothetic deformation [134], is a reasonable assumption and can be satisfied, at least approximately, by the appropriate selection of the initial points. The general requirement for the initial points is an even distribution. In areas of greater complexity, it may be necessary to include a denser distribution of initial points so that more accurate results can be derived. Deep sulcal areas have longer surface paths given the same Euclidean distance between the points. The initial point matching uses surface curvature, and the interpolation is based on geodesics. Therefore, our approach captures global and local surface shape, unlike other point matching algorithms.

17.6.3 Real Brain Pair Experiment

Figure 17.4 shows the automatic point matching results on a pair of brain surfaces (Sect. 17.6.1). Anatomic surface features are identified consistently due to the combined effect of the distance, normal and feature measures. The surface correspondence and reconstruction process for the atlas is shown in Fig. 17.5. In order to evaluate our approach, we also implemented a simple closest point method for correspondence finding. This method identifies the closest point on the individual as the corresponding point for the atlas. A comparison of these two approaches can be seen in Fig. 17.6, where the five points in the atlas correspond well by our geodesic method. The simple closest point method fails because it cannot compensate sufficiently for the relative deformation.

17 Methods for Nonrigid Image Registration   587

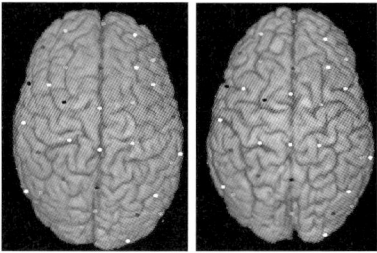

Fig. 17.4. Automatically matched initial points on the brain surface. *Left*: atlas surface with hand-labeled points (69); *right*: individual surface with corresponding points identified by our automatic point matching procedure (Sect. 17.6.1)

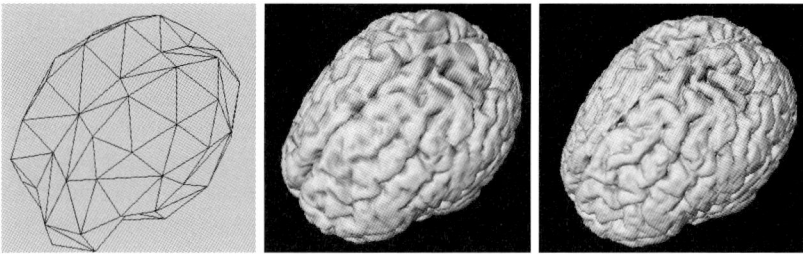

Fig. 17.5. Atlas surface correspondence and reconstruction by our hierarchical approach. *Left*: initial points and triangulation; *middle*: reconstructed surface after fourth iteration with normals extracted from the original surface; *right*: original atlas surface

17.6.4 Synthetic Warping Experiment

In order to further evaluate the methods, we define a known sinusoidal warp and apply it to the atlas brain image, generating a warped individual image to which both our algorithm and the simple closest point algorithm can be applied. The atlas initial points and triangulation are shown in Fig. 17.5. The initial points of the individual are derived from our matching procedure discussed above. We can calculate the distance between each pair of points and visualize the distance color map on the individual surface. We also have the known distance map (Fig. 17.7a) for comparison. Fig. 17.7 shows that the displacement pattern and mean square distance error for our method are much better than the simple closest point method. The simple closest point method tends to underestimate distances, while our approach results in more accurate distances.

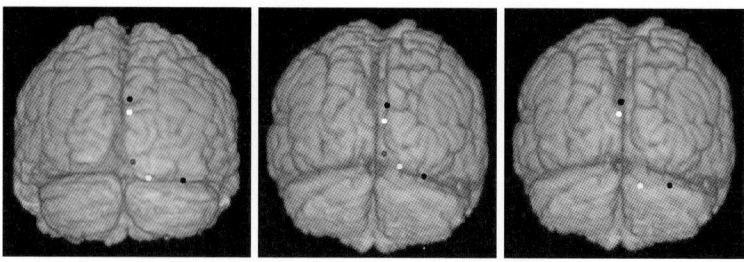

Fig. 17.6. Corresponding points comparison. *Left*: five points on the atlas posterior surface; *middle*: five corresponding points on the individual surface using shape-based method; *right*: five poorly corresponding points on the individual surface using the simple closest point method

17.7 Volumetric Warping

In this section, our goal is to incorporate statistical shape information into physical model registration and to develop a more accurate and robust algorithm. Without such information, such models tend to be underconstrained, yielding implausible matches. Our work here is most closely related to the work of Davatzikos [58, 59] and Christensen [30] described above. The algorithm we use is based on a physical model (linear elastic or viscous fluid) [50], a gray-level similarity measure and a consistency measure between corresponding boundary points. The statistical shape information is embedded in a separate boundary-finding process [38] applied to the individual. This method uses statistical point models with shape, and shape variation, generated from sets of examples by principal component analysis of the coordinate covariance matrix. The power of physical and statistical shape models are combined in our approach. For small deformations, our elastic model is appropriate and more efficient. For large deformations, we use our fluid model algorithm, which can track long-distance, nonlinear deformations.

17.7.1 Statistical Shape Information

The statistical shape information we use comes from corresponding boundary points. We have developed a boundary-finding method using a statistical shape model that also establishes correspondence, which has been described in detail [38]. Global shape parameters derived from the statistical variation of object boundary points in a training set are used to model the object [135]. A Bayesian formulation, based on this prior knowledge and the edge information of the input image, is used to find the object boundary with its subset points in correspondence with the point sets of boundaries in the training set. These points are then used in our nonrigid volumetric registration using a physical model.

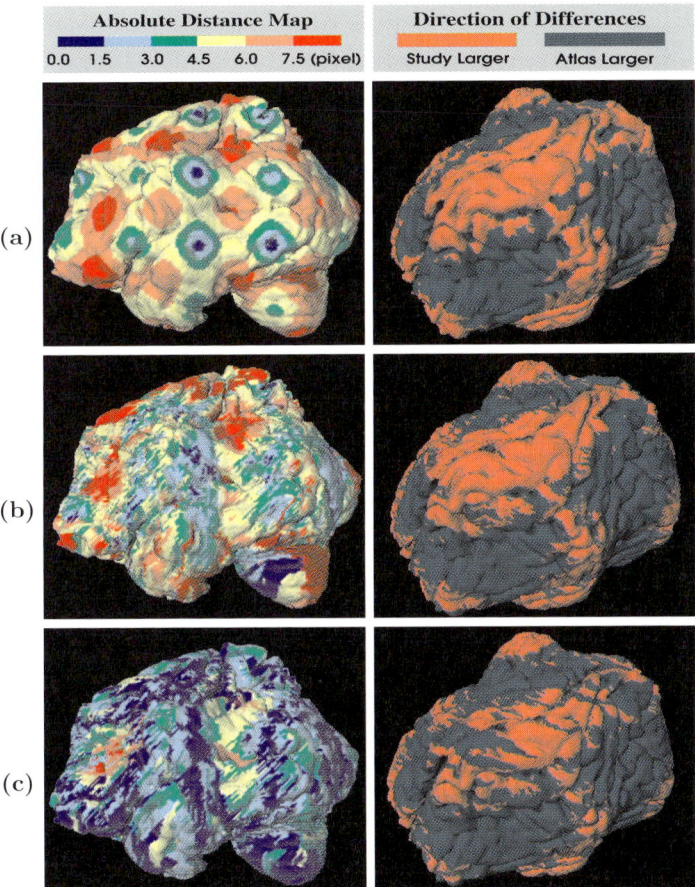

Fig. 17.7. Evaluation of different methods for the known warp. **a** True absolute distance map and direction of differences (*left* and *right* are two different views). **b** Results by our shape-based method (initial points from point-matching procedure) with absolute distance mean-square error 8.68 pixels. **c** Results by the simple closest point method with absolute distance mean-square error 16.03 pixels

17.7.2 Physical Model Integration

The physical models discussed previously (Sect. 17.5.1) are useful in nonrigid registration, but they are limited by themselves because they are too generic. Instead, boundary information can be used to guide the deformation in a way governed by structure in the image. In order to constrain solutions to more anatomically consistent deformations, we will start with the elastic and fluid physical models for intersubject deformation and then incorporate structural information from point models of boundary information.

We pose the registration problem as displacement estimation in a maximum a posteriori framework and derive corresponding forces [29, 50, 64]. As input to the problem, we have both the intensity image of the individual $I_s(\mathbf{w})$ and the boundary points corresponding to structure in the individual $\mathbf{b}_s(n) = (x_s(n), y_s(n))$, for $n = 1, 2 \cdots N$. The boundary points are derived from a separate boundary-finding process that uses statistical shape [38, 136]. We want to maximize:

$$\Pr(\mathbf{u}|I_s, \mathbf{b}_s) = \frac{\Pr(\mathbf{u}, I_s, \mathbf{b}_s)}{\Pr(I_s, \mathbf{b}_s)}. \quad (17.17)$$

Ignoring the denominator, which does not change with $\mathbf{u}$, and using Bayes' rule, our aim is to find:

$$\arg\max_{\mathbf{u}} \Pr(\mathbf{u}|I_s, \mathbf{b}_s)$$
$$= \arg\max_{\mathbf{u}} \Pr(\mathbf{b}_s|\mathbf{u}, I_s) \Pr(I_s|\mathbf{u}) \Pr(\mathbf{u}),$$
$$= \arg\max_{\mathbf{u}} \left[\ln \Pr(\mathbf{u}) + \ln \Pr(I_s|\mathbf{u}) + \ln \Pr(\mathbf{b}_s|\mathbf{u}) \right], \quad (17.18)$$

where we ignore the dependence of $\mathbf{b}_s$ on I_s (because $\mathbf{b}_s$ is obtained as a prior here and is not modified in this formulation), and take the logarithm.

The Bayesian posterior can be directly related to the PDE in Eq. (17.6) or Eqs. (17.7) and (17.8) based on the variational principle from which the PDE can be derived [29]. The first term in Eq. (17.18) corresponds to the transformation prior term, which gives high probability to transformations consistent with a physical model (elastic or fluid) and low probability to all other transformations.

The second term in Eq. (17.18) represents the likelihood that depends on the individual image. Let $I_a(\mathbf{w})$ be the intensity image of the atlas. We model the individual image as a Gaussian with mean given by the deformed atlas image, $I_a(\mathbf{w} - u(\mathbf{w}))$ [29] (since an Eulerian reference frame is used here, a mass particle instantaneously located at $\mathbf{w}$ originated from point $\mathbf{w} - \mathbf{u}(\mathbf{w})$). That is,

$$\ln \Pr(I_s|\mathbf{u}) = -\frac{1}{2\sigma_1^2} \int_\Omega [I_s(\mathbf{w}) - I_a(\mathbf{w} - u(\mathbf{w}))]^2 \, d\mathbf{w},$$

where σ_1 is the standard deviation of the Gaussian process.

The first body force $\mathbf{F}_1$ is the gradient of this likelihood term with respect to $\mathbf{u}$ at each $\mathbf{w}$ [29]:

$$\mathbf{F}_1(\mathbf{u}) = -k_1 \left[I_s(\mathbf{w}) - I_a(\mathbf{w} - u(\mathbf{w})) \right] \nabla I_a(\mathbf{w} - u(\mathbf{w})). \quad (17.19)$$

This force is the product of the difference in intensity between the individual (I_i) and the deformed atlas (I_a), and the gradient of the deformed atlas. The gradient term determines the directions of the local deformation forces applied to the atlas.

The last term of Eq. (17.18) incorporates structural information into the nonrigid registration framework. The extra constraint of corresponding boundary points is used as an additional matching criterion. The boundary point positions are the result of the deformation of the model to fit the data in ways consistent with the statistical shape models derived from the training set [38, 136]. Let $\mathbf{b}_a(n) = (x_a(n), y_a(n))$, for $n = 1, 2 \cdots N$, denote the atlas boundary points positions, which are known since we have full information about the atlas. We now model $\mathbf{b}_s$ as a Gaussian with mean given by the deformed atlas boundary positions, expressed as $\mathbf{b}_a(n) + \mathbf{u}(\mathbf{w})$, for pixels $\mathbf{w}$ at the deformed atlas boundary points. Then,

$$\ln \Pr(\mathbf{b}_s|\mathbf{u}) = -\frac{1}{2\sigma_2^2} \sum_{n=1}^{N} ||\mathbf{b}_s(n) - [\mathbf{b}_a(n) + \mathbf{u}(\mathbf{w})]||^2,$$

where σ_2 is again the standard deviation of the Gaussian process.

The second body force $\mathbf{F}_2$ is then the gradient of the above equation with respect to $\mathbf{u}$ for pixels $\mathbf{w}$ at the deformed atlas boundary points:

$$\mathbf{F}_2(\mathbf{u}) = k_2||\mathbf{b}_s(n) - [\mathbf{b}_a(n) + \mathbf{u}(\mathbf{w})]||, \qquad (17.20)$$

where $\mathbf{b}_a$ denotes the atlas boundary point positions, and $\mathbf{b}_s$ the individual boundary positions. The force is zero for pixels that are not on the deformed atlas boundary. From Eq. (17.20), we can see that the calculated displacements at the boundary points are constrained to match the vector difference of the corresponding atlas and individual boundary point positions. Using a weighted sum of these two forces will result in a match between shape features of the atlas and the individual, such as high-curvature points and important anatomical landmarks, as well as the intensity measure given by $\mathbf{F}_1(\mathbf{u})$.

We discretize these equations and employ Euler integration in time (Eq. 17.8) combined with successive overrelaxation (SOR) in the spatial domain (Eqs. 17.6 and 17.7).

17.7.3 Experimental Results

To evaluate the methodology, we quantify errors in the displacement field over the objects of interest and compare to either purely elastic [30] or purely fluid registration [29] based on our own implementation. Given a known warp, we can measure detailed displacement errors throughout the object. As before, we define a sinusoidally warped individual image. For a known nonrigid warp, the average (E_{oa}) and maximum (E_{om}) errors in the displacement vectors over the objects are used to measure accuracy. We also use the average error on the sparse boundary points, E_{qba}.

17.7.4 Real Image with Known Warping

In this experiment (Fig. 17.8), we apply a known sinusoidal warp to an MR sagittal brain image showing the corpus callosum. We compare our fluid

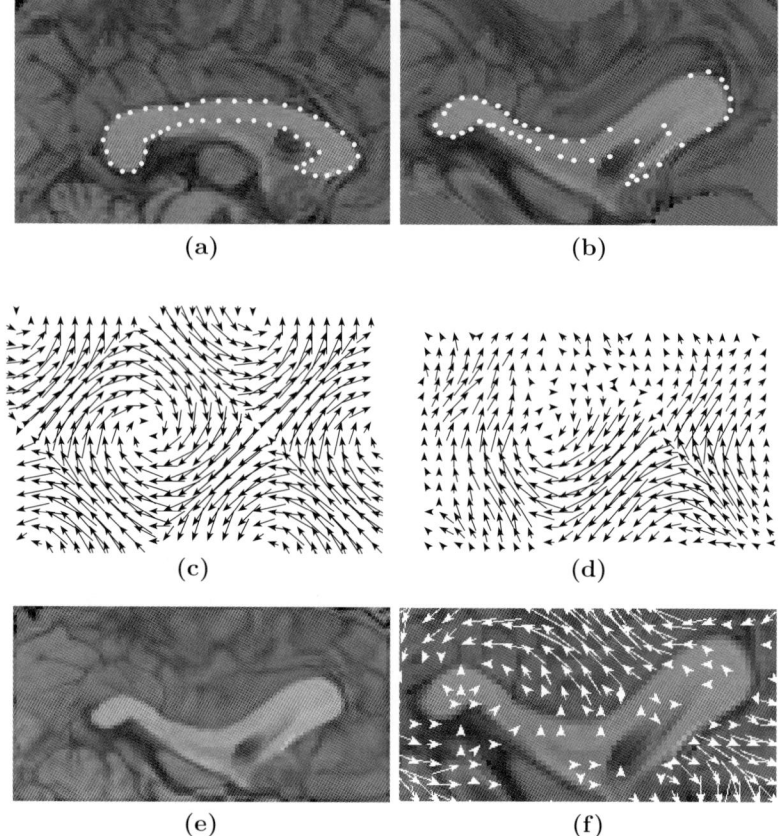

Fig. 17.8. MR sagittal corpus callosum image (100 × 64) example. **a** Atlas image with control points. **b** Individual image with control points. **c** True displacement vectors by sinusoidal warp. **d** Estimated vectors from our elastic method. **e** Deformed atlas by our elastic method. **f** Error vectors shown on individual image (cropped)

method, the purely fluid method and our elastic method, choosing the elasticity (μ) just large enough so that a homeomorphic map is ensured. The results (Fig. 17.8 and Table 17.1) show that our fluid method leads to a much better registration in the object of interest than the other methods, where the large deformation of the images cannot be tracked.

17.7.5 Real Atlas and Individual Images

Results of the method applied to an MR brain pair are shown in Fig. 17.9. These are 2D slices that roughly correspond from different brains for demonstration purposes. The control points of the individual image are derived from the shape model boundary finding algorithm [38]. We have found that the fi-

Table 17.1. Error measure for MR sagittal corpus callosum image with known warping (Fig. 17.8) showing the improvement of our fluid method. E_{oa}: average displacement error over corpus callosum; E_{om}: maximum displacement error over corpus callosum; E_{ba}: average displacement error on sparse boundary points. Note: The errors are in pixels; the percentages are with respect to the true average displacement

Method	E_{oa} (%)	E_{om}	E_{ba} (%)
Purely fluid	5.82 (63.2%)	13.82	5.91 (65.0%)
Our fluid	0.99 (11.4%)	2.29	0.89 (9.8%)
Our elastic	3.46 (37.6%)	10.54	3.32 (36.5%)

nal error of our methods is still much better than that of purely elastic or fluid methods. Note in particular the errors in the detailed matching by the purely elastic method shown in Fig. 17.9 compared with our approach.

17.8 Conclusion

Much work remains in the development of robust methods for nonrigid image registration, and much of this work will be tailored to the requirements of particular applications. We described two different approaches to the problem that take advantage of different available information. In the surface warping approach, we use local and global surface properties, while in the volumetric deformation method we use a combination of shape and intensity information. In general, the choice of the formulation in terms of the transformation, geometric structure, match metric and optimization method will depend on the requirements of the task, including the flexibility needed, the features available and the difficulty of the matching. A key aspect is, of course, the determination of correspondence, whether implicitly or explicitly. Ultimately, it is desirable to design a match metric that includes as much useful information as possible, a transformation tailored to the deformability needed and an efficient and reliable optimization. These requirements vary depending on the imaging source and problem domain. For particular applications, these elements need to be carefully chosen, and the best choice remains an open question in most areas.

References

1. Anuta, P.E. (1970) Spatial registration of multispectral and multitemporal digital imagery using fast Fourier transform techniques. *IEEE Trans. Geosci. Electron.* 8(4):353–368
2. Burr, D.J. (1981) A dynamic model for image registration. *Comp. Graphics Image Proc.* 15(2):102–112

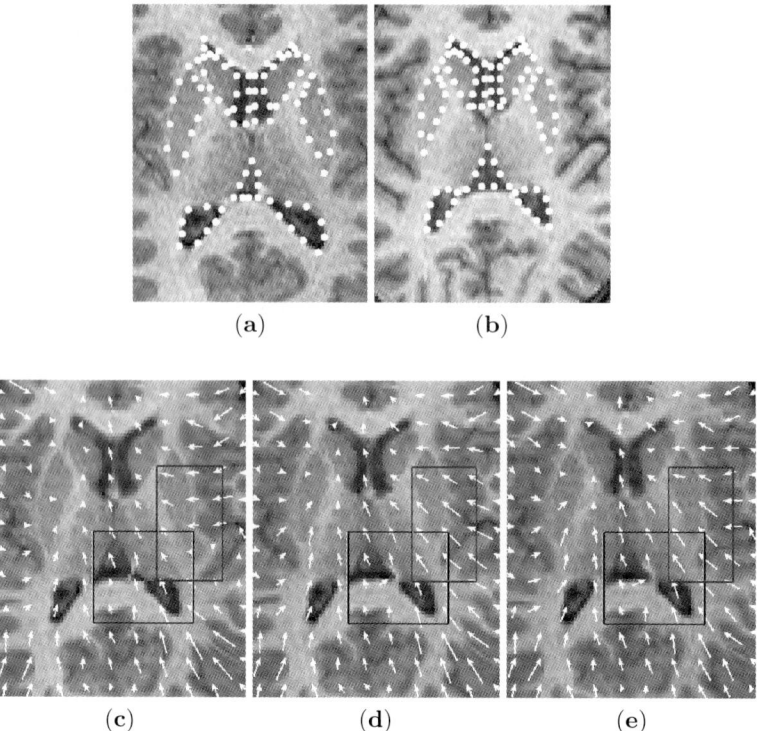

Fig. 17.9. MR axial brain images (80 × 100) and displacement vectors. **a** Atlas image with its control points. **b** Individual image with its control points derived from our boundary finding algorithm [38]. **c** Estimated vectors by the purely elastic method over their deformed atlas. **d** Our elastically estimated vectors over our elastically deformed atlas. **e** Our fluidly estimated vectors over our fluidly deformed atlas. compare the mappings of the ventricle corner (*lower square*) and the putamen (*upper rectangle*) showing the correct mapping by our algorithm

3. Bajcsy, R., Lieberson, R., Reivich, M. (1983) A computerized system for the elastic matching of deformed radiographic images to idealized atlas images. *J. Comp. Assisted Tomogr.* 7(4):618–625
4. Brown, L.G. (1992) A survey of image registration techniques. *ACM Comp. Surveys* 24(4):325–376
5. Maurer, C.R., Fitzpatrick, J.M. (1993) A review of medical image registration. In: Macuinas, R.J. (ed.) *Interactive Image-Guided Neurosurgery.* Am. Assoc. Neurological Surgeons, pp. 14–44
6. van den Elsen, P., Pol, E., Viergever, M. (1993) Medical image matching – A review with classification. *IEEE Eng. Med. Biol.* 12:26–39
7. Toga, A.W. (1999) *Brain Warping.* Academic, San Diego
8. Fitzpatrick, J.M., Hill, D.L.G., Maurer, C.R. (2000) Image registration. In: Sonka, M., Fitzpatrick, J.M. (eds.) *Handbook of Medical Imaging: Medical Image Processing and Analysis.* vol. 2. SPIE, Bellingham, WA

9. Bankman, I. (ed.) (2000) *Handbook of Medical Imaging: Processing and Analysis*. Academic, San Diego
10. Wolberg, G. (1990) *Digital Image Warping*. IEEE Comp. Soc., Los Alamitos, CA
11. Hajnal, J., Hawkes, D.J., Hill, D. (2001) *Medical Image Registration*. CRC Press, Boca Raton
12. Zitová, B., Flusser, J. (2003) Image registration methods: A survey. *Image and Vision Computing* 21:977–1000
13. Gee, J.C., Maintz, J.B.A., Vannier, M.W. (2003) *Biomedical Image Registration: 2nd International Workshop. LNCS, vol. 2717*. Springer, Berlin Heidelberg New York
14. Pernus, F., Kovacic, S., Stiehl, H.S., Viergever, M.A. (1999) *Biomedical Image Registration: 1st International Workshop*. Slovenian Pattern Recognition Society, Ljubljana, Slovenia
15. Goshtasby, A., Staib, L., Studholme, C., Terzopoulos, D. (2003) Nonrigid image registration: guest editors' introduction. *Comp. Vision and Image Understanding* 89(2-3):109–113
16. Pluim, J.P.W., Fitzpatrick, J.M. (2003) Image registration. *IEEE Trans. Med. Imaging* 22(11):1341–1343
17. Foley, J.D., Dam, A.V. (1982) *Fundamentals of Interactive Computer Graphics*. Addison-Wesley, Reading, MA
18. Secretta, G., Gregson, P.H. (1999) Automated registration of multimodal brain image sets using computer vision methods. *Comp. Biol. Med.* 29(5):333–359
19. Goshtasby, A. (1986) Piecewise linear mapping functions for image registration. *Pattern Recognition* 19(6):459–466
20. Boissonnat, J. (1984) Geometric structures for three-dimensional shape representation. *ACM Trans. Graphics* 3(4):266–286
21. Goshtasby, A. (1988) Image registration by local approximation methods. *Image and Vision Computing* 6(4):255–261
22. deBoor, C. (1978) *A Practical Guide to Splines*. Springer, New York
23. Sederberg, T.W., Parry, S.R. (1986) Free-form deformation of solid geometric models. *Comp. Graphics* 20(4):151–160
24. Rueckert, D., Sonoda, L., Hayes, C., Hill, D., Leach, M., Hawkes, D. (1999) Nonrigid registration using free-form deformations: application to breast MR images. *IEEE Trans. Med. Imaging* 18(8):712–721
25. Friston, K., Frith, C., Liddle, P., Frackowiak, R. (1991) Plastic transformation of PET images. *J. Comp. Assisted Tomogr.* 15(4):634–639
26. Jain, A.K., Zhong, Y., Lakshmanan, S. (1996) Object matching using deformable templates. *IEEE Trans. Pattern Anal. Machine Intell.* 18(3):267–278
27. Bajcsy, R., Kovačič, S. (1989) Multiresolution elastic matching. *Comp. Vision Graphics Image Proc.* 46:1–21
28. Bro-Nielson, M., Gramkow, C. (1996) Fast fluid registration of medical images. In: Höhne, K. (ed.) *Visualization Biomed. Comp. 1996. LNCS, vol. 1131*. Springer, Berlin Heidelberg New York, pp. 266–276
29. Christensen, G., Rabbitt, R., Miller, M. (1996) Deformable templates using large deformation kinematics. *IEEE Trans. Image Proc.* 5(10):1435–1447
30. Christensen, G.E., Miller, M.I., Vannier, M.W. (1996) Individualizing neuroanatomical atlases using a massively parallel computer. *Computer* 29:32–38

31. Davatzikos, C., Prince, J. (1994) Brain image registration based on curve mapping. In: *Proc. IEEE Workshop Biomedical Image Anal.* IEEE Comp. Soc., Los Alamitos, CA, pp. 245–254
32. Gee, J., Briquer, L.L., Barillot, C., Haynor, D. (1995) Probabilistic matching of brain images. In: Bizais, Y., Barillot, C., Paola, R.D. (eds.) *Information Proc. Med. Imaging.* Kluwer, Dordrecht, pp. 113–125
33. Miller, M., Christensen, G., Amit, Y., Grenander, U. (1993) Mathematical textbook of deformable neuroanatomies. *Proc. Natl. Acad. Sci. USA* 90:11944–11948
34. Cootes, T., Hill, A., Taylor, C., Haslam, J. (1993) The use of active shape models for locating structures in medical images. In: Barrett, H.H., Gmitro, A.F. (eds.) *Information Proc. Med. Imaging. LNCS, vol. 687.* Springer, Berlin Heidelberg New York, pp. 33–47
35. Staib, L.H., Duncan, J.S. (1992) Boundary finding with parametrically deformable models. *IEEE Trans. Pattern Anal. Machine Intell.* 14(11):1061–1075
36. Staib, L.H., Duncan, J.S. (1996) Model-based deformable surface finding for medical images. *IEEE Trans. Med. Imaging* 15(5):720–731
37. Székely, G., Kelemen, A., Brechbüler, C., Gerig, G. (1995) Segmentation of 3D objects from MRI volume data using constrained elastic deformations of flexible Fourier surface models. In: Ayache, N. (ed.) *Comp. Vision, Virtual Reality and Robotics in Med. LNCS, vol. 905.* Springer, Berlin Heidelberg New York, pp. 495–505
38. Wang, Y., Staib, L.H. (1998) Boundary finding with correspondence using statistical shape models. In: *Proc. Comp. Vision Pattern Recog.* IEEE Comp. Soc., Los Alamitos, CA, pp. 338–345
39. Ashburner, J., Neelin, P., Collins, D., Evans, A., Friston, K. (1997) Incorporating prior knowledge into image registration. *NeuroImage* 6:344–352
40. Thompson, P., Schwartz, C., Toga, A. (1996) High resolution random mesh algorithms for creating a probabilistic 3D surface atlas of the human brain. *NeuroImage* 3:19–34
41. Collins, D.L., Evans, A.C. (1996) Automatic 3D estimation of gross morphometric variability in human brain. *NeuroImage* 3(3):S129
42. Thompson, P., MacDonald, D., Mega, M., Holmes, C., Evans, A., Toga, A. (1997) Detection and mapping of abnormal brain structure with a probabilistic atlas of cortical surfaces. *J. Comp. Assisted Tomogr.* 21(4):567–581
43. Feldmar, J., Malandain, G., Declerck, J., Ayache, N. (1996) Extension of the ICP algorithm to non-rigid intensity-based registration of 3D volumes. In: *Proc. Workshop Math. Meth. Biomed. Image Anal.* IEEE Comp. Soc., Los Alamitos, CA, pp. 84–93
44. Thirion, J.P. (1998) Image matching as a diffusion process: An analogy with Maxwell's demons. *Med. Image Anal.* 2(3):243–260
45. Maintz, A., van den Elsen, P., Viergever, M. (1995) Comparison of feature-based matching of CT and MR brain images. In: Ayache, N. (ed.) *Comp. Vision, Virtual Reality and Robotics in Med. LNCS, vol. 905.* Springer, Berlin Heidelberg New York, pp. 219–228
46. D. Shen, C.D. (2001) HAMMER: Hierarchical attribute matching mechanism for elastic registration. In: *Proc. Workshop Math. Meth. Biomed. Image Anal.* IEEE Comp. Soc., Los Alamitos, CA
47. Rohr, K. (2000) On 3D differential operators for detecting point landmarks. *Image and Vision Computing* 15:219–233

48. Rohr, K. (2001) *Landmark-based Image Analysis*. Kluwer, Dordrecht
49. Rohr, K., Stiehl, H., Frantz, S., Hartkens, T. (2000) Performance characterization of landmark operators. In: Klette, R., et al. (eds.) *Performance Characterization in Computer Vision*. Kluwer, Dordrecht
50. Wang, Y., Staib, L.H. (1998) Elastic model based non-rigid registration incorporating statistical shape information. In: *Medical Image Computing and Computer-Assisted Intervention. LNCS, vol. 1496*. Springer, Berlin Heidelberg New York, pp. 1162–1173
51. Wang, Y., Staib, L.H. (1998) Integrated approaches to non-rigid registration in medical images. In: *IEEE Workshop on Applications of Computer Vision*. IEEE Comp. Soc., Los Alamitos, CA, pp. 102–108
52. Rangarajan, A., Chui, H., Bookstein, F.L. (1997) The softassign Procrustes matching algorithm. In: Duncan, J., Gindi, G. (eds.) *Information Proc. Med. Imaging. LNCS, vol. 1230*. Springer, Berlin Heidelberg New York, pp. 29–42
53. Rangarajan, A., Chui, H., Mjolsness, E., Pappu, S., Davachi, L., Goldman-Rakic, P., Duncan, J. (1997) A robust point-matching algorithm for autoradiograph alignment. *Med. Image Anal.* pp. 379–398
54. Besl, P.J., Mackay, N.D. (1992) A method for registration of 3-D shapes. *IEEE Trans. Pattern Anal. Machine Intell.* 14(2):239–256
55. Feldmar, J., Ayache, N. (1996) Rigid, affine and locally affine registration of free-form surfaces. *Int. J. Computer Vision* 18(2):99–199
56. Marais, P.C. (1998) *The Segmentation of Sparse MR Images*. PhD thesis, Oxford University, Oxford
57. Caunce, A., Taylor, C.J. (1999) Using local geometry to build 3D sulcal models. In: A. Kuba, M.v., Todd-Pokropek, A. (eds.) *Information Proc. Med. Imaging. LNCS, vol. 1613*. Springer, Berlin Heidelberg New York, pp. 196–209
58. Davatzikos, C. (1997) Spatial transformation and registration of brain images using elastically deformable models. *Comp. Vision and Image Understanding* 66(2):207–222
59. Davatzikos, C., Prince, J.L., Bryan, R.N. (1996) Image registration based on boundary mapping. *IEEE Trans. Med. Imaging* 15(1):112–115
60. Shi, P., Robinson, G., Chakraborty, A., Staib, L., Constable, R., Sinusas, A., Duncan, J. (1995) A unified framework to assess myocardial function from 4D images. In: Ayache, N. (ed.) *Comp. Vision, Virtual Reality and Robotics in Med. LNCS, vol. 905*. Springer, Berlin Heidelberg New York, pp. 327–340
61. Wang, Y., Peterson, B.S., Staib, L.H. (2000) Shape-based 3D surface correspondence using geodesics and local geometry. In: *Proc. Comp. Vision Pattern Recog.* vol. II. IEEE Comp. Soc., Los Alamitos, CA, pp. 644–651
62. McEachen, J., Duncan, J. (1997) Shape-based tracking of left ventricular wall motion. *IEEE Trans. Med. Imaging* 16(3):270–283
63. Tagare, H. (1999) Shape-based nonrigid correspondence with applications to heart motion analysis. *IEEE Trans. Med. Imaging* 18(7):570–579
64. Wang, Y., Staib, L.H. (2000) Physical model based non-rigid registration incorporating statistical shape information. *Med. Image Anal.* 4:7–20
65. Fleute, M., Lavallée, S. (1998) Building a complete surface model from sparse data using statistical shape models: Application to computer assisted knee surgery. In: *Medical Image Computing and Computer-Assisted Intervention. LNCS, vol. 1496*. Springer, Berlin Heidelberg New York, pp. 879–887

66. Davies, R., Twining, C., Cootes, T., Waterton, J., Taylor, C. (2002) A minimum description length approach to statistical shape modeling. *IEEE Trans. Med. Imaging* 21(5):525–537
67. Lopez, A., Lloret, D., Serrat, J. (1998) Creaseness measures for CT and MR image registration. In: *Proc. Comp. Vision Pattern Recog.* IEEE Comp. Soc., Los Alamitos, CA, pp. 694–699
68. van den Elsen, P. (1993) *Multimodality Matching of Brain Images*. PhD thesis, Utrecht University, Utrecht, The Netherlands ISBN 90-71546-02-0.
69. van den Elsen, P., Maintz, J., Pol, E., Viergever, M. (1992) Image fusion using geometrical features. In: Robb, R.A. (ed.) *Visualization Biomed. Comp. 1992, Proc. SPIE 1808*. SPIE, Bellingham, WA, pp. 172–186
70. Hellier, P., Barillot, C., Corouge, I., Gibaud, B., Le Goualher, G., Collins, D.L., Evans, A., Malandain, G., Ayache, N., Christensen, G.E., Johnson, H.J. (2003) Retrospective evaluation of inter-subject brain registration. *IEEE Trans. Med. Imaging* 22(9):1120–1130
71. Ashburner, J., Friston, K.J. (1999) Spatial normalization. *Human Brain Mapping* 7(4):254–266
72. Collins, D.L., Goualher, G.L., Evans, A.C. (1998) Non-linear cerebral registration with sulcal constraints. In: *Medical Image Computing and Computer-Assisted Intervention. LNCS, vol. 1496*. Springer, Berlin Heidelberg New York, pp. 974–984
73. Corouge, I., Barillot, C., Hellier, P., Toulouse, P., Gibaud, B. (2001) Non-linear local registration of functional data. In: Viergever, M.A., Dohi, T., Vannier, M. (eds.) *Medical Image Computing and Computer-Assisted Intervention. LNCS, vol. 2208*. Springer, Berlin Heidelberg New York, pp. 948–956
74. Pelizzari, C.A., Chen, G.T.Y., Spelbring, D.R., Weichselbaum, R.R., Chen, C.T. (1989) Accurate three-dimensional registration of CT, PET and MR images of the brain. *J. Comp. Assisted Tomogr.* 13(1):20–26
75. Borgefors, G. (1984) Distance transformations in arbitrary dimensions. *Comp. Vision Graphics Image Proc.* 27:321–345
76. Borgefors, G. (1988) Hierarchical chamfer matching: A parametric edge matching algorithm. *IEEE Trans. Pattern Anal. Machine Intell.* 10(6):849–865
77. Jiang, H.J., Robb, R.A., Holton, K.S. (1992) A new approach to 3-D registration of multimodality medical images by surface matching. In: Robb, R.A. (ed.) *Visualization Biomed. Comp. 1992, Proc. SPIE 1808*. SPIE, Bellingham, WA, pp. 196–213
78. Duncan, J.S., Owen, R.L., Staib, L.H., Anandan, P. (1991) Measurement of non-rigid motion in images using contour shape descriptors. In: *Proc. Comp. Vision Pattern Recog.* IEEE Comp. Soc., Los Alamitos, CA, pp. 318–324
79. Wang, Y., Peterson, B.S., Staib, L.H. (2003) 3D brain surface matching based on geodesics and local geometry. *Comp. Vision and Image Understanding* 89(2-3):252–271
80. Cover, T., Thomas, J. (1991) *Elements of Information Theory*. Wiley, New York
81. Collignon, A., Maes, F., Delaere, D., Vandermeulen, D., Suetens, P., Marchal, G. (1995) Automated multi-modality image registration based on information theory. In: Bizais, Y., Barillot, C., Paola, R.D. (eds.) *Information Proc. Med. Imaging*. Kluwer, Dordrecht, pp. 263–274

82. Maes, F., Collignon, A., Vandermeulen, D., Marchal, G., Suetens, P. (1997) Multimodality image registration by maximisation of mutual information. *IEEE Trans. Med. Imaging* 16(2):187–198
83. Studholme, C., Hill, D., Hawkes, D. (1996) Automated 3-D registration of MR and CT images of the head. *Med. Image Anal.* 1(2):163–175
84. Viola, P., Wells, W. (1995) Alignment by maximization of mutual information. In: *Proc. Fifth Int. Conf. Comp. Vision.* IEEE Comp. Soc., Los Alamitos, CA, pp. 16–23
85. Wells, W., Viola, P., Atsumi, H., Nakajima, S., Kikinis, R. (1996) Multi-modal volume registration by maximization of mutual information. *Med. Image Anal.* 1(1):35–52
86. Gaens, T., Maes, F., Vandermeulen, D., Suetens, P. (1998) Non-rigid multimodal image registration using mutual information. In: *Medical Image Computing and Computer-Assisted Intervention. LNCS, vol. 1496.* Springer, Berlin Heidelberg New York, pp. 1099–1106
87. Kim, B., Boes, J., Frey, K., Meyer, C. (1997) Mutual information for automated unwarping of rat brain autoradiographs. *NeuroImage* 5:31–40
88. Meyer, C., Boes, J., Kim, B., Bland, P. (1998) Evaluation of control point selection in automatic mutual information driven 3D warping. In: *Medical Image Computing and Computer-Assisted Intervention. LNCS, vol. 1496.* Springer, Berlin Heidelberg New York, pp. 944–951
89. Meyer, C.R., Boes, J.L., Kim, B., Bland, P.H., Zasadny, K.R., Kison, P.V., Koral, K., Frey, K.A., Wahl, R.L. (1997) Demonstration of accuracy and clinial versatility of mutual information for automatic multimodality image fusion using affine and thin-plate spline warped geometric deformations. *Med. Image Anal.* 1(3):195–206
90. Rueckert, D., Hayes, C., Studholme, C., Summers, P., Leach, M., Hawkes, D. (1998) Non-rigid registration of breast MR images using mutual information. In: *Medical Image Computing and Computer-Assisted Intervention. LNCS, vol. 1496.* Springer, Berlin Heidelberg New York, pp. 1144–1152
91. Pluim, J., Maintz, J., Viergever, M. (2000) Image registration by maximization of combined mutual information and gradient information. *IEEE Trans. Med. Imaging* 19(8):809–814
92. Pluim, J., Maintz, J., Viergever, M. (2003) Mutual-information-based registration of medical images: a survey. *IEEE Trans. Med. Imaging* 22(8):986–1004
93. Studholme, C., Hill, D.L.G., Hawkes, D.J. (1999) An overlap invariant entropy measure of 3D medical image alignment. *Pattern Recognition* 32:71–86
94. Press, W., Flannery, B., Teukolsky, S., Vetterling, W. (1986) *Numerical Recipes.* Cambridge University Press, Cambridge
95. Studholme, C., Hill, D., Hawkes, D. (1997) Automated three-dimensional registration of magnetic resonance and positron emission tomography brain images by multiresolution optimisation of voxel similarity measures. *Med. Phys.* 24(1):25–35
96. Staib, L.H., Lei, X. (1994) Intermodality 3D medical image registration with global search. In: *Proc. IEEE Workshop Biomedical Image Anal.* IEEE Comp. Soc., Los Alamitos, CA, pp. 225–234
97. Matsopoulos, G.K., Mouravliansky, N.A., Delibasis, K.K., Nikita, K.S. (1999) Automatic retinal image registration scheme using global optimization techniques. *IEEE Trans. Info. Tech. Biomed.* 3(1):47–60

98. Greminger, M., Nelson, B. (2003) Deformable object tracking using the boundary element method. In: *Proc. Comp. Vision Pattern Recog.* IEEE Comp. Soc., Los Alamitos, CA, pp. 289–294
99. Audette, M.A., Siddiqi, K., Ferrie, F.P., Peters, T.M. (2003) An integrated range-sensing, segmentation and registration framework for the characterization of intra-surgical brain deformations in image-guided surgery. *Comp. Vision and Image Understanding* 89(2–3):226–251
100. Lunn, K.E., Paulsen, K.D., Roberts, D.W., Kennedy, F.E., Hartov, A., Platenik, L.A. (2003) Nonrigid brain registration: synthesizing full volume deformation fields from model basis solutions constrained by partial volume intraoperative data. *Comp. Vision and Image Understanding* 89(2-3):299–317
101. Collins, D., Evans, A., Holmes, C., Peters, T. (1995) Automatic 3D segmentation of neuro-anatomical structures from MRI. In: Bizais, Y., Barillot, C., Paola, R.D. (eds.) *Information Proc. Med. Imaging.* Kluwer, Dordrecht, pp. 139–152
102. Declerck, J., Subsol, G., Thirion, J., Ayache, N. (1995) Automatic retrieval of anatomical structures in 3D medical images. In: Ayache, N. (ed.) *Comp. Vision, Virtual Reality and Robotics in Med. LNCS, vol. 905.* Springer, Berlin Heidelberg New York, pp. 153–162
103. Ojemann, J., Akbudak, E., Snyder, A., McKinstry, R., Raichle, M., Conturo, T. (1997) Anatomic localization and quantitative analysis of gradient refocused echo-planar fMRI susceptibility artifacts. *NeuroImage* 6:156–167
104. Farzaneh, F., Reidener, S., Pelc, N. (1990) Analysis of T_2 limitations and off-resonance effects on spatial resolution and artifacts in echo-planar imaging. *Mag. Res. Medicine* 14:123–139
105. Spencer, A.J.M. (1980) *Continuum Mechanics.* Longman, London
106. Davatzikos, C., Vaillant, M., Resnick, S., Prince, J., Letovsky, S., Bryan, R. (1996) A computerized approach for morphologic analysis of the corpus callosum. *J. Comp. Assisted Tomogr.* 20(1):88–97
107. Machado, A., Gee, J., Campos, M. (2000) A factor analytic approach to structural characterization. In: *Proc. Workshop Math. Meth. Biomed. Image Anal.* IEEE Comp. Soc., Los Alamitos, CA, pp. 219–223
108. Ashburner, J., Friston, K.J. (2000) Voxel-based morphometry–The methods. *NeuroImage* 11(6):805–821
109. Bookstein, F.L. (2001) "Voxel-based morphometry" should not be used with imperfectly registered images. *NeuroImage* 14(6):1454–1462
110. Malvern, L.E. (1969) *Introduction to the Mechanics of a Continuous Medium.* Prentice-Hall, Englewood Cliffs, NJ
111. Woods, R., Cherry, S., Mazziotta, J. (1992) Rapid automated algorithm for aligning and reslicing PET images. *J. Comp. Assisted Tomogr.* 16:620–633
112. Woods, R., Mazziotta, J., Cherry, S. (1993) MRI-PET registration with automated algorithm. *J. Comp. Assisted Tomogr.* 17:536–546
113. Friston, K., Ashburner, J., Poline, J.B., Frith, C.D., Heather, J.D., Frackowiak, R. (1995) Spatial registration and normalization of images. *Human Brain Mapping* 2:165–189
114. Thompson, P., Toga, A. (1996) A surface-based technique for warping three-dimensional images of the brain. *IEEE Trans. Med. Imaging* 15(4):402–417
115. Thompson, P., Toga, A. (1997) Detection, visualization and animation of abnormal anatomic structure with a deformable probabilistic brain atlas based on random vector field tranformations. *Med. Image Anal.* 1(4):271–294

116. Thompson, P.M., Mega, M.S., Narr, K.L., Sowell, E.R., Blanton, R.E., Toga, A.W. (2000) Brain image analysis and atlas construction. In: Sonka, M., Fitzpatrick, J.M. (eds.) *Handbook of Medical Imaging: Medical Image Processing and Analysis.* vol. 2. SPIE, Bellingham, WA
117. Collins, D., Peters, T., Dai, W., Evans, A. (1992) Model based segmentation of individual brain structures from MRI data. In: Robb, R.A. (ed.) *Visualization Biomed. Comp. 1992, Proc. SPIE 1808.* SPIE, Bellingham, WA, pp. 10–23
118. Collins, D., Goualher, G.L., Venugopal, R., Caramanos, Z., Evans, A., Barillot, C. (1996) Cortical constraints for non-linear cortical registration. In: Höhne, K. (ed.) *Visualization Biomed. Comp. 1996. LNCS, vol. 1131.* Springer, Berlin Heidelberg New York, pp. 307–316
119. Sandor, S., Leahy, R. (1997) Surface-based labeling of cortical anatomy using a deformable atlas. *IEEE Trans. Med. Imaging* 16(1):41–54
120. Chui, H., Rangarajan, A. (2003) A new point matching algorithm for non-rigid registration. *Comp. Vision and Image Understanding* 89(2–3):114–141
121. Papademetris, X., Jackowski, A.P., Schultz, R.T., Staib, L.H., Duncan, J.S. (2003) Computing 3D non-rigid brain registration using extended robust point matching for composite multisubject fMRI analysis. In: *Medical Image Computing and Computer-Assisted Intervention. LNCS, vol. 2879.* vol. II. Springer, Berlin Heidelberg New York, pp. 788–795
122. Lorenson, W., Cline, H. (1987) Marching cubes: A high resolution 3D surface construction algorithm. *Comp. Graphics* 21(4):163–169
123. Bookstein, F.L. (1997) Landmark methods for forms without landmarks: morphometrics of group differences in outline shape. *Med. Image Anal.* 1(3):225–243
124. ter Haar Romeny, B. (1994) *Geometry Driven Diffusion in Computer Vision.* Kluwer, Dordrecht
125. Zeng, X., Staib, L.H., Schultz, R.T., Duncan, J.S. (1998) Volumetric layer segmentation using coupled surfaces propagation. In: *Computer Vision and Pattern Recognition.* IEEE Comp. Soc., Los Alamitos, CA, pp. 708–715
126. Carmo, M.P.D. (1976) *Differential Geometry of Curves and Surfaces.* Prentice-Hall, New Jersey
127. Koenderink, J.J., van Doorn, A.J. (1992) Surface shape and curvature scales. *Image and Vision Computing* 10(8):557–565
128. Griffin, L.D. (1994) The intrinsic geometry of the cerebral cortex. *Journal of Theoretical Biology* 166(3):261–273
129. Khaneja, N., Miller, M., Grenander, U. (1998) Dynamic programming generation of curves on brain surfaces. *IEEE Trans. Pattern Anal. Machine Intell.* 20(11):1260–1265
130. Zeng, X., Staib, L.H., Schultz, R.T., Duncan, J.S. (1999) Segmentation and measurement of the cortex from 3D MR images using coupled surfaces propagation. *IEEE Trans. Med. Imaging* 18(10)
131. Brett, A.D., Hill, A., Taylor, C.J. (1999) A method of 3D surface correspondence and interpolation for merging shape examples. *Image and Vision Computing* 17(8):635–642
132. Kimmel, R., Sethian, J.A. (1998) Computing geodesic paths on manifolds. *Proc. Natl. Acad. Sci. USA* 95(15):8431–8435
133. Sethian, J.A. (1996) *Level Set Methods: Evolving Interfaces in Geometry, Fluid Mechanics, Computer Vision and Materials Science.* Cambridge University Press, Cambridge

134. Goldgof, D.B., Lee, H., Huang, T.S. (1988) Motion analysis of nonrigid surfaces. In: *Proc. Comp. Vision Pattern Recog.* IEEE Comp. Soc., Los Alamitos, CA, pp. 375–380
135. Cootes, T., Taylor, C. (1995) Combining point distribution models with shape models based on finite element analysis. *Image and Vision Computing* 13(5):403–410
136. Wang, Y., Staib, L.H. (2000) Statistical shape and smoothness models for boundary finding with correspondence. *IEEE Trans. Pattern Anal. Machine Intell.* 22(7):738–743

18

The Design of Implicit Functions for Computer Graphics

Alyn Rockwood

Mathematical and Computer Sciences
Colorado School of Mines
Golden, CO

18.1 Introduction

What we mean by an implicit function is a function $f : S \subseteq \Re^n \to \Re$ for a subset $S \subseteq \Re^n$. Often, $n = 3$ in the graphics context. The *zero set*, $f^{-1}(0)$, is called the *implicit model*. An important characteristic of implicit functions is that they characterize the space about the implicit model, that is $f(a) > 0$ can be seen as a quasi metric, giving proximity to the model. Implicit functions also conveniently partition space into regions outside, inside and on the surface corresponding to $f > 0, f < 0$ and $f = 0$. Finally, for a differentiable f, the surface normal of the implicit model is given simply by ∇f. This is needed in graphics for computing the lighting [2, 5]. Figure 18.1 shows a complex implicit model lighted using the gradient as a surface normal.

Parametric forms are the alternative way for modeling objects in graphics [3]. In three dimensions a parametric surface is given by $p : \Re^2 \to \Re^3$. The two domain variables parameterize the surface, yielding a useful local coordinate system for the surface that is utilized in tessellations and painting the surface with textures, among other applications. This typically makes the surface easier to render (polygonalize) and is a reason for the early popularity of parametric surfaces. As CPUs and graphics hardware become more powerful, the status is yielding to implicit techniques in many areas where the advantages of implicits can be used. We give a number of examples below, see also [2]. Initially, implicit models were used to represent canonical shapes that are common in engineering such as spheres, cylinders, cones and other quadrics, as well as torii. The unit sphere, for an obvious example, is given implicitly by

$$f(x) = x^2 + y^2 + z^2 - 1. \tag{18.1}$$

Superquadrics and blended surfaces were then added to this group. These are discusssed in Sect. 3. Bloomenthal [2] lists more interesting implicit models such as convolution surfaces, metaballs, alpha patches and others that are used

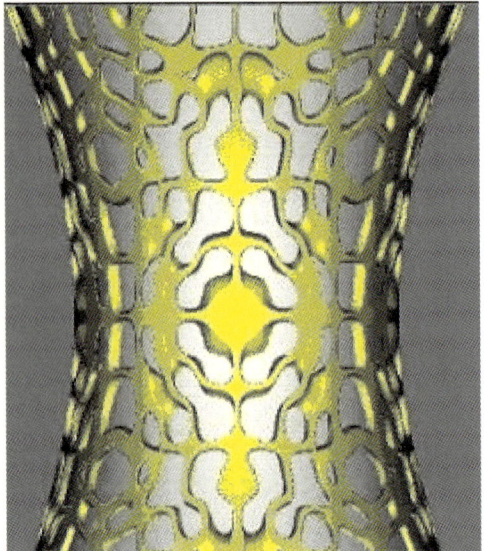

Fig. 18.1. A complex implicit model

in computer graphics. Two recent implicit methods are worth mentioning with some specifics. Level set methods introduced by Sethian [9] take a curve (on the left in Fig. 18.2), and build it into a surface. The cone-shaped surface shown on the right embeds the xy plane exactly where the curve sits. It is called the *level set function* and is an implicit function. The set is a zero set of the function.

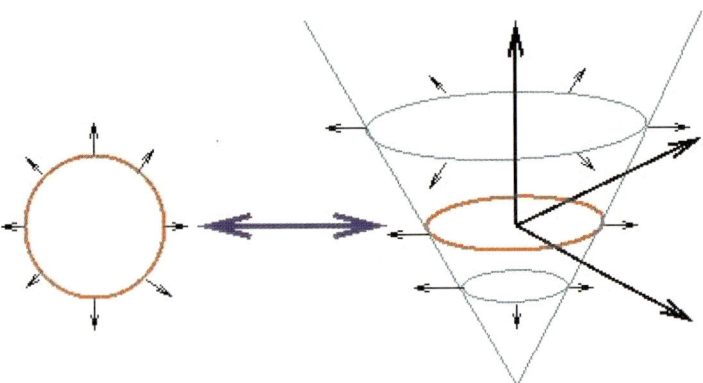

Fig. 18.2. The red curve is embedded into an implicit function

Quoting from http://math.berkeley.edu/~sethian/Flow_chart/chart.html: " The Osher-Sethian level set method tracks the motion of an object by embedding the object as the zero level set of the signed distance function. The motion of the interface is matched with the zero level set of the level set function, and the resulting initial value partial differential equation for the evolution of the level set function resembles a Hamilton-Jacobi equation. In this setting, curvatures and normals may be easily evaluated, topological changes occur in a natural manner, and the technique extends trivially to three dimensions. This equation is solved using entropy-satisfying schemes borrowed from the numerical solution of hyperbolic conservation laws, which produce the correct viscosity solution." As such it is a special case of implicit function in which the non-zero portion is constructed from entropy satisfying fluid flow schemes. Impressive results from these approaches have been obtained in fluid, fire and optical flow modeling, among others. We shall see other ways to construct implicit functions in Sects. 18.3 and 18.4. The second method, called *distance fields*, is closely related. Recent interest in distance fields is attested by several papers (see [4] and its references). They have shown wide versatility in computer graphics and related fields to the extent that they have even been suggested as a fundamental primitive for computer graphics, rivaling the polygon when appropriately organized as a sparse hierarchical data structure, called ADF for adaptively sampled distance field [4]. A distance field is often described with the popular notion that every point in space is associated with a "minimum" distance to the object. The notion derives both from natural experience and the classic definition of Hausdorf in which a distance from a point y in a metric space to a set A is given by

$$D(y, A) = \inf_{x \in A} D(x, y), \qquad (18.2)$$

where $D(x, y)$ is a standard metric satisfying

1. reflexivity, $D(x, y) = D(x, y)$,
2. positivity, $D(x, y) = 0$ if $x = y$, and
3. the triangle inequality rule, $D(x, z) \geq D(x, y) + D(y, z)$; see [1], p. 62.

For examples, the metric D may be (1) the Manhattan distance, i.e. the sum of absolute differences of components; (2) the Euclidean distance, i.e. the square root of the sum of squares of components; or (3) the infinity norm, i.e. the maximum of the absolute differences of components; known respectively as the l_1, l_2 and l_∞ norms. Figure 18.3 shows views of the Utah teapot, a set of 32 parametric Bezier patches, represented and rendered as a distance field from [4]. The field is computed by an intensive method that finds a nearest point on the teapot to any given point in space.

It is an example of how one can "implicitize" a parametric surface. One should notice the lack of polygonal artifacts, e.g. angular silhouettes and Mach banding in the highlights. Distances tend to yield much cleaner results. Unfortunately, the notion of minimum distance excludes many useful implicit

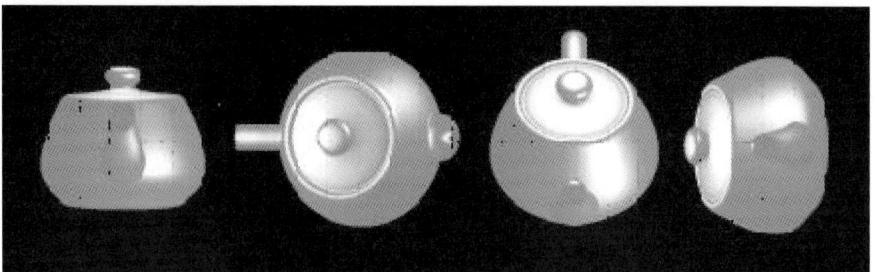

Fig. 18.3. Minimum distance teapot

models, for example, many zero sets of functions do not reference a minimum, nor do they need to designate a closest point on the object to define the field. Consider a simple ellipsoid defined by

$$f(x) = 20x^2 + y^2 + z^2 - 1. \tag{18.3}$$

For any point x in 3-space $f(x)$ represents a very usable distance to the flattened ellipsoid defined by $f(x) = 0$. It is an algebraic (non-Euclidean) distance that has no minimum or closest point in the calculation. Distance fields conveniently accommodate various operations such as Booleans, offsetting, collision detection, morphing, filleting or a number of rendering methods. But these can easily be carried over to other implicits. In Sect. 18.2 we classify some of the conspicuous types of implicit functions, and then in Sect. 18.3 we illustrate a few implicits that could also be embedded into a discrete distancelike field as in [4].

18.2 Types of Implicit Functions

The characterizations offered in this section will be useful in later developments of tools. We do not offer a comprehensive list of implicit function types, but concentrate on those that differentiate primarily on how they handle operations.

1. C^n. The function f is C^n continuous, where the most likely occurrences will be C^1, i.e. differentiable. Be aware, however, that differentiability of the function f alone does not guarantee a smooth zero set. The zero set may kink, or be non-manifold if the derivative is 0, which is a direct result of the implicit function theorem. Think of a zero set through a saddle point, for instance.
2. **Orientable**. We define a zero set to be orientable if its domain has an arbitrarily small neighborhood about any point of the zero set that contains points of both negative and positive distance. The set of negative points constitutes the *false side*, and the positive points constitute the

true side. For compact manifolds this corresponds to *inside* and *outside*, but in general this definition is quite different from the classical one. Our definition of orientability is far more forgiving than the traditional one.

3. **Attractor station**. If f is differentiable then ∇f induces a vector field, the gradient field of f. The *streamline* through a non-singular point x_0 is the set of points generated by

$$x_{i+1} = \pm \nabla f(xi) \Delta t + x_i, \Delta t \to 0, i = 0, 1, ... \qquad (18.4)$$

A zero set A is an *attractor station* of a region R if every streamline through a nonsingular point in R contains a point of A. Functionally it means that one can usually start at any nonsingular point in R and use Euler's method to generate a gradient descent (ascent) to arrive at A.

4. **Sloped**. There exists an arbitrarily small neighborhood around each point x of the function such that within the neighborhood there is a point x' with $f(x) > f(x')$. With other words, there is a path "down hill".

The different categories have no necessary dependencies with each other. The examples in the rest of the chapter demonstrate many independent combinations. The Weierstrass distance field (5) of Figure 18.2, for example, is sloped along the z-axes only, i.e. $\partial d / \partial z$ exists and is a constant, but none of the other partials exist in any other direction. Surprisingly too, it is orientable, despite the discontinuity.

18.3 Implicit Models

18.3.1 The Weierstrass function

The Weierstrass function given by

$$W(t) = \sum_k \alpha^k \cos(\beta^k t), \qquad (18.5)$$

for $k = 1, ..., \infty$, α real, β odd and $\alpha \beta > 1 + 3\pi/2$ is continuous everywhere, but differentiable nowhere. We must choose α less than 1 so the series converges. It is perhaps the first fractal described in the literature. We adapt Eq. (18.5) to define the *Weierstrass implicit function*

$$W_f(x) = z - \sum_k \alpha^k cos(\beta^k x)^k cos(\beta^k y), \qquad (18.6)$$

which is continuous and nondifferentiable in 3 dimensions. It's zero set, the *Weierstrass egg carton* is shown in Figure 18.4.

We call it an egg carton because of its shape that would hold eggs, but in its case, it would hold eggs of infinitely many sizes. We have combined it with an offset $W_f(x) - K$ for $y > 0$ to create a discontinuity in the field, which

Fig. 18.4. The split Weierstrass egg carton field

both illustrates that an implicit function need not be continuous, as well as showing its fractal like crosssection. As mentioned, the Weierstrass function 18.6 of Figure 18.4 is sloped along the z-axes only, i.e. $\partial d/\partial z$ exists and is a constant, but none of the other partials exist. The egg carton is not amenable to either analytic ray intersections, or triangulation, and it has no surface normals. Iterative ray intersections are costly and fallible to oversights.

Booleans and fillets

If f and g are two implicit functions with zero sets A and B, then the Boolean union, intersection and difference are given respectively by

$$A \cup B \equiv \{x|\min(f(x), g(x)) = 0\}, \tag{18.7}$$

$$A \cap B \equiv \{x|\max(f(x), g(x)) = 0\}, \tag{18.8}$$

$$A \backslash B \equiv \{x|\max(f(x), -g(x)) = 0\}. \tag{18.9}$$

The min/max functions typically create discontinuities of derivatives where $f(x) = g(x)$, i.e. nondifferentiable surfaces in the domain of the implicit function. Multiple Booleans are obtained by composition of the Boolean functions given by Eqs. (18.7)-(18.9) [8]. Figure 18.5 shows the union of three spheres configured to model the water molecule and displayed by ray tracing the zero set. The spheres are defined with the square root of the components squared; thus it is one of the rare cases in which the algebraic distance is equivalent to the Euclidean distance. The Boolean operators change that, of course. Let the spheres be given implicitly by functions $f(x)$, $g(x)$ and $h(x)$, where f and g are the smaller spheres and h the larger sphere. The new implicit function of the union is $\min(\min(f(x), g(x)), h(x))$, for example.

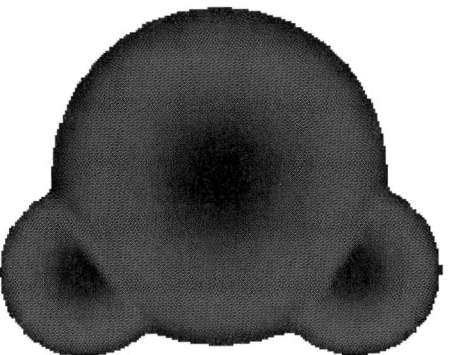

Fig. 18.5. H2O Boolean spheres model

In this case a nondifferentiability exhibits itself as creases where the spheres join, emanating into space as cones. To blend away the creases, we use an implicit blending form from CAD/CAM [8], which yields an implicit function

$$b(x) = 1 - \max\left(1 - \frac{f(x)}{R}, 0\right)^2 - 1 - \max\left(1 - \frac{g(x)}{R}, 0\right)^2$$
$$-1 - \max\left(1 - \frac{g(x)}{R}, 0\right)^2, \quad (18.10)$$

where R is the blending range chosen to determine the size of the fillet. Figure 18.6 shows the blended version of Fig. 18.5. It also has surfaces of nondifferentiability because of the max functions in Eq. (18.10). The function no longer exhibits creases in the zero set, however, which is shown to have continuous normals given by $\nabla b(x)$ wherever $b(x) = 0$.

Both distance fields for Figures 18.5 and 18.6 are orientable, sloped attractor stations. Even with the nondifferentiabilities, it is straightforward to march down viewing rays seeking sign differences to find intersections for "one bounce" ray tracing, which was done.

Thick surfaces

The blended union of Fig. 18.6 is extended in Fig. 18.7, first by adding more spheres of appropriate size and position, and second, by making it a *thick*

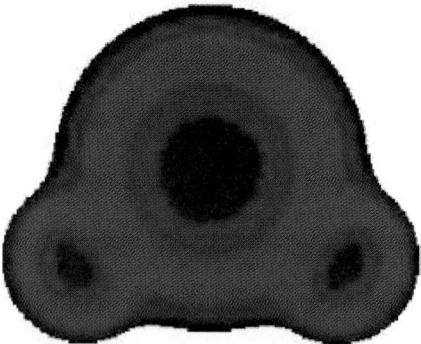

Fig. 18.6. The H2O molecule with blended creases

surface. The new function $m(x)$ can be used to define not just a zero set, but also a region about the zero set, given, for instance, by an *offset* $m(x) - K$, for K a constant. The so-called *thick or soft surface* is

$$B_K \equiv \{y | m(x) - K \geq y \geq b(x)\}. \tag{18.11}$$

The nicotine model is the same type as the water molecule, that is, it is sloped and orientable. Therefore, the same ray tracing algorithm that was used in Figs. 18.5 and 18.6 is employed, except that when the ray intersects the offset $b(x) - K$, a ray marching algorithm performs small incremental steps down the ray, while evaluating color transfer functions and accumulating the result with opacity; that is to say, a classic volume rendering routine is engaged at this point [5]. It continues until the ray emerges at the zero set. The ray is then traced again until it again intersects the soft surface and the marching method once again employed; and so forth until the ray exits the viewing window. In the figure, a 3D noise function governs the color transfer. Opacity is chosen so that the molecule is mostly translucent, except at silhouettes, where the larger traversal distance of the ray inside the offset region creates greater accumulation. The constant K is limited so that the offset does not contain any points of nondifferentiability.

18.3.2 The Color Gamut

The previous examples were based on implicit definitions directly or as compositions. Color gamuts are used in printing and photography to show the limits of the color devices. The color gamut is a case of a parametrically defined volume, in which the implicit function is defined as a function of the parameter space. Assume a mapping $p : (u, v, w) \rightarrow (x, y, z)$ that deforms space. Fix values of one of the u, v or w to define deformed surfaces in the range, i.e. images of the axial planes in parameter space. For the color gamut,

18 The Design of Implicit Functions for Computer Graphics

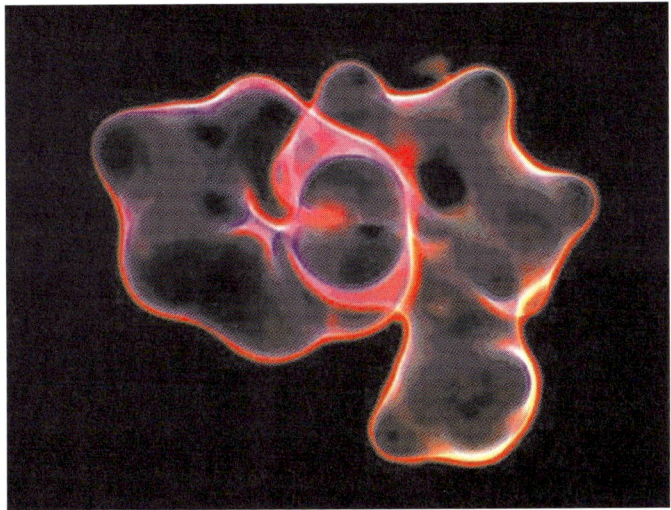

Fig. 18.7. Thick surface of blended unions - the nicotine molecule

p is the tri-linear blend of eight given points $x_i, y_i, z_i, i = 1, ..., 8$, that is, it maps the eight points of the unit cube in parameter space onto the eight given points in object space:

$$p(u, v, w) = (1 - w)[(1 - v)v]M[(1 - u)u]^\mathrm{T} + v[(1 - v)v]N[(1 - u)u]^\mathrm{T}, \quad (18.12)$$

where M and N are, respectively, matrices of the first and last four given points of the cube. The associated implicit function is formed by defining a function q in parameter space so that it is zero on the surface of the unit cube, negative inside and positive outside. Such a function is

$$q(u, v, w) = 2 \max \left[u - \frac{1}{2}, v - \frac{1}{2}, w - \frac{1}{2} \right] - 1. \quad (18.13)$$

The implicit function for the color gamut is then obtained as

$$f(x, y, z) = q(p^{-1}(x, y, z)). \quad (18.14)$$

Figure 18.8.a shows the surface of the gamut, and Figure 18.8.b shows an interior section of the gamut achieved as an offset value $f(x, y, z) - K$. That is, the color gamut is a legitimate volume, whose interior points are obtained as offsets of the implicit model. Notice that the interior points are less saturated than the pure color at the zero offset. What happens if we extrapolate beyond the hull of the gamut? Since the eight corner points are typically chosen as maximum values, it has little meaning in color theory to consider this, but visually the overflow values that result create an engaging image in Fig. 8c.

The form of Eq. (18.12) is a parametric tensor product, which can be generalized, for instance, to higher order Bezier or B-spline volumes [3] Rather than

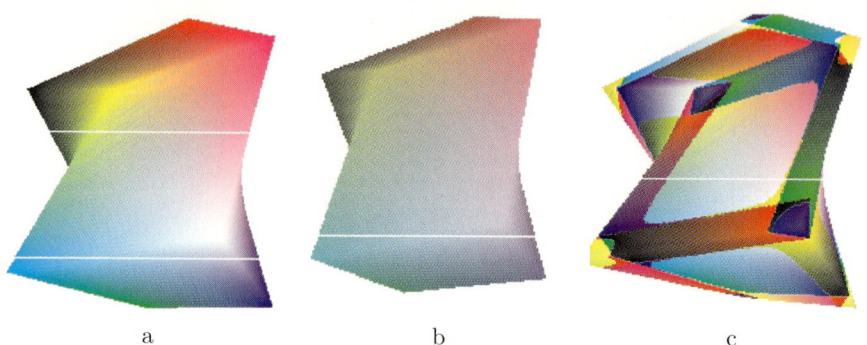

Fig. 18.8. a The (R,G,B) gamut. **b** Interior offset. **c** Extrapolate and overflow

actually performing the inverse of Eq. (18.14) in rendering, which is tantamount to solving a cubic equation, a forward scanning technique is employed. Parameters (u, v, w) are incremented through small step sizes, and substituted simultaneously into 18.12 and 18.13 to produce a cloud of points with their related distance values. As the points are produced the distance is eliminated if it is not within an epsilon tolerance of zero. The remaining points constitute a soft surface of the gamut. They are sorted into a z-buffer based on view. The color of each point is determined by using eight given (R,G,B) values in the matrices M and N in Formula Eq. (18.12), i.e. a color for each corner of the gamut. Clearly other color spaces could be used, e.g. CMY, CMYK, HSV and so forth [5]. The color gamut is another sloped, orientable model.

18.3.3 Equipotential fields

Potential fields can be formulated as distance fields by subtracting them from a constant threshold value. Hence the pervasive "r^2" fall-off in many physical fields is couched implicitly by

$$f(x) = T - \sum_i \frac{A_i}{\|x - x_i\|^2}, \tag{18.15}$$

where T is a threshold value, and the A_i determine the individual strength (charge) for each point x_i. This form can be used for depicting electrical, magnetic, light intensity and gravitational fields. Their implicit models represent surfaces of equipotential. Figures 18.9a,b show the equipotential surfaces of six arbitrary point sources with different values of T. As T increases in Fig. 18.9b the surface pulls in tighter. Different values for A_i account for the varying sizes of each blob.

In Fig. 18.9c variations of Eq. (18.15) generate potential versions of the cylinder, ellipsoid and superquadric by replacement in the denominator in Eq. (18.15 with) (respectively)

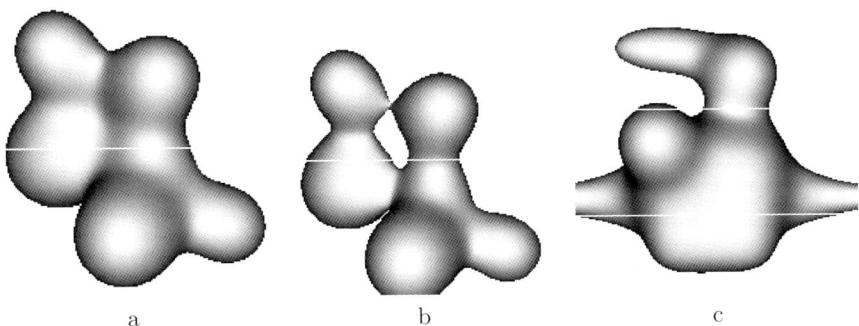

Fig. 18.9. a Equipotential. **b** Higher threshold. **c** Other potentials

$$(z - z_i)^2 + (y - y_i)^2 \tag{18.16}$$

$$0.1 * (x - x_i)^2 + (y - y_i)^2 + (z - z_i)^2 \tag{18.17}$$

$$(x - x_i)^4 + (y - y_i)^4 + (z - z_i)^2 \tag{18.18}$$

General cases are simple to extrapolate from the examples in Eqs. (18.16)-(18.18). These models are rendered in the same manner as the blends and fillets, previously described.

18.3.4 Skeletons

Any implicit function can generate a nonorientable model by changing the sign of either the inside or outside (negative or positive) part to match the other side. Squaring the implicit function is one way to accomplish this. Taking the absolute value is another. In both these cases, it can be shown that Boolean union (min) produces the union of the boundaries, for example, and without loss of generality

$$\partial A \cup \partial B \equiv x | \min\left(f(x)^2, g(x)^2\right) = 0, \tag{18.19}$$

where ∂A is the boundary of the set A. This is seen in Fig. 18.10a for two equipotential surfaces. Curiously, the max of the squares of the functions yields the same set

$$\partial A \cup \partial B \equiv x | \max\left(f(x)^2, g(x)^2\right) = 0. \tag{18.20}$$

In both cases we unite the zero sets of the squares. However if we take the sum of the sets, this time wlog using the absolute value to get the nonorientable version,

$$f(x) = |f(x)| + |g(x)|, \tag{18.21}$$

then we obtain $\partial A \cap \partial B$ as the implicit model displayed as a thick set in Fig. 18.10b.

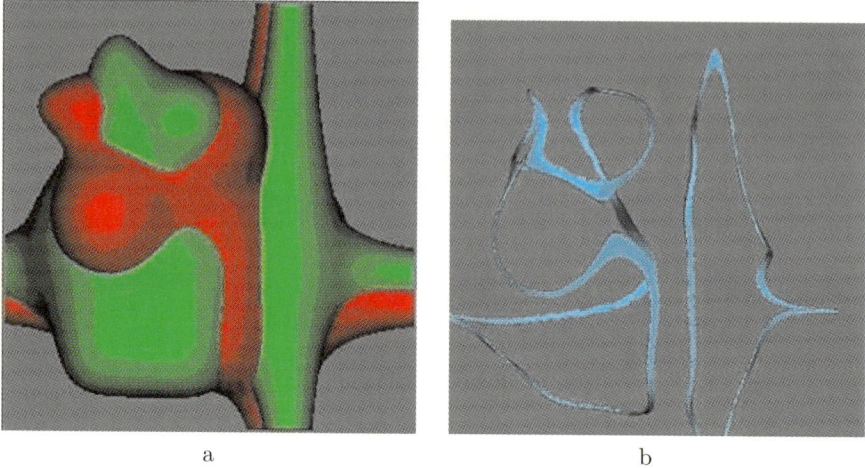

Fig. 18.10. a Union of boundaries of equipotential and **b** intersection of boundaries

18.3.5 Summary

The foregoing examples are intended not only as exhibits of implicit model versatility, but also as samples of a style of thinking with implicits that promotes and exploits its possibilities. It is also profitable background for understanding fundamental methods of dealing with implicits, including: texturing, field malleability, surface parameterizations and deformations. Even in the case of the parametric Bezier patches, the resulting set of voxels containing distances is a discrete distance field that *implicitizes* the object; any value of the distance field may be obtained through interpolation of the discrete points, and thus the implicit function, defined.

18.4 Re-tooling Implicit Models

Several points should be gleaned from the previous case studies. First, implicit functions can be generated in numerous ways, including algebraically, procedurally, as compositions and Booleans, and by interpolation of discrete data. Second, implicit models can be separated from the nonzero portion of the implicit functions such as by squaring; that is to say, there are many implicit functions that share the same zero set. The level set function where the domain is $n = 2$ provides a good visual device for thinking about implicits. The value $f(x)$ of the function at a point x may be thought of as a terrain height above or below the *water level* $f(x) = 0$, i.e. the zero set. The actual surface of the implicit model, the part displayed, is the *shoreline* of the water. If the water rises or recedes it creates a new shoreline, the offset. If the terrain contains sheer cliffs, it is not C^0; if it contains sharp ridges, peaks or valleys, it is not C^1. Any basins or bowls entirely above water level, or peaks entirely below water level disqualify it as an attractor station. These are traps that capture streamlines and keep them from passing through the shoreline. If it is nonorientable, then the shoreline is an infinitely thin moat with rising terrain on both sides. If one drops a ball onto a *sloped* terrain at any point, it will eventually find its way to water. We now carry some of these ideas over to operations on implicit models that widen our design and rendering facilities.

18.4.1 Three Dimensional Texturing

There are two ways of thinking about 3D texturing in this context. The first is the traditional graphics idea of embedding the object within a scalar "paint" field and coloring the object based on where it intersects the field [5], i.e. there exists a function $T : R^n \to C$, a color space, which determines the color for any point x in R^n. In our metaphor, T colors the terrain; in particular, the visible shoreline absorbs its color from the terrain it touches. Fig. 18.11 depicts the water molecule texture-mapped by decimating the Lambertian shaded model in Figure 18.4 according to whether

$$(x + y + z) \bmod \eta < \epsilon \qquad (18.22)$$

for some $\eta > \epsilon$. The simple "beat" pattern in each variable in 18.22 creates the characteristically observable Moiré texture on the surface of the implicit model.

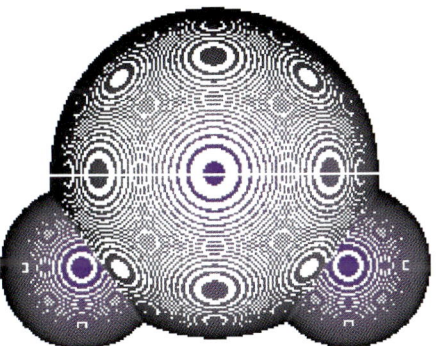

Fig. 18.11. A 3D texture on zero set

A second way to add detail to an existing implicit model is to define a so-called "shape texture" function $T(x)$ to add to the original shape:

$$f_{\texttt{new}}(x) = f_{\texttt{old}}(x) + T(x), \qquad (18.23)$$

which is usually has higher frequency components than the original implicit model. As an example, let

$$T(x) = \begin{cases} \cos(a\|x\|)\cos(A\|x\|) & \text{if } \|x\| < \frac{\pi}{2a} \\ 0 & \text{otherwise} \end{cases} \quad (18.24)$$

In Eq. (18.24) the texture has local support for $\|x\| < \pi/(2a)$. By adding a 3D translation the effect can be placed arbitrarily on the implicit model. In Fig. 18.12a the constant A is zero, which results in a simple cosine "bump" texture, i.e. the ball peen hammer strike is placed randomly on the zero set of cubic implicit model. The red line indicates a point at which the texture is added several times, creating a larger bump. In Fig. 18.12b the constant A is nonzero and yields the ripple effect. The line indicates an area where the ripples overlap, i.e. there are areas of support intersect. The result is a natural looking region where waves cross.

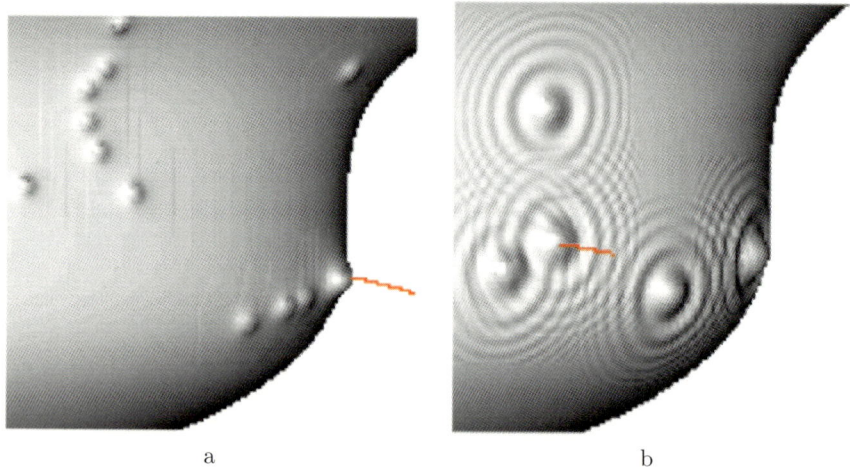

Fig. 18.12. Adding local texture effects **a** Ball peen hammer and **b** ripples

In Fig. 18.13 a globally supported, trivariate trigonometric combination is added as a texture to the base function (see Faberge egg code, Sect. 18.6.). The coefficients can be modified to increase the frequency of the texture function. It will serve as test functions for the next section. Notice that at the chosen frequency parts of the zero set detach, i.e. they are simply connected.

18.5 Regularizing Nonzero Space

Operations on implicit models such as Booleans, blending, shape texturing, equipotential surfaces, skeletons or local parameterization (described in the

18 The Design of Implicit Functions for Computer Graphics   617

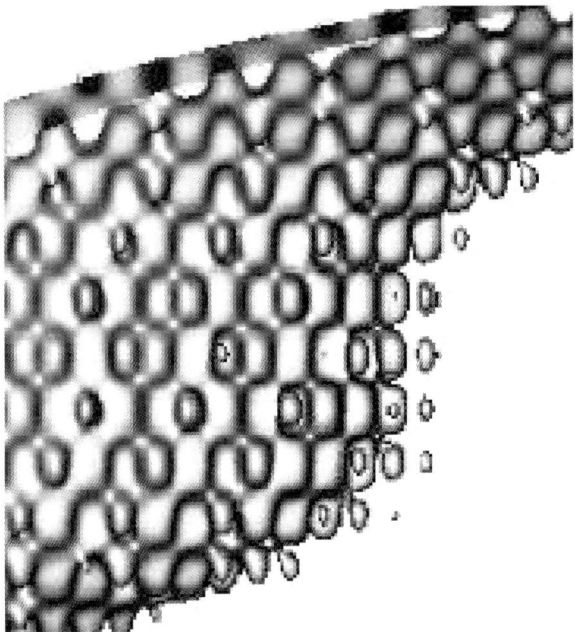

Fig. 18.13. A trigonometrically textured test function

next section) rely on smoothly varying space in the nonzero region in order to produce reasonably expected results. Many of the operations tend to degrade the space: thus Booleans and blending create nondifferentiabilities, shape texturing creates massive numbers of attractor basins off the zero surface. Some implicit functions like the Weierstrass function begin with undesirable characteristics. We seek a method that can redefine the nonzero space without affecting the implicit model itself. Generally, we want to create a new implicit function that possesses the desirable properties from Sect. 18.2. This will allow a wider range of operations on difficult functions, and more importantly, it will increase the reusability of implicit objects in subsequent operations. To understand what might happen, we visualize nonzero space by looking at the *streamlines* of the gradient field surrounding an implicit model. This is done by choosing a point x such that $f(x) \neq 0$, and then integrating along the steepest descent until the zero surface is reached. Using Euler integration, for instance, the streamline is generated by

$$x_{i+1} = h\nabla f(x_i), \qquad (18.25)$$

for given step size h. In Fig. 18.14 the streamlines begin at the corners and edge midpoints of a square that can be moved in space. The path of the streamlines indicates the shape of space. In Fig. 18.14a note that the streamlines contract, while in Fig. 18.14b there is an expansion, both depending on the starting position of the streamlines. Seen from a vector field perspective, it indicates that the gradient field has respectively: divergence less than, or greater than 1 in the regions tested. Also in Fig. 18.14b one can see the curl of the field as the streamlines on the top edge of the square twist as they converge to the zero surface.

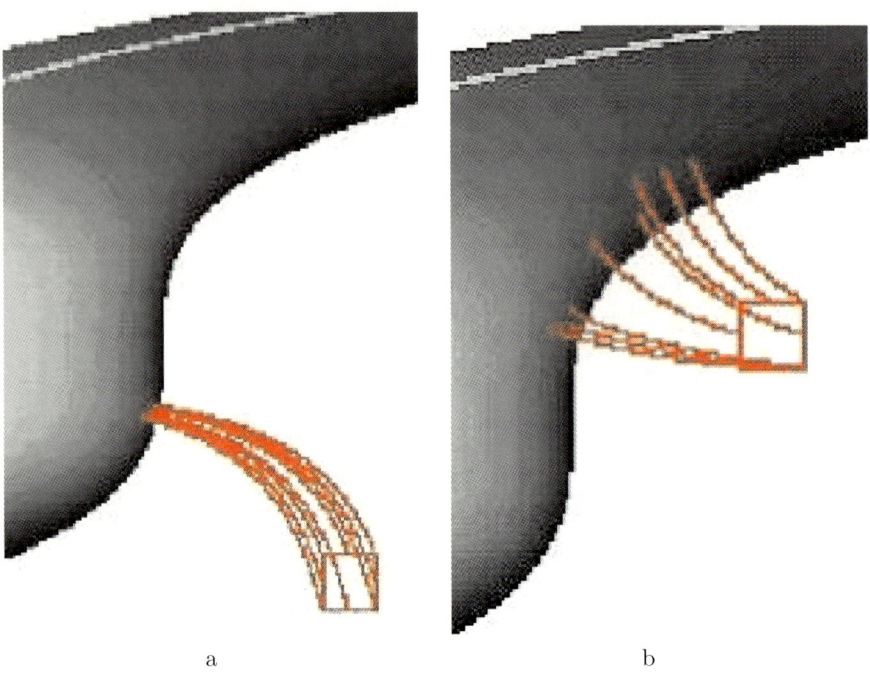

a b

Fig. 18.14. Visualizing the gradient field showing **a** contraction, **b** expansion

To demonstrate what can occur after operations, we calculate the streamlines for a textured function like Fig. 18.13, but with somewhat lower frequencies. Fig. 18.15 shows the tortuous path that results. The result for the original function in Fig. 18.13 was so tortured that it failed completely to compute and is not shown.

To resolve the problem of undesirable nonzero regions we propose a method that modifies this region by finding the minimal energy (regularization) function off the zero set, one that leaves the zero set unchanged. It is easiest to illustrate in one dimension lower. Consider the bivariate function shown in Fig. 18.16a.

The two peaks are defined as the zero set. The nonzero region is defined randomly by piecewise bilinear "twisted flats". This is as bad a nonzero region as could be imagined, or produced through any sequences of implicit operations. It is full of attractor basins and nondifferentiabilities. Our goal is to fix the two zero set points and the boundary, and then find the minimal energy

18 The Design of Implicit Functions for Computer Graphics   619

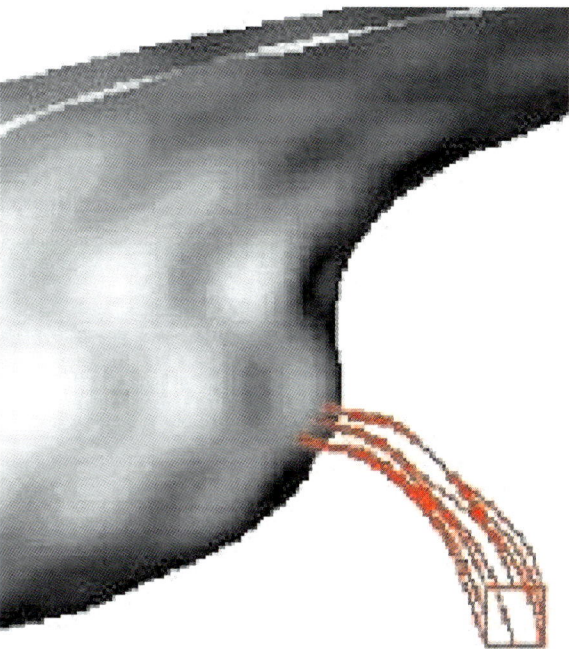

Fig. 18.15. Tortured streamline for textured surface

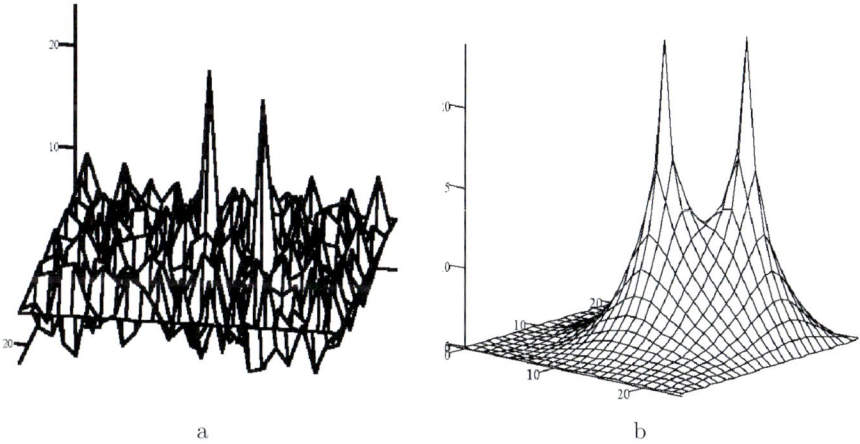

Fig. 18.16. a bivariate function with two zero-set points **b** the soap film surface

(soap film) surface that connects the fixed parts. Minimal energy surfaces are equivalent to minimal surface area, so our minimization algorithm uses the sum of the areas of the rectangles that tessellate the surface as a cost function. By varying a function value on an arbitrary rectangle vertex a new cost is computed, which can be entered to any of a number of optimization algorithms. We applied simulated annealing, but found that the energy-raising step was unnecessary; the process converged rapidly and robustly in a greedy fashion. The result is seen in Fig. 18.16b.

This is applied to 3D by minimizing the energy of the volume, i.e. changing the value of the function on a vertex of the cuboids that tessellate space and measuring the volume of the 4D cuboid (tesseract) given by the domain cuboid plus the function value. Using a simple greedy algorithm that accepted lower energy states and ignored others, the convergence was rapid and robust. To test the validity of our new nonzero regions we took the implicit models in Figs. 18.13 and 18.15, fixed their zero sets and a sufficient boundary, and then applied the energy minimization. We obtained results as seen in Fig. 18.17. Figure 18.17a shows streamlines for the implicit model of Fig. 18.15. It exhibits fewer variations and also contracts less than Fig. 18.14a. As a streamline it is always perpendicular to any contour $f(x) = K$, a constant, which includes the zero set. This is especially noteworthy in Fig. 18.17b where the streamlines are now possible to compute, unlike before, and are generally well behaved, making the twists into the zero set to maintain perpendicularity, only close to the surface.

There are other regularizations besides minimum energy that can be applied, such as minimization of divergence, curl or angles (conformality). The minimal energy predicate has sufficed for our needs, however. The first and immediate result of the method is the enhanced ability to parameterize the implicit model in a local area. We can imprint a parameterized object onto the surface of the implicit model by positioning it appropriately in space and for any point of the parameterized surface, follow its streamline to the zero surface, which inherits the parameter value. Figure 18.18 shows a radial sine function, the petal, parameterized by angle and length as it is mapped to the surface of two test functions.

18.6 Putting it together

We now use the previous tools to design some recognizable models. For those familiar with typical models in graphics it will be apparent that implicit models can readily attain complexity without large databases as with the more common parametric forms, e.g. Bezier, B-spline or polygon models. The first model illustrates.

18 The Design of Implicit Functions for Computer Graphics 621

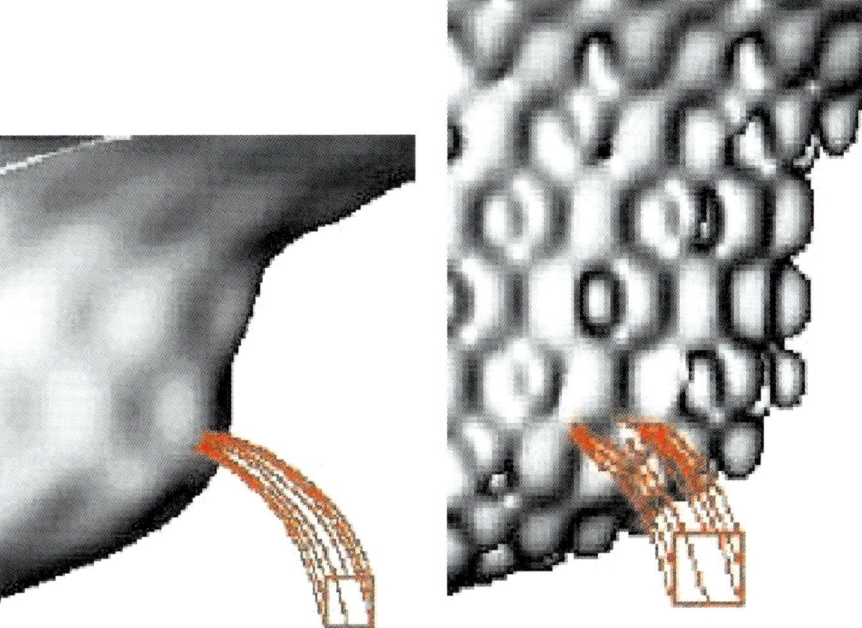

Fig. 18.17. Well-mannered streamlines on test objects

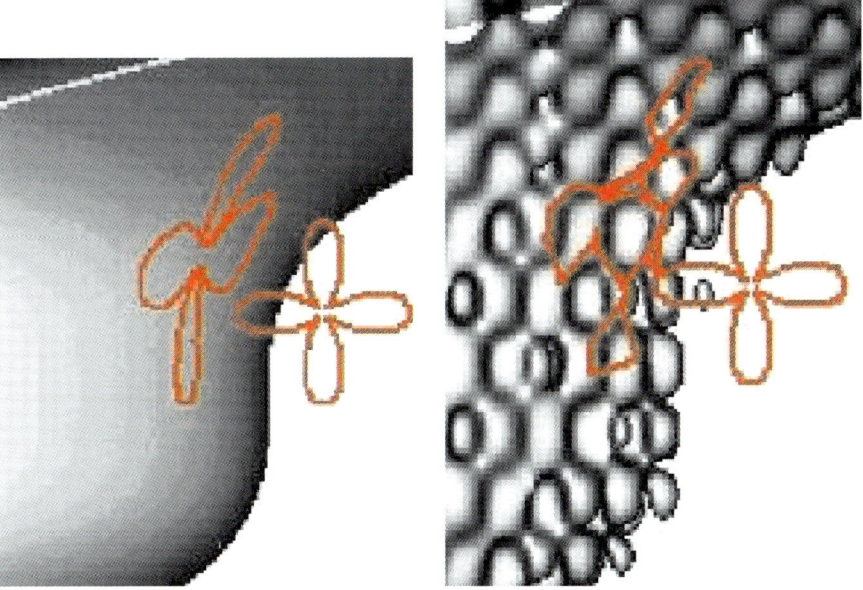

Fig. 18.18. Radial sine function parameterizing the implicit model

18.6.1 The Faberge Egg

Let $T(x, y, z)$ be a Fourier trigonometric combination like the "Berber weave" in Fig. 18.13. The actual form is seen in the code below. Next, create base shape with an implicit function $E(x, y, z)$, e.g. the ellipsoid $E(x, y, z) = x^2 + 4y^2 + 4z^2 - r$ of radius r. Then create an *outward textured shape* with $f_o(x, y, z, r) = E(x, y, z, r) + T(x, y, z)$. Create an inward textured shape with $f_i(x, y, z) = E(x, y, z) - T(x, y, z)$. Create the "filigree" version of the implicit function by taking the skeleton of the inwards and outward textured shapes:

$$g(x, y, z) = |f_o(x, y, z)| + |f_i(x, y, z)|. \qquad (18.26)$$

This is an interesting shape itself as seen in "gilding" of Fig. 18.19. Finally, create the "Faberge egg" as the zero set of the union of the filigree and the ellipsoid, *viz.* with the function

$$h(x, y, z, r) = \min[g(x, y, z), f(x, y, z)]. \qquad (18.27)$$

The color of the skeleton is chosen to be gold while the color of the ellipsoid is eggshell white, see Fig. 18.19. To emphasize its compact nature, the C++ code that defines the implicit model is given below:

```
{
  float Tex=  0.15*sin(x/0.1)*sin(y/0.1)*sin(z/0.1) +
      0.3*sin(x/0.05)*sin(y/0.05)*sin(z/0.05);
  ITag = 1;   // Set color gold
  E = pow(x,2)+2.0*pow(y,2)+2.0*pow(z,2)-1.0; // Ellipsoid
  float Fo = E+Tex;   // Add texture
  float Fi = E-Tex;   // Subtract Texture
  F = fabs(Fo)+fabs(Fi);  //Make skeleton
  float TempF;                     // Offset ellipsoid
  TempF = pow(x,2)+2.0*pow(y,2)+2.0*pow(z,2)-0.8;
  if (F > TempF)   //Create union
  {  F = TempF;
     ITag = 0;  // Set color eggshell
  }
}
```

We did not need the regularization algorithm to clean up nonzero space in modeling the egg. Any further operations, such as blending it to a base object, would require a regularization step, however, as the nonzero region is certainly as undesirable as Fig. 18.13. In the next models, this step is indispensable.

18.6.2 Two Broaches

We apply the same technique as in the Faberge egg to create a brooch, except that the ellipsoid of the egg is replaced by the superellipsoid of Eq. (18.18),

18 The Design of Implicit Functions for Computer Graphics

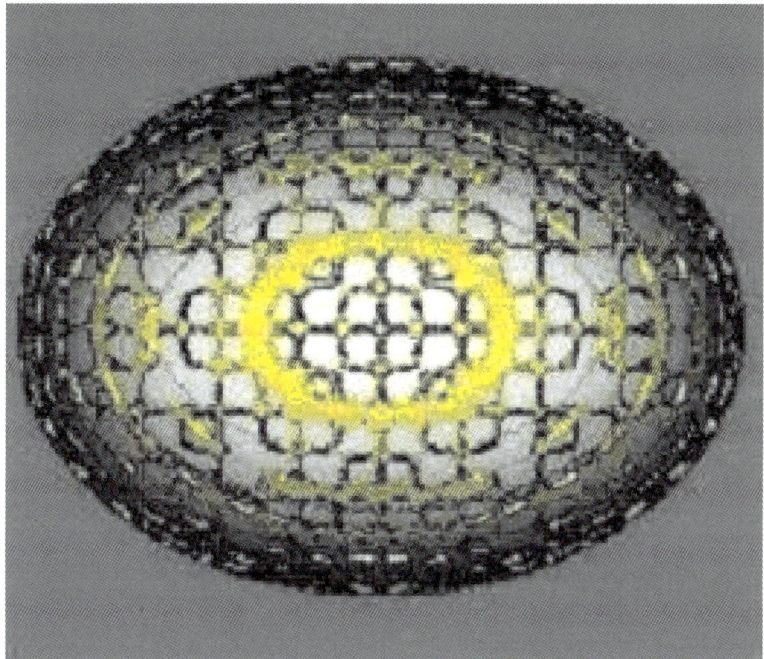

Fig. 18.19. The gilded Faberge egg

which gives it a flatter crosssection. The coefficients of the superellipsoid are adjusted so that the filigree skeleton disappears into the body at the border after the union. Because the next stages of design require a local parameterization via streamlines, the nonzero region of the implicit model is regularized by the soap film method. It is now possible to emboss and stencil some 2D patterns onto the brooch. In the first case (Fig. 18.20a), the petal of Fig. 18.18 is positioned in front of the brooch. By stepping at given parameter values around the petal, beginning points for the streamlines are generated. The streamlines are integrated with Eq. (18.25) until they reach the zero set. At the point where they reach the zero set a "ball peen" bump is added, as in Fig. 18.12. The difference between the three petals is the density of bumps, that is, the parameter step size at which the petal is evaluated. Figure 18.20b is simpler. A bitmap of a rose is positioned in front of the brooch. For each pixel in the bitmap, the associated streamline is followed to the surface, where that color is stenciled on the brooch. This, of course, brings up interesting questions about resampling as the samples of the bitmap expand or contract through the streamline mapping. That is beyond the scope of this exposition.

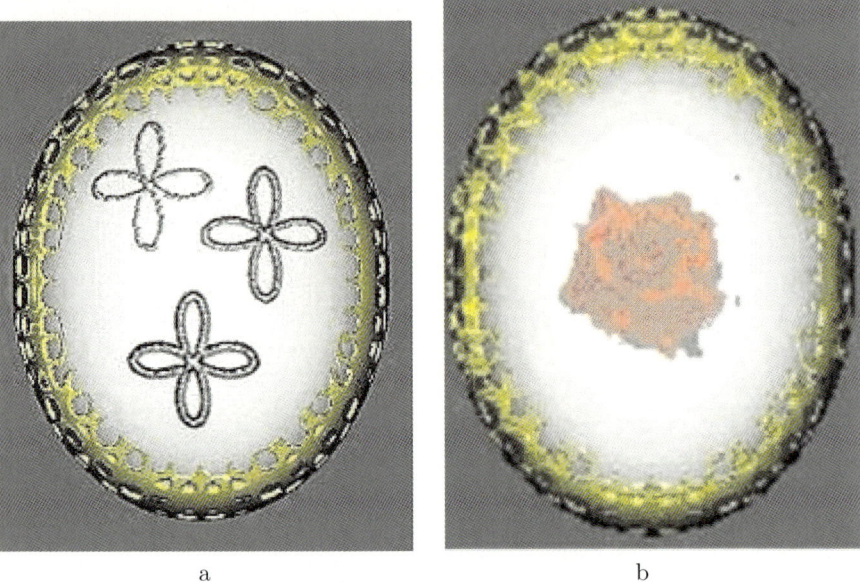

Fig. 18.20. A brooch with **a** embossed petals, and **b** stenciled bitmap of a rose

18.6.3 The Cocaine Molecule

Carefully sizing and placing spheres according to the chemical bonding of cocaine is shown in Fig. 18.21. It is then blended in exactly the same manner as H_2O and nicotine in Figures 18.5 and 18.6. Because of its complexity, however, it required a number of regularization steps so the blends could be sequenced without running over nondifferentiabilities in space and generating their attendant artifacts.

18.7 Summary and Future Development

Implicit functions can be defined in a variety of ways, depending on the object to be modeled. The different methods for defining implicit models often affect subsequent design options because they impinge on the nonzero regions of the implicit functions thus created. The regularization of this space by minimal energy methods enables a wider range of operations to be performed on a repeatable basis. One of the important and demonstrated characteristics of implicit function modeling is that it is capable of describing complex objects with a minimal database. Clearly, there are more investigations to be made into different impilict forms for shape, and into other tools for operating on them. The regularization methods mentioned that control divergence, curl and so on are yet to be implemented. We have, throughout the paper, used

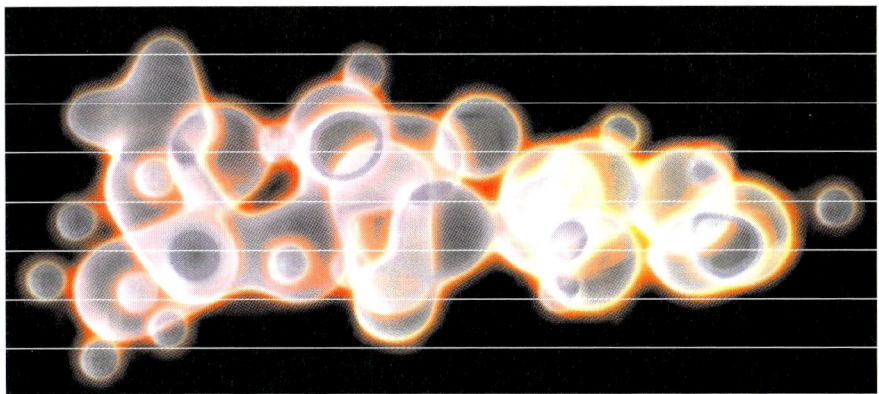

Fig. 18.21. The cocaine molecule, using blended spheres and a thick set

the classical definition of an implicit function as a real-valued mapping $\Re^n \to \Re$, the critical part of this behavior being the "real valued" portion, which allows the zero set assignment. With increasing attention being turned to non-Euclidean spaces in engineering and science, e.g. Clifford algebras, there is an interesting question of how general implicit functions could be used to model shape. In Hestenes geometric algebra [6, 7], for instance, one can easily image real valued functions of k-blades. What is the zero set of a mapping of bivectors to scalars? This occurs regularly through use of the pseudoscalar multiplication that creates dual forms for vectors, for example.

References

1. Bachman G. and Lawrence N. (1966) *Functional Analysis*, Academic Press, NY.
2. Bloomenthal J. (1997) *Introduction to Implicit Surfaces*, Morgan Kaufmann Publishers, San Francisco.
3. Farin G. (2002) *Curves and Surfaces for CAGD*, 5th ed., Morgan Kaufmann Publishers, San Francisco.
4. Frisken S. et al. (2000) Adaptively sampled distance fields: a general representation for shape for computer graphics. In:*SIGGRAPH 2000 Proc.*, July, pp. 249–257.
5. Foley J. and Van Dam A. (1982) *Fundamentals of Interactive Computer Graphics*, Addison-Wesley, NY.
6. Hestenes D (1986) *New Foundations for Classical Mechanics*, Kluwer Academic Publishing, Dordrecht.
7. Li H., Hestenes D. and Rockwood A. (2001) Generalized Homogeneous Coordinates for Computational Geometry. Somer, G. (ed.) In: *Geometric Computing with Clifford Algebras*. Springer-Verlag, Heidelberg, pp. 27–59.
8. Rockwood, A. (1989) The Displacement Method for Blending Surfaces in Solid Modeling. *ACM Transactions on Graphics, Special Issue on CAD/CAM*, vol.8, No.4, Oct., pp. 279–292.
9. Sethian J. (1996) *Level Set Methods: Evolving Interfaces in Geometry, Fluid Mechanics, Computer Vision and Materials Science*, Cambridge University Press.

Part VIII

Geometry and Robotics

19

Grassmann–Cayley Algebra and Robotics Applications

Neil L. White

Mathematics Dept., University of Florida, Gainesville, FL 32611-8105

19.1 Introduction

Grassmann–Cayley algebra is a means of writing expressions for geometric incidences in Euclidean or projective geometry, which gives a useful way to represent instantaneous kinematics and statics of robots. It is closely related to Grassmann algebra [9], and can be viewed as a special case of Clifford algebra [10]. It can also be described as exterior algebra with duality. However, all of these descriptions miss a crucial point, namely, that the formula for the meet operation provides a translation of geometric conditions into coordinate-free algebraic expressions. It even makes possible in some cases the reverse translation, called Cayley factorization. This can be very useful for analyzing critical positions of robots and several other applications, which we describe.

However, we begin with a more concrete version of this algebra, which involves Plücker coordinates.

19.2 Plücker Coordinates

Although we are primarily interested in three-dimensional space for our applications, our three-dimensional applications will also involve calculations in higher-dimensional space, so we start by considering d-dimensional space $\mathbb{R}^d$. If $x = (x_1, x_2, \ldots, x_d)$ is a point in $\mathbb{R}^d$, given in terms of the usual Cartesian coordinate system, then the *homogeneous coordinates* of x are obtained by adding a $(d+1)$st coordinate that is equal to 1. If λ is a nonzero scalar, then we also regard

$$\lambda(x_1, x_2, \ldots, x_d, 1) = (\lambda x_1, \lambda x_2, \ldots, \lambda x_d, \lambda)$$

as representing the same geometric point x. Furthermore, given any vector of $d+1$ real numbers, with last coordinate λ nonzero, we can divide by the scalar λ to recover the Cartesian coordinates of x.

If we also allow vectors of length $d+1$ whose last coordinate equals zero, which we regard as representing *points at infinity*, then we have the standard construction of *real d-dimensional projective space*. This construction is already very useful for robotics; for example, a revolute joint becomes a prismatic joint when its axis consists entirely of points at infinity. However, it is also very useful for a second reason. Let V denote the $(d+1)$-dimensional vector space in which all of these homogeneous coordinate vectors exist. Then subspaces of V correspond to points, lines, planes, and so on in $\mathbb{R}^d$, together with some points, lines, planes, and so on, which are composed entirely of points at infinity. A point, line, plane, and so on, in $\mathbb{R}^d$, which need not include the origin of $\mathbb{R}^d$, is referred to as an *affine* subspace, whereas the term *projective* subspace allows the subspaces at infinity, as well as the affine ones. This correspondence between projective (or affine) subspaces and vector subspaces of V is shown in Table 19.1. The crucial point here is that *all* affine

Table 19.1. Correspondence between projective subspaces and subspaces of V

points $\leftrightarrow$	1-dimensional subspaces of V
lines $\leftrightarrow$	2-dimensional subspaces of V
planes $\leftrightarrow$	3-dimensional subspaces of V
$k-1$ dimensional projective subspaces $\leftrightarrow$	k-dimensional subspaces of V

or projective subspaces are represented by subspaces of V, whereas if we use subspaces of the vector space $\mathbb{R}^d$, we get only lines, planes, and so on, through the origin.

Now let U be a $(k-1)$-dimensional affine subspace of $\mathbb{R}^d$, and U' the corresponding k-dimensional subspace of V. Let $u_1, u_2, \ldots, u_k$ be a basis of U'. It does not hurt to assume that each of these vectors has been chosen with $d+1$-st component equal to 1. We now form a matrix whose rows are these basis vectors:

$$M_U = \begin{bmatrix} u_{1,1} & u_{1,2} & \ldots & u_{1,d} & 1 \\ u_{2,1} & u_{2,2} & \ldots & u_{2,d} & 1 \\ & & \ldots & & \\ u_{k,1} & u_{k,2} & \ldots & u_{k,d} & 1 \end{bmatrix}. \tag{19.1}$$

Notice that the row space of M_U, which is just U', is uniquely determined by U.

Now choose any k columns of M_U, say the columns indexed by $j_1, j_2, \ldots, j_k$, with $1 \leq j_1 < j_2 < \ldots < j_k \leq d+1$. We then define the $j_1, j_2, \ldots, j_k$th *Plücker coordinate* of U, denoted by $P_{j_1, j_2, \ldots, j_k}$, to be the $k \times k$ determinant obtained from those k columns. Since there is one such Plücker coordinate for each choice of k of the columns, there are a total of

$$\binom{d+1}{k} = \frac{(d+1)!}{k!(d+1-k)!}$$

Plücker coordinates. We now define the *Plücker coordinate vector* P_U of U to be the vector we get by writing down all of these Plücker coordinates in some predetermined order.

Theorem 1. *An affine subspace U uniquely determines P_U up to scalar multiple. If P is a $\binom{d+1}{k}$-tuple of real numbers, and if $P = P_U$ for some affine (or projective) subspace U, then P uniquely determines U.*

For a proof, see Hodge and Pedoe [11], where Plücker coordinates are called Grassmann coordinates. If $P = P_U$ for some projective subspace U, we say that P is *decomposable*; there also exist indecomposable $\binom{d+1}{k}$-tuples, as will be seen in Section 6.

Example 1: Lines in $\mathbb{R}^3$.

We will adopt the mechanical engineer's convention of indexing the added fourth component by 0 instead of 4. Engineers also may write this added component first rather than last, though this really does not matter. We now write our Plücker coordinates in the order

$$P_L = (P_{0,1}, P_{0,2}, P_{0,3}, P_{2,3}, P_{3,1}, P_{1,2}),$$

where $P_{i,j} = -P_{j,i}$, and L is a line determined by points a and b (Fig. 19.1).

Our two points a and b give the 2×4 matrix

$$\begin{bmatrix} a_1 & a_2 & a_3 & 1 \\ b_1 & b_2 & b_3 & 1 \end{bmatrix},$$

and

$$P_L = (b_1 - a_1, b_2, a_2, b_3 - a_3, a_2 b_3 - a_3 b_2, a_3 b_1 - a_1 b_3, a_1 b_2 - a_2 b_1)$$

$$= (\mathbf{S}, \mathbf{r} \times \mathbf{S}),$$

where $\mathbf{S}$ is the first three components, the free vector from a to b, $\mathbf{r}$ is the coordinate vector of a (or any point on L), and the cross product $\mathbf{r} \times \mathbf{S}$, the last three coordinates of P_L, gives the moment of L about the origin.

Since P_L is determined only up to scalar multiple (indeed, a and b could be chosen to be any two points on L), mechanical engineers usually prefer to normalize P_L by choosing $\mathbf{S} \cdot \mathbf{S} = 1$, i.e., choosing a and b to have distance 1. Furthermore, we must have $\mathbf{S} \cdot (\mathbf{r} \times \mathbf{S}) = 0$, or $P_{0,1}P_{2,3} + P_{0,2}P_{3,1} + P_{0,3}P_{1,2} = 0$, and this condition, known as the Grassmann-Plücker relation, is the one equation that an arbitrary 6-tuple of real numbers must satisfy to be decomposable. These two conditions give us one way to see that there are four degrees of freedom in picking a line in 3-space.

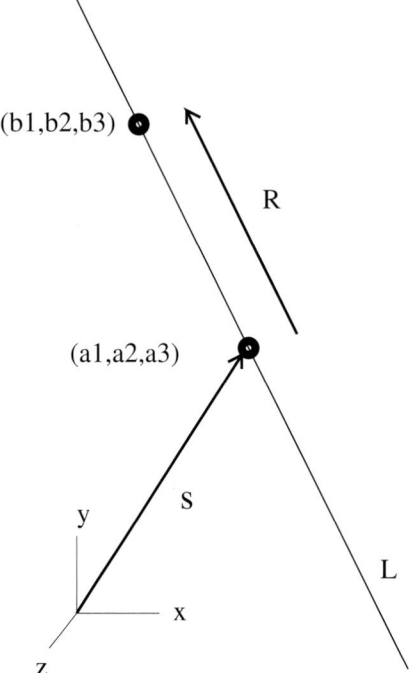

Fig. 19.1. A line in 3-space

The *point at infinity* on L corresponds to the homogeneous coordinate vector $(b_1 - a_1, b_2 - a_2, b_3 - a_3, 0)$, and its scalar multiples. The Plücker coordinate vector P_L can be computed from any two points on L, including the point at infinity. The same point at infinity lies on every line parallel to L, but nonparallel lines have distinct points at infinity (Fig. 19.2). Any 4-tuple $(x_1, x_2, x_3, 0) \neq (0, 0, 0, 0)$ represents a point at infinity, and any two such 4-tuples which are (nonzero) scalar multiples of each other represent the same point at infinity.

A *line at infinity* is the span in V of any two distinct points at infinity

$$\begin{bmatrix} x_1 & x_2 & x_3 & 0 \\ y_1 & y_2 & y_3 & 0 \end{bmatrix},$$

where these two rows are not scalar multiples of each other. Obviously, all points on such a line are points at infinity. If Π is an affine plane then all of the points at infinity on lines in Π comprise one line at infinity (Fig. 19.3). Now three-dimensional real projective space is $\mathbb{R}^3$ together with all of these points at infinity, with the latter regarded as constituting a single *plane at infinity*, which must therefore also contain all of the lines at infinity. Similar considerations apply to higher-dimensional projective spaces, though we will not provide details.

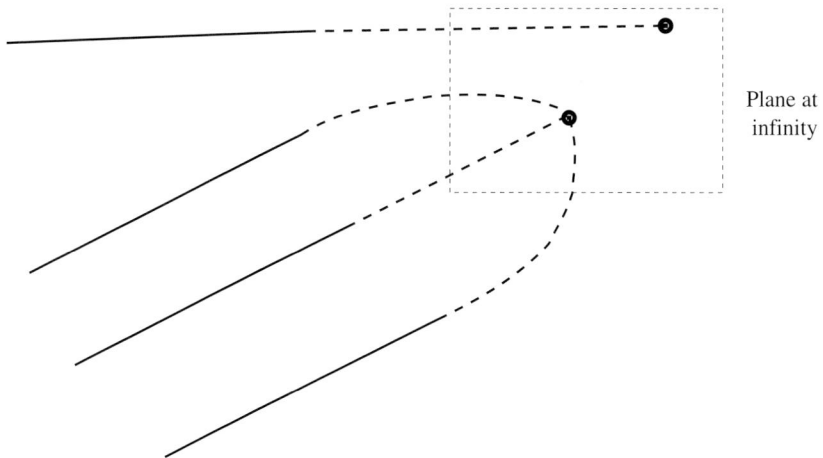

Fig. 19.2. Points at infinity

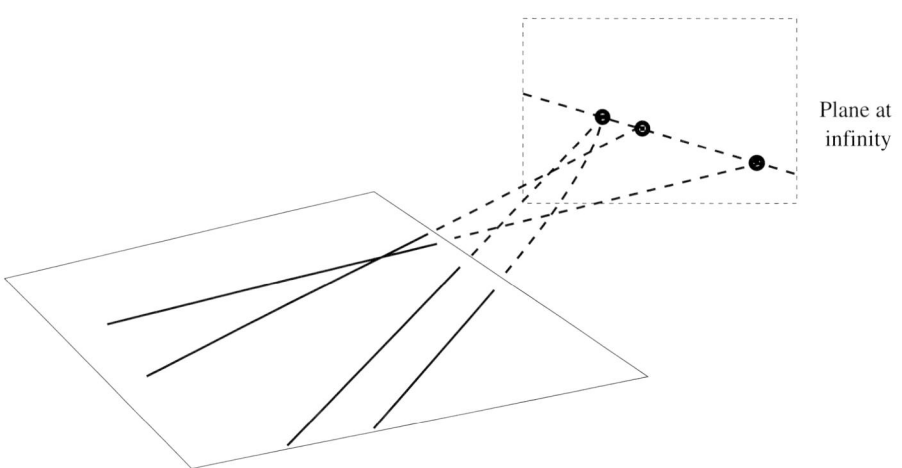

Fig. 19.3. A line at infinity

Example 2: Planes in $\mathbb{R}^3$.

Let Π be an affine plane determined by three affine points a, b, and c. The Plücker coordinates of Π are the 3×3 minors of

$$\begin{bmatrix} a_1 & a_2 & a_3 & 1 \\ b_1 & b_2 & b_3 & 1 \\ c_1 & c_2 & c_3 & 1 \end{bmatrix},$$

and an affine point (x, y, z) is on Π if and only if

$$\begin{bmatrix} a_1 & a_2 & a_3 & 1 \\ b_1 & b_2 & b_3 & 1 \\ c_1 & c_2 & c_3 & 1 \\ x & y & z & 1 \end{bmatrix} = 0.$$

By expanding this determinant by its last row, we see that this is true if and only if

$$P_{0,2,3}x - P_{0,1,3}y + P_{0,1,2}z - P_{1,2,3} = 0,$$

so we see that the Plücker coordinate vector may be written as

$$P_\Pi = (P_{0,2,3}, P_{0,3,1}, P_{0,1,2}, P_{1,2,3}) = (\mathbf{N}, \mathbf{r} \cdot \mathbf{N}),$$

where $\mathbf{N}$ is a normal to Π, and $\mathbf{r}$ is the coordinate vector of any point in Π. If we now normalize P_Π by dividing by the scalar $length(\mathbf{N})$, so that $\mathbf{N}$ now has length 1, then the last component $\mathbf{r} \cdot \mathbf{N}$ becomes the perpendicular distance from the origin to Π. Notice that the minus sign is absorbed in $P_{0,1,3}$ by writing $P_{0,3,1}$.

19.3 Dual Plücker Coordinates

Let U be a k-dimensional subspace of V, our $(d+1)$-dimensional vector space of homogeneous coordinates, and let $w_1, w_2, \ldots, w_{d+1-k}$ be $d+1-k$ linearly independent hyperplanes of V containing U. Each w_i is the solution set to a single homogeneous linear equation,

$$\Sigma_{j=1}^{d+1} w_{i,j} x_j = 0,$$

where, for a given i, the $w_{i,j}$ are defined only up to scalar multiple. However, if $\widetilde{M}_U$ is the $(d+1-k) \times (d+1)$ matrix with entries $w_{i,j}$, then the row space of $\widetilde{M}_U$ is uniquely determined by U. A *dual Plücker coordinate* of U is a $(d+1-k) \times (d+1-k)$ determinantal minor of $\widetilde{M}_U$, and the *dual Plücker coordinate vector* of U is the vector $\widetilde{P}_U$ of the dual Plücker coordinates in some predetermined order. Since

$$\binom{d+1}{d+1-k} = \binom{d+1}{k},$$

we see that the dual Plücker coordinate vector of U is the same length as the Plücker coordinate vector of U.

Theorem 2. *For U a subspace of dimension k of the $(d+1)$-dimensional vector space V, let P_U and $\widetilde{P}_U$ be a Plücker coordinate vector and dual Plücker coordinate vector, as above (recall that each is well defined only up to scalar multiple). Then there is a nonzero scalar ρ such that for every permutation $(j_1, j_2, \ldots, j_{d+1})$ of $(0, 1, 2, \ldots, d)$,*

$$\mathrm{sgn}(j_1, j_2, \ldots, j_{d+1})\widetilde{P}_{j_{k+1}, j_{k+2}, \ldots, j_{d+1}} = \rho P_{j_1, j_2, \ldots, j_k},$$

where sgn *denotes sign of a permutation.*

The proof is again in Hodge and Pedoe [11]. For lines in $\mathbb{R}^3$, this becomes

$$(\widetilde{P}_{2,3}, \widetilde{P}_{3,1}, \widetilde{P}_{1,2}, \widetilde{P}_{0,1}, \widetilde{P}_{0,2}, \widetilde{P}_{0,3}) = \rho(P_{0,1}, P_{0,2}, P_{0,3}, P_{2,3}, P_{3,1}, P_{1,2}),$$

where the signs are accomplished by the reversal of the indices $1, 3$. Thus we see that dual Plücker coordinates are essentially the same thing as Plücker coordinates, except for the order of the coordinates and perhaps some sign changes.

19.4 Basis Change

Suppose we wish to change the basis of V. The coordinates of a vector x in the new basis are found by taking a nonsingular $(d+1) \times (d+1)$ basis-change matrix S and computing Sx^{t}, that is, we are just effecting the linear transformation $x^{\mathrm{t}} \to Sx^{\mathrm{t}}$, where t denotes transpose. Now we wish to see how the Plücker coordinates of P_U change as a result. Let $V^{(k)}$ denote the $\binom{d+1}{k}$-dimensional vector space that is spanned by the Plücker coordinate vectors. Let $S^{l_1, l_2, \ldots, l_k}_{j_1, j_2, \ldots, j_k}$ denote the $k \times k$ determinantal minor of S consisting of rows $j_1, j_2, \ldots, j_k$ and columns $l_1, l_2, \ldots, l_k$, and let $S_{k \times k}$ denote the

$$\binom{d+1}{k} \times \binom{d+1}{k}$$

matrix of all these determinantal minors, with both rows and columns indexed in the same order we have indexed our Plücker coordinate vectors. The matrix $S_{k \times k}$ is called the k–**th compound matrix** of S.

Theorem 3. *(Hodge and Pedoe [11]). The linear transformation on $V^{(k)}$ given by $P_U^{\mathrm{t}} \to S_{k \times k} P_U^{\mathrm{t}}$ provides the Plücker coordinates of U in terms of the new basis. If S is nonsingular, then $S_{k \times k}$ is nonsingular.*

Thus the new Plücker coordinates are just $S_{k \times k}$ times the old Plücker coordinates, effectively, so a basis change in V effects nothing more than a basis change in $V^{(k)}$. In the next section, we will prefer to consider V as an *abstract* vector space, without a particular choice of basis. This theorem tells us that it does not hurt to do this, because we can then also think of P_U as a well-defined vector in another abstract vector space $V^{(k)}$. Then, any time we wish to work in terms of a specific basis of V, we just have to introduce the corresponding basis of $V^{(k)}$ according to the theorem.

Example 3: A rotation in $\mathbb{R}^3$.

Suppose we rotate the x, y, z-axes in $\mathbb{R}^3$ first by an angle Θ_1 in the y–z-plane, and then by an angle Θ_2 in the transformed xy-plane. Letting $c_1 = \cos\Theta_1, s_1 = \sin\Theta_1, c_2 = \cos\Theta_2, s_2 = \sin\Theta_2$, we get the following Euclidian basis change matrix:

$$\begin{bmatrix} c_2 & s_2 & 0 \\ -s_2 & c_2 & 0 \\ 0 & 0 & 1 \end{bmatrix} \begin{bmatrix} 1 & 0 & 0 \\ 0 & c_1 & s_1 \\ 0 & -s_1 & c_1 \end{bmatrix} = \begin{bmatrix} c_2 & s_2 c_1 & s_2 s_1 \\ -s_2 & c_2 c_1 & c_2 s_1 \\ 0 & -s_1 & c_1 \end{bmatrix}.$$

Now, to do the same thing in homogeneous coordinates, we simply need to extend our basis change matrix to

$$T = \begin{bmatrix} c_2 & s_2 c_1 & s_2 s_1 & 0 \\ -s_2 & c_2 c_1 & c_2 s_1 & 0 \\ 0 & -s_1 & c_1 & 0 \\ 0 & 0 & 0 & 1 \end{bmatrix}.$$

We should recall that if a vector represents the coordinates of a point in terms of the new axes, multiplying it by this basis change matrix would transform that vector to the coordinates of the point in terms of the old basis. We could, of course, use the inverse matrix to go the other way; in fact, as described above, we need to take $S = T^{-1}$.

We can now compute the second compound matrix, using the same order on our indices as in Example 1, namely, $01, 02, 03, 23, 31, 12$, for both rows and columns:

$$T_{2\times 2} = \begin{bmatrix} c_2 & s_2 c_1 & s_2 s_1 & 0 & 0 & 0 \\ -s_2 & c_2 c_1 & c_2 s_1 & 0 & 0 & 0 \\ 0 & -s_1 & c_1 & 0 & 0 & 0 \\ 0 & 0 & 0 & c_2 & s_2 c_1 & s_2 s_1 \\ 0 & 0 & 0 & -s_2 & c_2 c_1 & c_2 s_1 \\ 0 & 0 & 0 & 0 & -s_1 & c_1 \end{bmatrix}.$$

Notice that in multiplying P_L^t on the left by $T_{2\times 2}$, the upper left 3×3 submatrix is multiplying the free vector $\mathbf{S}$ of L by our original Euclidean basis change matrix, and also the lower right 3×3 submatrix is multiplying $\mathbf{r} \times \mathbf{s}$ by the same Euclidian basis change matrix. We could have predicted this, since a rotation is an orientation-preserving Euclidian isometry, and hence preserves the cross product.

19.5 Grassmann–Cayley Algebra; The Join Operation

Let V be an n-dimensional vector space over $\mathbb{R}$ (or any other field). We can let $n = d + 1$ if we are using homogeneous coordinate vectors for points in

$\mathbb{R}^d$, or for the d-dimensional projective space containing $\mathbb{R}^d$. Notice that we do not choose a basis of V, but rather think of V as an abstract vector space. Now let U be a k-dimensional subspace of V. Choose a basis $\{u_1, u_2, \ldots, u_k\}$ of U. Let A be the Plücker coordinate vector of U, but since we have not specified a basis in V, we also think of A as a vector in an abstract vector space $V^{(k)}$, of dimension $\binom{n}{k}$, with no particular basis specified. This is a symbolic approach to Plücker coordinates, although we can always return to using specific coordinates in applications, if we wish. The symbolic approach has several advantages: it is much easier to work with and easier to write down, and it is really much closer to the way we humans think about geometry than is any type of calculation using specific coordinates.

We denote the symbolic Plücker coordinate vector as

$$A = u_1 \vee u_2 \vee \ldots \vee u_k = u_1 u_2 \cdots u_k,$$

and refer to it as a *k-extensor* or *decomposable* (antisymmetric) *k-tensor*. There are also elements of $V^{(k)}$ that are not k-extensors; they are referred to as *indecomposable* (antisymmetric) *k-tensors*, and they can always be written as a linear combination of k-extensors. The general term *antisymmetric k-tensor* refers to either a decomposable or indecomposable antisymmetric k-tensor. The number k is called the *step* of the k-tensor, provided the k-tensor is not equal to 0. A sum of k-tensors of different steps is called simply a *tensor* (this sum can be realized in a bigger space $\Lambda(V)$, see below).

We can also write $A = u_1 u_2 \cdots u_k$ when $u_1, u_2, \ldots, u_k$ are linearly dependent; in that case $A = 0$ (independently of k). However, if $u_1, u_2, \ldots, u_k$ are linearly independent and a basis of U, we write $U = \overline{A}$, and say the U is the *support* of A. Two nonzero k-extensors are equal up to nonzero scalar multiple if and only if their supports are equal. We do not define the support of an indecomposable k-tensor, or of a tensor in general.

For example, a plane in affine or projective d-space corresponds to a three-dimensional subspace U of V. Let u_1, u_2, u_3 be a basis of U. Then U is the support of the 3-tensor $u_1 u_2 u_3$. In this way every subspace is represented by an extensor of the appropriate step.

The particular case of $k = n$ is of interest. If $A = u_1 u_2 \cdots u_n$ is an n-extensor, and if we expand each u_i in terms of fixed basis B of V, then for some fixed scalar μ depending only on B, $A = \mu \det(u_1, u_2, \ldots, u_n)$. We usually write $A = [u_1, u_2, \ldots, u_n]$ and refer to it as a *bracket*. The brackets form a subalgebra of the Grassmann–Cayley algebra, called the *bracket ring* or *bracket algebra*.

Next, we combine all of the vector spaces $V^{(k)}$ for various k into one big vector space:

$$\Lambda(V) = V^{(0)} \oplus V^{(1)} \oplus V^{(2)} \oplus \ldots \oplus V^{(n)}.$$

Here $V^{(1)}$ is just V, $V^{(0)}$ is just the real scalar field, and $V^{(n)}$ may also be thought of as another copy of the scalar field, though it is important for technical reasons to distinguish between scalars of step 0 and scalars of step

n. The brackets we think of as scalars of step n, although when we work with them on a symbolic level, they form a very complicated algebra. In particular, it is not easy to decide if two bracket expressions are equal, as this requires a *straightening algorithm* to put both expressions in normal form; see, for example, [22]. Since we also have another copy of the scalars in $V^{(0)}$, it is convenient to think of the brackets as scalars of step 0 at times. Tensors are now just arbitrary elements of $\Lambda(V)$. Notice that

$$\dim \Lambda(V) = \sum_{k=0}^{n} \binom{n}{k} = 2^n.$$

Now we wish to define the *join* operation on $\Lambda(V)$. Initially, we define only the join of two extensors. Let $A = a_1 \vee a_2 \vee \ldots a_j = a_1 a_2 \cdots a_j$ and $B = b_1 b_2 \cdots b_k$ be two extensors. Then their join, $A \vee B$, is defined by

$$A \vee B = a_1 \vee a_2 \vee \ldots a_j \vee b_1 \vee b_2 \vee \ldots \vee b_k = a_1 a_2 \cdots a_j b_1 b_2 \cdots b_k.$$

Thus $A \vee B$ is an extensor of step $k + l$, provided it is nonzero. It is nonzero precisely when $\{a_1, a_2, \ldots, a_j, b_1, b_2, \ldots, b_k\}$ is linearly independent. This can never be the case, for example, when $j + k > n$.

Theorem 4. *If $A = a_1 a_2 \cdots a_j$ and $B = b_1 b_2 \cdots b_k$ are nonzero extensors, then*

$$A \vee B \neq 0 \Leftrightarrow \overline{A} \cap \overline{B} = \{0\}$$
$$\Leftrightarrow \{a_1, a_2, \ldots, a_j, b_1, b_2, \ldots, b_k\} \quad \text{is linearly independent.}$$

If this is the case, then $\overline{A \vee B} = \overline{A} + \overline{B} = \mathrm{span}(\overline{A} \cup \overline{B})$. Furthermore,

$$A \vee B = (-1)^{jk} B \vee A.$$

This defines the join of two extensors. The above theorem shows that the join operation applied to two extensors results in an extensor whose support is the sum of the two supporting subspaces, $\overline{A} + \overline{B} = \{x + y : x \in \overline{A}, y \in \overline{B}\}$. This sum of two subspaces is the smallest subspace containing the two subspaces, and is sometimes called the geometric join of the two subspaces, written with the symbol $\vee$. We will avoid that notation here to prevent confusion between the geometric join of two subspaces and the algebraic join of two extensors.

Joins of tensors in general are defined by distributivity. Since a tensor may be written as a sum of extensors in more than one way, it is not completely trivial to assert that join is a well-defined operation on $\Lambda(V)$, but nevertheless, that is the case.

The vector space $\Lambda(V)$ together with the operation $\vee$ is an algebra, commonly known in mathematics as the *exterior algebra*, though mathematicians usually prefer to use the symbol $\wedge$ instead of $\vee$; here we prefer the latter because, as we have just seen, it corresponds to geometric join (or $\mathrm{span}(\overline{A} \cup \overline{B})$). Instead, we will use $\wedge$ for a second operation, called meet, which corresponds to geometric meet, and which we will introduce later, after looking at some applications of the join operation.

19.6 Application to Instantaneous Kinematics

We consider an instantaneous rotation of Euclidian 3-space about a line, so $d = 3$ and $n = d + 1 = 4$. Rotation about the line $\overline{ab}$ is represented by the 2-extensor $A = ab = a \vee b$, which lies in the six-dimensional vector space $V^{(2)}$ (as we have seen, the Plücker coordinate vector of a line is a 6-tuple of the form $(\mathbf{S}, \mathbf{r} \times \mathbf{S})$, where $\mathbf{S}$ is the vector from $\overline{a}$ to $\overline{b}$ and $\mathbf{r}$ is the coordinate vector of any point on the line). Let $\overline{p}$ be a point that is off the line $\overline{ab}$. Then $A \vee p = a \vee b \vee p$ is a 3-extensor whose support is the plane $\overline{abp}$. We saw in Example 2 that in the concrete form of Plücker coordinates, the first three components of $A \vee p$ give a normal vector to the plane (Fig. 19.4). Furthermore, it can be seen that the length of this (unnormalized) normal vector is proportional to the distance that $\overline{p}$ is from the line $\overline{A}$. However, the length of the instantaneous velocity vector of $\overline{p}$ is also proportional to the distance from $\overline{p}$ to the line $\overline{A}$. Thus there exists a constant ω such that the first three components of $\omega A \vee p$ are equal to the velocity vector at $\overline{p}$. In fact, we can always adjust the distance between a and b to make $\omega = 1$. Assuming that A has been so normalized, we call A the *center* or *axis* of the revolution. Another way to look at this is that we may adjust the distance between a and b, including reversing the positions of a and b if we wish to reverse the direction of rotation, so that the resulting rotation has angular velocity equal to any vector we choose that lies along our axis.

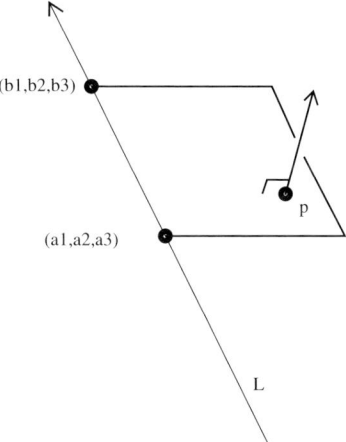

Fig. 19.4. Instantaneous rotation about an axis

If $\overline{A}$ is a line at infinity (thus both $\overline{a}$ and $\overline{b}$ are points at infinity), then A represents a translation in terms of Euclidean geometry. Indeed, using Plücker coordinates, $A \vee p$ is represented by the 3×3 minors of

$$\begin{bmatrix} a_1 & a_2 & a_3 & 0 \\ b_1 & b_2 & b_3 & 0 \\ p_1 & p_2 & p_3 & 1 \end{bmatrix}.$$

Now we see that the first three components of $A \vee p$, namely

$$P_{2,3,0} = a_2 b_3 - a_3 b_2, \; P_{3,1,0} = a_3 b_1 - a_1 b_3, \; P_{1,2,0} = a_1 b_2 - a_2 b_1,$$

are independent of p_1, p_2, p_3; thus every point p has the same velocity vector, and the result is a translation. As a result, in a very real sense, a translation is a rotation with an axis at infinity. This is fairly easy to visualize intuitively, as shown in Fig. 19.5. Imagine a point being rotated a fixed arc length l about an axis. Now keep the point's position fixed, but move the axis farther away. The point's trajectory is still a circular arc of length l, but with a larger radius. In the limit, as the axis moves away infinitely far, the trajectory becomes a straight line segment of length l. Since the axis is now infinitely far from all points in Euclidian space, all have the same direction of motion, and the result is a translation of Euclidian space.

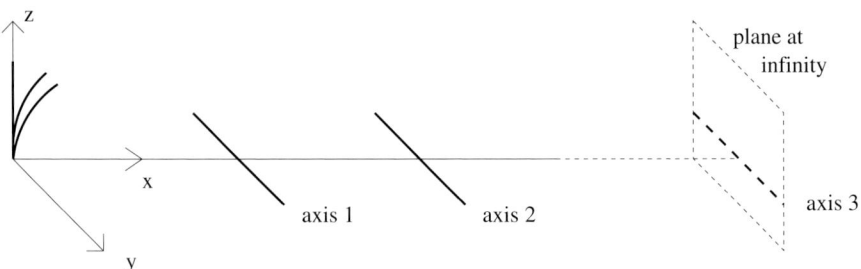

Fig. 19.5. Instantaneous translation as rotation about an axis at infinity

Consider now a robot arm with k revolute joints serially connected, which we consider to be in a certain position at time 0. Assume that all joints are activated simultaneously, each with a certain angular velocity. Let $A_1, A_2, \ldots, A_k$ be the centers of instantaneous rotation, each normalized as above to give the correct angular velocity. Then the net instantaneous motion of the end effector has center

$$t = A_1 + A_2 + \ldots + A_k,$$

see [15], which we call the **twist**. The actual instantaneous velocity vector of the end effector at time 0 is represented by $t \vee p$, where $\overline{p}$ is the position of the end effector. Notice that if $\overline{A}_1$ and $\overline{A}_2$ are skew lines, then $A_1 + A_2$ is indecomposable, and the motion represented by $A_1 + A_2$ is a screw motion. Implicit in this is that both rotations have nonzero angular velocities, for if $A_i = 0$, then $\overline{A}_i$ would not be a line, but would be undefined.

Any twist may be written as

$$t = (\omega, \mathbf{v}),$$

where ω and $\mathbf{v}$ are the angular and linear velocity, respectively, of the point of the end effector that instantaneously coincides with the origin (where we consider the end effector extended to include that origin, if necessary). See [21] for more details. We should also recall Chasles's Theorem [3], which states that any twist in $V^{(2)}$ can be expressed as a linear combination of two 2-extensors, one supported by a finite line L_1 and the other by a line at infinity L_2 that is uniquely determined as the intersection of the plane at infinity P_2 with a plane P_1 perpendicular to L_1 (Fig. 19.6). In other words, the corresponding instantaneous motion consists of a rotation about L_1 plus a translation in a direction parallel to L_1.

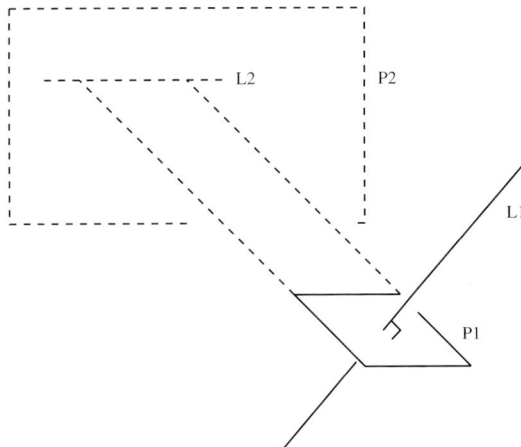

Fig. 19.6. Chasles's Theorem

More generally, if we wish to consider all possible instantaneous motions from the given starting position, we should consider all linear combinations of $A_1, A_2, \ldots, A_k$. This is the *twist space* of the robot arm in its initial position. Since each of $A_1, A_2, \ldots, A_k$ is a vector in $V^{(2)}$, we can think of the span of these k vectors as the support of an extensor in $\Lambda(V^{(2)})$, which we shall write as $A_1 \bigvee A_2 \bigvee \ldots \bigvee A_k$, where we use $\bigvee$, called a *superjoin*, to denote join in $\Lambda(V^{(2)})$ as opposed to join $\vee$ in $\Lambda(V)$. Notice that since $V^{(2)}$ is six-dimensional, $A_1 \bigvee A_2$ is the join of two vectors in that six-dimensional space. Since $\binom{6}{2} = \frac{6!}{4!2!} = 15$, $A_1 \bigvee A_2$ is a vector in a 15-dimensional vector space, $(V^{(2)})^{(2)}$.

As we have seen, a single revolute joint may be represented by a 2-extensor as its center, and a prismatic joint, which allows a unique translation (in its initial position), may be represented by a 2-extensor at infinity. Likewise, a screw joint, which allows a unique screw motion, would be represented by an indecomposable 2-tensor. Spherical, cylindrical, and planar joints, which have more than one degree of freedom, may be represented by combinations of the simple joints above; for example, a spherical joint is represented by three revolute joints with axes all intersecting in one point but not all three axes coplanar. See Hunt [12], Table 1.1, for details and figures.

Suppose a three-dimensional robot arm has six simple joints (after replacing joints of more than one degree of freedom by combinations of simple joints), say $A_1, A_2, \ldots, A_6$. Some of the A_i are allowed to be indecomposable. Then the robot arm has full mobility if

$$\overline{A_1 \bigvee A_2 \vee \ldots \bigvee A_6} = V^{(2)},$$

in other words, $A_1, A_2, \ldots, A_6$ span all of $V^{(2)}$. A critical configuration is any configuration in which $A_1 \bigvee A_2 \vee \ldots \bigvee A_6 = 0$. Figure 19.7 shows a robot arm with full mobility in the illustrated position. Since it has six simple joints, full mobility means the six joints have Plücker coordinate vectors that are linearly independent, and hence are NOT in a critical configuration.

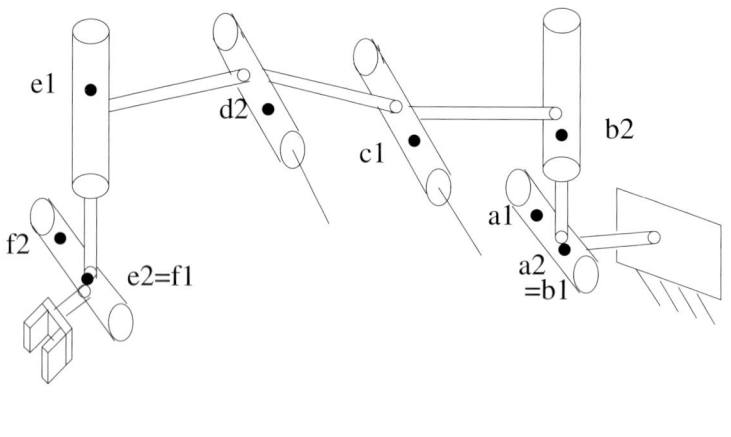

Fig. 19.7. A robot arm of six joints

Much of the theory of screw systems and of line geometry (see Hunt [12]) can be expressed in terms of the Grassmann–Cayley Algebra of $V^{(2)}$. For example, a cylindroid may be defined by a 2-system of screws, which may be represented as an general two-dimensional subspace of $V^{(2)}$, and thus as the support of $A_1 \bigvee A_2$ for some 2-tensors A_1 and A_2. Here we must exclude the degenerate case where A_1 and A_2 are extensors supporting two intersecting lines, as well as excluding any case where A_1 and A_2 are linearly dependent.

A fundamental concept of line geometry is the line complex. A line complex may be defined as the set of all lines whose six Plücker coordinates satisfy a given homogeneous linear equation. Thus a line complex is represented by a five-dimensional subspace of $V^{(2)}$, but one is primarily interested in the lines represented by extensors in that subspace, rather than the screws represented by indecomposable 2-tensors. Thus a line complex represented by $A_1 \vee A_2 \vee \ldots \vee A_5$ would be the collection of all 2-extensors $A = a_1 \vee a_2$ such that $A \vee A_1 \vee \ldots \vee A_5 = 0$. However, $A_1, A_2, \ldots, A_5$ themselves could be either decomposable or indecomposable. A **line congruence** is similar, except that it is based on a four-dimensional subspace of $V^{(2)}$. For more on these concepts and their application to robotics, see [18].

19.7 The Meet Operation

Now we define the *meet* operation on $\Lambda(V)$. Let $A = a_1 a_2 \cdots a_j$ and $B = b_1 b_2 \cdots b_k$ be two extensors with $j + k \geq n$. Then

$$A \wedge B = \frac{1}{(n-k)!(j+k-n)!} \times \qquad (19.2)$$

$$\sum_\sigma \operatorname{sgn}(\sigma)[a_{\sigma(1)}, a_{\sigma(2)}, \ldots, a_{\sigma(n-k)}, b_1, b_2, \ldots, b_k] a_{\sigma(n-k+1)} \cdots a_{\sigma(j)}, \qquad (19.3)$$

where the sum is over all permutations σ of $\{1, 2, \ldots, j\}$. Notice that this formula can be thought of as a linear combination of extensors of step $j+k-n$ (the a's outside the bracket), with brackets as scalar coefficients. In fact, this $(j + k - n)$-tensor turns out always to be a $(j + k - n)$-extensor when A and B are extensors. However, we may also extend the meet operation to take the meet of any two tensors by assuming distributivity.

Actually, this formula is correct when we are working over the field $\mathbb{R}$ of real numbers, or any field of characteristic 0 (besides $\mathbb{R}$, this includes the field of rational numbers and the field of complex numbers). If a permutation is applied just to the a's inside the bracket, because of the antisymmetry of the bracket and of the sign in front of the bracket, terms are reinforced, and the same happens for permutations just of the a's outside the bracket. Thus an equivalent formula eliminates the numerical coefficient in front of the summation, and sums only over permutations with

$$\sigma(1) < \sigma(2) < \ldots < \sigma(n-k) \quad \text{and} \quad \sigma(n-k+1) < \sigma(n-k+2) < \ldots \sigma(j).$$

These permutations are called *shuffles* of the $(n-k, j+k-n)$-split of A. The shuffle formula as opposed to the formula summing over all permutations is essential when working over fields of finite characteristic, but is not necessary when working over $\mathbb{R}$.

Theorem 5. *[5] If A and B are a j-tensor and a k-tensor (resp.), then*

$$A \wedge B = (-1)^{(n-j)(n-k)} B \wedge A.$$

Theorem 6. *If A and B are nonzero extensors, then*

$$A \wedge B \neq 0 \Leftrightarrow \overline{A} \cup \overline{B} \text{ spans } V.$$

If this is the case, then $A \wedge B$ is a $(j+k-n)$-extensor, and $\overline{A \wedge B} = \overline{A} \cap \overline{B}$.

Example 5.

In Euclidian 3-space ($n = 4$), let us take extensors $a_1 a_2 a_3$ and $b_1 b_2$, where we assume that $\overline{a_1 a_2 a_3}$ and $\overline{b_1 b_2}$ are a plane and a line not contained in the plane. Thus $\overline{a_1 a_2 a_3} \cap \overline{b_1 b_2}$ is a point, which is a point at infinity in the case that $\overline{b_1 b_2}$ is parallel to $\overline{a_1 a_2 a_3}$. Indeed,

$$a_1 a_2 a_3 \wedge b_1 b_2 = \frac{1}{2} \sum_\sigma \text{sgn}(\sigma)[a_{\sigma(1)}, a_{\sigma(2)}, b_1, b_2] a_{\sigma(3)}$$

$$= [a_1, a_2, b_1, b_2] a_3 - [a_1, a_3, b_1, b_2] a_2 + [a_2, a_3, b_1, b_2] a_1,$$

using Eq. (19.2). Thus $a_1 a_2 a_3 \wedge b_1 b_2$ is a linear combination of the three vectors a_1, a_2, and a_3, and hence is supported by some point on the plane $\overline{a_1 a_2 a_3}$. But we also see by Theorem 5 that

$$a_1 a_2 a_3 \wedge b_1 b_2 = b_1 b_2 \wedge a_1 a_2 a_3 = [b_1, a_2, a_3, a_4] b_2 - [b_2, a_1, a_2, a_3] b_1.$$

Thus $a_1 a_2 a_3 \wedge b_1 b_2$ is also supported by a point on the line $\overline{b_1 b_2}$ hence its support must be precisely the point of intersection $\overline{a_1 a_2 a_3} \cap \overline{b_1 b_2}$.

The *Grassmann–Cayley algebra* is the vector space $\Lambda(V)$ together with the operations $\vee$ and $\wedge$. These two operations are both associative, distributive over addition, distributive over scalar multiplication (i.e., $\alpha(A \vee B) = (\alpha A) \vee B = A \vee (\alpha B)$, and similarly for $\wedge$), and anticommutative in the sense described in the theorems above. The *Grassmann algebra* is the same algebra plus the usual dot product on V. However, Grassmann did not know our formulas for the $\wedge$ involving shuffles or all permutations, which considerably restricted the usefulness of his algebra. These formulae for $\wedge$ were discovered (in their full generality) by Rota (see [2, 5, 20], where the Grassmann–Cayley algebra is called the Cayley algebra or double algebra). We should also mention that Grassmann–Cayley algebra can also be realized as a subalgebra of a certain kind of Clifford algebra, see [10].

Grassmann thought of $\vee$ and $\wedge$ as dual operations. If V is an n-dimensional vector space over $\mathbb{R}$, then the *dual vector space* V^* is the collection of linear transformations from V to $\mathbb{R}$ (or to any one-dimensional vector space over $\mathbb{R}$). Each vector in V^* corresponds to a hyperplane (or $(n-1)$-dimensional subspace), and likewise each vector in V can be associated with a hyperplane in V^*, or better yet, with a *covector*, or $(n-1)$-extensor in $(V^*)^{(n-1)}$, whose support is that hyperplane. This association can be extended to an invertible linear transformation from V to $(V^*)^{(n-1)}$ and from $V^{(n-1)}$ to V^*, and indeed from all of $\Lambda(V)$ to $\Lambda(V^*)$, which maps $V^{(k)}$ to $(V^*)^{(n-k)}$ for each k. This map interchanges $\vee$ on $\Lambda(V)$ with $\wedge$ on $\Lambda(V^*)$. Since V and V^* are both vector spaces of dimension n over $\mathbb{R}$, they are effectively interchangeable, so we have the following theorem.

Theorem 7. *The operations $\vee$ and $\wedge$ on $\Lambda(V)$ are dual operations. That is, if we prove any theorem about the Grassmann–Cayley algebra, we immediately get another equivalent theorem by duality: interchange $\vee$ with $\wedge$, vectors with covectors, step k with step $n - k$, while at the same time interchanging the geometric content: affine subspaces of dimension j are interchanged with affine subspaces of dimension $n - 1 - j$, and containment is reversed.*

There are some issues with global signs that we are glossing over, although we have already seen them in the more concrete case of Plücker coordinates. By global signs, we mean signs in front of the entire Grassmann–Cayley expression. However, the geometric validity of any theorem is normally independent of global signs or other scalar multiples, so this is not a big deal. On the contrary, *relative* signs between individual terms in a Grassmann–Cayley algebraic expression or a bracket expression are crucial.

19.8 Application to Statics

The next two sections follow Staffetti [21]. The result of a force $\mathbf{f} = (f_1, f_2, f_3)$ applied at the point $\overline{p}$ of a rigid body M may be represented by the join

$$F = p \vee f = (\mathbf{f}, \overline{p} \times \mathbf{f})$$

of the projective points (or homogeneous 4-tuples) $p = (p1, p2, p3, 1)$ and $f = (f_1, f_2, f_3, 0)$. A force is a free vector and is represented by the point at infinity common to all of the lines parallel to the vector. However, the effect of the force on a point of a rigid body is a line-bound vector, represented by the 2-extensor of that line. Notice that the magnitude of the force may be adjusted by taking scalar multiples of f, although that does not change the point at infinity $\overline{f}$. If two forces $F_1 = p \vee f$ and $F_2 = q \vee g$ with $f = -g$ are applied at two distinct points $\overline{p}$ and $\overline{q}$ of M, the resultant

$$G = F_1 + F_2 = p \vee f + q \vee (-f) = (p - q) \vee f$$

is called a *couple*. Note that the couple is nonzero only if the line $\overline{pq}$ is not parallel to the force vector $\mathbf{f}$. Since $p - q = (p_1 - q_1, p_2 - q_2, p_3 - q_3, 0)$ represents a point at infinity, this shows that we can conceptualize a couple as resulting from the force $\mathbf{f}$ acting at a point at infinity on the body M. The couple is said to act in the plane determined by $\overline{p}, \overline{q}$, and $\mathbf{f}$.

In general, if two or more forces or couples are applied to a rigid body, the resultant is typically neither a force nor a couple, but a **wrench**, which is represented by a 2-tensor, rather than a 2-extensor. This representing 2-tensor is gotten by summing the 2-extensors representing the individual forces or couples. We may always write a wrench as

$$w = (\mathbf{f}, \mathbf{m}),$$

where $\mathbf{f}$ and $\mathbf{m}$ are the force and the moment, respectively, that must be applied at the origin to equal w. Poinsot's Theorem says that every wrench can be rewritten as a sum of a single force acting along a certain line, plus a single couple, with the couple acting in a plane orthogonal to the line of the force (Fig. 19.8).

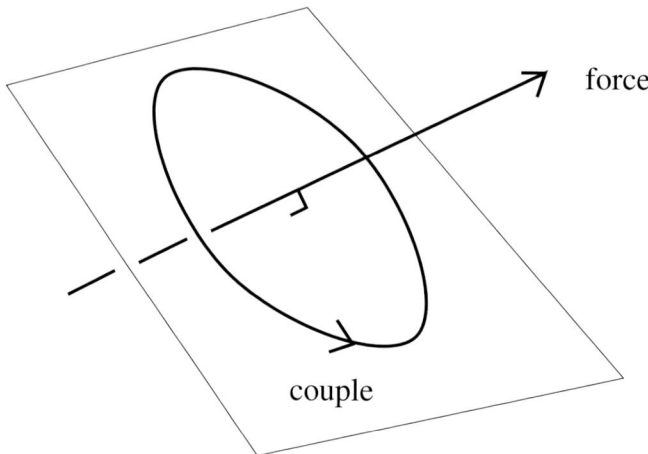

Fig. 19.8. Poinsot's Theorem

By now, the reader will no doubt have noticed that the representation of wrenches by 2-tensors is very analogous to the representation of twists by 2-tensors, with the special case of rotation corresponding to force, translation to couple, etc. In fact, for reasons that will become clear shortly, it is best to think of twists as 2-tensors in $V^{(2)}$ and wrenches as 2-tensors in $(V^*)^{(2)}$, where V^* is the dual vector space of V, that is, the vector space of all linear transformations from V to $\mathbb{R}$. Since each such linear transformation has a kernel that is a hyperplane, this really means that in the concrete interpretation of Plücker coordinates, we are really representing the wrenches by dual Plücker coordinates. Furthermore, in our case of $n = 4$, this really just means that we are interchanging the first three Plücker coordinates with the last three, so we now write

$$w = (\mathbf{m}, \mathbf{f}).$$

With this convention, the idea of reciprocity of a twist and a wrench has a very simple presentation. Let $t \in V^{(2)}$ and $w \in (V^*)^{(2)}$ be a twist and a wrench, respectively. Then

$$t \cdot w = t^{\mathrm{t}} w = \omega \cdot \mathbf{m} + \mathbf{v} \cdot \mathbf{f}$$

is the instantaneous power generated by t against w (see [16] or [21] for details). We say that t and w are **reciprocal** when $t^{\mathrm{t}}w = 0$. This formula really expresses the duality between V and V^*, in the following sense. If $T \subseteq V^{(2)}$ is the twist space of a partially constrained rigid body M, and $W \subseteq (V^*)^{(2)}$ is the vector space of all wrenches that are reciprocal to every twist in T, then T and W are dual spaces in a sense that directly reflects the duality of the Grassmann–Cayley algebra. This will become more apparent in the next section.

19.9 Application to Series–Parallel Mechanisms

Let C and D be robot arms, each attached to the ground at one end and having an end effector at the other end. The *series connection* of C and D is obtained by attaching the ground link of D rigidly to the end effector of C, where we assume that D was already in position to make this attachment, so that the twist and wrench spaces of D do not change in making this connection. The *parallel connection* of C and D is obtained by attaching the ungrounded end of each to a common end effector. Common robot arms have six joints connected in series, but other structures, such as a Stewart platform, have their joints connected in parallel. A *series–parallel robot* is any robot created by a sequence of such series and parallel connections. Let us consider the twist and wrench spaces of a series–parallel robot.

Let T_C and T_D be the twist spaces of C and D, both of which spaces are subspaces of $V^{(2)}$, and let A_C and A_D be the extensors in $\Lambda(V^{(2)})$ that are supported by those twist spaces. Then the twist space of their series connection is just the sum of the two subspaces, and hence is the support of $A_C \bigvee A_D$, provided this superjoin is nonzero. If that superjoin is zero, this means that the two twist spaces have nonzero intersection, that is, there is some particular twist that can be realized by both C and D. Usually one can eliminate this intersection; for example, if both C and D are arms with simple joints connected in series, then at least one of the simple joints is dependent on the others, and should be eliminated for purposes of the calculation. Of course, in this case, the combined arm was in a critical configuration.

Similarly, if W_C and W_D are the wrench spaces of C and D, and B_C and B_D are the corresponding extensors in $\Lambda((V^*)^{(2)})$, then the wrench space of their series connection is the intersection of the two wrench spaces, and hence is the support of $B_C \bigwedge B_D$ if this supermeet is not zero. Again, if the supermeet is zero, there are ways to get around this by adding artificial wrenches to cause $W_C \cup W_D$ to span $(V^*)^{(2)}$. We are using $\bigwedge$ for the meet in $\Lambda((V^*)^{(2)})$, as well as in $\Lambda(V^{(2)})$, to distinguish it from the meet in $\Lambda(V)$ or in $\Lambda(V^*)$.

Furthermore, the above two paragraphs remain true if we replace series connection by parallel connection, provided we also interchange $\bigvee$ and $\bigwedge$.

Thus the twist space of the parallel connection of C and D is the support of $A_C \bigwedge A_D$ if the super meet is not zero, and the wrench space of the parallel connection of C and D is the support of $B_C \bigvee B_D$ if the superjoin is not zero.

Similar considerations apply to series–parallel robots, by applying the above ideas to one step of the construction at a time. There is also the possibility of computing the twist and wrench spaces of complicated robots which are not constructable by successive series and parallel constructions. This is done by Delta–Wye transformations; see [8] or [21].

19.10 An Example of a Robot Arm

Consider the example of a robot arm with six revolute joints, as shown in Fig. 19.7. This is modified from a robot considered for the US space shuttle. The large cylinders represent revolute joints, and the thin cylinders represent links. We wish to find the critical configurations of the arm. We choose two points on each joint axis, and consider the six joint centers $a_1a_2, b_1b_2, c_1c_2, d_1d_2, e_1e_2, f_1f_2$. Notice that where possible we have chosen the points determining the axes to coincide, namely $a_2 = b_1, e_2 = f_1$, and $c_2 = d_1$, where the last of these is a point at infinity. Since the robot arm consists of the six joints in series, the twist space is $a_1a_2 \bigvee a_2b_2 \bigvee c_1c_2 \bigvee c_2d_2 \bigvee e_1e_2 \bigvee e_2f_2$. Since these six centers are in $V(2)$, a six-dimensional space, the twist space is all of $V^{(2)}$ precisely when the superjoin of these six centers is nonzero, in other words, when the Plücker coordinate vectors of the six axes are linearly independent. Then the arm has full mobility and is not in a critical configuration.

So far, none of this is surprising or new. It is well known that the criticality of a six-joint arm depends just on the linear dependence of the six Plücker coordinate vectors. What is surprising and fairly new, however, is that we can use ideas from the Grassmann–Cayley algebra to learn the geometry of the critical configurations. The superjoin of the six vectors in $V^{(2)}$ amounts to computing the determinant of the six Plücker coordinate vectors (up to some constant scalar). This calculation is just the bracket in the Grassmann–Cayley algebra $\Lambda(V^{(2)})$, which we call the *superbracket* $[[a_1a_2, b_1b_2, c_1c_2, d_1d_2, e_1e_2, f_1f_2]]$, to distinguish it from the ordinary bracket of four points in V.

The superbracket of 6 general 2-extensors can also be considered as determined by the 12 points selected on the axes. Then it is possible to prove using the theory of projective invariants that the superbracket can be written in terms of ordinary brackets involving those 12 points. This has been worked out in [17], where the expression for our robot arm reduces to

$$[[a_1a_2, a_2b_2, c_1c_2, c_2d_1, e_1e_2, e_2f_2]]$$
$$= -\sum_\sigma [a_1, a_2, b_2, \sigma(c_1)][a_2, c_2, d_2, e_2][\sigma(c_2), e_1, e_2, f_2]$$

$$+ \sum_\tau [a_1, a_2, b_2, \tau(c_2)][a_2, c_1, c_2, e_2][\tau(d_2), e_1, e_2, f_2]$$
$$= \sum_\rho [\rho(c_1), a_1, a_2, b_2][\rho(d_2), a_2, c_2, e_2][\rho(c_2), e_1, e_2, f_2],$$

where each summation is over all permutations of the points involved (the sums over σ and τ each involve two terms, and the same four terms are in the sum over ρ, since two of the six terms in that sum are zero by virtue of repeated elements).

Now we can recognize the last line as the expansion into brackets of the Grassmann–Cayley expression

$$c_1 d_2 c_2 \wedge a_1 a_2 b_2 \wedge a_2 c_2 e_2 \wedge e_1 e_2 f_2.$$

Thus we can recognize that the geometric conditions corresponding to criticality are precisely those that make this Grassmann–Cayley expression equal to 0, namely:

1. one or more of the planes $\overline{c_1 c_2 d_2}, \overline{a_1 a_2 b_2}, \overline{a_2 c_2 e_2}, \overline{e_1 e_2 f_2}$ is degenerate, or
2. the four planes have nonempty intersection.

Notice that in the robot arm under consideration, none of the degeneracies of type 1 can occur. To the author's knowledge, condition 2 was not known for such robot arms previous to the Grassmann–Cayley analysis.

19.11 Cayley Factorization

Consider the following diagram. We have seen that we can translate projective or affine geometric incidence conditions into the Grassmann–Cayley algebra, and, furthermore, it is easy to translate backwards if we have a **simple** Grassmann–Cayley expression, that is, one involving only joins, meets, and scalar multiplication. However, the backwards translation is much more difficult if addition (or subtraction) in $\Lambda(V)$ is involved.

(1) Projective geometry
     $\updownarrow$
(2) Grassmann–Cayley algebra
     $\downarrow$ $\uparrow$ Cayley factorization
(3) Bracket algebra
     $\downarrow$
(4) Coordinate algebra

The translation from Grassmann–Cayley algebra to bracket algebra is straightforward using our formula for expanding the meet, provided the Grassmann–Cayley expression involved is of step 0 or n. Fortunately, this is the case in most applications of interest.

The translation from bracket algebra to ordinary coordinate algebra is trivial: just introduce a fixed basis of V so that each bracket becomes a determinant, and write out the usual formula for each determinant in terms of the coordinates of the vectors involved. The backwards translation from coordinate algebra to bracket algebra is possible, though computationally expensive, provided the given coordinate expression has some expression in terms of determinants (such expressions are known as **projective invariants**). See Sturmfels [22] for an algorithm.

A very interesting problem is the translation from the bracket algebra to the Grassmann–Cayley algebra. In other words, given an expression Q in brackets, does there exist, and if so, can we find, a simple Grassmann–Cayley expression that expands to Q? We call this process **Cayley factorization**. By the way, if we allowed the Grassmann–Cayley expression to use addition, the process becomes both very easy and useless. We saw an example of Cayley factorization in the previous section, and it immediately gave us the desired geometric interpretation of criticality. No practical algorithm is known for this problem in the general case.

The importance of Cayley factorization arises from the following "philosophical" point of view. It is best for many purposes to avoid level 4, and to work instead on levels 2 and 3. This is because the expressions in those two levels are symbolic and coordinate-free, and are therefore much closer to the way we humans really think about the problems. A practical algorithm for Cayley factorization would greatly facilitate this approach. The final section shows two more applications closely related to robotics that use this coordinate-free approach, and that illustrate the use of Cayley factorization.

19.12 Rigidity of Frameworks

Consider a d-dimensional *bar framework*, by which we mean a collection of rigid bars connected to each other only at their endpoints, by spherical joints. An instantaneous motion of such a framework is an assignment of velocity vectors to each joint such that the lengths of the bars are instantaneously preserved. A framework is **rigid** if the only instantaneous motions are the rigid-body Euclidian motions in $\mathbb{R}^d$. We represent the framework by a graph, with vertices of the graph representing the joints, and edges of the graph representing the bars. Given such a graph, we can also represent the graph by a framework, by choosing a point in $\mathbb{R}^d$ for each vertex, and then joining the appropriate points by bars of length exactly equal to the distance between the points involved. A graph G is called **generically isostatic** (for bar frameworks) in $\mathbb{R}^d$ if there exists a realization of it in $\mathbb{R}^d$ that is rigid, but it is minimally rigid in the sense that the removal of any one bar would make it nonrigid. An example in $\mathbb{R}^2$ is shown in Fig. 19.9, which has six vertices and nine edges. It is not difficult to show that a generically isostatic graph in $\mathbb{R}^2$

with v vertices always has exactly $2v - 3$ edges; if it is in $\mathbb{R}^3$ it must have $3v - 6$ edges.

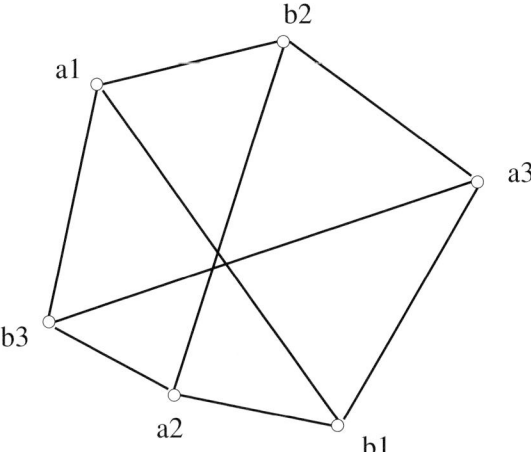

Fig. 19.9. A bar framework in the plane

A key question is whether we can describe geometrically the realizations of a generically isostatic graph G that are not rigid. It can be proved that such geometric conditions must be projectively invariant, and therefore describable by a bracket condition; that is, there must be a bracket expression C_G involving the vertices of the graph that is zero precisely for the nonrigid (i.e., critical) realizations of G. In fact, [26] not only proves these facts, but gives an algorithm to construct this bracket condition C_G, called the **pure condition** of G. Then it becomes a question of Cayley factorization of C_G to discover the desired geometric condition. In the example of Fig. 19.9, we find (using the algorithm and the bracket algebra straightening algorithm) that

$$C_G = [a_1, a_2, a_3][a_1, b_2, b_3][b_1, a_2, b_3][b_1, b_2, a_3]$$
$$- [b_1, b_2, b_3][b_1, a_2, a_3][a_1, b_2, a_3][a_1, a_2, b_3].$$

For a more complete description of how to compute the pure condition, and an accessible summary of how it is done for this example, see [24] or [25].

This bracket expression is not easy to Cayley factor. However, it was recognized as the bracket condition obtained by expanding the Grassmann–Cayley expression

$$C_G = (a_1 b_2 \wedge a_2 b_1) \vee (a_1 b_3 \wedge a_3 b_1) \vee (a_2 b_3 \wedge a_3 b_2).$$

Now we can see geometrically that $C_G = 0$ precisely when each pair of opposite sides of the hexagon in Fig. 19.9 is intersected to give a point (in the projective plane) and these three points are collinear. By Pascal's Theorem, this is in turn equivalent to the six joints lying on a common conic (a circle, ellipse, parabola, hyperbola, or possibly a degenerate conic, which is just two lines). Thus by Cayley factorization, we have the result that a realization of G is rigid in the plane if and only if the six joints do not lie on a common conic (possibly degenerate). Incidentally, further degenerate cases, such as two of the joints being coincident or two of the bars lying on the same line, all fall under the case of all six joints lying on a degenerate conic of two lines.

A second kind of framework we will consider is a **bar-and-body framework**. This consists of a number of rigid bodies in $\mathbb{R}^d$, connected by some rigid bars, with each bar connected at its endpoints to two of the bodies, using spherical joints. The bodies themselves may be either d-dimensional or $(d-1)$-dimensional. Now an instantaneous motion is an assignment of a twist to each body, in such a way as to preserve instantaneously the length of each bar. Again, the framework is rigid if the only instantaneous motions are those obtained from a single rigid-body motion, which is then applied to all of the bodies (i.e., a Euclidian motion of the entire framework). Again, we can represent the framework by a graph, using vertices for the bodies and edges for the bars. A graph is again called **generically isostatic** if some realization of it as a bar-and-body framework in $\mathbb{R}^d$ is minimally rigid. Incidentally, the size and shape of the bodies is essentially irrelevant— it is really just the geometry of the bars that controls the rigidity here. Such a graph with v vertices must have exactly $6(v-1)$ edges to be generically isostatic for bar-and-body frameworks in $\mathbb{R}^3$, and $3(v-1)$ in $\mathbb{R}^2$.

As in the case of bar frameworks, there is a pure condition for bar-and-body frameworks, and an algorithm to compute it, in [27]. One major difference, however, is that the brackets contain bars, not just joints of the framework. In fact, each bracket is really a superbracket (see Sect. 19.10). As a simple example, consider Fig. 19.10. There are three rigid bodies in the plane, connected by six bars. The pure condition is

$$C_G = [[a,b,c]][[d,e,f]] - [[a,b,d]][[c,e,f]], \qquad (19.4)$$

which can be easily Cayley factored as

$$C_G = ab \bigwedge cd \bigwedge ef.$$

How is this super-Grassmann–Cayley expression to be interpreted geometrically? First, let us note that since we are working in $\mathbb{R}^2$, V is three-dimensional, as is $V^{(2)}$. In fact, it is very natural to think of $V^{(2)}$ as V^*, the dual vector space of V, because each vector in $V^{(2)}$ is supported by a line, which is a hyperplane in V. Thus the join ab (or really $a \bigvee b$ in $\Lambda(V^{(2)})$) corresponds to $a \wedge b$ in $\Lambda(V)$, where we are now thinking of a and b as 2-extensors in $\Lambda(V)$. Thus the support of $a \bigvee b$ is the point of intersection of the (lines determined by) the bars a and b. Thus the critical configurations are all those in which the three points of intersection are collinear. This result has the following kinematic interpretation. The intersection of the lines of the bars a and b is the center of relative motion of the two bodies connected by a and b, and similarly for the other pairs of bars connecting a given pair of bodies. A theorem of Aronhold [1] states that if three rigid bodies are in relative motion, then the three centers of relative motion must be collinear, a result we have just verified (even for instantaneous motions) using our techniques of pure condition and Cayley factorization.

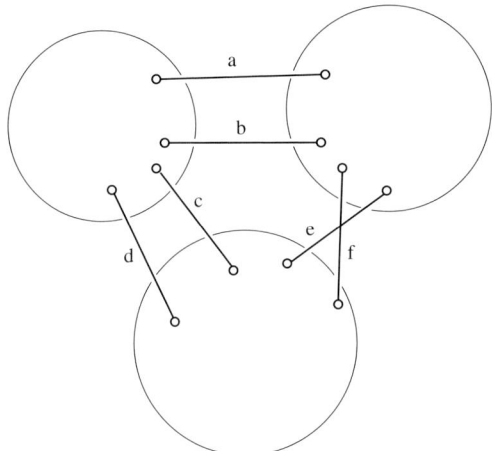

Fig. 19.10. A bar-and-body framework in the plane

An interesting feature of the pure condition of a bar-and-body framework is that every bar appears only once in each term of the pure condition. For example, in the framework of Fig. 19.10, we saw that the pure condition in Eq. 19.4 is a difference of two terms, each of which is a product of two brackets, with each bar occuring exactly once in the brackets of a term. This situation is described by saying that the pure condition of a bar-and-body framework is *multilinear*, and in the multilinear case, Cayley factorization is much easier than in the general case. In fact, [23] gives an algorithm to Cayley factor a multilinear bracket expression (or announce its impossibility if that is the case), although the algorithm is effective in practice only up to about 20 bars. No algorithm is known for the nonmultilinear case of Cayley factorization, short of literally expanding all possible Grassmann–Cayley expressions (although even then, comparing the resulting expansion to the given bracket expression is nontrivial, requiring the straightening algorithm to reduce both expressions to a standard form). Nevertheless, attempts at providing such an algorithm in certain situations has met with some success, see [13] and [14].

Let us briefly see how the multilinear Cayley factorization algorithm works. Suppose we wish to Cayley factor the bracket polynomial

$$P = [a,b,c][d,e,f] + [a,b,d][c,e,f].$$

We see that P is indeed a multilinear bracket polynomial on six points in the plane. The algorithm first checks to see which pairs of points have the property that P is antisymmetric in that pair. For example, to see if P is antisymmetric in a,b, we take P and subtract from it the result of interchanging a and b in P. We now use the straightening algorithm to check whether this difference is 0 (although in this case, this should be obvious, because each of the two terms in P is antisymmetric in a and b). It turns out that the only such pairs are the two obvious ones, a,b and also e,f. This already is enough to see that P cannot be Cayley factored. If there were a Cayley factorization, c would have to appear in some join with at least one other point, and then P would be antisymmetric in c and that other point. If all points had been involved in antisymmetric pairs, then the algorithm would compute all maximal sets of points that are pairwise antisymmetric. These are candidates for joins of points that are involved in meets in a Cayley factorization of P. The algorithm then checks for antisymmetry in the sense of Theorem 5. If even one such antisymmetry is found, it is then possible to reduce the number of points involved in P and continue inductively. For complete details, see [23].

Another nice example of the use of the pure condition and Cayley factorization to determine the critical configurations of a framework is given in [6]. They consider an octahedral manipulator, which is just a Stewart platform, with both the platform and base considered as rigid bodies in $\mathbb{R}^3$, with six bars connecting them. It is easy to see that the critical configurations of the Stewart platform as a robotic structure are the same as those of the bar-and-body framework, since in both cases it is precisely the configurations where the superbracket of the six bars is zero, that is, the six bars have dependent Plücker coordinates.

Thinking of the last example as a Stewart platform raises the following question: if a single bar in a generically isostatic bar-and-body framework is replaced by a linear actuator and then actuated, can the resulting instantaneous motion be computed? The answer is yes, and it is easily gotten from the pure condition, as shown in [27]. Let us see how this works in the example in Fig. 19.10. Let us actuate the bar a by a unit velocity (instantaneously), and ask for the relative motion between the two bodies connected by a. If we think of the first body as fixed to the ground, then this gives the instantaneous motion of the other body. We just have to take the pure condition in Eq. 19.4 and factor out a. Recalling that the superbrackets in that equation are really superjoins of step 3, we get

$$m = [[d,e,f]]b \bigvee c - [[c,e,f]]b \bigvee d.$$

Recalling the duality explained above, $b \bigvee c$ is a point on the line b, as is $b \bigvee d$. Thus the center of relative motion m is a particular linear combination of those two points.

The analogous problem for a bar framework is also of interest, and is currently being studied by the author.

19.13 Other Applications

Other applications of the Grassmann–Cayley algebra include computer vision [7], proving geometry theorems [20], and even automatic geometry theorem-proving [4, 19].

19.14 Conclusion

The Grassmann–Cayley algebra is a symbolic approach to Plücker coordinates, which is very useful for analyzing the instantaneous kinematics and statics of robots, as well as determining the geometry of the critical configurations. The latter involves the process of Cayley factorization, which has met with some success, but is not well understood in general.

References

1. S. Aronhold (1872) Gründzuge der kinematischen Geometrie. *Verh. d. Ver. z. Beförderung des Gewerbefleiss in Preussen* 51:129–155
2. M. Barnabei, A. Brini, G.-C. Rota (1985) On the exterior calculus of invariant theory. *J. Algebra* 96:120–160
3. M. Chasles (1830) Note sur les propriétés générales du système de deux corps semblables entr'eux et placés d'une manière quelconque dans l'espace; et sur le déplacement fini ou infiniment petit d'un corps solide libre. *Bulletin des Sciences Mathematiques, Astronomiques, Physiques, et Chimiques* 14:321–326
4. H. Crapo, J. Richter-Gebert (1995) Automatic proving of geometric theorems. In: N. White (ed.) *Invariant Methods in Discrete and Computational Geometry*. Kluwer, Dordrecht, pp. 167-196
5. P. Doubilet, G.-C. Rota, J. Stein (1974) On the foundations of combinatorial theory (IX): Combinatorial methods in invariant theory. *Stud. Appl. Math.* 53:185–216
6. D. Downing, A. Samuel, K. Hunt (2002) Identification of the special configuration of the octahedral manipulator using the pure condition. *International J. Robotics Research* 21:147–159
7. O. Faugeras, T. Papadopoulo (1998) Grassmann–Cayley algebra for modelling systems of cameras and the algebraic equations of the manifold of trifocal tensors. *Phil. Trans. R. Soc. London, Series A*, 356:1123–1152
8. T. Feo, J. Provan (2003) Delta-Wye transformation and the efficient reduction of two-terminal planar graphs. *Operations Research*, 41:572-582
9. H. Grassmann (1911) *Gesammelte Mathematische und Physikalische Werke*. Teubner, Leipzig

10. D. Hestenes, G. Sobczyk (1984) *Clifford Algebra to Geometric Calculus:a Unified Language for Mathematics and Physics.* D. Reidel, Dordrecht
11. W. Hodge, D. Pedoe (1968) *Methods of Algebraic Geometry*, vols. 1 and 2. Cambridge Univ. Press, Cambridge, UK
12. K. Hunt (1978) *Kinematic Geometry of Mechanisms.* Clarendon Press, Oxford
13. H. Li, Y. Wu (2003) Automated short proof generation for projective geometric theorems with Cayley and bracket algebras, I. Incidence geometry. *J. Symbolic Computation* 36:717-762
14. H. Li, Y. Wu (2003) Automated short proof generation for projective geometric theorems with Cayley and bracket algebras, II. Conic geometry. *J. Symbolic Computation* 36:763-809
15. H. Lipkin, J. Duffy (1982) Analysis of industrial robots via the theory of screws. In: Proc. Internat. Symp. Industrial Robots, Paris
16. J. McCarthy (2000) *Geometric Design of Linkages.* Springer, Berlin Heidelberg New York
17. T. McMillan, N. White (1991) The dotted straightening algorithm. *J. Symbolic Comput.* 11:471–482
18. J.-P. Merlet (1989) Singular configurations of parallel manipulators and Grassmann geometry. *International J. Robotics Research* 8:45–56
19. J. Richter-Gebert (1995) Mechanical theorem proving in projective geometry. *Ann. Math. Artif. Intell.* 13:139–172
20. G.-C. Rota, J. Stein (1976) Applications of Cayley algebras. In: *Colloquio Internazionale sulle Teorie Combinatorie.* Accademia Nazionale dei Lincei, Rome, pp. 71-97
21. E. Staffetti (2004) Kinestatic analysis of robot manipulators using the Grassmann–Cayley algebra. *IEEE Trans. Robotics and Automation* 20:200-210
22. B. Sturmfels (1993) *Algorithms in Invariant Theory.* Springer, Berlin Heidelberg New York
23. N. White (1991) Multilinear Cayley factorization. *J. Symbolic Comput.*, 11:421–438
24. N. White (1994) Grassmann–Cayley algebra and robotics. *J. Intell. Robot. Syst.* 11:91–107
25. N. White (1995) A tutorial on Grassmann-Cayley algebra. In: N. White (ed.) *Invariant Methods in Discrete and Computational Geometry.* Kluwer, Dordrecht, pp. 93-106
26. N. White, W. Whiteley (1983) The algebraic geometry of stresses in frameworks. *SIAM J. Algebraic Discrete Meth.* 4:481–511
27. N. White, W. Whiteley (1987) The algebraic geometry of motions in bar-and-body frameworks. *SIAM J. Algebraic Discrete Meth.* 8:1–32

20

Clifford Algebra and Robot Dynamics

J. M. Selig

Faculty of Business, Computing and Information Systems
London South Bank University,
Borough Road
London SE1 0AA, UK
seligjm@lsbu.ac.uk

20.1 Introduction

In this chapter the classical mechanics of rigid bodies will be investigated using a novel Clifford algebra representation. A key point is that this algebra contains elements representing the velocities, momenta and inertias of rigid bodies. Modern approaches to classical mechanics treat velocities and momenta as dual objects. That is, the velocities are vectors and the momenta are linear functionals on the vectors. Hence velocities and momenta are fundamentally different types of object. There is a pairing, or evaluation map that produces a scalar energy given a velocity and a momenta. The mapping which turns a velocity into a momenta is provided by the inertia. The inertia is also represented by a element in the Clifford algebra. The approach taken has some similarities to recent attempts to represent linear algebra using Clifford algebra [11].

The other key ingredient here is the action of the group of rigid body motions or proper Euclidian group $SE(3)$. The position and orientation of a rigid body is given by setting a 'home' position for the body and then specifying the rigid motion which transforms the home configuration into the current one. The elements of this group can also be found as elements of the Clifford algebra. The action of this group on the velocities, momenta and inertias is extremely important and turns out to be quite simple in this representation.

This allows a very succinct formulation of the equations of motion for a rigid body. This material is reviewed in Sect. 20.2. In Sect. 20.3 the methods are extended to look at the dynamics of serial robots. These can be considered as mechanisms composed of several rigid bodies connected by simple joints. First, a fairly simple derivation of the equations of motion is given. Then, the Lagrangian and Hamiltonian mechanics of these robots are introduced. These are used to find the equations of motion again, but this time in terms of the joint variables and their derivatives. In particular, simple formulas are given for the elements of the generalised mass matrix, and the centrifugal and Coriolis terms.

In Sect. 20.4 parallel machines are considered. Here the robot is considered as a star-structured mechanism with constraints on the leaf links. Hence, the representation of constraints by a linear system of constraint wrenches which are dual to a system of screws of freedom is introduced. Finally, a method to find the equations of motion for a Stewart platform is sketched.

The Clifford algebra used here was introduced in [7], where more details on the application to a single rigid body can be found. More detail on the dynamics of serial robots can be found in [5, Chap. 12] and on parallel robots in [9]. To begin with the Clifford algebra will be introduced.

20.2 The Clifford algebra $C\ell(0, 6, 2)$

20.2.1 Rigid Transformations

The group of proper rigid body motions in three dimensions is usually denoted $SE(3)$. Its elements are sometimes called finite screws. This is because all elements of the group, except the pure translations, are helical motions consisting of a rotation about an axis followed by a translation along that axis. In the kinematics literature the Lie algebra of the group (denoted $se(3)$) are usually called twists or infinitesimal screws. The term screw will also be used here to denote a Lie algebra element.

It has been known for a long time that both finite and infinitesimal screws can be represented using the Clifford algebra $C\ell(0, 3, 1)$. This is the basis for Study's dual quaternions, see [12] and also [4]. Here, we have three generators which square to -1 and one which squares to 0. Say, $e_1^2 = e_2^2 = e_3^2 = -1$, and $e^2 = 0$.

In this algebra we can represent rotations by elements of the form

$$r = \cos\frac{\theta}{2} + \sin\frac{\theta}{2}(u_x e_2 e_3 + u_y e_3 e_1 + u_z e_1 e_2). \tag{20.1}$$

The angle of rotation is θ, and $(u_x, u_y, u_z)^T$ is a unit vector in the direction of the rotation axis. This is simply the representation of rotations by quaternions, in a slightly different guise; just replace $e_2 e_3 \mapsto i$, $e_3 e_1 \mapsto j$ and $e_1 e_2 \mapsto k$. So, this is a double-valued representation, or rather a representation of the covering group Spin(3).

Translations can be represented in this algebra by elements of the form,

$$t = 1 + \frac{1}{2}(t_x e_1 e + t_y e_2 e + t_z e_3 e). \tag{20.2}$$

Since rigid motions are combinations of rotations and translations $g = tr$, they will have the general form

$$g = \alpha_0 + \alpha_1 e_2 e_3 + \alpha_2 e_3 e_1 + \alpha_3 e_1 e_2 + \beta_0 e_1 e_2 e_3$$
$$+ \beta_1 e_1 e + \beta_2 e_2 e + \beta_3 e_2 e + \beta_3 e_3 e \quad (20.3)$$

where the αs and βs are real coefficients. The coefficients are not completely arbitrary, for such an element to be a rigid body motion it must satisfy the condition

$$gg^* = 1 \quad (20.4)$$

where $*$ denotes the Clifford conjugate. In terms of the coefficients this gives just two relations

$$\alpha_0^2 + \alpha_1^2 + \alpha_2^2 + \alpha_3^2 = 1, \quad \text{and} \quad \alpha_0 \beta_0 + \alpha_1 \beta_1 + \alpha_2 \beta_2 + \alpha_3 \beta_3 = 0. \quad (20.5)$$

The Lie algebra of the group, the twists or screws, are represented by arbitrary elements of grade 2,

$$\mathbf{s} = \omega_x e_2 e_3 + \omega_y e_3 e_1 + \omega_z e_1 e_2 + v_x e_1 e + v_y e_2 e + v_z e_3 e. \quad (20.6)$$

As usual, $(\omega_x, \omega_y, \omega_z)^\mathrm{T}$ is the angular velocity vector of the body and $(v_x, v_y, v_z)^\mathrm{T}$ is the linear velocity vector of the body. The Lie bracket of a pair of screws $\mathbf{s}_1$ and $\mathbf{s}_2$ is simply their commutator,

$$[\mathbf{s}_1, \mathbf{s}_2] = \frac{1}{2}\left(\mathbf{s}_1 \mathbf{s}_2 - \mathbf{s}_2 \mathbf{s}_1\right). \quad (20.7)$$

The adjoint representation of the group on its Lie algebra is given in the Clifford algebra by the conjugation

$$\mathrm{Ad}(g)\mathbf{s} = g\mathbf{s}g^*. \quad (20.8)$$

Next, the exponential map from the Lie algebra to the group is consistent with the Clifford algebra, given a Lie algebra element $\mathbf{z}$ we have

$$g = e^{\frac{1}{2}\mathbf{z}}. \quad (20.9)$$

The factor $\frac{1}{2}$ here is due to the fact that we are working in the double cover of $SE(3)$.

Now the adjoint representation of the Lie algebra on itself can be found. This is done by differentiating the adjoint action of the group at the identity.

$$\mathrm{ad}(\mathbf{s}_1)\mathbf{s}_2 = \frac{\mathrm{d}}{\mathrm{d}t}\left. e^{\frac{t}{2}\mathbf{s}_1} \mathbf{s}_2 e^{-\frac{t}{2}\mathbf{s}_1}\right|_{t=0} = \frac{1}{2}\left(\mathbf{s}_1 \mathbf{s}_2 - \mathbf{s}_2 \mathbf{s}_1\right). \quad (20.10)$$

That is, the adjoint action of the Lie algebra is simply the Lie bracket, $\mathrm{ad}(\mathbf{s}_1)\mathbf{s}_2 = [\mathbf{s}_1, \mathbf{s}_2]$.

The velocity screw comes from the time derivative of the exponential. This is simple if the motion of the body is a uniform motion about a constant screw, $g = e^{\frac{t}{2}\mathbf{s}_0}$. As above, the derivative of this is,

$$\frac{d}{dt}g = \frac{1}{2}\mathbf{s}_0 g. \tag{20.11}$$

When the motion is more general, $g = e^{\frac{1}{2}\mathbf{z}(t)}$, then it is harder to show but nevertheless true that

$$\frac{d}{dt}g = \frac{1}{2}\mathbf{s}g, \tag{20.12}$$

where $\mathbf{s}$ is the velocity screw of the rigid body. The relationship between $\mathbf{s}$ and $\mathbf{z}$ is given by

$$\mathbf{s} = \sum_{i=0}^{\infty} \frac{1}{(i+1)!}\operatorname{ad}^i(\mathbf{z})\frac{d}{dt}\mathbf{z}, \tag{20.13}$$

where $\operatorname{ad}^i(\mathbf{z})\mathbf{y} = [\mathbf{z},\operatorname{ad}^{i-1}(\mathbf{z})\mathbf{y}]$, and $\operatorname{ad}^1(\mathbf{z})\mathbf{y} = \operatorname{ad}(\mathbf{z})\mathbf{y} = [\mathbf{z},\mathbf{y}]$, see [1].

20.2.2 Momenta and Inertia

In order to include momenta and inertias the algebra can be 'doubled', that is, the Clifford algebra $C\ell(0,6,2)$ can be used. The generators can be labelled, e_1, e_2, e_3, e and a_1, a_2, a_3, a, and assume that $a_i^2 = e_j^2 = -1$ and $a^2 = e^2 = 0$.

In this algebra the coscrews, that is elements of the dual to the Lie algebra, are represented by grade-2 elements of the form

$$\mathcal{P} = p_x a_2 a_3 + p_y a_3 a_1 + p_z a_1 a_2 + l_x a_1 a + l_y a_2 a + l_z a_3 a \tag{20.14}$$

A combination of a force and a torque applied to a rigid body is also a coscrew. A three-dimensional force $\mathbf{F}$ and torque $\boldsymbol{\tau}$ would be represented as

$$\mathcal{W} = F_x a_2 a_3 + F_y a_3 a_1 + F_z a_1 a_2 + \tau_x a_1 a + \tau_y a_2 a + \tau_z a_3 a. \tag{20.15}$$

The action of the group of rigid body motions on these coscrews is exactly the same as the action on the screws but using as instead of es. So, let us introduce a new operation, an involution, which exchanges as for es. If we write this with an overbar then $\bar{a}_i = e_i$ and $\bar{e}_i = a_i$.

Now the group action on momenta can be written

$$\mathcal{P} \longrightarrow \bar{g}\mathcal{P}\bar{g}^*, \tag{20.16}$$

where g would be the corresponding group element that acts on the screws. This is the coadjoint representation of the group.

Notice that in fact the same transformation law could be used for the screws and coscrew, $\mathbf{s} \to (g\bar{g})\mathbf{s}(g\bar{g})^*$; since $\bar{g}$ commutes with $\mathbf{s}$ and $\mathcal{P} \to (g\bar{g})\mathcal{P}(g\bar{g})^*$, since g commutes with $\mathcal{P}$.

We will need the coadjoint action of the Lie algebra on its dual space. This can be found from the group action on the coscrews given above. Let us suppose that a coscrew is subject to a uniform motion about a screw. The relevant group elements can be written as exponentials, $g = e^{t\mathbf{s}/2}$, where t is time. Hence, as a function of time the coscrews is given by,

$$\mathcal{P}(t) = e^{\frac{t}{2}\bar{s}}\mathcal{P}e^{-\frac{t}{2}\bar{s}}. \tag{20.17}$$

Now if we differentiate this with respect to t and then set $t = 0$, we get the coadjoint action of the screw $\mathbf{s}$ on the coscrew $\mathcal{P}$,

$$\{\mathbf{s}, \mathcal{P}\} = \frac{1}{2}(\bar{s}\mathcal{P} - \mathcal{P}\bar{s}). \tag{20.18}$$

Note that the standard notation for this action is given by curly brackets as above.

Next we look at the inertia operator. The inertia operator here will be a combination of the usual 3×3 inertia matrix, the mass and centre of gravity of the body. In screw theory, this can be represented using a 6×6 symmetric matrix. A diagonal inertia matrix can be represented as

$$N = d_x a_1 a e_1 e + d_y a_2 a e_2 e + d_z a_3 a e_3 e + m a_2 a_3 e_2 e_3$$
$$+ m a_3 a_1 e_3 e_1 + m a_1 a_2 e_1 e_2. \tag{20.19}$$

Under a rigid body motion this transforms according to

$$N \longrightarrow (g\bar{g})N(g\bar{g})^*. \tag{20.20}$$

It is well known that any inertia matrix can be diagonalised using a rigid motion, hence we can find the Clifford algebra element representing any inertia matrix.

As an example, consider how the diagonal inertia above transforms under a translation of t_x in the x-direction. For this transformation we have

$$g = 1 + \frac{1}{2}t_x e_1 e \tag{20.21}$$

and hence,

$$g\bar{g} = \left(1 + \frac{1}{2}t_x e_1 e\right)\left(1 + \frac{1}{2}t_x a_1 a\right) = 1 + \frac{1}{2}t_x a_1 a + \frac{1}{2}t_x e_1 e + \frac{1}{4}t_x^2 a_1 a e_1 e \tag{20.22}$$

For hand calculation it is probably better to perform the multiplications by g and $\bar{g}$ separately. The result is

$$g\bar{g}N\bar{g}^*g^* = d_x a_1 a e_1 e + (d_y + mt_x^2)a_2 a e_2 e + (d_z + mt_x^2)a_3 a e_3 e$$
$$+ mt_x a_3 a e_3 e_1 - mt_x a_2 a e_1 e_2$$
$$+ mt_x a_3 a_1 e_3 e - mt_x a_1 a_2 e_2 e$$
$$+ m a_2 a_3 e_2 e_3 + m a_3 a_1 e_3 e_1 + m a_1 a_2 e_1 e_2. \tag{20.23}$$

Compare this with the corresponding 6×6 inertia matrix,

$$\begin{bmatrix} d_x & 0 & 0 & 0 & 0 & 0 \\ 0 & d_y + mt_x^2 & 0 & 0 & 0 & -mt_x \\ 0 & 0 & d_z + mt_x^2 & 0 & mt_x & 0 \\ 0 & 0 & 0 & m & 0 & 0 \\ 0 & 0 & mt_x & 0 & m & 0 \\ 0 & -mt_x & 0 & 0 & 0 & m \end{bmatrix}. \tag{20.24}$$

The inertia matrix is always a symmetric matrix; in the Clifford algebra this corresponds to the property $\bar{N} = N$.

Finally in this section we look at some invariants under this group action. The pseudoscalar in the algebra $C\ell(0,3,1)$ is simply $E = e_1 e_2 e_3 e$, which is clearly invariant under the action $E = gEg^*$. As an element of $C\ell(0,6,2)$ it commutes with $\bar{g}$ and hence

$$E = (g\bar{g})E(g\bar{g})^*. \tag{20.25}$$

By a similar argument, the element $A = \bar{E} = a_1 a_2 a_3 a$ is also invariant.

$$A = (g\bar{g})A(g\bar{g})^*. \tag{20.26}$$

The third invariant we will need is the following grade-4 element:

$$Q_0 = a_2 a_3 e_1 e + a_3 a_1 e_2 e + a_1 a_2 e_3 e + a_1 a e_2 e_3 + a_2 a e_3 e_1 + a_3 a e_1 e_2. \tag{20.27}$$

It is straightforward to verify that this does indeed satisfy the relation

$$Q_0 = (g\bar{g})Q_0(g\bar{g})^*. \tag{20.28}$$

Notice also that $\bar{Q}_0 = Q_0$.

Finally, there are a couple of neat equations which tie some of these ideas together,

$$Q_0 \wedge \mathbf{s} = \bar{\mathbf{s}} E, \quad \text{and} \quad \bar{\mathbf{s}} \wedge Q_0 = A\mathbf{s}. \tag{20.29}$$

These relations are simple to verify by direct computation.

20.2.3 Operation and Products

Here we look at how to combine the objects that were introduced above. The theory of exterior or Grassmann products ($\wedge$) in Clifford algebras is well established, see [3], for example. However, we will also need the dual of this operation. In a nondegenerate Clifford algebra this could be done simply using the unit pseudoscalar, that is the element $e_1 e_2 \cdots e_n$ with maximum grade. The algebra used here is degenerate since some of the generators square to zero, so this operation has to be introduced explicitly.

The construction is borrowed directly from the theory of Grassmann–Cayley algebras, see [13]. The operation we need is called the shuffle product

$\vee$ and is defined as follows. Let $b = b_1 \wedge b_2 \wedge \cdots \wedge b_j$ and $c = c_1 \wedge c_2 \wedge \cdots \wedge c_k$ in a general Clifford algebra, with $j + k \geq n$ the dimension of the algebra. Then,

$$b \vee c = \sum_\sigma \text{sign}(\sigma) \det(b_{\sigma(1)}, \ldots, b_{\sigma(n-k)}, c_1, \ldots, c_k) b_{\sigma(n-k+1)} \wedge \cdots \wedge b_{\sigma(j)}. \tag{20.30}$$

The sum is taken over all permutations σ of $1, 2, \ldots, j$ such that $\sigma(1) < \sigma(2) < \cdots < \sigma(n-k)$, and $\sigma(n-k+1) < \sigma(n-k+2) < \cdots < \sigma(j)$.

Each b_i can be written as a sum of basis elements

$$b_i = b_{i1} e_1 + b_{i2} e_2 + \cdots + b_{in} e_n. \tag{20.31}$$

So the determinant in the above definition is the determinant of the matrix whose columns are the coefficients $a_{\sigma(1)i}, a_{\sigma(2)i}, \ldots, b_{\sigma(k)i}$. The shuffle product is then extended to the entire Clifford algebra by demanding that it distributes over addition. It is also possible to show that it is associative.

As an example, consider the Clifford algebra $C\ell(0, 6, 2)$. Before doing any calculations an order for the generators must be fixed. So assume that the standard order of the generators is given by

$$a_1, a_2, a_3, a, e_1, e_2, e_3, e. \tag{20.32}$$

Now consider $A \vee E = (a_1 \wedge a_2 \wedge a_3 \wedge a) \vee (e_1 \wedge e_2 \wedge e_3 \wedge e)$, for orthogonal generators we can confuse the Clifford and exterior products. From the definition above we have that

$$A \vee E = \det(a_1, a_2, a_3, a, e_1, e_2, e_3, e) = 1. \tag{20.33}$$

As another example, suppose we need to compute $(a_1 a_2 a_3 a) \vee (a_3 a_1 e_1 e_2 e_3 e)$:

$$(a_1 a_2 a_3 a) \vee (a_3 a_1 e_1 e_2 e_3 e) = \det(a_1, a_2, a_3, a_1, e_1, e_2, e_3, e) a_3 a + \cdots$$
$$- \det(a_2, a, a_3, a_1, e_1, e_2, e_3, e) a_1 a_3 + \cdots$$
$$+ \det(a_3, a, a_3, a_1, e_1, e_2, e_3, e) a_1 a_2 \tag{20.34}$$

Clearly only the middle term on the right-hand side is nonzero so we have the result

$$(a_1 a_2 a_3 a) \vee (a_3 a_1 e_1 e_2 e_3 e) = a_3 a_1. \tag{20.35}$$

Now it is possible to write the evaluation map of a coscrew on a screw. Recall that the coscrews are dual to the screws and hence can be thought of as linear functionals on the screws. The evaluation map can be written in several possible forms,

$$\mathcal{P}(\mathbf{s}) = \mathcal{P} \vee (Q_0 \wedge \mathbf{s}) = (\mathcal{P} \wedge Q_0) \vee \mathbf{s}, \tag{20.36}$$

using the invariant element Q_0 described above. Notice that this is very similar to traditional screw theory. There, velocities and momenta would both be represented by screws and the evaluation map would be given by the reciprocal product.

Using Eq. (20.29), the evaluation map can also be written in the following forms:
$$\mathcal{P}(\mathbf{s}) = \mathcal{P} \vee \bar{\mathbf{s}} E = A \bar{\mathcal{P}} \vee \mathbf{s}. \tag{20.37}$$

The elements $\mathcal{P}$ and $\bar{\mathbf{s}}$ have grade 2 and contain only as; likewise the elements $\bar{\mathcal{P}}$ and $\mathbf{s}$ have grade 2 and contain only es. From these facts and the definition of the shuffle product, given by (20.30), it is easy to see that we can rearrange the previous equations to get yet another couple of versions of the evaluation map,
$$\mathcal{P}(\mathbf{s}) = \bar{\mathbf{s}} \vee \mathcal{P} E = A \mathbf{s} \vee \bar{\mathcal{P}}. \tag{20.38}$$

Next, the inertia operator operates on velocities and turns them into momenta. This operation can be written using the A invariant defined above,
$$\mathcal{P} = A \vee (N \wedge \mathbf{s}). \tag{20.39}$$

Finally, the kinetic energy of a rigid body is given by evaluating its momentum on its velocity, and dividing by 2. In the Clifford algebra this can be written
$$E_k = \frac{1}{2}\mathcal{P} \vee (Q_0 \wedge \mathbf{s}) = \frac{1}{2} A \vee (Q_0 \wedge \mathbf{s}) \vee (N \wedge \mathbf{s}). \tag{20.40}$$

From Eq. (20.29) we have
$$A \vee (Q_0 \wedge \mathbf{s}) = A \vee \bar{\mathbf{s}} E = \bar{\mathbf{s}}. \tag{20.41}$$

So the kinetic energy can be written neatly as
$$E_k = \frac{1}{2}\bar{\mathbf{s}} \vee (N \wedge \mathbf{s}). \tag{20.42}$$

Notice that, because N is symmetric, this combination is symmetric. That is, for a pair of screws $\mathbf{s}_1, \mathbf{s}_2$ we have that
$$\bar{\mathbf{s}}_1 \vee (N \wedge \mathbf{s}_2) = \bar{\mathbf{s}}_2 \vee (N \wedge \mathbf{s}_1). \tag{20.43}$$

20.2.4 Equations of Motion for a Single Body

The equations of motion can be derived very simply now. According to Newton's second law, the force applied to the body is equal to the rate of change of momentum, which in our notation becomes
$$\frac{d}{dt}\mathcal{P} = \mathcal{W}, \tag{20.44}$$

where $\mathcal{W}$ is the applied wrench. The momentum is given by $\mathcal{P} = A \vee (N \wedge \mathbf{s})$, as we saw above. The derivatives of the factors here are simple to find, A is an invariant so its derivative is zero, and the screw $\mathbf{s}$ is the unknown here so we simply write $\frac{d}{dt}\mathbf{s} = \dot{\mathbf{s}}$. The derivative of the inertia matrix can be found by differentiating the group action

$$\frac{d}{dt}N = \frac{1}{2}(\mathbf{s}N - N\mathbf{s}) + \frac{1}{2}(\bar{\mathbf{s}}N - N\bar{\mathbf{s}}). \quad (20.45)$$

A straightforward computation reveals that $(\mathbf{s}N - N\mathbf{s}) \wedge \mathbf{s} = 0$. So when we combine these results to form the derivative of the momenta we get

$$\frac{d}{dt}\mathcal{P} = A \vee (N \wedge \dot{\mathbf{s}}) + \frac{1}{2}A \vee \big((\bar{\mathbf{s}}N - N\bar{\mathbf{s}}) \wedge \mathbf{s}\big). \quad (20.46)$$

The second term on the right-hand side here is

$$\frac{1}{2}\big(\bar{\mathbf{s}}(A \vee (N \wedge \mathbf{s})) - (A \vee (N \wedge \mathbf{s}))\bar{\mathbf{s}}\big) = \{\mathbf{s}, A \vee (N \wedge \mathbf{s})\} = \{\mathbf{s}, \mathcal{P}\}. \quad (20.47)$$

So the equation of motion can be written as

$$A \vee (N \wedge \dot{\mathbf{s}}) + \{\mathbf{s}, A \vee (N \wedge \mathbf{s})\} = \mathcal{W}. \quad (20.48)$$

We can tidy our Clifford equation of motion a little by introducing the invariant element $E = e_1 e_2 e_3 e$ (Sect. 20.2.2). Suppose that $c = x_a E$, where x_a is an element of the algebra that contains a_is but no e_is, then it is not hard to see that $A \vee c = A \vee x_a E = x_a$. Hence, $E(A \vee c) = Ex_a = c$; remember that E is a product of an even number of e_is and so commutes with any x_a. Taking the Clifford product of Eq. (20.48) with E results in

$$N \wedge \dot{\mathbf{s}} + \frac{1}{2}\big(\bar{\mathbf{s}}(N \wedge \mathbf{s}) - (N \wedge \mathbf{s})\bar{\mathbf{s}}\big) = \mathcal{W}E. \quad (20.49)$$

Finally, since $\mathbf{s}$ and $\bar{\mathbf{s}}$ commute, we may write the equation of motion as

$$N \wedge \dot{\mathbf{s}} + \frac{1}{2}(\bar{\mathbf{s}}N - N\bar{\mathbf{s}}) \wedge \mathbf{s} = \mathcal{W}E. \quad (20.50)$$

Notice that this is a relation among elements of the algebra with degree six. However, all the terms will have the general form $a_i a_j E$.

20.3 Serial Robots

A serial robot consists of several rigid links joined in series by one-degree-of-freedom joints, (Fig. 20.1). For industrial applications it is most common to use six links, this is so that the robot's end effector, or gripper, will have six degrees of freedom. However, several commercial robots have only four or five joints, for example, most spray-painting robots have only five joints because the final rotation about the axis of the spray nozzle is not important. Usually the joints are rotary joints, called revolute joints by mechanical engineers. These can be driven easily by electric motors. More rarely, prismatic or sliding joints are used.

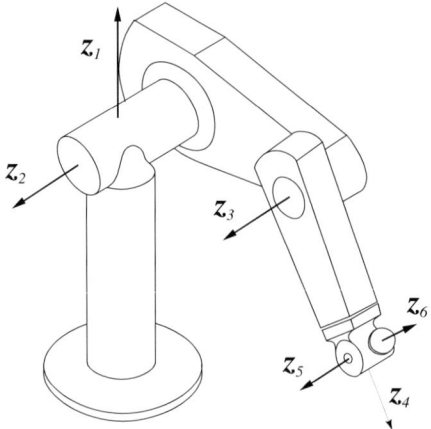

Fig. 20.1. A six-joint industrial robot arm

20.3.1 Equations of Motion

The simplest way to derive the equations of motion is to copy the methods of elementary mechanics. That is to say, we can write an equation of motion for each rigid link and then manipulate these equations to remove the unknown reaction forces and torques at the joints. The equations of motion for each link are simply

$$N_i \wedge \dot{\mathbf{s}}_i + \frac{1}{2}\bigl(\bar{\mathbf{s}}_i(N_i \wedge \mathbf{s}_i) - (N_i \wedge \mathbf{s}_i)\bar{\mathbf{s}}_i\bigr) = \mathcal{W}_i E, \qquad i = 1, 2, \ldots, 6. \quad (20.51)$$

The subscript i here refers to the ith link. It will be assumed that the robot has six links.

The wrench acting on each link can be written as a sum of five terms

$$\mathcal{W}_i = \mathcal{T}_i - \mathcal{T}_{i+1} + \mathcal{R}_i - \mathcal{R}_{i+1} + \mathcal{G}_i, \qquad (20.52)$$

here $\mathcal{T}_i$ is the torque due to the ith motor, $\mathcal{R}_i$ is the reaction wrench at the ith joint and $\mathcal{G}_i$ is the wrench due to gravity (Fig. 20.2). At the last link, however, there are only three wrenches:

$$\mathcal{W}_6 = \mathcal{T}_6 + \mathcal{R}_6 + \mathcal{G}_6. \qquad (20.53)$$

So if we add the equations of motion cumulatively from the ith to the last, we get

$$\sum_{j=i}^{6} \left(N_j \wedge \dot{\mathbf{s}}_j + \frac{1}{2}\bigl(\bar{\mathbf{s}}_j(N_j \wedge \mathbf{s}_j) - (N_j \wedge \mathbf{s}_j)\bar{\mathbf{s}}_j\bigr) - \mathcal{G}_j E \right) = \mathcal{T}_i E + \mathcal{R}_i E, \quad (20.54)$$

where $i = 1, 2, \ldots, 6$. To get rid of the reaction wrenches we pair the equations with the joint screws; that is, we can use the evaluation map. Recall from

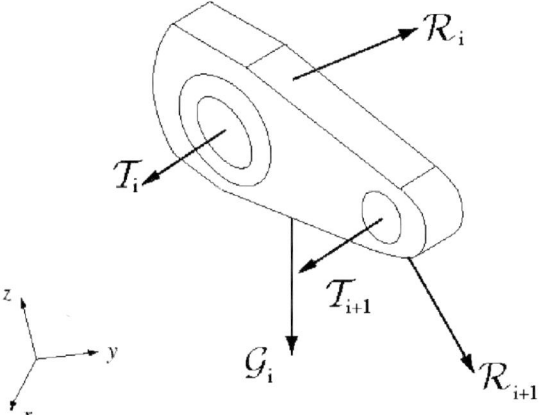

Fig. 20.2. Force/torque diagram for a single link

Sect. 20.2.3 that evaluating a momentum coscrew on a velocity screw gives the kinetic energy, up to a constant factor, see Eq. (20.42). Here we evaluate a wrench on a velocity screw, and this gives the power or rate of work. By the principle of virtual work the reaction wrenches can do no work on the joint screw, so $\mathcal{R}_i(\mathbf{z}_i) = 0$, where $\mathbf{z}_i$ is the ith joint screw. On the other hand, pairing the wrench due to the motor with this screw gives the amplitude of the torque provided by the motor, $\mathcal{T}_i(\mathbf{z}_i) = \tau_i$. The amplitude τ_i here is a scalar. We can use the expression $\mathcal{P}(\mathbf{s}) = \bar{\mathbf{s}} \vee PE$, from Eq. (20.38) to substitute for the evaluation map. So the equations of motion for our robot become

$$\sum_{j=i}^{6} \left(\bar{\mathbf{z}}_i \vee (N_j \wedge \dot{\mathbf{s}}_j) + \frac{1}{2} \bar{\mathbf{z}}_i \vee \left(\bar{\mathbf{s}}_j (N_j \wedge \mathbf{s}_j) - (N_j \wedge \mathbf{s}_j) \bar{\mathbf{s}}_j \right) - \bar{\mathbf{z}}_i \vee \mathcal{G}_j E \right) = \tau_i, \quad (20.55)$$

where again $i = 1, 2, \ldots, 6$.

There are a couple of ways in which this can be tidied. First, the middle terms on the left can be simplified to

$$\frac{1}{2} \bar{\mathbf{z}}_i \vee \left(\bar{\mathbf{s}}_j (N_j \wedge \mathbf{s}_j) - (N_j \wedge \mathbf{s}_j) \bar{\mathbf{s}}_j \right) = \frac{1}{2} (\bar{\mathbf{z}}_i \bar{\mathbf{s}}_j - \bar{\mathbf{s}}_j \bar{\mathbf{z}}_i) \vee (N_j \wedge \mathbf{s}_j). \quad (20.56)$$

This follows from the symmetry properties of triple products. The second simplification involves the introduction of the 'gravity screw'. Assuming that gravity acts in the minus z-direction, we set $\mathbf{g} = -g e_3 e$ with g the acceleration due to gravity (9.81 m/s^2). The wrench due to gravity on the ith link is now given by

$$\mathcal{G}_i E = N_j \wedge \mathbf{g}. \quad (20.57)$$

This is easily verified using a diagonal inertia matrix; in such a case the origin of the coordinate system would correspond to the centre of mass of the link. And in these coordinates the gravity wrench would be a pure force with no moment. The action of the group of rigid motions can then be used to show that the relation is true in general.

The equations of motion for the robot can now be written in the simple form

$$\sum_{j=i}^{6}\left(\bar{\mathbf{z}}_i \vee (N_j \wedge (\dot{\mathbf{s}}_j - \mathbf{g})) + \frac{1}{2}(\bar{\mathbf{z}}_i\bar{\mathbf{s}}_j - \bar{\mathbf{s}}_j\bar{\mathbf{z}}_i) \vee (N_j \wedge \mathbf{s}_j)\right) = \tau_i, \quad (20.58)$$

and as usual by now, $i = 1, 2, \ldots, 6$.

Although this is rather neat, it is not that useful as it stands. The equations involve the velocity screws $\mathbf{s}_i$ of the links. In most robot applications it is more useful to have the equations in terms of the joint variables. For a revolute joint this would be the angle turned by the joint. Rather than develop the above equations another approach will be illustrated in the next section.

20.3.2 Lagrangian and Hamiltonian Methods

In the interests of brevity the weight of the links will be ignored in this section. Hence the Lagrangian of the robot is simply its total kinetic energy

$$L = \frac{1}{2}\sum_{i=1}^{6} \bar{\mathbf{s}}_i \vee (N_i \wedge \mathbf{s}_i). \quad (20.59)$$

In order to develop the Euler–Lagrange equations we need to know about the kinematics of the robot. Suppose that the joint screws of the robot in the home position are given by $\mathbf{z}_1^0, \mathbf{z}_2^0, \ldots \mathbf{z}_6^0$. The position of the i-link is given by the rigid transformation g_i, which moves the link from its home position to its current one. This can be written as a product of exponentials

$$g_i = e^{\frac{1}{2}\theta_1 \mathbf{z}_1^0} e^{\frac{1}{2}\theta_2 \mathbf{z}_2^0} \cdots e^{\frac{1}{2}\theta_i \mathbf{z}_i^0}. \quad (20.60)$$

Here θ_i represents the joint variable for the ith joint. If the the joint is revolute then this will be an angle, while for prismatic joints it is a length. In either case the variable will be zero in the home configuration of the robot. To understand this formula, imagine moving the robot from its home position to the current position by moving each joint in turn, beginning with the one nearest the last link and then working towards the ground link.

The velocity screw of the ith link is given by differentiating with respect to time and then translating back to the identity $\mathbf{s}_i = \dot{g}_i g_i^{-1}$. Here the inverse of a group element is given by the Clifford conjugate; hence we have,

$$\mathbf{s}_i = \dot{g}_i g_i^* = \frac{1}{2}\dot{\theta}_1 \mathbf{z}_1^0 + \frac{1}{2}\dot{\theta}_2 g_1 \mathbf{z}_2^0 g_1^* + \cdots + \frac{1}{2}\dot{\theta}_i g_{i-1} \mathbf{z}_i^0 g_{i-1}^*, \quad i = 1, 2, \ldots 6. \quad (20.61)$$

Now the current position of the ith joint screw, relative to its home position, is given by the adjoint action:

$$\mathbf{z}_i = g_{i-1}\mathbf{z}_i^0 g_{i-1}^*, \quad i = 1, 2, \ldots 6. \tag{20.62}$$

So, the velocity screw of the the ith joint can be written as

$$\mathbf{s}_i = \frac{1}{2}(\dot{\theta}_1 \mathbf{z}_1 + \dot{\theta}_2 \mathbf{z}_2 + \cdots + \dot{\theta}_i \mathbf{z}_i), \quad i = 1, 2, \ldots 6. \tag{20.63}$$

In terms of the joint screws the Lagrangian can be written as

$$L = \frac{1}{2} A_{ij} \dot{\theta}_i \dot{\theta}_j \tag{20.64}$$

here, and in the following, the summation convention is adopted; it is assumed that repeated indices are to be summed over, unless otherwise stated. In robotics the object A_{ij} is known as the 'generalised mass matrix' of the robot. From the above we can write the elements of this matrix as

$$A_{ij} = \sum_{l=\max(i,j)}^{6} \bar{\mathbf{z}}_i \vee (N_l \wedge \mathbf{z}_j), \quad i, j = 1, 2, \ldots 6. \tag{20.65}$$

The Euler–Lagrange equations are

$$\frac{\mathrm{d}}{\mathrm{d}t}\left(\frac{\partial L}{\partial \dot{\theta}_i}\right) - \frac{\partial L}{\partial \theta_i} = \tau_i, \quad i = 1, 2, \ldots 6. \tag{20.66}$$

This is often written as

$$A_{ij}\ddot{\theta}_j + B_{ijk}\dot{\theta}_j \dot{\theta}_k = \tau_i, \quad i = 1, 2, \ldots 6. \tag{20.67}$$

The B_{ijk} terms are the Christoffel symbols of the metric determined by the kinetic energy; in robotics they are usually called the centrifugal and Coriolis terms, see [10, Chap.6], for example. It is not hard to see that

$$B_{ijk}\dot{\theta}_j \dot{\theta}_k = \frac{\partial A_{ij}}{\partial \theta_k}\dot{\theta}_j \dot{\theta}_k - \frac{1}{2}\frac{\partial A_{jk}}{\partial \theta_i}\dot{\theta}_j \dot{\theta}_k, \quad i = 1, 2, \ldots 6, \tag{20.68}$$

using the chain rule to differentiate with respect to t. This is usually written as the more symmetrical formula

$$B_{ijk}\dot{\theta}_j \dot{\theta}_k = \frac{1}{2}\left(\frac{\partial A_{ij}}{\partial \theta_k} + \frac{\partial A_{ki}}{\partial \theta_j} - \frac{\partial A_{jk}}{\partial \theta_i}\right)\dot{\theta}_j \dot{\theta}_k, \quad i = 1, 2, \ldots 6. \tag{20.69}$$

This follows because A_{ij} is symmetric, so we can exchange i and j, then since we are summing over j and k these index variables may be relabelled. Now we can set

$$B_{ijk} = \frac{1}{2}\left(\frac{\partial A_{ij}}{\partial \theta_k} + \frac{\partial A_{ki}}{\partial \theta_j} - \frac{\partial A_{jk}}{\partial \theta_i}\right), \quad i, j, k = 1, 2, \ldots 6. \tag{20.70}$$

This is not the only solution for B_{ijk} since we only know Eq. (20.69) above, but it is a reasonable choice.

To proceed we need the derivatives of the joint screws and link inertias with respect to the joint angles. Recall that in Eq. (20.62) the forward kinematics of the joint screws were given as $\mathbf{z}_i = g_{i-1}\mathbf{z}_i^0 g_{i-1}^*$. So, differentiating with respect to a joint angle above this joint in the serial chain gives nothing

$$\frac{\partial}{\partial \theta_i}\mathbf{z}_j = 0, \qquad 1 \le j \le i \le 6. \tag{20.71}$$

On the other hand, if the joint angle we are moving is below the joint in the chain, using Eq. (20.62) we have,

$$\frac{\partial}{\partial \theta_i}\mathbf{z}_j = \frac{\partial}{\partial \theta_i}(g_{j-1}\mathbf{z}_j^0 g_{j-1}^*), \qquad 1 \le i < j \le 6$$

$$= \frac{1}{2}g_{i-1}\mathbf{z}^0 g_{i-1}^* g_{j-1}\mathbf{z}_j^0 g_{j-1}^* - \frac{1}{2}g_{j-1}\mathbf{z}_j^0 g_{j-1}^* g_{i-1}\mathbf{z}_i^0 g_{i-1}^*$$

$$= \frac{1}{2}(\mathbf{z}_i\mathbf{z}_j - \mathbf{z}_j\mathbf{z}_i). \tag{20.72}$$

The derivative of $\bar{\mathbf{z}}_i$ can be found by 'barring' the above result. For the inertias we have that $N_i = g_i \bar{g}_i N_i^0 \bar{g}_i^* g_i^*$, where N_i^0 is the inertia of the ith link in the home position of the robot. So, by a similar argument to the one for the joint screws we have that

$$\frac{\partial}{\partial \theta_i}N_j = \begin{cases} \frac{1}{2}(\mathbf{z}_i N_j - N_j \mathbf{z}_i) + \frac{1}{2}(\bar{\mathbf{z}}_i N_j - N_j \bar{\mathbf{z}}_i), & if\ 1 \le i \le j \le 6 \\ 0, & otherwise \end{cases} \tag{20.73}$$

A relation can be derived here which will help simplify the derivatives in a moment. Consider the scalar $\bar{\mathbf{s}}_1 \vee (N \wedge \mathbf{s}_2)$. There are many ways to verify that this is indeed a scalar, perhaps the simplest is to observe that it is invariant under an arbitrary rigid body transformation. See Sect. 20.2.2 for the relations specifying how $\bar{\mathbf{s}}_1$, $\mathbf{s}_2$ and N transform under rigid body motions.

Now using those transformation properties we have

$$\bar{\mathbf{s}}_1 \vee (N \wedge \mathbf{s}_2) = \bar{g}\bar{\mathbf{s}}_1\bar{g}^* \vee (g\bar{g}N\bar{g}^*g^* \wedge g\mathbf{s}_2 g^*). \tag{20.74}$$

Now suppose that $g = e^{\frac{1}{2}t\mathbf{s}_3}$ and differentiate the above with respect to t

$$0 = \frac{1}{2}(\bar{\mathbf{s}}_3\bar{\mathbf{s}}_1 - \bar{\mathbf{s}}_1\bar{\mathbf{s}}_3) \vee (N \wedge \mathbf{s}_2)$$

$$+ \frac{1}{2}\bar{\mathbf{s}}_1 \vee ((\mathbf{s}_3 N - N\mathbf{s}_3) \wedge \mathbf{s}_2 + (\bar{\mathbf{s}}_3 N - N\bar{\mathbf{s}}_3) \wedge \mathbf{s}_2)$$

$$+ \frac{1}{2}\bar{\mathbf{s}}_1 \vee (N \wedge (\mathbf{s}_3\mathbf{s}_2 - \mathbf{s}_2\mathbf{s}_3)). \tag{20.75}$$

Rearranging this and using the symmetry of the pairing as given in Eq. (20.43), we get the relation

$$\frac{1}{2}\bar{\mathbf{s}}_1 \vee \left((\mathbf{s}_3 N - N\mathbf{s}_3) \wedge \mathbf{s}_2 + (\bar{\mathbf{s}}_3 N - N\bar{\mathbf{s}}_3) \wedge \mathbf{s}_2\right) =$$

$$\frac{1}{2}(\bar{\mathbf{s}}_1\bar{\mathbf{s}}_3 - \bar{\mathbf{s}}_3\bar{\mathbf{s}}_1) \vee (N \wedge \mathbf{s}_2) + \frac{1}{2}(\bar{\mathbf{s}}_2\bar{\mathbf{s}}_3 - \bar{\mathbf{s}}_3\bar{\mathbf{s}}_2) \vee (N \wedge \mathbf{s}_1). \quad (20.76)$$

Now returning to the problem of differentiating A_{ij} (see Eq. (20.65)), since the mass matrix is symmetric, we have only three cases to consider. That is, we will assume that $i \leq j$. The simplest case to consider is when $i \leq j < k$:

$$\frac{\partial A_{ij}}{\partial \theta_k} = \frac{1}{2} \sum_{l=k}^{6} \left((\bar{\mathbf{z}}_i\bar{\mathbf{z}}_k - \bar{\mathbf{z}}_k\bar{\mathbf{z}}_i) \vee (N_l \wedge \mathbf{z}_j) + (\bar{\mathbf{z}}_j\bar{\mathbf{z}}_k - \bar{\mathbf{z}}_k\bar{\mathbf{z}}_j) \vee (N_l \wedge \mathbf{z}_i)\right). \quad (20.77)$$

When $i < k \leq j$ we get some cancellation of terms using Eq. (20.76),

$$\frac{\partial A_{ij}}{\partial \theta_k} = \frac{1}{2} \sum_{l=j}^{6} (\bar{\mathbf{z}}_i\bar{\mathbf{z}}_k - \bar{\mathbf{z}}_k\bar{\mathbf{z}}_i) \vee (N_l \wedge \mathbf{z}_j). \quad (20.78)$$

Finally, if $k \leq i \leq j$ using Eq. (20.76) all the terms cancel

$$\frac{\partial A_{ij}}{\partial \theta_k} = 0. \quad (20.79)$$

At last we can substitute these results into Eq. (20.70). Analysing the results it is clear that there are only two different formulas depending on whether $j < k$ or not. These results can be combined into the single formula:

$$B_{ijk} = \frac{1}{4} \sum_{l=\max(i,j,k)}^{6} \left((\bar{\mathbf{z}}_i\bar{\mathbf{z}}_j - \bar{\mathbf{z}}_j\bar{\mathbf{z}}_i) \vee (N_l \wedge \mathbf{z}_k)\right.$$
$$\left. \pm (\bar{\mathbf{z}}_i\bar{\mathbf{z}}_k - \bar{\mathbf{z}}_k\bar{\mathbf{z}}_i) \vee (N_l \wedge \mathbf{z}_j) + (\bar{\mathbf{z}}_i\bar{\mathbf{z}}_k - \bar{\mathbf{z}}_k\bar{\mathbf{z}}_i) \vee (N_l \wedge \mathbf{z}_j)\right) (20.80)$$

The plus sign should be taken for the second term here when $j < k$ and the minus sign otherwise. Notice that this ensures that B_{ijk} is invariant with respect to the interchange of the last two indices j, k.

Finally, in this section a little can be said about the Hamiltonian mechanics of these machines. First, we compute the momentum conjugate to the joint variables, that is

$$\pi_i = \frac{\partial L}{\partial \dot{\theta}_i}, \quad i = 1, 2, \ldots 6. \quad (20.81)$$

Differentiating Eq. (20.63), it is easy to see that

$$\frac{\partial}{\partial \dot{\theta}_i} \mathbf{s}_j = \begin{cases} \frac{1}{2}\mathbf{z}_i, & if\ i \geq j \\ 0, & if\ i < j \end{cases} \quad (20.82)$$

So the momentum is obtained by differentiating Eq. (20.59)

$$\pi_i = \frac{1}{4} \sum_{j=i}^{6} \left(\bar{\mathbf{z}}_i \vee (N_j \wedge \mathbf{s}_j) + \bar{\mathbf{s}}_j \vee (N_j \wedge \mathbf{z}_i) \right), \qquad i = 1, 2, \ldots 6. \tag{20.83}$$

Using the symmetry of the pairing in Eq. (20.43) this becomes

$$\pi_i = \frac{1}{2} \sum_{j=i}^{6} \bar{\mathbf{z}}_i \vee (N_j \wedge \mathbf{s}_j), \qquad i = 1, 2, \ldots 6. \tag{20.84}$$

Notice that this is equivalent to,

$$\pi_i = \frac{1}{2} A_{ij} \dot{\theta}_j. \tag{20.85}$$

The Hamiltonian function h is equal to the Lagrangian here as we are ignoring the weight of the links, hence

$$h = L = \frac{1}{2} A_{ij} \dot{\theta}_i \dot{\theta}_j = \dot{\theta}_i \pi_i = 2 (A^{-1})_{ij} \pi_i \pi_j, \tag{20.86}$$

where summation is assumed on repeated indices. Rather than attempt to differentiate this expression we can rearrange the Euler–Lagrange equations above to yield Hamilton's equations

$$\dot{\theta}_i = 2 (A^{-1})_{ij} \pi_j, \tag{20.87}$$

$$\dot{\pi}_i = \frac{\partial A_{jk}}{\partial \theta_i} \dot{\theta}_j \dot{\theta}_k + \tau_i. \tag{20.88}$$

Here the partial derivatives are as in the case of Lagrangian mechanics, that is, with $\dot{\theta}_i$s held constant, rather than the Hamiltonian case were π_i would be constant. So the terms $\partial A_{jk}/\partial \theta_i$ are as we have calculated in Eqns. (20.77), (20.78) and (20.79). Hence if we want to investigate whether or not θ_i is an ignorable coordinate it is these terms that must compare with zero.

20.4 Parallel Robots

Parallel robots are usually complex mechanisms with many links and joints. Hence a description of the dynamics of these devices is necessarily complicated and contains a large number of equations of motion. Rather than give a complete account here we simply look at a single important example, the Stewart platform. This device consists of a platform connected to the ground by six legs (Fig. 20.3). Each of these legs consists of a hydraulic ram, providing the actuation, and two spherical joints at either end. The spherical joints are passive, that is, they are not driven. They connect the base to the leg at the top end, and join the leg to ground at the bottom. This robot was originally designed as an aircraft simulator but more recently has found application as a machine tool.

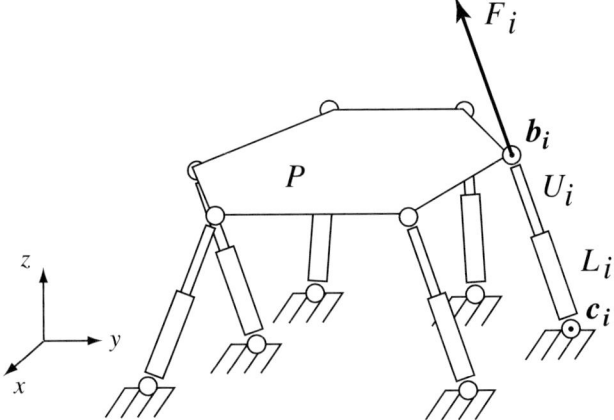

Fig. 20.3. The Stewart platform

20.4.1 Constrained Dynamics

We begin here with some generalities on the dynamics of constrained rigid bodies. Consider a single rigid body moving subject to some constraint; here we are usually thinking of some constraint on the position of the body. Constraints on the velocity of the body could presumably be handled in a similar fashion. Suppose, for example, that a point on the body is constrained to be fixed. For example, this might happen if the body is connected to a spherical joint, the centre of the joint will be the fixed point. For the moment we can take the fixed point as the origin of our coordinates. Now the possible velocities that the body can have are just rotation about the fixed point. Another way to say this is that the allowed velocities for the body are linear combinations of the three screws

$$\mathbf{s}_x = e_2 e_3, \quad \mathbf{s}_y = e_3 e_1, \quad \mathbf{s}_z = e_1 e_2. \tag{20.89}$$

These will sometimes be called 'screws of freedom'. Dual to these are 'constraint wrenches', which are wrenches that annihilate the freedom screws. In our case,

$$\mathcal{W}_x = a_2 a_3, \quad \mathcal{W}_y = a_3 a_1, \quad \mathcal{W}_z = a_1 a_2. \tag{20.90}$$

Clearly the pairing between any screw of freedom and wrench of constraint vanishes,

$$\mathcal{W}_\alpha \vee (Q_0 \wedge \mathbf{s}_\beta) = 0, \qquad \alpha, \beta = x, y, z. \tag{20.91}$$

This is usually referred to as the principle of virtual work; the constraint wrenches cannot do work on the freedom screws. This leads to the interpretation of the constraint wrenches as the forces and torques acting on the body to maintain the constraints. Hence, using Eq. (20.50), we may write the equations of motion of the body as

$$N \wedge \dot{\mathbf{s}} + \frac{1}{2}(\bar{\mathbf{s}}N - N\bar{\mathbf{s}}) \wedge \mathbf{s} = \mathcal{W}_a E + \lambda_x \mathcal{W}_x E + \lambda_y \mathcal{W}_y E + \lambda_z \mathcal{W}_z E. \quad (20.92)$$

Here $\mathcal{W}_a$ is the external wrench applied to the rigid body, and the λ_αs are the amplitudes of the constraint wrenches. This approach is clearly equivalent to the standard method of using Lagrange multipliers. The λ_αs are three new variables, and hence we need three more equations to make the system fully determinate. These extra equations are provided by the principle of virtual work

$$\mathcal{W}_x \vee (Q_0 \wedge \mathbf{s}) = 0, \quad \mathcal{W}_y \vee (Q_0 \wedge \mathbf{s}) = 0, \quad \mathcal{W}_z \vee (Q_0 \wedge \mathbf{s}) = 0. \quad (20.93)$$

More detail on this approach to the dynamics of constrained robots can be found in [8].

20.4.2 The Stewart Platform

The Stewart platform contains 13 rigid bodies and 18 joints (see Fig. 20.3), so we are only going to be able to sketch a method for finding its equations of motion here. The rigid links consist of a platform which will be labelled P, and six legs each consisting of an upper part U_i and a lower part L_i connected by a prismatic joint. The prismatic joints correspond to the hydraulic rams. We will label the force delivered to the ith leg by F_i. As mentioned above, the other two joints on each leg are passive spherical joints. The joints at the bottom of each leg, the ones that connect the legs to the ground, will be considered as constraints here. At each joint we will have constraint wrenches of the form

$$\begin{aligned}
\mathcal{W}_{ix} &= a_2 a_3 + c_{iz} a_2 a - c_{iy} a_3 a, \\
\mathcal{W}_{iy} &= a_3 a_1 - c_{iz} a_1 a + c_{ix} a_3 a, \quad (20.94) \\
\mathcal{W}_{iz} &= a_1 a_2 + c_{iy} a_1 a - c_{ix} a_2 a,
\end{aligned}$$

where $\mathbf{c}_i$ is the position vector of the ith joint centre.

The upper spherical joint on each leg is nominally a 3-degree-of-freedom joint. However, it is clear that the leg can rotate about a line joining its two spherical joints without affecting the position of the platform. Such motion is usually called a passive or internal freedom. These motions are undesirable, and in practice steps are taken to stop such motions. For instance, sometimes the spherical 'ball-and-socket' joint is replaced by a pair of revolute joints whose axes meet at right angles. This is a universal or Hooke's joint (see Fig. 20.4). Alternatively, the spherical joint can be 'keyed' to prevent rotation about the leg axis. In either case we may represent the two possible motions of the joint by a pair of joint screws $\mathbf{u}_i$ and $\mathbf{v}_i$ with corresponding joint variables θ_i and ϕ_i, respectively, (see Fig. 20.4). The joint variable for the prismatic joints will be labelled d_i, and the corresponding joint screws will be

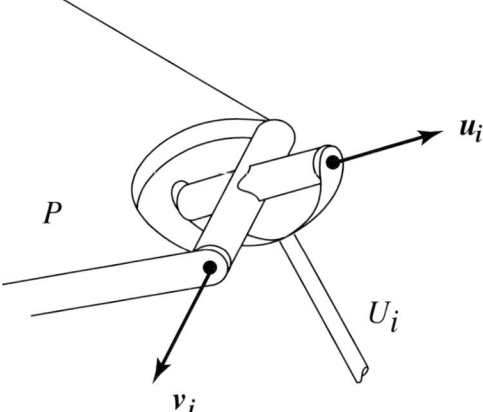

Fig. 20.4. Detail of an upper passive joint shown as a Hooke's joint

$$\mathbf{l}_i = \frac{1}{M}\big((b_{ix} - c_{ix})e_1 e + (b_{iy} - c_{iy})e_2 e + (b_{iz} - c_{iz})e_3 e\big), \quad (20.95)$$

where $M = \sqrt{(b_{ix} - c_{ix})^2 + (b_{iy} - c_{iy})^2 + (b_{iz} - c_{iz})^2}$. Notice that if the joint centres are given in the algebra by the grade-3 elements

$$\mathbf{b}_i = e_1 e_2 e_3 + b_{ix} e_2 e_3 e + b_{iy} e_3 e_1 e + b_{iz} e_1 e_2 e, \quad (20.96)$$

and

$$\mathbf{c}_i = e_1 e_2 e_3 + c_{ix} e_2 e_3 e + c_{iy} e_3 e_1 e + c_{iz} e_1 e_2 e, \quad (20.97)$$

as in [6], then we can write

$$\mathbf{l}_i = \frac{1}{2M}(\mathbf{b}_i \mathbf{c}_i - \mathbf{c}_i \mathbf{b}_i). \quad (20.98)$$

This can be verified by direct computation.

For each rigid body we can write an equation of motion. For the lower leg links we can eliminate the reaction wrenches at the prismatic joint by pairing with the joint screw to get

$$\bar{\mathbf{l}}_i \vee (N_{Li} \wedge \dot{\mathbf{s}}_{Li}) + \frac{1}{2}(\bar{\mathbf{l}}_i \bar{\mathbf{s}}_{Li} - \bar{\mathbf{s}}_{Li} \bar{\mathbf{l}}_i) \vee (N_{Li} \wedge \mathbf{s}_{Li}) =$$
$$F_i + \bar{\mathbf{l}}_i \vee (\lambda_{ix} \mathcal{W}_{ix} + \lambda_{iy} \mathcal{W}_{iy} + \lambda_{iz} \mathcal{W}_{iz}) E. \quad (20.99)$$

The velocity screw of a lower leg can be written

$$\mathbf{s}_{Li} = \mathbf{s}_P + \dot{\theta}_i \mathbf{u}_i + \dot{\phi}_i \mathbf{v}_i + \dot{d}_i \mathbf{l}_i, \quad (20.100)$$

with $\mathbf{s}_P$ the velocity of the platform.

For each upper leg we get a pair of equations by pairing with the two joint screws. First, however, we add the equations for the lower leg to get rid of the reactions at the prismatic joint. The two equations are

$$\bar{\mathbf{u}}_i \vee (N_{Ui} \wedge \dot{\mathbf{s}}_{Ui}) + \frac{1}{2}(\bar{\mathbf{u}}_i \bar{\mathbf{s}}_{Ui} - \bar{\mathbf{s}}_{Ui}\bar{\mathbf{u}}_i) \vee (N_{Ui} \wedge \mathbf{s}_{Ui})$$

$$+ \bar{\mathbf{u}}_i \vee (N_{Li} \wedge \dot{\mathbf{s}}_{Li}) + \frac{1}{2}(\bar{\mathbf{u}}_i \bar{\mathbf{s}}_{Li} - \bar{\mathbf{s}}_{Li}\bar{\mathbf{u}}_i) \vee (N_{Li} \wedge \mathbf{s}_{Li}) =$$

$$\bar{\mathbf{u}}_i \vee (\lambda_{ix}\mathcal{W}_{ix} + \lambda_{iy}\mathcal{W}_{iy} + \lambda_{iz}\mathcal{W}_{iz})E, \quad (20.101)$$

and

$$\bar{\mathbf{v}}_i \vee (N_{Ui} \wedge \dot{\mathbf{s}}_{Ui}) + \frac{1}{2}(\bar{\mathbf{v}}_i \bar{\mathbf{s}}_{Ui} - \bar{\mathbf{s}}_{Ui}\bar{\mathbf{v}}_i) \vee (N_{Ui} \wedge \mathbf{s}_{Ui})$$

$$+ \bar{\mathbf{v}}_i \vee (N_{Li} \wedge \dot{\mathbf{s}}_{Li}) + \frac{1}{2}(\bar{\mathbf{v}}_i \bar{\mathbf{s}}_{Li} - \bar{\mathbf{s}}_{Li}\bar{\mathbf{v}}_i) \vee (N_{Li} \wedge \mathbf{s}_{Li}) =$$

$$\bar{\mathbf{v}}_i \vee (\lambda_{ix}\mathcal{W}_{ix} + \lambda_{iy}\mathcal{W}_{iy} + \lambda_{iz}\mathcal{W}_{iz})E, \quad (20.102)$$

where the velocity of an upper leg link is

$$\mathbf{s}_{Ui} = \mathbf{s}_P + \dot{\theta}_i \mathbf{u}_i + \dot{\phi}_i \mathbf{v}_i. \quad (20.103)$$

So far we have $3 \times 6 = 18$ scalar equations. The final equation is for the platform itself. Apart from the reactions at the upper joints there are no forces of torques acting on the platform, so we do not need to pair the equation with any screw. However, we do need to add all the equations for all the links to remove all the reaction wrenches. The result is a single wrench equation

$$N_P \wedge \dot{\mathbf{s}}_P + \frac{1}{2}(\bar{\mathbf{s}}_P N_P - N_P \bar{\mathbf{s}}_P) \wedge \mathbf{s}_P$$

$$+ \sum_{i=1}^{6} N_{Ui} \wedge \dot{\mathbf{s}}_{Ui} + \frac{1}{2}(\bar{\mathbf{s}}_{Ui}N_{Ui} - N_{Ui}\bar{\mathbf{s}}_{Ui}) \wedge \mathbf{s}_{Ui}$$

$$+ l \sum_{i=1}^{6} N_{Li} \wedge \dot{\mathbf{s}}_{Li} + \frac{1}{2}(\bar{\mathbf{s}}_{Li}N_{Li} - N_{Li}\bar{\mathbf{s}}_{Li}) \wedge \mathbf{s}_{Li} =$$

$$\sum_{i=1}^{6}(\lambda_{ix}\mathcal{W}_{ix} + \lambda_{iy}\mathcal{W}_{iy} + \lambda_{iz}\mathcal{W}_{iz})E. \quad (20.104)$$

So we now have effectively $18 + 6 = 24$ equations. The unknowns are the 6 components of $\mathbf{s}_p$, the 18 joint variables θ_i, ϕ_i and d_i, and the 18 Lagrange multipliers, λ_{ix}, λ_{iy}, λ_{iz}. The remaining equations are given by the principle of virtual work,

$$\mathcal{W}_{ix} \vee (Q_0 \wedge \mathbf{s}_{Li}) = 0, \ \mathcal{W}_{iy} \vee (Q_0 \wedge \mathbf{s}_{Li}) = 0, \ \mathcal{W}_{iz} \vee (Q_0 \wedge \mathbf{s}_{Li}) = 0, \quad (20.105)$$

where $i = 1, 2, \ldots, 6$.

Finally, here we note that by choosing suitable combinations of the equations it is possible to eliminate the Lagrange multipliers. Further, the kinematics of the machine can be used to write the joint variables and their derivatives in terms of the velocity screw of the platform and its derivative. So it is possible to find equations relating the forces at the prismatic joints to the platform's velocity and acceleration. See [9] for further details.

20.5 Conclusions

In an earlier paper [7], the Clifford algebra $C\ell(0, 6, 2)$ we introduced as a vehicle for rigid body dynamics computations with the hope that it would be possible to extend its applicability to robots. This aim has been fully realised here.

There are several different formalisms for robot dynamics. Of course they are all equivalent to Newton's laws, but they have been investigated mainly for their computational efficiency. This is because a major use for the equations of motion is in the control of robot manipulators. The Clifford algebra formalism is not presented here for its computational efficiency, rather for its symbolic power. Notice that the inertia of a rigid body, velocities, wrenches, the positions of points and many other geometric and dynamical entities are all represented in the same algebra by elements of different grades. This means that there is a uniformity to the way that these objects are treated.

So although, in terms of robot dynamics, the results presented here are not particularly new the methods used are very different from traditional approaches to the subject. There have been other attempts to formulate rigid body dynamics using Clifford algebra, notably Hestenes. In [2], Hestenes uses $C\ell(4, 1)$. This algebra contains a pair of distinguished grade one elements which square to 0. This allows him to define screws and wrenches but it is not clear, at least not to this author, how the mass and inertia terms are handled. In this work the inertia of the rigid body is represented by an honest element of the algebra. Essentially a representation of the group $SE(3)$ has been found in the algebra $C\ell(0, 6, 2)$. This representation is isomorphic to the symmetric square of the adjoint representation. The algebra also contains copies of the adjoint and coadjoint representations as well as many others. It has been suggested that $C\ell(7, 1)$ and/or $C\ell(4, 4)$ could also be used to represent rigid body dynamics. This seems unlikely since these algebras will suffer from the same problems as $C\ell(4, 1)$; the two distinguished grade-one elements which square to 0 neither commute nor anticommute in these algebras. This makes it difficult to see how we can combine a screw and a wrench to produce an invariant scalar.

The Lagrangian and Hamiltonian mechanics of parallel mechanisms has not been investigated here. There would not appear to be any fundamental difficulties in applying these theories using the Clifford algebra. The simple methods used here seem to be adequate for the examples presented. There may, however, be more complex mechanisms where there is advantage in using more advanced methods.

It is perhaps a little surprising that although the Clifford, exterior and shuffle products have all been used, the contraction does not seem to appear [3]. The contraction seems to be related quite closely to the distance and angle in the underlying Euclidian space. Perhaps some of the formulas derived here can be simplified using the contraction.

References

1. Hausdorff F. (1906) Die Symbolische exponential formel in den grupen theorie. *Berichte de Sächichen Akademie de Wissenschaften (Math Phys Klasse)* **58**:19–48
2. Hestenes D. (1999) *New Foundations for Classical Mechanics, 2nd edn.* Reidel, Dordrecht
3. Lounesto P. (2001) *Clifford Algebras and Spinors, 2nd edn.* Cambridge University Press, Cambridge
4. Porteous I.R. (1981) *Topological Geometry, 2nd edn.* Cambridge University Press, Cambridge
5. Selig J.M. (1996) *Geometrical Methods in Robotics.* Springer, Berlin Heidelberg New York
6. Selig J.M. (2000) Clifford algebra of points, lines and planes. *Robotica* **18**:545–556
7. Selig J.M., Bayro-Corrochano E. Rigid Body Dynamics using Clifford Algebra to appear *Phil. Trans. Royal Soc.*
8. Selig J.M., McAree P.R. (1999) Constrained robot dynamics. I. Serial robots with end-effector constraints, *Journal of Robotic Systems* **16**(9):471–486
9. Selig J.M., McAree P.R. (1999) Constrained robot dynamics. II. Parallel machines, *Journal of Robotic Systems* **16**(9):487–498
10. Spong M.W., Vidyasagar M. (1989) *Robot Dynamics and Control.* Wiley, New York
11. Sobczyk G. (2001) Universal Geometric Algebra. In: Bayro-Corrochano E., Sobczyk G. (eds.) *Geometric Algebra with Applications in Science and Engineering*, Birkhäuser, Boston, pp. 18–41
12. Study E. (1891) von den Bewegungen und Umlegungen. *Math. Ann.* **39**:441–566
13. White N. (1994) Grassmann–Cayley algebra and robotics. *J. Intell. Robot Syst.* **11**:97–107

21

Geometric Methods for Multirobot Optimal Motion Planning

Calin Belta[1] and Vijay Kumar[2]

[1] Mechanical Engineering and Mechanics, Drexel University, Philadelphia, PA
 calin@drexel.edu
[2] GRASP Laboratory, University of Pennsylvania, Philadelphia, PA,
 kumar@grasp.cis.upenn.edu

21.1 Introduction

As a result of technological advances in control techniques for single vehicles and the explosion in computation and communication capabilities, the interest in cooperative robotics has dramatically increased in the last few years. The research in the field of control and coordination for multiple robots is currently progressing in areas like automated highway systems [34], formation flight control [2], unmanned underwater vehicles [29], satellite clustering [22], exploration [7], surveillance [15], search and rescue, mapping of unknown or partially known environments, distributed manipulation [21], and transportation of large objects [31].

There are roughly three approaches to multivehicle coordination reported in literature: leader following, behavioral methods, and virtual structure techniques. In leader following, some robots are designated as leaders, while others are followers [10]. In behavior-based control [1] several desired behaviors are prescribed for each agent, the final control being derived by weighting the relative importance of each behavior. In the virtual structure approach, the entire formation is treated as a rigid body [13, 19, 24]. Desired motion is assigned to the virtual structure that traces out trajectories for each member of the formation to follow.

Virtual structures, as rigid bodies, evolve on the Lie group of all translations and orientations in 3D, $SE(3)$. The problem of finding a smooth interpolating curve is well understood in Euclidean spaces [14], but it is not clear how these techniques can be generalized to curved spaces. There are two main issues that need to be addressed. First, it is desired that the computational scheme be independent of the description of the space and invariant with respect to the choice of the coordinate systems used to describe the motion. Second, the smoothness properties and the optimality of the trajectories need to be considered. Shoemake [28] proposed a scheme for interpolating rotations with Bezier curves based on the spherical analog of the de Casteljau algorithm. This idea was extended by Park and Ravani [25] to spatial motions. Another class of methods is based on the representation of Bezier curves with Bernstein polynomials. Ge and Ravani [16] used the dual unit quaternion representation of $SE(3)$ and subsequently applied Euclidean methods to interpolate in this space. Srinivasan [30] and Jütler [18] propose the use of spatial rational B-splines for interpolation. Marthinsen [20] suggests the use of Hermite interpolation and the use of truncated inverse of the differential of the exponential mapping and the truncated Baker–Campbell–Hausdorff formula to simplify the construction of interpolation polynomials. The advantage of these methods is that they produce rational curves. It is worth noting that all these works (with the exception of [25]) use a particular parameterization of the group and do not discuss the invariance of their methods. In contrast, Noakes et al. [23] derived the necessary conditions for cubic splines on general manifolds without using a coordinate chart. These results are extended in [9] to the dynamic interpolation problem. Necessary conditions for higher-order splines are derived in Camarinha et al. [8]. A coordinate-free formulation of the variational approach was used to generate shortest paths and minimum acceleration and jerk trajectories on $SO(3)$ and $SE(3)$ in [36]. However, analytical solutions are available only in the simplest of cases, and the procedure for solving optimal motions, in general, is computationally intensive. If optimality is sacrificed, it is possible to generate bi-invariant trajectories for interpolation and approximation using the exponential map on the Lie algebra [35]. While the solutions are of closed form, the resulting trajectories have no optimality properties.

Most of the existing works on motion planning and control of virtual structures use formation graphs, whose nodes capture the individual agent kinematics or dynamics, and edges represent interagent constraints that must be satisfied [10, 32, 33]. The notions of graph rigidity, minimally rigid graphs, and node augmentation are studied and applied to formations by Olfati-Saber and Murray [24] and Eren and Morse [13]. Stabilization of a formation at a given rigid configuration is formulated in terms of a structural potential function [24] or a formation function [12]. An alternative to constructing structural potential functions induced by formation graphs and a relaxation to the rigidity constraint is to use biologically inspired *artificial potential functions*, as Leonard and Fiorelli suggest in [19]. Along different lines, a geometric formulation of feasibility on formation graphs is given by Tabuada et al. [32].

We first describe a method to generate smooth trajectories for a rigid body with specified boundary conditions. The method involves two key steps: (1) the generation of optimal trajectories in $GA^+(n)$, a subgroup of the affine group in $\mathbb{R}^n$; (2) the projection of the trajectories onto $SE(3)$, the Lie group of rigid body displacements. The overall procedure is invariant with respect to both the local coordinates on the manifold and the choice of the inertial frame. The benefits of the method are threefold. First, it is possible to apply any of the variety of well-known, efficient techniques to generate optimal curves on $GA^+(n)$. Second, the method yields approximations to optimal solutions for general choices of Riemannian metrics on $SE(3)$. Third, from a computational point of view, the method we propose is less expensive than traditional methods.

These results are then extended to generate motion plans for fully actuated robots required to maintain a rigid structure. The fundamental idea is based on the definition of a kinetic energy metric in the configuration space of the team. We decompose the kinetic energy into two terms: the first corresponds to the motion of a rigid structure, and the second to motions that violate the rigidity constraint. The first set of motions can be associated to orbits of the Euclidean group, $SE(3)$ or $SE(2)$. The second corresponds to velocity vectors that are orthogonal to the first. The kinetic energy metric is "shaped" by assigning different weights to each contribution. This idea of a "decomposition" and a subsequent "modification" is related to the methodology of controlled Lagrangians described in [6]. The geodesic flow for this modified metric is derived, and trajectories of the individual robots are generated. When the weights are biased toward the rigid body motion, the obtained trajectories correspond to optimal rigid body motions in 3D space ($SE(3)$) or in the plane ($SE(2)$). Other choices of weights lead to the special cases of the robots moving toward each other or each individual robot traversing its own optimal path.

The remainder of this chapter is organized as follows. Section 21.2 is a short overview of the differential geometry tools that are used in this work. Section 21.3 describes a computationally efficient, left-invariant method for generating smooth trajectories for a moving rigid body with specified boundary conditions. Smooth trajectories for a set of mobile robots satisfying constraints on relative positions are generated in Sect. 21.4. The paper concludes with final remarks and directions of future work in Sect. 21.5.

21.2 The Geometry of Rigid Body Motion

This section is a short review of the mathematical tools that are used in this chapter. The reader interested in a more detailed description is referred to [11].

21.2.1 Matrix Lie Groups and Rigid Motion

Let $GL^+(n)$ denote the set of all $n \times n$ real matrices with positive determinant:

$$GL^+(n) = \{M \mid M \in \mathbb{R}^{n \times n}, \det M > 0\}. \tag{21.1}$$

$SO(n)$ is a subset of GL^+, defined as

$$SO(n) = \{R \mid R \in GL^+(n), RR^{\mathrm{T}} = I\}. \tag{21.2}$$

Let
$$GA^+(n) = \left\{ B \mid B = \begin{bmatrix} M & d \\ 0 & 1 \end{bmatrix}, M \in GL^+(n), d \in \mathbb{R}^n \right\}, \quad (21.3)$$

and
$$SE(n) = \left\{ A \mid A = \begin{bmatrix} R & d \\ 0 & 1 \end{bmatrix}, R \in SO(n), d \in \mathbb{R}^n \right\}. \quad (21.4)$$

$GL^+(n)$, $SO(n)$, $GA^+(n)$, and $SE(n)$ have the structure of a group under matrix multiplication. Moreover, matrix multiplication and inversion are both smooth operations, which make all $GL^+(n)$, $SO(n)$, $GA^+(n)$, and $SE(n)$ Lie groups [11].

$GL^+(n)$ and $GA^+(n)$ are subgroups of the general linear group $GL(n)$ (the set of all nonsingular $n \times n$ matrices) and of the affine group $GA(n) = GL(n) \times \mathbb{R}^n$, respectively. $SO(n)$ is referred to as the special orthogonal group or the rotation group on $\mathbb{R}^n$. The special Euclidean group $SE(n)$ is the set of all rigid displacements in $\mathbb{R}^n$.

Special consideration will be given to $SO(3)$ and $SE(3)$. Consider a rigid body moving in free space. Assume any inertial reference frame $\{F\}$ fixed in space and a frame $\{M\}$ fixed to the body at point O' (Fig. 21.1). At each instance the configuration (position and orientation) of the rigid body can be described by a homogeneous transformation matrix, $A \in SE(3)$, corresponding to the displacement from frame $\{F\}$ to frame $\{M\}$.

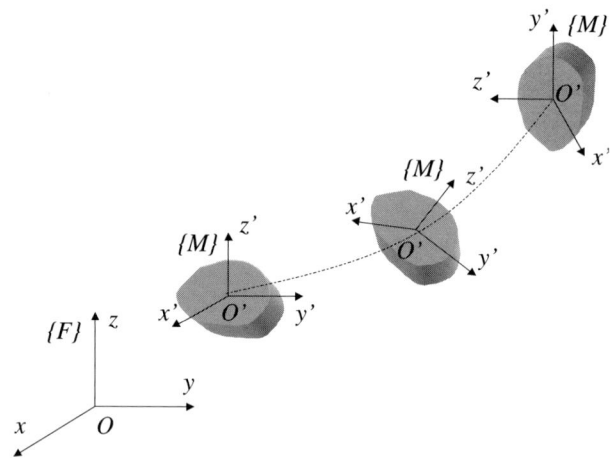

Fig. 21.1. The inertial (fixed) frame and the moving frame attached to the rigid body

On any Lie group the tangent space at the group identity has the structure of a Lie algebra. The Lie algebras of $SO(3)$ and $SE(3)$, denoted by $so(3)$ and $se(3)$, respectively, are given by:

$$so(3) = \{\hat{\omega} \mid \hat{\omega} \in \mathbb{R}^{3\times 3},\ \hat{\omega}^{\mathrm{T}} = -\hat{\omega}\}, \tag{21.5}$$

$$se(3) = \left\{ S = \begin{bmatrix} \hat{\omega} & v \\ 0 & 0 \end{bmatrix} \,\middle|\, \hat{\omega} \in so(3),\ v \in \mathbb{R}^{3} \right\}, \tag{21.6}$$

where $\hat{\ }$ is the skew-symmetric operator.

Given a curve

$$A(t): [-a, a] \to SE(3),\quad A(t) = \begin{bmatrix} R(t) & d(t) \\ 0 & 1 \end{bmatrix},$$

an element $S(t)$ of the Lie algebra $se(3)$ can be identified with the tangent vector $\dot{A}(t)$ at an arbitrary point t by:

$$S(t) = A^{-1}(t)\dot{A}(t) = \begin{bmatrix} \hat{\omega}(t) & R^{\mathrm{T}}d \\ 0 & 0 \end{bmatrix}, \tag{21.7}$$

where $\hat{\omega}(t) = R(t)^{\mathrm{T}}\dot{R}(t)$ is the corresponding element from $so(3)$.

A curve on $SE(3)$ physically represents a motion of the rigid body. If $\{\omega(t), v(t)\}$ is the vector pair corresponding to $S(t)$, then ω physically corresponds to the angular velocity of the rigid body, while v is the linear velocity of the origin O' of the frame $\{M\}$, both expressed in the frame $\{M\}$. In kinematics, elements of this form are called twists, and $se(3)$ thus corresponds to the space of twists. The twist $S(t)$ computed from Eq. (21.7) does not depend on the choice of the inertial frame $\{F\}$ and is therefore called left invariant.

The standard basis for the vector space $so(3)$ is:

$$L_1^o = \hat{e}_1,\ L_2^o = \hat{e}_2,\ L_3^o = \hat{e}_3, \tag{21.8}$$

where

$$e_1 = \begin{bmatrix} 1 & 0 & 0 \end{bmatrix}^{\mathrm{T}},\ e_2 = \begin{bmatrix} 0 & 1 & 0 \end{bmatrix}^{\mathrm{T}},\ e_3 = \begin{bmatrix} 0 & 0 & 1 \end{bmatrix}^{\mathrm{T}}.$$

L_1^o, L_2^o, and L_3^o represent instantaneous rotations about the Cartesian axes x, y, and z, respectively. The components of a $\hat{\omega} \in so(3)$ in this basis are given precisely by the angular velocity vector ω.

The standard basis for $se(3)$ is:

$$L_1 = \begin{bmatrix} L_1^o & 0 \\ 0 & 0 \end{bmatrix}\ L_2 = \begin{bmatrix} L_2^o & 0 \\ 0 & 0 \end{bmatrix}\ L_3 = \begin{bmatrix} L_3^o & 0 \\ 0 & 0 \end{bmatrix}$$
$$L_4 = \begin{bmatrix} 0 & e_1 \\ 0 & 0 \end{bmatrix}\ L_5 = \begin{bmatrix} 0 & e_2 \\ 0 & 0 \end{bmatrix}\ L_6 = \begin{bmatrix} 0 & e_3 \\ 0 & 0 \end{bmatrix} \tag{21.9}$$

The twists L_4, L_5, and L_6 represent instantaneous translations along the Cartesian axes x, y, and z, respectively. The components of a twist $S \in se(3)$ in this basis are given precisely by the velocity vector pair $s := \{\omega, v\} \in \mathbb{R}^6$.

21.2.2 Riemannian Metrics on Lie Groups

If a smoothly varying, positive definite, bilinear, symmetric form $<.,.>$ is defined on the tangent space at each point on the manifold, such a form is called a Riemannian metric and the manifold is Riemannian [11]. On an n-dimensional manifold, the metric is locally characterized by a $n \times n$ matrix of $\mathcal{C}^\infty$ functions $\tilde{g}_{ij} = <X_i, X_j>$, where X_i are basis vector fields. If the basis vector fields can be defined globally, then the matrix $[\tilde{g}_{ij}]$ completely defines the metric.

On $SE(3)$ (on any Lie group), an inner product on the Lie algebra can be extended to a Riemannian metric over the manifold using left (or right) translation. To see this, consider the inner product of two elements $S_1, S_2 \in se(3)$ defined by

$$<S_1, S_2>_I = s_1^T \tilde{G} s_2, \qquad (21.10)$$

where s_1 and s_2 are the 6×1 vectors of components of S_1 and S_2 with respect to some basis, and G is a positive definite matrix. If V_1 and V_2 are tangent vectors at an arbitrary group element $A \in SE(3)$, the inner product $<V_1, V_2>_A$ in the tangent space $T_A SE(3)$ can be defined by:

$$<V_1, V_2>_A = <A^{-1}V_1, A^{-1}V_2>_I. \qquad (21.11)$$

The metric satisfying the above equation is said to be *left invariant* [11]. Right invariance is defined similarly. A metric is *bi-invariant* if it is both left and right invariant.

21.2.3 Geodesics and Minimum Acceleration Curves

Any motion of a rigid body is described by a smooth curve $A(t) \in SE(3)$. The velocity is the tangent vector to the curve $V(t) = \frac{dA}{dt}(t)$.

An *affine connection* on $SE(3)$ is a map that assigns to each pair of C^∞ vector fields X and Y on $SE(3)$ another C^∞ vector field $\nabla_X Y$, which is bilinear in X and Y and, for any smooth real function f on $SE(3)$ satisfies $\nabla_{fX} Y = f \nabla_X Y$ and $\nabla_X fY = f \nabla_X Y + X(f)Y$.

The *Christoffel symbols* Γ^i_{jk} of the connection at a point $A \in SE(3)$ are defined by $\nabla_{\bar{L}_j} \bar{L}_k = \Gamma^i_{jk} \bar{L}_i$, where $\bar{L}_1, \ldots, \bar{L}_6$ is the basis in $T_A SE(3)$ and the summation is understood.

If $A(t)$ is a curve and X is a vector field, the *covariant derivative* of X along A is defined by $DX/dt = \nabla_{\dot{A}(t)} X$. Vector field X is said to be *autoparallel* along A if $DX/dt = 0$. A curve A is a *geodesic* if $\dot{A}$ is autoparallel along A. An equivalent characterization of a geodesic is the following set of equations:

$$\ddot{a}^i + \Gamma^i_{jk} \dot{a}^j \dot{a}^k = 0, \qquad (21.12)$$

where a_i, $i = 1, \ldots, 6$ is an arbitrary set of local coordinates on $SE(3)$. Geodesics are also minimum length curves. The length of a curve $A(t)$ between the points $A(a)$ and $A(b)$ is defined to be $\mathbf{L}(A) = \int_a^b <V,V>^{\frac{1}{2}} dt$, where $V = \frac{dA(t)}{dt}$. It can be shown [11] that if there exists a curve that minimizes the functional L, this curve also minimizes the so-called *energy functional*:

$$\mathbf{E}(A) = \int_a^b <V,V> dt. \tag{21.13}$$

For a manifold with a Riemannian (or pseudo-Riemannian) metric, there exists a unique symmetric connection that is compatible with the metric [11]. Given a connection, the acceleration and higher derivatives of the velocity can be defined. The acceleration $\mathcal{A}(t)$ is the covariant derivative of the velocity along the curve $\mathcal{A} = \frac{D}{dt}\left(\frac{dA}{dt}\right) = \nabla_V V$. Minimum acceleration curves are defined as curves minimizing the square of the L^2 norm of the acceleration:

$$\mathbf{A}(A) = \int_a^b <\nabla_V V, \nabla_V V> dt, \tag{21.14}$$

where $V(t) = \frac{dA(t)}{dt}$, $A(t)$ is a curve on the manifold, and ∇ is the unique symmetric connection compatible with the given metric. The initial and final point as well as the initial and final velocity for the motion are prescribed.

21.2.4 The Kinetic Energy Metric

A metric that is attractive for trajectory planning can be obtained by considering the dynamic properties of the rigid body. The kinetic energy of a rigid body is a scalar that does not depend on the choice of the inertial reference frame. It thus defines a left-invariant metric. If the body-fixed reference frame is attached at the centroid the matrix $\tilde{G}$ as in Eq. (21.10)

$$\tilde{G} = \frac{1}{2}\begin{bmatrix} G & 0 \\ 0 & mI \end{bmatrix}, \tag{21.15}$$

where m is the mass of the rigid body, and G is the inertia matrix of the body about the body frame $\{M\}$. If $\{\omega, v\} \in se(3)$ is the vector pair associated with some velocity vector V, the norm of the vector V assumes the familiar expression of the kinetic energy:

$$<V,V> = \frac{1}{2}\omega^T G\omega + \frac{1}{2}mv^T v. \tag{21.16}$$

In [36] it was proved that a geodesic $A(t)$ on $SE(3)$ equipped with metric (21.15) is described by

$$\frac{d\omega}{dt} = -G^{-1}(\omega \times (G\omega)), \tag{21.17}$$

$$\ddot{d} = 0. \tag{21.18}$$

If $G = \alpha I$, an analytical expression for the geodesic passing through

$$A(0) = (R(0), d(0)), \quad A(1) = (R(1), d(1)) \tag{21.19}$$

at $t = 0$ and $t = 1$, respectively, is given by [36]

$$A(t) = (R(t), d(t)) \in SE(3), \tag{21.20}$$

where

$$R(t) = R(0)\exp(\hat{\omega}_0 t), \tag{21.21}$$
$$\hat{\omega}_0 = \log(R(0)^\mathsf{T} R(1)), \tag{21.22}$$

and

$$d(t) = (d(1) - d(0))t + d(0) \tag{21.23}$$

In the case when $G \neq \alpha I$, there is no closed-form expression for the corresponding geodesic, and numerical methods should be employed.

If $G = \alpha I$, the differential equations to be satisfied by a minimum acceleration curve are [36]:

$$\omega^{(3)} + \omega \times \ddot{\omega} = 0 \tag{21.24}$$
$$d^{(4)} = 0, \tag{21.25}$$

As observed in [23], Eq. (21.24) can be integrated to obtain $\omega^{(2)} + \omega \times \dot{\omega} = $ constant. However, this equation cannot be further integrated analytically for arbitrary boundary conditions. In [36] it is shown that for special choice of the initial and final velocities, minimum acceleration curves are reparameterized geodesics. If $G \neq \alpha I$ in metric (21.15), the differential equations to be satisfied by the minimum acceleration curves are difficult to derive and are not suited for numerical integration.

21.3 An SVD-Based Method for Interpolation on $SE(3)$

In this section it is shown that there is a simple way of defining a left- or right-invariant metric on $SO(n)$ ($SE(n)$) by introducing an appropriate constant metric in $GL^+(n)$ ($GA^+(n)$). Defining a metric (i.e., the kinetic energy) at the Lie algebra $so(n)$ (or $se(n)$) and extending it through left (right) translations is equivalent to inheriting the appropriate metric at each point from the ambient manifold.

21.3.1 Riemannian Metrics on $SO(n)$ and $SE(n)$

3.1.1 A Metric in $GL^+(n)$

Let W be a symmetric positive definite $n \times n$ matrix. For any $M \in GL^+(n)$ and any $X, Y \in T_M GL^+(n)$, define

$$< X, Y >_{GL^+} = \text{Tr}(X^T Y W) = \text{Tr}(W X^T Y) = \text{Tr}(Y W X^T). \quad (21.26)$$

By definition, form (21.26) is the same at all points in $GL^+(n)$. It is easy to see that Eq. (21.26) is a Riemannian metric on $GL^+(n)$ when W is symmetric and positive definite. The following interesting result is proved in [4]:

Proposition 1. *The metric given by Eq. (21.26) defined on $GL^+(n)$ is left invariant when restricted to $SO(n)$. The restriction on $SO(n)$ is bi-invariant if $W = \alpha I$, $\alpha > 0$, where I is the $n \times n$ identity matrix.*

Remark 1. If right invariance on $SO(n)$ is desired (and left invariance is not needed), we can define

$$<< X, Y >>_{GL^+} = \text{Tr}(XY^T W) = \text{Tr}(Y^T W X) = \text{Tr}(W X Y^T).$$

Similarly, metric $<<, >>_{GL^+}$ will be right invariant on $SO(n)$ for W symmetric and positive definite and bi-invariant if $W = \alpha I$.

3.1.2 The Induced Metric on $SO(3)$

Let $R \in SO(3)$, $X, Y \in T_R SO(3)$, and $R_x(t), R_y(t)$ the corresponding local flows so that

$$X = \dot{R}_x(t), \ Y = \dot{R}_y(t), \ R_x(t) = R_y(t) = R.$$

The metric inherited from $GL^+(3)$ can be written as:

$$< X, Y >_{SO} = < X, Y >_{GL^+} = \text{Tr}(\dot{R}_x^T(t) \dot{R}_y(t) W) =$$
$$= \text{Tr}(\dot{R}_x^T(t) R R^T \dot{R}_y(t) W) = \text{Tr}(\hat{\omega}_x^T \hat{\omega}_y W),$$

where $\hat{\omega}_x = R_x(t)^T \dot{R}_x(t)$ and $\hat{\omega}_y = R_y(t)^T \dot{R}_y(t)$ are the corresponding twists from the Lie algebra $so(3)$. If we write the above relation using the vector form of the twists, some elementary algebra leads to:

$$< X, Y >_{SO} = \omega_x^T G \omega_y, \quad (21.27)$$

where

$$G = \text{Tr}(W) I_3 - W \quad (21.28)$$

is the matrix of the metric on $SO(3)$ as defined by Eq. (21.10). A different but equivalent way of arriving at the expression of G as in Eq. (21.28) would be defining the metric in $so(3)$ i.e., at identity of $SO(3)$) as being the one inherited from $T_I GL^+(3)$: $g_{ij} = \text{Tr}(L_i^{o T} L_j^o W)$, $i, j = 1, 2, 3$ (L_1^o, L_2^o, L_3^o is the basis in $so(3)$). Left-translating this metric throughout the manifold is equivalent to inheriting the metric at each three-dimensional tangent space of $SO(3)$ from the corresponding nine-dimensional tangent space of $GL^+(3)$.

Using Eq. (21.28), it is easy to verify that the metric W on $GL^+(3)$ and the induced metric G on $SO(3)$ share the following properties:

- G is symmetric if and only if W is symmetric.
- If W is positive definite, then G is positive definite.
- If G is positive definite, then W is positive definite if and only if the eigenvalues of G satisfy the triangle inequality.

In the particular case when $W = \alpha I$, $\alpha > 0$, from Eq. (21.28), we have $G = 2\alpha I$, which is the standard bi-invariant metric on $SO(3)$. This is consistent with the second assertion in Proposition 1. For $\alpha = 1$, metric (21.26) induces the well-known Frobenius matrix norm on $GL^+(3)$ [17].

The quadratic form $\omega^T G \omega$ associated with metric (21.27) can be interpreted as the (rotational) kinetic energy. Consequently, $2G$ can be thought of as the inertia matrix of a rigid body with respect to a certain choice of the body frame $\{M\}$. The triangle inequality restriction on the eigenvalues of G therefore simply states that the principal moments of inertia of a rigid body satisfy the triangle inequality, which, by definition, is true for any rigid body. Therefore, for an arbitrarily shaped rigid body with inertia matrix $2G$, we can formulate a (positive definite) metric (21.26) in the ambient manifold $GL^+(3)$ with matrix

$$W = \frac{1}{2}\text{Tr}(G)I_3 - G. \tag{21.29}$$

Thus Eq. (21.29) gives us a formula for constructing an ambient metric space that is compatible with the given metric structure of $SO(3)$.

3.1.3 A Metric in $GA^+(n)$

Let

$$\tilde{W} = \begin{bmatrix} W & a \\ a^T & w \end{bmatrix} \tag{21.30}$$

be a symmetric positive definite $(n+1) \times (n+1)$ matrix, where W is the matrix of metric (21.26), $a \in \mathbb{R}^n$, and $w \in \mathbb{R}$. Let X and Y be two vectors from the tangent space at an arbitrary point of $GA^+(n)$ (X and Y are $(n+1) \times (n+1)$ matrices with all entries of the last row equal to zero). A quadratic form

$$<X, Y>_{GA^+} = \text{Tr}(X^T Y \tilde{W}) \tag{21.31}$$

is symmetric and positive definite if and only if $\tilde{W}$ is symmetric and positive definite.

3.1.4 The Induced Metric in $SE(3)$

We can get a left-invariant metric on $SE(n)$ by letting $SE(n)$ inherit the metric $<.>_{GA^+}$ given by Eq. (21.31) from $GA^+(n)$.

Let A be an arbitrary element from $SE(3)$. Let X, Y be two vectors from $T_A SE(3)$, and $A_x(t), A_y(t)$ the corresponding local flows so that

$$X = \dot{A}_x(t), \ Y = \dot{A}_y(t), \ A_x(t) = A_y(t) = A.$$

Let
$$A_i(t) = \begin{bmatrix} R_i(t) & d_i(t) \\ 0 & 1 \end{bmatrix}, \ i \in \{x, y\}$$

and the corresponding twists at time t:

$$S_i = A_i^{-1}(t)\dot{A}_i(t) = \begin{bmatrix} \hat{\omega}_i & v_i \\ 0 & 0 \end{bmatrix}, \ i \in \{x, y\}.$$

The metric inherited from $GA^+(3)$ can be written as:

$$<X, Y>_{SE} = <X, Y>_{GA^+} = \text{Tr}(\dot{A}_x^T(t)\dot{A}_y(t)\tilde{W}) = \text{Tr}(S_x^T A^T A S_y \tilde{W}).$$

Now using the orthogonality of the rotational part of A and the special form of the twist matrices, a straightforward calculation leads to the result:

$$<X, Y>_{SE} = \text{Tr}(S_x^T S_y \tilde{W}) = \text{Tr}(\hat{\omega}_x^T \hat{\omega}_y W) + \text{Tr}(\hat{\omega}_x^T v_y a^T) + v_x^T \hat{\omega}_y a + v_x^T v_y w$$

If G is the matrix of the metric in $SO(3)$ induced by $GL^+(3)$, then

$$<X, Y>_{SE} = \begin{bmatrix} \omega_x^T & v_x^T \end{bmatrix} \tilde{G} \begin{bmatrix} \omega_y \\ v_y \end{bmatrix}, \ \tilde{G} = \begin{bmatrix} G & \hat{a} \\ -\hat{a} & wI_3 \end{bmatrix}, \quad (21.32)$$

and G is given by Eq. (21.28).

The metric given by Eq. (21.32) is left invariant since the matrix $\tilde{G}$ of this metric in the left invariant basis vector field is constant. Also, if $\tilde{W}$ is symmetric and positive definite, then $\tilde{G}$ given by Eq. (21.32) is symmetric and positive definite.

The quadratic form $s^T \tilde{G} s$ associated with metric (21.32) can be interpreted as being the kinetic energy of a moving (rotating and translating) rigid body, where w is twice the mass m of the rigid body. If the body fixed frame $\{M\}$ is placed at the centroid of the body, then $a = 0$. Moreover, if $\{M\}$ is aligned with the principal axes of the body, then $G = \frac{1}{2}H$, where H is the diagonal inertia matrix of the body. In the most general case, when the frame $\{M\}$ is displaced by some (R_0, d_0) from the centroid and the orientation parallel with the principal axes, we have [36]:

$$G = R_0^T H R_0 - m R_0^T \hat{d}_0 R_0, \ a = -m R_0 d_0.$$

21.3.2 Projection on $SO(n)$

We can use the norm induced by metric (21.26) to define the distance between elements in $GL^+(3)$. Using this distance, for a given $M \in GL^+(3)$, we define *the projection* of M on $SO(3)$ as being the closest $R \in SO(3)$ with respect to the metric from Eq. (21.26). The solution of the projection problem is derived for the general case of $GL^+(n)$ [4]:

Proposition 2. *Let $M \in GL^+(n)$ and U, Σ, V the singular value decomposition of MW (i.e., $MW = U\Sigma V^T$). Then the projection of M on $SO(n)$ with respect to metric (21.26) is given by $R = UV^T$.*

It is easy to see that the distance between M and R in metric (21.26) is given by $\text{Tr}(W^{-1}V\Sigma^2 V^T) + \text{Tr}(W) - 2\text{Tr}(\Sigma)$. For the particular case when $W = I_3$, the distance becomes $\sum_{i=1}^{n}(\sigma_i - 1)^2$, which is the standard way of describing how far a matrix is from being orthogonal.

The question we ask is what happens with the solution to the projection problem when the manifold $GL^+(n)$ is acted upon by the group $SO(n)$. The answer is given below and the proof in [4].

Proposition 3. *The solution to the projection problem on $SO(n)$ is left invariant under actions of elements from $SO(n)$. If $W = \alpha I_3$, the solution is bi-invariant.*

For the case $W = I$, it is worthwhile to note that other projection methods do not exhibit bi-invariance. For instance, it is customary to find the projection $R \in SO(n)$ by applying a Gram–Schmidt procedure (QR decomposition). In this case it is easy to see that the solution is left invariant, but in general it is not right invariant.

21.3.3 Projection on $SE(n)$

Similar to the previous section, if a metric of the form given in Eq. (21.31) is defined on $GA^+(n)$ with the matrix of the metric given by Eq. (21.30), we can find the corresponding projection on $SE(n)$. We consider the case $a = 0$, which corresponds to a body frame $\{M\}$ fixed at the centroid of the body.

Proposition 4. *Let $B \in GA^+(n)$ with the following block partition:*

$$B = \begin{bmatrix} B_1 & B_2 \\ 0 & 1 \end{bmatrix}, \quad B_1 \in GL^+(n), \ B_2 \in \mathbb{R}^n,$$

and U, Σ, V be the singular value decomposition of $B_1 W$. Then the projection of B on $SE(n)$ is given by

$$A = \begin{bmatrix} UV^T & B_2 \\ 0 & 1 \end{bmatrix} \in SE(n).$$

The proof is given in [4]. Similar to the $SO(n)$ case, the projection on $SE(n)$ exhibits interesting invariance properties.

Proposition 5. *The solution to the projection problem on $SE(n)$ is left invariant under actions of elements from $SE(n)$. In the special case when $W = \alpha I$, the projection is bi-invariant under rotations.*

21.3.4 The Projection Method

Based on the results we presented so far, we can outline a method to generate an interpolating curve $A(t) \in SE(3)$, $t \in [0, 1]$ while satisfying the boundary conditions:
$$A(0), A(1), \dot{A}(0), \dot{A}(1), \ldots, A^{(m)}(0), A^{(m)}(1),$$
where the superscript $(\cdot)^{(m)}$ denotes the mth derivative. The projection procedure consists of two steps:

- **Step 1:** Generating the optimal curve $B(t)$ in the ambient manifold $GA^+(3)$ that satisfies the boundary conditions, and
- **Step 2:** Projecting $B(t)$ from step 1 onto $A(t) \in SE(3)$.

Because the metric we defined on $GA^+(3)$ is the same at all points, the corresponding Christoffel symbols are all zero. Consequently, the optimal curves in the ambient manifold assume simple analytical forms. For example, geodesics are straight lines, minimum acceleration curves are cubic polynomial curves, and minimum jerk curves are fifth-order polynomial curves in $GA^+(3)$, all parameterized by time. Therefore, in step 1 the following curve is constructed in $GA^+(3)$:
$$B(t) = B_0 + B_1 t + \ldots + B_{2m-1} t^{2m-1},$$
where the coefficients B_i $i = 1, \ldots, 2m-1$ are linear functions Γ_i of the input data:
$$B_i = \Gamma_i \left(A(0), A(1), \dot{A}(0), \dot{A}(1), \ldots, A^{(m)}(0), A^{(m)}(1) \right).$$

Step 2 consists of a singular value decomposition (SVD) decomposition weighted by the matrix W as described in Proposition 4 to produce the curve $A(t)$. Using the linearity of Γ_i and Proposition 5, we can prove:

Proposition 6. *The projection method on $SE(3)$ is left invariant, i.e., the generated trajectories are independent of the choice of the inertial frame $\{F\}$.*

Because of the linearity on the boundary conditions of the curve in the ambient manifold, the first step is always bi-invariant, i.e., invariant to arbitrary displacements in both the inertial frame $\{F\}$ and the body frame $\{M\}$. The invariance properties of the overall method are, therefore, dictated by the second step. According to Proposition 5, the procedure is bi-invariant with respect only to rotations of $\{F\}$ in the particular case of $W = \alpha I$. In the most general case, i.e., for arbitrary choices of W, the method is left invariant to arbitrary displacements of the inertial frame.

21.3.5 Geodesics and Minimum Acceleration Curves

Consider a rigid body with inertial properties described by inertia matrix G (in a frame placed at the centroid) and mass m. As shown in Sect. 21.2.4, the kinetic energy of the body can be written in terms of a product metric given in Eq. (21.15). Given two boundary conditions for the pose $A(0) = (R(0), d(0))$ at $t = 0$ and $A(1) = (R(1), d(1))$ at $t = 1$, the translational part of the geodesic interpolant is simply the linear interpolant. The rotational part is constructed by numerically solving a boundary value problem consisting of end values $R(0)$ and $R(1)$ and the system of differential equations (21.17) augmented by the expressions of the time derivatives of some chosen coordinates on $SO(3)$ (exponential coordinates, Euler angles, quaternions) [3]. The relaxation or the shooting method are among the most popular [26]. For $G = \alpha I$, the interpolating minimum acceleration curve for the same position end conditions and velocity boundary conditions $\dot{R}(0)$, $\dot{d}(0)$, $\dot{R}(1)$, $\dot{d}(1)$ has a cubic translational part. The interpolating rotation can be found by solving a boundary value problem consisting of $R(0)$, $\dot{R}(0)$, $R(1)$, $\dot{R}(1)$ and 12 differential equations: Eq. (21.24) and the derivatives of the parameterization.

If the projection method described above is used, an approximate geodesic for the metric given in Eq. (21.15) and the same boundary conditions is given by

$$d(t) = d(0) + (d(1) - d(0))t, \quad R(t) = U(t)V^\mathrm{T}(t),$$

with U and V determined from the weighted SVD

$$M(t)W = U(t)\Sigma(t)V^\mathrm{T}(t),$$

where

$$M(t) = R(0) + (R(1) - R(0))t, \quad W = \frac{1}{2}\mathrm{Tr}(G)I_3 - G.$$

Similarly, an approximate minimum acceleration curve can be constructed as

$$d(t) = d_0 + d_1 t + d_2 t^2 + d_3 t^3, \quad R(t) = U(t)V^\mathrm{T}(t),$$

where

$$d_0 = d(0), \ d_1 = \dot{d}(0), \ d_2 = -3d(0) + 3d(1) - 2\dot{d}(0) - \dot{d}(1),$$

$$d_3 = 2d(0) - 2d(1) + \dot{d}(0) + \dot{d}(1),$$

$$M(t) = M_0 + M_1 t + M_2 t^2 + M_3 t^3,$$

$$M_0 = R(0), \ M_1 = \dot{R}(0), \ M_2 = -3R(0) + 3R(1) - 2\dot{R}(0) - \dot{R}(1),$$

$$M_3 = 2R(0) - 2R(1) + \dot{R}(0) + \dot{R}(1)$$

For geodesics on $SO(3)$ with Euclidean metric, we prove [4] that the projection of the geodesic from $GL^+(3)$ and the true geodesic on $SO(3)$ follow the same path but with different parameterizations. However, one can reparameterize the geodesic from $GL^+(3)$ so that it projects to the exact geodesic on $SO(3)$. We also show that uniqueness of projected geodesics and minimum acceleration curves is guaranteed under reasonable assumptions on the amount of rotation and the magnitude of the end velocities. In [3], we show that, from a computational point of view, it is much less expensive to generate interpolating motion using the projection method as opposed to the relaxation method. Specifically, if M is the number of uniformly distributed time points in $[0, 1]$, then the number of flops required by the projection method in $GL^+(n)$ is of order $O(n^3 M)$. On the other hand, the number of flops required by the relaxation method for generating solution at M mesh points of a system of N differential equations with boundary conditions is of order $O(M^3 N^3)$. For example, generating geodesics on $SO(3)$ at $M = 100$ time points involves millions of flops by the relaxation method, while only thousands by the projection method.

21.3.6 Simulation Results

In this section, we generate motion for a homogeneous parallelepipedic rigid body We assume that the body frame $\{M\}$ is placed at the center of mass and aligned with the principal axes of the body. Let a, b, and c be the lengths of the body along its x, y, and z axes respectively, and m the mass of the body. For visualization, a small square is drawn on one of its faces and the center of the parallelepiped is shown starred.

The matrix G of metric $<,>_{SO}$ is given by

$$G = \begin{bmatrix} \frac{m}{24}(b^2 + c^2) & 0 & 0 \\ 0 & \frac{m}{24}(a^2 + c^2) & 0 \\ 0 & 0 & \frac{m}{24}(a^2 + b^2) \end{bmatrix}. \tag{21.33}$$

True and projected minimum acceleration motions for a cubic rigid body with $a = b = c = 2$ and $m = 12$ are given in Fig. 21.2 for comparison. Note that for this case $G = \alpha I$ with $\alpha = 4$. Geodesics for the same boundary conditions and a parallelepipedic body with $a = c = 2$, $b = 10$ and $m = 12$ are given in Fig. 21.3.

As seen in Figs. 21.2 and 21.3, even though the total displacement between the initial and final positions on $SO(3)$ is large (rotation angle of $\pi\sqrt{14}/6$), there is no noticeable difference between the true and the projected motions.

21.4 Optimal Motion Generation for Groups of Robots

This section presents a method for generating smooth trajectories for a set of mobile robots satisfying constraints on relative positions. It is shown that, given two end configurations of the set of robots, by tuning one parameter, the user can choose an interpolating trajectory from a continuum of curves varying from the trajectory corresponding to maintaining a rigid formation to trajectories that allow the formation to change and the robots to reconfigure while moving.

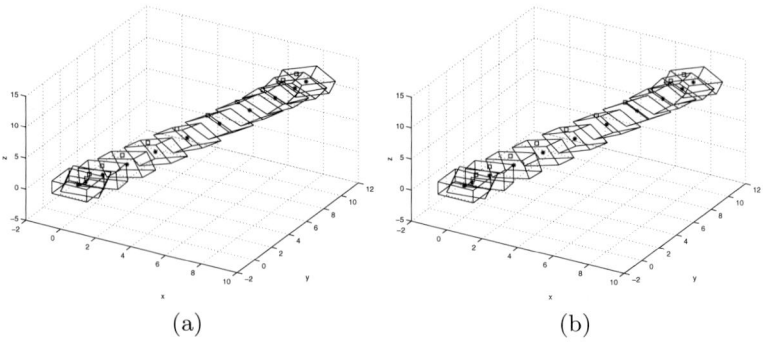

Fig. 21.2. a,b. Minimum acceleration motion for a cube in free space: **a** relaxation method, **b** projection method

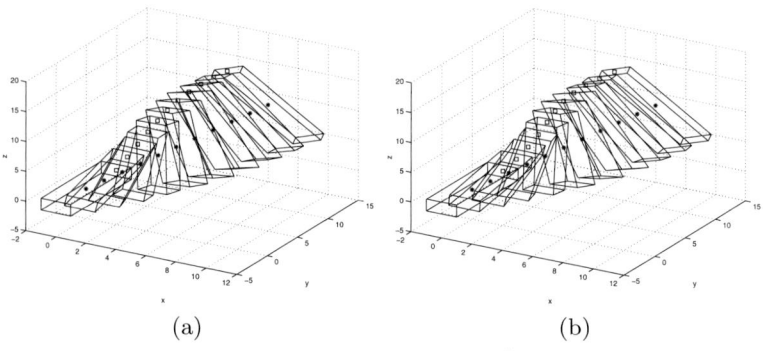

Fig. 21.3. a,b. Geodesics for a parallelepipedic body: **a** relaxation method, **b** projection method

21.4.1 Problem Statement and Notation

Consider N robots moving (rotating and translating) in 3D space with respect to an inertial frame $\{F\}$. We choose a reference point on each robot at its center of mass O_i. A moving frame $\{M_i\}$ is attached to each robot at O_i (see Fig. 21.4).

Robot i has mass m_i and matrix of inertia H_i with respect to frame $\{M_i\}$. Let $R_i \in SO(3)$ denote the rotation of $\{M_i\}$ in $\{F\}$ and $q_i \in \mathbb{R}^3$ the position vector of O_i in $\{F\}$. Let ω_i denote the expression in $\{M_i\}$ of the angular velocity of $\{M_i\}$ with respect to $\{F\}$. The *formation* is defined by the reference points O_i. The moving formation is called *rigid* if the relative distance between any of the points O_i is maintained constant. Sometimes it is also useful to define a *formation frame* $\{M\}$, attached at some virtual point O' and with pose $(R, d) \in SE(3)$ in $\{F\}$. Let q_i^0 denote the position vectors of O_i in $\{M\}$.

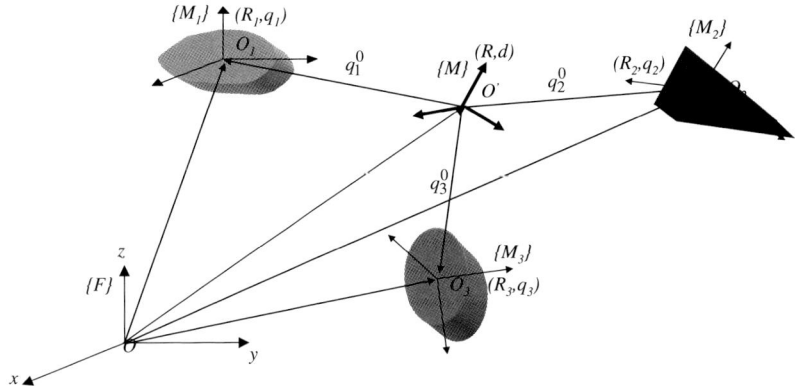

Fig. 21.4. A set of $N = 3$ robots

The configuration space is the $6N$-dimensional manifold, $SE(3) \times \ldots \times SE(3)$, given by the poses of each robot. Given two configurations at times $t = 0$ and $t = 1$ respectively, the goal is to generate smooth interpolating motion for each robot so that the total kinetic energy is minimized.

The kinetic energy T of the system of robots is the sum of the individual energies. Since the frames $\{M_i\}$ were placed at the centroids O_i of the robots, T can be written as the sum of the total rotational energy T_r and the total translational energy T_t in the form:

$$T = T_r + T_t, \quad T_r = \frac{1}{2} \sum_{i=1}^{N} (\omega_i^T H_i \omega_i), \quad T_t = \frac{1}{2} \sum_{i=1}^{N} (m_i \dot{q}_i^T \dot{q}_i). \tag{21.34}$$

Since our definition of a formation only involves the reference points O_i, a formation requirement will only constrain the q_i's from the above equation. Therefore, as a result of the decomposition in Eq. (21.34), minimizing the total energy is equivalent to solving $N + 1$ independent optimization subproblems:

$$\min_{\sigma_i} \int_0^1 \omega_i^T H_i \omega_i \, dt, \ i = 1, \ldots, N, \tag{21.35}$$

$$\min_{q_i = 1, \ldots, N} \int_0^1 T_t \, dt \tag{21.36}$$

where σ_i is some parameterization of the rotation of $\{M_i\}$ in $\{F\}$, i.e., some local coordinates on $SO(3)$. The solutions to Eq. (21.35) are given by N geodesics on $SO(3)$ with left-invariant metrics with matrices H_i. A relaxation method [26] or the projection method described in Section 21.3 can be used to generate the solution. An example is given in Sect. 21.4.4.

The main focus in this section is solving the problem given by Eq. (21.36) while satisfying constraints on the positions of the reference points O_i that may be imposed by the requirements on the task. Thus the configuration space we are interested in is just the $3N$-dimensional $Q = \{q|q = (q_1,\ldots,q_N)\}$ that collects all the position vectors of the chosen reference points. Maintaining a rigid formation (a virtual structure) imposes constraints on the configuration space, Q, and these constraints may be relaxed as necessary.

21.4.2 The Rigidity Constraint: Virtual Structures

The group of N robots is said to form a *virtual structure* if the relative distance between any of the reference points O_i is maintained constant. Let $q = [q_1^T,\ldots,q_N^T]^T$ denote an arbitrary point in Q. For an arbitrary pair of reference points with position vectors q_i and q_j, $i,j = 1,\ldots,N$, $i < j$, the constraints can be written as:

$$(q_i - q_j)^T(q_i - q_j) = constant, \tag{21.37}$$

or, by differentiation:

$$(q_i - q_j)^T \dot{q}_i - (q_i - q_j)^T \dot{q}_j = 0.$$

By lifting this constraint to the configuration manifold Q, the coordinates of the corresponding differential one form can be written as a $1 \times 3N$ row vector:

$$\omega_{ij} = \begin{bmatrix} 0 \ldots 0 \ (q_i - q_j)^T \ 0 \ldots 0 \ -(q_i - q_j)^T \ 0 \ldots 0 \end{bmatrix}.$$

The nonzero 1×3 blocks in the above matrix are in positions i and j, respectively. If we consider all $(N-1)N/2$ possible constraints, we can construct the codistribution ω_R as the span of all the corresponding covectors:

$$\omega_R = \text{span}\{\omega_{ij},\ i,j = 1,\ldots,N,\ i < n\}.$$

It is obvious that not all the $(N-1)N/2$ covectors (constraints) are independent. To insure rigidity, it is necessary and sufficient to impose $2N - 3$ constraints of the type (21.37) in plane, while in 3D the number is $3N - 6$. By simple inspection, it is easy to prove that the annihilating distribution of ω_R ($\omega_R(\Delta_R) = 0$) is:

$$\Delta_R = \text{Range}(A(q)),\ A(q) = \begin{bmatrix} -\hat{q}_1 & I_3 \\ \ldots & \ldots \\ -\hat{q}_N & I_3 \end{bmatrix}. \tag{21.38}$$

Therefore, by lifting each constraint to the configuration manifold Q, the virtual structure (rigidity) constraint can be written as

$$\dot{q} \in \Delta_R(q). \tag{21.39}$$

If q_i are not all contained in any proper hyperplane of $\mathbb{R}^d$ ($d = 2,3$), it can be proved [27] that the distribution Δ_R is regular, and, therefore integrable, since involutivity is always guaranteed. The distribution $\Delta_R(q)$ determines a foliation of Q with leaves given by orbits of $SE(3)$. Indeed, assume $q(0) = q^0$ and $\dot{q}(0) \in \Delta_R(q^0)$. Then, the rigidity constraint in Eq. (21.39) is satisfied for all $t \geq 0$ if and only if

$$q_i(t) = d(t) + R(t)q_i^0, \ i = 1, \ldots, N, \qquad (21.40)$$

where $(R(t), d(t))$ is a trajectory of the left-invariant control system

$$\dot{g}(t) = gS \qquad (21.41)$$

starting from $R(0) = I_3$, $d(0) = 0$.

Note that, under the rigidity assumption in Eq. (21.39), the coordinates r of the expansion of $\dot{q} \in \Delta_R(q)$ along the columns of $A(q)$, i.e., $\dot{q} = A(q)r$, are exactly the components of the left-invariant twist of a virtual structure formed by $(q_1, \ldots, q_N)$ and $\{F\}$ at that instant.

Also, if Eq. (21.39) is satisfied, then s from Eq. (21.41) is the left-invariant twist of a moving rigid structure formed by $(q_1^0, \ldots, q_N^0)$ and $\{M\}$ and for which the mobile frame $\{M\}$ was coincident with $\{F\}$ at $t = 0$. The pose of the moving frame $\{M\}$ in $\{F\}$ is $g = (R, d)$. Moreover, we have

$$\dot{q}_i = R[-\hat{q}_i^0 \ I]s. \qquad (21.42)$$

It follows that motion planning (control) problems for a set of N robots in 3D required to maintain a rigid formation can be reduced to motion planning (control) problems for a left-invariant control system on $SE(3)$.

21.4.3 Motion Decomposition: Rigid vs. Nonrigid

We first define a metric $<,>$ in the position configuration space, which is the same at all points $q \in Q$:

$$<V_q^1, V_q^2> = V_q^{1\mathrm{T}} M V_q^2, \qquad (21.43)$$

$$V_q = \dot{q} \in T_qQ, \quad M = \frac{1}{2}\mathrm{diag}\{m_1 I_3, \ldots, m_N I_3\}.$$

Metric (21.43) is called the *kinetic energy metric* because its induced norm ($V_q^1 = V_q^2 = \dot{q}$) assumes the familiar expression of the kinetic energy of the system $1/2 \sum_{i=1}^N m_i \dot{q}_i^\mathrm{T} \dot{q}_i$. If no restrictions are imposed on Q, the geodesic between $q(0) = q^0$ and $q(1) = q^1$ for metric (21.43) is obviously a straight line uniformly parameterized in time interpolating between q^0 and q^1 in Q.

At each point q in the configuration space Q, $\Delta_R(q)$ locally describes the set of all rigid body motion directions. The orthogonal complement to $\Delta_R(q)$, $\Delta_{NR}(q)$, will be the set of all directions violating the rigid body constraints. [3]

---

[3] In [6], the tangent space at q to the orbit of $SE(3)$ is called the vertical space at q, Ver_q, and its orthogonal complement is the horizontal space at $q \in Q$, Hor_q.

For an arbitrary tangent vector $V_q \in T_qQ$, let $\mathbf{R}V_q$ denote the projection onto Δ_R and $\mathbf{NR}V_q$ denote the projection onto Δ_{NR}.

Using the metric in Eq. (21.43), the orthogonal complement of the "rigid" distribution $\Delta_R(q)$ is the "nonrigid" distribution

$$\Delta_{NR}(q) = \text{Null}(A(q)^T M). \tag{21.44}$$

Let $B(q)$ denote a matrix whose columns are a basis of $\Delta_{NR}(q)$.

Let ψ denote the components of the projection in this basis: $\mathbf{NR}V_q = B(q)\psi$. Therefore, the velocity at point q can be written as:

$$V_q = \mathbf{R}V_q + \mathbf{NR}V_q = A(q)r + B(q)\psi. \tag{21.45}$$

Then, for any $V_q^1, V_q^2 \in T_qQ$, we have:

$$<V_q^1, V_q^2> = V_q^{1^T} M V_q^2 = <\mathbf{NR}V_q^1, \mathbf{NR}V_q^2> + <\mathbf{R}V_q^1, \mathbf{R}V_q^2>$$

because both $A^T M B$ and $B^T M A$ are zero from Eq. (21.44). Also, note that

$$r = (A^T M A)^{-1} A^T M V, \ \psi = (B^T M B)^{-1} B^T M V, \tag{21.46}$$

where the explicit dependence of A and B on q was omitted for simplicity. Therefore, the translational kinetic energy (which is the square of the norm induced by metric (21.43)) becomes:

$$T_t(q, \dot{q}) = \dot{q}^T M \dot{q} = r^T A^T M A r + \psi^T B^T M B \psi. \tag{21.47}$$

In (21.47), $r^T A^T M A r$ captures the energy of the motion of the system of particles as a rigid body, while the remaining part $\psi^T B^T M B \psi$ is the energy of the motion that violates the rigid body restrictions. For example, in the obvious case of a system of $N = 2$ particles, the first part corresponds to the motion of the two particles connected by a rigid massless rod, while the second part would correspond to motion along the line connecting the two bodies.

21.4.4 Motion Generation for Rigid Formations

In this section, we will assume that the robots are required to move in *rigid formation*, i.e., the distances between any two reference points O_i are preserved, or, equivalently, the reference points form a rigid polyhedron.

In our geometric framework, the rigid body requirement means restricting the trajectory $q(t) \in Q$ to be a $SE(3)$-orbit, or equivalently, $\mathbf{NR}\dot{q} = 0$ or $\dot{q} \in \Delta_R(q)$, for all q.

In this case, one can imagine a body frame $\{M\}$ moving with the virtual structure determined by the O_i's. Initially ($t = 0$), the frame $\{M\}$ is coincident with $\{F\}$ and $q(0) = q^0$. The position vector of O_i in $\{M\}$ is constant during this motion and equal to q_i^0.

Using Eq. (21.42), the kinetic energy T_t becomes:

$$T_t = s^{\mathrm{T}} \mathcal{M} s, \quad \mathcal{M} = A(q^0)^{\mathrm{T}} M A(q^0), \tag{21.48}$$

where $s \in se(3)$ is the instantaneous twist of the virtual structure.

Therefore, if the set of robots is required to move while maintaining a constant shape q^0, the optimization problem is reduced from dimension $6N$ to dimension $3N + 6$, and consists of solving for N geodesics on $SO(3)$ with metrics H_i (individual rotations) and one geodesic on the $SE(3)$ of the virtual structure with left invariant metric $\mathcal{M}$ as in Eq. (21.48).

4.4.1 Example: Five Identical Robots in 3D Space

For illustration, we consider five identical parallelepipedic robots with dimensions a, b, c and masses $m_i = m$, $i = 1, \ldots, 5$ required to move in formation while minimizing energy. The virtual structure is a pyramid with a square base of side l and height h. The body frames and the formation frame are placed at the center of mass and aligned with the principal axis. As outlined in the previous section, generating optimal motion for this group of robots reduces to generating five geodesics on the $SO(3)$ of each robot with left-invariant metric $H_i = G$ as in Eq. (21.33), $i = 1, \ldots, 5$ and one geodesic on the $SE(3)$ of the virtual structure endowed with a left-invariant metric with matrix

$$\tilde{G} = \frac{m}{2} \begin{bmatrix} 2l^2 & 0 & 0 & 0 \\ 0 & l^2 + \frac{4h^2}{3} & 0 & 0 \\ 0 & 0 & l^2 + \frac{4h^2}{3} & 0 \\ 0 & 0 & 0 & 3I_3 \end{bmatrix}$$

The resulting motion is presented in Fig. 21.5 for numerical values $a = c = 2$, $b = 10$, $m = 12$, $h = 20$, and $l = 10$. The projection method presented in Sect. 21.3 was used to generate the interpolating motions.

21.4.5 Motion Generation by Kinetic Energy Shaping

By shaping the kinetic energy, we mean smoothly changing the corresponding metric (21.43) at $T_q Q$ so that motion along some specific directions is allowed while motion along some other directions is penalized. The new metric will no longer be constant: the Christoffel symbols of the corresponding symmetric connection will be nonzero. The associated geodesic flow gives optimal motion.

In this work, the original metric (21.43) is shaped by putting different weights on the terms corresponding to the rigid and nonrigid motions:

$$<V_q^1, V_q^2>_\alpha = \alpha <\mathbf{N}\mathbf{R}V_q^1, \mathbf{N}\mathbf{R}V_q^2> + (1-\alpha) <\mathbf{R}V_q^1, \mathbf{R}V_q^2>. \tag{21.49}$$

Using Eq. (21.46) to go back to the original coordinates, we get the modified metric in the form:

$$<V_q^1, V_q^2>_\alpha = V_q^{1\mathrm{T}} M_\alpha(q) V_q^2, \tag{21.50}$$

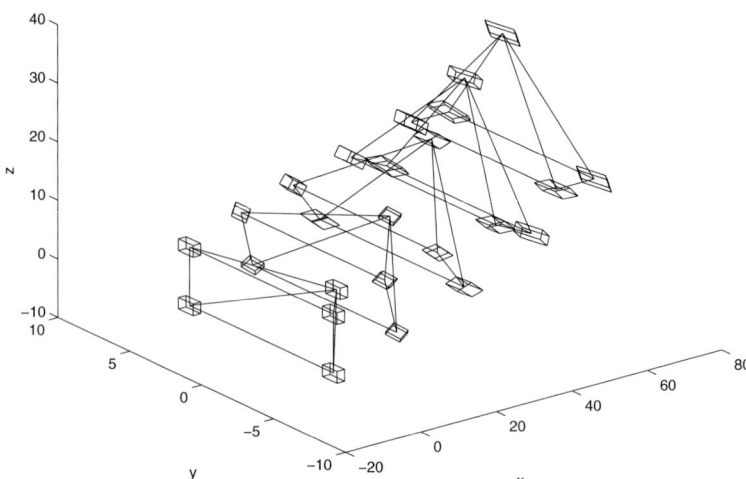

Fig. 21.5. Optimal motion for five identical robots required to maintain a rigid formation

where the new matrix of the metric is now dependent on the artificially introduced parameter α and the point on the manifold $q \in Q$:

$$M_\alpha(q) = \alpha M A (A^\mathrm{T} M A)^{-\mathrm{T}} A^\mathrm{T} M + (1-\alpha) M B (B^\mathrm{T} M B)^{-\mathrm{T}} B^\mathrm{T} M. \qquad (21.51)$$

The influence of the parameter α can be best seen by examining the significance of α taking on the values of 0, 0.5, and 1. As α tends to 0, the preferred motions will be ones where robots cluster together through much of the duration of the trajectory, thus minimizing the *rigid body energy* consumption. As α approaches 0.5, the motions degenerate toward uncoordinated, independent motions. As α tends to 1, the preferred motions are ones where the robots stay in rigid formation through most of the trajectory, thus minimizing the energy associated with *deforming* the formation.

We use the geodesic flow of metric (21.50) to produce smooth interpolating motion between two given configurations:

$$q^0 = q(0), \quad q^1 = q(1) \in \mathbb{R}^{3N}. \qquad (21.52)$$

To simplify the notation, let x_i, $i = 1, \ldots, 3N$ denote the coordinates $q_i \in \mathbb{R}^3$, $i = 1, \ldots, N$ on the configuration manifold Q. In this coordinates, the geodesic flow is described by the following differential equations [11]:

$$\ddot{x}_i + \sum_{j,k} \Gamma^i_{jk} \dot{x}_j \dot{x}_k = 0, \ i = 1, \ldots, 3N, \qquad (21.53)$$

where Γ^k_{ij} are the Christoffel symbols of the unique symmetric connection associated to metric (21.50):

$$\Gamma_{ij}^k = \frac{1}{2} \sum_h \left(\frac{\partial m_{hj}}{\partial x^i} + \frac{\partial m_{ih}}{\partial x^j} - \frac{\partial m_{ij}}{\partial x^h} \right) m^{hk}, \qquad (21.54)$$

for m_{ij} and m^{ij} elements of M_α and M_α^{-1}, respectively.

Because $\alpha = 0$ and $\alpha = 1$ make the metric singular, Eq. (21.54) can only be used for $0 < \alpha < 1$.

4.5.1 Example: Two Bodies in Plane

Consider two bodies of masses m_1 and m_2 moving in the x-y plane. The configuration space is $Q = R^4$ with coordinates $q = [x_1, y_1, x_2, y_2]^T$. The A and B matrices describing $\Delta_R(q)$ and $\Delta_{NR}(q)$ as in Eqns. (21.38) and (21.44) are:

$$A = \begin{bmatrix} -y_1 & 1 & 0 \\ x_1 & 0 & 1 \\ -y_2 & 1 & 0 \\ x_2 & 0 & 1 \end{bmatrix}, \quad B = \begin{bmatrix} \frac{m_2(x_2-x_1)}{m_1(y_1-y_2)} \\ -\frac{m_2}{m_1} \\ \frac{x_1-x_2}{y_1-y_2} \\ 1 \end{bmatrix}.$$

The 64 Christoffel symbols $\Gamma^k = (\Gamma_{ij}^k)_{ij}$ of the connection associated with the modified metric at $q \in Q$ become:

$$\Gamma^1 = \frac{2(1-2\alpha)}{\alpha} \frac{m_2}{m_1 + m_2} \frac{d_x}{(d_x^2 + d_y^2)^2} \Gamma,$$

$$\Gamma^2 = \frac{2(1-2\alpha)}{\alpha} \frac{m_2}{m_1 + m_2} \frac{d_y}{(d_x^2 + d_y^2)^2} \Gamma,$$

$$\Gamma^3 = -\frac{2(1-2\alpha)}{\alpha} \frac{m_1}{m_1 + m_2} \frac{d_x}{(d_x^2 + d_y^2)^2} \Gamma,$$

$$\Gamma^4 = -\frac{2(1-2\alpha)}{\alpha} \frac{m_1}{m_1 + m_2} \frac{d_y}{(d_x^2 + d_y^2)^2} \Gamma,$$

where

$$\Gamma = \begin{bmatrix} -d_y^2 & d_x d_y & d_y^2 & -d_x d_y \\ d_x d_y & -d_x^2 & -d_x d_y & d_x^2 \\ d_y^2 & -d_x d_y & -d_y^2 & d_x d_y \\ -d_x d_y & d_x^2 & d_x d_y & -d_x^2 \end{bmatrix},$$

and $d_x = x_1 - x_2$, $d_y = y_1 - y_2$. It can be easily seen that, as expected, all Christoffel symbols are zero if $\alpha = 0.5$. Also, the actual masses of the robots are not relevant, only the ratio m_1/m_2 is important.

In this example, we assume $m_2 = 2m_1$ and the boundary conditions:

$$q^0 = \begin{bmatrix} 1 \\ 0 \\ -0.5 \\ 0 \end{bmatrix}, \quad q^1 = \begin{bmatrix} 3 - \frac{\sqrt{2}}{2} \\ -\frac{\sqrt{2}}{2} \\ 3 + \frac{\sqrt{2}}{4} \\ \frac{\sqrt{2}}{4} \end{bmatrix},$$

which correspond to a rigid body displacement so that we can compare our results to the optimal motion corresponding to a rigid body.

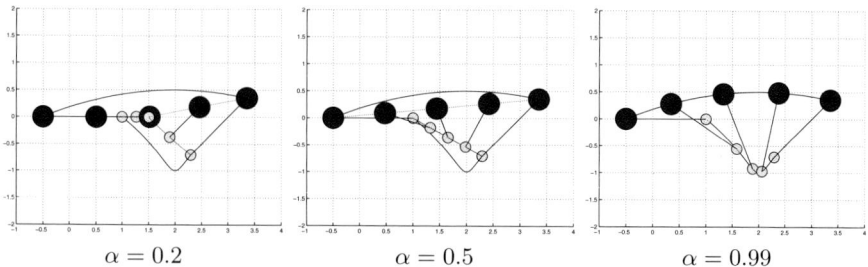

Fig. 21.6. Three interpolating motions for a set of two planar robots as geodesics of a modified metric defined in the configuration space

If the structure was assumed rigid, then the optimal motion is described by uniform rectilinear translation of the center of mass between $(0,0)$ and $(3,0)$ and uniform rotation between 0 and $3\pi/4$ around $-z$ placed at the center of mass. The corresponding trajectories of the robots are drawn in solid line in all the pictures in Fig. 21.6. It can be easily seen that there is no difference between the optimal motion of the virtual structure solved on $SE(2)$ and the geodesic flow of the modified metric with $\alpha = 0.99$ (Fig. 21.6, right). If $\alpha = 0.5$, all bodies move in straight line as expected (Fig. 21.6, middle). For $\alpha = 0.2$, the bodies go toward each other first, and then split apart to attain the final positions (Fig. 21.6, left).

4.5.2 Example: Three Bodies in Plane

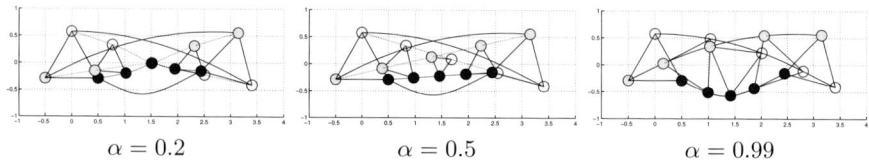

Fig. 21.7. Three interpolating motions for a set of three planar robots as geodesics of a modified metric defined in the configuration space

The calculation of the trajectories for three bodies moving in the plane is simplified by assuming that the robots are identical, and, without loss of generality, we assume $m_1 = m_2 = m_3 = 1$. The rigid and the nonrigid spaces at a generic configuration

$$q = [x_1, y_1, x_2, y_2, x_3, y_3]^T \in Q = \mathbb{R}^6$$

are given by

$$\Delta_R = \text{Range}(A), \quad A = \begin{bmatrix} -y_1 & 1 & 0 \\ x_1 & 0 & 1 \\ -y_2 & 1 & 0 \\ x_2 & 0 & 1 \\ -y_3 & 1 & 0 \\ x_3 & 0 & 1 \end{bmatrix},$$

$$\Delta_{NR} = \text{Range}(B), \quad B = \begin{bmatrix} \frac{x_3-x_1}{y_1-y_2} & \frac{y_2-y_3}{y_1-y_2} & \frac{x_2-x_1}{y_1-y_2} \\ -1 & 0 & -1 \\ \frac{x_1-x_3}{y_1-y_2} & \frac{y_3-y_1}{y_1-y_2} & \frac{x_1-x_2}{y_1-y_2} \\ 0 & 0 & 1 \\ 0 & 1 & 0 \\ 1 & 0 & 0 \end{bmatrix}.$$

For simplicity, we omit the expressions of the modified metric and of the Christoffel symbols. The simulation scenario resembles the one in Sect. 21.4.5: the end poses correspond to a rigid structure consisting of an equilateral triangle with sides equal to 1. The optimal trajectory solved on $SE(2)$ corresponds to rectilinear uniform motion of the center of mass (line between (0,0) and (3,0) in Fig. 21.7) and uniform rotation from angle 0 to $3\pi/4$ around axis $-z$. The resulting motion of each robot is shown solid, while the actual trajectory for the corresponding value of α is shown dashed. First note for $\alpha = 0.99$ the trajectories are basically identical with the optimal traces produced by the virtual structure, as expected. In the case $\alpha = 0.5$ the bodies move in straight line (corresponding to the unmodified metric). The tendency to cluster as α decreases is seen for $\alpha = 0.2$. Note also that because of our choice $m_1 = m_2 = m_3$, the geometry of the equilateral triangle is preserved for all values of α, it only scales down when α decreases from 1.

21.5 Conclusion

In this chapter, we survey the problem of generating interpolating motion for groups of robots required to maintain a rigid formation, or virtual structure. Since energy consumption is an important issue, especially for deep space formations, motion planning for virtual structures is often accomplished by posing the problem as an optimization problem. For example, satellite formation reconfiguration demands a fuel-optimal trajectory to preserve mission life and is constrained by the limited thrust available. Also, it is desired that the generated trajectories be independent of a chosen reference frame.

Virtual structures, as rigid bodies, evolve on the Lie group of all translations and orientations in 3D, $SE(3)$. The first part of this chapter is concerned with optimally interpolating trajectories on $SE(3)$. The second part investigates the rigidity constraint for a team of robots and shows how individual motion plans can be constructed so that the overall energy of the formation is minimized. The methodology and results are organized around two main issues: optimality and invariance of the generated trajectories. The price one has to pay to achieve these is, of course, a large amount of computation. We expect that these methods will only find applications in areas where the number of agents is small and fuel consumption is critical, as in satellite reconfiguration.

When a large number of agents is required to be coordinated and controlled, some level of abstraction is necessary, dependent on the imposed task. The virtual structure approach, which can be seen as an abstraction, is not appropriate in many applications, including passing of a tunnel and obstacle avoidance. Preliminary results on developing a general control method for large groups of inexpensive agents based on abstractions are presented in [5].

References

1. T. Balch, R. Arkin (1998) Behavior-based formation control for multi-robotic teams. *IEEE Transactions on Robotics and Automation* 14:926–934
2. R. W. Beard, F. Y. Hadaegh (1998) Constellation templates: an approach to autonomous formation flying. In: *World Automation Congress*, pp. 177.1–177.6, Anchorage, Alaska
3. C. Belta, V. Kumar (2002) Euclidean metrics for motion generation on $SE(3)$. *Journal of Mechanical Engineering Science Part C* 216:47–61
4. C. Belta, V. Kumar (2002) An SVD-based projection method for interpolation on se(3). *IEEE Trans. on Robotics and Automation*, 18:334–345
5. C. Belta, V. Kumar (2002). Towards abstraction and control for large groups of robots. In: *Control Problems in Robotics, Springer, Berlin Heidelberg New York*, pp. 169–182
6. A.M. Bloch, N.E. Leonard, J.E. Marsden (2000). Controlled lagrangians and the stabilization of mechanical systems 1: the first matching theorem. *IEEE Transactions on Automatic Control*, 45:2253–2270
7. W. Burgard, M. Moors, D. Fox, R. Simmons, S. Thrun (2000) Collaborative multi-robot exploration. In: *Proc. IEEE Int. Conf. Robot. Automat.*, pp. 476–481, San Francisco, CA
8. M. Camarinha, F. Silva Leite, P. Crouch (1995) Splines of class C^k on non-Euclidean spaces. *IMA J. Math. Control Inform.*, 12:399–410
9. P. Crouch, F. Silva Leite (1995) The dynamic interpolation problem: on Riemannian manifolds, Lie groups, and symmetric spaces. *J. Dynam. Control Systems*, 1:177–202
10. J. Desai, J. P. Ostrowski, V. Kumar (1998) Controlling formations of multiple mobile robots. In: *Proc. IEEE Int. Conf. Robot. Automat.*, pp. 2864–2869, Leuven, Belgium
11. M. P. do Carmo (1992) *Riemannian geometry*. Birkhauser, Boston
12. M. Egerstedt, X. Hu (2001) Formation constrained multi-agent control. *IEEE Trans. Robotics and Automation*, 17:947–951

13. T. Eren, P. N. Belhumeur, B. D. O. Anderson, A. S. Morse (2002) A framework for maintaining formations based on rigidity. In: *IFAC World Congress*, pp. 2752–2757, Barcelona, Spain
14. G. E. Farin (1992) *Curves and surfaces for computer aided geometric design : a practical guide, 3rd edn*, Academic Press, Boston
15. J. Feddema, D. Schoenwald (2001) Decentralized control of cooperative robotic vehicles. In: *Proc. SPIE Vol. 4364, Aerosense*, Orlando, Florida
16. Q. J. Ge, B. Ravani (1994) Computer aided geometric design of motion interpolants. *ASME Journal of Mechanical Design*, 116:756–762
17. G. H. Golub, C. F. van Loan (1989) *Matrix computations*. The Johns Hopkins University Press, Baltimore
18. B. Juttler, M. Wagner (1996) Computer aided design with rational b-spline motions. *ASME J. of Mechanical Design*, 118:193–201
19. N. E. Leonard, E. Fiorelli (2001) Virtual leaders, artificial potentials, and coordinated control of groups. In: *40th IEEE Conference on Decision and Control*, pp. 2968 – 2973, Orlando, FL
20. A. Marthinsen (1999) Interpolation in Lie groups and homogeneous spaces. *SIAM J. Numer. Anal.*, 37:269–285
21. M. Mataric, M. Nilsson, K. Simsarian (1995) Cooperative multi-robot box pushing. In *IEEE/RSJ International Conf. on Intelligent Robots and Systems*, pp. 556–561, Pittsburgh, PA
22. C. R. McInnes (1995) Autonomous ring formation for a planar constellation of satellites. *AIAA Journal of Guidance Control and Dynamics*, 18:1215–1217
23. L. Noakes, G. Heinzinger, B. Paden (1989) Cubic splines on curved spaces. *IMA J. of Math. Control & Information*, 6:465–473
24. R. Olfati-Saber, R. M. Murray (2002) Distributed cooperative control of multiple vehicle formations using structural potential functions. In: *IFAC World Congress*, pp. 346–352, Barcelona, Spain
25. F. C. Park, B. Ravani (1994) Bezier curves on Riemannian manifolds and Lie groups with kinematics applications. *ASME Journal of Mechanical Design*, 117:36–40
26. W. H. Press, S. A. Teukolsky, W. T. Vetterling, B. P. Flannery (1998) *Numerical Recipes in C*. Cambridge University Press, Cambridge, UK
27. B. Roth (1981) Rigid and flexible frameworks. *American Mathematical Monthly*, 88:6–21
28. K. Shoemake (1985) Animating rotation with quaternion curves. *ACM Siggraph*, 19:245–254
29. T. R. Smith, H. Hanssmann, N. E. Leonard (2001) Orientation control of multiple underwater vehicles. In: *40th IEEE Conference on Decision and Control*, pp. 4598–4603, Orlando, FL
30. L. Srinivasan, Q. J. Ge (1998) Fine tuning of rational b-spline motions. *ASME Journal of Mechanical Design*, 120:46–51
31. T. Sugar, V. Kumar (2000) Control and coordination of multiple mobile robots in manipulation and material handling tasks. In: P. Corke and J. Trevelyan (eds.), *Experimental Robotics VI: Lecture Notes in Control and Information Sciences, vol.* 250, Springer, Berlin Heidelberg New York, pp. 15–24
32. P. Tabuada, G. J. Pappas, P. Lima (2001) Feasible formations of multi-agent systems. In: *American Control Conference*, Arlington, VA

33. Herbert G. Tanner, George J. Pappas, Vijay Kumar (2002) Input-to-state stability on formation graphs. In: *Proceedings of the IEEE International Conference on Decision and Control*, pp. 2439–2444, Las Vegas, NV
34. P. Varaiya (1993) Smart cars on smart roads: problems of control. *IEEE Transactions on Automatic Control*, 38:195–207
35. M. Žefran, V. Kumar (1998) Interpolation schemes for rigid body motions. *Computer-Aided Design*, 30:179–189
36. M. Žefran, V. Kumar, C. Croke (1995) On the generation of smooth three-dimensional rigid body motions. *IEEE Transactions on Robotics and Automation*, 14:579–589

Part IX

Reaching and Motion Planning

The Computation of Reachable Surfaces for a Specified Set of Spatial Displacements

J. Michael McCarthy[1] and Hai-Jun Su[2]

[1] Department of Mechanical and Aerospace Engineering, University of California, Irvine, `jmmccart@uci.edu`
[2] Robotics and Automation Laboratory, University of California, Irvine, `suh@eng.uci.edu`

22.1 Overview

In this chapter, we consider the geometry of configurations of points generated by arbitrary sets of positions of a rigid body. The goal is to find the points in a moving body that lie on specific algebraic surfaces. This means we seek the solutions to sets of algebraic equations that become more complex as the number of parameters that define each surface increases.

The problem originates with Schoenflies [1], who sought points that remain in a specific configuration for a given set of spatial displacements. Burmester [2] applied this idea to planar mechanism design by seeking the points in a planar moving body that remain on a circle. These *Burmester points* form the joints of planar articulated chain that guides the body through the specified positions. His result was a graphical solution to a set of five quadratic equations in five unknown parameters. See Hartenberg and Denavit [3] and Sandor and Erdman [4] for analytical solutions to these design equations.

Chen and Roth [5] generalized Burmester's approach by seeking points and lines in a moving body that take positions on surfaces associated with the articulated chains used to build robot manipulators–also see Suh and Radcliffe [6] or McCarthy [7]. A discussion of general serial chain robots can be found in Craig [8] or Tsai [9]. Our focus is on two-jointed chains that support a spherical wrist. The center of this wrist traces a surface which is said to be "reachable" by the chain. Considering the various ways of assembling these articulated chains, we obtain seven reachable algebraic surfaces, and the problem reduces to computing the dimensions of these chains from a set of polynomial equations.

The complexity of this problem increases with the number of dimensional parameters and the degree of the surface. The total degree of the polynomial systems that we consider range from 32 for the simplest case to over 4×10^6 for the most complex. The equations have significant internal structure, so it is possible to use a linear product decomposition (Bernshtein[10]; Morgan et al. [11]) to provide a better bound on the number of solutions, which range from 10 to over 800,000. Where possible we have used polynomial elimination to verify the number of roots (Nielsen and Roth [12]; Husty [13]), but in the more complex examples we use polynomial homotopy algorithms (Tsai and Morgan [14], Verschelde and Haegemans [15]) to determine the number of roots by numerical experiments. The more difficult cases required adapting the software POLSYS_PLP (Wise et al. [16]) for parallel computation, renamed POLSYS_GLP (Su et al. [17]).

Our results are summarized in Table 22.3, which compares the total degree of each polynomial system, the bound obtained using the linear product structure of these polynomials, and the number of solutions obtained either analytically or using the homotopy algorithms.

22.2 Introduction

An important problem in geometry is determining an algebraic surface that passes through a given set of points. If the surface is defined by a single polynomial equation, then we can evaluate it on each of the given points to obtain a linear set of equations in the unknown coefficients of the surface.

For example, a sphere of radius R with center $\mathbf{B} = (u, v, w)$ is defined so a general point $\mathbf{P} = (X, Y, Z)$ satisfies the equation

$$\mathcal{S}: \quad (X-u)^2 + (Y-v)^2 + (Z-w)^2 - R^2 = 0. \tag{22.1}$$

Given four points $\mathbf{P}^i, i = 1, \ldots, 4$, we can evaluate this equation on each point to obtain four equations $\mathcal{S}(\mathbf{P}_i) = 0, i = 1, \ldots, 4$, which are linear in the coefficients of $\mathcal{S}$. These equations are easily solved to determine the four parameters $\mathbf{B} = (u, v, w)$ and R. This is described by saying that four points define a sphere.

We generalize this problem by considering the points $\mathbf{P}^i$ to be images of a single point $\mathbf{p} = (x, y, z)$ in a moving body transformed by a specified set of spatial displacements T_i. If we include the coordinates of $\mathbf{p}$ as part of the problem, we find that seven spatial displacements completely define the sphere and point, such that the sphere passes through all seven positions of the point (Chen and Roth [5]).

This generalized problem finds a practical application in the design of articulated serial chains. In particular, the TS chain shown in Fig. 22.1 has a gimbal joint (T-joint) at its shoulder, a ball joint (S-joint) at the wrist of its gripper, and no elbow joint (McCarthy [7]). Thus, the wrist center $\mathbf{P}$ moves on a sphere about the center $\mathbf{B}$ of the gimbal joint and traces the reachable surface of the chain. Given seven arbitrarily specified positions, we can compute the points $\mathbf{P}$ and $\mathbf{B}$ that define a sphere and obtain a TS chain that can reach each position. In what follows, we introduce homogeneous transforms that define the spatial positions of a moving body and form the

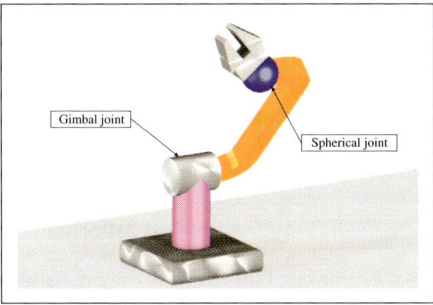

Fig. 22.1. The TS serial constraints the wrist center to the surface of a sphere

data for our generalized problem. We then enumerate the reachable surfaces and associated serial chains and, finally, focus on characterizing and solving the polynomial systems that define these surfaces.

22.3 Spatial Displacements

The position of a body in space is defined by attaching a frame M and determining the 3×1 translation vector $\mathbf{d}$ that locates the origin of M relative to a ground frame F, and by determining the 3×3 rotation matrix $[A]$ that defines the orientation of M relative F. This matrix and vector combine to transform the coordinates $\mathbf{p} = (x, y, z)$ of a point in the moving frame M to $\mathbf{P} = (X, Y, Z)$ in the fixed frame F, by the formula

$$\mathbf{P} = T(\mathbf{p}) = [A]\mathbf{p} + \mathbf{d}. \tag{22.2}$$

This transformation has the property that distances between points measured in M are preserved in F, and is called a *spatial displacement*. For a more detailed presentation, see Suh and Radcliffe [6], Bottema and Roth [18], or McCarthy [19].

Spatial displacements as in Eq. (22.2) are not linear operators, because $T(\mathbf{x} + \mathbf{y}) \neq T(\mathbf{x}) + T(\mathbf{y})$, due to the translation term. It is easy to adjust for this inhomogeneity by adding a fourth component to our position vectors that always equals 1. The result is the rotation matrix $[A]$ and translation vector $\mathbf{d}$ combine to form the 4×4 homogeneous transform $[T] = [A, \mathbf{d}]$, so we have

$$\begin{bmatrix} \mathbf{X} \\ 1 \end{bmatrix} = \begin{bmatrix} A & \mathbf{d} \\ 000 & 1 \end{bmatrix} \begin{bmatrix} \mathbf{x} \\ 1 \end{bmatrix}. \tag{22.3}$$

which we write as

$$\mathbf{X} = [T]\mathbf{x}. \tag{22.4}$$

For convenience we do not distinguish homogeneous point coordinates. Therefore, our vectors have three components, and we add a fourth component of 1 when appropriate for the use of 4×4 transforms.

22.3.1 Dual Quaternions

For every homogeneous transform $[T] = [A, \mathbf{d}]$, we can determine an invariant line $L(t) = \mathbf{C} + t\mathbf{S}$, which we also write as the Plücker vector $\mathsf{S} = (\mathbf{S}, \mathbf{C} \times \mathbf{S})$. The direction $\mathbf{S}$ of this line is the rotation axis of the matrix $[A]$, which means it satisfies the relation $[I - A]\mathbf{S} = 0$. The point $\mathbf{C}$ is obtained by solving the equation

$$[I - A]\mathbf{C} = \mathbf{d} - (\mathbf{d} \cdot \mathbf{S})\mathbf{S}, \tag{22.5}$$

where we assume $|\mathbf{S}| = 1$. See McCarthy [19, 7] for an explicit formula for $\mathbf{C}$. The displacement $[T]$ defines the position of M relative to F as the result of a rotation around $\mathbf{S}$ by the angle ϕ given by the complex eigenvalues of $[A]$, followed by a slide k along $\mathbf{S}$ where $k = \mathbf{S} \cdot \mathbf{d}$. This is the well-known result that every spatial displacement is equivalent to a rotation around and translation along a fixed line called the *screw axis* of the displacement (Bottema and Roth [18]).

The parameters ϕ, k and $\mathsf{S} = (\mathbf{S}, \mathbf{C} \times \mathbf{S})$ that define $[T]$ can be assembled into an eight dimensional vector known as a *dual quaternion*. To define a dual quaternion, we introduce the dual number $\hat{a} = a + \epsilon a^\circ$, where $\epsilon^2 = 0$. Two dual numbers add componentwise and multiply by the rule

$$\hat{a}\hat{b} = (a + \epsilon a^\circ)(b + \epsilon b^\circ) = ab + \epsilon(a^\circ b + ab^\circ). \tag{22.6}$$

This formalism allows us to define the dual angle $\hat{\phi} = \phi + \epsilon k$, such that

$$\sin\frac{\hat{\phi}}{2} = \sin\frac{\phi}{2} + \epsilon\frac{k}{2}\cos\frac{\phi}{2}, \text{ and } \cos\frac{\hat{\phi}}{2} = \cos\frac{\phi}{2} - \epsilon\frac{k}{2}\sin\frac{\phi}{2}. \tag{22.7}$$

A rotation of ϕ and slide k around and along the line S defines the dual quaternion

$$\hat{Q} = \sin\frac{\hat{\phi}}{2}\hat{\mathsf{S}} + \cos\frac{\hat{\phi}}{2}, \tag{22.8}$$

where $\hat{\mathsf{S}} = \mathbf{S} + \epsilon \mathbf{C} \times \mathbf{S}$ is the dual vector form of the Plücker coordinates of the line. The set of dual quaternions forms a vector space known as a *Clifford algebra* (McCarthy [19]). The components of a dual quaternion $\hat{Q} = (q_1, q_2, q_3, q_4) + \epsilon(q_1^\circ, q_2^\circ, q_3^\circ, q_4^\circ) = \mathbf{q} + \epsilon\mathbf{q}^\circ$ can be assembled into an eight-dimensional vector $\hat{Q} = (\mathbf{q}, \mathbf{q}^\circ)$, and it is easy to verify that these component satisfy the constraints,

$$\mathbf{q} \cdot \mathbf{q} = 1, \text{ and } \mathbf{q} \cdot \mathbf{q}^\circ = 0. \tag{22.9}$$

Thus, dual quaternions provide a coordinate representation of the set of spatial displacements as an algebraic submanifold in $\mathbf{R}^8$ that is called its *image space* by Ravani [20]; also see Ravani and Roth [21]. A slight variation of this was called *soma space* by Study (Bottema and Roth [18]). The elements of the rotation matrix and translation vector in a homogeneous transform can be defined in terms of the components of the associated dual quaternion, so we have $[T(\hat{\mathbf{Q}})] = [A(\mathbf{q}), \mathbf{d}(\mathbf{q}, \mathbf{q}^\circ)]$. The result is a mapping of the constraint equation of the reachable surface to an algebraic manifold in the space of dual quaternions, called a *constraint manifold*. Our problem now becomes that of fitting this constraint manifold to the dual quaternion points that represent the specified displacements.

22.4 Articulated Serial Chains

A serial chain is a sequence of rigid links connected by joints that forms the skeleton of a mechanical system. One end is connected to ground while the other end forms the workpiece, or end effector. In general, each joint of the chain allows either pure rotation about, or a linear slide along, the joint axis, termed a revolute or prismatic joint, respectively. The number of these joints defines the *degree of freedom* of the chain.

Table 22.1. The five basic joints

Joint	Diagram	Symbol	DOF
Revolute		R	1
Prismatic		P	1
Cylindric		C	2
Universal		T	2
Spherical		S	3

Revolute and prismatic joints combine to form three additional joints. A gimbal, or universal, joint, is formed by two revolute joints with axes that intersect at right angles–we denote this joint by T. The combination of a revolute and a prismatic joint such that their axes are parallel forms a cylindric joint, denoted by a C. Finally, a three revolute chain constructed so the joint axes are concurrent is called a spherical, or ball, joint, denoted by S. See Table 22.1 for descriptions of these five basic joints. In order to define reachable surfaces, we consider only articulated chains for which the end effector is supported by a spherical wrist that allows full orientation of the gripper about its center, **P**. This means the reachable surface is traced by **P** with the movement of two remaining joints. The possibilities are simply the combinations of R and P for the first two joints, that is, RRS, RPS, PRS, and PPS. The surfaces reachable by the wrist centers of these chains are, respectively, the general torus, the circular hyperboloid, the elliptic cylinder, and the plane.

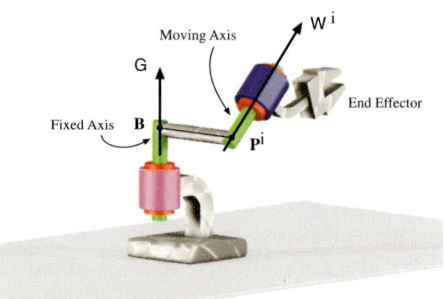

Fig. 22.2. The RR serial chain formed by a sequence of two revolute joints

We obtain additional reachable surfaces by specializing the dimensional parameters that characterize the first two joints. In particular, the RR chain Fig. 22.2, has two defining parameters: the distance ρ between the joint axes along their common normal line, and the angle α between them measured around this common normal. For $\alpha = \frac{\pi}{2}$ we have the "right RRS" chain that traces a circular torus. The case $\alpha = 0$ yields the "parallel RRS" chain, which constrains its wrist center to a plane, and is therefore equivalent to the PPS chain. If the parameter $\rho = 0$, then the surface is part of a sphere, and fills the sphere for $\alpha = \frac{\pi}{2}$, which characterizes the TS chain. For RP and PR chains, only the angle α is important because the P joint ensures that all points travel on lines parallel to its direction. We can identify the special cases of the RPS and PRS chains for which this angle is $\alpha = 0$. Both cases become the CS chain that traces a circular cylinder. If this angle is $\alpha = \frac{\pi}{2}$, called a "right RPS" or "right PRS" chain, then the surface is again a plane equivalent to that traced by the PPS chain. Finally, all PP chains are essentially the same as long the directions of the two joints are not parallel, so that some component of movement perpendicular to the first prismatic joint is available by sliding along the second joint.

The result is seven articulated chains and the associated algebraic surfaces that are reachable by their wrist centers (Table 22.2). These surfaces define seven constraint manifolds in the Clifford algebra of dual quaternions. In what follows, we determine the number of solutions to the polynomial equations that define these reachable surfaces. This is equivalent to determining the number of constraint manifolds for a particular chain that pass through a given set of displacements.

Table 22.2. The basic serial chains and their associated reachable surfaces

Case	Chain	Angle	Length	Surface
1	PPS	–	–	Plane
2	TS	$\frac{\pi}{2}$	0	Sphere
3	CS	0	–	Circular cylinder
4	RPS	α	–	Circular hyperboloid
5	PRS	α	–	Elliptic cylinder
6	Right RRS	$\frac{\pi}{2}$	ρ	Circular torus
7	RRS	α	ρ	General torus

22.5 Linear Product Decomposition

The fundamental theorem of algebra states that the number of roots of a polynomial is equal to or less than its degree, which is the integer value of its highest power–equality is obtained if roots are counted with the appropriate multiplicity. This has been generalized to Bezout's theorem, which states that the number of roots of a system of polynomials is less than or equal to the product of the degrees of the individual polynomials, called the *total degree* of the system.

In the problems that we consider in this paper, the total degree overestimates the number of roots in the polynomial system $\mathcal{P}(\mathbf{z})$ by a significant amount. Morgan et al. [11] show that a "generic" system of polynomials that includes every monomial of a particular system of polynomials will have as many or more solutions as any polynomial system obtained by specifying values for the coefficients. This provides the foundation for the construction of the *linear product decomposition* of a system of polynomials.

The linear product decomposition of a system of polynomials is a polynomial system that includes all of the monomials of the original system, but in which each polynomial is formed by the product of linear combinations of the variables.

Let $\langle x, y, 1 \rangle$ represent the set of linear combinations of parameters x, y, and 1, which means a typical term is $\alpha x + \beta y + \gamma \in \langle x, y, 1 \rangle$, where α, β, and γ are arbitrary constants. Using this notation, we define the product of $\langle x, y, 1 \rangle \langle u, v, 1 \rangle$ as the set of linear combinations of the product of the elements of the two sets, that is

$$\langle x, y, 1 \rangle \langle u, v, 1 \rangle = \langle xu, xv, yu, yv, x, y, u, v, 1 \rangle. \tag{22.10}$$

This product commutes, which means $\langle x \rangle \langle y \rangle = \langle y \rangle \langle x \rangle$, and it distributes over unions, such that $\langle x \rangle \langle y \rangle \cup \langle x \rangle \langle z \rangle = \langle x \rangle (\langle y \rangle \cup \langle z \rangle) = \langle x \rangle \langle y, z \rangle$. Furthermore, we represent repeated factors using exponents, so $\langle x, y, 1 \rangle \langle x, y, 1 \rangle = \langle x, y, 1 \rangle^2$.

In order to illustrate a linear product decomposition, we consider our example in Eq. (22.1) in more detail. Write these polynomials in vector form to obtain

$$(\mathbf{P}^i - \mathbf{B}) \cdot (\mathbf{P}^i - \mathbf{B}) = R^2, \quad i = 1, \ldots, 7, \tag{22.11}$$

where the dot denotes the vector dot product. Now subtract the first equation from the rest in order to eliminate R^2. This reduces the problem to six equations in the unknowns $\mathbf{z} = (x, y, z, u, v, w)$, given by

$$\mathcal{S}_j(\mathbf{z}) = (\mathbf{P}^{j+1} \cdot \mathbf{P}^{j+1} - \mathbf{P}^1 \cdot \mathbf{P}^1) - 2\mathbf{B} \cdot (\mathbf{P}^{j+1} - \mathbf{P}^1) = 0, \quad j = 1, \ldots, 6. \tag{22.12}$$

We now focus attention on the monomials formed by the unknown parameters.

Recall that $\mathbf{P}^i = [A_i]\mathbf{p} + \mathbf{d}_i$, where $[A_i]$ and $\mathbf{d}_i$ are known, so it is easy to see that

$$2\mathbf{B} \cdot (\mathbf{P}^{j+1} - \mathbf{P}^1) \in \langle u, v, w \rangle \langle x, y, z, 1 \rangle. \tag{22.13}$$

It is also possible to compute

$$\mathbf{P}^{j+1} \cdot \mathbf{P}^{j+1} - \mathbf{P}^1 \cdot \mathbf{P}^1 = 2\mathbf{d}_{j+1} \cdot [A_{j+1}]\mathbf{p} - 2\mathbf{d}_1 \cdot [A_1]\mathbf{p} + \mathbf{d}_{j+1}^2 - \mathbf{d}_1^2 \in \langle x, y, z, 1 \rangle. \tag{22.14}$$

Each of the equations in (22.12) has the same monomial structure given by

$$\langle x, y, z, 1 \rangle \cup \langle u, v, w \rangle \langle x, y, z, 1 \rangle \subset \langle x, y, z, 1 \rangle \langle u, v, w, 1 \rangle. \tag{22.15}$$

From this we see that a generic set of polynomials, which contains our system as a special case, can be formed as products of linear factors, that is

$$Q(\mathbf{z}) = \begin{bmatrix} (a_1 x + b_1 y + c_1 z + d_1)(e_1 u + f_1 v + g_1 w + h_1) \\ \vdots \\ (a_6 x + b_6 y + c_6 z + d_6)(e_6 u + f_6 v + g_6 w + h_6) \end{bmatrix} = 0, \tag{22.16}$$

where the coefficients are known constants. This structure is called the *linear product decomposition* (LPD) of the polynomial system.

Solutions to the LPD of a set of polynomials are easily determined by assembling all combinations of factors, one from each equation, that can be set to zero and solved for the unknown parameters (Wampler[22]). In our example, select the factors $a_i x + b_i y + c_i z + d_i = 0$ from three of the six equations, and combine with the three factors $e_i u + f_i v + g_i w + h_i = 0$ from the remaining equations. A solution of this set of six linear equations is a root of Eq. (22.16). Thus, we find that this system has $\binom{6}{3} = 20$ solutions, which matches the known result for Eq. (22.12). For the problems we consider, the LPD provides a bound on the number of solutions that is significantly less than the total degree.

In the following sections, we consider each reachable surface in turn. We derive a defining polynomial system, evaluate its total degree, compute its LPD bound, and then determine the roots of a generic problem to find the number of articulated chains that reach a specified set of displacements.

22.6 The Plane

The PPS serial chain has the property that the wrist center $\mathbf{P}$ is constrained to lie on a plane (Fig. 22.3). A point $\mathbf{P} = (X, Y, Z)$ lies on a plane with the surface normal $\mathbf{G} = (a, b, c)$ if it satisfies the equation

$$aX + bY + cZ - d = \mathbf{G} \cdot \mathbf{P} - d = 0. \tag{22.17}$$

The parameter d is the product of the magnitude $|G|$ and the signed normal distance to the plane.

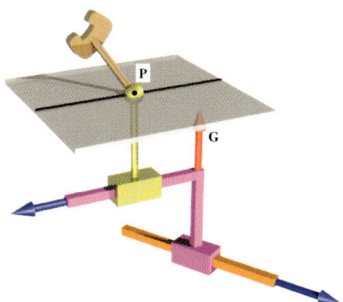

Fig. 22.3. A plane as traced by a point at the wrist center of a PPS serial chain

Given a set of spatial displacements, $\hat{Q}_i$, $i = 1, \ldots, n$, we have the images $\mathbf{P}^i = [T(\hat{Q}_i)]\mathbf{p}$ of a single point $\mathbf{p}$ in the moving frame M. We seek both the plane $P : (\mathbf{G}, d)$ and the point $\mathbf{p} = (x, y, z)$, so each $\mathbf{P}^i$ lies on this plane. There are seven parameters in this problem, however, the components of $\mathbf{G}$ are not independent because only its direction is important to defining the plane, not its magnitude. A convenient way to constrain this magnitude is to choose an vector $\mathbf{m}$ and a scalar e, and require that $\mathbf{m} \cdot \mathbf{G} = e$. Add to this six equations obtained by evaluating Eq. (22.17) on six arbitrary displacements, given by

$$\mathbf{G} \cdot \mathbf{P}^i - d = 0, \quad i = 1, \ldots, 6. \tag{22.18}$$

Subtract the first of these equations from the remaining to eliminate d, and the result is the polynomial system

$$\begin{aligned} \mathcal{P}_j : \quad & \mathbf{G} \cdot (\mathbf{P}^{j+1} - \mathbf{P}^1) = 0, j = 1, \ldots, 5, \\ \mathcal{C} : \quad & \mathbf{m} \cdot \mathbf{G} - 1 = 0. \end{aligned} \tag{22.19}$$

This is a set of five quadratic equations and one linear equation in the six unknowns $\mathbf{z} = (a, b, c, x, y, z)$. The total degree of this system of polynomials is $2^5 = 32$, which means that for six arbitrary displacements there are most 32 points in the moving body that have six positions on a plane.

The linear combinations of monomials that contain the plane equations (22.19) are given by

$$\mathcal{P}_j \in \langle a, b, c \rangle \langle x, y, z, 1 \rangle |_j = 0, \quad j = 1, \ldots, 5,$$
$$\mathcal{C} \in \langle a, b, c, 1 \rangle = 0. \qquad (22.20)$$

This is the LPD of the polynomial system. The root count for this LPD is given by the combinations of linear factors that can be set to zero and solved for the unknown parameters. In this case, we have $\binom{5}{2} = 10$ roots, which means that there are at most 10 points in the moving body that have 6 positions on a plane.

In this case, direct elimination of the parameters (Raghavan 1995[23]) yields a univariate polynomial of degree 10, which shows that this LPD bound is exact. So far, our study of example problems has not yielded more than four real solutions.

Once the plane P and point $\mathbf{p}$ are defined, then it is possible to determine a PPS chain, a parallel RRS, or right RPS chain that guides this point through the specified positions.

22.7 The Sphere

We now return to our opening example of a point $\mathbf{P} = (X, Y, Z)$ constrained to lie on a sphere of radius R around the point $\mathbf{B} = (u, v, w)$ (Fig. 22.4). This means its coordinates satisfy the equation

$$(X - u)^2 + (Y - v)^2 + (Z - w)^2 - R^2 = (\mathbf{P} - \mathbf{B})^2 - R^2 = 0. \qquad (22.21)$$

We now consider $\mathbf{P}^i$ to be the image of a point $\mathbf{p} = (x, y, z)$ in a moving frame M that takes positions in space defined by the displacements $\hat{\mathbf{Q}}_i$, $i = 1, \ldots, n$. See Innocenti [24], Liao and McCarthy [25] and Raghavan [26].

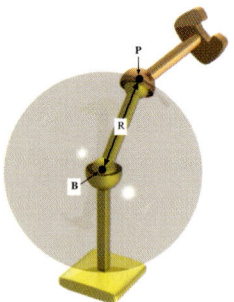

Fig. 22.4. A sphere traced by a point at the wrist center of a TS serial chain

This problem has seven parameters, therefore we can evaluate Eq. (22.21) on $n = 7$ displacements. We reduced these equations to the set of six quadratic polynomials,

$$\mathcal{S}_j : (\mathbf{P}^{j+1^2} - \mathbf{P}^{1^2}) - 2\mathbf{B} \cdot (\mathbf{P}^{j+1} - \mathbf{P}^1) = 0, j = 1, \ldots, 6. \quad (22.22)$$

This system has total degree of $2^6 = 64$.

We have already seen that the system Eq. (22.22) has the LPD

$$\mathcal{S}_j \in \langle x, y, z, 1\rangle\langle u, v, w, 1\rangle|_j = 0, \quad j = 1, \ldots, 6. \quad (22.23)$$

From this we can compute the LPD bound $\binom{6}{3} = 20$. Parameter elimination yields an univariate polynomial of degree 20, so we see that this bound is exact. Innocenti [24] presents an example that results in 20 real roots; also see Wampler et al. [27].

The conclusion is that given seven arbitrary spatial positions there can be as many as 20 points in the moving body that have positions lying on a sphere. For each real point, it is possible to determine an associated TS chain.

22.8 The Circular Cylinder

In order to define the equation of a circular cylinder, let the line $L(t) = \mathbf{B} + t\mathbf{G}$ be its axis. A general point $\mathbf{P}$ on the cylinder lies on a circle about the point $\mathbf{Q}$ closest to it on the axis $L(t)$ (Fig. 22.5). Introduce the unit vectors $\mathbf{u}$ and $\mathbf{v}$ along $\mathbf{G}$ and the radius R of the cylinder, respectively, so we have

$$\mathbf{P} - \mathbf{B} = d\mathbf{u} + R\mathbf{v}, \quad (22.24)$$

where d is the distance from $\mathbf{B}$ to $\mathbf{Q}$. Compute the cross product of this equation with $\mathbf{G}$, in order to cancel d before squaring both sides. The result is

$$((\mathbf{P} - \mathbf{B}) \times \mathbf{G})^2 = R^2 \mathbf{G}^2. \quad (22.25)$$

In this calculation we use the fact that $(\mathbf{v} \times \mathbf{G})^2 = \mathbf{G}^2$.

Another version of the equation of the cylinder is obtained by substituting $d = (\mathbf{P} - \mathbf{B}) \cdot \mathbf{u}$ into Eq. (22.24) and squaring both sides to obtain

$$(\mathbf{P} - \mathbf{B})^2 - ((\mathbf{P} - \mathbf{B}) \cdot \mathbf{G})^2 \frac{1}{\mathbf{G} \cdot \mathbf{G}} - R^2 = 0. \quad (22.26)$$

Notice that we allow $\mathbf{G}$ to have an arbitrary magnitude. This form of the cylinder is related to the equation of the circular hyperboloid, which is discussed in Sect. 22.9.

Equation (22.25) has ten parameters, the radius R and three each in the vectors $\mathbf{P} = (X, Y, Z)$, $\mathbf{B} = (u, v, w)$, and $\mathbf{G} = (a, b, c)$. However, because only the direction of $\mathbf{G}$ is important to the definition of the cylinder, its three

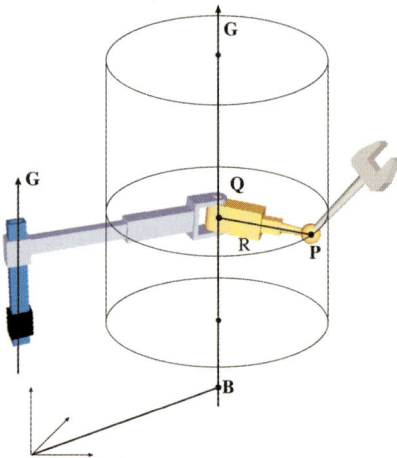

Fig. 22.5. The circular cylinder reachable by a CS serial chain

components are not independent. We set the magnitude of $\mathbf{G}$ as we did above for the equation of the plane. Choose an arbitrary vector $\mathbf{m}$ and scalar e and require the components of $\mathbf{G}$ to satisfy the constraint,

$$\mathcal{C}_1: \quad \mathbf{G} \cdot \mathbf{m} - e = 0. \tag{22.27}$$

The components of the point $\mathbf{B}$ are also not independent, but for a different reason. It is because any point on the line $L(t)$ can be selected as the reference point $\mathbf{B}$. We identify this point by requiring $\mathbf{B}$ to lie on a specific plane $U: (\mathbf{n}, f)$, that is

$$\mathcal{C}_2: \quad \mathbf{B} \cdot \mathbf{n} - f = 0, \tag{22.28}$$

where $\mathbf{n}$ and f are chosen arbitrarily to avoid the possibility that the line $L(t)$ may lie on U. Eight more polynomials are obtained by evaluating Eq. (22.26) with $\mathbf{P}$ specified as the image of $\mathbf{p} = (x, y, z)$ for eight spatial displacements, that is, $\mathbf{P}^i = [T(\hat{\mathbf{Q}}_i)]\mathbf{p}, i = 1, \ldots, 8$. The result is

$$((\mathbf{P}^i - \mathbf{B}) \times \mathbf{G})^2 - R^2 \mathbf{G}^2 = 0, i = 1, \ldots, 8. \tag{22.29}$$

Subtract the first equation from the remaining to eliminate R

$$((\mathbf{P}^{j+1} - \mathbf{B}) \times \mathbf{G})^2 - ((\mathbf{P}^1 - \mathbf{B}) \times \mathbf{G})^2 = 0, \quad j = 1, \ldots, 7. \tag{22.30}$$

Expand the terms in this equation to obtain the system of polynomials

$$\mathcal{P}_i: \quad (\mathbf{P}^{j+1} \times \mathbf{G})^2 - (\mathbf{P}^1 \times \mathbf{G})^2 - 2((\mathbf{P}^{j+1} - \mathbf{P}^1) \times \mathbf{G}) \cdot (\mathbf{B} \times \mathbf{G}) = 0,$$
$$j = 1, \ldots, 7,$$
$$\mathcal{C}_1: \quad \mathbf{G} \cdot \mathbf{m} - e = 0, \text{ and } \mathcal{C}_2: \quad \mathbf{B} \cdot \mathbf{n} - f = 0. \tag{22.31}$$

This is a set of seven polynomials of degree four and two linear equations. The total degree is $4^7 = 16{,}384$. See Nielsen and Roth [12] and Su et al. [28] for additional details about this problem. We now consider the monomial structure of the polynomial system given by Eq. (22.31). The polynomials $\mathcal{P}_i$ are linear combinations of monomials in the set generated by

$$(\langle x,y,z,1\rangle\langle a,b,c\rangle)^2 \cup \langle x,y,z,1\rangle\langle a,b,c\rangle\langle u,v,w\rangle\langle a,b,c\rangle. \tag{22.32}$$

The products commute, $\langle a\rangle\langle b\rangle = \langle b\rangle\langle a\rangle$, and they distribute over unions, $\langle a\rangle\langle b\rangle \cup \langle a\rangle\langle c\rangle = \langle a\rangle(\langle b\rangle \cup \langle c\rangle) = \langle a\rangle\langle b,c\rangle$, therefore Eq. (22.32) becomes

$$\langle a,b,c\rangle^2(\langle x,y,z,1\rangle^2 \cup \langle x,y,z,1\rangle\langle u,v,w\rangle), \tag{22.33}$$

which can be written as

$$\langle a,b,c\rangle^2\langle x,y,z,1\rangle\langle x,y,z,u,v,w,1\rangle. \tag{22.34}$$

This shows that the polynomial system Eq. (22.31) has the monomial structure,

$$\begin{aligned}
&\mathcal{P}_j \in \langle a,b,c\rangle^2\langle x,y,z,1\rangle\langle x,y,z,u,v,w,1\rangle|_j = 0, \quad j=1,\ldots,7, \\
&\mathcal{C}_1 \in \langle u,v,w,1\rangle = 0, \quad \text{and} \\
&\mathcal{C}_2 \in \langle a,b,c,1\rangle = 0.
\end{aligned} \tag{22.35}$$

In order to estimate the number of roots we see that to specify $\mathbf{G}=(a,b,c)$ we must combine $\mathcal{C}_2$ with two terms $\langle a,b,c\rangle$ from the seven polynomials $\mathcal{P}_i$. Because this term is squared, the number of choices is increased by a factor of $2^2 = 4$. For the remainder of the parameters, we can choose from zero to three of the terms $\langle x,y,z,1\rangle$ from the remaining five polynomials to define $\mathbf{p}=(x,y,z)$. The third terms in what is left and $\mathcal{C}_1$ define the remaining parameters. This yields the LPD bound of

$$2^2 \binom{7}{2} \sum_{i=0}^{3} \binom{5}{i} = 2{,}184, \tag{22.36}$$

which is much reduced from the total degree of 16,384. For polynomial systems with a large number of roots, elimination is notattractive, but we may find all solutions using polynomial continuation. We used the software PHC (Verschelde[29]) and POLSYS_PLP (Watson et al.[30]) to compute the roots for random test cases and determined the exact root count for this problem to be 804. Clearly, there is more structure in this system of polynomials than what is shown in the linear product decomposition.

Thus, we find that for eight arbitrary spatial positions we can find as many as 804 points in the moving body, each of which has all eight positions on a circular cylinder. For each of these points, we can determine an associated CS chain.

22.9 The Circular Hyperboloid

A circular hyperboloid is generated by rotating one line around another so that every point on the moving line traces a circle around the fixed line, which is the axis of the hyperboloid (Fig. 22.6). Of all of these circles there is one with the smallest radius, R, and its center $\mathbf{B} = (u, v, w)$ is the center of the hyperboloid. Let $\mathbf{G} = (a, b, c)$ be the direction of the axis, and denote its Plücker coordinates as $\mathsf{G} = (\mathbf{G}, \mathbf{B} \times \mathbf{G})$. A unit vector $\mathbf{N}$ perpendicular to $\mathbf{G}$ though $\mathbf{B}$ is the common normal between the axis G and one of the generated lines H. The generator is located at the distance R along $\mathbf{N}$, and lies at an angle α around $\mathbf{N}$ relative to the axis G.

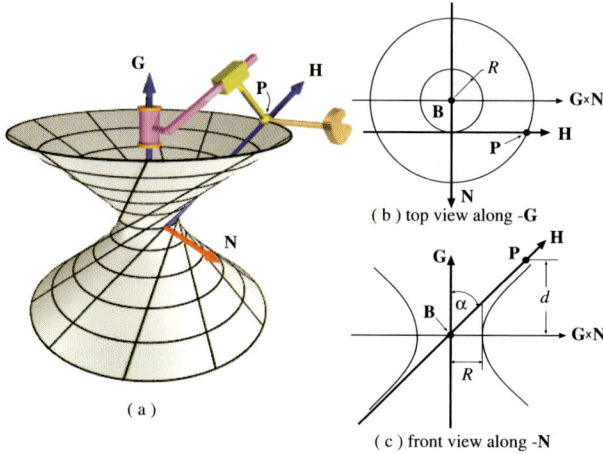

Fig. 22.6. The circular hyperboloid traced by the wrist center of an RPS serial chain

If point $\mathbf{P}$ is a point on the generator H, then its d measured along the axis G from $\mathbf{G}$ is given by

$$d = \frac{(\mathbf{P} - \mathbf{B}) \cdot \mathbf{G}}{\sqrt{\mathbf{G} \cdot \mathbf{G}}}. \tag{22.37}$$

Notice that we are not assuming that $\mathbf{G}$ is a unit vector. The magnitude of $\mathbf{P} - \mathbf{B}$ is now computed to be

$$(\mathbf{P} - \mathbf{B})^2 = R^2 + d^2 + (d \tan \alpha)^2. \tag{22.38}$$

Substitute d into this equation to obtain the equation of a circular hyperboloid

$$(\mathbf{P} - \mathbf{B})^2 - ((\mathbf{P} - \mathbf{B}) \cdot \mathbf{G})^2 \left(\frac{1 + \tan^2 \alpha}{\mathbf{G} \cdot \mathbf{G}} \right) - R^2 = 0. \tag{22.39}$$

When $\alpha = 0$, this equation becomes the equation of a cylinder presented in Sect. 22.8.

Figure 22.6a shows the RPS chain associated with the circular hyperboloid. The R-joint axis is $\mathbf{G}$, and its P-joint axis is in the direction α measured around the common normal. The point $\mathbf{P}$ is the center of the S-joint, and lies at the distance R in the direction $\mathbf{N}$ of the common normal. Expand Eq. (22.39) and collect terms to obtain

$$k_0 \mathbf{P} \cdot \mathbf{P} + 2\mathbf{K} \cdot \mathbf{P} - (\mathbf{P} \cdot \mathbf{G})^2 - \zeta = 0, \qquad (22.40)$$

where we have introduce the parameters k_0, $\mathbf{K} = (k_1, k_2, k_3)$, and ζ defined by

$$k_0 = \frac{\mathbf{G} \cdot \mathbf{G}}{1 + \tan^2 \alpha}, \quad \mathbf{K} = (\mathbf{B} \cdot \mathbf{G})\mathbf{G} - k_0 \mathbf{B}, \quad \zeta = (\mathbf{B} \cdot \mathbf{G})^2 - k_0 \mathbf{B} \cdot \mathbf{B} + k_0 R^2. \quad (22.41)$$

Given values for ζ, k_0, $\mathbf{K}$, and $\mathbf{G}$, we can compute $\mathbf{B}$ by solving the linear equations

$$\begin{bmatrix} k_1 \\ k_2 \\ k_3 \end{bmatrix} = \begin{bmatrix} a^2 - k_0 & ab & ac \\ ab & b^2 - k_0 & ac \\ ac & bc & c^2 - k_0 \end{bmatrix} \begin{bmatrix} u \\ v \\ w \end{bmatrix}. \qquad (22.42)$$

Then the length and twist parameters, R and α, are obtained from the formulas

$$\alpha = \arccos\left(\sqrt{\frac{k_0}{\mathbf{G} \cdot \mathbf{G}}}\right), \quad R = \sqrt{\frac{\zeta - (\mathbf{B} \cdot \mathbf{G})^2 + k_0 \mathbf{B} \cdot \mathbf{B}}{k_0}}. \qquad (22.43)$$

Thus, the 11 dimensional parameters ζ, k_0, $\mathbf{K}$, $\mathbf{G}$, and $\mathbf{P}$ define a circular hyperboloid.

As we have done previously, we set the length of $\mathbf{G}$ by choosing an arbitrary vector $\mathbf{m}$ and scalar e to define the constraint

$$\mathcal{C}: \quad \mathbf{G} \cdot \mathbf{m} - e = 0. \qquad (22.44)$$

This means given ten arbitrary displacements $\hat{\mathbf{Q}}_i$, we can map a point $\mathbf{p} = (x, y, z)$ to its displaced positions $\mathbf{P}^i = [T(\hat{\mathbf{Q}}_i)]\mathbf{p}$, $i = 1, \ldots, 10$. Evaluating the equation of the hyperboloid on these ten points, we obtain

$$k_0 \mathbf{P}^i \cdot \mathbf{P}^i + 2\mathbf{K} \cdot \mathbf{P}^i - (\mathbf{P}^i \cdot \mathbf{G})^2 - \zeta = 0, \quad i = 1, \ldots, 10. \qquad (22.45)$$

Subtract the first of these equation from the remaining in order to eliminate ζ and define the system of polynomials

$$\mathcal{H}_j: \quad k_0(\mathbf{P}^{j+1^2} - \mathbf{P}^{1^2}) + 2\mathbf{K} \cdot (\mathbf{P}^{j+1} - \mathbf{P}^1) - (\mathbf{P}^{j+1} \cdot \mathbf{G})^2 + (\mathbf{P}^1 \cdot \mathbf{G})^2 = 0,$$
$$j = 1, \ldots, 9,$$
$$\mathcal{C}: \quad \mathbf{G} \cdot \mathbf{m} - e = 0. \qquad (22.46)$$

This is a system of nine fourth-degree polynomials $\mathcal{H}_j$ and one linear equation $\mathcal{C}$, which has a total degree of $4^9 = 262{,}144$. See Nielsen and Roth [12] and Kim and Tsai [31] for other formulations of this problem.

A better bound on the number of solutions can be obtained by considering the monomial structure of the these equations. Recall that the term $\mathbf{P}^{j+1^2} - \mathbf{P}^{1^2}$ is linear in x,y, and z, because the quadratic terms cancel, see Eq. (22.14). This means the polynomials $\mathcal{H}_j$ have the monomial structure

$$\mathcal{H}_j \in \langle k_0\rangle\langle x,y,z,1\rangle \cup \langle k_1,k_2,k_3\rangle\langle x,y,z,1\rangle \cup (\langle x,y,z,1\rangle\langle a,b,c\rangle)^2. \quad (22.47)$$

This simplifies to yield the linear product decomposition for the system (22.46) as

$$\begin{aligned}\mathcal{H}_j &\in \langle a,b,c\rangle^2\langle x,y,z,1\rangle\langle x,y,z,k_0,k_1,k_2,k_3,1\rangle|_j, \quad j=1,\ldots,9,\\ \mathcal{C} &\in \langle a,b,c,1\rangle.\end{aligned} \quad (22.48)$$

This structure allows us to count the number of roots from the number of admissible sets of linear equations that yield solutions for the unknown parameters. In this case we obtain

$$LPD = 2^2 \binom{9}{2} \sum_{j=0}^{3} \binom{7}{j} = 9{,}216. \quad (22.49)$$

The POLSYS_PLP algorithm yielded a generic root count of 1024 in a calculation that took approximately 24h on a single 2.4-GHz PC (384 paths per processor-hour). The parallel version POLSYS_GLP was run on eight 64-bit processors of UCI's Beowulf cluster, and required 30min (2304 paths per processor-hour).

This particular problem has a structure that is convenient for polyhedral homotopy algorithms. Using this setting for PHC, we obtained the same solutions in 24min on a single processor by tracking only 1024 paths. See Gao et al. [32, 33] for a discussion of polyhedral homotopy methods. Thus, we see that there are as many as 1024 points, each of which lies on a circular hyperboloid when displaced through ten arbitrarily specified spatial displacements. For each real root we can find an associated RPS chain.

22.10 The Elliptic Cylinder

An elliptic cylinder is generated by a circle that has its center swept along a line $L(t) = \mathbf{B} + t\mathbf{S}_1$ such that the vector through the center normal to the plane of the circle maintains a constant direction $\mathbf{S}_2$ at an angle α relative to the direction $\mathbf{S}_1$ of $L(t)$ (Fig. 22.7). The major axis of the elliptic cross-section is the radius R of the circle and the minor axis is $R\cos\alpha$. This surface is generated by the wrist center of a PRS chain that has its P-joint aligned with the axis $L(t)$ and its R-joint positions so its axis is along $\mathbf{S}_2$.

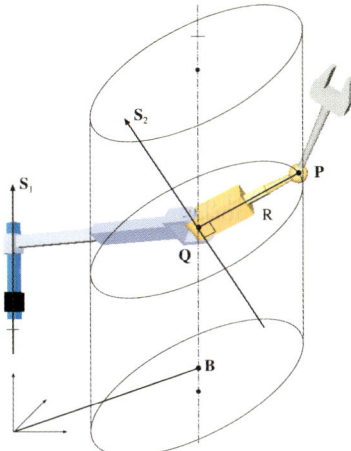

Fig. 22.7. The elliptic cylinder reachable by a PRS serial chain

Consider a general point on the cylinder $\mathbf{P}$, and let $\mathbf{Q}$ be the center of the circle. The point $\mathbf{Q}$ moves along the axis $L(t)$, which has the Plücker coordinates $\mathsf{S}_1 = (\mathbf{S}_1, \mathbf{B} \times \mathbf{S}_1)$. The distance from the reference point $\mathbf{B}$ to $\mathbf{Q}$ is denoted d. These definitions allow us to express the location of $\mathbf{P}$ relative to $\mathbf{B}$ as

$$\mathbf{P} - \mathbf{B} = d\mathbf{S}_1 + R\mathbf{u}, \tag{22.50}$$

where $\mathbf{u}$ is a unit vector in the direction $\mathbf{S}_1 \times \mathbf{S}_2$. Compute the cross product with $\mathbf{S}_1$ to eliminate d, and the cross product with $\mathbf{S}_2$ to obtain

$$\mathbf{S}_2 \times ((\mathbf{P} - \mathbf{B}) \times \mathbf{S}_1) = R(\mathbf{S}_2 \cdot \mathbf{S}_1)\mathbf{u}. \tag{22.51}$$

The magnitude of this vector identity yields our equation of the elliptic cylinder

$$\left(\mathbf{S}_2 \times ((\mathbf{P} - \mathbf{B}) \times \mathbf{S}_1)\right)^2 = R^2(\mathbf{S}_1 \cdot \mathbf{S}_2)^2. \tag{22.52}$$

This equation has 13 dimensional parameters: the radius R, three each for the directions $\mathbf{S}_1$, $\mathbf{S}_2$, and the points $\mathbf{P}$ and $\mathbf{B}$. Notice that if $\mathbf{S}_1 = \mathbf{S}_2 = \mathbf{G}$ this simplifies to the equation of a circular cylinder.

There are actually only ten independent parameters in Eq. (22.52) and we can determine three additional linear constraints. First, note that it is the directions of $\mathbf{S}_1$ and $\mathbf{S}_2$ that matter, not their magnitude. We specify these magnitudes by introducing two arbitrary planes $V_k : (\mathbf{m}_k, e_k), k = 1, 2$. In general, the lines through the origin parallel to $\mathbf{S}_i$ must intersect these planes, respectively. We select these points of intersection to be $\mathbf{S}_i$; that is, we require

$$\mathcal{C}_k: \quad \mathbf{m}_k \cdot \mathbf{S}_k - e_k = 0, \quad k = 1, 2. \tag{22.53}$$

Next, we note that any point along the line S_1 can serve as the reference point $\mathbf{B}$ for the axis of the cylinder. We determine $\mathbf{B}$ by specifying an arbitrary plane

$U : (\mathbf{n}, f)$. In general, the line G must intersect this plane, and we select this point at $\mathbf{B}$. Thus, $\mathbf{B}$ satisfies the linear equation

$$\mathcal{C}_3: \quad \mathbf{n} \cdot \mathbf{B} - f = 0. \tag{22.54}$$

Notice that $\mathbf{n}$ is the unit normal to the plane and f the directed distance from the origin to the plane.

We now consider the images of a point $\mathbf{p} = (x, y, z)$ generated by ten spatial displacements, that is $\mathbf{P}^i = [T(\hat{\mathbf{Q}}_i)]\mathbf{p}, i = 1, \ldots, 10$. Evaluate the equation of the elliptic cylinder on these ten points to obtain

$$\left(\mathbf{S}_2 \times ((\mathbf{P}^i - \mathbf{B}) \times \mathbf{S}_1)\right)^2 - R^2(\mathbf{S}_1 \cdot \mathbf{S}_2)^2 = 0, \quad i = 1, \ldots, 10. \tag{22.55}$$

Subtract the first of these equations from the remaining to obtain the system of polynomials

$$\begin{aligned}
\mathcal{E}_j &: \quad (\mathbf{S}_2 \times ((\mathbf{P}^{j+1} - \mathbf{B}) \times \mathbf{S}_1))^2 - (\mathbf{S}_2 \times ((\mathbf{P}^1 - \mathbf{B}) \times \mathbf{S}_1))^2 = 0, \, j = 1, \ldots, 9, \\
\mathcal{C}_k &: \quad \mathbf{m}_k \cdot \mathbf{S}_k - e_k = 0, \, k = 1, 2, \\
\mathcal{C}_3 &: \quad \mathbf{n} \cdot \mathbf{B} - f = 0.
\end{aligned} \tag{22.56}$$

The result is nine polynomials of degree six and three linear equations. The total degree of this polynomial system is $6^9 = 10{,}077{,}696$.

The total degree of this system can be reduced as follows. Expand the triple product

$$\begin{aligned}
\mathbf{S}_2 \times ((\mathbf{P} - \mathbf{B}) \times \mathbf{S}_1) &= (\mathbf{S}_1 \cdot \mathbf{S}_2)(\mathbf{P} - \mathbf{B}) - ((\mathbf{P} - \mathbf{B}) \cdot \mathbf{S}_2)\mathbf{S}_1 \\
&= (\mathbf{S}_1 \cdot \mathbf{S}_2)(\mathbf{P} - (\mathbf{P} \cdot \mathbf{K})\mathbf{S}_1 + \mathbf{Q}),
\end{aligned} \tag{22.57}$$

where

$$\mathbf{K} = \frac{\mathbf{S}_2}{\mathbf{S}_1 \cdot \mathbf{S}_2}, \quad \text{and} \quad \mathbf{Q} = (\mathbf{B} \cdot \mathbf{K})\mathbf{S}_1 - \mathbf{B}. \tag{22.58}$$

Add to this the constraints

$$\mathbf{S}_1 \cdot \mathbf{S}_1 = 1, \quad \mathbf{K} \cdot \mathbf{S}_1 = 1, \quad \text{and} \quad \mathbf{Q} \cdot \mathbf{K} = 0. \tag{22.59}$$

This combines with the other constraints to reduce the degree of these polynomials to four, so we have

$$\begin{aligned}
(\mathbf{P} - (\mathbf{P} \cdot \mathbf{K})\mathbf{S}_1 + \mathbf{Q})^2 = \\
\mathbf{P}^2 + (\mathbf{P} \cdot \mathbf{K})^2 + \mathbf{Q}^2 - 2(\mathbf{P} \cdot \mathbf{S}_1)(\mathbf{P} \cdot \mathbf{K}) + 2\mathbf{P} \cdot \mathbf{Q} - 2(\mathbf{P} \cdot \mathbf{K})(\mathbf{Q} \cdot \mathbf{S}_1).
\end{aligned} \tag{22.60}$$

The result is a new version of the polynomial system

$$\begin{aligned}
\mathcal{E}'_j &: (\mathbf{P}^{j+1} - (\mathbf{P}^{j+1} \cdot \mathbf{K})\mathbf{S}_1 + \mathbf{Q})^2 - (\mathbf{P}^1 - (\mathbf{P}^1 \cdot \mathbf{K})\mathbf{S}_1 + \mathbf{Q})^2 = 0, \\
&\qquad\qquad j = 1, \ldots, 9, \\
\mathcal{C}'_1 &: \mathbf{S}_1 \cdot \mathbf{S}_1 - 1 = 0, \\
\mathcal{C}'_2 &: \mathbf{K} \cdot \mathbf{S}_1 - 1 = 0, \\
\mathcal{C}'_3 &: \mathbf{Q} \cdot \mathbf{K} = 0,
\end{aligned} \tag{22.61}$$

which has the total degree $(2^3)(4^9) = 2{,}097{,}152$.

As we have done previously, we examine the monomial structure of the equations $\mathcal{E}'_j$. Let $\mathsf{S}_1 = (a,b,c)$, $\mathbf{K} = (k_1, k_2, k_3)$, and $\mathbf{Q} = (q_1, q_2, q_3)$, and recall that the quadratic terms in $\mathbf{P}^{j+1^2} - \mathbf{P}^{1^2}$ cancel, as does the term $\mathbf{Q}^2$. Thus, the polynomials $\mathcal{E}'_j$ have the monomial structure

$$\langle x,y,z,1\rangle \cup \langle x,y,z,1\rangle^2 \langle k_1,k_2,k_3\rangle^2 \cup \langle x,y,z,1\rangle^2 \langle k_1,k_2,k_3\rangle\langle a,b,c\rangle$$
$$\cup \langle x,y,z,1\rangle\langle q_1,q_2,q_3\rangle \cup \langle x,y,z,1\rangle\langle k_1,k_2,k_3\rangle\langle a,b,c\rangle\langle q_1,q_2,q_3\rangle. \quad (22.62)$$

This leads to the linear product decomposition for this polynomial system

$$\mathcal{E}'_j \in \langle x,y,z,1\rangle\langle x,y,z,q_1,q_2,q_3,1\rangle\langle k_1,k_2,k_3,1\rangle\langle k_1,k_2,k_3,a,b,c,1\rangle|_j = 0,$$
$$j = 1,\ldots,9,$$

$$\mathcal{C}'_1 \in \langle a,b,c,1\rangle^2 = 0,$$
$$\mathcal{C}'_2 \in \langle k_1,k_2,k_3,1\rangle\langle a,b,c,1\rangle = 0,$$
$$\mathcal{C}'_3 \in \langle k_1,k_2,k_3,1\rangle\langle q_1,q_2,q_3,1\rangle = 0. \quad (22.63)$$

The LPD bound for this system is 247,968 which is large.

This system was solved using our parallelized POLSYS_GLP on 128 nodes of the Blue Horizon supercomputer at the San Diego Supercomputer Center. The result was 18,120 solutions in almost 33min. Each node of Blue Horizon has eight processors, so this corresponds to 563 cpu hours, or approximately 440 paths per processor-hour. These 18,120 real and complex solutions require further study to evaluate the associated PRS chains.

22.11 The Circular Torus

A circular torus is generated by sweeping a circle around an axis so its center traces a second circle. Let the axis be $L(t) = \mathbf{B} + t\mathbf{G}$, with Plücker coordinates $\mathsf{G} = (\mathbf{G}, \mathbf{B} \times \mathbf{G})$ (Fig. 22.8). Introduce a unit vector $\mathbf{v}$ perpendicular to this axis so the center of the generating circle is given by $\mathbf{Q} - \mathbf{B} = \rho\mathbf{v}$. Now define $\mathbf{u}$ to be the unit vector in the direction $\mathbf{G}$, then a point $\mathbf{P}$ on the torus is defined by the vector equation

$$\mathbf{P} - \mathbf{B} = \rho\mathbf{v} + R(\cos\phi\,\mathbf{v} + \sin\phi\,\mathbf{u}), \quad (22.64)$$

where ϕ is the angle measured from $\mathbf{v}$ to the radius vector of the generating circle.

An algebraic equation of the torus is obtained from Eq. (22.64) by first computing the magnitude

$$(\mathbf{P} - \mathbf{B})^2 = \rho^2 + R^2 + 2\rho R\cos\phi. \quad (22.65)$$

Next compute the dot product with $\mathbf{u}$, to obtain

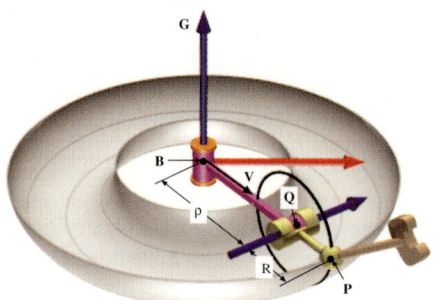

Fig. 22.8. The circular torus traced by the wrist center of a right RRS serial chain

$$(\mathbf{P} - \mathbf{B}) \cdot \mathbf{u} = R \sin \phi. \tag{22.66}$$

Finally, eliminate $\cos \phi$ and $\sin \phi$ from these equations, and the result is

$$\mathbf{G}^2((\mathbf{P} - \mathbf{B})^2 - \rho^2 - R^2)^2 + 4\rho^2((\mathbf{P} - \mathbf{B}) \cdot \mathbf{G})^2 = 4\rho^2 \mathbf{G}^2 R^2. \tag{22.67}$$

This is the equation of a circular torus. It has 11 parameters, the scalars ρ and R, and the three vectors $\mathbf{G}$, $\mathbf{P}$, and $\mathbf{B}$.

In contrast to what we have done previously, here we set the magnitude of $\mathbf{G}$ to a constant, in order to simplify the polynomial (22.67),

$$\mathcal{G} : \mathbf{G} \cdot \mathbf{G} = 1. \tag{22.68}$$

Unfortunately, this doubles the number of solutions since $-\mathbf{G}$ and $\mathbf{G}$ define the same torus, however, it reduces this polynomial from degree sixth to degree four.

Let $\hat{\mathbf{Q}}_i, i = 1, \ldots, 10$ be a specified set of displacements, so we have the ten positions $\mathbf{P}^i = [T(\hat{\mathbf{Q}}_i)]\mathbf{p}$ of a point $\mathbf{p} = (x, y, z)$ that is fixed in the moving frame M. Evaluating Eq. (22.67) on these points, we obtain the polynomial system

$$\begin{aligned} \mathcal{T}_i : \quad & ((\mathbf{P}^i - \mathbf{B})^2 - \rho^2 - R^2)^2 + 4\rho^2((\mathbf{P}^i - \mathbf{B}) \cdot \mathbf{G})^2 \\ & - 4\rho^2 R^2 = 0, \ i = 1, \ldots, 10, \\ \mathcal{G} : \quad & \mathbf{G} \cdot \mathbf{G} - 1 = 0. \end{aligned} \tag{22.69}$$

The total degree of this system is $2(4^{10}) = 2{,}097{,}152$.

In order to simplify the polynomials $\mathcal{T}_i$ we introduce the parameters

$$\mathbf{H} = 2\rho \mathbf{G} \quad \text{and} \quad k_1 = \mathbf{B}^2 - \rho^2 - R^2, \tag{22.70}$$

which yields the identity

$$4\rho^2 R^2 = \mathbf{H}^2 \left(\mathbf{B}^2 - \frac{\mathbf{H}^2}{4} - k_1 \right). \tag{22.71}$$

Substitute these relations into $\mathcal{T}_i$ to obtain

$$\mathcal{T}_i': \quad ((\mathbf{P}^i)^2 - 2\mathbf{P}^i \cdot \mathbf{B} + k_1)^2 + ((\mathbf{P}^i - \mathbf{B}) \cdot \mathbf{H})^2 - \mathbf{H}^2 \left(\mathbf{B}^2 - \frac{\mathbf{H}^2}{4} - k_1 \right) = 0,$$
$$i = 1, \ldots, 10, \quad (22.72)$$

It is difficult to find a simplified formulation for these equations, even if we subtract the first equation from the remaining in order to cancel terms.

Expanding the polynomial $\mathcal{T}_i'$ and examining each of the terms, we can identify the linear product decomposition

$$\mathcal{T}_i' \in \langle x, y, z, h_1, h_2, h_3, 1 \rangle^2 \langle x, y, z, h_1, h_2, h_3, u, v, w, k_1, 1 \rangle^2. \quad (22.73)$$

This allows us to compute the LPD bound on the number of roots as

$$\text{LPD} = 2^{10} \sum_{j=0}^{6} \binom{10}{j} = 868{,}352. \quad (22.74)$$

Our POLSYS_GLP algorithm tracked these homotopy paths in 72 min on 128 nodes of the Blue Horizon supercomputer. This means the over 800,000 paths were tracked on 1024 processors at a rate of approximately 707 paths per processor-hour. We obtained 94,622 real and complex solutions for a random set of specified displacements. However, this problem needs further study to provide an efficient way to evaluate and sort the large number of right RRS chains.

22.12 The General Torus

A general torus is defined by sweeping a circle that has a general orientation in space about an arbitrary axis (Fig. 22.9). Let $\mathsf{S}_1 = (\mathbf{S}_1, \mathbf{B} \times \mathbf{S}_1)$ be the Plücker coordinates of the line that forms the axis of the torus, and $\mathsf{S}_2 = (\mathbf{S}_2, \mathbf{Q} \times \mathbf{S}_2)$ define the line through the center of the circle that is perpendicular to the plane of the circle. These two lines define a common normal N, and we choose its intersection with S_1 and S_2 to be the reference points $\mathbf{B}$ and $\mathbf{Q}$, respectively. The normal angle and distance between these lines around and along their common normal are denoted α and ρ. Finally, we identify the center of the circle as lying a distance d along S_2 measured from $\mathbf{Q}$.

In this derivation, we constrain $\mathbf{S}_1$ and $\mathbf{S}_2$ to be unit vectors, in order to reduce the degree of the resulting equation. This allows us to define the unit vector in the common normal direction as $\mathbf{n} = (\mathbf{S}_1 \times \mathbf{S}_2)/\sin\alpha$, so we obtain a general point $\mathbf{P}$ on the torus from the vector equation,

$$\mathbf{P} - \mathbf{B} = \rho \mathbf{n} + d\mathbf{S}_2 + R(\cos\phi \mathbf{n} + \sin\phi(\mathbf{S}_2 \times \mathbf{n})). \quad (22.75)$$

The algebraic equation for the torus is obtained by first computing

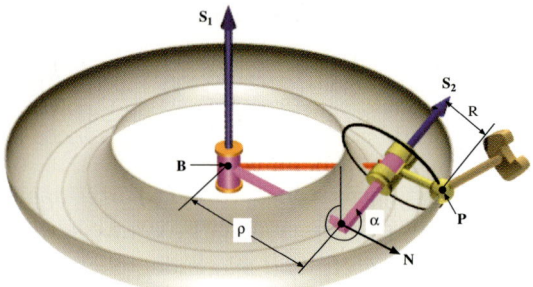

Fig. 22.9. The general torus reachable by the wrist center of an RRS serial chain

$$(\mathbf{P} - \mathbf{B})^2 = \rho^2 + d^2 + R^2 + 2\rho R \cos\phi, \quad (22.76)$$

and

$$(\mathbf{P} - \mathbf{B}) \cdot (\mathbf{v} \times \mathbf{n}) = R \sin\phi. \quad (22.77)$$

Notice that $\mathbf{S}_2 \times \mathbf{n}$ is

$$\mathbf{S}_2 \times \frac{\mathbf{S}_1 \times \mathbf{S}_2}{\sin\alpha} = \frac{1}{\sin\alpha}(\mathbf{S}_1 - \cos\alpha \mathbf{S}_2). \quad (22.78)$$

Now, eliminate ϕ between these two equations to obtain

$$((\mathbf{P}-\mathbf{B})^2 - \rho^2 - d^2 - R^2)^2 + \frac{4\rho^2}{\sin^2\alpha}((\mathbf{P}-\mathbf{B})\cdot\mathbf{S}_1 - d\cos\alpha)^2 - 4\rho^2 R^2 = 0. \quad (22.79)$$

This equation has the four scalar parameters ρ, α, d, and R, and the three vector parameters $\mathbf{P}$, $\mathbf{B}$, and $\mathbf{S}_1$. These 13 parameters combine with the constraint that $|\mathbf{S}_1| = 1$, to yield 12 independent parameters.

In order to simplify the use of Eq. (22.79), we introduce the parameters

$$k_1 = \mathbf{B} \cdot \mathbf{B} - \rho^2 - R^2 - d^2,$$
$$k_2 = (\mathbf{B} \cdot \mathbf{S}_1 + d\cos\alpha)\frac{2\rho}{\sin\alpha},$$
$$k_3 = 4\rho^2 R^2,$$
$$\mathbf{H} = \frac{2\rho}{\sin\alpha}\mathbf{S}_1. \quad (22.80)$$

These parameters allow us to write Eq. (22.79) in the form

$$(\mathbf{P}^2 - 2\mathbf{P}\cdot\mathbf{B} + k_1)^2 + (\mathbf{P}\cdot\mathbf{H} - k_2)^2 - k_3 = 0. \quad (22.81)$$

This is a quartic polynomial in the three scalars $k_i, i = 1, 2, 3$, and three vectors $\mathbf{P}$, $\mathbf{B}$, and $\mathbf{H}$.

Given a set of displacements $\hat{\mathbf{Q}}_i, i = 1, \ldots, 12$, we evaluate Eq. (22.81) on the points $\mathbf{P}^i = [T(\hat{\mathbf{Q}}_i)]\mathbf{p}, i = 1, \ldots, 12$. Subtract the first of these equations from the remaining to cancel k_3 and obtain

$$\mathcal{GT}_j : (\mathbf{P}^{j+1^2} - 2\mathbf{P}^{j+1} \cdot \mathbf{B} + k_1)^2 - (\mathbf{P}^{1^2} - 2\mathbf{P}^1 \cdot \mathbf{B} + k_1)^2$$
$$+ (\mathbf{P}^{j+1} \cdot \mathbf{H} - k_2)^2 + (\mathbf{P}^1 \cdot \mathbf{H} - k_2)^2 = 0, \quad j = 1, \ldots, 11. \quad (22.82)$$

The total degree of this system of polynomials is $4^{11} = 4,194,304$.

We can refine the estimate of the number of roots of this polynomial system by using the linear product decomposition. Expanding the polynomial $\mathcal{GT}_j$, we obtain the terms

$$\mathbf{P}^{j+1^4} - \mathbf{P}^{1^4} \in \langle x, y, z, 1 \rangle^3,$$
$$(2\mathbf{P}^{j+1} \cdot \mathbf{B})^2 - (2\mathbf{P}^1 \cdot \mathbf{B})^2 \in \langle x, y, z, 1 \rangle^2 \langle u, v, w \rangle^2,$$
$$-4\mathbf{P}^{j+1^2}(\mathbf{P}^{j+1} \cdot \mathbf{B}) + 4\mathbf{P}^{1^2}(\mathbf{P}^1 \cdot \mathbf{B}) \in \langle x, y, z, 1 \rangle^3 \langle u, v, w \rangle,$$
$$2k_1(\mathbf{P}^{j+1^2} - \mathbf{P}^{1^2} - 2\mathbf{P}^{j+1} \cdot \mathbf{B} + 2\mathbf{P}^1 \cdot \mathbf{B}) \in \langle x, y, z, 1 \rangle \langle u, v, w, 1 \rangle \langle k_1 \rangle,$$
$$(\mathbf{P}^{j+1} \cdot \mathbf{H})^2 - (\mathbf{P}^1 \cdot \mathbf{H})^2 \in \langle x, y, z, 1 \rangle^2 \langle h_1, h_2, h_3 \rangle^2$$
$$-2k_2(\mathbf{P}^{j+1} \cdot \mathbf{H} - \mathbf{P}^1 \cdot \mathbf{H}) \in \langle x, y, z, 1 \rangle \langle h_1, h_2, h_3 \rangle \langle k_2 \rangle \quad (22.83)$$

Notice that the quartic terms in the first expression cancel. We combine these monomials into the LPD

$$\mathcal{GT}_j : \langle x, y, z, 1 \rangle^2 \langle u, v, w, h_1, h_2, h_3, 1 \rangle \langle x, y, z, u, v, w, h_1, h_2, h_3, k_1, k_2, 1 \rangle |_j,$$
$$j = 1, \ldots, 11. \quad (22.84)$$

This allows us to compute the LPD bound of 448,702.

Our parallel POLSYS_GLP algorithm computed 42,786 solutions in 42min using 128 nodes of Blue Horizon. This is approximately 626 paths per processor-hour. Each real solution can be used to design an RRS chain to reach the specified displacements. The distribution and utility of these solutions requires further study.

22.13 Generalization of the Problem

Each of the polynomial systems that we have studied is an algebraic manifold in the Clifford algebra of dual quaternion coordinates. These constraint manifolds (McCarthy[19]) lie in what Ravani and Roth [21] call the *image space* of spatial displacements, or what Study [34] called *soma space*; also see Blaschke [35]. Thus, our results can be viewed as determining the parameters of a constraint manifold such that it passes through given set of points $\hat{\mathbf{Q}}_i$ in the image space of spatial displacements. Ravani introduced this approach to the synthesis of articulated chains.

While our focus has been limited to chains that have algebraic constraint equations, it is possible to extend these ideas to manifolds parameterized by the joint variables of an articulated chain by means of its kinematics equations.

The physical dimensions of the chain appear in these equations as parameters that define the manifold reachable by its end effector in the image space of spatial displacements. The joint variables introduce extra parameters that must be eliminated as part of the solution process, but it does allow us to consider articulated chains that do not have a spherical wrist.

Tsai and Roth [36] used the geometry of the screw triangle, which is a geometric representation of the kinematics equations, to formulate design equations for serial chains with two and three joints; also see [37]. In Tsai and Roth [38], they solved these equations to obtain a complete solution for the design of a spatial RR chain, (also see Perez and McCarthy [39] and Mavroidis et al. [40]).

Lee and Mavroidis [41, 42] and Lee et al. [43] formulated and solved the kinematics equations for the spatial RRR and PRR chains that reach a specified set of spatial displacements. Perez and McCarthy [44] formulated this problem directly in dual quaternion coordinates, and obtained explicit parameterized manifolds in the image space. Their "dual quaternion synthesis" technique has been applied to the four degree of freedom RPRP serial chain (Perez and McCarthy [45]).

As we have seen the complexity of these problems increases with the number of dimensional variables that define the manifold, or articulated chain. The general 5R spatial chain, which has at least 20 variables, 4 for each of the 5 joint axes, appears to be the most challenging case. More research is needed to determine the number of solutions to many of these parameterized constraint manifold fitting problems.

Table 22.3. Summary of the total degree, LPD bound, and number of solutions of the polynomial equations that define each reachable surface

Case	Surface	Total degree	LPD bound	Number of roots
1	Plane	32	10	10
2	Sphere	64	20	20
3	Circular cylinder	16,384	2,184	804
4	Circular hyperboloid	262,144	9,216	1,024
5	Elliptic cylinder	2,097,152	247,968	18,120
6	Circular torus	2,097,152	868,352	94,622
7	General torus	4,194,304	448,702	42,786

22.14 Conclusion

In this chapter we have examined the geometric problem of fitting an algebraic surface to points generated by a set of spatial displacements. We focus on seven surfaces that are associated with articulated chains that have spherical wrists. Table 22.3 lists the number of solutions for the polynomial systems for each of these reachable surfaces.

The complexity of this problem increases with the number of dimensional parameters that define the surface. In particular, the elliptic cylinder, circular torus, and general torus have 10, 11, and 12 parameters, respectively, and have solutions that number in the tens of thousands. This large number of solutions is remarkable and requires further study to identify those that are useful for practical applications.

This problem can be generalized to fitting the parameterized manifold defined by the kinematics equations of an articulated chain to a set of spatial displacements. Furthermore, rather than focus on finding a finite set of solutions to our design equations, we can specify fewer task positions and obtain a curve, surface, or submanifold of chains. These results have practical application in the design of devices that provide controlled spatial movement.

References

1. Schoenflies A. (1886) *Geometrie der Bewegung in Synthetischer Darstellung.* Leipzig, Germany. (See also the French translation: *La Géométrie du Movement*, Paris, 1983)
2. Burmester L. (1886) *Lehrbuch der Kinematik.* Verlag Von Arthur Felix, Leipzig
3. Hartenberg R., Denavit J. (1964) *Kinematic Synthesis of Linkages.* McGraw-Hill, New York
4. Sandor G.N., Erdman A.G. (1984) *Advanced Mechanism Design: Analysis and Synthesis, vol. 2.* Prentice-Hall, Englewood Cliffs, NJ
5. Chen P., Roth B. (1967) Design equations for finitely and infinitesimally separated position synthesis of binary link and combined link chains, *ASME J. Engineering for Industry* 91:209-219
6. Suh C. H., Radcliffe C.W. (1978) *Kinematics and Mechanism Design.* Wiley, New York
7. McCarthy J.M. (2000) *Geometric Design of Linkages.* Springer-Verlag, New York
8. Craig J.J. (1989) *Introduction to Robotics, Mechanics and Control*, Addison Wesley, MA
9. Tsai L.W. (1999) *Robot Analysis: The Mechanics of Serial and Parallel Manipulators.* John Wiley and Sons, New York
10. Bernshtein D.N. (1975) The number of roots of a system of equations. *Functional Anal. Appl.* 9(3):183–185
11. Morgan A.P, Sommese, A.J., Wampler, C.W. (1995) A product-decomposition bound for Bezout numbers. *SIAM J. of Numerical Analysis*,32(4):1308-1325
12. Nielsen J., Roth B. (1995) Elimination methods for spatial synthesis. Merlet J.P., Ravani B. (eds.) *Computational Kinematics, Solid Mechanics and Its Applications*, vol.40 Kluwer, Dordrecht, pp. 51–62
13. Husty M. L. (1996) An algorithm for solving the direct kinematics of general Stewart–Gough platforms. *Mech. Mach. Theory*, 31(4):365–380

14. Tsai, L.W., Morgan A.P. (1985) Solving the kinematics of the most general six- and five-degree-of-freedom manipulatorsby continuation methods. *ASME J. Mech. Trans. Automation Design*, 107:189–200
15. Verschelde J, Haegemans A. (1993) The GBQ-Algorithm for constructing start systems of homotopies for polynomial systems. *SIAM J. Numerical Analysis*, 30(2):583-594
16. Wise S.M., Sommese A.J., Watson L.T. (2000) Algorithm 801: POLSYS_PLP: A partitioned linear product homotopy code for solving polynomial systems of equations. *ACM Trans. Math. Software* 26(1):176–200
17. Su H.-J., McCarthy J.M., Watson, L.T. (2004) Generalized linear product homotopy algorithms and the computation of reachable Surfaces. *ASME Journal of Computers and Information Science and Engineering*, 4(3)
18. Bottema O., Roth B. (1979) *Theoretical Kinematics*, North Holland Press, NY
19. McCarthy J.M. (1990)*An Introduction to Theoretical Kinematics*,MIT Press, Cambridge, MA
20. Ravani B., Roth B. (1984) Mappings of spatial kinematics.*ASME J. of Mechanisms, Transmissions, and Automation in Design*,106(3):341–347
21. Ravani B., Roth B. (1983) Motion synthesis using kinematic mapping.*ASME J. of Mechanisms, Transmissions, and Automation in Design*,105(3):460–467
22. Wampler C. (1994) An efficientsStart system for multi-homogeneous polynomial continuation. *Numerical Mathematics*, 66:517-523
23. Raghavan M., Roth B. (1995) Solving polynomial systems for the kinematic analysis and synthesis of mechanisms and robotmanipulators. *ASME J. of Mechanical Design*, 117(B):71Đ79
24. Innocenti C. (1995) Polynomial solution of the spatial Burmester problem", *ASME J. Mech. Design* 117(1)
25. Liao Q.Z., McCarthy J.M. (2001) On the seven position synthesis of a 5-SS platform linkage.*ASME J. Mechanical Design*, 123(1):74-79
26. Raghavan M. (2002) Suspension Mmchanism synthesis for linear toe curves. *Proc. Des. Eng. Tech.Conf.* paper no. DETC2002/MECH-34305, Sept. 29–Oct. 2, Montreal, Canada
27. Wampler C.W., Morgan A.P., Sommese, A.J. (1990) Numerical continuation methods for solving polynomial systems arising in kinematics. *ASME Journal of Mechanical Design*, 112(1):59-68
28. Su H.-J., Wampler C., McCarthy J.M. (2003) Geometric design of cylindric PRS serial chains. *ASME Journal of Mechanical Design*, 126(2):269-277
29. Verschelde J. (1999) Algorithm 795: PHCpack: A general purpose solver for polynomial systems by homotopy continuation*ACM Transactions on Mathematical Software*, 25(2):251–276. Software available at http://www.math.uic.edu/~jan
30. Watson L.T., Sosonkina M., Melville R.C., Morgan A.P., Walker,H.F. (1997) Algorithm 777: HOMPACK90: A suite of Fortran 90 codes for globallyconvergent homotopy algorithms. *ACM Trans. Math. Software* 23, 514-549
31. Kim H.S., Tsai, L.W. (2002) Kinematic synthesis of spatial 3-RPS parallel manipulators. *Proc. ASME Des. Eng. Tech.Conf.* paper no. DETC2002/MECH-34302, Sept. 29–Oct. 2, Montreal, Canada
32. Gao T., Li T.Y., Wang X. (1999) Finding all isolated zeros of polynomial systems in C^n via stable mixed volumes. *J. Symbolic Comput.* 28(1-2):187–211
33. Gao T., Li T.Y., Wu, M. (2003) MixedVol: A software package for mixed volume computation. submitted to *ACM Transactions on Math. Sofware*, August

34. Study E. (1912) *Sitzungsberichte der Berliner Mathematischen Gesellschaft*, 104. Sitzung, 12 Dec. pp. 36-60
35. Blaschke W. (1960) *Kinematik and Quaternionen*. VEB, Berlin
36. Tsai L.W., Roth B. (1972) Design of dyads with helical, cylindrical, spherical, revolute and prismatic joints," *Mechanism and Machine Theory*, 7:591–598
37. Tsai L.W. (1972) *Design of open loop chains for rigid body guidance*, Ph.D. Thesis, Department of Mechanical Engineering, StanfordUniversity
38. Tsai L.W., Roth B. (1973) A note on the design of revolute–revolute cranks," *Mechanismand Machine Theory,* 8:23–31
39. Perez, A, and McCarthy, J.M., 2000, "Dimensional synthesis of Bennett linkages," *ASME Journal of Mechanical Design*, 125(1):98-104, March 2003
40. Mavroidis C., Lee E., Alam M. (2001) A new polynomial solution to the geometric design problem of spatial RR robot manipulators using theDenavit-Hartenberg parameters," *ASME J. Mechanical Design*, 123(1):58-67
41. Lee E., Mavroidis D. (2002) Solving the geometric design problem of spatial 3R robot manipulators using polynomial homotopy continuation.*ASME J. Mechanical Design*, 124(4):652-661
42. Lee E., Mavroidis D. (2002) Geometric design of spatial PRR manipulators using polynomial elimination techniques.*Proc. ASME 2002 Design Eng. Tech. Conf.*, paper no. DETC2002/MECH-34314, Sept. 29–Oct. 2, Montreal, Canada
43. Lee E., Mavroidis C., Merlet J.P. (2002) Five precision points synthesis of spatial RRR manipulators using interval analysis. *Proc. ASME 2002 Design Eng. Tech. Conf.*, paper no. DETC2002/MECH-34272, Sept. 29–Oct. 2, Montreal, Canada
44. Perez A., McCarthy J.M. (2002) Dual quaternion synthesis of constrained robots. *Advances in Robot Kinematics*, (J.Lenarcic and F. Thomas, eds.) Kluwer, Dordrecht, pp.443–454, Barcelona, Spain, June 24–29
45. Perez A., McCarthy J.M. (2003) "Dual quaternion synthesis of constrained robotic systems," *ASME Journal of Mechanical Design*, 126(3):425-435

Planning Collision-Free Paths Using Probabilistic Roadmaps

Seth Hutchinson[1] and Peter Leven[2]

[1] University of Illinois at Urbana-Champaign, Urbana, Illinois seth@uiuc.edu
[2] Hewlett-Packard, San Diego, California p.leven@computer.org

Planning collision-free paths is one of the central research problems that confronts intelligent robotics. In its simplest form, the path planning problem is to determine a path in the configuration space that moves the robot from an initial configuration to a goal configuration, such that the robot never contacts any object in its environment. Even this most basic problem is computationally intractable, and at present, the best known algorithms for its solution require time that grows exponentially with the dimension of the robot's configuration space [14, 36].

Because of the inherent complexity of the planning problem, in recent years a number of probabilistic approaches have been developed [5, 19, 27, 50, 54]. These algorithms work by constructing a set of randomly generated sample configurations, and connecting these samples using local planning algorithms that require only moderate computation. The result is a graph in the configuration space that is referred to as a *probabilistic roadmap* (PRM). These approaches sacrifice completeness for computational efficiency, but one can often derive limiting properties of the algorithms, such as probabilistic completeness (i.e., a guarantee that with probability one the algorithm will find an existing solution as the number of samples approaches infinity).

Most of the PRM planners require significant preprocessing to construct the roadmap. The idea that the cost of planning will be amortized over many planning episodes provides a justification for spending extensive amounts of time during this preprocessing stage, provided the resulting representation can be used to generate plans very quickly during a query stage. Thus, these planners use a two-stage approach. During the preprocessing stage, the planner generates a set of vertices that correspond to random configurations in the configuration space, connects these vertices using a simple, local path planner to form a roadmap, and, if necessary, uses a subsequent sampling stage to enhance the roadmap. During the second, on-line stage, planning is reduced to query processing, in which the initial and final configurations are connected to the roadmap, and the augmented roadmap is searched for a feasible path.

These planners tend to be easy to implement, but there are many design choices that affect overall performance, both in terms of the required computation and in terms of the success of the planner in finding paths in complicated environments. In this chapter, we investigate a number of these design choices. We begin with a brief review of the lineage of these planners, followed by a brief overview of how they function. Following this, we investigate several specific aspects of the planning process, including sample generation, rejection-based importance sampling techniques, transforming samples to improve coverage of the configuration space, connecting samples to form a graph, and enhancing the roadmap to improve connectivity.

23.1 The Evolution of Sampling-Based Methods

The earliest work in path planning produced exact algorithms (for example, [14, 52]) and methods that build an approximate representation of the full volume of configuration space (for example, [13, 23, 45]). In the former case, the best known algorithms have exponential complexity and require exact descriptions of both the robot and its environment, whereas in the latter case, the size of the representation of configuration space grows exponentially in the dimension of the configuration space.

The fact that real robots rarely have an exact description of the environment, coupled with a desire for real-time planning, led to the development of potential field approaches [22, 32, 34]. The idea of potential field approaches is to construct a scalar function over the configuration space that represents the goal region as the global minimum in the field and the obstacles as local maxima. Path planning is then reduced to following the gradient of the potential function until the goal is reached. The advantage of this approach is that the potential functions are easy to compute, making the planner fast. Unfortunately, it is difficult to create potential functions with a single global minimum at the goal; therefore, these planners are easily trapped by local minima.

The problems of local minima in potential field planners led to the development of randomized planning [8]. In this approach, when a local minimum is detected, a random motion is performed to try to escape the local minimum. Planning then can be considered a graph search, where the vertices of the graph are the sequence of local minima encountered when searching for the goal.

The randomized motion planners proved effective for a large range of problems, but required extensive computation time for some robots in certain environments [21, 31]. This limitation, together with the idea that a robot will operate in the same environment for a long period of time, led to the development of the probabilistic roadmap (PRM) planners [27, 50]. As mentioned above, these planners use a preprocessing stage to create the PRM, and plans are generated at run time in the query stage.

The two stages of the probabilistic roadmap planners can be described as follows. In the preprocessing stage, a roadmap is constructed in the free configuration space. The vertices of the roadmap are created by some random sampling scheme, and pairs of vertices are connected using a simple local planner. After construction, this roadmap may contain more than one connected component, in which case an enhancement operation may be performed to try to connect the different components together. In the second stage, planning queries are performed. For each planning query, the initial and goal configurations are connected to the roadmap and the resulting augmented roadmap is searched for a path.

The PRM planners that use this two-stage processing mechanism are targeted toward environments in which the obstacles are stationary and their positions are known in advance. For environments for which this assumption does not hold, single-query variants were developed that build the roadmap as they search for a path [21, 35, 53]. PRM type approaches have also been used for sensor-based exploration of unknown environments. For example, [48] describes a robot equipped with a skin sensor to explore the environment using a lazy-PRM approach.

There have also been a number of sampling-based approaches to path planning in changing environments. In some cases, planners have execution times that make it feasible to directly use them in some kinds of changing environments with no modifications. This is the case, for example, for the Ariadne's Clew algorithm reported in [9, 46]. The Ariadne's Clew algorithm operates by generating landmarks (during an exploration phase) and then connecting them to the existing roadmap (the search phase). Variations of this algorithm can be obtained by varying the search phase and by using different optimization criteria to select candidate landmarks [2, 47]. The idea of incrementally expanding a roadmap for single-query planning has also been used in [21] and [35]. In both of these, roadmaps are grown from both the initial and goal configurations until they can be connected, though the details for expanding the roadmap differ. The incremental expansion method in [21] has also been used in a dynamic environment [33]. In [53] an adaptable approach that uses multiple local planners is described. At run time, characteristics of the problem are used to determine which (combination of) local planners will be most effective.

23.2 An Overview of the Approach

Sampling-based approaches construct a roadmap that represents a set of paths in the free configuration space of the robot. This roadmap itself is represented as a graph $G = (V, E)$ in which the vertices correspond to free configurations of the robot and edges correspond to free paths between these configurations. Constructing the roadmap consists of generating the vertices and edges in G. The typical procedure is illustrated in Fig. 23.1.

```
1.   V ← ∅, E ← ∅
2.   while |V| < N
3.       v ← GenerateSample()
4.       if Reject(v)
5.           goto 3
6.       endif
7.       v ← Transform(v)
8.       forall v' ∈ Neighborhood(v)
9.           if e ← Connect(v, v')
10.              E ← E ∪ {e}
11.          endif
12.          V ← V ∪ {v}
13.      endfor
14.  end while
```

Fig. 23.1. Algorithm to generate a sample-based roadmap in the configuration space

The set of candidate configurations generated by step 3 can be created using a random number generator (perhaps the most common approach), or using a deterministic algorithm that generates a sequence of configurations that satisfy user-specified criteria. The earliest of the sampling-based approaches used the former approach [27, 50]. Using this approach, a number of interesting properties can be proven using the randomized aspect of the underlying algorithm (see, e.g., [25, 29, 30, 51]). In the latter case, samples are typically generated so that they satisfy some sort of uniformity criteria. The simplest approach is to use a uniform grid to generate the samples, but more-sophisticated approaches have recently been introduced into the motion planning literature (see, e.g., [12, 38]). Each of these approaches is described below in Sect. 23.3.

Generating samples (step 3) is nearly always done using a simple algorithm that does not take into account any features that are specific to the current motion planning problem. For example, random number generators generate independent samples, distributed uniformly on some transformed version of configuration space (typically the unit cube in n dimensions), and do not consider the geometry of the obstacles or even the topology of the configuration space. The advantage is that samples can be generated rapidly. However, the quality of the resulting set of samples may be low. For example, samples may lie in collision configurations, or they may fail to cover the "interesting" parts of the configuration space. In order to obtain a good set of samples, most all approaches generate a very large number of samples using step 3, large enough that some subset of those samples will have the desired properties. Since this set will often contain samples that have little value for motion planning, step

4 is used to reject samples, thus producing a roadmap of reasonable size that possesses the desired properties.

Many criteria have been proposed for the rejection in step 4. Indeed, this has been one of the major research thrusts in recent years by PRM researchers. The most conspicuous reason to reject a sample is that it lies in the configuration space obstacle region (however, some methods retain these samples and transform them into free configurations in step 7). Other reasons to reject samples include redundancy (e.g., if the existing samples already cover a particular region of the configuration space), location (e.g., some algorithms reject most samples that do not lie near the boundary of the configuration space obstacle region), or properties of the local geometry of the configuration space (e.g., rejecting samples for which manipulability of the robot arm is low). We describe a number of these approaches in Sect. 23.4.

An alternative to rejecting samples is to transform them so that they obtain more desirable properties (step 7). For example, if a sample lies within the configuration space obstacle region, it can be moved until it lies just beyond an obstacle boundary. The resulting transformed sample will lie near the obstacle boundary, which can be beneficial in certain applications. Other examples of this approach include pushing samples toward the medial axis of the free configuration space, and the method used to construct rapidly-exploring random trees (RRTs). These and other methods to transform samples are described in Sect. 23.5.

Once a set of vertices has been generated, pairs of vertices are connected by edges that represent free paths in the configuration space (steps 8-11). This is done by identifying candidate pairs of vertices, and then using a local path planner to find the free path between the corresponding configurations. Candidate pairs are typically chosen using a k-nearest neighbor scheme, sometimes enforcing the restriction that the resulting graph be acyclic. Determining the nearest neighbors for a vertex relies on a distance function, and a number of these have been explored in the literature. For the local planning algorithm, a simple straight-line planner is often used. We discuss these and related issues in Sect. 23.6.

The algorithm shown in Fig. 23.1 constructs a roadmap; however, the roadmaps constructed using this algorithm often contain multiple components, even when the free configuration space consists of a single connected component. For this reason, many planners use a final enhancement step to add vertices to the roadmap in an effort to add edges that connect distinct connected components. This is often done by generating new vertices by densely sampling "difficult" regions of the configuration space. We describe enhancement strategies in Sect. 23.7.

After enhancement, a planner can use the roadmap to generate plans very quickly. This is done by using a local planner to connect the initial and final configuration to the roadmap, then searching the augmented roadmap for a path that connects these two new vertices. In some cases, an additional step is used to smooth the resulting path. In Fig. 23.2 the entire process is

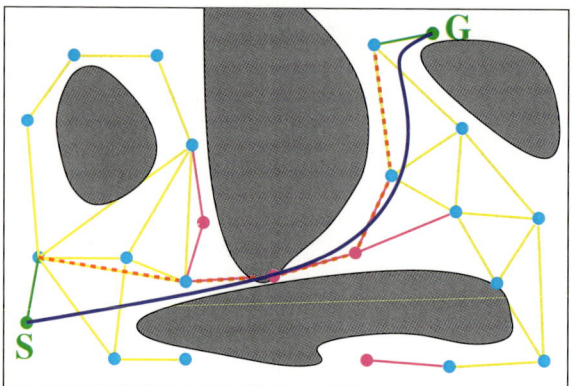

Fig. 23.2. Planning using sample-based roadmaps

illustrated. In the figure, the lightly shaded vertices are added by the algorithm of Fig. 23.1, and darkly shaded vertices are added during enhancement. The vertices labeled S and G are the initial and goal configurations. The dashed line is the resulting path in the roadmap, and the dark line is a smoothed version of the path.

23.3 Generating Sample Configurations

The easiest way to generate sample configurations is to invoke a random number generator. Since most readily available random number generators return a scalar sample from the uniform distribution on the unit interval, we will here define our sample space X to be the unit d-cube, i.e., $X = [0, 1]^d$. A sample p from X can be generated by merely invoking the random number generator d times, once for each coordinate of p. Once a sample has been generated in this way, it can be transformed to the appropriate space, since the configuration space for most robots is not the unit d-cube. Since this transformation is straightforward, we restrict our attention here to sampling $X = [0, 1]^d$.

Recently, a number of deterministic alternatives to random sampling were introduced [12, 38]. These alternatives aim to optimize various properties of the distribution of the samples on X. Before introducing some of these alternatives, we will briefly describe two ways to evaluate the quality of a set of samples on X. A more comprehensive introduction to these ideas can be found in [37].

Let P be a set of point samples, and N be the number of points in P. One way to evaluate the quality of the samples in P is to assess how "uniformly" the points in P cover the space. This is done with respect to a specific collection of subsets of X, called a *range space*, denoted by $\mathcal{R}$. Let $\mathcal{R}$ be the set of all axis-aligned rectangular subsets of X. Since P contains N points, the fraction of samples contained in $R \in \mathcal{R}$ is given by the ratio

$$\frac{|P \cap R|}{N}.$$

If we define μ to be the measure (or volume) of a set, we can compare the relative volume of R to the relative portion of samples that lie in R,

$$\left| \mu(R) - \frac{|P \cap R|}{N} \right|,$$

since X is the unit cube and thus $\mu(X) = 1$. If we take the supremum of this difference over all $R \in \mathcal{R}$ we obtain the concept of *discrepancy*.

Definition. The *discrepancy* of point set P with respect to range space $\mathcal{R}$ over some space X is defined as

$$D(P, \mathcal{R}) = \sup_{R \in \mathcal{R}} \left| \mu(R) - \frac{|P \cap R|}{N} \right|.$$

It is not necessary to take $\mathcal{R}$ as the subset of axis-aligned rectangles, but this choice gives an intuitive understanding of discrepancy. Another common choice is to take $\mathcal{R}$ as the set of d-balls, i.e., for each $R \in \mathcal{R}$ we have $R = \{x' \mid \|x - x'\| < \epsilon\}$, for some point x and radius $\epsilon > 0$.

While discrepancy provides a measure of how uniformly points are distributed over the space X, *dispersion* provides a measure of the largest portion of X that contains no points in P. For a given metric ρ, the distance between a point $x \in X$ and a point $p \in P$ is given by $\rho(x, p)$. Thus,

$$\min_{p \in P} \rho(x, p)$$

gives the distance from x to the nearest point in P. If we take ρ to be the Euclidean metric, this gives the largest empty ball centered on x. If we then take the minimization over all points in X, we obtain the size of the largest empty ball in X. This is exactly the concept of dispersion.

Definition. The *dispersion* δ of point set P with respect to the metric ρ is given by

$$\delta(P, \rho) = \sup_{x \in X} \min_{p \in P} \rho(x, p).$$

If we instead use the L_∞ norm, $\rho(x, p) = \max_i |x_i - p_i|$, the dispersion gives the size of the largest empty axis-aligned rectangular subset of X. In this case, we obtain a relationship between discrepancy and dispersion:

$$\delta(P, \rho) \leq D(P, \mathcal{R})^{\frac{1}{d}}$$

in which d is the dimension of X. Thus, in this case, if P has low discrepancy, it will also have low dispersion, since the latter is bounded from above by the former.

An important result found by Sukharev gives a bound on the number of samples required to achieve a given dispersion. In particular, the *Sukharev sampling criterion* states that when ρ is taken as the L_∞ norm, a set P of N samples on the d-dimensional unit cube will have

$$\delta(P, \rho) \geq \frac{1}{2 \lfloor N^{\frac{1}{d}} \rfloor}.$$

So, to achieve a given dispersion value, say δ^*, since N must be an integer, we have

$$\delta^* \geq \frac{1}{2 \lfloor N^{\frac{1}{d}} \rfloor} \to N \geq \left(\frac{1}{2\delta^*}\right)^d,$$

i.e., the number of samples required to achieve a desired dispersion grows exponentially with the dimension of the space. In some sense, this result implies that to minimize dispersion, sampling on a regular grid will yield results that are as good as possible.

Now that we have quantitative measures for the quality of a set of samples, we describe some common ways to generate samples. For the case of $X = [0, 1]$ the *van der Corput sequence* gives a set of samples that minimizes both dispersion and discrepancy. The nth sample in the sequence is generated as follows. Let $a_i \in \{0, 1\}$ be the coefficients that define the binary representation of n,

$$n = \sum_i a_i 2^i = a_0 + a_1 2 + a_2 2^2 \cdots.$$

The nth element of the van der Corput sequence $\Phi(n)$ is defined as

$$\Phi(n) = \sum_i a_i 2^{-(i+1)} = a_0 2^{-1} + a_1 2^{-2} \cdots.$$

Fig. 23.3a shows the first 16 elements of a van der Corput sequence. Figure 23.4 shows how the values of discrepancy and dispersion vary as samples are added to the van der Corput sequence.

The van der Corput sequence can only be used to sample the real line. The *Halton sequence* generalizes the van der Corput sequence to d dimensions. Let $\{b_i\}$ define a set of d relatively prime integers, e.g., $b_1 = 2$, $b_2 = 3$, $b_3 = 5$, $b_4 = 7 \cdots$. The integer n has a representation in base b_j given by

$$n = \sum_i a_{ij} b_j^i, \quad a_{ij} \in \{0, 1 \cdots b_j\},$$

and we define $\Phi_{b_j}(n)$ as

$$\Phi_{b_j}(n) = \sum a_{ij} b_j^{-(i+1)}.$$

n	n binary	$\Phi(n)$ binary	$\Phi(n)$	n	$\Phi_2(n)$	$\Phi_1(n)$
0	0	0.0	0	0	0	0
1	1	0.1	1/2	1	1/3	1/2
2	10	0.01	1/4	2	2/3	1/4
3	11	0.11	3/4	3	1/9	3/4
4	100	0.001	1/8	4	4/9	1/8
5	101	0.101	5/8	5	7/9	5/8
6	110	0.011	3/8	6	2/9	3/8
7	111	0.111	7/8	7	5/9	7/8
8	1000	0.0001	1/16	8	8/9	1/16
9	1001	0.1001	9/16	9	1/27	9/16
10	1010	0.0101	5/16	10	10/27	5/16
11	1011	0.1101	13/16	11	19/27	13/16
12	1100	0.0011	3/16	12	4/27	3/16
13	1101	0.1011	11/16	13	13/27	11/16
14	1110	0.0111	7/16	14	22/27	7/16
15	1111	0.1111	15/16	15	7/27	15/16

(a)   (b)

Fig. 23.3. a van der Corput sequence. **b** Halton sequence for $d = 2$

N	2	4	8	16
Discrepancy	1/2	1/4	1/8	1/16
Dispersion	1/2	1/4	1/8	1/16

Fig. 23.4. Discrepancy and dispersion for increasing values on N for the van der Corput sequence

The nth sample is then defined by the coordinates $p_n = (\Phi_{b_1}(n), \Phi_{b_2}(n) \cdots \Phi_{b_d}(n))$. Fig. 23.3b shows the first 16 elements of a Halton sequence for $b_1 = 2, b_2 = 3$.

When the range space $\mathcal{R}$ is the set of axis-aligned rectangular subsets of X, the discrepancy for the Halton sequence is bounded by

$$D(P, \mathcal{R}) \leq O\left(\frac{\log^d N}{N}\right).$$

When the range space $\mathcal{R}$ is the set of d-balls, the discrepancy is bounded by

$$D(P, \mathcal{R}) \leq O\left(N^{-\frac{(d+1)}{2}}\right).$$

One difficulty in generating sequences with low discrepancy is that the number of samples N is not specified a priori. Thus, samples are generated incrementally, and this must be done in a manner that maintains (or improves) discrepancy. The van der Corput sequence achieves this by sampling progressively smaller subintervals of the unit interval. The best asymptotic discrepancy that can be attained by such a sequence (i.e., a sequence for which

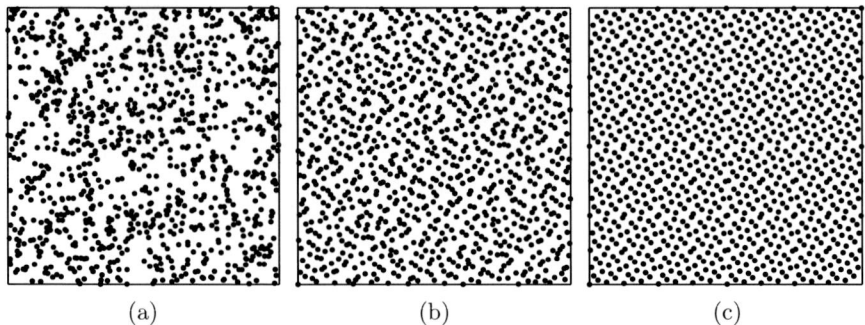

Fig. 23.5. These figures shows 1024 samples generated in the plane using (**a**) a random number generator, (**b**) a Halton sequence, (**c**) a Hammersley sequence

N is not specified a priori) is $O(N^{-1} \log^d N)$. When N is known in advance, we can do better. In this case, the best asymptotic discrepancy that can be attained is $O(N^{-1} \log^{d-1} N)$. In either case, the best asymptotic dispersion that can be attained is $O(N^{-1/d})$.

When N is specified, a *Hammersley sequence* (sometimes called a Hammersley point set, since the number of points is known and finite) achieves the best possible asymptotic discrepancy. The nth point in a Hammersley sequence is obtained by using the first $d-1$ coordinates of a point in the Halton sequence, with the ratio n/N as the first coordinate,

$$p_n = (n/N, \Phi_{b_1}(n), \Phi_{b_2}(n) \cdots \Phi_{b_{d-1}}(n)), \qquad n = 0, \ldots, N-1.$$

Figure 23.5 shows point sets generated using random number generator (Fig. 23.5a), a Halton sequence (Fig. 23.5b), and a Hammersley sequence (Fig. 23.5c). Each point set contains 1024 points.

23.4 Sample Rejection

As described in section 23.2, many sample-based methods for path planning use a rejection step, in which samples that do not have desired properties are discarded. This essentially amounts to *importance sampling*, an idea that is well known in the statistics and numerical integration literature, since the idea is to concentrate samples in the more important regions of the configuration space. Samples can be rejected using either a deterministic or probabilistic algorithm. The simplest deterministic case is to reject sample configurations that lie in the configuration space obstacle region (i.e., sample configurations for which the robot contacts some obstacle). Probabilistic rejection schemes are more popular, owing in part to asymptotic properties that can be proven for such methods. Below we describe a number of probabilistic rejection schemes. We describe in some detail an approach that rejects

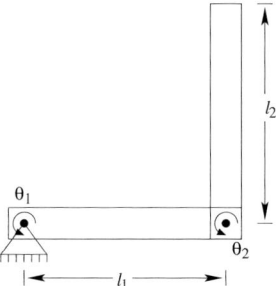

Fig. 23.6. A two-link planar robot arm

samples based on manipulability. Following this, we give brief descriptions of several other schemes.

23.4.1 Manipulability-based sampling

In [44], we introduced a method for biasing the sampling during the vertex-generation stage used to build a PRM. Our method is based on *manipulability* [55], an intrinsic property of robot arms, which measures an arm's freedom to move in all directions. Our rationale for this approach is that in regions of the configuration space where manipulability is high, the robot has great dexterity, and therefore relatively fewer samples should be required in these areas. Conversely, regions in which the manipulability is low tend to be near (or to include) singular configurations of the arm, where the range of possible motions is reduced; therefore such regions should be sampled more densely. Another interpretation of this is that for regions of the configuration space where manipulability is low, large joint motions correspond to small workspace motions. Thus, in these regions, traversing small paths in the workspace requires traversing relatively longer paths in the configuration space, consequently increasing the chance that such a path would intersect the configuration space obstacle region.

Let $J(q)$ denote the manipulator Jacobian matrix (i.e., the matrix that relates velocities of the end effector to joint velocities). For a redundant arm (e.g., an arm with more than six joints for a 3D workspace) the manipulability in configuration q is given by

$$\omega(q) = \sqrt{\det J(q)J^T(q)}.$$

Consider the robot shown in Fig. 23.6 as an example. The manipulability for this robot is $\omega = l_1 l_2 |\sin \theta_2|$, where l_1 and l_2 are the lengths of the two links. The configuration shown in Fig. 23.6 corresponds to one of the configurations at which the manipulability is highest for this robot. For this robot, the manipulability does not depend on the position of the first joint.

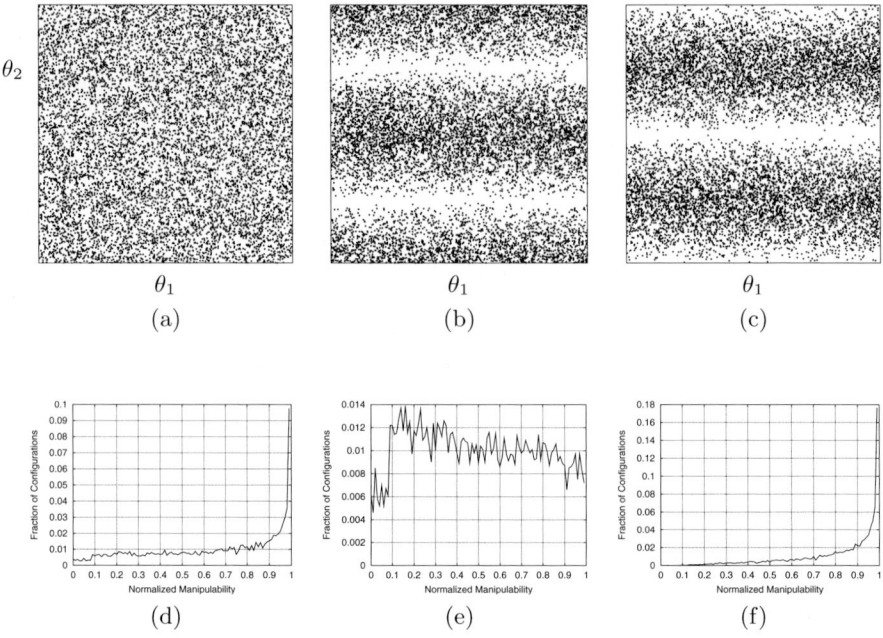

Fig. 23.7. a-c. Sample distributions for a two-joint planar robot: (a) uniform, (b) higher density in regions of low manipulability, (c) higher density in regions of high manipulability. **d-f.** Histograms for the sample distributions shown in a–c: (d) uniform, (e) higher density in regions of low manipulability, (f) higher density in regions of high manipulability

Three different sample distributions for this robot are shown in Fig. 23.7. As shown in the figure, concentrating the sampling in regions of low manipulability results in more samples near $\theta_2 = -\pi$, 0, and π at the bottom, middle, and top of the views of configuration space, respectively (Fig. 23.7b); sampling in regions of high manipulability results in more samples near $\theta_2 = -\pi/2$ and $\pi/2$ (Fig. 23.7c).

In order to bias sampling based on manipulability, we can use an approximation of the cumulative density function (CDF) for manipulability. If we treat manipulability as a random quantity, denoted by the random variable Ω, with probability density function p_Ω, the CDF is given by

$$P_\Omega(\omega) = \int_0^\omega p_\Omega(t) \mathrm{d}t.$$

We compute a discrete representation of P_Ω as follows. First, we create a discrete approximation to p_Ω. This is done by sampling the configuration space of the robot uniformly at random and computing the manipulability for each sample configuration. We exclude from this computation any configuration in which the robot collides with itself. We then create a histogram of the manipulability values that have been computed. We normalize the number in each bucket of the histogram and create the approximation to P_Ω from these normalized values.

The CDF for manipulability P_Ω can be used to drive a a rejection-based approach to bias the sampling of the configuration space. For each sample, we use the following procedure. First, a candidate sample q_c is generated using uniform random sampling of the configuration space. If q_c is a self-collision configuration, it is rejected. If q_c is not rejected, we compute the manipulability $\omega(q_c)$. We reject q_c with probability $P_\Omega(\omega(q_c))$. This approach was used to generate the sample distribution in Fig. 23.7b. In Fig. 23.7c, we use $P_\Omega(\omega(q_c))$ as the probability of acceptance. Many rejection-based approaches to importance sampling use a CDF in this way to reject samples with finite probability (rather than deterministically deciding whether or not to reject a sample).

One shortcoming of the manipulability measure for our purposes is that it does not reflect joint limits. When the robot is near a joint limit, its movement is restricted. In an effort to include samples near joint limits we can adopt the following convention: at configurations in which some joint is near a limit, the manipulability is defined to be zero. The nearness of a joint to its limit is a parameter of the sampling algorithm.

Example manipulability histograms are shown in Fig. 23.8a, c, and e. As can be seen in the figure, the manipulability histograms tend to be unimodal, and quite smooth. Figures 23.8b, d, and f show the histogram of manipulability of vertices that are selected using biased sampling. Note that sampling biased toward low manipulability has a tendency to shift the histogram to the left, while sampling biased toward high manipulability has a tendency to shift the histogram to the right. In these figures, the plots labeled "Not filtered" correspond to sampling the manipulability of the robot without filtering out samples in which the robot is in self-collision; the plots labeled "Filtered" exclude such samples. In all cases, 10 million samples were evaluated for manipulability. In addition, the gnuplot "csplines" function was used to smooth the plots.

To evaluate sampling biased by manipulability, we used a modified form of the original PRM planner for planar fixed-based articulated robots described in [24]. In particular, we added a function to the preprocessing phase to compute whether to reject a configuration based on its manipulability. We further modified the planner to adjust the order in which tests are applied to a random sample of the configuration space to determine whether to accept a sample. For each random sample, we test first whether the robot is in self-collision, then we apply the manipulability bias criterion, and last test the sample for collision between the robot and the obstacles. If the sample passes all tests, it is added to the roadmap. The remainder of the preprocessing phase continues as described in [24].

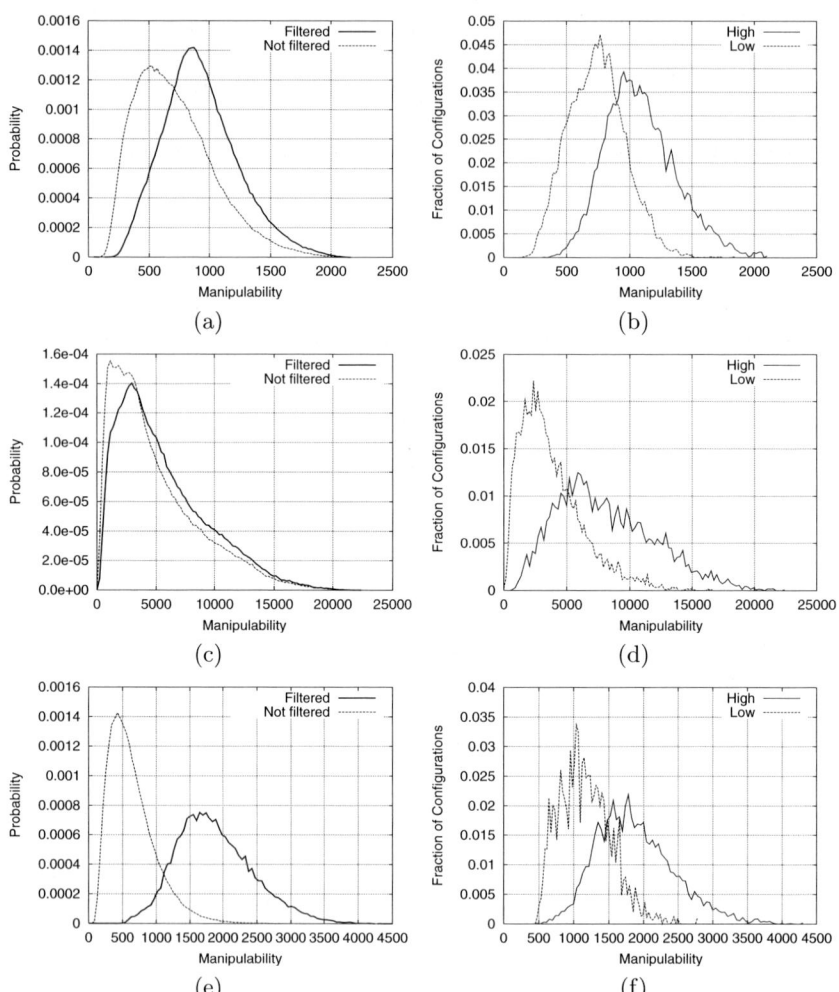

Fig. 23.8. (**a**) The histogram for the manipulability of a planar robot with six joints. (**b**) The histogram for the manipulability of the roadmap vertices for this robot. (**c**) The histogram for the manipulability of a robot with six joints in a 3D workspace. (**d**) The histogram for the manipulability of the roadmap vertices for this robot. (**e**) The histogram for the manipulability of a planar robot with 20 joints. (**f**) The histogram for the manipulability of the roadmap vertices for this robot

To evaluate the planner, we performed a similar set of experiments to those described in [24]: for each set of parameters, we generate 40 roadmaps and then test whether 8 test configurations (Fig. 23.9), can be connected to the roadmap. As a baseline, we include the results using unbiased sampling.

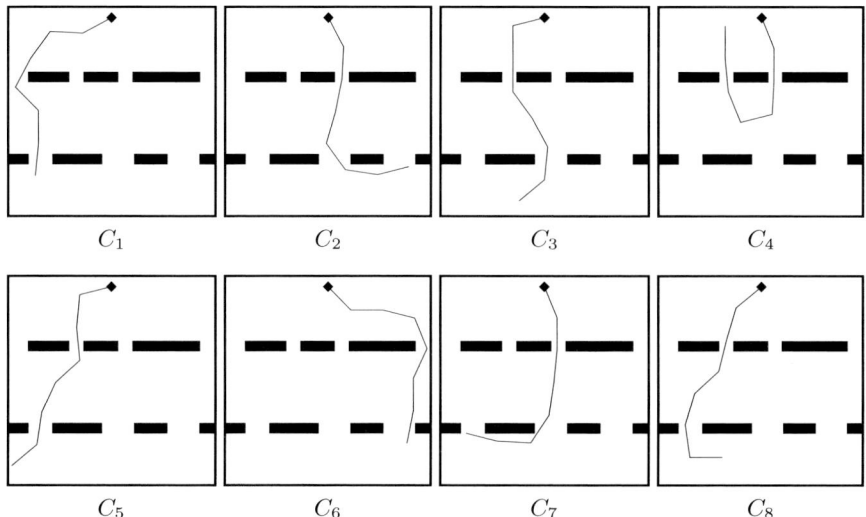

Fig. 23.9. Eight configurations of a 7-revolute-joint fixed-base robot

An explanation for the labels on the tables in Fig. 23.10 is as follows: The columns marked "Nodes" represents the target number of vertices for the roadmap after preprocessing, with "N" vertices generated during random sampling and "M" vertices generated during enhancement. The columns labeled "Rejected" list the number of vertices that failed a test: manipulability bias ("Manip."), or robot collision with an obstacle ("Obstacle"). The next three columns show three more statistics for the preprocessing phase. The column labeled "Avg. size" lists the average size of the largest connected component in the roadmap after preprocessing. The column labeled "Avg. comps." lists the average number of components in the roadmap after preprocessing, and the column labeled "Avg. time" lists the average processing time required by the preprocessing phase. The last columns show the success rate over the 40 roadmaps of connecting the configurations shown in Fig. 23.9.

We begin by noting that some of our results in Fig. 23.10a are slightly better than those originally reported in [24]. This can be attributed to improvements in computing power since those early results were published. It can be seen in Figs. 23.10b and c that the importance sampling scheme is significantly more selective than unbiased approaches. The manipulability-based rejection criterion rejects 2 to 3 times the number of vertices as are rejected due to collision with obstacles. Thus, one can see from these tables the trade-off between efficacy in vertex selection and the amount of computation required to construct the PRM.

By comparing Tables 23.10a and c, it can be seen that using manipulability-biased sampling *without* enhancement produces PRMs that are nearly as effective as those that are produced by unbiased sampling *with* enhancement

Nodes		Rejected	Avg.	Avg.	Avg.	Connection success rate (%)							
N	M	obstacle	size	comps.	time	C_1	C_2	C_3	C_4	C_5	C_6	C_7	C_8
800	400	43799	911	59	9.604	100	63	60	60	60	100	65	58
1000	500	54253	1253	52	12.978	100	78	78	78	78	100	78	78
1200	600	65137	1584	48	16.559	100	88	88	88	90	100	88	88
1400	700	75763	1916	43	20.338	100	98	90	98	93	100	98	90
1600	800	86218	2240	42	24.157	100	98	100	98	100	100	98	100
1800	900	97039	2534	40	28.108	100	100	98	100	98	100	100	98
2000	1000	108022	2862	37	32.126	100	100	100	100	100	100	100	100
2200	1100	118841	3144	35	36.323	100	100	100	100	100	100	100	100
2400	1200	129942	3456	34	40.626	100	100	100	100	100	100	100	100
2600	1300	140305	3747	32	44.903	100	100	100	100	100	100	100	100
3000	1500	161483	4359	31	53.779	100	100	100	100	100	100	100	100

(a) Results for unbiased sampling with enhancement

Nodes		Rejected	Avg.	Avg.	Avg.	Connection success rate (%)							
N	M	manip.	size	comps.	time	C_1	C_2	C_3	C_4	C_5	C_6	C_7	C_8
800	400	96440	970	43	10.793	98	75	85	75	85	100	75	85
1000	500	74035	1366	37	14.509	100	95	98	95	98	100	95	98
1200	600	88094	1651	36	18.473	100	95	98	95	98	100	95	98
1400	700	103999	1955	36	22.697	100	95	100	95	100	100	95	100
1600	800	118832	2275	33	26.885	100	100	100	100	100	100	100	100
1800	900	134298	2567	31	31.233	100	100	100	100	100	100	100	100
2000	1000	148752	2868	28	35.729	100	100	100	100	100	100	100	100
2200	1100	163294	3156	28	40.162	100	100	100	100	100	100	100	100
2400	1200	177390	3463	27	44.795	100	100	100	100	100	100	100	100
2600	1300	192478	3758	25	49.453	100	100	100	100	100	100	100	100
3000	1500	222769	4348	25	59.231	100	100	100	100	100	100	100	100

(b) Sampling biased toward low manipulability with enhancement

Nodes		Rejected	Avg.	Avg.	Avg.	Connection success rate (%)							
N	M	manip.	size	comps.	time	C_1	C_2	C_3	C_4	C_5	C_6	C_7	C_8
1200	0	144614	605	136	10.558	80	48	50	48	50	60	48	50
1500	0	181626	913	135	14.482	93	48	63	48	63	80	48	63
1800	0	217943	1367	142	18.474	98	65	83	65	83	98	65	83
2100	0	253770	1647	148	22.638	100	68	85	68	85	98	68	85
2400	0	290852	2085	150	26.852	100	85	95	85	95	100	85	95
2700	0	326285	2472	152	31.224	100	98	98	98	98	100	98	98
3000	0	364893	2709	156	35.763	100	90	100	90	100	100	90	100
3300	0	399917	3092	157	40.265	100	100	100	100	100	100	100	100
3600	0	436321	3385	159	44.754	100	100	100	100	100	100	100	100
3900	0	473301	3678	163	49.450	100	100	100	100	100	100	100	100
4500	0	546276	4258	170	59.148	100	100	100	100	100	100	100	100

(c) Sampling bias toward lower manipulability without enhancement

Fig. 23.10. Comparison of results using unbiased sampling and sampling biased by manipulability

(we describe enhancement in section 23.7). This indicates that it may be possible to drive PRM enhancement using primarily intrinsic properties of the robot arm, as opposed to properties that are specific to the obstacles in a given workspace. This opens the door for new representations that can be constructed for arbitrary workspaces, as in some of our related work [42, 43].

From this single set of experiments, one should not draw the conclusion that biasing samples toward regions of low manipulability will always lead to improved performance. Indeed, in the experiments presented here, we have chosen an environment with many small passages, and we use planning problems that often require the robot to operate near singularities (e.g., when the robot must "stretch" to reach a goal). While we believe that for these kinds of environments our approach will lead to performance improvements, it seems equally clear that little would be gained by applying our approach in sparsely populated environments.

Furthermore, even though the results shown here for biasing toward low manipulability are quite good, it should be noted that the results for biasing toward high manipulability are also reasonably good. For results presented in [44], biasing toward higher manipulability gave roadmaps that were as good or better than the traditional PRM for about half of the problems, performing more than 10% worse than the traditional approach only 18% of the time. From this, we surmise that there may be environments for which biasing toward higher manipulability would be more appropriate. This can be justified intuitively by noting that vertices in regions of the configuration space in which manipulability is high have the potential to be connected to many configurations, possibly generating roadmaps with higher connectivity.

23.4.2 The Visibility Roadmap

The rejection scheme discussed above rejects a configuration based on intrinsic properties of the robot evaluated at that configuration. In some sense, this is a first-order rejection scheme, since relationships between sample configurations are not considered. The approach introduced in [49] rejects sample configurations based on their relationship to the current roadmap. The end result, called the *visibility roadmap*, is typically much smaller than the traditional PRM, requires much less computation to build, and effectively covers the free configuration space.

Let $V(q)$, called the visibility domain of q, be the set of all configurations that can be reached from configuration q by the local planner. When the local planner is a straight-line planner in the configuration space, $V(q)$ is exactly the set of configurations that are visible from q, which we refer to as the *guard* for $V(q)$, since a guard located at this configuration would see all of $V(q)$. Thus, we say that the configurations in $V(q)$ are visible from q. A configuration is referred to as a *connection* if it lies in the visibility domains of more than one guard, i.e., q is a connection if $q \in V(q_i) \cap V(q_j), i \neq j$.

The visibility PRM consists of vertices that are either guards or connections, and edges that represent local paths between the vertices. Let s be the number of guards. The set of guards $q_1 \ldots q_s$ is such that $\cup V(q_i)$ covers the free configuration space and $q_j \notin V(q_i)$ for $i \neq j$, i.e., every free configuration is visible to at least one of the guards, and no two guards are visible to one another. For any two guards q_i, q_j whose visibility domains intersect, a connection q' is added, and the local planner is used to construct paths from q_i to q' and from q' to q_j.

It is easy to see that a roadmap constructed in this way has connectivity consistent with the connectivity of the free configuration space relative to the local planner. More specifically, for configurations q_init and q_goal if there exists a sequence of local paths in the free configuration space that connect these two configurations, then there exist guards q_i and q_j in the visibility roadmap such that $q_\text{init} \in V(q_i)$ and $q_\text{goal} \in V(q_j)$, and with q_i and q_j in the same connected component of the roadmap.

The algorithm to construct the visibility roadmap is a straightforward implementation of the definition given above. A free configuration q is generated, and this vertex is added to the roadmap if (a) it is not visible from any existing configuration in the roadmap (it becomes a new guard), or (b) it is visible from a pair of configurations q_i and q_j such that q_i and q_j do not lie in a single connected component of the existing roadmap (it becomes a new connection). If q is visible only from configurations in a single existing connected component of the roadmap, it is rejected.

Although the visibility PRM described above has connectivity consistent with that of the free configuration space, it is not necessarily the case that a planner will be able to successfully construct such a visibility PRM. For example, if $V(q_i) \cap V(q_j)$ is very small, there is a likelihood that no randomly generated q will lie in this intersection, and therefore there will be no connection for guards q_i and q_j. However, as noted in [49], since the size of the intersection will typically be small only for well-chosen q_i and q_j, the probability of failing to generate connections for this reason is small. The narrow passage problem also poses difficulties for this planner, since, if a free corridor in the configuration space is narrow, it is unlikely that randomly generated samples will fall in the corridor, or that samples at opposite ends of the corridor will be visible to one another.

In spite of these limitations, the visibility PRM has proven effective for a wide range of difficult path planning problems. The visibility PRM is typically much smaller than a traditional PRM. Further, since it rejects a large fraction of candidate vertices, the amount of computation required for local path planning (which requires collision checking along the local path) is greatly reduced. Experimental results given in [49] show examples for which the visibility PRM requires less than 10% of the computation time required to construct the traditional PRM.

23.4.3 Gaussian Sampling

The two approaches described above (manipulability and visibility) do not explicitly take into account the geometry of the obstacle region in the configuration space (although this geometry is, of course, implicitly taken into account in the notion of visibility). Since difficult motion planning problems tend to require the robot to move in close proximity to obstacles, it would seem reasonable to concentrate the sample configurations in the roadmap near the boundary of the configuration space obstacle region. This is the motivation for the Gaussian sampling strategy introduced in [11].

One way to find samples that lie near the obstacle boundary is to generate pairs of samples, discarding any pair that lies completely in the free configuration space or completely in the configuration space obstacle region. Any pair that is not discarded will be such that one sample lies in the free configuration space and the other lies in the configuration space obstacle region. The former is a sample that lies near the obstacle boundary, provided the samples are not too far apart.

In [11], the method is implemented via a Gaussian sampling strategy. Let D be a zero-mean Gaussian random variable that denotes distance, with probability density function given by

$$\phi_D(d) = \frac{1}{\sqrt{2\pi}\sigma} e^{-\frac{d^2}{2\sigma^2}}.$$

A pair of samples is generated as follows. First, generate q_1 using any of the sampling methods described in Sect. 23.3. Second, generate a sample d from ϕ_D. Next, generate a second sample q_2 at a distance d from q_1. If either of q_1 or q_2 is in the free configuration space while the other lies in the configuration space obstacle regions, add it to the roadmap. This is illustrated in Fig. 23.11. The value of σ^2 determines how close to the obstacle boundary the samples will lie.

For the experiments reported in [11], it was observed that σ^2 should be chosen so that most configurations lie closer to the obstacle than the maximum length of the robot. It is also noted in [11] that special care must be taken when dealing with the dimensions of the configuration space that correspond to rotational degrees of freedom. This method has been shown to be effective for a number of difficult motion planning problems.

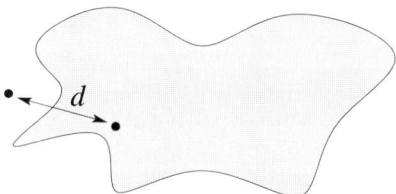

Fig. 23.11. Gaussian sampling

23.5 Sample Transformation

The methods described above reject samples that do not satisfy certain criteria. An alternative to rejecting such a sample is to transform it so that the transformed sample does satisfy those criteria. A number of methods to achieve this have been proposed in the literature. Here we describe four approaches: pushing samples toward the boundary of the obstacle region, pushing samples toward the medial axis of the free space, bouncing samples off of obstacle boundaries using random walks, and pulling samples from the existing roadmap toward randomly drawn samples to construct rapidly-exploring random trees (RRTs).

23.5.1 Pushing Samples to the Obstacle Boundary

The planners described in [5, 7] attempt to place samples near the boundaries of obstacles. The motivation, like that for the Gaussian sampling method described above, is that difficult planning problems often require the robot to move in close proximity to obstacles, and therefore regions of the configuration space that lie near obstacle boundaries should be sampled more densely.

The approach described in [5, 7] is to first identify samples that lie inside the configuration space obstacle, and to then "push" these samples to the obstacle boundary. The method is conceptually straightforward. Generate samples until a sample q is found to lie within the configuration space obstacle. This can be determined by existing collision-checking algorithms. Choose a set of directions $\{v_1, v_2 \cdots v_m\}$, and for each direction v_i a binary search can be used to find the boundary of the configuration space obstacle along v_i from q. This obstacle-based PRM (OBPRM) method is illustrated in Fig. 23.12.

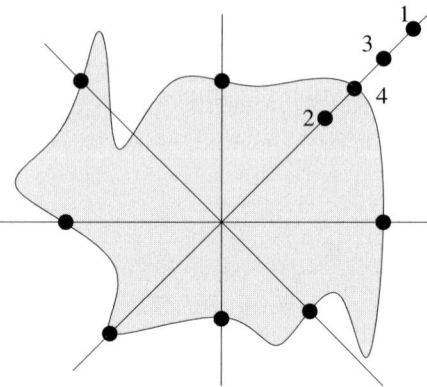

Fig. 23.12. OBPRM approach to finding samples near the boundary of the configuration space obstacle

This sampling strategy has been investigated in conjunction with a number of local planners. It has been shown to be effective for planning problems in cluttered environments, and for configuration spaces that contain narrow passages.

23.5.2 Pushing Samples to the Medial Axis

An alternative to placing samples near the obstacle boundaries is to sample near the medial axis of free configuration space or free workspace [16, 18, 54]. This approach is inspired, to some extent, by retraction approaches to path planning, in which the free configuration space is retracted onto a one-dimensional subset. If this can be done successfully, samples will be found in narrow corridors in the free configuration space, as well as in large regions of the free configuration space.

The medial axis of the free configuration space is the set of points that are equidistant from multiple distinct boundary points in the obstacle region of the free configuration space. For the 2D case, in a polygonal environment the medial axis is just the generalized Voronoi diagram. An example is shown in Fig. 23.13. It is generally not feasible to compute explicitly the medial axis of the free configuration space, but it can be possible to transform randomly drawn sample configurations so that the transformed samples lie on the medial axis.

The case of a polyhedral robot moving in a 3D workspace populated by polyhedral obstacles is considered in [54]. For a given sample configuration q, a sample on the medial axis q' can be generated as follows. For the case when q lies in the free configuration space, apply a pure translation to the configuration until the robot in the workspace is equidistant from two distinct obstacle points in the workspace. This translation is along the line segment that defines the minimum distance between the robot and the nearest obstacle. Note that this does not require any distance computations in the configuration space, since only workspace distances to obstacles are used in the computation. For the case when q is a collision configuration, first apply the shortest pure translation to q that will free the robot from collision, then proceed as above. In some cases, the shortest translation to free a collision configuration will cause the robot to contact workspace obstacles at more than a single unique point. These configurations are discarded by the sampling algorithm of [54].

This method has also been applied to narrow corridor problems and has proven effective in a number of experiments. However, it is not clear that the method will scale well to problems in highly cluttered environments, since the calculations for determining nearest contact configurations will become the dominant factor in the computation.

23.5.3 Random Walks

A popular way to transform an existing sample in the configuration space is to execute a random walk from that sample. This method has its roots in the randomized potential field planner (RPP) describe in [8]. The goal of this original planner was to use random walks to escape local minima in a potential field. The approach has subsequently been adapted to generate vertices in a graph embedded in the free configuration space. A number of

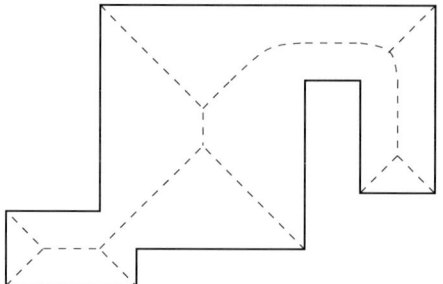

Fig. 23.13. The medial axis of a polygonal region

these approaches are referred to as landmark-based approaches (the sample configurations being the landmarks), but they are very similar in spirit to the PRM method.

The Ariadne's Clew algorithm employs this approach (see, e.g., [2, 3, 4, 56]). Some versions of the algorithm use an exploration stage that generates Manhattan paths in the configuration space (Fig. 23.14). A Manhattan path consists of straight line segments that meet at orthogonal junctions. In practice, these Manhattan paths are typically generated by genetic algorithms, which are amenable to problem of generating Manhattan paths, since there is a small set of discrete choices (the Manhattan directions) at each junction.

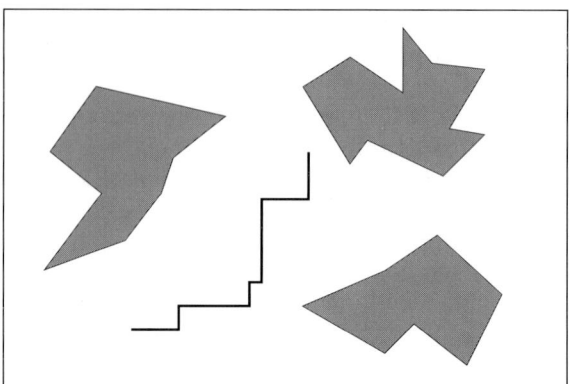

Fig. 23.14. Manhattan paths

The samples found at termination of the Manhattan paths can be improved by "bouncing" the path off of obstacle boundaries when collisions occur. In particular, at a collision point, the path is retraced to a previous point on the path and a new direction is chosen. Figure 23.15 illustrates the process. The first figure on the left shows an initial path. From the collision point (indicated by the dark bar), the path is retraced, and a new segment is added going downward in the figure. This new segment collides with an obstacle, as shown in the second figure, so it is retraced and new segments are added, leading to the collision shown in the third figure. The third figure shows the same path after two bounces. The final figure shows a valid path obtained after a final bounce.

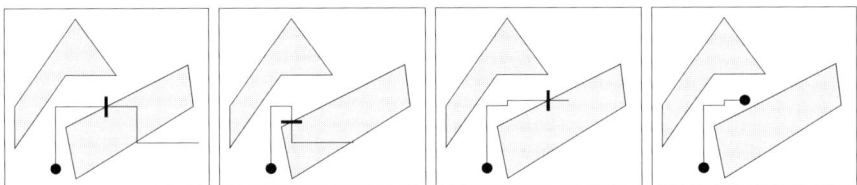

Fig. 23.15. Using Manhattan paths to bounce off of obstacles

23.5.4 Rapidly-Exploring Random Trees (RRTs)

Like the Ariadne's clew algorithms, rapidly-exploring random trees (RRTs) [35] use an existing tree in the configuration space to bias sample generation. The basic algorithm is defined recursively. At the ith iteration, generate a sample q using any of the methods described in section 23.3. Let q_{near} be the vertex in the existing tree that is nearest to q. The sample q_i to be added to the tree is generated by taking a small step from q_{near} toward q.

Construction of RRTs requires a distance metric on the configuration space to determine the vertex q_{near} and a local planner to construct a path from q_{near} to q_i. Since the destination configuration, q_i is not explicitly specified (it is merely the configuration that terminates the short path from q_{near} toward q), there is considerable flexibility in the implementation of the local planner. This, in turn, allows the application of RRT algorithms to more difficult planing problems, such as the kinodynamic planning problem.

For the problem of kinodynamic planning (which considers constraints on robot velocity as well as configuration) samples can be generated by sampling from the space of control inputs, and applying the sample control at the vertex q_{near} in the existing roadmap [39, 40]. In this case, the local planner is implemented by applying the chosen control input to the dynamical model of the robot system. A similar idea is described in [33].

A number of important theoretical results have been derived for the performance of RRTs [40]. For holonomic planning, the distribution of RRT vertices converges to the sampling distribution (typically uniform). This result leads to the probabilistic completeness of the RRT algorithm. For q_{init} and q_{goal} in a single connected component of the configuration space, the probability that an RRT constructed from q_{init} will find a path to q_{goal} approaches one as the number of RRT vertices approaches infinity.

RRTs have been applied to a wide variety of motion planning problems, including holonomic and nonholonomic mobile robots and autonomous space craft. These problems include cases for which robot motion is governed by a complex system of dynamics equations. They have also been adapted to perform bidirectional search (RRTs are grown simultaneously from both q_{init} and q_{goal}).

23.6 Connecting the Vertices

Once a sample configuration has been generated, it must be connected to the existing roadmap. For some of the methods described above (e.g., Ariadne's clew and RRTs), this connection is taken care of in the process of generating the sample. In most cases, however, sample configurations are generated without consideration of the existing roadmap, and connection to the roadmap must be handled explicitly.

Connecting a new vertex to the existing roadmap involves two operations: determining a set of candidate neighbors for the new vertex and generating local plans to connect the new vertex to those neighbors. For the latter issue, any local planning method may be used, and a simple straight-line planner in the configuration space is a typical choice. For the former issue, a distance function (not always a metric) is used to define the set of k nearest neighbors to the new configuration. In the remainder of this section, we discuss a few distance functions that have been used for this purpose.

The distance function provides a measure of the difficulty the local planner is likely to have when attempting to connect two configurations. An ideal distance function would be the swept volume in the workspace of the trajectory connecting the two configurations, since intuitively trajectories with larger swept volumes are more likely to be blocked by obstacles in the environment. Unfortunately, as noted by others [6, 31], this distance function is very expensive to compute; therefore, most PRM methods use approximations based solely on the two configurations that are to be connected.

Several distance functions have been defined on the configuration space of the robot. These distance functions typically treat the configuration space as a Cartesian space and define the distance function accordingly. For example, the Euclidean distance is used in a few planners [9, 7, 17]. The l_1 norm has also been used [41]. A problem with these distance functions is that they weight the configuration parameters equally, when some parameters may have a larger effect than others. A solution to that problem is to add weights to the different configuration parameters, and this has been used in a number of planners [5, 40, 51, 10].

Workspace distance functions attempt to measure the motion of the robot in the workspace. One method that has been used for articulated robots is to take the 2-norm of the Euclidean distances between the joint positions in the workspace [27]. Two distance functions for rigid objects in 3D that

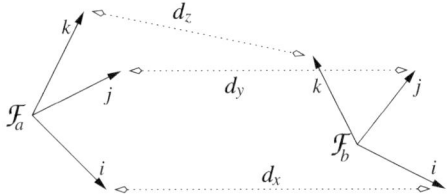

Fig. 23.16. Illustration of d_x, d_y, and d_z

were tested in [6] are the distance between the center of mass of the object at two configurations and the maximum distance between any vertex of the bounding box at one configuration and its corresponding vertex at the other configuration. For objects with flexible surfaces, a workspace-defined distance function is the sum of the translation distance, scaled rotation, and maximum displacement of a control point [26]. Another distance function is defined as the sum of the distances between unit vectors of the coordinate frame of the end-effector of the robot. This distance function, illustrated in Fig. 23.16, has been used for manipulation planning [2] and for inverse kinematics of redundant robots [1]. Another variant on a workspace distance function is defined for nonholonomic robots in 2D workspaces in [51].

Table 23.1. Four distance functions from the literature that we have investigated

2-norm in C-space:	$\mathcal{D}_2^C(q,q') = \|q' - q\| = \left[\sum_{i=1}^n (q'_i - q_i)^2\right]^{\frac{1}{2}}$
∞-norm in C-space:	$\mathcal{D}_\infty^C(q,q') = \max_n |q'_i - q_i|$
2-norm in workspace:	$\mathcal{D}_2^W(q,q') = \left[\sum_{p \in \mathcal{A}} \|p(q') - p(q)\|^2\right]^{\frac{1}{2}}$
∞-norm in workspace:	$\mathcal{D}_\infty^W(q,q') = \max_{p \in \mathcal{A}} \|p(q') - p(q)\|$

Table 23.1 shows four distance functions. For the equations in this table, the robot has n joints, q and q' are the two configurations corresponding to different vertices in the roadmap, q_i refers to the configuration of the ith joint, and $p(q)$ refers to the workspace reference point p of the set of reference points $\mathcal{A}$ at configuration q. Versions of $\mathcal{D}_\infty^W$ and $\mathcal{D}_2^W$ were also used in [31].

Figure 23.17 shows the roadmaps that result from using these four distance functions with a collection of 50 vertices. As can be seen in the figure, each of the roadmaps is somewhat different from the others, particularly as the distance between vertices increases. The advantage of the roadmaps generated using the workspace distance function is that the links of the robot are relatively closer together in the workspace; therefore, the volume swept by the

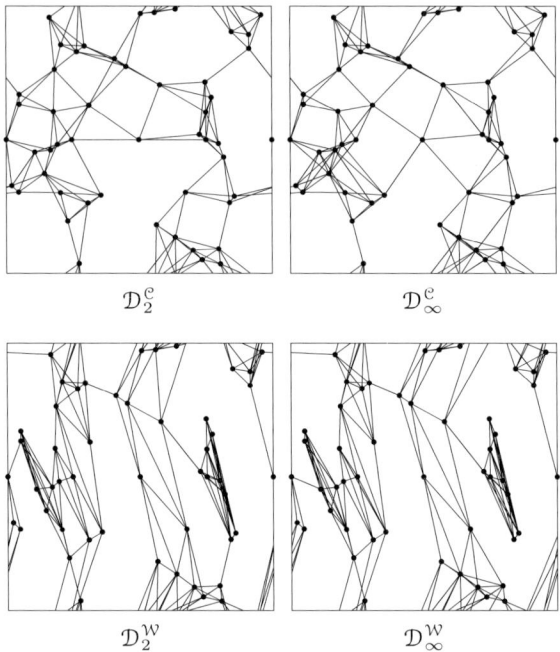

Fig. 23.17. The roadmaps that result from using the four distance functions

robot as it traverses a path is, on average, smaller than the volume swept by paths defined by the configuration space distance functions.

For some methods of constructing the roadmap, there is no explicit distance function. Some of these methods involve incremental expansion of the roadmap, and the distance function is implicit in how new samples are generated around a vertex. Connections are then attempted between all new samples and the chosen vertex. The tree expansion method described in [21] uses this approach. In addition, tree expansion methods that derive new vertices by integrating the motion equations with an input automatically connect the new vertices with the old, and no distance function is needed in this case [33]. Other methods that do not use a distance function when constructing the roadmap are those that attempt to connect every pair of vertices [20]. This approach is particularly useful when analyzing the performance of the roadmap [25].

23.7 Enhancement

As mentioned above, the initial roadmap often contains multiple connected components, even when the free configuration space contains a single connected component. Because of this, it is typical to use an enhancement phase of roadmap construction, during which vertices and edges are added to the roadmap in an attempt to connect disjoint components. The basic idea is to identify vertices in the existing roadmap that are near problem areas (e.g., narrow corridors in the free configuration space), and to expand the roadmap from these vertices.

In the original work on PRM planners, a vertex was determined to be good candidate for expansion when the local planner required many attempts to connect it to the existing roadmap. More precisely, for vertex q, let $n(q)$ be the number of times that the planner attempted to connect q to the roadmap, and let $f(q)$ be the number of failures to connect q to the existing roadmap. To make this more precise, define the failure ratio $r_f(q)$ as

$$r_f(q) = \frac{f(q)}{n(q)+1}.$$

By normalizing this ratio, we obtain a measure that can be used as a probability,

$$w(q) = \frac{r_f(q)}{\sum r_f(q)}.$$

It is now possible to use $w(q)$ in an importance sampling scheme to select vertices for expansion. In particular, we expand vertex q with probability $w(q)$. To do this, a random number generator can be used to generate a sample x from the uniform density on the unit interval. If $x < w(q)$, expand q.

Once a vertex has been selected for expansion, most any planning algorithm can be used to generate paths from that vertex, in an attempt to find connections to vertices in the existing roadmap. As an example, in [28] a randomized potential field planner [8] is used to generate paths from candidate vertices. In [19] paths are created by using a random walk that bounces off of obstacle boundaries (Fig. 23.18).

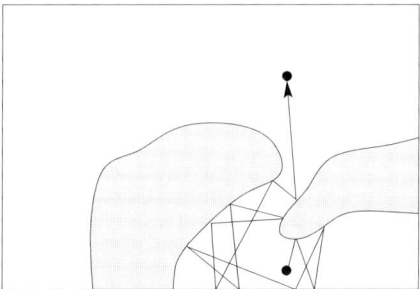

Fig. 23.18. A random walk that bounces off of obstacle boundaries

23.8 Conclusions

In the early 1990s, probabilistic roadmap approaches were introduced in the robot motion planning literature. Since that time, there has been an explosion in their use and development. PRM planners tend to be easy to implement, but there are many design choices, and these choices have considerable impact on the overall performance of the planner (see, e.g., [15] for a comparative study of approaches). In this chapter we have attempted to describe the general algorithm and to present a discussion of these design choices.

At the present time, PRM planners are able to solve a large range of motion planning problems; however, many problems remain. In particular, problems associated with narrow corridors in the free configuration space continue to be difficult for these planners. Further, the relationship between the geometry of the workspace (both obstacles and robots) and the geometry of the free configuration space is not yet well understood, making a thorough analysis of these methods difficult. At present, the asymptotic performance of these algorithms has been fairly well characterized, but it remains an open problem to determine how well a given PRM algorithm will perform for a specific workspace.

References

1. J.M. Ahuactzin and K. Gupta. The kinematic roadmap: A motion planning based global approach for inverse kinematics of redundant robots. *IEEE Transactions on Robotics and Automation*, 15(4):653–669, August 1999.
2. J.M. Ahuactzin, K. Gupta, and E. Mazer. Manipulation planning for redundant robots: A practical approach. *International Journal of Robotics Research*, 17(7):731–747, July 1998.
3. J.M. Ahuactzin, E. Mazer, and P. Bessiere. Fondements mathematiques d'algorithme "Fil d'Ariane". *Revue d'Intelligence Artificielle*, 9(1):7–34, 1995.
4. J.M Ahuactzin, E.-G. Talbi, P. Bessiere, and E. Mazer. Using genetic algorithms for robot motion planning. In *European Conference on Artificial Intelligence*, pp. 671–5, 1992.
5. N.M. Amato, O.B. Bayazit, L.K. Dale, C. Jones, and D. Vallejo. OBPRM: An obstacle-based PRM for 3D workspaces. In *Proceedings of Workshop on Algorithmic Foundations of Robotics*, pp. 155–168, 1998.
6. N.M. Amato, O.B. Bayazit, L.K. Dale, C. Jones, and D. Vallejo. Choosing good distance metrics and local planners for probabilistic roadmap methods. *IEEE Transactions on Robotics and Automation*, 16(4):442–447, August 2000.
7. N.M. Amato and Y. Wu. A randomized roadmap method for path and manipulation planning. In *Proceedings of IEEE Conference on Robotics and Automation*, volume 1, pp. 113–120, 1996.
8. J. Barraquand and J.-C. Latombe. Robot motion planning: A distributed representation approach. *International Journal of Robotics Research*, 10(6):628–649, December 1991.

9. P. Bessiere, J.M. Ahuactzin, E.-G. Talbi, and E. Mazer. The "Ariadne's clew" algorithm: Global planning with local methods. In *Proceedings of Workshop on Algorithmic Foundations of Robotics*, pp. 39–47, 1994.
10. R. Bohlin and L.E. Kavraki. Path planning using lazy PRM. In *Proceedings of IEEE Conference on Robotics and Automation*, pp. 521–528, 2000.
11. V. Boor, N. H. Overmars, and A. F. van der Stappen. The Gaussian sampling strategy for probabilistic roadmap planners. In *Proceedings of IEEE Conference on Robotics and Automation*, pp. 1018–1023, 1999.
12. M. S. Branicky, S. M. LaValle, K. Olson, and L. Yang. Quasi-randomized path planning. In *Proc. IEEE Int'l Conf. on Robotics and Automation*, pp. 1481–1487, 2001.
13. R. Brooks and T. Lozano-Pérez. A subdivision algorithm in configuration space for findpath with rotation. In *International Joint Conference on Artificial Intelligence*, pp. 799–806, 1983.
14. J. F. Canny. *The Complexity of Robot Motion Planning*. MIT Press, Cambridge, MA, 1988.
15. R. Geraerts and M. H. Overmars. A comparative study of probabilistic roadmap planners. In *Proceedings of Workshop on Algorithmic Foundations of Robotics*, pp. 43–57, 2002.
16. L.J. Guibas, C. Holleman, and L.E. Kavraki. A probabilistic roadmap planner for flexible objects with a workspace medial-axis-based sampling approach. In *Proceedings of IEEE/RSJ Conference on Intelligent Robots and Systems*, pp. 254–259, 1999.
17. L. Han and N.M. Amato. A kinematics-based probabilistic roadmap method for closed chain systems. In *Proceedings of Workshop on Algorithmic Foundations of Robotics*, 2000.
18. C. Holleman and L.E. Kavraki. A framework for using the workspace medial axis in PRM planners. In *Proceedings of IEEE Conference on Robotics and Automation*, pp. 1408–1413, 2000.
19. T. Horsch, F. Schwarz, and H. Tolle. Motion planning with many degrees of freedom — random reflections at c-space obstacles. In *Proceedings of IEEE Conference on Robotics and Automation*, pp. 3318–3323, 1994.
20. D. Hsu, L.E. Kavraki, J.-C. Latombe, and R. Motwani. Capturing the connectivity of high-dimensional geometric spaces by parallelizable random sampling techniques. In P. M. Pardalos and S. Rajasekaran, editors, *Advances in Randomized Parallel Computing*, pp. 159–182. Kluwer Academic Publishers, 1999.
21. D. Hsu, J.-C. Latombe, and R. Motwani. Path planning in expansive configuration spaces. *International Journal of Computational Geometry and Applications*, 9(4 & 5):495–512, 1999.
22. Y.K. Hwang and N. Ahuja. Path planning using a potential field representation. Technical Report UILU-ENG-8-2251, University of Illinois, October 1988.
23. S. Kambhampati and L.S. Davis. Multiresolution path planning for mobile robots. *IEEE Journal of Robotics and Automation*, 2(3):135–145, September 1986.
24. L.E. Kavraki. *Random Networks in Configuration Space for Fast Path Planning*. PhD thesis, Stanford University, Stanford, CA, 1994.
25. L.E. Kavraki, M. N. Kolountzakis, and J.-C. Latombe. Analysis of probabilistic roadmaps for path planning. In *Proceedings of IEEE Conference on Robotics and Automation*, volume 4, pp. 3020–3025, 1996.

26. L.E. Kavraki, F. Lamiraux, and C. Holleman. Towards planning for elastic objects. In *Proceedings of Workshop on Algorithmic Foundations of Robotics*, 1998.
27. L.E. Kavraki and J.-C. Latombe. Randomized preprocessing of configuration space for fast path planning. In *Proceedings of IEEE Conference on Robotics and Automation*, volume 3, pp. 2138–2145, 1994.
28. L.E. Kavraki and J.-C. Latombe. Probabilistic roadmaps for robot path planning. In K. Gupta and P. del Pobil, editors, *Practical Motion Planning in Robotics: Current Approaches and Future Directions*, pp. 33–53. John Wiley & Sons LTD, 1998.
29. L.E. Kavraki, J.-C. Latombe, R. Motwani, and P. Raghavan. Randomized query processing in robot motion planning. In *Proceedings of the ACM Symposium on Theory of Computing*, pp. 353–362, 1995.
30. L.E. Kavraki, J.-C. Latombe, R. Motwani, and P. Raghavan. Randomized query processing in robot path planning. *Journal of Computer and System Science*, 57(1):50–60, August 1998.
31. L.E. Kavraki, P. Švestka, J.-C. Latombe, and M.H. Overmars. Probabilistic roadmaps for path planning in high-dimensional configuration spaces. *IEEE Transactions on Robotics and Automation*, 12(4):566–580, August 1996.
32. O. Khatib. Real-time obstacle avoidance for manipulators and mobile robots. *International Journal of Robotics Research*, 5(1):90–98, 1986.
33. R. Kindel, D. Hsu, J.-C. Latombe, and S. Rock. Kinodynamic motion planning amidst moving obstacles. In *Proceedings of IEEE Conference on Robotics and Automation*, pp. 537–543, 2000.
34. D.E. Koditschek. Robot planning and control via potential functions. In *The Robotics Review 1*, pp. 349–367. MIT Press, 1989.
35. J.J. Kuffner, Jr. and S.M. LaValle. RRT-connect: An efficient approach to single-query path planning. In *Proceedings of IEEE Conference on Robotics and Automation*, pp. 995–1001, 2000.
36. J. C. Latombe. *Robot Motion Planning*. Kluwer Academic Publishers, Boston, 1991.
37. S. M. LaValle. *Planning Algorithms*. [Online], 1999-2003. Available at http://msl.cs.uiuc.edu/planning/.
38. S. M. LaValle and M. S. Branicky. On the relationship between classical grid search and probabilistic roadmaps. In *Proceedings of Workshop on Algorithmic Foundations of Robotics*, 2002.
39. S.M. LaValle and J.J. Kuffner, Jr. Randomized kinodynamic planning. In *Proceedings of IEEE Conference on Robotics and Automation*, pp. 473–479, 1999.
40. S.M. LaValle and J.J. Kuffner, Jr. Rapidly-exploring random trees: Progress and prospects. In *Proceedings of Workshop on Algorithmic Foundations of Robotics*, 2000.
41. S.M. LaValle, J.H. Yakey, and L.E. Kavraki. A probabilistic roadmap approach for systems with closed kinematic chains. In *Proceedings of IEEE Conference on Robotics and Automation*, pp. 1671–1676, 1999.
42. P. Leven and S. Hutchinson. Toward real-time path planning in changing environments. In *Proceedings of Workshop on Algorithmic Foundations of Robotics*, 2000.
43. P. Leven and S. Hutchinson. Real-time path planning in changing environments. *International Journal of Robotics Research*, 21(12):999–1030, December 2002.

44. P. Leven and S. Hutchinson. Using manipulability to bias sampling during the construction of probabilistic roadmaps. *IEEE Transactions on Robotics and Automation*, 19(6), December 2003.
45. T. Lozano-Pérez. Spatial planning: A configuration space approach. *IEEE Transactions on Computers*, February 1983.
46. E. Mazer, J.M. Ahuactzin, and P. Bessiere. The Ariadne's clew algorithm. *Journal of Artificial Intelligence Research*, 9:295–316, 1998.
47. A. McLean and I. Mazon. Incremental roadmaps and global path planning in evolving industrial environments. In *Proceedings of IEEE Conference on Robotics and Automation*, pp. 101–107, 1996.
48. M. Mehrandezh and K. Gupta. Simultaneous path planning and free space exploration with skin sensor. In *Proceedings of IEEE Conference on Robotics and Automation*, pp. 3838–3843, 2002.
49. C. Nissoux, T. Simeon, and J-P. Laumond. Visibility based probabilistic roadmaps. In *Proceedings of IEEE/RSJ Conference on Intelligent Robots and Systems*, pp. 1316–1321, 1999.
50. M.H. Overmars and P. Švestka. A probabilistic learning approach to motion planning. In *Proceedings of Workshop on Algorithmic Foundations of Robotics*, pp. 19–37, 1994.
51. M.H. Overmars and P. Švestka. A paradigm for probabilistic path planning. Technical Report UU-CS-1995-22, Utrecht University, March 1995.
52. J. T. Schwartz, M. Sharir, and J. Hopcroft, editors. *Planning, Geometry, and Complexity of Robot Motion*. Ablex, Norwood, NJ, 1987.
53. D. Vallejo, C. Jones, and N.M. Amato. An adaptive framework for 'single shot' motion planning. Technical Report TR99-024, Department of Computer Science, Texas A&M University, College Station, TX, October 1999.
54. S.A. Wilmarth, N.M. Amato, and P.F. Stiller. Motion planning for a rigid body using random networks on the medial axis of the free space. In *Proceedings of ACM Symposium on Computational Geometry*, pp. 173–180, 1999.
55. T. Yoshikawa. Manipulability of robotic mechanisms. *International Journal of Robotics Research*, 4(2):3–9, April 1985.
56. Y. Yu and K. Gupta. Sensor-based roadmaps for motion planning for articulated robots in unknown environments: Some experiments with an eye-in-hand system. In *Proceedings of IEEE/RSJ Conference on Intelligent Robots and Systems*, 1999.

Index

absolute conic, 312
additive equation, 26
adjoint representation, 659, 668, 677
affine
    camera, 319
    piece, 311
    tensor, 326
affine connection, 684
affine plane, 428
    3D, 430
    for incidence relations, 432
affine subspace, 630
affine transformation, 312, 572, 573, 576, 577, 581
Akaike information criterion (AIC), 470
Akaike, H., 470
algebra
    Grassmann, 644
    Grassmann–Cayley, 644
    lattice algebra, 97, 108–110, 113, 122
    linear algebra, 100, 110
    linear algebraic operations, 110
    minimax algebra, 100
algebraic error, 513
algebras of
    Clifford, 406
    Gibbs, 406
    Grassmann, 406
amacrine cell, 5
ambiguity, 270, 278, 279, 282, 293, 297
    intersection, 297
    resection, 297
    structure and motion, 297, 298, 300

angle meter, 270
Ariadne's clew algorithm, 758
articulated chain, 709, 731
artificial potential fields, 738
aspect ratio, 315
associative memory, 101, 123
    autoassociative, 101, 102, 105, 123
    correlation, 101, 102
    Hopfield, 101, 123
    linear, 101
    matrix, 100, 109
    morphological, 97, 100–102, 105–110, 122, 123
astigmatism, 515
asymptotic analysis, 466
asymptotic efficiency, 472
attack, 233
autocalibration, 341

back-projection, 521
backpropagation, 77
ball tensor, 542
ball voting field, 545
bar framework, 650
batch learning, 76
Bayesian, 472
    predictive distribution, 85, 90
beacon, 270
Beltrami flow, 211
Bezout's theorem, 715
bias, 480
    bilinear form, 505
    mean, 504
    spherical normalization, 506

variance, 504
bifocal constraint, 328
bifocal tensor, 328
bilinear constraint, 328
bilinear form, 505
bilocal operator, 180
bin widths, 175
biochemical reaction, 26
bipolar cell, 4
bivector, 407
blade, 407
blur
  discrimination, 13
  perception, 13
blurring, 178
blurry isophotes, 182
blurry region, 181
body-eye calibration, 435
Boolean operations, 606, 608, 613
boundary conditions, 701
boundary encoding, 7
boundary value problem, 692
bracket, 637
bracket ring, 637
bundle adjustment, 338
Burmester points, 709

calibrated trilinear tensor, 275
camera
  calibrated, 315
  calibration, 320
  centre, 314
  constant, 314
  equation, 315
  matrix, 315
  one-dimensional, 271
  state, 271
  uncalibrated, 315
canonical correlation analysis (CCA), 142
  regularized (RCCA), 146
cardinal depth conflict, 388
Carlsson
  duality, 281, 289
Casorati curvature, 196
Cayley factorizatioin, 650
Cayley–Klein geometries, 183, 184
cell
  K, 5
  M, 5
  P, 5
center of revolution, 639
central limit theorem, 472
chaos-based block cipher, 243
chaos-based image encryption scheme, 242
chaos-based method, 236
chaos-based stream cipher, 253
chaos (definition), 239
chaotic pseudorandom number generators (CPRNG), 254
Christoffel symbol, 684, 691, 699–701, 703
cipher, 232
cipherkey, 232
ciphertext, 232
circles of the first kind, 187
circles of the second kind, 187
Clifford
  conjugate, 668
  conjugation, 412
  planes, 173, 190
  product, 407
Clifford algebra, 657–678, 712
  dual quaternions, 714, 731
coadjoint representation, 660, 661, 677
codimension, 467, 474
codistribution, 696
coin-flipping model, 71, 79
collineation, 350
color images, 210
color space, 211, 217
common dividend of lowest grade, 409
compositionality, 384
compound matrix, 635
conditioning, 530
cone-type singular structure, 74
configuration space distance functions, 760
conformal geometry, 417
conformal mappings, 187
conformal split, 414, 416
conformal transformation, 424
confusion, 243
conic, 311, 477
consistency, 471, 476
constraint
  bilinear, 355

Index 771

trilinear, 274, 275, 301
constraint manifold, 713, 731
constraint manifold, algebraic, 714
constraint manifold, parameterized, 732
contraction product, 678
contrast, 12
   constancy, 12
contravariant index, 326
contravariant tensor, 349
coordinate homogeneous, 629
correlation, 575, 582
correspondence, 571, 574–578, 582, 583, 586–588, 593
couple, 645
covariance matrix, 464
   estimated, 511
   singular, 499, 500
covariant derivative, 684
covariant index, 326
covector, 644
Cramer–Rao lower bound (CRLB), 472
critical configuration, 642
cross entropy error, 79
cryptosystem, 231
curse of dimensionality, 136
curvature, 575, 584–586, 591
curvature flow, 205
curve evolution, 205
curve saliency, 542
curvedness, 196

Data Encryption Standard (DES), 231
data space, 467
decomposition
   eigenvalue, 133
   singular value, 134
   spectral, 133
deep structure, 175
degeneracy, 474
degeneracy detection, 474
degree of freedom, 713
dendritic computing, 109–112, 122
dendritic structure, 110–120
dense vote, 544, 547, 557
densification, 557, 562
depth, 315
   , cardinal, 388
   , ordinal, 388, 393
differential invariants of images, 194

differential operators, 180
differentiation of images, 181, 193
diffusion, 203, 243
   anisotropic diffusion, 207
   anisotropic nonlinear diffusion, 208
   inverse diffusion, 219
   isotropic nonlinear diffusion, 207
   linear diffusion, 205, 221
   nonlinear diffusion, 222
   orientation diffusion, 218
diffusion equation, 178
diffusivity, 207
dilation, 427
dilator, 427
dimensionality reduction, 140
direct linear transformation (DLT), 320
directed distance, 431
discrepancy, 743
dispersion, 743
distance fields, 605–607, 609, 612, 614
distance in image space, 185
distribution, 696–698
   bilinear form, 505
divergence, 470
divisors of zero, 186
dual, 514
   number, 712
dual number
   plane, 186, 187
dual projective space, 352
dual quaternion, 658, 712, 714
dual variables, 135
dual vector space, 644
dual vectors, 131
duality, 134, 135, 280, 300, 302, 310, 408
   Carlsson, 281, 289
dynamic
   alignment problem, 351
   viewing, 21

edge, 5, 465
   blur, 7
   blur discrimination, 13
   blur perception, 13
   sharpening, 13
   sharpness, 7
edge detection, 465
edge operators, 174
edges, 174

edginess, 174
efficiency, 472
egomotion, 384–386, 393
eigenvalue, 132
eigenvalue decomposition, 133
eigenvector, 132
eight-point algorithm, 332
Eikonal equation, 226, 585
elastic, 571, 573, 580–582, 588–594
ensemble, 461
epipolar constraint, 324
epipolar equation, 467, 478
epipolar line, 324
epipole, 322
equilibration, 484
equipotential surface, 612, 613, 616
equivalence
   homogeneous entities, 501
   uncertain homogeneous entities, 503
equivalence relation, 74
ergodicity, 239, 241
error propagation, 498
error propagation, 533
   implicit, 511
errors-in-variables model, 487
estimating function, 487
Euclidian differential invariants, 172
Euler–Lagrange equations, 668, 669, 672
evaluation map, 657, 663, 664, 666, 667
excitation, 12
   center, 12
   one-to-one, 12
exponential map, 659, 660, 668
extensor, 637
exterior algebra, 638
exterior product, 662, 678
extrinsic parameters, 316

factorization, 336, 337
feature, 36, 38, 39, 47, 48, 57
feature detectors, 174
feature space, 211, 217
features, 181
feedback, 12
fillet, 606, 608
filtering, 33, 39
first-order voting, 549
first-order voting fields, 550

Fisher discriminant analysis (FDA), 156
Fisher information matrix, 72, 472
Fisher metric, 72
fixed point, 100, 103–105
fluid, 573, 580, 581, 588–593
focal length, 314, 315
focal point, 313–316
focus of expansion, 393
foliation, 697
framework plane and parallax, 350
Fubini–Study, 38, 54
functional magnetic resonance imaging, 571, 578
fundamental matrix, 324, 467, 478, 521
fundamental numerical scheme (FNS), 480, 481
fundamental second-order stick voting field, 544

Galilean group, 184
gamma transformations, 172
Gauss–Helmert model, 508, 512, 529
Gaussian curvature, 48
Gaussian kernel, 178
Gaussian noise model, 71, 78
Gaussian sampling, 754
generalization, 178
generalization error, 75
   expected, 85
generalized function, 500
generalized homogeneous coordinates, 415
generalized inverse, 469
generalized 3D baker map, 246
generically isostatic, 650, 652
geodesic, 583, 585, 586, 684–686, 692, 693, 695, 697
geometric
   algebra, 515
      conformal, 531
   constraints, 524
   constructions, 515
   entities, representation of uncertainty, 515
   mappings, 520
   relations, 514
   testing uncertain relations, 527
geometric AIC (G-AIC), 470
geometric algebra, 406

geometric fitting, 467
geometric inference, 461
geometric MDL (G-MDL), 471
geometric model, 467
geometric model selection, 470
geometric product, 407
  of two multivectors, 410
geometrical loci, 181
germ, 284–289, 292
Gold sequence, 255
grade involution, 412
gradient descent learning
  standard, 76
  stochastic, 77
Gram matrix, 135, 136
Grassmann–Cayley algebra, 499, 662
Grassmann–Plücker relation, 631
group of motions, 183, 184
group of movements in image space, 189

Halton sequence, 744
Hamilton relations, 412
Hamiltonian mechanics, 657, 671–672, 678
heat equation, 204
  affine heat equation, 206
  geometric heat equation, 205, 206
HEIV, 481, 507, 512
  heteroscedastic errors-in-variables, 482
Hesse distance, 432
heteroscedastic, 487
hierarchical structures, 73
hills and dales, 196
histogram, 175
histogram valued images, 175
Hodgkin–Huxley equation, 25
homogeneous
  vectors, equivalence, 501
homogeneous coordinates, 308, 414
homogeneous spaces, 183
homogeneous transform, 711
homogenous vectors, uncertain, 498
homography, 311, 521
homography tensor, 349, 355, 361
homotopy algorithm, 710, 724
horizontal cell, 5
horosphere, 415
hyperplane, 415, 422

hypersphere, 417
hypothesis testing, 513

ideal point, 306
image deep structure, 173
image denoising, 217
image noise, 465
image processing, 171, 172, 183, 199
image space, 172, 173, 177, 183
image space, spatial displacements, 731, 732
images, 171
image encryption algorithm, 236
implicit function, 603
implicit model, 603
implicit texturing, 615, 616
importance sampling, 746
incidence operators, 409
inertia, 657, 661–662, 664, 665, 677
inertia matrix, 689, 691
infinite homography, 324
information
  dynamics, 38
  Fisher, 39, 50, 51
  maximal, 54
  sparse, 36
information geometry, 69
inhibition, 12
  many-to-one, 12
  shunting, 12
  surround, 12
inner product, 407
inner scale, 173, 174
intensity domain, 177, 183
intensity gradients, 173
international data encryption algorithm (IDEA), 231
interpolation, 691, 692
intersection, 274, 277, 282, 283, 285, 297, 332
  ambiguity, 298
intrinsic parameters, 315
invariant elements, 662, 664, 665
inverse kinematics, 440
  of a pan-tilt unit, 440
inverse optics, 3
inversion, 424, 425
involution, 427, 660
isogonal conjugate point, 279

iterative factorization, 337

join, 409, 429, 638
joint, 713
   ball joint (S), 710
   cylindric (C), 713
   gimbal (T), 710
   prismatic (P), 713
   revolute (R), 713
   spherical (S), 713
   universal (T), 713
   variables, 731
joint space, 467
junction saliency, 542

Kerckhoff's principle, 232
kernel, 106–109, 136, 223
   Gaussian kernel, 223
   one-dimensional kernel, 224
   short time kernel, 225
   function, 136
   matrix, 135, 136
   matrix, centered, 139
   methods, 135
   methods (KM), 135
   trick, 138
kernelizing, 136
kinetic energy, 657, 664, 667–669
Kullback–Leibler
   divergence, 72, 75
   information, 470

Lagrange multipliers, 674, 676, 677
Lagrangian mechanics, 657, 668–671, 678
Laplacian, 178
laser scanner, 270
laser-guided vehicle (LGV), 270
lateral geniculate nucleus (LGN), 5
lattice, 97, 98
   bounded lattice-ordered group (blog), 97, 99, 110
   complete, 98
   computation, 113
   dependency, 103, 105
   distributive, 98, 99
   dual of, 98
   independence, 105, 108
   independent pattern, 109

lattice-ordered group (ℓ-group), 97–100
lattice-ordered semigroup (ℓ-semigroup), 99
semilattice, 98
semilattice-ordered group ($s\ell$-group), 98, 100
semilattice-ordered semigroup ($s\ell$-semigroup), 99, 110
sublattice, 98
theory, 98, 100, 105
law of large numbers, 472
leaky-integrator model, 26
least squares, 479
least-squares regression, 137
least-squares solution, 479
   LS solution, 479
lens formula, 313
level set, 205, 604
level set function, 604
Lie algebra, 659, 660, 682, 687
Lie bracket, 659
Lie group, 682
likelihood, 468
likelihood function, 50
line at infinity, 306, 632
line complex, 642
line geometry, 642
linear functional, 657
linear geometric fitting, 477
linear product decomposition, 709, 715, 716
   circular cylinder, 721
   circular hyperboloid, 724
   circular torus, 729
   elliptic cylinder, 727
   general torus, 731
   plane, 718
   sphere, 719
local disorder, 175
local jet, 181
local operators, 174
log-polar coordinates, 390
Lorentz–Cauchy, 37, 40
   LC, 40, 44
Lyapunov exponent, 241

magnetic resonance imaging, 571, 572, 578, 591–594

Mahalanobis distance, 468
manipulability-based sampling, 747
maps, 5
    color, 5
    direction of motion, 5
    orientation, 5
    retinotopic, 5
    spatial frequency, 5, 9
match metric, 571–574, 593
matching, 555, 561
mathematical morphology, 97
maximum likelihood estimation, 468, 471
maximum likelihood estimator, 468
maximum-likelihood estimation, 509
maximum-likelihood estimator, 84
MDA multiple discriminant analysis, 159
MDL, 470
mean curvature flow, 209
medial axis, 757
meet, 410, 643
membrane equation, 25
meter, 270
metric, 217, 684
    bi-invariant, 684, 688
    kinetic energy, 685, 689, 692, 697
    left invariant, 684, 685, 687, 688, 695, 699
metric duality in image space, 191
metric shaping, 699
minimal cases, 336
minimal energy surface, 620
minimal representation, 107–109
minimum acceleration curve, 685, 686, 691–693
minimum description length, 470
Minkowski plane, 414
missing data, 338
mixing, 241
ML estimator, 468
MLE, 84, 87, 468, 471
MLP, 69
model, 466, 467, 471
model selection, 470, 472
modulus constraint, 343
moment, 431, 497
monofocal tensor, 328
Moore–Penrose generalized inverse, 469

Moore–Penrose pseudo inverse, 469
motion
    clustering, 385
motion clustering, 387, 392
motion deblurring, 19
motion estimation, 433
    ambiguity, 386
    line-based, 434
    point-based, 433
motion segmentation, 383
motion valley, 387
motor, 430
movements in image space, 183
multilayer perceptron, 69, 70
    stochastic, 70
multilinear constraint, 349
multilocal geometry, 175
multiple discriminant analysis, 159
multiplicative equation, 25
multivector, 407
    dual of a, 408
    homogeneous, 407
multiview analysis, 350
mutual information, 575, 576

natural gradient learning, 77
    adaptive, 78
navigation, 438, 455
Nernst potential, 25
neural network
    adaptive logic, 97
    architecture of, 114
    artificial, 97, 100, 109, 110, 112, 113, 123
    biological, 110
    feedforward, 97, 109, 122, 123
    fuzzy lattice, 97
    fuzzy min-max, 119
    hybrid morphological-rank-linear, 97
    min-max, 97
    morphological, 97, 100, 109, 110, 122
    morphological perceptron, 97, 110, 112–114, 121
    perceptron, 110, 111, 113, 114, 121
    radial basis function, 117
    recurrent, 101
    regularization, 97
    shared-weight, 97
neuromanifold, 69, 72

Newton's formula, 314
Neyman–Scott problem, 487
noise, 100, 106, 109, 123
   dilative, 105–107
   erosive, 105–107
   noisy pattern, 105, 107–109, 123
   random, 106–109
noise level, 464
nonrigid transformation, 571–574, 576–583, 588, 589, 591, 593
normalization, 530
   spherical, bias, 506
normalized covariance matrix, 464
nuisance parameter, 487
null basis, 413
null cone, 416
null space, 134
null vectors, 416, 428
numerical schemes, 203, 219
   explicit, 221
   finite difference, 220
   implicit, 221
   additive operator splitting (AOS), 222
   alternating direction implicit (ADI), 222
   locally one dimensional (LOD), 223

object manipulation, 440
occlusion, 384, 388
occlusion filling, 393, 394
occlusion-motion conflict, 388
occlusions, 392
occlusions and ordinal depth, 393, 394
omnidirectional vision, 448
   conformal unified model, 449
   for robot navigation, 455
   using conformal geometric algebra, 448
online learning, 77
operator
   bounded, 42
   Harris, 462
   intertwining, 42
   momentum, 45
   position, 45
   rotation, 46
   scaling, 46
   SUSAN, 462

   translation, 46
optical axis, 313
optical flow, 384, 386, 392
optical ray, 317
ordinal depth conflict, 388, 393
origin, 414
orthographic camera, 318
outer product, 407
outer scale, 173, 174
over-realizable scenario, 83

parallel connection, 647
parallel points, 185
parameter space, 467
partial differential equations, 203
partial least squares, 147
path planning, 737
pathways, 5
   action, 5
   dorsal, 5
   magnocellular, 5
   parvocellular, 5
   perception, 5
   ventral, 5
   what, 5
   where, 5
pencil, 310
phase, 17
   feed-forward dominant, 18
   feedback dominant, 18
   reset, 17
phase correlation, 389
photoreceptor, 4
pinhole camera, 314
pinhole camera model, 352
pixels, 172
Plücker
   constraint, 514, 530, 531
   coordinates, 514, 522, 529, 532
   matrix, 519, 525
   matrix, dual, 519, 525
Plücker coodinate
   vector, 631
Plücker coordinate
   dual, 634
plaintext, 232
plane at infinity, 632
plate tensor, 542
plate voting field, 546

plateau, 81, 92
PLS
　EZ-, 150
　Partial Least Squares, 147
　Regression-, 152
point at infinity, 414, 428, 630, 632
　ideal point, 306
point isogonal conjugate, 279
point operator, 179, 193
polarity vector, 549
polynomial systems, 709, 711, 721, 731, 733
primal variables, 135
prime, 270, 283, 285, 286, 288–290, 292–295
principal component analysis (PCA), 140
principal point, 315
prior
　Jeffreys', 91
　uniform, 91
PRM, 737
probabilistic roadmap, 737
product
　inner, 408
　inner general definition, 410
　outer, 408
progressive coding, 257
projection, 414, 689, 690, 692, 695
projection matrix, 469, 479, 521
projective geometry, 306
　uncertain reasoning, 499
projective invariant, 650
projective line, 306
projective plane, 306, 352
projective space, 38, 54, 57, 306, 352, 630
projective subspace, 630
projective transformation, 297, 311
pseudo inverse, 469
pseudoscalar, 408
pure isotropic rotations, 193
pure isotropic shifts, 193

quadrifocal constraint, 330
quadrifocal tensor, 331
quadrilinear constraint, 330
quadrilinear tensor, 280
quaternion, 413

quaternions, 658

rank, 467
ransac, 341
rapidly-exploring random trees (RRTs), 759
RCCA, 146
re-entrant, 12
reachable surface, 709, 710, 713–716, 732
reachable surface, circular cylinder, 719
reachable surface, circular hyperboloid, 714, 722
reachable surface, circular torus, 727
reachable surface, elliptic cylinder, 714, 724
reachable surface, general torus, 714, 729
reachable surface, plane, 714, 717
reachable surface, sphere, 718
reasoning under uncertainty, 499
reciprocity, 646
reconstruction, 438
reflection, 424, 425
reflector, 270
region of interest, 174
registration, 571–583, 588, 591
regularization, 618, 620, 622, 624
rejection, 414
relation sphere and hyperplane, 422
relations
　geometric, 514
renormalization, 507
　method, 483
representation
　geometric elements, 514
　uncertainty, 497
representation theory, 369
resection, 282, 283, 285, 297, 331
　ambiguity, 298
residual, 470
residual sum of squares, 470
resolution, 33, 38, 48, 173
retinal ganglion cell, 4
retino-cortical dynamics (RECOD) model, 8
retinotopic map, 5
reversion, 412
ridge regression, 137

ridges and ruts, 198
Riemannian, 72
    manifold, 72
    metric, 51, 57, 72
rigid, 650
rigid body motion, 657–660, 670
rigid transformation, 571, 572, 575, 577, 581, 583
rigidity constraint, 696, 697
robot dynamics, 657–678
robot kinematics, 668
robot manipulator, 709
    following a spherical path, 445
    grasping an object, 446
    touching a point, 442
robot navigation, 455
rotation, 426
rotor, 412, 427, 430

saliency decay function, 543
scale ambiguity, 316
scale space, 175, 178, 204
scaled orthographic camera, 319
screw axis, 712
screw system, 642
second-order tensor, 541
security analysis, 260
segmented stationary points, 375
semiparametric model, 487
sensitivity to initial condition, 239
serial chain, 713
    CS, 714, 720, 721
    parallel RRS, 714, 718
    PPS, 714, 717, 718
    PRR, 732
    PRS, 714, 727
    right RRS, 714
    RPRP, 732
    RPS, 714, 718, 722
    RR, 714, 732
    RRR, 732
    RRS, 728–731
    RRS , 714
    TS, 710, 714, 718, 719
series connection, 647
series–parallel robot, 647
set of equivalent points, 74
set partitioning in hierarchical trees (SPIHT), 256

seven-point algorithm, 333
shape index, 196
shuffle, 643
shuffle product, 662–664, 678
shunting equation, 25
similarities of the first kind, 193
similarities of the second kind, 193
similarity
    transformation, 273, 284, 312
singular structure, 74
singular value, 134
singular value decomposition, 350, 690, 691
singular vector, 134
singularity, 74
singularity problem, 83
six-point algorithm, 333
skew, 315
soma space, 712, 731
space conic, 508
sparse vote, 544
spatial displacement, 711
spatial frequency, 5
spatial relations, 514
    testing, 513
special Euclidean group, 682
special orthogonal group, 682
spectral
    algorithms, 139
    clustering, 160
    decomposition, 133
spheres, 418
spherical normalization
    bias, 506
spherical wrist, 709, 733
standard 2D baker map, 245
statistical estimation, 466
step, 637
stereo, 385, 388, 397
stereographic projection, 417
Stewart platform, 654, 658, 672, 674, 677
stick tensor, 542, 543
stick voting field, 544
stochastic model, 471
stochastic model selection, 472
straightening algorithm, 638
structural parameter, 487
structure and motion, 270

Index   779

problem, 321, 331
structure from motion, 369, 386, 388
structure of a point, 174
superbracket, 648
superjoin, 641
support, 637
support vector machines, 164
surface saliency, 542
surveying, 270
synapse, 5
synaptic development, 7
synaptic pattern, 7

tangent space, 469
tensor
   antisymmetric, 637
   calculus, 326
   constraint, 357
   homography, 361
   of the dynamic 3D-to-3D alignment
      problem, 363
   symmetric, 358, 366
   to distinguish dynamic and stationary
      points, 363
   decomposable antisymmetric, 637
   indecomposable antisymmetric, 637
   quadrilinear, 280
   trilinear, 276, 277, 297, 298, 300, 302
tensor voting 4D, 560
tensorial transfer, 330
testing
   spatial relations, 495, 513
   uncertain geometric relations, 527
test standard, 255
texture, 214
top-down, 35, 36, 38
topological group, 42
topological transitivity, 239
total least squares, 479, 507
total variation, 207
training error, 76
   expected, 85
transformation
   projective, 297
   similarity, 273, 284
transient regime, 21
translation, 424, 426
translator, 426, 430
transversion, 426

transversor, 426
triangulation, 494, 508
trifocal constraint, 329
trifocal tensor, 329
trilinear
   constraint, 329
   constraint, 274, 275, 301
   tensor, 276, 277, 297, 298, 300, 302
trivector, 407, 412
twist, 640, 683, 689
   left invariant, 683, 697
   space, 641
two-stage encoding, 471

uncalibrated image sequence, 321
uncertain
   equivalence, 503
   homogeneous vectors, 498, 503
uncertainty
   3D line transformation, 532
   geometric entities, 515
   propagation, 498
   representation, 497
uncertainty ellipse, 464
unitary
   E(2), 44
   operators, 41
   representation, 41

van der Corput sequence, 744
variance components, 509
versor, 424
   for
      dilation, 427
      inversion, 425
      reflection, 426
      rotation, 427
      translation, 426
      transversion, 426
   representation, 424
video, 216
video stabilization, 389
visibility roadmap, 753
visual system, 3
volumetric data, 216
voting fields, 544, 546

wavelet, 39, 42, 43
Weierstrass function, 608, 617